# MODELS FOR CARBONATE STRATIGRAPHY FROM MIOCENE REEF COMPLEXES OF MEDITERRANEAN REGIONS

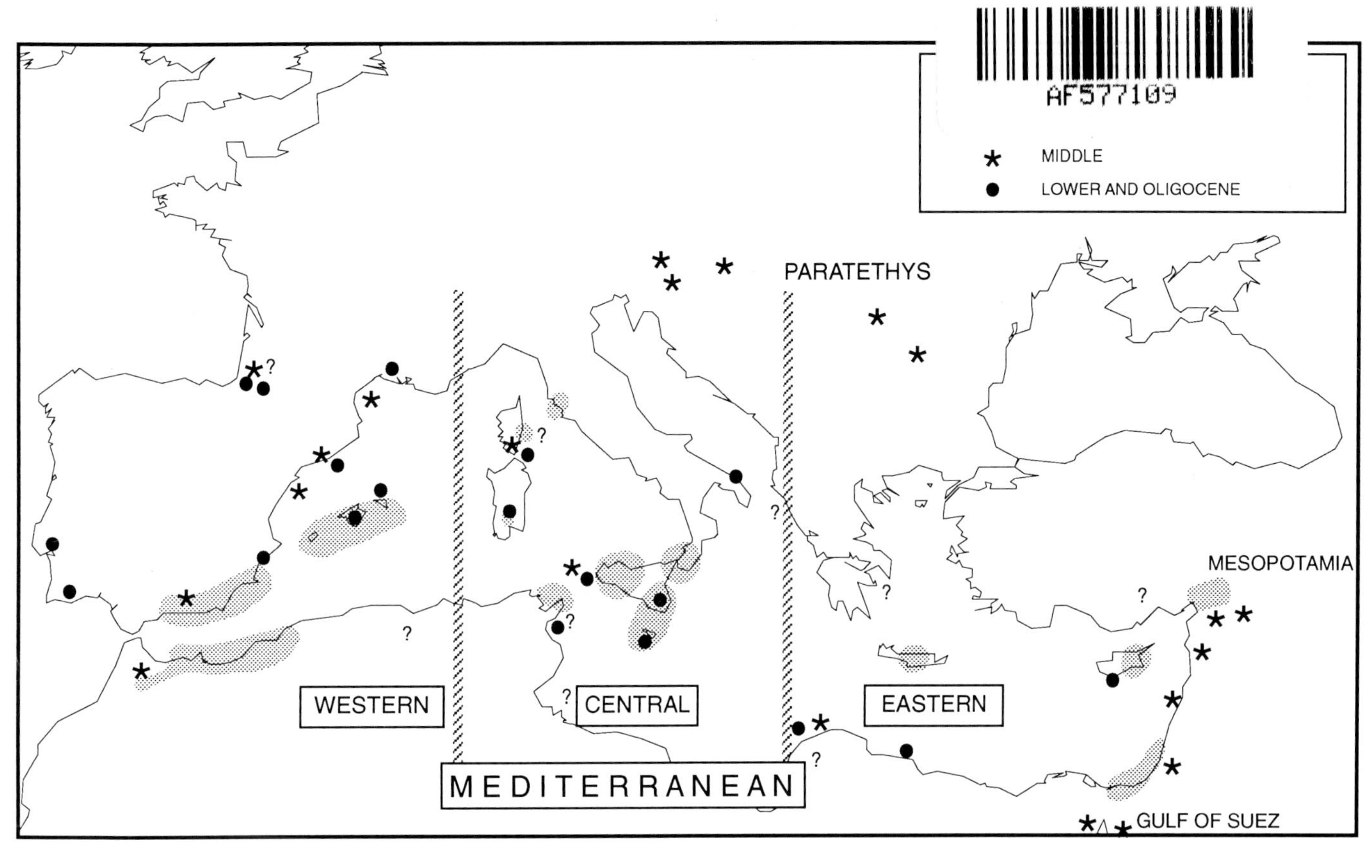



*Edited by:*

*Evan K. Franseen, Kansas Geological Survey, University of Kansas, Lawrence, Kansas*

*Mateu Esteban, Carbonates International Ltd, Esporles, Mallorca, Spain*

*William C. Ward, Department of Geology and Geophysics, University of New Orleans, Louisiana*

*and Jean-Marie Rouchy, Laboratoire de Geologie, Museum National D'Histoire Naturelle, Paris, France*





*Peter A. Scholle, Editor of Special Publications*
*Concepts in Sedimentology and Paleontology Volume 5*

*Tulsa, Oklahoma, U.S.A.* *April, 1996*

A Publication of
SEPM (Society for Sedimentary Geology)

ISBN 1-56576-033-6

1731 E. 71st Street
Tulsa, Oklahoma 74136-5108
Printed in the United States of America

# TABLE OF CONTENTS

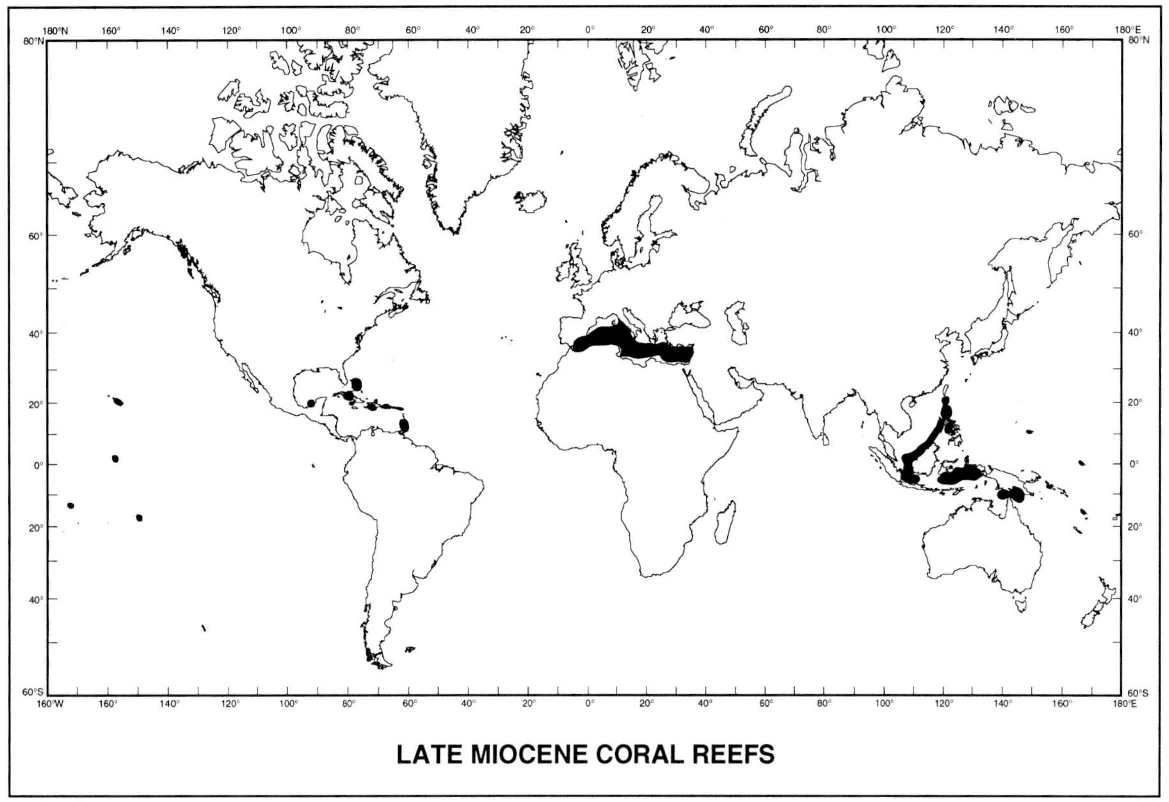


FIG. 1.—(A) Global distribution of Early to Middle Miocene coral reefs and hermatypic coral mounds. (B) Global distribution of Modern coral reefs and zone with hermatypic corals (dashed line). (C) Global distribution of Late Miocene coral reefs.

generating considerable interest, discussion and direction for recent and ongoing studies. Because the 1988 symposium clearly showed that there was a great deal of interest in the well-exposed Miocene reefs of the Mediterranean regions, our editorial team was stimulated to produce a volume that focuses on Miocene reefs of those regions and shows their value as models for carbonate stratigraphy.

The 21 revised and updated papers in the resultant volume, Models for Carbonate Stratigraphy from Miocene Reef Complexes of Mediterranean Regions, are a culmination of over 20 years of research in Mediterranean regions, and they represent our understanding of Miocene carbonate complexes in those areas at this point in time. However, many problems remain to be solved. One of the most significant problems that is currently receiving much attention is regional correlation within the uncertainties of the available chronostratigraphic charts (Fig. 2). We hope this current volume will provide a useful basis on which to continue to build in ongoing and future studies of the Miocene carbonate complexes from these fascinating regions.

The volume is divided into two major sections: (1) Regional Reviews and (2) Detailed Studies. The Regional Reviews section contains six papers. The first paper is a synthesis of general trends and facies models of the entire Mediterranean area. The next four papers describe general characteristics of four Mediterranean regions divided into: the western Mediterranean; the central Mediterranean; the eastern Mediterranean, Red Sea and Mesopotamian basin regions; and the Paratethys region. The final paper in this section describes Miocene reefs bordering the nearby Atlantic Ocean and provides a comparison for the Miocene reefs in the Mediterranean regions.

The Detailed Studies section contains fifteen papers. These papers describe specific details of exceptionally exposed carbonate complexes from the various regions of the Mediterranean. They are arranged by general geographical setting starting with carbonate complexes in the western Mediterranean region, then with examples from the central Mediterranean region and ending with those in the eastern Mediterranean, Paratethys and Red Sea regions. Many of the papers in this section describe Upper Miocene carbonate complexes. This is partly due to the spectacular exposures of the Upper Miocene part of the section and also partly due to the attention that has been given to this part of the section to gain a better understanding of the nature of the "Messinian Salinity Crisis" evaporite event in the Mediterranean basin. In addition to highlighting Miocene reefs of the Mediterranean regions as important models for carbonate stratigraphy, we hope that this volume will serve to stimulate interest and additional research on the Lower and Middle Miocene carbonate complexes of the Mediterranean regions.

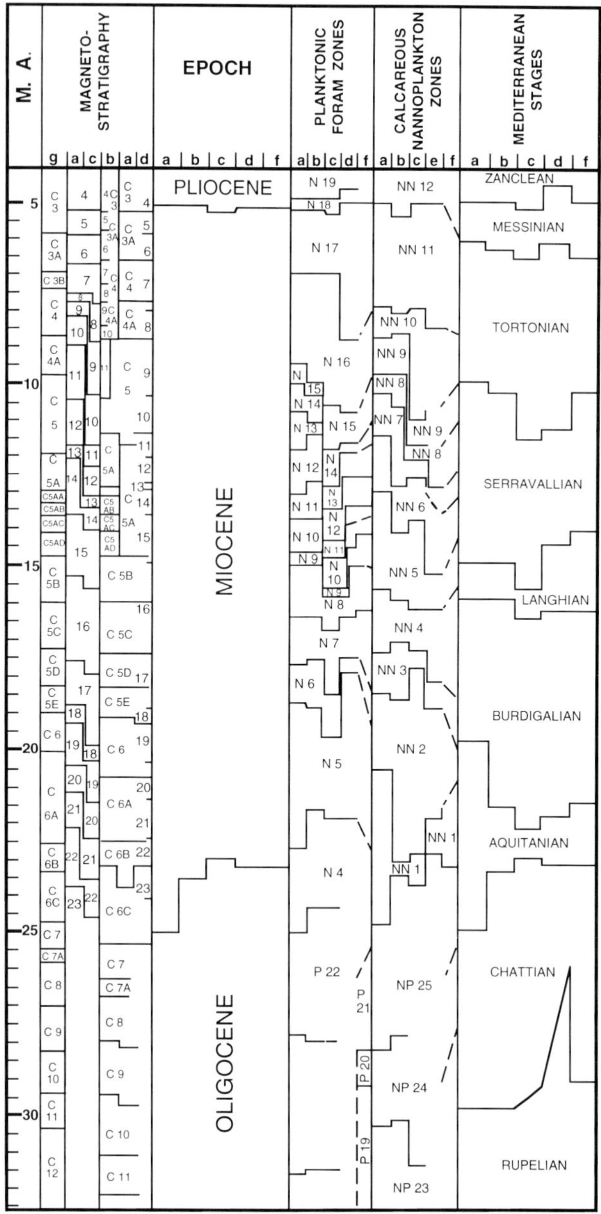


FIG. 2.—Simplified Correlation chart modified from Berger et al. (1990). a- Haq et al. (1987), b- Berggren et al. (1985), c- Steininger et al. (1985), d- Steininger et al. (1990), e- Vass et al. (1987), f- Harland et al. (1990), g- Cande and Kent (1995).

REFERENCES

BERGER, J. P., ENGESSER, B., BARVIN, V., BOLLIGERT, T., KELLER, B., AND WEIDMANN, M., 1990, Correlative chart of the European Oligocene and Miocene: Fribourg, Institut de Géologie, Unversité de Fribourg, 1 table.

BERGGREN, W. A., KENT, D. V., FLYNN, J. J., AND VAN COUVERING, J. A., 1985, Cenozoic geochronology: Geological Society of America Bulletin, v. 96, p. 1407-1418.

CANDE, S. C., AND KENT, D. V., 1995, Revised calibration of the geomagnetic polarity timescale for the Late Cretaceous and Cenozoic: Journal Geophysical Research, v. 100, p. 6093-6096.

HARLAND, W. B., ARMSTRONG, R. L., COX, A. V., GRAIG, L. E., SMITH, A. G., AND SMITH, D. G., 1990, A Geologic Time Scale 1989: Cambridge, Cambridge University Press, 263 p.

HAQ, B. U., HARDENBOL, J. AND VAIL, P. R., 1987, Chronology of fluctuating sea levels since the Triassic: Science, v. 235, p. 1156-1166.

STEININGER, F. F., SENES, J., KLEEMANN, K., AND RÖGL, F., 1985, Neogene of the Mediterranean Tethys and the Paratethys, stratigraphic correlation tables and sediment distribution maps: Institute of Paleontology of Vienna, v. 1, p. 472, and v. 2, p. 504.

STEININGER, F. F., BERNOR, R. L., AND FAHLBUSCH, V., 1990, European Neogene marine/continental chronologic correlations, *in* Lindsay, E. H., Fahlbusch, V., and Mein, P., eds., European Neogene Mammal Chronology: New York, Plenum Press, p. 155-285.

VASS., D., REPCOK, I., BALOGH, K., AND HALMAI, J., 1987, Revised radiometric time-scale for the Central Paratethyan Neogene: Annalses Instituti Geologici Publici Hungaricici, v. LXX, p. 423-434.

ACKNOWLEDGMENTS

We appreciate the efforts of Dana S. Ulmer-Scholle, SEPM Technical Editor, in expediting the publication of this volume. We are grateful to the people listed below for their time and expertise in reviewing the papers included in this volume. Finally, we thank the authors for their efforts, cooperation, and patience in the preparation and completion of the manuscripts.

REVIEWERS

R. H. Benson, Smithsonian Institution
D. Bosence, Royal Holloway, University of London
J. C. Braga, Universidad de Granada
B. Buchbinder, Geological Survey of Israel
N. Budd, University of Iowa
B. Cahuzac, Université Bordeaux I
F. Calvet, Universitat de Barcelona
G. Carannante, Università di Napoli
M. Colgan, Mobil Oil Company
M. Coniglio, University of Waterloo
K. J. Cunningham, University of Miami
C. Dabrio, Universidad de Madrid
R. H. Goldstein, University of Kansas
R. Handford, Consultant, Denver
N. P. James, Queens University
J. Jimenez, Universitat de Barcelona
C. Jordan, Consultant, Missouri
R. Loucks, Arco Oil Company
C. Mankiewicz, Beloit College
J. Martin, Universidad de Granada
W. Meyers, SUNY-StonyBrook
F. Orszag-Sperber, Université de Paris-Orsay
E. Oswald, Exxon Production Research Company
M. Pedley, The University of Hull
A. Pisera, Instytut Paleobiologii, Warszawa
L. Pomar, Universitat de les Illes Balears
E. G. Purdy, PetroQuest, London
R. Riding, University of Wales College of Cardiff
J.-P. Saint Martin, Université de Marseille

C. Santisteban, Universidad de Valencia
J. F. Sarg, Mobil Oil Company
W. Schlager, Vrije Universitat, Amsterdam
A. Simo, University of Wisconsin-Madison
L. Simone, Università di Napoli
F. Steininger, University of Vienna
S. Q. Sun, Carbonates International, London
C. Taberner, Institut Jaume Almera, Barcelona
B. Ward, Texas Bureau of Economic Geology
I. Zamareño, Institut Jaume Almera, Barcelona
M. Ziegler, Consultant, Binningen, Switzerland

# Part I:
# REGIONAL REVIEWS

# AN OVERVIEW OF MIOCENE REEFS FROM MEDITERRANEAN AREAS: GENERAL TRENDS AND FACIES MODELS

MATEU ESTEBAN
*Carbonates International Ltd, Vilanova 70, E-07190 Esporles, Mallorca, Spain*

ABSTRACT: Miocene carbonates in the Mediterranean are dominated by organic buildups of rhodalgal and coral-reef facies with local stromatolitic mounds, ahermatypic coral mounds and oyster banks and occur in a wide variety of tectonic settings and substrates. Regional chronostratigraphic correlation is in a state of flux, but it appears that coral reef development was extensive during the climatic optimum of the Chattian-Aquitanian, Langhian and Late Tortonian-Messinian times corresponding to global 2nd-order highstands or supercycles of relative sea level. The coral reef provinces of the Mediterranean reflect the transition between Early Miocene open-oceanic, humid-tropical conditions and Late Miocene land-locked, semi-arid and marginally subtropical environments. This is interpreted as a reflection of the global cooling trend and the northward displacement of the European plate, but also the increasing involvement of the rising Alpine foldbelts in controlling the climatic trends. The evolution of the connecting seaways with the Atlantic and Indo-Pacific domains are critical in the development of the Mediterranean Miocene carbonates.

Most facies models in the region are based on superb outcrops of Upper Miocene carbonate complexes, which are largely applicable to lesser known Lower and Middle Miocene carbonates. Narrow platforms with fringing reefs are predominant; lagoonal facies are poorly developed and commonly with variable amounts of terrigenous mixing. Extensive carbonate platforms with barrier reefs and lagoons occur in Oligocene-Lower Miocene carbonates but are very scarce or ephemeral in Upper Miocene platforms. The best outcrops show excellent preservation of depositional morphologies (platform slopes, reef buttresses and spur-and-grooves, reefal patches, skeletal sand bodies and lobes) and allow detailed paleogeographic reconstructions. Depositional sequences of different orders of magnitude display a basic stacking pattern consisting of vertical aggradation, progradation and offlaping (downstepping); faithfully reflecting inferred relative sea-level oscillations. Coral diversity decreased from Early to Late Miocene times; part of the Messinian coral reefs are essentially monogeneric (*Porites*, one of the main reef builders in all Miocene times). There are also Messinian reefs with 3-5 coral species; variations in diversity reflect local conditions rather than age or basin-wide events. The largest Upper Miocene reef complexes tend to be monogeneric and show good vertical zonation in colonial morphologies. However, these vertical zonations cannot be generalized for an entire basin or even a single reef complex.

Upper Miocene coral reefs developed before, during and after the repeated deposition of basinal evaporite units and marine marls, resulting in complex wedge-on-wedge geometries of difficult correlation. Some of the Messinian coral reefs in the western Mediterranean exhibit peculiar features: exuberant monogeneric coral branches coated by cyanobacterial crusts, locally associated with giant stromatolitic domes, algal blooms and diatomitic marls. The peculiar look of some Messinian reefs is interpreted to be a result of the influx of cold, nutrient-rich Atlantic waters and their interaction with dense, warmer Mediterranean waters with a tendency to eutrophic, stressed marine conditions. These features are considered part of the scenario referred to as the Messinian crises, leading to major salinity variations and evaporite deposition in the basin.

Miocene carbonates have a marked cyclicity of different orders of magnitude, particularly well recorded on Upper Miocene platforms. Rhodalgal carbonate ramps are common in the transgressive sections, and the coral reefs predominate during highstands. However, there are times (Burdigalian, Serravallian) or regions where depositional sequences are completely dominated by rhodalgal facies, with high-frequency cycles of rhodolith sizes, variations in large benthic foraminifers, bryozoans or encrusting forams. This predominance of rhodoalgal facies is considered a result of upwellings or the influx of cooler, nutrient-rich waters preventing the development of coral reefs, and could correspond to estuarine type of circulation in the basinal areas. The presence of coral reefs is considered as evidence of protected areas from cool, nutrient-rich waters as well as effective discharge of surface waters into partly restricted basins with predominant lagoonal circulation. Cyclicity in Miocene carbonates also is expressed in stromatolitic and oolitic units and diagenetic patterns (cementation).

Overall, the rhodalgal carbonates are, volumetrically, the most important facies in the Miocene Mediterranean area but have been less studied than the spectacular coral reefs. However, the main problem detected in this review is the need for accurate timing and correlation of events across the Mediterranean regions to understand the observed patterns in Miocene deposition.

## INTRODUCTION

The presence of Miocene reef-building corals in many Mediterranean areas was well recorded by paleontologists at the end of the last century. Chevalier (1961) synthesized and updated this classic knowledge for the western and central Mediterranean. Until the mid 1970's, these Miocene reefs received little attention, mostly as a brief mention in regional paleontologic and biostratigraphic monographs. Petrologic and sedimentologic studies were initiated in the 1970's (Permanyer and Esteban, 1973; Montenat, 1975; Dabrio, 1974, 1975; Buchbinder, 1975a, b, 1979; Pedley, 1976, 1979; Esteban and Permanyer, 1977; Alvarez et al., 1977; Pagnier, 1977; Armstrong et al., 1977; Esteban et al., 1977a, b; Catalano, 1979; Esteban, 1979, 1980a; Rouchy, 1979), mostly for the Upper Miocene and for a few Middle Miocene reefs. This early phase of work was characterized by general facies descriptions with emphasis on depositional geometries and the significance of these reefs in relation to the Messinian and other evaporitic events widely discussed in the literature. Esteban (1979) summarized the available observations and proposed interpretations, models and working hypotheses for the Upper Miocene reefs of the western and central Mediterranean. Since then, continued work by different schools has advanced considerably the understanding of the Upper

Models for Carbonate Stratigraphy from Miocene Reef Complexes of Mediterranean Regions,
SEPM Concepts in Sedimentology and Paleontology #5, Copyright © 1996,
SEPM (Society for Sedimentary Geology), ISBN 1-56576-033-6, p. 3-53.

Miocene reefs in the western and central Mediterranean (Bossio et al., 1978, 1981; Santisteban, 1981; Rouchy, 1982a, b; Rouchy et al., 1982a, 1986; Esteban, 1980a, b; Esteban and Giner, 1980; Esteban et al., 1982a, b; Buxton and Pedley, 1989; Grasso and Pedley, 1988, 1989; Saint Martin, 1990; Saint Martin and Rouchy, 1990; Pomar et al., 1983; Braga and Martin, 1988; Dabrio and Polo, 1988; Martin et al., 1989; Martin and Braga, 1990; Rouchy and Saint Martin, 1992) and the Middle Miocene reefs in the Red Sea (Rouchy, 1979, 1982a, b; Rouchy et al., 1982b, 1983; Purser et al., 1987; Monty et al., 1987; James et al., 1988; Montenat et al., 1988; Burchette, 1988). A third phase of study has just started, using sequence stratigraphic concepts, modern petrologic techniques, isotopic geochemistry and very detailed outcrop studies (Franseen, 1989; Goldstein et al., 1990; Pomar, 1991, 1993; Pomar and Ward, 1994; Oswald, 1992; Follows, 1992; Saint Martin et al., 1992; Buchbinder et al., 1993; Franseen and Mankiewicz, 1991; Franseen et al., 1993; Cornée et al., 1994; and papers in this volume).

This paper attempts to summarize and organize the large volume of information now available in the Mediterranean area with the aim of facilitating future work in the region and contributing to the development of new working hypotheses. For this purpose, this paper reviews the types, settings and paleogeographic trends of Miocene reefs, with discussion of facies models for Upper Miocene reefs of the western and central Mediterranean and their applicability to other Miocene reefs in the region. It is also an introduction to the regional syntheses of the Miocene reefs in Atlantic Europe, western, central and eastern Mediterranean and Paratethys and provides the framework to the selected case histories presented in this volume.

## TYPES OF MIOCENE REEFS

Four major types of Miocene reefs, classified by the predominant frame builder, can be recognized in the Mediterranean area:

1. Hermatypic coral reefs are commonly associated with red algae, molluscs, bryozoans, benthic forams, locally green algae, hydrozoans and ahermatypic corals. Fringing reefs, patch reefs, coral thickets and mounds predominate, but there are also some extensive barrier reefs and terraced erosion platforms. *Porites* and *Tarbellastraea* are the dominant reef builders through Miocene times and to a lesser extent, *Stylophora, Heliastraea, Siderastraea, Acropora* and *Platygyra.* Coral diversity can be relatively high (more than 15 coral species present), moderate (5-15), low (2-5 species) or very low (1 or 2 species dominant or exclusive as significant builders). The highest diversity occurs in the Lower Miocene reefs of the Aquitaine Basin (96 coral species in Chevalier, 1961) and Provence (45 hermatypic coral species in Chevalier, 1961), but a number of these species are now under revision (Chaix, unpubl. data). The size of reef buildups varies from 2-m-thick mounds to 70-m-thick reef cores in prograding complexes 2-20 km in width and up to more than 300 m in thickness; size of the reefs is independent of the diversity of corals.

2. Ahermatypic coral mounds and thickets, with or without the presence of hermatypic coral species, are not studied in detail; a good example of ahermatypic coral mound framework is exposed in the quarries of Zebegeny (Mid Miocene, Hungary), with dominant *Orbicella,* presence of four more ahermatypic species and few *Porites,* reaching a thickness of 30-40 m (pers. observ.). The Lower Miocene thickets of Torino-Casale (northern Italy) contain 50 to more than 100 ahermatypic species and 10-30 hermatypic corals (Chevalier, 1961) in terrigenous marls with red algae, bryozoans, milleporids, gastropods and *Heterostegina.* These mixed ahermatypic coral mounds appear to occur as part of a belt around the northern limits of the hermatypic coral reef distribution.

3. Red algal mounds, biostromes and pavements make up the rhodalgal units. The general term 'rhodalgal unit' (Carannante et al., 1988) refers to sedimentary accumulations with any type of red algae as a predominant constituent and without major development of coral reefs. Red algae can occur as rhodoliths, *in situ* frameworks (branching, crustose, massive) or bioclastic debris, with variable and diverse amounts of bryozoans, echinoderms, ahermatypic corals, vermetids, serpulids, encrusting forams, molluscs, brachiopods and large benthic forams. Glauconite, phosphatized grains and crusts, terrigenous, carbonate and siliciclastic grains, planktonic forams and fish debris are locally present to abundant. Hermatypic corals can be locally present but do not form large reef frameworks. Rhodalgal units also occur as the substratum for hermatypic coral reefs and may appear alternating cyclically with them. Rhodalgal units, mostly as well-developed ramps with mounds up to 50-150 m thick, are the most abundant and extensive Miocene organic buildup in the Mediterranean and adjacent areas;

4. Stromatolite mounds and blankets, with stromatolitic heads up to 5-8 m in diameter locally, form thick mounds (up to 20-30 m thick) in the Upper Miocene carbonates of the Mediterranean and the Middle Miocene carbonates of the Red Sea (Esteban and Giner, 1977a, b; Rouchy, 1979, 1982a, b). These stromatolites are commonly intercalated with oolitic or skeletal grainstones with cerithid gastropods; locally they are associated with *Porites* thickets and patch reefs. Stromatolites can develop as thick mounds on shelf margins or coating carbonate slopes; thinner developments occur on flat-lying shelf interiors and basin floors. Many or most of these stromatolites are considered to have formed in marine environments unfavorable to coral reefs and rhodalgal carbonates. Dense micrite coatings considered to be marine stromatolites occur as important contributions to some Upper Miocene coral reef frameworks (Riding et al., 1991a, b).

There are other types of locally conspicuous reefal buildups in Miocene deposits. For instance, in-situ monospecific oyster "mounds" are up to 4 m thick and 100 m long in the Lower Miocene sections of eastern Spain and also in the Upper Miocene sections of southeastern Spain (Santisteban, 1981). Extensive oyster blankets (up to several km$^2$) occur in the Tortonian units of Algeria and Morocco (Saint Martin, 1990). At a much smaller scale, there are also examples of organic buildups of bryozoans, serpulids, vermetids and balanids, individually or in association (Rouchy et al., 1986; Pisera, 1985 and this volume).

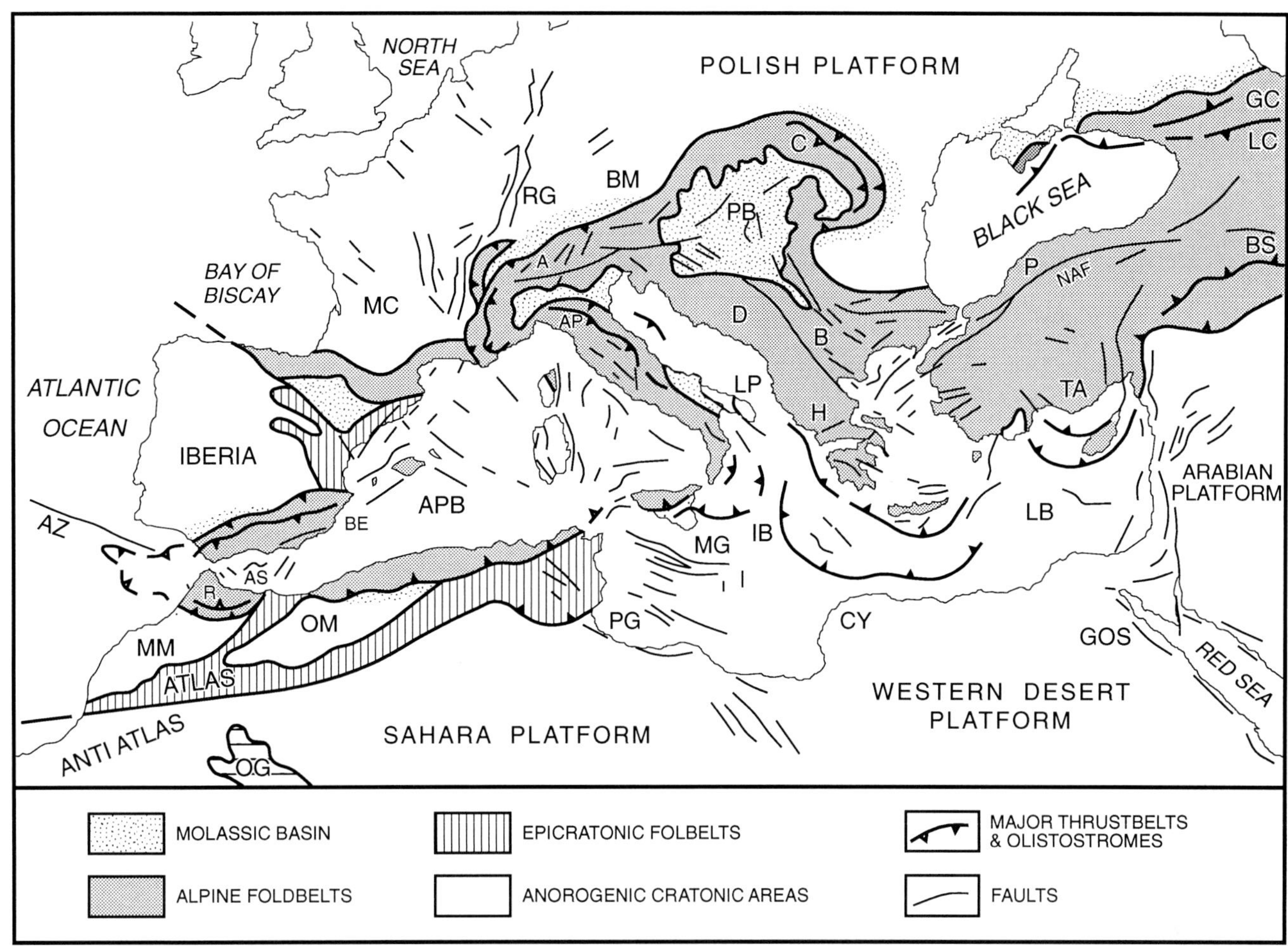


FIG. 1.—Major structural units of the Mediterranean regions (extensively modified from Ziegler, 1988). AZ: Azores Fault; MM: Moroccan Meseta; OM: Oran Meseta; R: Rif; B: Betics; OG: Ougarta trough; APB: Algerian-Provençal Basin; MC: Massif Central; RG: Rhine Graben; A: Alps; BM: Bohemian Massif; AP: Apennines; MG: Malta Graben; PG: Pelagian Shelf; IB: lonian Basin; LP: Apulia; D: Dinarides; PB: Pannonian Basin; C: Carpathians; H: Hellenides; B: Balcans; CY: Cyrenaica Plateau; LB: Levantine Basin; TA: Taurides; NAF: North Anatolian Fault; P: Pontides; LC: Lesser Caucasus; GC: Great Caucasus; BS: Bitlis Suture; GOS: Gulf of Suez.

Nevertheless, these are volumetrically less important than the four reef types mentioned above and in general are not considered in this review.

## GEOLOGIC SETTING OF MIOCENE REEFS

### *Structural Trends and Evolution*

The complexity of the structural trends and evolution of the Mediterranean regions (Fig. 1) result from the interaction of the Eurasian plate with the African and Arabian plates and involves a large number of intermediate microplates (e.g., Alboran, Ionian, Aegean, Levantine, Caspian, Adriatic, Turkish, Van, Iran). General reviews of Mediterranean geology are presented in Biju-Duval et al. (1977), Lemoine (1978), Rehault et al. (1984), Dewey et al. (1973, 1989), Dercourt et al. (1986), Ziegler (1988) and many others. An extensive review and correlation of the complex Miocene stratigraphy is presented in Steininger et al. (1985); Alvinerie et al. (1992) offer a synthesis of the Miocene connections with the northeastern Atlantic. Miocene carbonates (including coral reefs) of different ages occur across all these plates and can be grouped into provinces (Fig. 2) in different structural settings:

A. Mesozoic and Tertiary Alpine foldbelts, with Miocene carbonates developed pre-, syn- and post-thrusting (e.g., Mallorca, central and western Sicily, Provence, Languedoc, Alacant, Murcia, Adana, Chelif and Rif);

B. Epicratonic and intracratonic domains folded and thrusted during the Alpine orogeny (e.g., eastern Tunisia and Moldavian platform);

C. Ophiolite remnants of oceanic crust (i.e., Rosignano in Tuscany, Cyprus);

D. Variscan or older continental crust reworked in the Alpine orogeny (e.g., Calabria, southern France);

E. Areas affected by Mid Tertiary (Miocene) metamorphism (i.e., Melilla, Crete) and Upper Tertiary volcanic rocks (e.g., Almería);

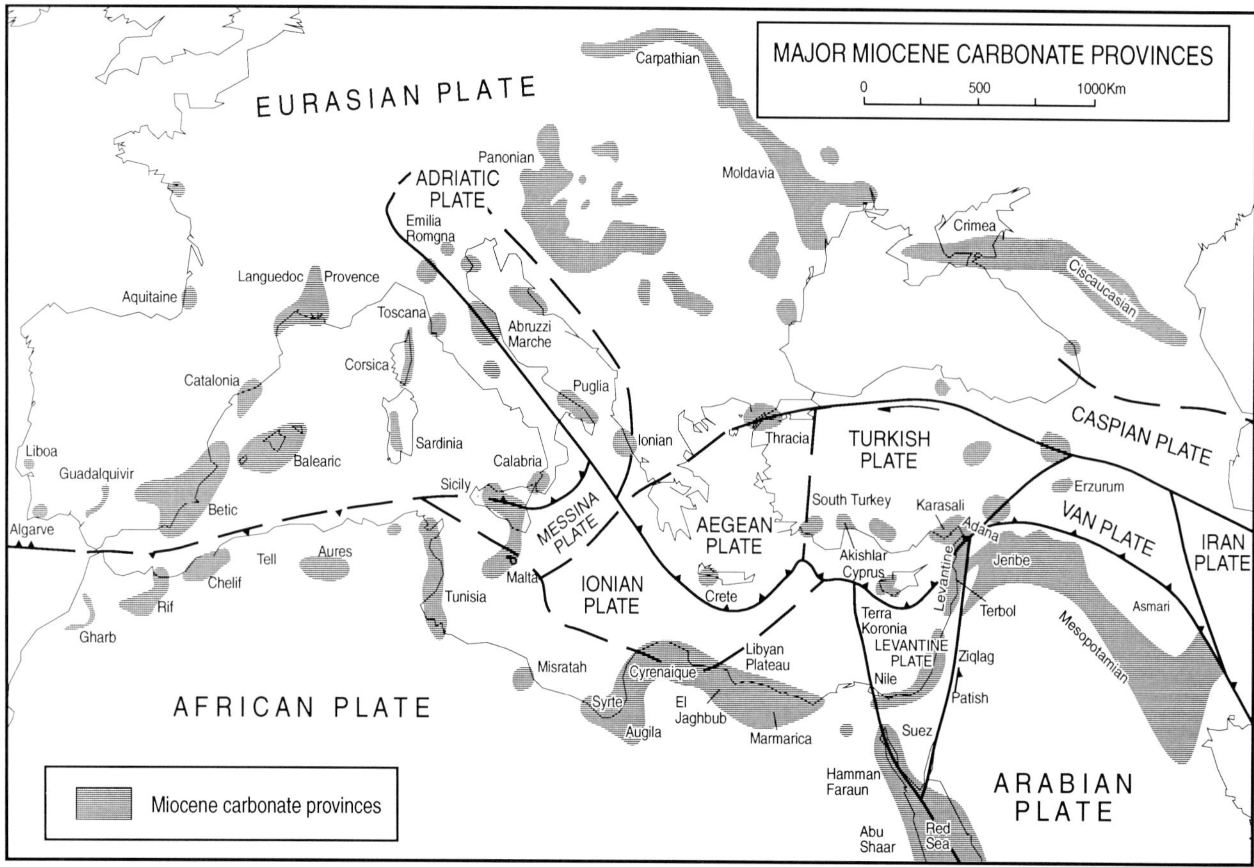


FIG. 2.—Main onshore Miocene carbonate provinces of the Mediterranean (extensively modified from Steininger et al., 1985) with present-day outline of plate boundaries (simplified from Dewey et al., 1973)

F. Tertiary rift grabens (e.g., Sardinia, Gulf of Valencia, Malta, Lampedusa, Gulf of Suez, Red Sea); and

G. Stable foreland (e.g., southeastern Sicily, Israel, Egypt's Western Desert, Cyrenaica).

The Mediterranean regions have suffered a complex history of deformation. A large number of Miocene reefs occur in areas intensely folded in Cretaceous and Paleogene times and can be considered as post-orogenic successor basins (southeastern Spain, Catalan Ranges, Leithakalk in the Pannonian Basin, southern Turkey, Rif and Chelif). However, most were affected to varying degrees by several phases of block faulting, which was particularly intense in Mid Miocene, Early Tortonian, Messinian and Mid Pliocene times. Gravitational tectonics are especially important in deforming Miocene reefs in central Sicily. Miocene folding affects reefs in southeastern Turkey, Cyprus, central and south Italy, south Sicily and Mallorca. Reactivation during the Pliocene times of localized Triassic diapirs intensely deformed some Upper Miocene reefs (e.g., Guercif Basin). In spite of the intensity of the Late Miocene-Pliocene block faulting and subsidence, a large number of Upper Miocene reefs can be considered as essentially post-orogenic, without evidence of major structural tilts. Depositional geometries and slopes of Upper Miocene platforms are well preserved over extensive areas of Upper Miocene reefs in Mallorca, Santa Pola, Almería, Algeria (Djebel Murdjadjo), Melilla, and Malta and in the Middle Miocene reefs of the Red Sea.

### *Reef Substrates and Associated Lithologies*

The morphology and lithology of Miocene reef substrates (and associated rhodalgal units) also show considerable variation. Substrates consist of older Miocene carbonates (Mallorca, Israel), Mesozoic and Tertiary carbonates (Murcia, Alacant, Almería, Saïss in Morocco), Neogene volcanic rocks (Cabo de Gata), granitic rocks (Calabria), ophiolites (Livorno) and metamorphics (Melilla, Kebdana, Almería). These substrates commonly show terraced erosion surfaces interpreted as marine planation surfaces (Esteban and Giner, 1980; Vallés-Roca, 1986; Bossio et al., 1981; Franseen, 1989; Calvet et al., this volume).

Miocene terrigenous sediments (siliciclastics, carbonate clastics, volcanoclastics) occur as the substratum for a large number of Miocene reefs (Santisteban, 1981; Hayward, 1982; Santisteban and Taberner, 1988; Hayward et al., this volume;

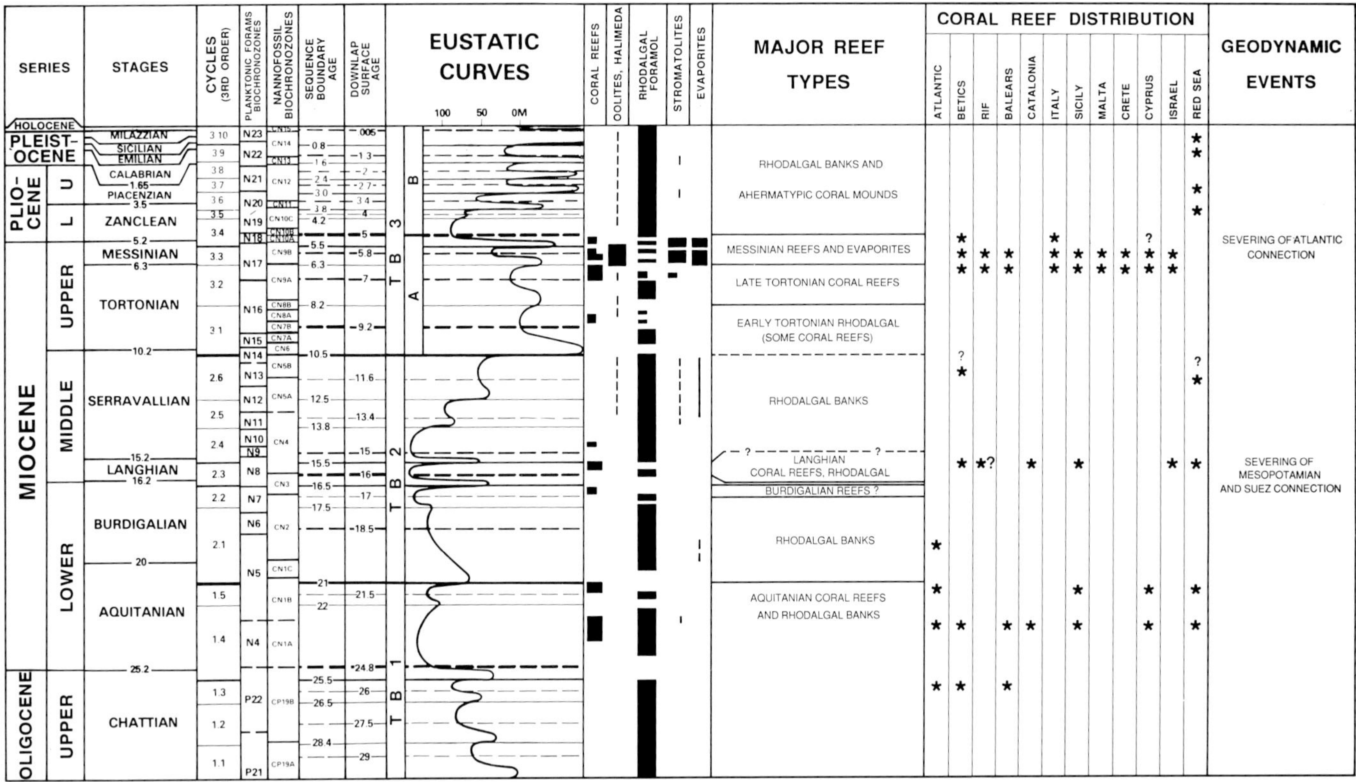


FIG. 3.—Stratigraphy of the major reef types and associated lithologies of the Miocene Mediterranean and their correlation with the stratigraphic terminology, sequence boundaries and eustatic curve of Haq et al. (1987).

Friebe, 1990). These terrigenous clastic deposits are commonly interpreted as deltaic and delta-fan deposits, and the reefs were developed on abandoned channel margins, stream-mouth bars and sand shoals. Some Miocene reefs developed in areas of repeated terrigenous sedimentation, resulting in alternating layers of coral reefs intercalated with wedges of terrigenous rocks. Moreover, some Upper Miocene *Porites* coral reefs in growth position have a fine-grained siliciclastic matrix (mud and silt), suggesting active coral growth at times of siliciclastic deposition (Livorno- Bossio et al., 1981; central Sicily- Grasso and Pedley, 1989; Almería, Martin et al., 1989; Dabrio 1974, 1975; Algeria and Morocco, Saint Martin, 1990). *Porites* is well known for its ability to withstand high levels of fine-grained clastic sedimentation (Hubbard and Pocock, 1972).

Miocene coral reefs were formed just before the deposition of Miocene evaporite units during Early and Mid Miocene times in the Mesopotamian basin, Mid Miocene times in the Gulf of Suez, the Red Sea and Pannonian basins and Messinian time in the Mediterranean (see Rouchy, 1982a, b, for a review). However, there are also coral reefs post-dating evaporitic units in the Messinian strata of Fortuna Basin (Santisteban, 1981; Müller, 1986) and Tuscany (Bossio, et al., this volume), and in the Lower-Middle Miocene deposits of the Gulf of Suez (Purser et al., 1987; Montenat et al., 1988). The marked similarity of all these occurrences suggests a general facies model: the classic wedge-on-wedge relationship (cyclic and reciprocal sedimentation of Wilson, 1975), with a vertical evolution from normal-marine to hypersaline conditions. Very commonly, evaporites occur offlapping and onlapping a truncated and karstified coral reef. A review of the Messinian coral reef-evaporite relationships will be presented later in this paper

Some Miocene coral reefs may have developed in marginal areas of the Moroccan basins under the intermittent influx of continental fresh waters (suggested by Saint Martin, 1990, p. 204, on the basis of ostracod and benthic-foram assemblages in the area of Sefrou). These settings with marked brackish-marine character also could be common in the Paratethys, but we are unaware of adequate documentation of such occurrences. *Porites,* one of the main reef builders in the Miocene times, is resistant to wide variations (19-48%) of marine salinities (Downing, 1985; Braithwaite, 1971; Kinsman, 1964).

### *Geologic Controls on the Setting of Miocene Reefs*

In view of the wide variety of structural and lithologic settings of the Mediterranean Miocene reefs, the basic similarities of facies patterns and reef geometries are quite remarkable. This suggests the dominance of regional paleoceanographic controls overriding local structural and lithological trends. Significantly, coral-reef events in the Mediterranean (Fig. 3) appear synchronous; most of them are concentrated in the Aquitanian, Langhian (and latest Burdigalian?) and Late Tortonian-Messinian intervals. This pattern reflects a relationship with 2nd-order eustatic sea-level highstands of Haq et al. (1987). One exception is the absence of coral reefs during the Pliocene highstand, probably due to the low temperatures of the marine waters in the Mediterranean (but coral reefs related to the Indian Ocean are present in the Pliocene of the Red Sea). In contrast, rhodalgal carbonates occur throughout the Neogene and Pleistocene sections.

There are still major problems in the biostratigraphy and correlation of Mediterranean Miocene reefs. The stratigraphic terminology of Haq et al. (1987) is followed in this paper, while noting the important inconsistencies with other stratigraphic charts (Steininger et al., 1990; Berggren et al., 1985a, b; Berger et al., 1990; Harland et al., 1990; see also the Introduction to this volume). To avoid confusion in stratigraphic correlation, it is essential to refer to the foraminifer zones and to consider the distinctive usage of the stratigraphic stages by different authors. Correlation problems with the Paratethys sections are particularly serious; local factors such as water chemistry or amount of connection with open oceans could have played a more important role than in the Mediterranean (Pisera, this volume). It is not the purpose of this paper to discuss Miocene biostratigraphy but to confirm the importance of coral reef development in the highstands of the three 2nd-order cycles of the Miocene times.

## PALEOGEOGRAPHIC TRENDS AND REEF EVENTS

The Miocene history of the Mediterranean Tethys reflects the relative movements of the Eurasian and African-Arabian plates, with the progressive narrowing and severing of extensive Paleogene deepwater seaways that openly communicated with the Atlantic and Indian oceans. Steininger et al. (1985) and Rögl and Steininger (1983, 1984) have provided an extensive paleogeographic review which has been adopted in the following paragraphs (Figs. 4-8). Also, Alvinerie et al. (1992) synthesized the Neogene history of the Atlantic side of southeastern Europe and the connections with the Mediterranean. During most of Paleogene time, Europe was an epicontinental sea with vast emerging archipelagos corresponding to the rising Alpine system. The intensification of these orogenic movements during the Late Eocene-Oligocene times gave rise to extensive emergent areas in central Europe, with scattered restricted and lacustrine basins, and produced the major paleogeographic trends that dominated Miocene time (Fig.1). Rögl and Steininger (1984) distinguish: (1) the Atlantic-Boreal bioprovince of the coasts of western and northern Europe, that during highstands of sea level extended across central and eastern Europe, (2) the Mediterranean bioprovince, a relict of the Mesozoic Tethys, covering southern Europe, northern Africa and the Middle East, and (3) the Paratethys bioprovince, a northern branch of the Tethys created by the rise of the Alps, Balkans and Pontids and with major connections with the Indo-Pacific realm (also with the Mediterranean and Atlantic-Boreal bioprovinces). A large Anatolian-Balkan continent separated the Paratethys from the Mediterranean during Miocene time.

Miocene sedimentary history comprises three supercycles in the sense of Haq et al. (1987; Fig. 3): TB1-Chattian and most of Aquitanian (21-30 my), TB2-Uppermost Aquitanian, Burdigalian, Langhian and Serravallian (21-10.5 my) and TB3-from Tortonian to Recent. However, in the Mediterranean it seems more accurate to distinguish between a Late Miocene cycle (A in Fig. 3) and a Plio-Quaternary cycle (B in Fig. 3) separated by the evaporitic drawdown of the Messinian event. These are here informally termed as early, middle and late supercycles. The following sections summarize the key paleogeographic trends and main features of the reefs of the Miocene supercycles.

### *Early Supercycle (TB 1)*

A global climatic minimum and sea-level drop occurred at the Rupelian-Chattian boundary (30 Ma) related to plate movements in the Pacific realm (Hallam, 1981). The first land bridge between the European archipelago and Asia was established by the rise of the Ural mountains. Tropical to subtropical climates returned to the Mediterranean by the Late Oligocene with the invasion (Earliest Aquitanian?) of large benthic forams *(Miogypsina, Miogypsinoides, Lepidocyclina, Operculina, Heterostegina, Austrotrillina* and *Cycloclypeus;* Adams, 1976, 1981; Adams et al., 1983; Drooger, 1979; Steininger et al., 1976) that reached as far as the North Sea and southwestern France (Cahuzac and Poignant, 1988, 1990). Wide and deep seaways existed between the Mediterranean, Atlantic, Paratethys and the Indo-Pacific realms. Extensive rhodalgal carbonates developed on the flanks of these seaways, mostly in Mesopotamia (Asmari and Euphrates limestones of Iran, Iraq and Syria) and the Mediterranean (Adana Bay in Turkey, Lower Coralline Limestone of Malta). Reefs with the highest coral diversity of the entire Miocene time occurred in numerous localities (Cyprus, Libya, Puglia, Malta, Sicily, southeastern France and eastern Spain; Fig. 4). There are mentions of barrier reef development (Chevalier, 1961), but, other than extensive paleontological lists, little detailed documentation is available from the Mediterranean Aquitanian reefs. Evaporitic conditions appeared (Aquitanian?, Adams et al., 1983) in the shallower parts of the Mesopotamian seaway and on the eastern Carpathian flank (Fig. 4), but most of Mediterranean regions had open oceanic conditions in many respects similar to the present southeastern Asia reefs.

It is important to realize that there are major differences in the precise stratigraphy and absolute ages of the Lower Cycle. Haq et al. (1987) consider the N4 as well as a good part of the N5 as Aquitanian age . In contrast, Steininger et al. (1990) limit the Aquitanian times to the N4 only and consider the lower part of the N4 as latest Chattian. Absolute ages may differ as much as 2 my.

### *Middle Supercycle (TB 2)*

The collision of the Arabian and Turkish plates interrupted the marine connection with the Indo-Pacific area around the

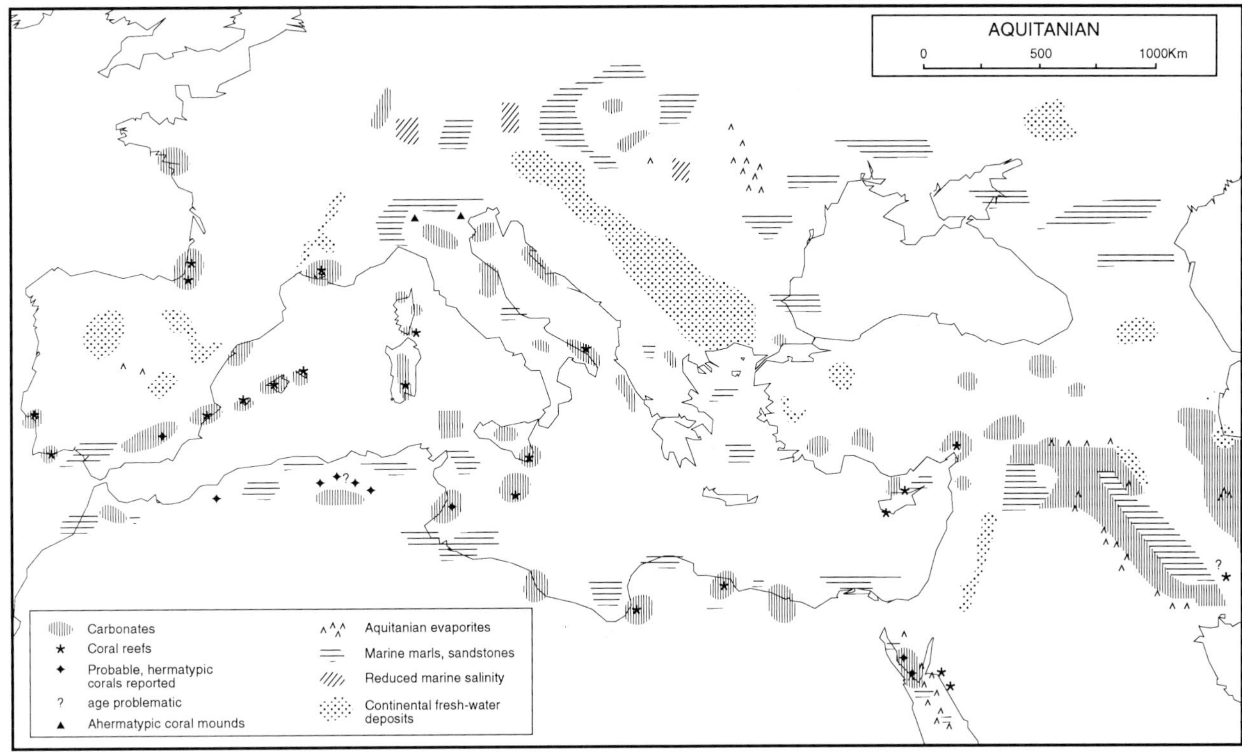


FIG. 4.—Distribution of Aquitanian lithofacies and reef types in the Mediterranean. Extensively modified from data in Steininger et al. (1985).

Aquitanian-Burdigalian boundary (20-21 Ma). The Lower Fars evaporite was deposited extensively over the Mesopotamian straits, and intermittent land bridges existed between Arabia and Anatolia (Fig. 5). The eastern Paratethys connection with the Indo-Pacific was also severed with a consequent change to brackish marine conditions. The central Paratethys remained fully marine because of its connections with the western Mediterranean along the short-lived seaway of the western Alps foredeep and the Rhone valley and from the Atlantic-Boreal province along the Rhine graben. This major geodynamic event (Aquitanian-Burdigalian boundary) coincided with a global cooling, intensification of Antarctic glaciation and a sea-level drop (Barron and Keller, 1982).

This climatic deterioration (cooling) continued during the Early Burdigalian times. Carbonates were almost exclusively of rhodalgal and foramol facies (sensu Lees and Buller, 1972; Carannante et al., 1988; no coral reefs), with reduced areal extent compared to the Aquitanian or Langhian carbonates (Figs. 4, 6). Nebelsick (1989) provided an example of the temperate water carbonate facies that developed in Mediterranean Europe during the Burdigalian times. Basinal areas contain abundant organic-rich marls and diatomitic deposits.

The climate warmed by Late Burdigalian and mostly during Langhian time. Extensive carbonate platforms include numerous coral reefs (Fig. 6) with relatively moderate coral diversity in the Red Sea (Purser et al., this volume), Israel (Buchbinder et al., 1993; this volume), Egypt's Western Desert, eastern Spain (Permanyer and Esteban, 1973), southeastern France (Chevalier, 1961), Corsica (Orszag-Sperber and Pilot, 1976) and central Paratethys (Pisera, this volume). It appears that the Paratethyan basin recovered from the Early Miocene "temperate" conditions, while on the Atlantic side of Europe the Langhian coral reef zone was clearly displaced to the south: Madeira (Lietz and Schwarzbach, 1970, Wijsman-Best and Boekschoten, 1982), and Morocco (pers. observ.).

The best-known Langhian coral reefs occur in the Red Sea (Coniglio et al., this volume; Purser et al., this volume), Paratethys (Pisera, this volume) and Catalonia (Permanyer and Esteban, 1973; Esteban and Permanyer, 1977; Alvarez et al., 1977; Permanyer, 1990; Cabrera et al., 1991). These Langhian coral reefs are dominated by *Stylophora, Porites* and *Tarbellastraea*, and locally reach considerable size (50- to 70-m-thick coral framework, progradation of 1-2 km). In most of these regions, there are two distinctive coral-reef units (Langhian and uppermost Burdigalian?) separated by an unconformity. Portugal, Atlantic France, Vienna basin, northern Hungary and Bulgaria probably represent the boreal limit of the Langhian coral reefs, with mixed ahermatypic hermatypic coral mounds of exceptionally high diversity (Chevalier, 1961).

Climatic conditions deteriorated (cooled) during the Serravallian, with impoverished sub-tropical to temperate faunas in most of the Mediterranean. Some authors, as Rögl et al.

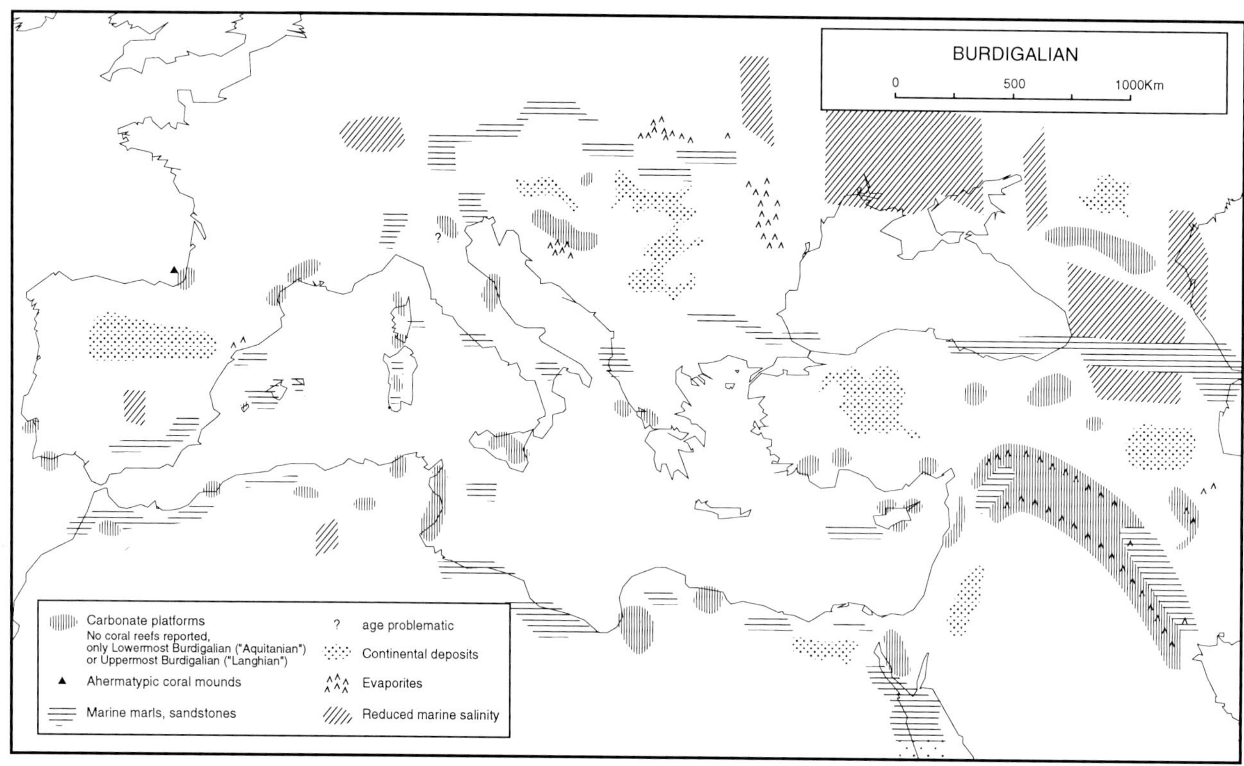


FIG. 5.—Distribution of Burdigalian lithofacies and reef types in the Mediterranean. Extensively modified from data in Steininger et al. (1985).

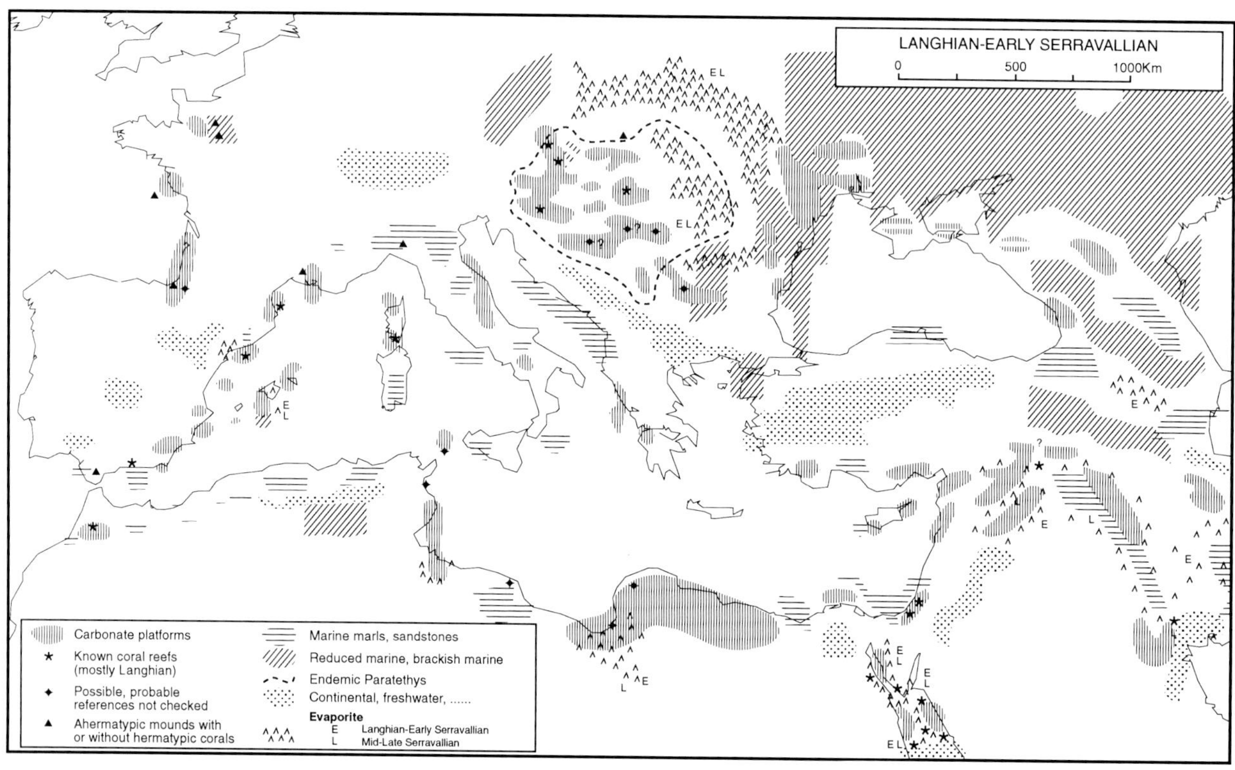


FIG. 6.—Distribution of Langhian lithofacies and reef types in the Mediterranean. Extensively modified from data in Steininger et al. (1985).

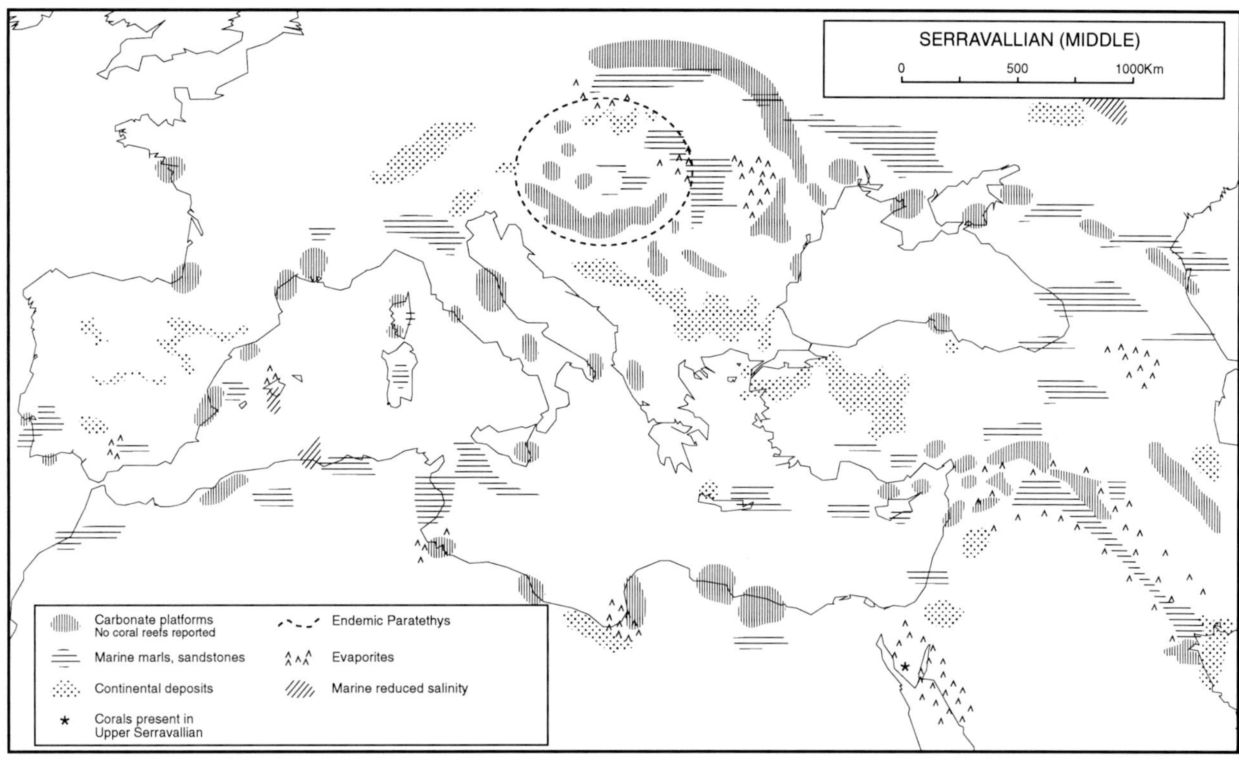


FIG. 7.—Distribution of Serravallian lithofacies and reef types in the Mediterranean. Extensively modified from data in Steininger et al. (1985).

(1978) and Demarcq (1984, 1985), refer to this time as the "Serravallian crisis." There is a reported occurrence of a rhodalgal Serravallian carbonate with abundant coral fauna in the Gulf of Suez (Scott and McGovean, 1985). If this can be corroborated as a true coral reef, it would be one of the rare Serravallian reef occurrences.

Important salinity crises represented by up to several hundred (locally few thousand) meters of evaporites occurred during the Middle Supercycle in central Paratethys, the Red Sea, Gulf of Suez and Mesopotamia but also reached as far away as the marginal intramontane basins in Barcelona, Mallorca, eastern Tunisia and Sirte basin (Fig. 7). Facies patterns and relationships are remarkably similar to those of the Messinian evaporites (Rouchy, 1982a, b).

The stratigraphy of the Middle Cycle is also controversial. Zone N8 was traditionally considered as part of the late Burdigalian age, but now most of the zone is placed in the Langhian age by Haq et al. (1987) and Steininger et al. (1990). The Red Sea reefs are considered as Late Burdigalian age by Montenat et al. (1988) and Purser et al. (1990), but are dated as Langhian age in Evans (1988), James et al. (1988), Burchette (1988) and Smale et al. (1988). Zone N9 is considered by Haq et al (1987) as the base of the Serravallian age, but Steininger et al. (1990) include it at the top of the Langhian age. Moreover, the base of the Tortonian age is coincident with the base of N15 (CN 7A) in Haq et al. (1987), but it is considered to be younger in Steininger et al. (1990; base of CN 7B).

### *Upper Miocene Supercycle (TB 3A)*

During the latest Serravallian times, a major global sea-level fall (10.5 Ma) coincided with the uplift of the Arabian and Turkish plates and marked the beginning of the main post-orogenic phase in many fold belts (Betics, Rif, Zagros). This major event produced the definitive severing of the Mesopotamian connection to the Indo-Pacific realm, recorded by extensive red beds and evaporites of the Upper Fars units (Fig. 8). Similar types of deposits occurred in the Gulf of Suez-Red Sea, and the area became independent from the Mediterranean. The Mediterranean basin was semi-isolated from the world oceans; the only connection was in the west to the Atlantic, across the Betic and Rif Straits (Esteban et al., this volume). These straits consisted of complex archipelagos with shallow sills and allowed normal-marine communication during the global highstands of sea level. Well-developed coral reefs flourished in the Mediterranean part of the Betic and Rif Straits (Granada, Saïss, Almería, Guercif, Murcia, Alacant, Melilla, Oran, Chelif) and extended to the north to the Balearic Islands, Sardinia and Tuscany. The northern limit of the coral-reef zone suffered a marked shift to the south (Figs. 8, 9) along Italy and re-appeared in Calabria, Sicily, Malta and continued to Crete, southern Turkey, Cyprus and Israel. Small coral reefs occurred during the Early Tortonian

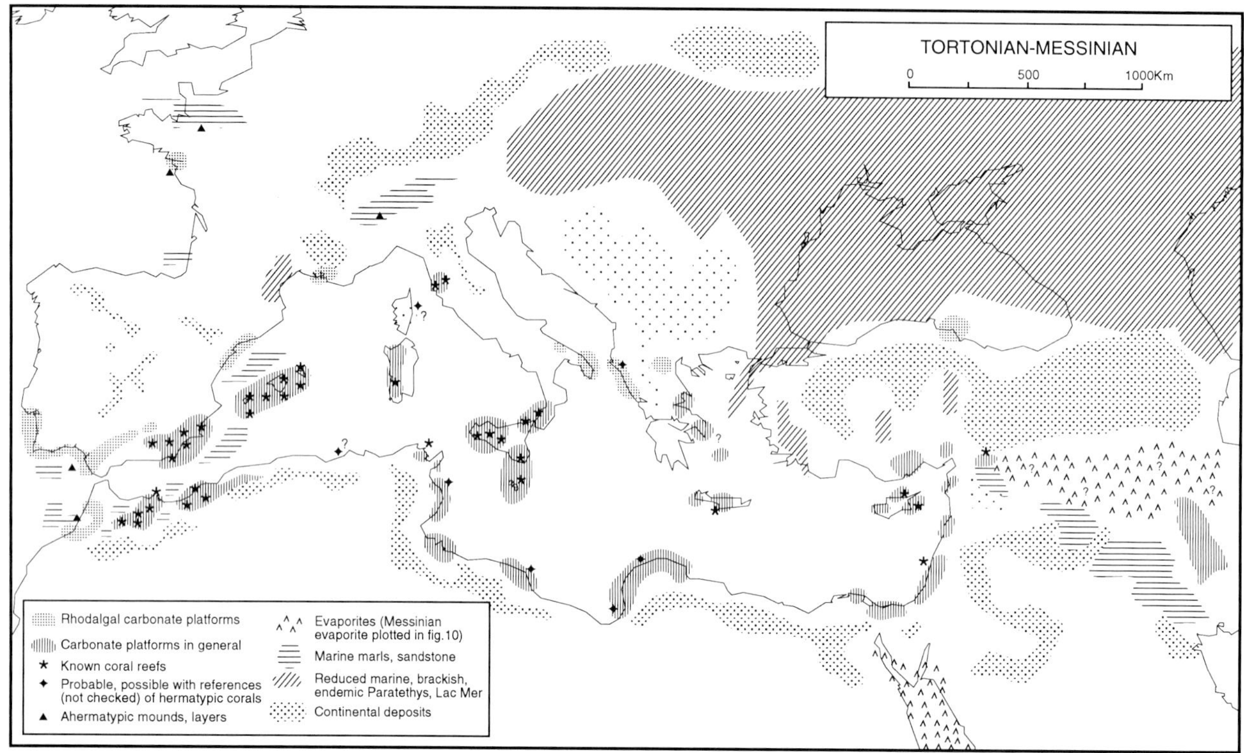


FIG. 8.—Distribution of Tortonian-Messinian lithofacies and reef types in the Mediterranean. Extensively modified from data in Steininger et al. (1985).

times, but these were better developed during the Late Tortonian and the Messinian times.

During Messinian time, the communication with the Atlantic was repeatedly reduced by the combined effect of tectonic uplifts, lowering sea levels, sediment accumulation and gravitational and diapiric adjustments. This resulted in the deposition of several thick evaporitic units intercalated between marine sediments (as well as with restricted marine, brackish and hypersaline sediments) over extensive areas of the Mediterranean (Fig. 10). A review of the widely debated Messinian events (Hsü et al., 1973, 1977; Adams et al., 1977; Ryan et al., 1973; Cita, 1982; Rouchy, 1982a, b; Cita and McKenzie, 1986; Benson et al., 1991) is beyond the purpose of this paper, but two key points are retained here: (1) the Mediterranean base level of erosion was repeatedly lowered, as documented by extensive and well-developed subaerial erosion surfaces and incised valleys and (2) different types of evaporites as well as brackish and lacustrine sediments occurred over vast Mediterranean regions. During the Messinian period, there were times of major evaporitic drawdown of the Mediterranean with numerous land bridges (across Betic and Rif Straits, Sicily Straits, Suez Straits, Balearics to central Mediterranean, etc.) separated by evaporitic seas; it is unlikely that there was complete desiccation of the entire Mediterranean basin.

The Paratethys basins became isolated from the Mediterranean and the Indo-Pacific and are characterized by facies of marine salinities evolving into brackish and freshwater associations. These facies contain markedly endemic fossils (Lac Mer biofacies) and during the Late Messinian times expanded over most of the Mediterranean basins up to southeastern Spain during times of minor influx of Atlantic waters.

The detailed stratigraphy of the Tortonian-Messinian cycle also shows minor differences of interpretation amongst authors. These are particularly important in the precise correlation of the well studied coral reef events and will be discussed below.

### *Pliocene-Quaternary Supercycle (TB 3B)*

The Early Pliocene global highstand resulted in an abrupt marine transgression from the Atlantic, coinciding with the formation of the Gibraltar Strait by block faulting and regional uplift into the terrestrial domain of the Betic and Rif corridors. The Pliocene transgression, with extensive deposits of hemipelagic and pelagic facies (*Globigerina* oozes), reached the Ponto-Caspian region, in marked contrast with the predominantly shallow-water (hypersaline to fresh) Messinian deposits. No hermatypic coral reefs occur in the Mediterranean Plio-Quaternary Supercycle, but rare and small zooxanthellate corals are recorded in some Pliocene sediments (Chevalier, 1961) and in the Holocene (*Madracis, Oculina, Cladocora,* as reported by

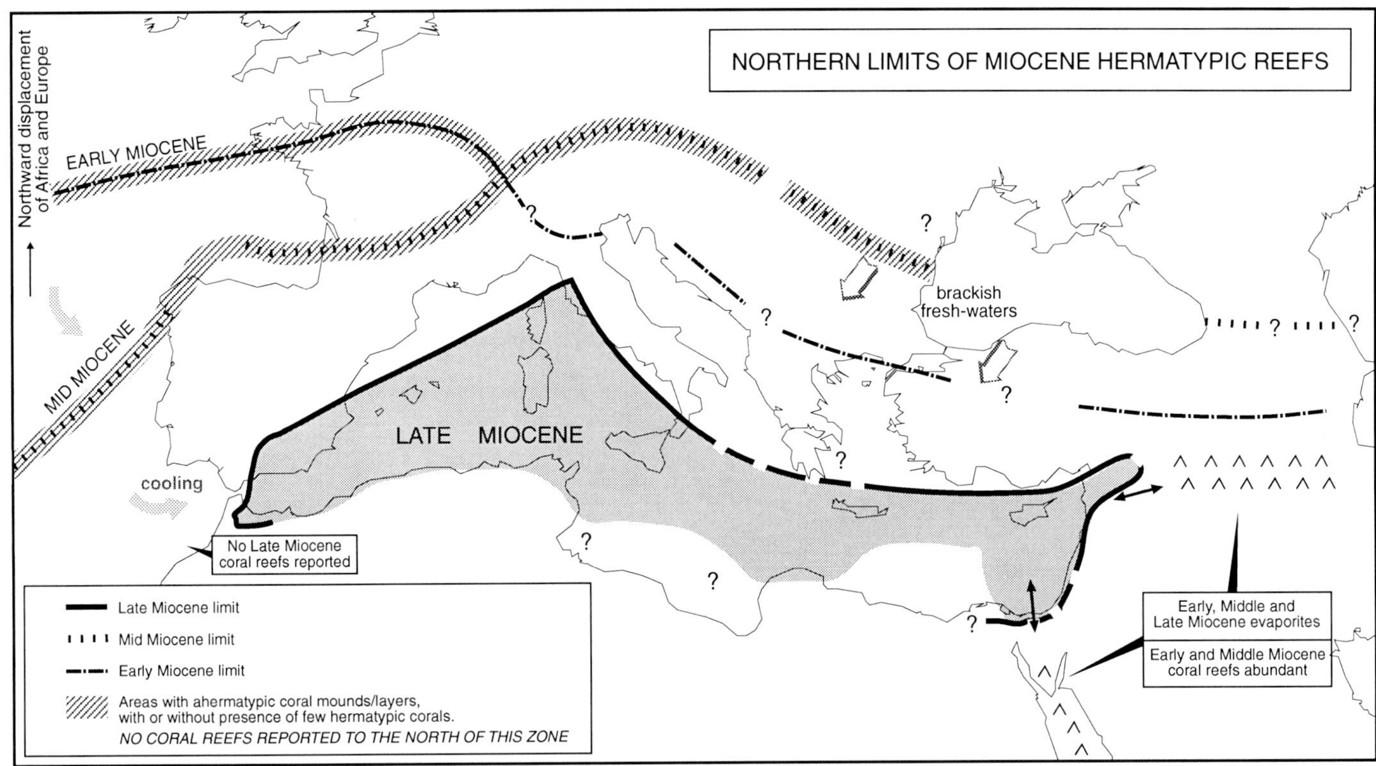


FIG. 9.—Boundaries of the coral reef regions during Early, Middle and Late Miocene times in the Mediterranean.

Zibrownius, 1980). The Mediterranean Plio-Quaternary "reefs" are reduced to ahermatypic coral mounds and extensive rhodalgal ramps (Fig. 3). In contrast, the Gulf of Suez-Red Sea was isolated from the Mediterranean and well-connected to the Indian Ocean; coral reefs flourished during the Plio-Quaternary Supercycle.

### *Lessons from the Modern Mediterranean Sea*

The Mediterranean today is a small zonal ocean almost completely severed from the world ocean with complex peculiar features (Hopkins, 1985): (1) its thermohaline circulation is driven by evaporation, not by polar cooling; (2) continental meteorology plays a most important control on Mediterranean oceanography; and (3) the total loss of vertical stratification during certain winter circulations leads to one of the most extensive vertical circulation patterns known to occur anywhere in the world's oceans. The Mediterranean, as its name implies, is situated within a continental land mass; the complex mountain systems behind the shores of the Mediterranean impose prevailing wind directions (Flos, 1985). This results in one of the world's most cyclogenetically active centers (Fig. 11). Very strong winds and gales are frequent in the winter-spring of the Gulf of Lions and northern part of the Adriatic and Aegean Seas. As a consequence, these strong winds and low pressures favor seasonal vertical mixing, and nutrient-rich waters extensively affect coastal areas (continental shelves are very narrow). Another enriching mechanism is the undercurrent created by major river discharge (more abundant in the northern Mediterranean and mostly during winter and spring). Strictly speaking, it is difficult to distinguish the enriching mechanisms in the coastal areas of the Mediterranean: seasonal mixing, or the 'pump' effect of atmospheric depressions and upwelling (Flos, 1985). Coastal upwelling by advection is not particularly intense in the Mediterranean but has been observed in southern France and eastern Spain, with strong winds parallel to the coast or from land. In any event, some local shallow areas of the northern Mediterranean are frequently affected by waters richer in nutrients than the rest of the basin.

It is likely that a similar situation existed during Late Miocene time. Most of the Alpine topography was already erected, and the connection to the Atlantic was reduced to one or two narrow straits (Esteban et al., this volume). The distribution of Upper Miocene coral reefs shows important gaps in northeastern Spain, southern France and northern Adriatic sea (Figs. 8, 12), but Upper Miocene rhodalgal carbonates are extensively developed. Although some of these areas have major erosional truncations (Messinian unconformity) or had emergent conditions prior to Messinian time, it is here proposed that the observed coral reef distribution could reflect a general paleoenvironmental trend. Coral reef progradation is predominantly to the south or southeast; coral reefs facing the northern winds are less common or poorly developed. It could be assumed that unfavorable conditions for reef growth existed in the coastal areas of northeastern Spain and southern France during Late Miocene time; these conditions could be related to the frequent

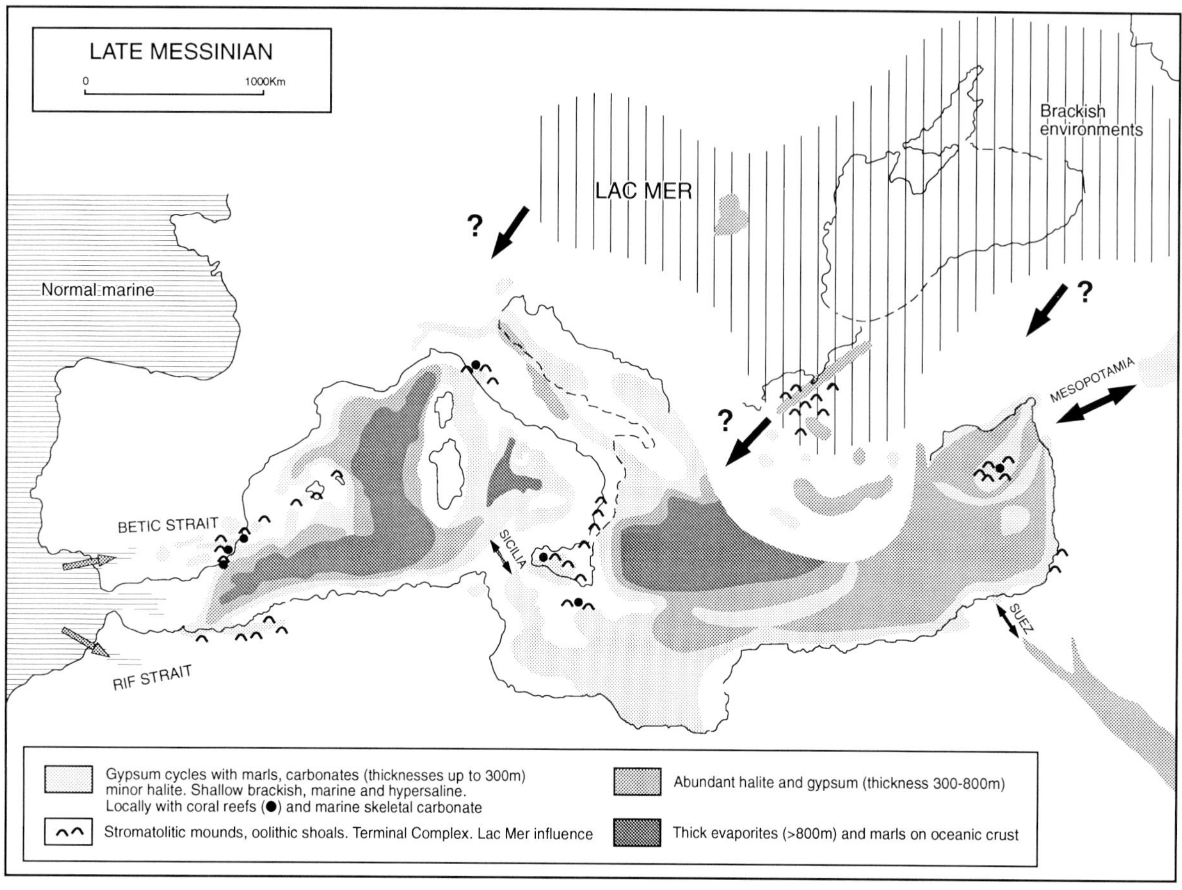


FIG. 10.—Distribution of the Messinian evaporites in the Mediterranean (extensively modified from Ziegler, 1988; Rouchy, 1982a, b; Biju-Duval and Montadert, 1976; Mauffret, 1976; and others).

exposure to cold winds and nutrient-rich waters under mechanisms similar to those occurring today in the Mediterranean.

The land-locked Late Miocene Mediterranean was in contrast with the more open oceanic conditions that dominated during the Early Miocene period. During Middle Miocene time, these oceanic conditions were somewhat reduced by intense phases of the Alpine orogeny (Pedley, this volume), resulting in complex paleoceanographic patterns distinct from those today and during Late Miocene time. Probably, the overall paleoclimatic conditions also followed a parallel trend. The humid, tropical Early Miocene Mediterranean evolved into a drier sub-tropical to warm temperate Late Miocene Mediterranean (Sun and Esteban, 1994).

## CORAL REEF PROVINCES

The main carbonate outcrops in the Mediterranean (Fig. 2) can be grouped into provinces and regions (Fig. 12). The relatively well-studied western region is very complex (Esteban et al., this volume) and includes the Betic, Balearic, Rif and Tell provinces (mostly Upper Miocene reefs with few Lower and Mid Miocene reefs). These areas show the most exuberant reef building for the entire Mediterranean and adjacent areas, reflected in the predominance of papers in this volume. The Early Miocene reefs of the Aquitaine, Lisboa and Algarve provinces are Atlantic-related (Cahuzac and Chaix, this volume). The Mid Miocene reefs of Languedoc, Catalonia (Calvet at al., this volume) and Corsica define a Provençal reef province, probably already established during Early Miocene deposition. Corsica and Sardinia are part of the central Mediterranean region but were located closer to the western region during Early Miocene time.

The central Mediterranean region (Pedley, this volume) was well defined during Late Miocene time as an area of narrowing connections between the eastern and western Mediterranean.

FIG. 11.—Present-day patterns of Mediterranean oceanography and relation to Upper Miocene reef distribution. (A) Intensity, frequency and direction of strong winds (Air Ministry, Meteorological Office, 1936) and distribution pattern of Atlantic zooplankton in the Mediterranean (Hopkins, 1985). (B) Distribution of Upper Tortonian and Messinian coral reefs with predominant direction of progradation.

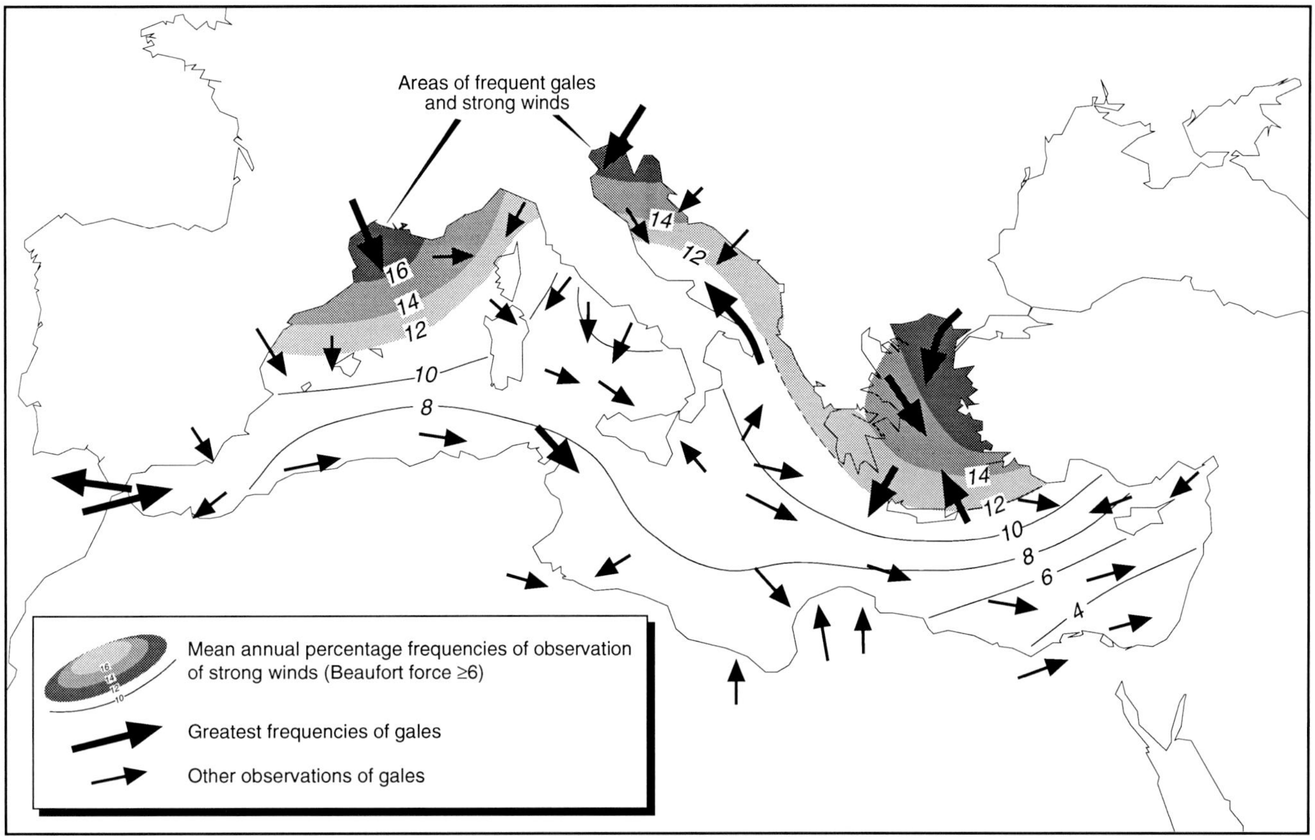

Areas of frequent gales and strong winds

16 14 12 10 8 6 4

Mean annual percentage frequencies of observation of strong winds (Beaufort force ≥6)

Greatest frequencies of gales

Other observations of gales

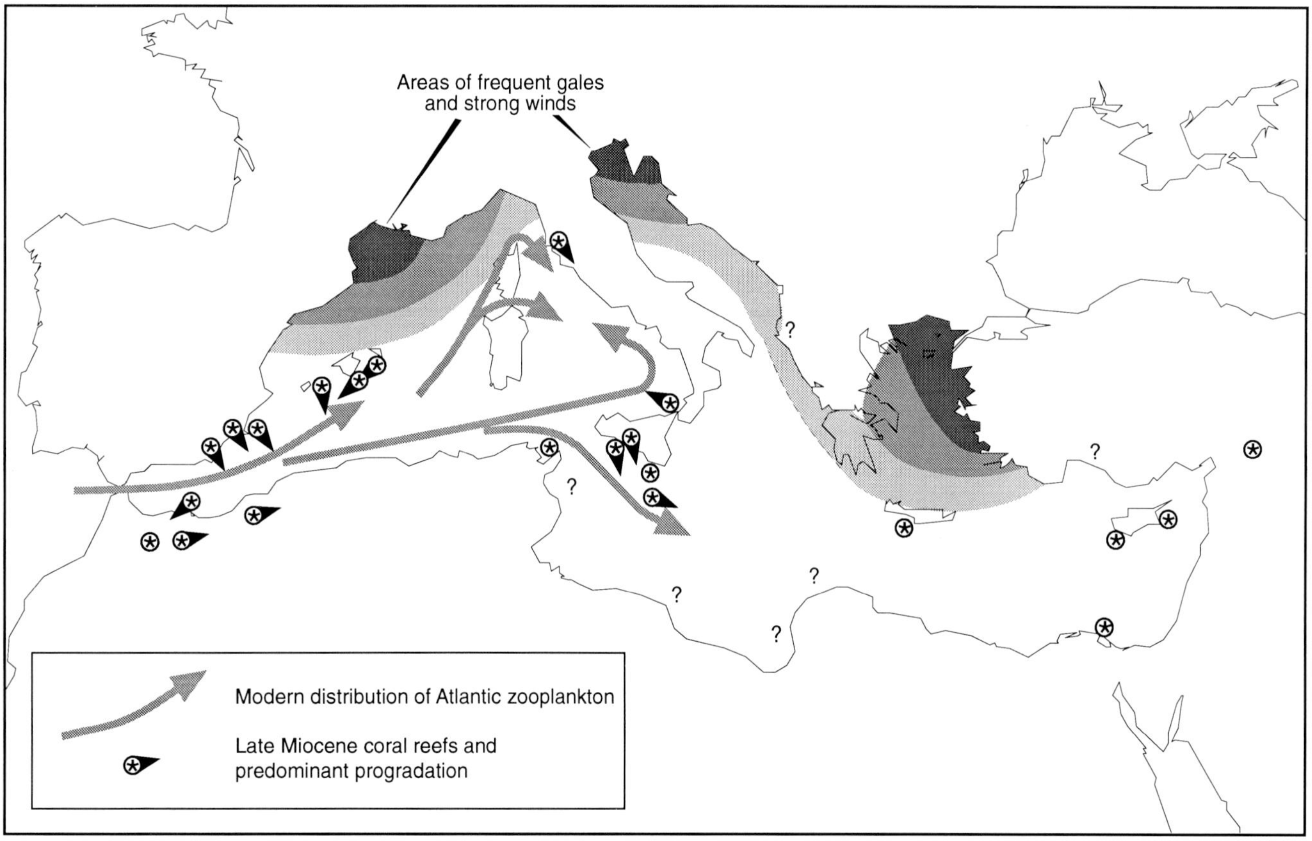

Areas of frequent gales and strong winds

Modern distribution of Atlantic zooplankton

Late Miocene coral reefs and predominant progradation

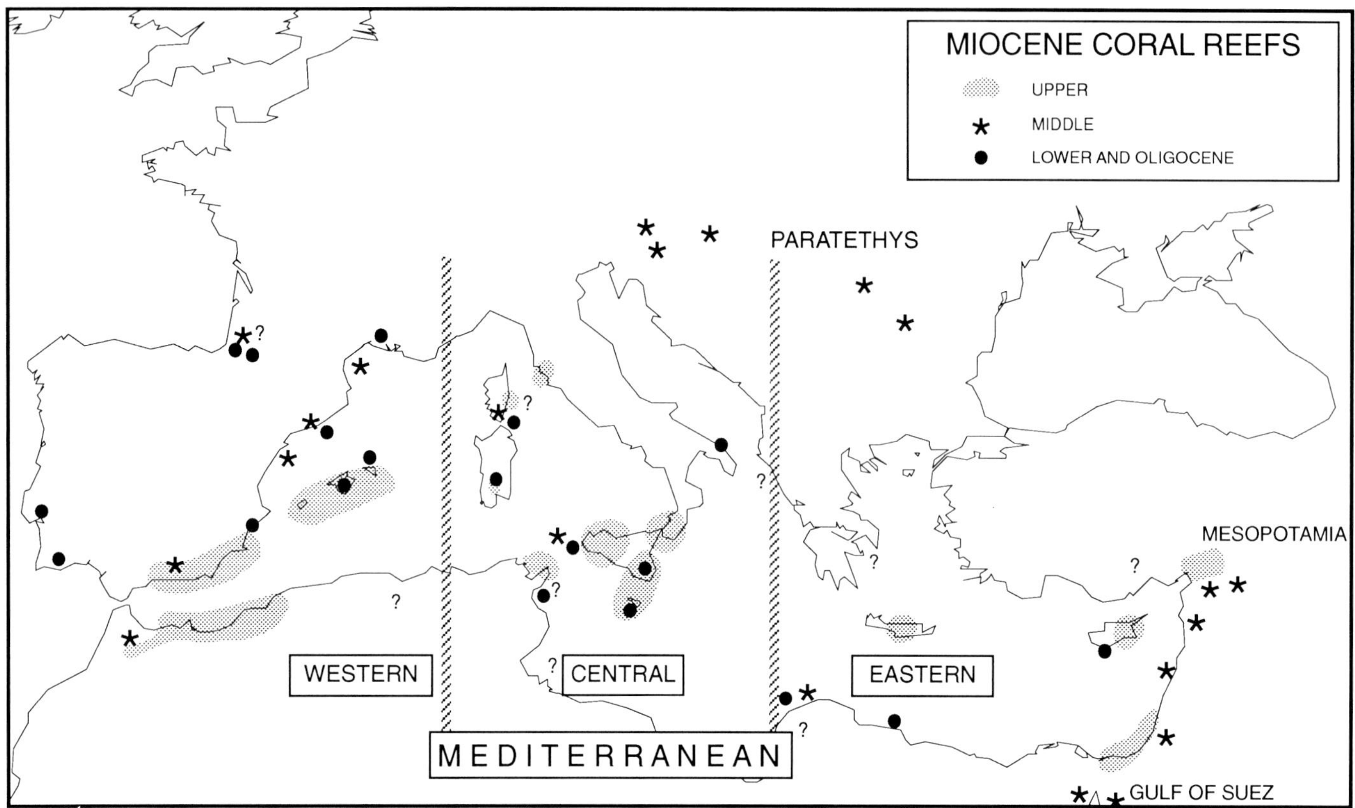


FIG. 12.—The Miocene coral reefs in the Mediterranean regions.

The central region contains the Upper Miocene reef provinces of Toscana (Bossio et al., this volume), Calabria, Sicily-Malta (Pedley, this volume) Corsica and Tunisia and the Lower-Mid Miocene rhodalgal platforms of central Appennines (Carannante and Simone, this volume), Emilia, Romagna and Abruzzi. We lack adequate stratigraphic and facies descriptions for a number of Miocene carbonates in Adriatic basins and the Libyan coast. Nevertheless, one of the most striking features of the central region is the absence of major coral reefs in the Lower-Mid Miocene record, in contrast with their occurrences in the western and eastern Mediterranean regions and also with the importance of the Oligocene reefs in Italy. Similarly, Upper Miocene coral reefs are less developed than in the western Mediterranean.

The Paratethys region (Pisera, this volume) shows extensive development of Mid Miocene carbonate reefs (including some coral reefs) and had major connections with the Indian Ocean region to the southeastern across the Ciscaucasian province and, to a lesser extent, with the Adriatic provinces of the Central Mediterranean region. Paratethyan regional stratigraphy is complicated by the interaction of brackish marine, hypersaline and freshwater deposition with eastern marine influxes. Red algae, bryozoans, vermetids, serpulids and stromatolites are the main reef builders in this region. The influence of Paratethyan waters into the Mediterranean is particularly well recorded with characteristic faunal assemblages in the Uppermost Miocene sections.

The eastern Mediterranean region (Buchbinder, this volume) has extensive Middle Miocene reefs in the Ziqlag (Buchbinder and Martinotti, this volume), Jeribe and Mesopotamian provinces and Upper Miocene reefs in Crete, Cyprus (Follows et al., this volume), Israel, Turkey and Libyan plateau. The distribution of coral reefs in the eastern Mediterranean appears to have been at least partially influenced by major terrigenous discharge from the Nile. Other Lower Miocene reefs occur in southern Turkey and Cyprus (Follows, 1992). The eastern region connected with the Mesopotamian basins during the Early Miocene times and developed extensive carbonate banks and locally reefs; little documentation is available about this area. The eastern region also connected with well-studied Lower and Middle Miocene reef provinces of the Gulf of Suez and Red Sea (Purser et al., this volume; Coniglio et al., this volume); this was a Miocene Mediterranean province until the flooding by the Indian Ocean in Pliocene times.

## UPPER MIOCENE FACIES MODEL

It is not surprising that most of the studies of Miocene reefs in the Mediterranean are of Late Miocene age. The excellent quality and accessibility of many of the outcrops, particularly in the western Mediterranean, provides an unrivalled opportunity for description of facies and geometric relationships, the most solid support for "realistic" facies models. Examples such as the well-studied Cap Blanc, Fortuna, Níjar, La Molata and Melilla reefs show detailed facies relationships that do not rely on the

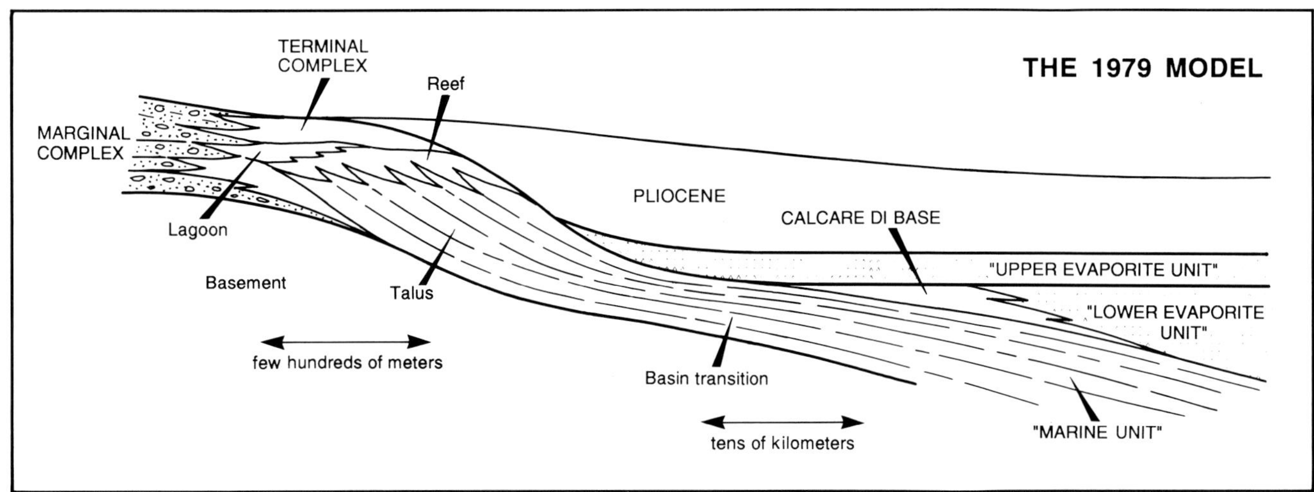


FIG. 13.—The generalized model of Esteban (1979) for the Messinian reefs in the western Mediterranean and the relationships with the Messinian Evaporite Units.

application of an "ideal" synthetic scheme based on modern coral reefs. Saint Martin and Thomassin (1989) stress the problems of "transferring" ecology and facies zonation from the "ideal" modern coral reef to Miocene reefs. The best of Upper Miocene outcrops clearly provide the "observation" of facies and geometric relationships that can be organized into descriptions and interpretations as barrier, fringing and patch reef associations, with well-preserved talus slopes, reef walls with spur-and-grooves or buttresses, reef flats or crests and back reef lagoons, without the need of applying ideal models from modern reefs. Esteban and Giner (1980) and Dabrio et al. (1981) noted that in some localities present-day geomorphology closely follows the depositional Miocene reef morphologies (archipelagos, paleovalleys, buttresses and slopes are all beautifully preserved in the present-day landscape with only minor Plio-Pleistocene modifications).

Many of the Upper Miocene outcrops in the western Mediterranean are of the same calibre as the classic outcrops of the Capitan Reef or the Dolomites, where facies models have been used and refined for over four decades and continue to teach important new lessons. This section will review facies models of Upper Miocene reefs, using the basic model of Esteban (1979; "the 1979 model," Fig. 13) with modifications proposed by subsequent workers. The 1979 model summarized collective work since 1974 in eastern Spain (Barcelona, Mallorca, Alacant, Murcia, Almería), Morocco (Melilla, Guercif, Taza, Saïss), Italy (Toscana, Calabria) and western Sicily (Esteban et al., 1977a, b; 1978; Esteban, 1978, 1979; Esteban and Giner, 1980; Dabrio et al., 1981; Bossio et al., 1978; Catalano and Esteban, 1978). These publications also recognized that Miocene reefs were far more extensively developed than previously documented in the literature. Subsequent work by numerous authors indicates that the general Upper Miocene reef model is also applicable to Lower and Middle Miocene reefs.

### *Types of Reef Assemblages*

The majority of the Upper Miocene reefs in western Mediterranean are fringing reefs (Esteban 1979) with small (few hundred meters wide), poorly developed or absent lagoons. These fringing reefs developed directly on basement (volcanic, metamorphic or older Cenozoic or Mesozoic sedimentary rocks), commonly with thin intercalated clastic layers and following linear trends. There are also complex fringing reef morphologies on active Upper Miocene coastal fans (fan deltas, delta platforms, interdistributary mouth bars, beaches, ...), representing the Marginal Terrigenous Complex (Fig. 13). These reefs consist of layers and laterally discontinuous patches and arcuate trends, exhibiting a wide variety of sizes and geometries, (some with small lagoons) and adapted to the morphology of the terrigenous sedimentary bodies. Excellent sedimentological characterization of these environments is provided by Santisteban (1981), Santisteban and Taberner (1983), also Dabrio and Polo (1988), Braga and Martin (1988), Dabrio (1989, 1990), Martin et al. (1989), Grasso et al. (1982), Pedley (1983), Grasso and Pedley (1988, 1989), Braga et al. (1990) and Saint Martin (1990). A wide variety of morphologies result from this dynamic interaction of coral reef growth and terrigenous siliciclastics (thickets, mounds, layers, "pinnacles"); these are also controlled by the profile and lithology of the substrate (Fig. 14; Santisteban, 1981; Saint Martin, 1990). It should be stressed that in the early work, the term "patch reef" was used descriptively to refer to laterally discontinuous, patchy coral framework in shallow-water areas but not necessarily in a back-reef lagoon.

True barrier reefs with extensive lagoonal systems (at least 1-2 km wide) are the exception rather than the norm (e.g., Mallorca). Pomar et al. (1983) and Pomar (1991) show that lagoonal units are not always present in the 5-km-long progradational reef complex; there are repeated intercalations of fringing reef episodes. Terms such as fringing or barrier reefs are difficult to

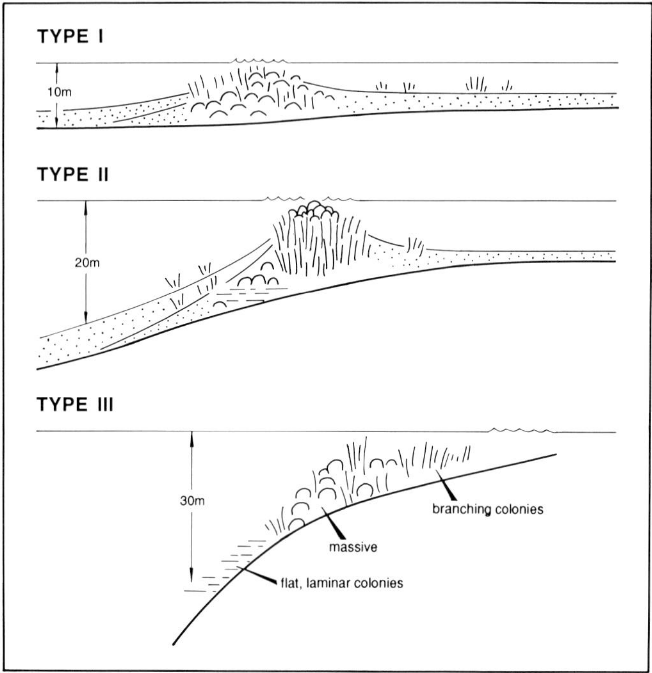


FIG. 14.—Three major types of reef profile in the Upper Tortonian-Messinian Fortuna Basin (southeastern Spain). After Santisteban (1981).

apply when referring to entire reef complexes with long histories of progradation. For this reason, some authors (Saint Martin, 1990) dislike the term barrier reef and prefer the less specific term "platform reef complex." The Llucmajor platform (Mallorca) is the widest (20 km) and thickest (300 m) of these platform reef complexes in the Mediterranean (Pomar et al., 1983; Pomar, 1991, 1993).

There were different types of ramp settings with Miocene carbonates dominated by rhodalgal facies associated with large benthic forams and local sigmoidal mounds of rhodolithic carbonates (mostly in the Burdigalian, Lower Tortonian, some Langhian and Upper Tortonian; Bosence, this volume; Obrador et al., 1992; Buxton and Pedley, 1989). In these settings, coral reefs, if present at all, tended to be small, discontinuous patches on the inner ramp; there is no evidence of vertical pinnacle development on the outer ramp.

## *Reef Facies*

*Lagoonal facies.—*

Lagoonal carbonate facies, where present, are characterized (Esteban, 1979) by foraminifer-rich shoals (miliolids, peneroplids and alveolinids), oolitic-grapestone bars, red algae, shell beds, coral knobs and patch reefs *(Porites, Tarbellastraea* and *Siderastraea),* mud-supported facies with gastropods, stromatolites and traces of mangrove roots. Locally, there are concentrations of winnowed planktonic forams (Santisteban, 1981). Where properly defined, lagoonal carbonates grade laterally into frontal reef core (reef wall) and are typically subhorizontal and well bedded. In some localities, the calcareous lagoonal sediment grades landward into terrigenous-siliciclastic deposits (shales, silts, sands and conglomerates, up to a few hundred meters thick) of the Marginal Terrigenous Complex (lacustrine, fan-delta, fluviodeltaic-plains and alluvial-fan facies). Mixed coral reef and terrigenous coastal facies are common in southeastern Spain (Santisteban, 1981; Dabrio and Polo, 1987, 1988; Braga et al., 1990; Dabrio, 1990), Sicily (Catalano and Esteban, 1978; Catalano, 1979) and Tuscany (Bossio et al., 1978, 1981). The Marginal Complex, however, may grade directly into the reef wall or into talus-slope facies and turbidites; where absent, the coral reef directly overlies the volcanic, metamorphic or older Cenozoic and Mesozoic substrate.

In some localities, mostly with mixed carbonate-clastic sedimentation (fringing reefs on fan-delta lobes, channels, bars of the Marginal Complex, carbonate ramps), there are obvious "patch-reef" geometries (laterally discrete) without evidence of a coeval frontal barrier reef (Catalano and Esteban, 1978; Catalano, 1979; Bossio et al., 1978, 1981; Santisteban, 1981; Saint Martin, 1990). As stressed by Saint Martin (1990), these should not be called "back-reef lagoons."

*Reef-core facies.—*

The reef-core facies is characterized by in-situ coral framework grading laterally into bioclastic beds. In the reef front, the reef-core facies (reef wall) grades seaward into reef-slope facies. Many of these reef-core facies appear in present-day cliffs and steep slopes, which are controlled by the depositional morphology with only minor modifications by Miocene to Quaternary erosion. Coral knobs and patch reefs lack a well-developed slope but may have small flank deposits. This discussion, however, is mostly concerned with the marginal reef-core facies (also called reef front). Many reef-front facies can be traced laterally for many 10's of km (Fig. 15), faithfully reflecting a complex archipelago paleogeography (partially preserved in present-day topography) and a history of sea-level oscillations (e.g., Esteban and Giner, 1980; Dabrio et al., 1981; Pomar, 1991, 1993; Pomar and Ward, 1994; Franseen et al., 1993; Franseen and Goldstein, this volume).

The reef-core framework (commonly up to 20-30 m thick) is predominantly constructed by *Porites,* with or without the association of other frame-building corals *(Tarbellastraea, Acanthastraea, Paleoplesiastraea, Siderastraea).* Santisteban (1981) mentioned an exceptional 100-m-thick reef framework in El Desastre resulting from the vertical stacking of five reef events. Red algae, bryozoans, encrusting forams, serpulids, vermetids, micritic (stromatolitic in Pedley, 1979; Riding et al., 1991b) and fibrous marine cements contribute in variable amounts to the binding of the reef. Framestones, bafflestones, bindstones and rudstones are common depositional textures. Echinoderms, bivalves, gastropods, benthic forams and crustaceans are common skeletal grains in a wide variety of mud-supported and grain-supported textures. Large benthic forams tend to be smaller and less abundant than in the rhodalgal facies. Organic boring (molluscs, sponges, worms and algae) is very conspicuous in the upper part of the reef wall and in the lagoon. Dascycladacean algae are rare but have been cited by Fois

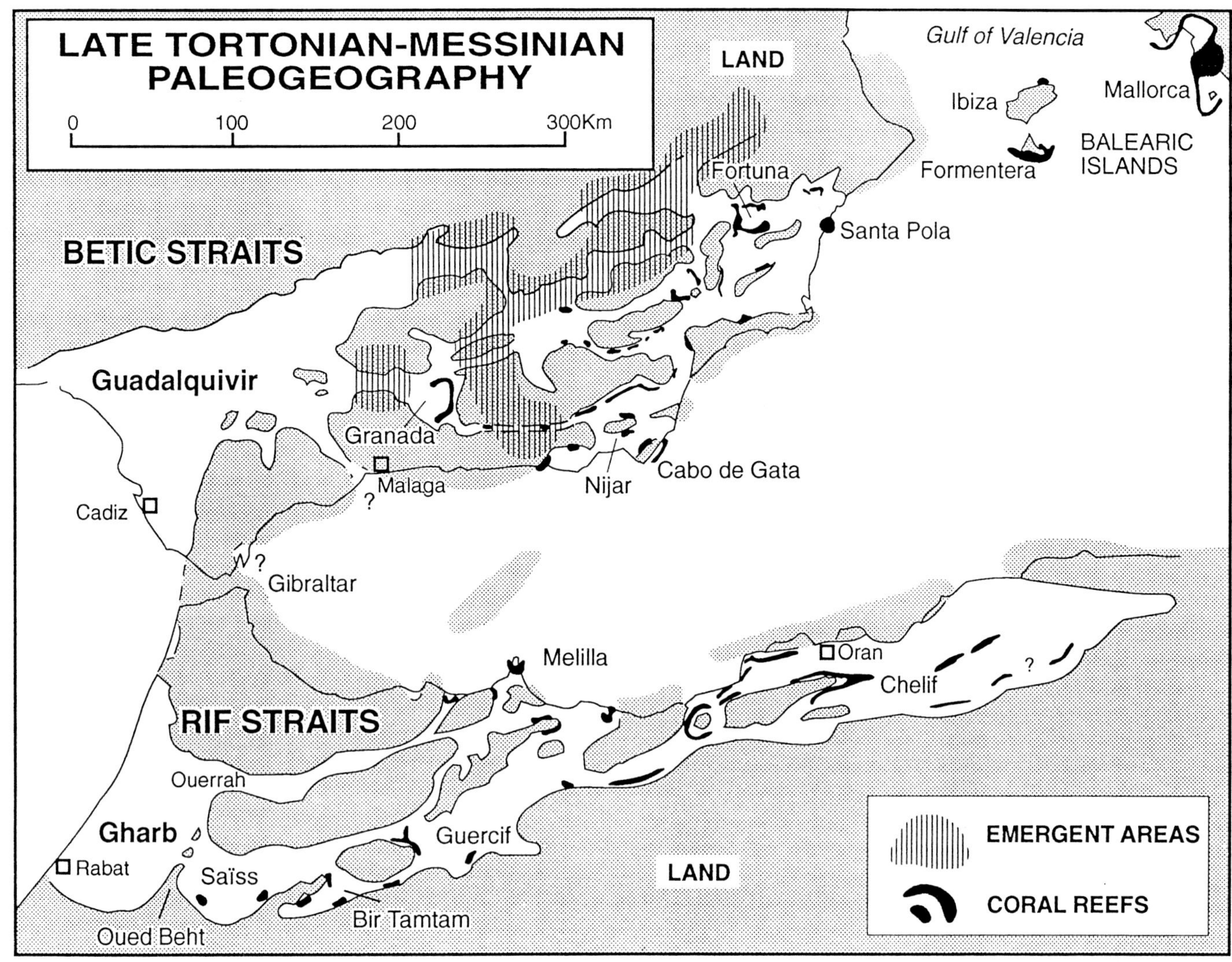


FIG. 15.—Paleogeographic trends of the Upper Tortonian-Messinian coral reefs in the Betic-Rif straits (modified from Esteban, 1979; Dabrio et al., 1981; Santisteban, 1981; Rouchy et al., 1986; Saint Martin, pers. commun., and unpubl. observ.). This synthetic scheme presents significant variations during Tortonian-Messinian times; the emergent areas repeatedly interrupted the marine connection along the Betic and Rif Straits.

(1990), Franseen and Goldstein (this volume) and Riding et al. (1991b). Addicott et al. (1977) mentioned some milleporid reefs in Almería but this has not been confirmed. Pagnier (1977) was the first to mentioned the presence of conspicuous vermetid reefs. Ben-Moussa et al. (1989), Chaix et al. (1986), Saint Martin (1984, 1990), Saint Martín and Chaix, 1981, 1984; Saint Martin et al. (1985), Martin et al. (1989) and Braga and Martín (1988) present a detailed paleontological and paleoecological analysis of the Upper Miocene reef community, listing more than 300 species in total.

The reef-core framework can be massive or formed by juxtaposed thickets, layers or mounds (Fig. 16). Clean, large-scale exposures show a complex internal architecture of units of different orders of magnitude (4th- to 6th-order depositional sequences or subsequences) with basic sigmoid geometries and truncated wedges (locally oblique tangential to parallel; Franseen and Goldstein, this volume) stacked in different progradational patterns separated by well-developed terraced erosional surfaces (Pomar, 1991). Other examples of this complex internal architecture are presented in Dabrio et al. (1981), Pomar et al. (1983), Rouchy (1982a, his Fig. 58) and in this volume. Field observations in the best outcrops of Late Tortonian-Messinian reefs show unequivocal terraced erosional geometries (Pomar 1991; Franseen, 1989; Vallés-Roca, 1986; Esteban and Giner 1980; Buchbinder and Martinotti, this volume) that compare in many aspects with Quaternary reef terraces.

Commonly, where the reef core is interbedded with terrigenous clastics, it appears stratified (Santisteban, 1981; Saint Martin, 1984, 1990) with surfaces of sediment bypass interrupting reef growth (Fig. 16). Braga and Martin (1988) and Martin et al. (1989) present a detailed analysis of high-frequency cycles with coral succession *(Porites-Platygyra-Tarbellastraea*-red algae) controlled by clastic input. Spur-and-grooves and larger-scale buttresses (Fig. 17) can be seen in the best outcrops in Mallorca, Fortuna and Níjar (Esteban et al., 1977a, b; Esteban, 1979; Esteban and Giner, 1977a, b; Santisteban, 1981; Pomar et

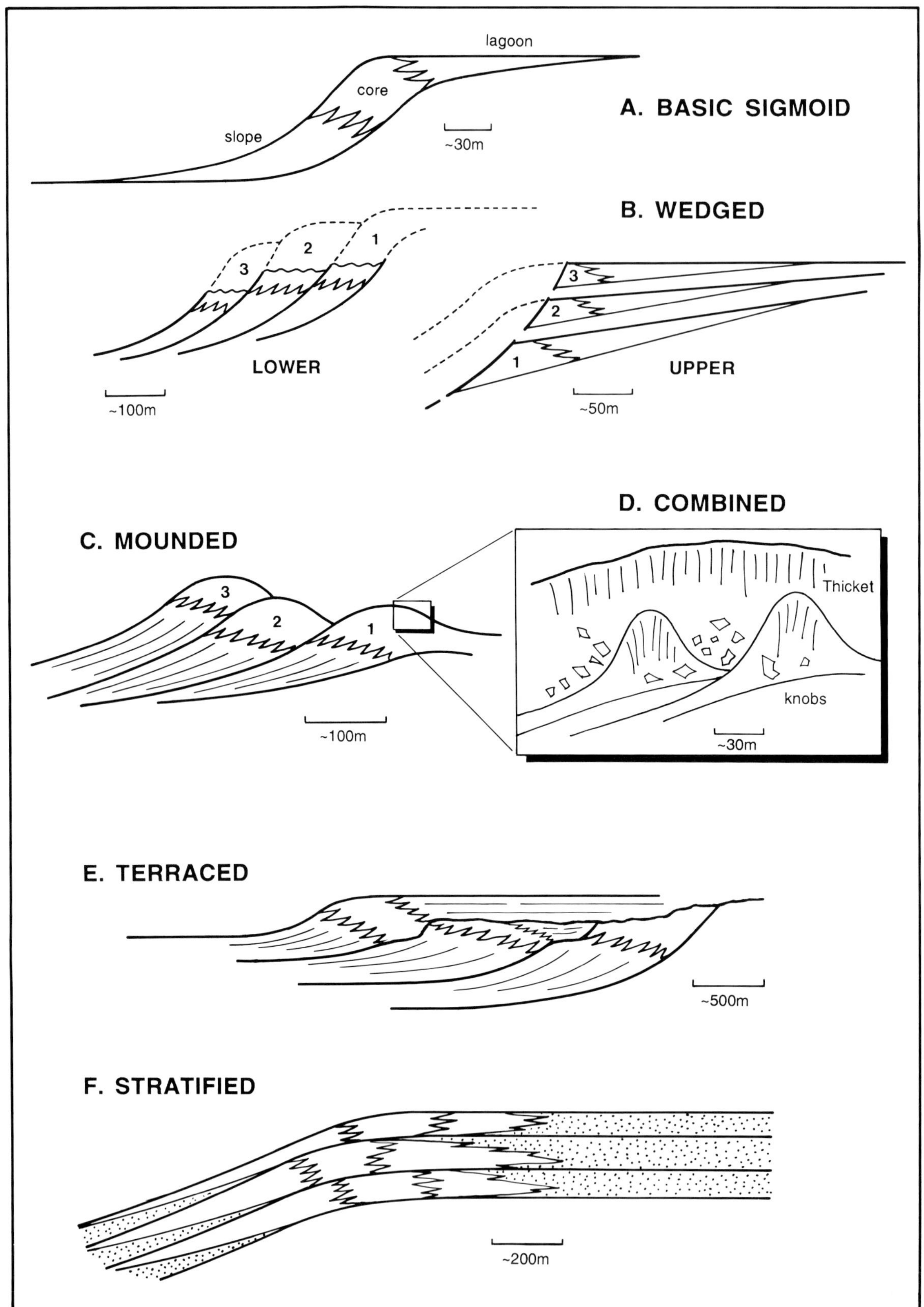


FIG. 16.—Common geometries of coral reef complexes.

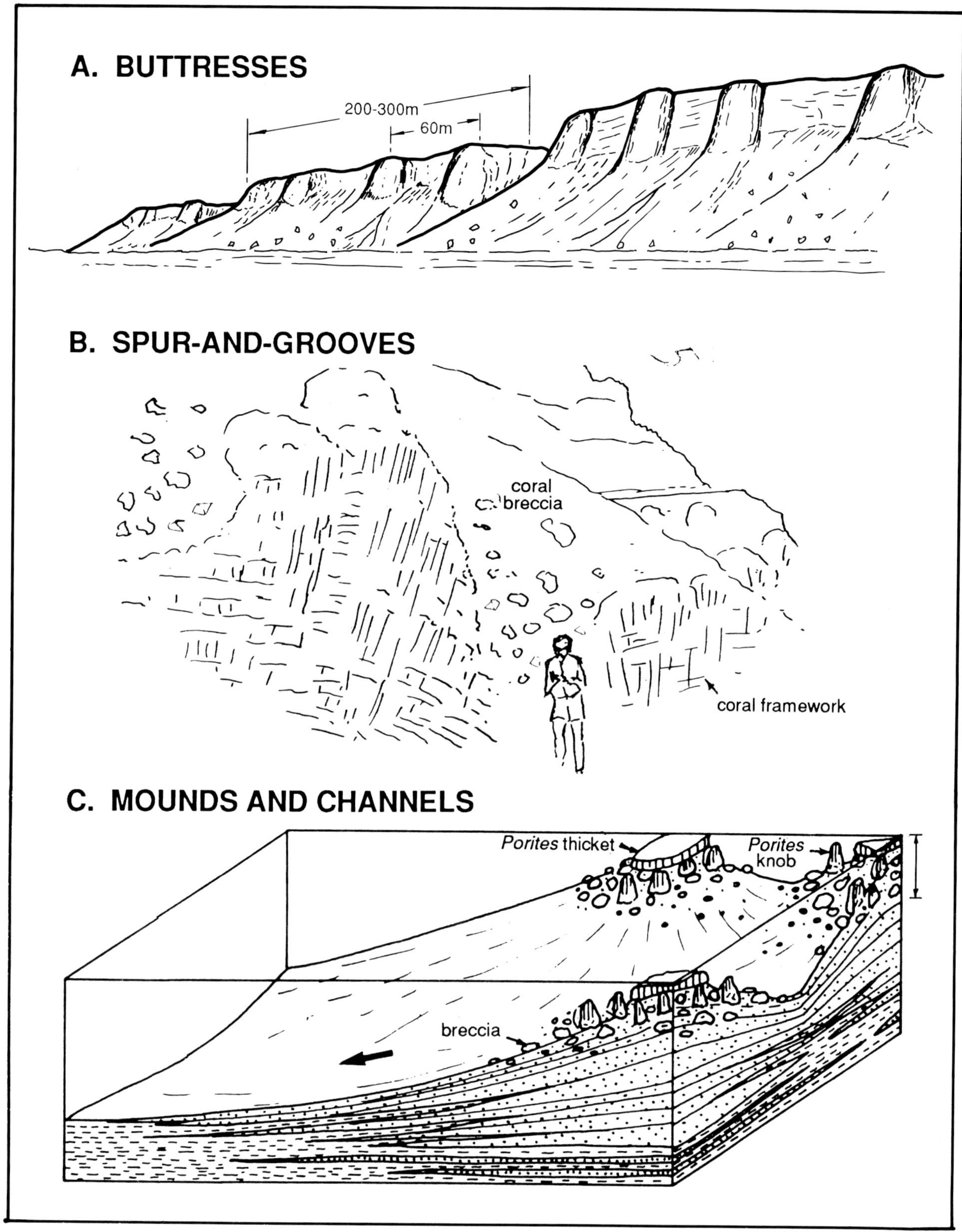


FIG. 17.—(A) Buttresses in the reef front of Santa Pola (southeastern Spain), after Esteban et al., 1977b). (B) Spur-and-grooves in the Cap Blanc section. (C) Mounds and channels in Níjar, after Dabrio et al. (1981). Scale: 50 m.

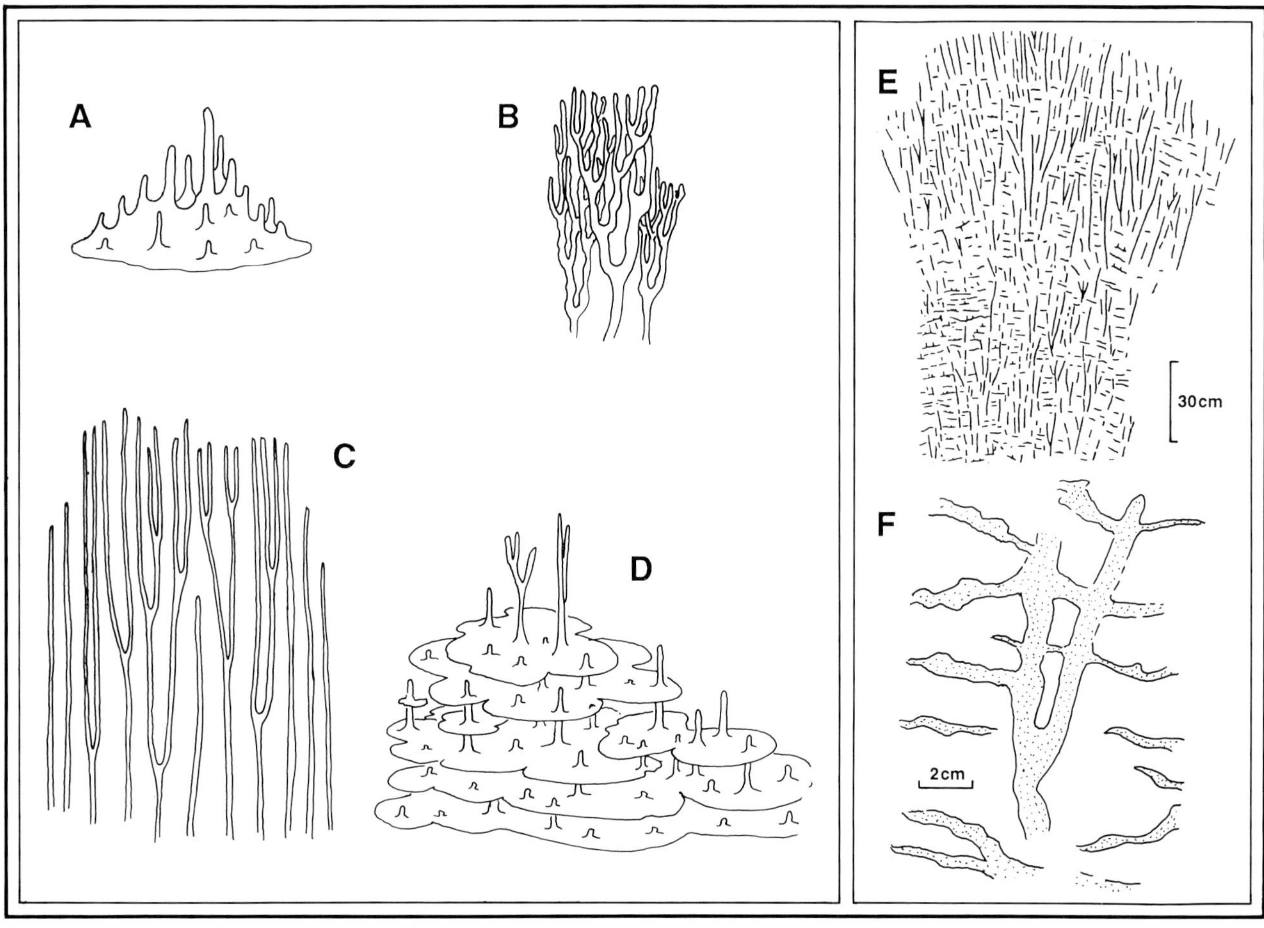


FIG. 18.—Some of the typical colonial morphologies of *Porites* in the Upper Miocene coral reefs of the Mediterranean (modified from Esteban, 1979; Esteban and Giner, 1977; Esteban et al., 1977b). (A) Elliptical colony with vertical fingers, (B) branching fingers, (C) long sticks, (D) laminar, disc-like colonies with vertical sticks, and (E, F) field sketches.

al., 1983, 1985; Figs. 185 and 186 in James, 1983) and control the transport of large amounts of sediments and coral breccia down the upper slope. However, some of these features are considered as erosional rather than constructional (Esteban and Giner, 1977a, b; Santisteban, 1981) and some could be controlled by syndepositional dilation fractures along the platform margin.

On the basis of coral diversity, Esteban (1978,1979) distinguished two types of Upper Miocene reefs: Type A with up to 5-15 coral species (normally only about 2-4 corals are the main frame builders) associated with a relatively rich community of echinoids, molluscs, red algae, bryozoans, and foraminifers and Type B with one dominant coral (locally two) that is commonly but not exclusively *Porites* (locally *Tarbellastraea)* with stick morphologies and laminar colonies at the base (Fig. 18). Other corals, when present, are much less abundant, and the associated community is much more impoverished and with smaller size of specimens in relation to the type A reefs. There is an important intermediate type AB reef with one (locally two) dominant corals but with associated communities similar to type A (e.g., the Cap Blanc reef of Mallorca). For this reason, some authors offer a more detailed classification with three reef types (Rouchy et al., 1986).

Esteban (1979) assumed that Type A and B reefs were related to different ages: Type A, Early Messinian or Tortonian ages and Type B, Messinian age. Santisteban (1981) clearly disproved this generalized age interpretation of coral diversity by demonstrating reversals in the general Late Miocene trend of decreasing diversity. He proposed that sustained or repeated terrigenous influx in large amounts (but below lethal levels) would favor higher coral diversity. However, very large amounts of terrigenous clastics decrease the diversity to one coral *(Porites)* or to rhodalgal sandy carbonates without corals (Rodríguez-Perea, 1989; Saint Martin, this volume). Other authors (as Bossio et al., 1981; Saint Martin, 1984, 1990; Rouchy et al., 1986) have recognized the weakness of the 1979 age hypothesis, but there is general agreement regarding the presence of the two reef types: oligotypic to monogeneric reefs (Type B) and taxonomically more diversified reefs (Type A). The problem is still what determines (local vs. Mediterranean-wide factors) the appear-

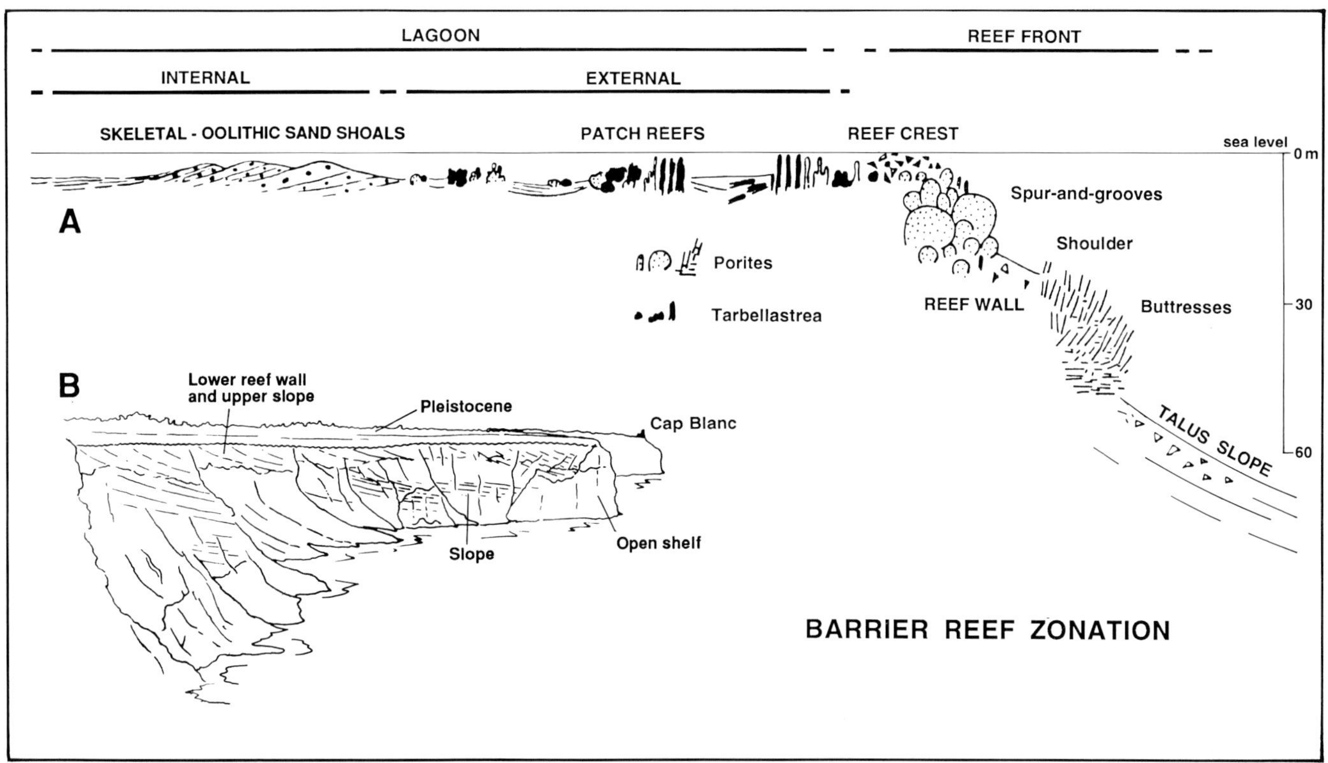


FIG. 19.—(A) Reef facies zonation in the Upper Miocene Cap Blanc coral reef complex (Mallorca, Balearic Islands) after Esteban (1979). (B) Sketch from Santisteban (1981) of the 90-m-high seacliffs west of the Cap Blanc, showing well developed progradation of the coral-reef slopes

ance of each reef type. This topic of coral diversity also will be discussed below in relation to the peculiarities of some Type B reefs.

Saint Martin (1990) summarized the variety of Upper Miocene reefs in Morocco and Algeria into four major groups: Group 1, plurigeneric stratified or layered reefs (fringing, submerged highs) without rimmed platforms and slopes; Group 2, plurigeneric fringing-reefs with steep slopes; Group 3, platform patches and lenses with dominant *Porites* and *Tarbellastraea* and abundant red algae; and Group 4, monogeneric *Porites* reefs with marked downstepping progradation and accumulations of abundant *Halimeda.*

*Zonation of colonial morphologies.—*

On the basis of the magnificent outcrops of Cap Blanc of Mallorca, Esteban (1979) and Pomar et al. (1983) presented a basic zonation of the reef wall (Fig. 19): (i) rear zone — coral heads, pillars and columns grading into the lagoon; (ii) reef flat or crest — coral breccia, red algae, flat-coral encrustations and truncated coral heads; (iii) shallow reef wall (up to -10 m) — massive coral head zone; (iv) middle reef wall (up to -20 m) — commonly with long, thin coral sticks, and locally including sediment-rich pockets; and (v) deep reef wall (up to -30 m) — predominantly laminar corals. In all these zones, *Porites* is the main coral, and locally, *Tarbellastraea* and minor amounts of *Siderastraea.* These zones correspond to a single progradational event (Pomar, 1991). The paleodepth zonation also is reflected by red algae associations (Bosence, pers. commun., 1993). Subsequent field work and numerous geologists visiting the Cap Blanc outcrops have corroborated this basic zonation. However, there are indications of variations in the zonation along the 6-km-long prograding cross section (Pomar, 1991). It has to be stressed that this zonation is only a general pattern, probably applicable to subsequences (4th- to 6th-order sequences) with a dominant aggrading stacking in clean-water settings.

Many Upper Miocene reef walls have a more simplified zonation, consisting of long, branching *Porites* (locally *Tarbellastraea)* sticks with laminar, horizontal expansions gradually predominant at the base, forming bushes, thickets and wedges (Esteban et al., 1977b, 1978, 1982a, b; Esteban and Giner, 1980; Dabrio et al., 1981). Many of these outcrops were truncated at the top and overlain by other carbonate units; consequently, the depositional depths of this type of colonial morphology could not be established. However, Riding et al. (1991b) interpreted the presence of reef-crest facies (massive coral heads) on top of some of these branching thickets, thus implying the local preservation of a complete profile. The excellent outcrops of the Fortuna Basin show numerous variations in zonation patterns (Santisteban, 1981, Figs. 14, 20), apparently also controlled by local rates of sediment supply, subsidence, substrate morphology and relative sea-level oscillations. The main variation in the steep reef walls is the presence

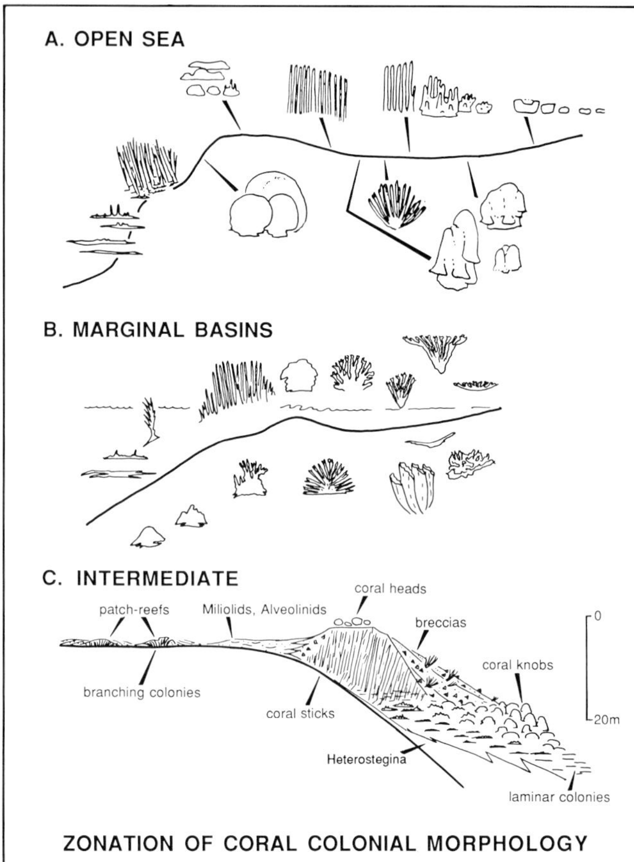


FIG. 20.—Zonation of colonial morphology in the Upper Tortonian-Messinian coral reefs of the Fortuna Basin (southeastern Spain) after Santisteban (1981). (A) Open sea, deep-water margins, (B) restricted, shallow margins, and (C) example of intermediate coral zonation (modified from Santisteban, 1981).

of a deep massive-coral-head zone below the *Porites* branches, also shown by Saint Martin (1990, p. 104).

Reefs in mixed terrigenous environments show a wider variety of colonial morphologies (blades, conical-branching aggregates, botryoidal heads, and many more), previously recognized by Bossio et al. (1978, 1981), Grasso and Pedley (1988, 1989) and Saint Martin (1990). In addition, Santisteban (1981) and Martin et al. (1989) mentioned the presence of microatolls. It is quite difficult to establish general zonations in these mixed terrigenous settings, other than a general landward trend toward smaller colony size, predominance of branching forms and the development of laminar skirts in the lower parts of branching and massive corals.

It is not the purpose of this review to analyze the environmental implications of the coral morphologies. Saint Martin (1990) and Grasso and Pedley (1988, 1989) offer a good account of the controls on coral morphology. There is good agreement that the patterns observed in the Upper Miocene are consistent with those in modern reefs. The stick morphology responds to the need for fast growth rates (to compete for space, to avoid burial and to maximize light availability). Esteban et al. (1977b) and Esteban (1979) noted that the typical thin branches of *Porites* would be quite fragile if affected by waves and the rarity of piles of broken branches is remarkable. The coral framework should be less fragile where coral branches have laminar bridging expansions. It can be assumed that support was provided by sediment packed around the branches, and the 2- to 4-cm-thick red-algal and micritic crusts were common in many of the *Porites* frameworks. Lower light intensity could explain the laminar expansions of *Porites* and *Tarbellastraea* in the lower part of the reef wall (as in Barnes, 1973). Many laminar coral colonies show distinctive growth lines (seasonal?) at the base. Zonations in colonial morphology can be best seen in low-coral-diversity reefs; genotypic and phenotypic variations in morphology in higher-diversity reefs tend to obscure the zonation. In summary, the field geometries of the best Upper Miocene outcrops offer the possibility of determining quite accurately the absolute and relative paleodepths of the different coral assemblages, providing a useful control for local field work. This zonation , howerver, cannot be generalized for the entire basin or province, or even for a single reef complex.

*Reef-slope facies (fore reef).—*

The presence of prograding reef-talus slopes in the Upper Miocene platforms of western Mediterranean was first recognized by Esteban et al. (1977a, b). Reef slopes were characterized by foreset-like clinoforms grading upslope to reef-wall facies and horizontal lagoonal beds grading downslope to flat, fine-grained deeper-marine carbonates and marls. These early descriptions recognized the presence of layers with abundant *Halimeda,* red algae, molluscs, bryozoans, coral breccias, vermetids and serpulids on the reef slopes, including some skeletal grains supposedly derived from the lagoon (miliolids, *Borelis* and peneroplids). In many areas, the reef slopes were previously considered in the local literature to be tectonically deformed shallow-marine deposits. The reef slopes of Níjar (southeastern Spain) were described as a model (Fig. 21) for other areas (Dabrio et al., 1981). The talus slope beds are thicker and coarser in the upper part with up to 20-30° dips (proximal talus) and become thinner and finer grained with gentle seaward dips in the lower part (distal talus). Observed field geometries indicate up to 200 m of relief, commonly about 60-80 m. Santisteban (1980, 1981) mentions more than 400 m of relief adjacent to El Desastre. There are a few references to steeper (40-50°) reef slopes, but they seem to involve early lithification and/or erosional events. The proximal talus is predominantly grain-supported and contains abundant skeletal grains and coral breccias derived from the reef wall and, to a lesser extent, from the lagoonal areas. The distal talus is mud-supported with abundant grain-supported layers; skeletal grains tend to be smaller, more micritized and contain variable amounts of planktonic foraminifers. Slumps and convoluted beds are common. The best exposures also show channel and lobe morphologies in the reef slopes (Santisteban, 1981). Large carbonate blocks detached from the reef wall and proximal talus can be found up to more than 2 km away from the source area (Melilla, Carboneras)

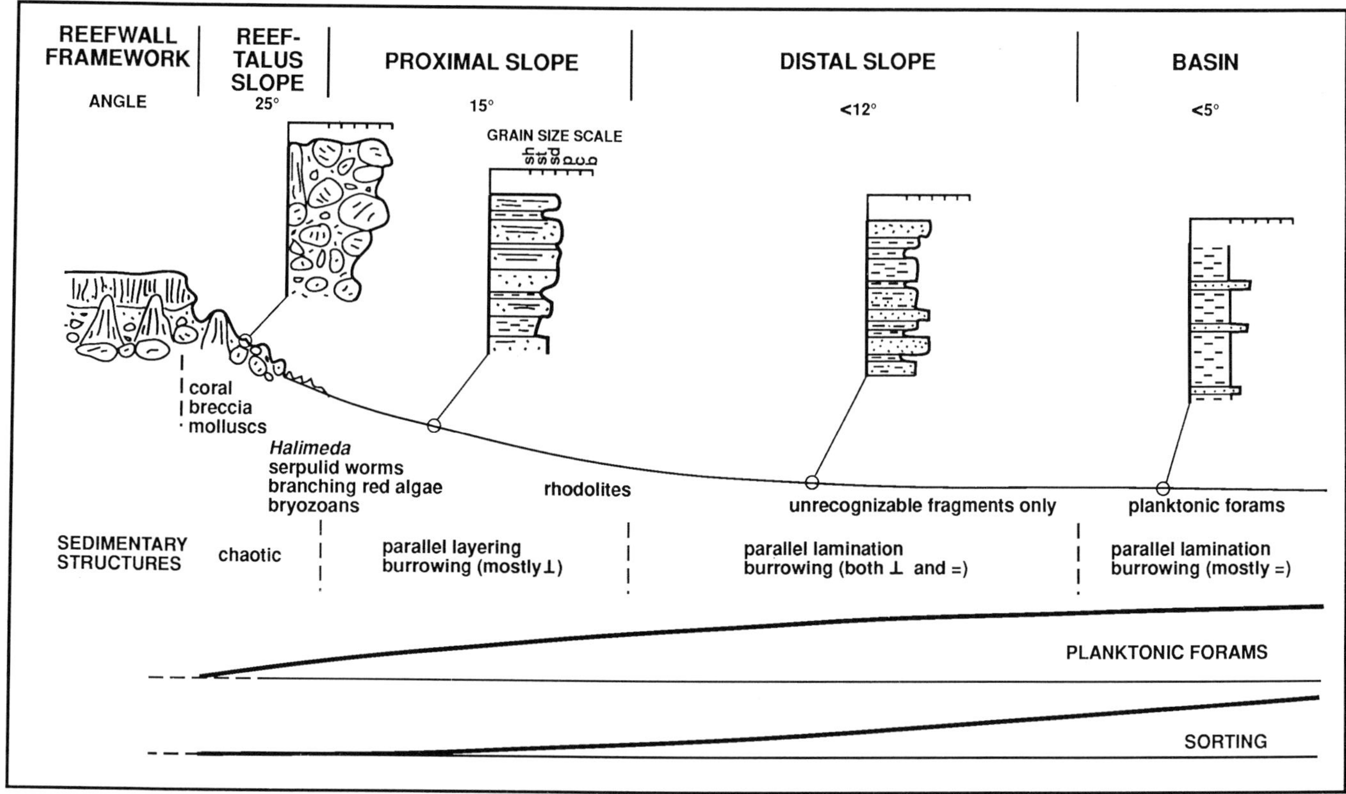


FIG. 21.—Facies model of the reef slopes of Níjar (southeastern Spain) showing distribution of organisms and sedimentary structures (from Dabrio et al., 1981).

and can be confused with patch reefs (Esteban and Giner, 1980). Where slumped blocks are in upright position, the distinction is even more difficult. A rich and varied fauna including ahermatypic corals, gorgonians and hydrozoans is recorded in these slope breccias (e.g., Carboneras- Bordet et al., 1982; La Chapelle, 1988; Barrier et al., 1990). The distal slopes pass basinward, very gradually, into basinal calcisiltites marls and turbiditic sands.

Mankiewicz (1987, this volume), Franseen (1989), Franseen and Mankiewicz (1991), Franseen et al. (1993) and Franseen and Goldstein (this volume) offer excellent sedimentologic and petrologic descriptions of reef talus facies and geometries and distinguish several depositional sequences previously lumped together in the slope facies by the 1979 model. Rouchy (1980), Rouchy et al. (1986) and Saint Martin and Rouchy (1990) also distinguished a rhodalgal unit at the base of the progradational coral reefs. In Mallorca and Menorca, extensive rhodalgal biostromes appear intercalated in the coral reef slopes. All this recent work shows the complexity in facies patterns of the reef talus, pointing to high-frequency cyclicity. Mankiewicz (1987, 1988) studied the distinctive layers with high concentrations of *Halimeda* and suggested the correlation with diatomaceous cycles in the basin. This is supported, at least in part, by Rouchy (1979,1982a, b), Martín and Braga (1990), Buxton and Pedley (1989), Saint Martin (1990) and Grasso et al. (1990), who also see the diatomaceous deposits as laterally equivalent in part to the rhodalgal unit or to the lower part of the coral reef framework.

The spectacular progradational character of the reef cores and slopes allows detailed high-frequency sequence stratigraphy with patterns that can be recognized in seismic lines (Pomar, 1993). Pomar (1991, 1993), Pomar and Ward (1994) and Pomer et al. (this volume) show the complexity of this progradation pattern (aggradation, progradation in highstand, downstepping, progradation in lowstand) and the need for considering the high-frequency cycles (4th-, 5th- and 6th-order) of sea level to understand the progradation patterns observed in the outcrops. Retrogradation with relative sea-level rise was first recognized by Santisteban (1981) but seems to be relatively uncommon.

One of the most intriguing geometries is the large-scale (1-2 km) downward shift (downstepping, "offlapping" in the sense of Swain, 1949) of the progradational units involving monogeneric reef frameworks and reef slopes. This was first recognized by Esteban et al. (1977a, b) and described by Dabrio et al. (1981; see also Fig. 184 in James, 1983) and interpreted to be the result of a gradual seal-level lowering by evaporation prior to evaporite deposition in the basin. Esteban and Giner (1980), Santisteban (1981), Mankiewicz (1987), Franseen (1989), Ott d'Estevou (1980), Bernet Rollande et al. (1980), Rouchy et al. (1982a, b), Saint Martin et al. (1983), Saint Martin (1990) and many others also have documented the generalized occurrence of downstepping geometries in reefs preceeding deposition of

Messinian evaporites. As noted by Schlager and Pomar (pers. communs., 1992), the very high rate of uplift of the Betic-Rif foldbelt probably started in Late Miocene times and could have played a role in the large-scale downstepping. In fact, large-scale downstepping in the Upper Miocene platform slopes is not extensively recorded outside the Betic-Rif foldbelt (Bossio et al., this volume, as a possible exception). Large-scale, downstepping (offlapping) progradational geometries are uncommon in carbonates (Kendall and Schlager, 1981), nevertheless downstepping of units prograding about 1-2 km seems to be a common part of the normal cycles of progradation-aggradation in the 20-km-wide Llucmajor platform in Mallorca without leading to evaporite deposition and without uplift during the Late Miocene times (Pomar, 1991, 1993; Pomar and Ward, 1994). Recently, Cornée et al. (1994) questioned the syndepositional character of the large-scale downstepping of some Messinian reefs, arguing that the apparent offlap is in fact the result of post-Miocene structural tilt. However, evidence is not conclusive (Franseen and Goldstein, this volume), and the 1- to 2-km offlapping cycles in Mallorca, associated with horizontal lagoonal beds, cannot be related to post-Miocene structural tilt. Most of the downstepping reefs appear related to fast-growing, oligotypic coral reefs with peculiar features suggesting deteriorating marine environments (Esteban, 1979). Furthermore, the offlapping pattern does not affect other Upper Miocene sediment types suggesting a syndepositional origin (evaporitic drawdown and/or Late Miocene uplift) rather than a post-Miocene structural tilt. Most authors have recognized minor tilts which may accentuate the offlapping pattern but seem insufficient to produce it.

*The Terminal Carbonate Complex.—*

The Terminal Carbonate Complex (TCC or TC) was defined by Esteban and collaborators during early field work in Mallorca and Almería as a distinctive mappable unit considered to be a depositional sequence different from the main reef complex (Fig. 12), bounded by two major unconformities, both with evidence of paleokarst and paleosols and onlapping expansively the truncated Upper Miocene coral reef complex or older rock sections (Esteban, 1979; Esteban and Giner, 1977, 1980; Dabrio et al., 1981). The Terminal Complex includes characteristic lithologies and was considered as a lateral equivalent (or post-dating) to the Messinian evaporites. Other authors documented the presence of the Terminal Complex and TC-like facies and geometries in many Mediterranean localities (Bossio et al., 1978; Catalano and Esteban, 1978; Pedley, 1979, 1983; Rouchy, 1979, 1982a, b, 1986; Grasso et al., 1982; Rouchy et al., 1982a, b; 1986; Pomar et al., 1983; Simó and Giner, 1983; Saint Martin and Rouchy, 1986, 1990; Pagnier, 1977; Rouchy and Pierre, 1979; Bernet-Rollande et al., 1980; Monty et al., 1980; Saint Martin, 1990; Rouchy and Saint Martin, 1992), although not always using this informal terminology. In fact, a literature review has traced references to this characteristic lithofacies as far back as Hermite's (1879) work in Mallorca. Montenat (1973a, b, 1977) described extensive occurrences of TC-type facies in southeastern Spain, but coined the term "Miocene Terminal" to refer to all Upper Tortonian-Messinian deposits. Incidentally, the TC is also widespread in the eastern Mediterranean, where blocks of this rock type are in the ancient wall of the city of Troy (Esteban, unpubl. data).

As originally defined (Esteban, 1979), the Terminal Complex is characterized by cyclic, alternating layers of oolite shoals, large stromatolites, serpulid reefs, oysters, fresh and brackish water limestones and marls, evaporitic carbonates and breccias, evaporites and *Porites* patch reefs (Type B). Stromatolites and oolites are very common lithologies, but the key aspect of the Terminal Complex is its marked cyclic character (from freshwater to normal marine to hypersaline) with erosion surfaces bounding the cycles (3-15 cycles, 0.5-5 m each). The Terminal Carbonate Complex is up to 150 m thick, but rarely exceeds 30 m and is intensively eroded by the pre-Pliocene erosion surface. In many field localities, the TC is a very distinctive unit with marked lithologic contrast to the truncated top of the reef complex. Dabrio et al. (1985) and Fornós and Pomar (1983) studied the well-preserved sedimentary structures (current and wave ripples, trough lamination) of these oolitic bars. The TC grades landward into the Marginal Terrigenous Complex, and commonly includes brackish and lacustrine facies in its upper part. In Melilla and southeastern Spain, the typical TC with predominantly carbonate lithology is unconformably overlain by a siliciclastic-rich TC (Cunningham et al., 1994).

Catalano et al. (1976), Di Stefano and Catalano (1976), Montenat (1977), Catalano and Esteban (1978) and Catalano (1979) observed that the Terminal Complex grades laterally and interfingers into basinal units attributed to the Upper Evaporite (Fig. 22). As a consequence, the 1979 model considered that the Terminal Carbonate Complex in the Mediterranean was the lateral, upslope equivalent of the Upper Messinian Evaporite, while the "Calcare di Base" of the Italian authors, an evaporitic carbonate, was considered as the lateral equivalent of the Lower Evaporite (Decima and Wezel, 1971; Decima et al., 1988). The transition between stromatolitic facies and Messinian evaporites also was documented by Delfaud and Revert (1974), Rouchy (1980), Rouchy et al. (1986), Montenat (1977), Santisteban (1981), Montenat et al. (1980, 1987), Monty (1981), Rouchy and Monty (1981), Ortí-Cabo and Shearman (1977), Saint Martin and Rouchy (1990) and Rouchy and Saint Martin (1992). Some authors only recognize the TC as a carbonate unit overlying a evaporitic unit in outcrops (Riding et al., 1991a), although this is not necessarily the last evaporite of the Messinian evaporite unit.

Santisteban (1981) in Fortuna, Bossio et al (1978, 1981) in Tuscany, Grasso et al. (1982), Pedley (1983, 1987a, b) and Grasso and Pedley (1989) in Malta and southeastern Sicily encountered difficulty in applying the simplistic 1979 concept of the Terminal Complex. In these areas, there are at least three mappable evaporitic units; each one of them separated from the other by marine to brackish water deposits and including numerous stromatolitic layers. Also, it was documented that some of the coral reefs intercalated between the evaporitic units were of

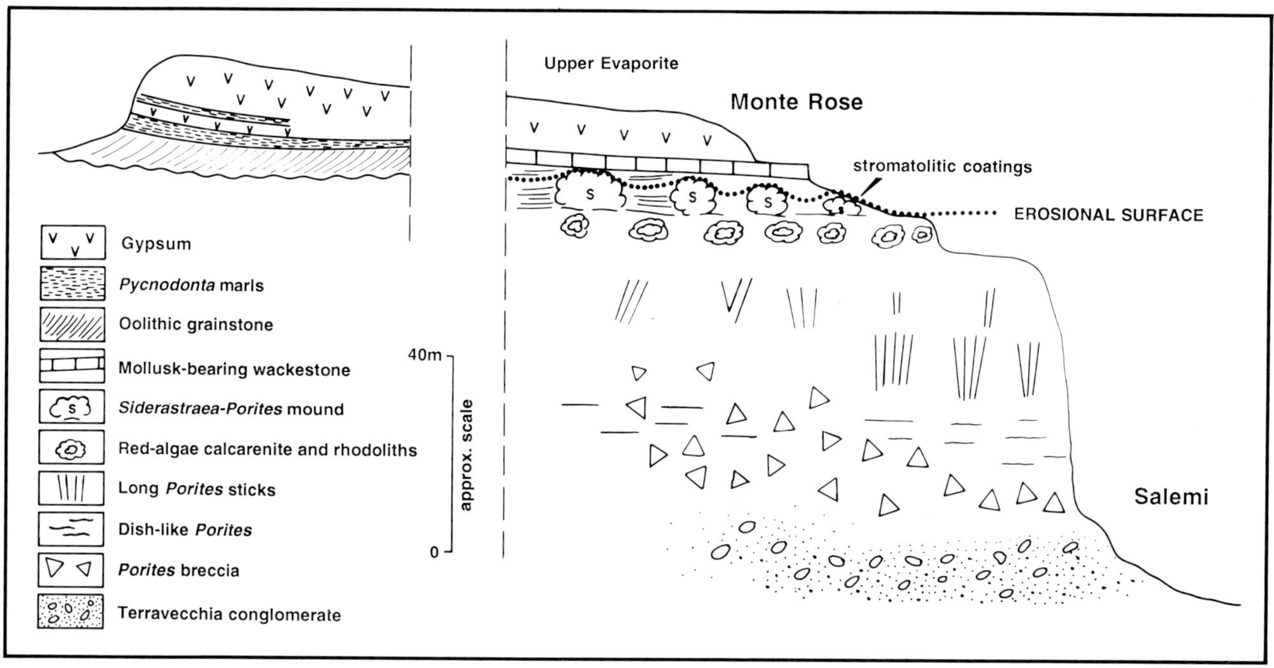


FIG. 22.—The Messinian reef of Salemi (western Sicily) overlain by transitional beds between the Terminal Complex and the Upper Evaporite (after Catalano, 1979; Catalano and Esteban, 1978).

relatively high diversity and well developed, practically indistinguishable from older Miocene reefs. The working hypothesis of Esteban (1979) proposed that only Type B reefs would be present in the TC. This was an oversimplification; it is now clear that the TC also includes intercalations of Type-A reefs. Furthermore, some of the lagoonal back-reef lithofacies resemble the stromatolites and oolites of the TC and also occur on truncated coral reefs; their distinction can be problematic (e.g., in Mallorca). In essence, a satisfactory definition of the TC should include its relationship (documented or assumed) with Messinian evaporites; either grading into, interfingering with or post-dating a basinal evaporitic unit (not only the Upper Evaporite). Some basins may have 2 or 3 Terminal Complexes (Bossio et al., this volume; Cunningham et al., 1994); the lithofacies alone or the marked erosional truncation at the base are not sufficient criteria.

In Santa Pola, southeastern Spain, Esteban and Giner (1977), Esteban et al. (1977b) and Esteban (1979) erroneously considered the entire structure as part of the TC, and Maurin et al. (1980) demonstrated the importance of a karstified subaerial exposure surface separating the truncated reef complex from the stromatolitic TC. However, continued work in the area (also Vallés-Roca, 1986; Calvet et al., this volume; Saint Martin, 1993) confirmed that the stromatolitic-oolitic TC contains several intercalations of *Porites* patch reefs. Montenat et al. (1980, 1987) also mentioned small buildups of *Porites-Tarbellastraea* intercalated in the stromatolites near Santa Pola. The presence of *Porites* reefs in the TC is also confirmed in Almería (Esteban and Giner, 1980; Dabrio and Martin, 1978; Dabrio et al., 1981, 1985; Franseen, 1989; Riding et al., 1991a), Fortuna (Santisteban, 1981; Müller, 1986; Müller and Hsü, 1987), Sicily (Esteban et al., 1982a, b; Grasso et al., 1982) and Tuscany (Bossio et al., 1981). Small *Porites* reefs occur on a erosion surface truncating a gypsum unit in Cyprus, but its interpretation is unclear; they could be slumped blocks (Orszag-Sperber and Rouchy, 1979). Except for some possible isolated coral colonies or knobs, Rouchy (1982a, b), Saint Martin (1990), Saint Martin and Rouchy (1990) and Rouchy and Saint Martin (1992) did not recognize the presence of true coral reefs intercalated between stromatolites and oolites of the TC (post-dating Messinian evaporite).

The Terminal Complex (TC) remains a useful mappable unit grading into, interfingering with or pre-, or post-dating a Messinian evaporite unit and showing strong or extreme cyclic character involving freshwater, marine and hypersaline coastal facies. Stromatolites and oolites are typical lithologies, but not always present. These coastal carbonate facies are commonly difficult to date paleontologically as Messinian age; they show important influences or affinities with "Lac Mer"-Paratethys type of biofacies *(Congeria, Melanopsis, Hydrobia, Cyprideis, Ammonia,* limmocardids), more definitive in the upper part, where lacustrine marls, conglomerates and sands are common. In many localities, the TC is severely karstified and onlapped by well-defined Lower Pliocene open-marine sediments. On top of the truncated reef platform and slopes, the lower limit of the Terminal Complex is a major unconformity, locally with terraced erosion surfaces (Fig. 23). Esteban and Giner (1980), Vallés-Roca (1986), Franseen (1989) and Calvet et al. (this

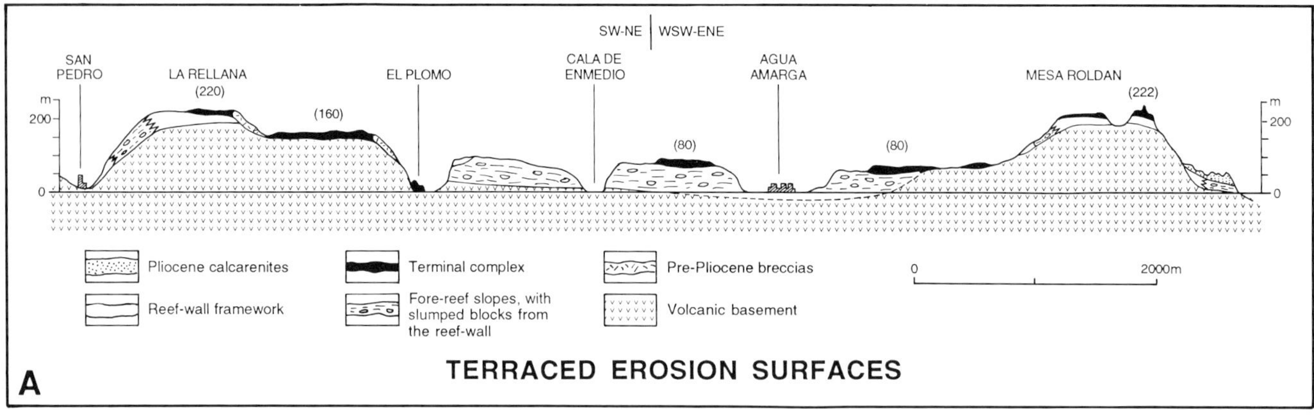


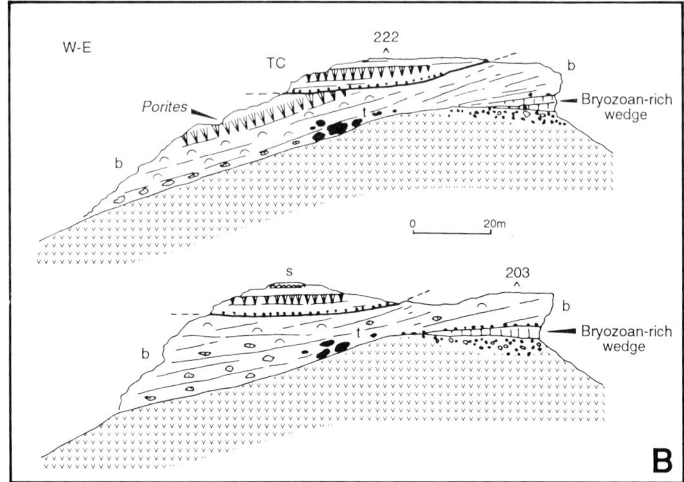


FIG. 23.—Terraced erosion surfaces onlapped by the Terminal Complex. (A) Cabo de Gata (Esteban and Giner, 1980). (B) Mesa Roldán (Esteban and Giner, 1977); v: volcanic basement with conglomerates; b: coral reef slopes with *Halimeda* sands, rhodoliths and slumped blocks of *Tarbellastraea* (t), grading into lower reef wall with in-place *Porites;* TC: Terminal Complex, with *Porites,* oolitic-skeletal sands and stromatolites (s). Notice the presence of *Porites* patch reefs intercalated in the Terminal Complex.

volume) interpreted these surfaces as successive coastal marine planation surfaces and sea cliffs related to the onlapping TC sequence. On the distal reef slopes and basinal areas, the base of the TC commonly shows little evidence of erosion and appears offlapping and interfingering with the underlying distal reef slopes and restricted basinal marine marls (Montenat et al., 1980; Van de Poel, 1991). Some authors believe that some of the TC units are Early-Middle Messinian in age (based on data summarized by Saint Martin, 1990; Grasso et al., 1982, 1990; Pedley, 1987a, b, this volume; Van de Poel, 1991; Pedley and Grasso, 1993; magnetostratigraphy of Cunningham et al., 1994; and $^{87}Sr/^{86}Sr$ data of Oswald, pers. commun.) and could be correlative with the Lower Evaporite. According to the present definition, the age of TC units could range from earliest to latest Messinian times.

Units in many aspects similar to the Terminal Complex (including relationship with basinal evaporites) also are known in the Early-Mid Miocene reefs of the Red Sea (Rouchy, 1982b; Monty et al., 1987; James et al., 1988; Burchette, 1988; Purser et al., 1987, 1990; Montenat et al., 1988) and in reefs of other ages, such as the Priabonian (Late Eocene) of Catalonia (Salas, 1979; Salas and Esteban, 1980; Puigdefábregas, 1975), the Silurian of the Michigan Basin (Mesolella et al., 1974; Huh et al., 1977; Gill, 1977, 1985) and the Devonian of Saskatchewan (Maiklem, 1971). This indicates a basic common pattern of association of reefs and evaporites. It is not surprising that the lateral relationships of reefs-evaporites and TC-like lithologies have engendered numerous discussions.

### *Reef-Evaporite Relationships*

Supporting many early observations (Montenat, 1973a, b, 1977; Dronkert and Pagnier, 1977), the 1979 model recognized a classic wedge-on-wedge relationship (reciprocal sedimentation in Wilson, 1975) between the coral reef complex (Upper Tortonian-Lower Messinian) and the Messinian evaporite (Fig. 13), with a major discontinuity and erosion surface separating the two units. Esteban and Giner (1977a, b, 1980) and Esteban (1978, 1979) interpreted the lateral facies transition between the Terminal Complex and the Upper Evaporite as multiple, stacked

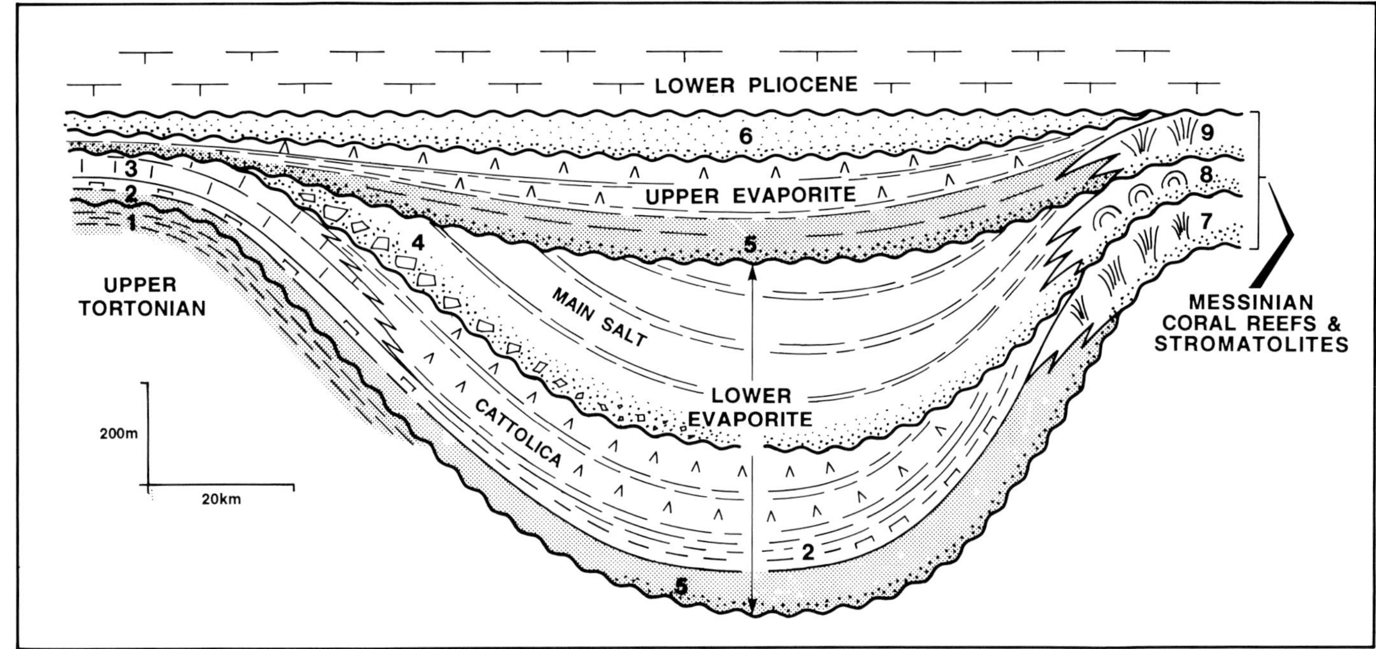


FIG. 24.—Stratigraphy of the Messinian Evaporite in Sicily (modified from Decima and Wezel, 1971) and the hypothetic correlation with the Upper Miocene coral reefs in Sicily, Italy and southeastern Spain. (1) Uppermost Tortonian marine marls. (2) Lowermost Messinian marls and diatomites (Tripoli). (3) Calcare di Base, calcitized evaporites and stromatolites. (4) Anhydritic basal breccia and gypsum-turbidites. (5) Gypsarenites. (6) Arenazzolo, lacustrine sands and marls. (7) Lower Messinian coral reefs. (8) Mid Messinian stromatolites and rhodalgal carbonates. (9) Upper Messinian coral reefs and stromatolites.

wedge-on-wedge relationships between carbonates and evaporites separated by erosional surfaces. This review indicates that reef-evaporite relationships are more complex than previously thought; Santisteban and Dawans (1985) show an interpretation of these complex relationships.

The Lower Messinian coral reefs appear well dated micropaleontologically (Martín and Braga, 1990; Saint Martin and Rouchy, 1990; Benson et al., 1991; Bossio et al., this volume) and occur on the flanks of older reefs considered as Late Tortonian in age. Some well-developed coral reefs were erroneously considered to post-date the Messinian Evaporite (Garrido-Megías, 1985; Dabrio et al., 1985), but Martín and Braga (1990) demonstrated that these reefs were pre-evaporitic. Some uncertainty exists regarding the age of the base of the reef complex; in some areas it could represent the latest Tortonian (Fini-Tortonian).

Only the terminations (up to few hundreds of meters) of the Mediterranean Messinian Evaporite (1,500-3,500 m) crops out along the coastal areas. The classic Upper and Lower Evaporite subdivision of Decima and Wezel (1971) serves for most seismic work in the region, but it is insufficient to accommodate current field observations (Fig. 24). At least three major evaporitic units have been documented in Fortuna, Sicily and Tuscany, most of which include basal diatomitic deposits, resedimented and turbiditic (evaporite) horizons, and stromatolites. They are correlative with erosion surfaces truncating coastal carbonate cycles (including coral reefs). The Lower Evaporite in central Sicily can be further subdivided into the Cattolica Gypsum and the Main Salt by a horizon with breccias and gypsoturbidites. The Upper Evaporite is clearly transgressive over the Lower Evaporite and part of the Lower Messinian coral reefs (western Sicily). Each evaporitic unit consists of 6 to more than 30 cycles of marine (normal, oligotypic, dwarfed faunas) marls, diatomites, turbidites and gypsoturbidites; the Upper Evaporite also has marine cycles at the base, but salinity is progressively reduced toward freshwater in the upper part (Lac Mer, Paratethyan affinities in faunal assemblages). The rhythmic character of the Messinian evaporites has been emphasized since Nesteroff and Glaçon (1975). Troelstra et al. (1980), Rouchy (1982a, b) and Van de Poel (1991) provide micropaleontological documentation and comparisons with other localities. Benson et al. (1991) propose that the base of the type section of the Tripoli in Sicily roughly corresponds to the Tortonian-Messinian boundary (also McKenzie et al., 1979; Chamley et al., 1986; Müller and Hsü, 1987). This is considered the onset of the Messinian crisis represented by "deteriorating" cyclic marine conditions occurring before evaporite deposition. However, basin-wide correlations should take into account the fact that younger or older evaporite units with basal diatomitic layers may occur locally in marginal, partly restricted basins.

Field observations by numerous authors are here interpreted to indicate that there are two types of packages of evaporites and associated restricted marine carbonates in the Terminal Complex (Fig. 25). The lower package is in offlapping continuity with the toe of the slope of the last coral reefs (typically downstepping with abundant *Halimeda* and monogeneric reef cores) and is best developed in basinal areas. This package is characterized by high-frequency cycles of increasing salinity and biotic restriction (restricted marine to stromatolitic to evapor-

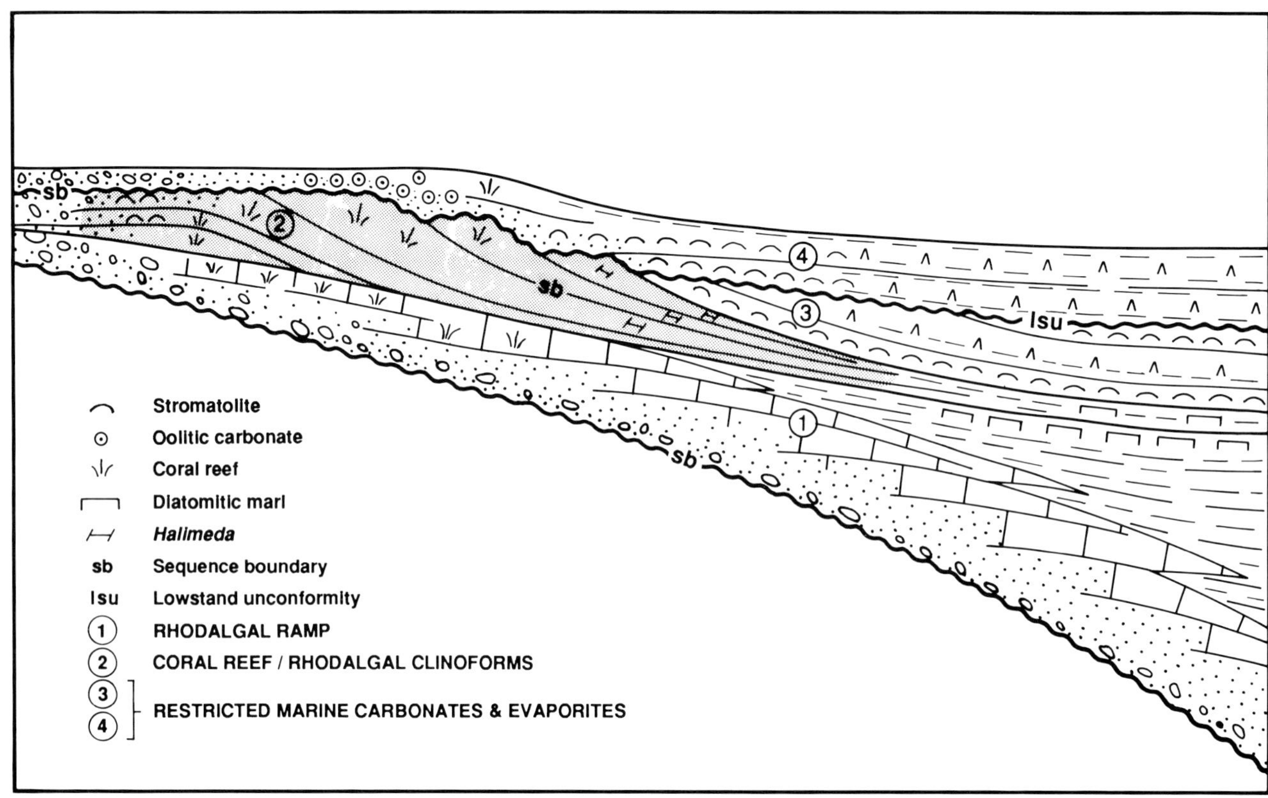


FIG. 25.—Reef-evaporite relationships

ite) and include intercalations of restricted marine marls with oligotypic planktonic forams and diatomites. The upper package is onlapping a major erosional surface truncating the offlapping lower package and coral reef. Gypsarenites, conglomerates and breccia blocks of different lithologies are particularly abundant at the base of the upper package. High-frequency cycles, each of them with upward-decreasing salinity, characterize the upper package (evaporite or minor erosion surface to stromatolite to restricted marine to shallow, open marine); influx of continental waters are locally recorded and become predominant in the uppermost cycles of the upper evaporitic package (Rouchy and Pierre, 1979; Pierre, 1982; Pierre and Rouchy, 1990).

The main problem of the Messinian carbonates in the Mediterranean is the short amount of time (about 1 my) to account for thick evaporites and intercalated marine deposits, at least four major erosional surfaces, and complex stratigraphic relationships (Fig. 24). Haq et al. (1987) show a single 3rd-order cycle corresponding approximately to most of the Messinian times; several authors have attempted to correlate the general Messinian stratigraphy to the global 3rd-order cycle (Thunell et al., 1987; Mankiewicz, 1987; Franseen, 1989; Saint Martin and Rouchy, 1990; Franseen et al., 1993). To better accommodate the stratigraphic packages defined by numerous authors, it is proposed here to consider the fourth-order depositional sequences rather than the standard 3rd-order depositional sequences; an approach already suggested for some of the sequences by Franseen and Mankiewicz (1991) and Franseen et al. (1993). Figure 26 presents a first attempt of correlation between Vail's global 3rd-order cycles and the western Mediterranean (eastern Betics) 4th-order cycles of relative sea level. The magnitudes of these 4th-order cycles vary in different localities, and the correlation of events is still problematic. However, 4th-order sequence boundaries are believed to be synchronous and correlatable at the Mediterranean scale. The character of each Messinian 4th-order carbonate sequence seems to be largely dependent on the local conditions of subsidence, sediment supply and accomodation. Available data suggest different ages for the large-scale downstepping Messinian reefs. Although there is good general agreement of the local sea-level curves (3rd-4th order) for all the Mediterranean (as in Santisteban, 1981; Franseen and Mankiewicz, 1991; Franseen et al., 1993) and Atlantic Miocene sediments (Esteban, unpubl. data) with Vail's eustatic curve the Messinian units present a significant departure. This is interpreted as a reflection of the isolation of the Mediterranean and the repeated evaporitic drawdown of sea level. It would seem that the global Messinian highstand (TB3.3) supplied the necessary Atlantic water to produce the thick Lower Evaporite unit in the semi-isolated Mediterranean.

The correlation hypothesis of Figure 26 suggests that the Upper Evaporite in the western Mediterranean still contains marine intercalations, but these are less prominent in the eastern

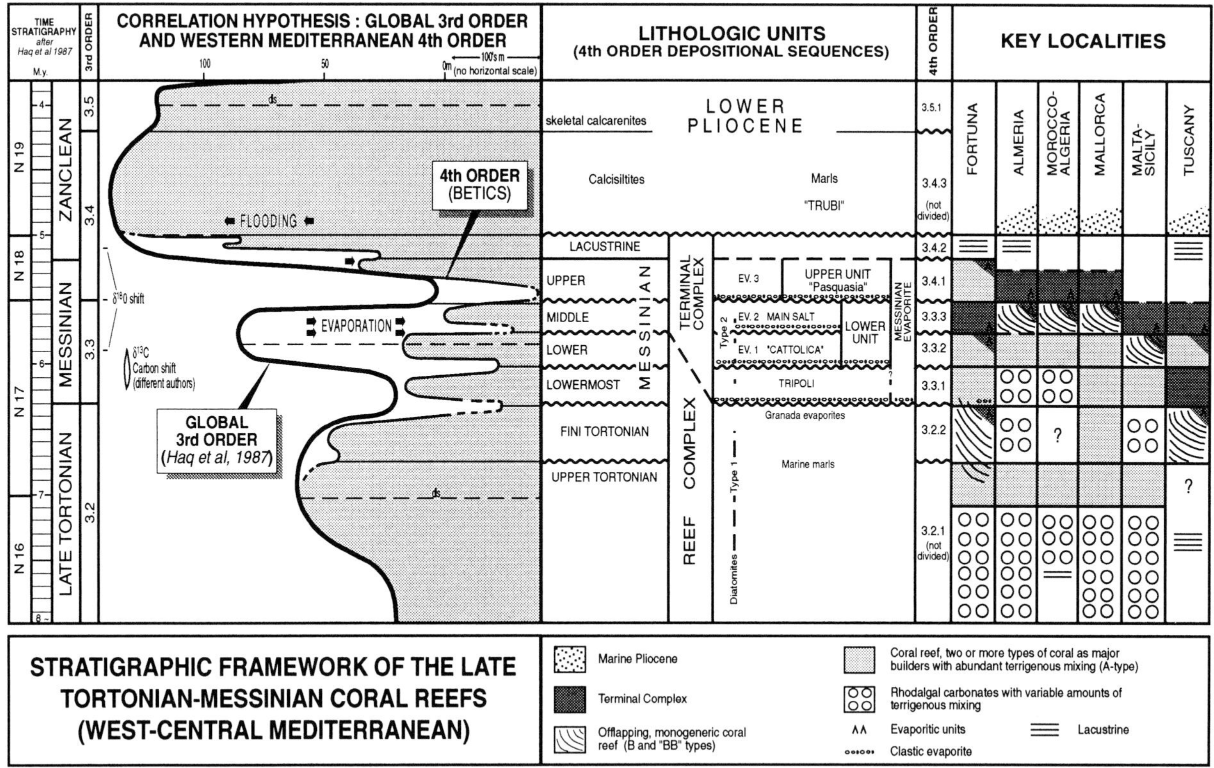


FIG. 26.—Fourth-order Upper Miocene depositional sequences and the corresponding cycles of the relative sea level in the Betic province of the western Mediterranean, as compared with the 3rd-order sequence stratigraphy of Haq et al. (1987).

Mediterranean. This hypothesis is consistent with Müller and Hsü (1987). Alternatively, Benson et al. (1991) and Van de Poel (1991) propose that there is no record of marine deposits related to the Upper Evaporite; in this case, the Terminal Complex should correspond to a markedly transgressive 4th-order sequence within the TB3.3. This requires a shift of the sea-level curve to include the Fini-Tortonian cycle as part of the Early Messinian times (as indicated by Saint Martin and Rouchy, 1990). These alternative hypotheses have to be refined and evaluated by further field and biostratigraphic work.

Recently, Gautier et al. (1994) propose that the entire salinity crisis (meaning the two main evaporite bodies in the Mediterranean basin) occurred later and was shorter than previously thought (Benson et al., 1991), corresponding entirely to the first reverse chron of the Gilbert Epoch and a duration of about 0.4 Ma (from 5.3 to 5.7 Ma). The base of the Gilbert Epoch was previously considered as the limit of the Lower-Upper Evaporite. New stratigraphic data consider the Messinian period older and longer (5.3-6.9 or 7.2 Ma), allowing more time for the development of complex architecture of the Messinian carbonates and associated unconformities. The Messinian stratigraphy is in a state of flux, and the correlation presented here (Fig. 26) is only an attempt. The main problem may derive from the presence of older evaporite units in marginal, restricted basins without correlation with the main evaporite units in the Sicilian reference sections in the central Mediterranean (already indicated by Rouchy, 1982a).

The significance of the Messinian evaporites, popularly referred to as the expression of the "Messinian salinity crisis or crises" or "Messinian events," continues to inspire intense discussion. Extensive and thick deposits of evaporites (locally 2-4,000 m thick) also occur in Lower, Middle and other Upper Miocene units in the Mediterranean, Paratethys, Mesopotamia and the Red Sea (Figs. 4 to 8). It is remarkable that only the Mediterranean Messinian evaporites have generated such intensive and polemic attention; perhaps this is because of the use of terms such as "crisis". While popular usage equates the terms "salinity crisis" and "Messinian events" to massive evaporite precipitation, it is well known that the Messinian evaporite unit also contains many cyclic intercalations of marine to freshwater deposits. Messinian events have to be considered not only as a time of salinity crises but also not only a time of hypersalinity. Indeed, Rouchy (1986) considers the Messinian evaporites the result of repeated phases (not a single event) of concentration of marine waters in a fragile equilibrium with increasing freshwater influences and controlled by a wide variety of factors.

### *Peculiarities of Some Messinian Reefs in the Mediterranean*

With the aim of drawing attention to the peculiarities of some of the Upper Miocene reefs in the Mediterranean, Esteban et al. (1977a, b; 1978) and Esteban (1978, 1979) pointed out the presence of so-called "aberrant" features, mainly those considered to be Early Messinian in age and of low taxonomic diversity (type B). These features were not encountered in: (a) the "normal" Upper Miocene higher diversity (type A) in the Medi-

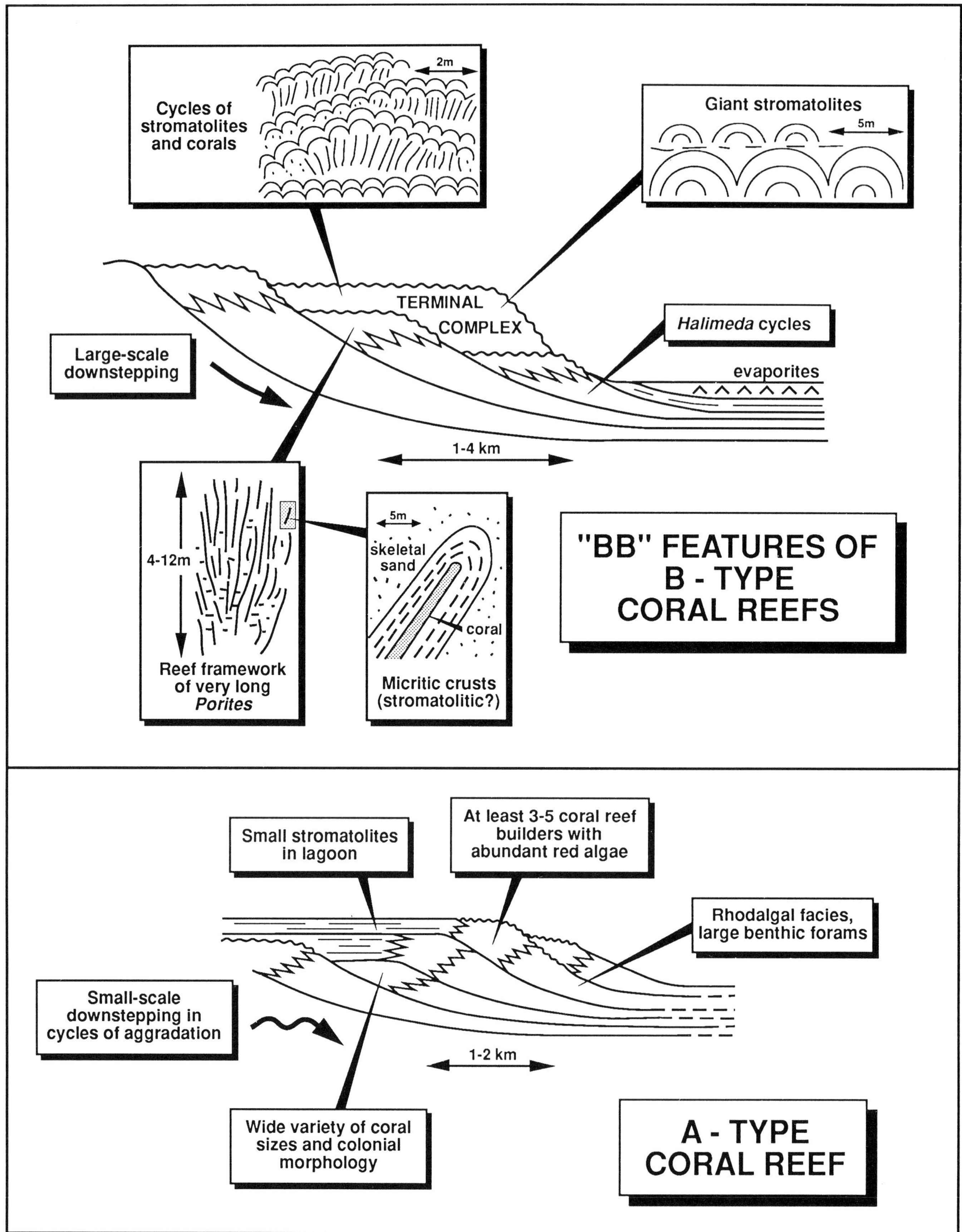


FIG. 27.—The peculiar "BB" features of some Messinian coral reefs in the western Mediterranean and their comparison with Tortonian-Lower Messinian A-type coral reefs.

terranean, (b) older Miocene reefs in the Mediterranean and (c) most other reefs in the Mediterranean and elsewhere. While terms such as "aberrant" and "normal" are not the most adequate for natural phenomena, the point was to stress the peculiar or uncommon features of these reefs. This objective was fully accomplished, and it generated much attention and intense discussion. Probably one of the factors that generated controversy was the interpreted relationship of these features with the prelude to the Messinian "crises" (see above).

*The hypothesis.—*

Some of the oligotypic (type B) reefs show distinctive peculiar features that will be here collectively referred to as BB features and reefs (Fig. 27) in order to avoid the polemic term "aberrant". Other authors use terms such as "spectacular," "impressive," "original," "special" and "uncommon." Expanding from Esteban (1979), the peculiar BB features essentially consist of the following:

1. The very large size of these oligotypic *(Porites,* locally with *Tarbellastraea)* reefs with very impoverished associated community; progradations of more than 3 km and thicknesses of more than 200 m are not uncommon. The point is the rarity of similar-sized monogeneric coral reefs in the geologic record.

2. These Messinian reefs depart from the general trend of progressive decrease in coral reef diversity and associated reef size, from the maximum in the Oligocene and disappearance in the Early Pliocene times (Wells, 1956; Heckel, 1974; Chevalier, 1961). There seems to be an inverse relationship: higher diversity-smaller reef size, and vice versa.

3. The impressive colonial morphologies of the thin (1-3 cm) and very long *Porites* (locally also Tarbellastraea*)* branches, commonly displayed by these Messinian reefs. Individual *Porites* sticks are 40-60 cm long, but if branching is included, individual sticks can be traced for 4 m and probably for much more (at least 7 m in Santisteban, 1981). The synoptic relief (actual height of the colony on the sediment bottom) is difficult to evaluate, but it is likely to have been much less. While the growth strategy appears as standard (variations of light availability), the size of the colonies makes them special; they are uncommon in any other coral reef. Esteban et al. (1977a, b and subsequent papers) noted that physically stressed modern corals do not show the exuberant colonial morphologies of these Messinian reefs. These Messinian corals are even more peculiar when considering the interpretation of the associated micritic crusts as marine stromatolites (Pedley, 1979, 1983; Riding et al., 1991b). In some coral frameworks, these micrite crusts are the dominant rock type (30-80% of the reef rock).

4. The presence in the TC of gigantic (up to 8 m in diameter) domal stromatolites is uncommon after Early Paleozoic time, although large stromatolitic mounds have been described in normal, slightly restricted, high-energy marine environments on the margins of the Bahamian platforms (Dill et al., 1986; Awramik and Riding, 1988), and Riding et al. (1991a) believe that they are similar to the Messinian stromatolites. However, most of the Messinian domes are characterized by the abundance of huge accumulations of monogeneric gastropods (cerithids) and predominance of oolitic-peloidal grains; only in few localities there are minor, scattered and scarce marine borings, red algae coatings and small coral knobs and fragments. Although some authors refer to them as "normal" marine, it has to be recognized that, at best, this is a very impoverished, low-diversity assemblage suggesting generalized marine eutrophic conditions. Evaporitic molds are locally present but scarce; lateral transitions between stromatolites and evaporites have been observed (see above), but all these observations only support local evaporitic-marine environments rather than generalized hypersaline settings. Perhaps the important point is that all these giant stromatolites are unexpected, surprising, uncommon and difficult to explain with our present models.

These BB features were considered "peculiar" by Esteban et al. (1977a, b) and Esteban (1978, 1979), who interpreted them as a response to deteriorating environmental conditions (namely temperature and salinity variations, increase in concentration of nutrients or organic matter) leading to the Messinian "crises" (rather than "crisis"). Terms such as "salinity-?" variations (not hypersalinity) were used. Esteban et al. (1977a, b) pointed out that BB features were also present in some coral reefs in the Terminal Complex and discussed the significance of the survival of corals after extensive evaporite deposition.

Esteban et al. (1977a, b) recognized a global cooling trend and a decrease in coral diversity during Miocene time (as demonstrated by Chevalier, 1961, 1977) and referred to the Late Miocene Mediterranean as a "coral refuge" with connections or influences from colder, adjacent Atlantic waters without coral reefs (Figs. 9, 15). Esteban (1978, 1979) speculated about the possibility of physical disturbance introduced by entering cold Atlantic waters in reducing the taxonomic diversity of coral reefs (as a retrogressive succession in a relict mature community). This could be one of the factors eliminating the more delicate, aggressive and ecologically adapted members of the reef community and allowing un-checked growth of *Porites* without competitors. Esteban et al. (1977a, b, 1978) and Esteban (1979) also speculated on the contribution of other factors such as organic matter and nutrients. In any event, these BB Messinian reefs cannot be compared to the typical tropical reefs, nor can they be compared with the modern sub-tropical to temperate reef occurrences.

*The evaluation.—*

Santisteban (1981) proved that relatively higher-diversity reefs (type A) occurred sandwiched between and after evaporitic units, whereas low-diversity BB reefs occurred clearly before the onset of Messinian evaporite deposition. Bossio et al. (1981, this volume) support this pattern. Santisteban (1981), Saint Martin (1984) and Pedley (1987a, b) suggested that variations in coral diversity appear related to local factors (bottom topography, subsidence, restriction and sediment supply) rather than age or Messinian events. They concluded, therefore, that BB features could not have been related to the sharp onset of the Messinian "crisis" (in the sense of synchronous, Mediterranean-wide evaporite deposition). This is a significant departure from one of the assumptions of the 1979 model: the isochron value of the BB reefs.

With respect to the peculiar colonial morphology of *Porites,* Santisteban (1981) and Pedley (1983) attributed the very long coral sticks to fast rates of relative sea-level rise in areas with active slope sedimentation, rather than to any environmental condition associated with Messinian events. This interpretation is possible and also was considered in Esteban (1979), but it does not solve the main problem. Why do we not see similar coral morphologies and sizes in either synchronous, older or younger coral reefs in other places? Why should a very fast rise in relative sea level be peculiar to the Late Miocene Mediterranean?

Saint Martin (1990) discovered a small (few meters thick) but significant coral reef south of Meknes, Morocco in the most western part of the Rif corridor (Fig. 15) and stressed its similarity with reefs on the Mediterranean coasts of Morocco and Algeria. However, the area around Meknes (Saïss Basin) is separated from the Atlantic embayment of the Gharb basin by the submerged sill of the Oued Beth (Wernli, 1987; Esteban, unpubl. report). Moreover, there is no record (surface or subsurface) of coral reefs in the Tortonian and Messinian of the Gharb Basin or the Guadalquivir Basin (Fig. 15). Obviously, the Atlantic waters in these areas were too cold (or too rich in nutrients) to allow coral reef growth.

The peculiar "aberrant" (BB) reefs occur clearly before the onset of evaporite deposition; hypersalinity could not have been the cause of any aberrant feature or stress in the markedly stenohaline coral reef community (echinoderms, *Halimeda,* red algae). Esteban et al. (1977a, b) and Esteban (1978, 1979) never suggested that these "aberrant" reefs grew in hypersaline conditions; it was only assumed that they were somehow connected to a series of environmental factors in deteriorating marine conditions that preceded extensive evaporite deposition. Salinity (not hypersalinity) variations were considered, but Esteban (1978, 1979) emphasized the possible role of cooler waters of Atlantic origin and (since Esteban et al., 1977a, b) the possible role of increasing nutrient levels in disrupting advanced stages of ecologic succession as a possible explanation of the "aberrant" features. Confusion probably has arisen from the discussion of the survival of corals in the cycles of the Terminal Complex (Esteban et al., 1977a, b; Esteban, 1978, 1979), which clearly involved the re-establishment of "normal-marine" conditions after extensive evaporite deposition in the basin.

The basinal marls deposited during the growth of the BB reefs contain numerous intercalations of laminated diatomitic deposits (Troelstra et al., 1980; Rouchy et al., 1986; Rouchy, 1988; Saint Martin, 1990; Saint Martin and Rouchy, 1990; Van de Poel, 1991). According to Meulenkamp et al. (1979), these basinal marls precede the evaporites and represent "aberrant" bottom conditions, and Troelstra et al. (1980) assumed an increase in marine salinities and decrease in water temperature to be the cause of extremely low diversity of abundant planktonic forams. Montenat et al. (1980) and Van de Poel (1991) emphasized the oxygen-deficiency to explain the dwarf faunas, dominance of *Buliminacea* and abundance of well preserved fish and metahaline environments. At the base of the diatomitic units, the planktonic fauna is more diversified and presents Atlantic affinities (type 1 of Rouchy, 1982a, b). Towards the top, the marls are more oligotypic, reflecting more euxinic, eutrophic conditions (type 2 of Rouchy, 1982a, b). These basinal marls are very similar to the marls intercalated in the BB coral reef slopes and in between the evaporite layers in their lower part.

Rouchy et al. (1986, p. 344) concluded that increasing confinement and restriction led to the deposition of diatomitic laminites as a response to increasing "salinity" and caused the extinction of corals (other than *Porites)* prior to evaporite deposition. Charrière and Saint Martin (1989), Saint Martin (1990) and Moissette and Saint Martin (1990) observe that monogeneric reefs with peculiar BB growth forms occur in areas adjacent to diatomitic deposits and consider the role of local "upwelling" (and the associated cooler, high-nutrient waters) in reducing diversity and increasing growth rates. Diatomitic deposits are more abundant in the western part of the western Mediterranean (Rouchy, 1982a, b), in the general area with common monogeneric reefs. However, this relationship is far from clear in local basins (i.e., Tuscany, Melilla). Poorly diversified coral reefs tend to occur in basins with diatomites, but the opposite is not true. Diversified reefs do occur in basins with diatomites (Rosignano Limestone; Bossio et al., this volume). Martin and Braga (1990) and Dabrio et al. (1985) also reported lateral transition and downlapping of prograding *Halimeda* slopes of BB reefs onto diatomites. In any case, many authors (see above) consider the deposition of the Tripoli diatomites as marking the onset of the Messinian events (salinity crises). Pedley (this volume) speculates on the possibility of spill-over from adjacent diatomitic embayments, rather than direct upwelling, to explain the BB features. All data and interpretations strongly support the hypothesis of Esteban (1978, 1979); confinement and increasing marine salinity is already part of the Messinian event (before the deposition of the evaporite), as indicated by Rouchy (1986, p. 517, 519).

Many of the reef frameworks with BB features also show thick (2-4 cm), poorly laminated micritic crusts extensively coating the corals. These crusts contain variable amounts of encrusting forams and red algae. The importance of these crusts as the volumetrically predominant rock type in the reef framework volume was first pointed out by Lloyd C. Pray (pers. commun., 1979). They may form up to 40-70% of the reef rock (Dabrio et al., 1981). Although some possible algal filaments were recognized and the stromatolite interpretation was considered (Esteban et al., 1977b, p. 4.25), these crusts were eventually interpreted as submarine micritic cements (Dabrio et al., 1981), but Pedley (1979, 1983) insisted on a subtidal stromatolite interpretation. Their similarity to micritic coatings in the Pleistocene *Acropora cervicornis* frameworks of Northern Jamaica (Esteban, unpubl. data) and the Holocene micritic cements in Discovery Bay (Land and Goreau, 1970) favored the micritic cement interpretation. Jones and Hunter (1991) show *Acropora* branches from the Late Pleistocene reefs of the Caribbean with a community replacement sequence (succession) remarkably similar to the Messinian crusts on *Porites*. Dead coral branches were encrusted by a succession of red algae, foraminifers, serpulids and stromatolites, reflecting regressive conditions (Jones and Hunter, 1991). Similar succession could also be explained by increasing nutrient levels and decreasing temperatures, without the need of shallower and quieter conditions. Riding et al. (1991b) affirmed them as marine stromatolites. If this is the case, one could presume the presence of high nutrient levels and algal blooms periodically affecting these peculiar (although marine) BB reefs for a sufficiently short period of time to allow recovery of coral growth. It could also be possible that the very long coral sticks were a response to this microbial infestation. These micritic crusts may coincide with important accumulations of *Halimeda* in the BB reef slopes (Mankiewicz, 1987, 1988). In essence, micritic crusts, locally the most outstanding contributor to the reef rock volume, are suggestive of cyanobacterial (also heterotropic bacteria, Pedley, pers. commun.) populations affecting the coral reefs. In any event, these comparisons illustrate that micritic crusts in coral reefs correspond to "special" marine conditions not favorable to coral growth. Increased awareness of these crusts on modern coral reefs and their relation to microbial infestations will assist in accurate interpretations of the Messinian environments.

Riding et al. (1991b) and Martín and Braga (1994) insist in considering as "normal" marine the large stromatolitic domes on the basis of their local association with minor amounts of marine fossils (see above). The discussion is semantic to a great extent. The Messinian evaporite units contain abundant marine intercalations, and the evaporite rock itself contains scattered marine fossils. The fact that some cyclic stromatolitic sections interfinger with cyclic evaporite sections does not support a hypersaline origin of the stromatolites (see above references), but the assemblages in the stromatolites do not support a "normal" marine setting either. The presence of stromatolitic coatings on the coral framework or the replacement of the coral reef by stromatolitic mounds suggest massive microbial infestations, sickness and death of the live coral. These microbial infestations may occur in marine environments but are not normal for coral reefs. The study of extensively infested modern reefs in the Caribbean could provide insights in the evaluation of the Messinian reef crisis.

In conclusion, this discussion confirms that the development of BB reefs was apparently related to "special" marine environmental conditions (water temperatures, salinities, nutrients) that can be seen as part of the complex Messinian events (Esteban, 1979). Most of the objections to the 1979 model can be accommodated by avoiding colorful terms such as "aberrant" and "crisis" and considering that the Messinian events started with the deposition of basinal diatomitic marls before basinal evaporite cycles. Hypersalinty is not and was not implied in the interpretation of the "special" character of the Messinian reefs. However, these "special" Messinian conditions were not synchronous all over the Messinian Mediterranean, and there are Messinian coral reefs without BB features. It is quite likely that there is a chain of inter-related environmental factors (water circulation, upwellings, temperature, nutrients, salinity) that explains the "connection" between the peculiar "BB" reefs, the diatomites and the evaporites.

*The synthesis.—*

The above paragraphs suggest a relationship between: (i) the entry of colder Atlantic waters (Esteban, 1979) that could have increased nutrient levels, also involved variations in salinity and organic matter (Esteban, 1978, 1979), (ii) upwellings, type-1 diatomites and their correlation to rhodalgal units, as convincingly demonstrated by Rouchy (1982a, b; 1988) and Saint Martin (1990), (iii) change of environmental conditions with aggradation and progradation of coral reefs ending with downstepping, *Halimeda* concentrations and BB features; this suggests a trend towards evaporation (but still mostly marine!), stagnation, euxinification, ending with type-2 diatomites, (iv) increased evaporation, subaerial exposure of the margins, and stromatolites, diatomites and evaporite-marl cycles in the basin, and (v) onlap of the Terminal Complex on the truncated coral reef complex. Figure 28 reflects this evolution of coral reef types.

Coral reefs with the peculiar "BB" features are all Messinian; they appear at times of environmental change preluding the onset of local cycles of evaporite deposition in the basin (Messinian events), but not all the Messinian reefs show "BB" features. The Esteban (1988) model (see below) suggests that the "BB" features may occur in the latest part of any prograding coral reef during relative sea-level fall before the deposition of basinal evaporites. In this way, the absence of "BB" features in some of the reefs preceding or sandwiched between evaporitic units (Santisteban, 1981, Fortuna; Rouchy, 1982a, b, Lorca) could be explained as the non-deposition or erosion of the latest part of the prograding reef complex.

Benson et al. (1991) finally demonstrated the influx of the "cold" Atlantic waters into the Mediterranean postulated by Esteban (1979). They documented the influx of Atlantic psychrospheric and temperate mesopelagic faunas in the Rif Straits at the beginning of Messinian time (before the Tripoli) and during the entire Early Messinian time till about 5 Ma. Benson et al. (1991) suggest that the paleoceanographic evolution since Mid Miocene time led to an acceleration and reduction in thickness of the North Atlantic Gyre in the Late Tortonian. Seasonal upwelling off the Atlantic coast of Morocco was "siphoned" into the Mediterranean as a consequence of the water-budget deficit in the Mediterranean (and reduction or severance of the discharge waters into the Atlantic). The model of Benson et al. (1991) provides a consistent explanation for the presence of nutrient-rich temperate waters of Atlantic origin and the development of BB reefs. However, the presence of type-A coral reefs in the Messinian units and the cyclicity of coral and algal (red algal, cyanobacterial, green algal?) events suggest that the "siphon" pulsated rather than flowed steadily. The main objection to the Benson et al. (1991) model is the nature and location of the Betic Strait, further discussion is provided below

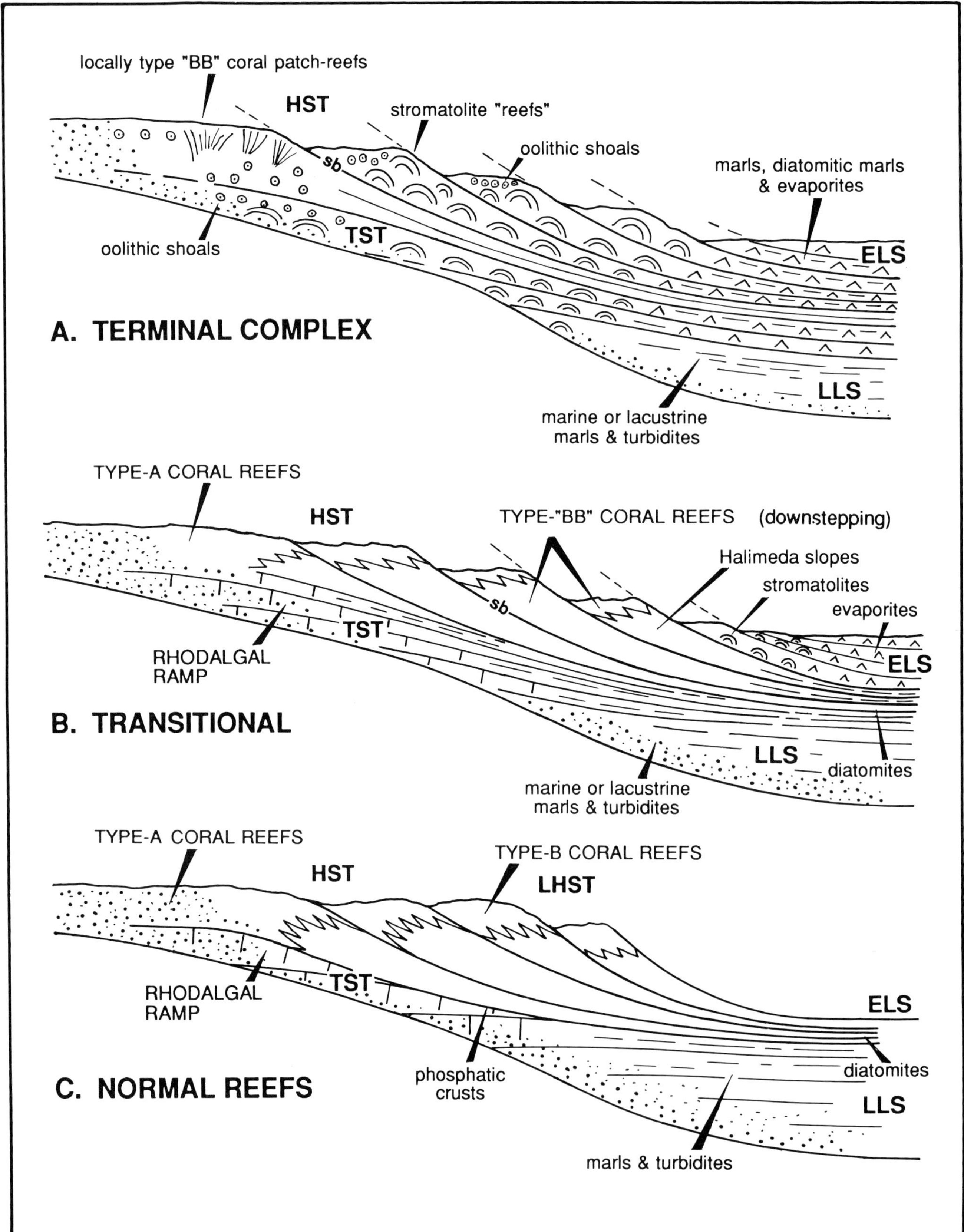


FIG. 28.—Three main types of facies models of Upper Miocene reefs in the Mediterranean. (A) Terminal Complex, with stromatolitic reefs, oolite shoals and, locally, coral reefs (marine-evaporitic cycles). (B) Normal marine reefs evolving into evaporitic conditions. (C) Normal marine coralgal reef complex. ELSW: Early lowstand wedge. LLSW: Late lowstand edge. TST: Transgressive systems tract. HST: Highstand systems tract.

and in Esteban et al. (this volume).

Since Margalef (1968), it has been recognized that the negative influence of high-nutrient levels on coral reef communities leads to extensive disruption, algal infestation, bioerosion and extinction (also Hallock and Schlager, 1986; Hallock, 1988). The case of the Messinian reefs is also peculiar in that, although algal infestation is documented, evidence of increased bioerosion is scarce; in fact, organic borings are more abundant and varied in type A reefs than in types B and BB. More work is required to understand the response of reef communities to the influx of waters with high-nutrient levels.

*Further questions.—*

Why are there no reports of "BB" features in the pre-evaporitic Middle Miocene reefs of Paratethys, Gulf of Suez-Red Sea or in the Mid-Lower Miocene reefs of the Mesopotamian basin? We do not know, but several suggestions can be formulated. The Red Sea-Gulf of Suez is different in the sense of its connection to the extensive Mid Miocene reef provinces in the Mediterranean. In contrast, the Upper Miocene reefs in the Mediterranean represent the northern limit of the hermatypic reefs and do not have counterparts in the Atlantic neighborhood. This could possibly suggest a relationship with migration and radiation patterns. Micritic crusts are mentioned but seem to be less developed. Infra-gypsum diatomitic deposits are present but less developed in the Red Sea-Gulf of Suez area than in the Mediterranean (Rouchy, 1982a, b); stromatolites are very abundant, and intra-gypsum diatomites are even more abundant than in the Mediterranean (Rouchy et al., 1995). These observations suggest differences in blue-green algae populations and nutrient levels. Finally, the Red Sea-Gulf of Suez basin is not connected with an extensive fresh-brackish water basin such as the Late Miocene Paratethys.

Within this context, the lack of reported "BB" reefs in the Pannonian basin is less explicable. There the northern limit of hermatypic reefs was directly connected with extensive fresh to brackish water bodies, very similar to the Late Miocene setting in the Mediterranean. Diatomitic marls and stromatolites are present but not abundant, and giant domes have not been reported. Little is known of the Mesopotamian Mid Miocene reefs that could help this discussion. The Upper Eocene units of the Ebro Basin (northeastern Spain) contain coral reefs, TC-like geometries and facies (stromatolites and oolites), basinal evaporites and laminated marls in a pattern that closely resembles the Messinian model (Salas, 1979; Salas and Esteban, 1980). However, these reefs show a high diversity and standard colonial morphologies. Similarly, no special features have been reported in the Paleozoic reefs associated with evaporites of the North America craton, though laminated marls and some TC-like facies appear to be present. During the Early Burdigalian times, extensive diatomitic deposits occurred in many Mediterranean areas with abundant rhodalgal carbonates and without major coral reef development. The clue may be in the type of currents and upwellings in the basin.

In summary, it is suggested that final stages of reef progradation may present "peculiar" BB features (monogeneric buildups with fast growth rates, very long coral branches, association with giant stromatolites and stromatolite coatings) when preceding deposition of basinal evaporites at times of high-nutrient levels in basins possibly adjacent to major freshwater bodies (?) and disconnected from major reef provinces (?). The above examples show that this is insufficient; something is still missing. The special BB features are definitely related to the concept of a Messinian event, but not all the Messinian reefs show BB features, and Messinian-like events in basins of different ages show no record of any peculiar BB features.

## CYCLICITY IN MIOCENE CARBONATES

Miocene carbonate lithologies show a marked cyclicity at different orders of magnitude. Esteban (1988) reviewed the available data on Miocene reef platforms from the Mediterranean in a preliminary attempt at developing an "ideal" Miocene reef model reflecting this marked cyclicity: the 1988 model (Fig. 29). Esteban et al. (1992), Pomar (1993), Pomar and Ward (1994) and Franseen et al. (this volume) provide further discussion of the cyclicity and sequence stratigraphy of the magnificent outcrops of the Cap Blanc (Mallorca); these studies have shown important implications to all Miocene reefs in the Mediterranean region. Franseen (1989), Franseen and Mankiewicz (1991), Franseen et al. (1993) and Franseen and Goldstein (this volumne) analyze the sequence stratigraphy of the outcrops in Almeria. The following discussion summarizes the main aspects and preliminary interpretations of the cyclicity of the Miocene carbonates.

### *The Rhodalgal Units*

Esteban (1988) interpreted the well-known rhodalgal carbonates in the Mediterranean as a transgressive system tract of 3rd-order depositional sequences. However in Melilla, Rouchy et al. (1986), Saint Martin and Rouchy (1990) and Saint Martin et al. (1991) distinguished a basal terrigenous-rich transgressive rhodalgal unit and younger progradational rhodalgal sigmoids as a highstand deposits. Mankiewicz (1987, this volume), Franseen et al. (1988), Franseen (1989) and Franseen and Mankiewicz (1991) also consider the rhodalgal unit without in-place corals as a separate 3rd-order depositional sequence, well differentiated from the overlying coral reefs, but these are here re-interpreted as 4th-order sequences within a 3rd-order transgressive system tract.

Pedley (1976), Bosence and Pedley (1982) and Saint Martin (1984) mentioned an antagonistic relationship between the presence of rhodalgal and coral reef facies that is commonly in evidence at the top of the rhodalgal unit. This also can be seen in small-scale (0.5-2 m) sequences of rhodalgal pavements and coral, of coral debris and *Halimeda* pavements (Fig. 29) and of rhodolithic units and large benthic forams (Fig. 30). Rouchy (1979, 1980, 1982a, b) indicated the relationship between rhodalgal facies, upwellings and basinal diatomitic deposits.

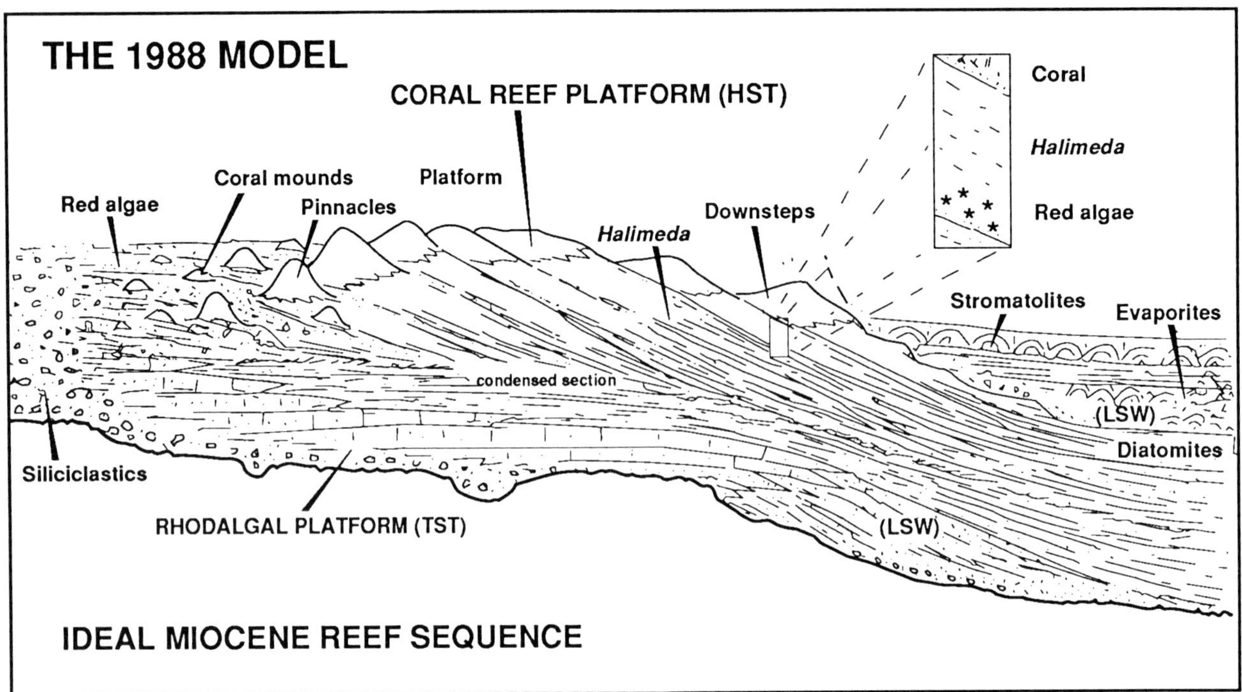


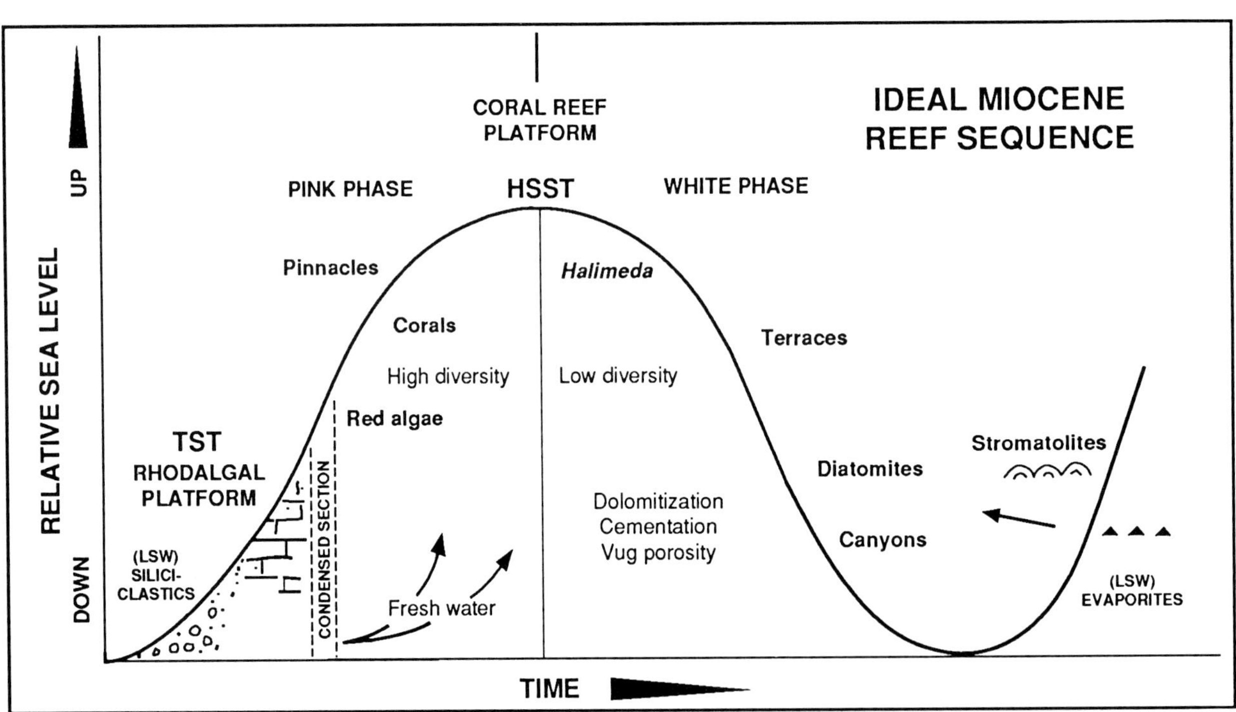


Fig. 29.—The ideal Miocene reef model as a 3rd-order depositional sequence (from Esteban 1988). LSW: Lowstand wedge. TST: Transgressive systems tract. HST: Highstand systems tract.

Carannante et al. (1988) also proposed that upwellings favored rhodalgal units and displaced coral reefs. On a much larger scale, this antipathetic relationship is also apparent in the 2nd-order Miocene stratigraphy (Fig. 3).

When coral reefs are present (2nd-order highstands, Fig. 3), a well-developed rhodalgal unit is characteristic of the transgressive system tracts of 3rd-order depositional sequences and excludes major coral-reef development (Figs. 25, 29). Thinner

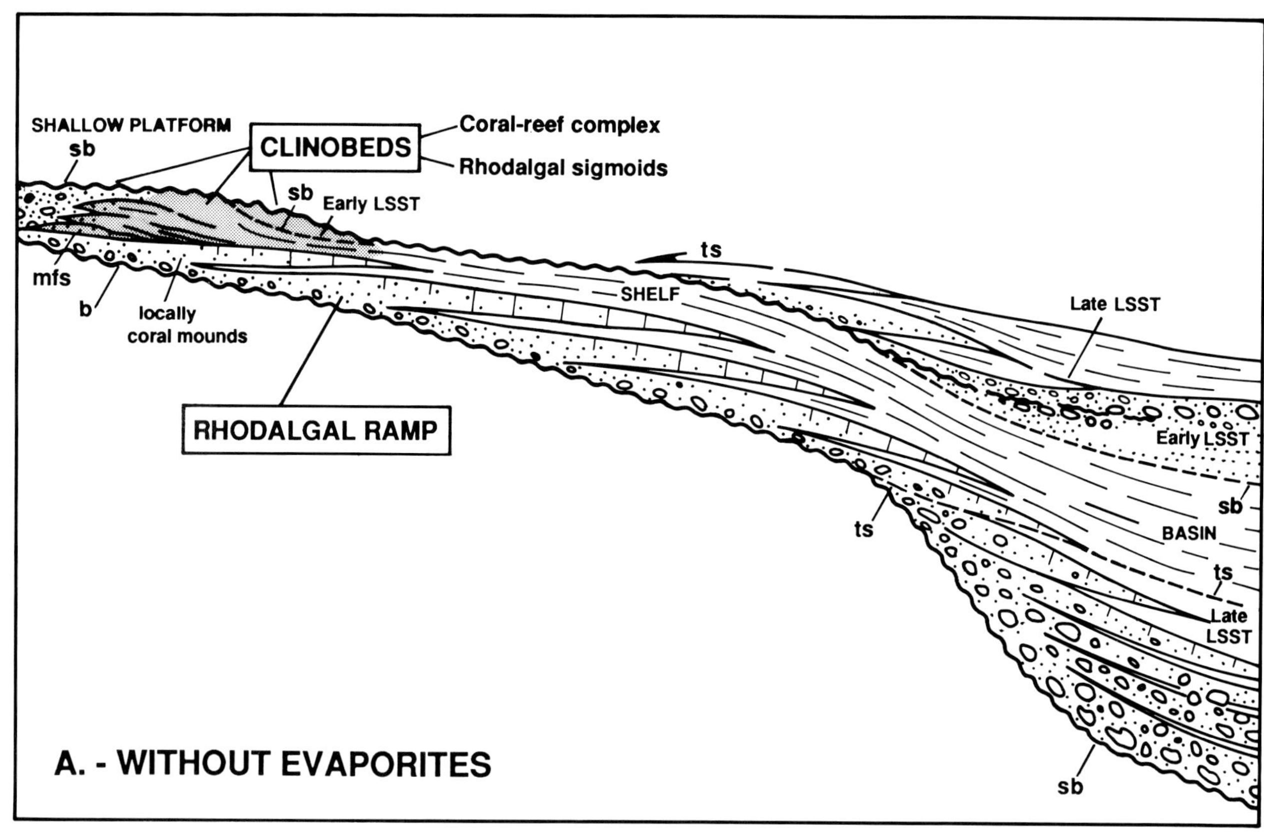


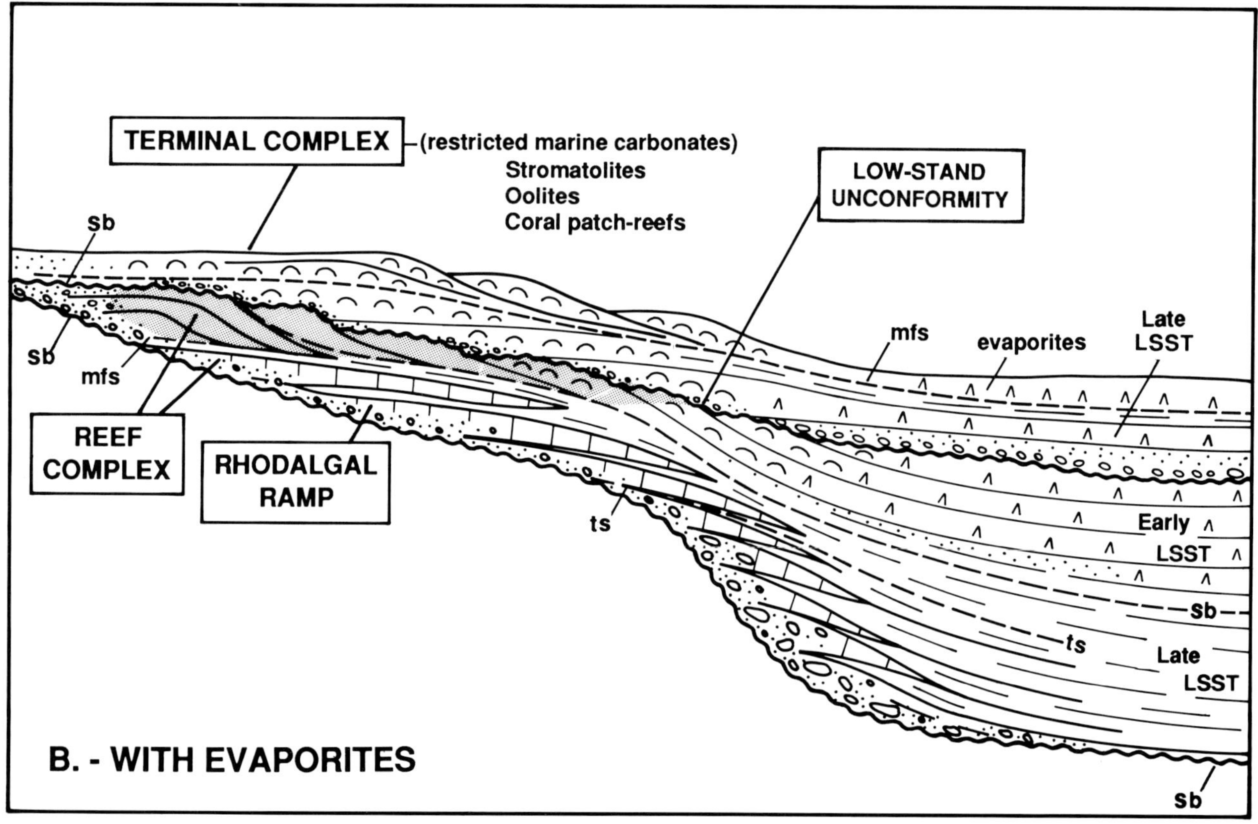


FIG. 30.—Cycles of rhodoliths and benthic forams in the Middle Miocene carbonates of northeastern Spain.

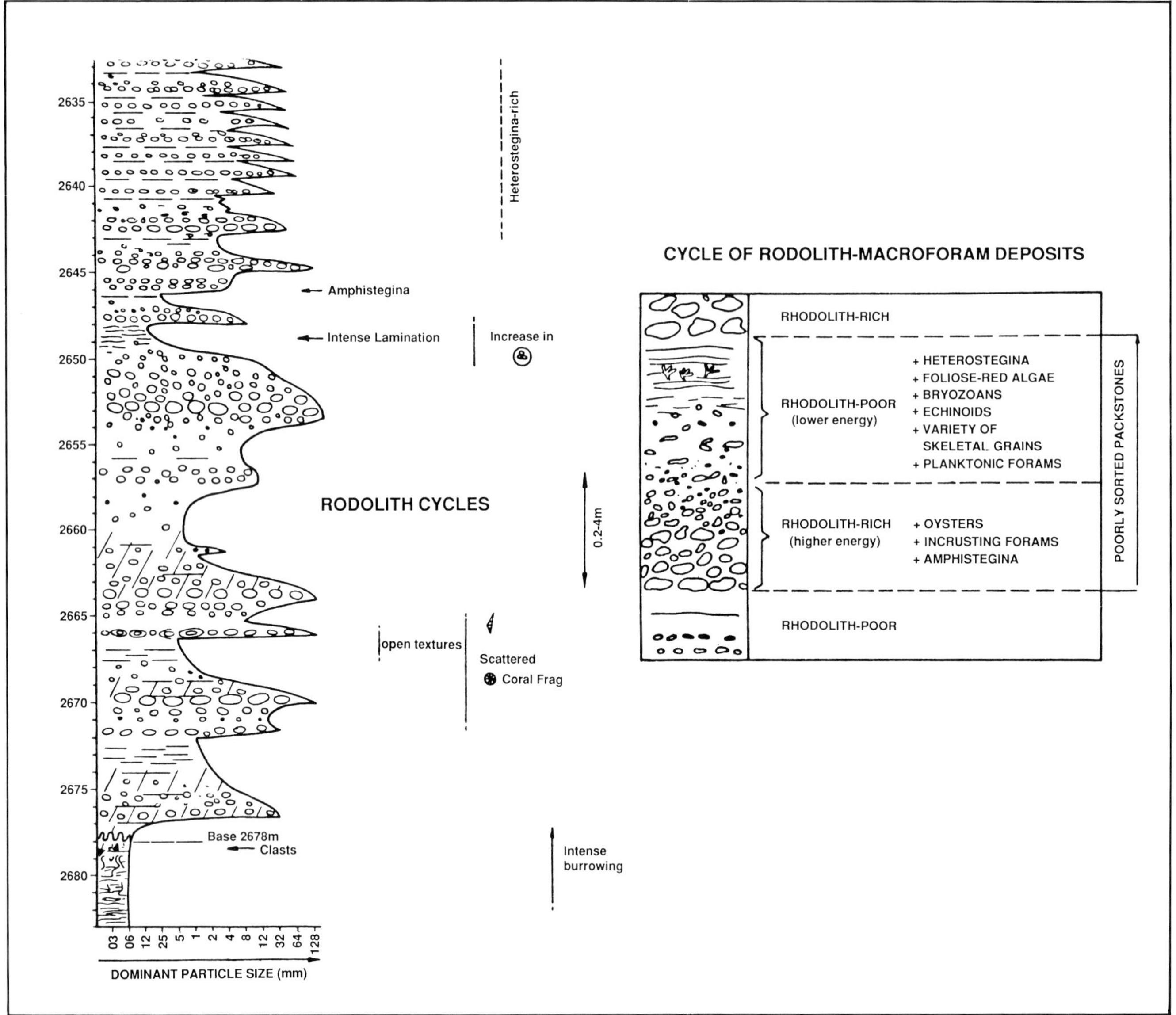


FIG. 31.—General facies model of the Miocene carbonates in the Mediterranean. (A) Without evaporites and (B) with evaporites. ts: transgressive surface; sb: sequence boundary; mfs: maximum flooding surface.

rhodalgal units also appear as the open-shelf equivalent of lowstand and offlapping units of the higher-frequency cycles (4th, 5th, 6th) intercalated in prograding reef complexes at times of reduced sediment production from the coral reef. When coral reefs are absent in 2nd- and 3rd-order sequences, rhodalgal units dominate all the system tracts and significant lithologic variations (large-scale sigmoidal clinobeds of rhodolithic facies; mollusc, foram, or echinoderm-rich layers).

The rhodalgal ramp facies is the most extensively developed Miocene carbonate of the Mediterranean (Figs. 3-8). There are regions (i.e., central Appennines, Paratethys) and time intervals (Burdigalian, Serravallian) where and when Miocene carbonates are completely dominated by rhodalgal facies (and variations). Hypothetically, this could represent the location or time of influence of cooler, nutrient-rich waters or persistent upwellings. Expanding the concepts of Lees and Buller (1972), Lees (1974, 1975), Franseen (1989) and Carannante et al. (1988), we could consider the alternations of foramol (red algae, mollusks, echinoids, benthic forams), chlorozoan (coral reefs) and chloralgal (*Halimeda,* benthic forams) accumulations as representative of cyclic sedimentation.

### *The Coralgal Reefs*

Coral reefs are characteristic of the 3rd-order highstand system tracts (HST) and the early lowstand (offlapping; Figs. 29, 31) during the 2nd-order highstands (supercycles of Fig. 3). Locally, coral reefs also occur in the late transgressive system tracts (LTST) of the 3rd-order depositional sequences. Retrograding or backstepping coral reef sequences of high frequency (4th-, 5th-order) has been detected associated with terrigenous deposits in the Fortuna Basin (Santisteban, 1981) and probably also in Algeria (Saint Martin, 1990), but it is rather uncommon. The late TST and early part of HST show relatively high diversity coral-reef mounds, pinnacles, layers and patches (type A reefs), commonly still mixed with abundant rhodalgal facies (Figs. 29, 31). Large molluscs, oysters and bryozoans are typically abundant. The late TST and early HST are characterized by the presence of mixed carbonate-terrigenous environments (in fan deltas, deltas and beaches, Marginal Terrigenous Complex) and associated red-algal pavements and oysters (Braga et al., this volume) as stabilizing communities. Terrigenous intercalations decrease gradually towards the younger prograding beds. Where vertical aggradation was dominant, coral mounds and pinnacles occur as the base of the prograding reef complex, and extensive back-reef lagoons with patch reefs and knobs can be found (as in Mallorca). In the later HST, terrigenous intercalations are uncommon and coral pinnacles and mounds appear amalgamated with increasing development of prograding carbonate slopes and decrease of coral diversity (and variety of colonial morphology) and associated fauna. Horizontal progradation is dominant, lagoons are very reduced, absent or eroded; the coral reefs are of fringing type. Downstepping (offlapping) of the coral-reef (Fig. 31B) results from fall in sea level and is considered as part of the early lowstand system tract (ELST) in Esteban et al. (1992) or a new systems tract in Pomar and Ward (1994). Major subaerial exposure and partial truncation of the preceding HST reefs occurs landward of the offlapping reefs (Pomar, 1991). The offlapping reefs seem to correspond to a marked decrease in diversity, whereas the successive rising sea-level cycle commonly appears to show an increase in diversity in the new TST reefs. Local tectonic uplift during reef progradation may accentuate the offlapping pattern (see above discussion). Some of the offlapping reefs in restricted marine basins preceded basinal evaporite deposition.

### *The Stromatolitic Units*

Large stromatolitic mounds or "reefs," up to 20-30 m in thickness formed by laterally linked hemispheroids individually up to 8 m in diameter, occur in the Messinian carbonates of the Mediterranean and in the Middle Miocene carbonates of the Red Sea. Much smaller stromatolitic units occur in the Aquitanian units of La Nerthe (southern France) and in the marine lagoons of the Upper Miocene platforms of Mallorca, but those are not discussed here.

The largest stromatolitic mounds appear to occur in late TST and early HST (Fig. 31) in association with oolitic shoals and intercalated with *Porites* patch reefs (i.e., Santa Pola, Fig. 32) or with siliciclastic conglomerates and sands (Cariatiz, SE Spain; Tuscany). Thin red-alga coatings and mollusc borings within the stromatolite domes are locally present (Riding et al., 1991a) but are very scarce; in contrast, large accumulations of cerithid gastropods are common. This suggests a very restricted, eutrophic marine environment for the largest stromatolites. These units show shallowing-upwards cycles with stromatolites at the base and capped by coral patch reefs (Fig. 32). The transgressive phases of each high-frequency cycle were not favorable to normal marine carbonates

Evidence of evaporite molds and possible evaporite-collapse breccia (partly silicified) is very scarce and appears reduced to smaller stromatolites on offlapping layers in more basinal settings (early lowstand, Fig. 31B) grading and interfingering with evaporites. There are also stromatolites in freshwater environments, as in Tuscany and northern Morocco (unpubl. data), locally described as lacustrine carbonates and travertines. In essence, Miocene stromatolitic units seem to occur in a wide range of environments, although they are best developed in TST and EHST with evidence of very restricted, eutrophic marine environments unfavorable to coral reef development.

### *Diagenetic Patterns*

The diagenetic patterns of the Mediterranean Miocene carbonates are not as extensively studied as their stratigraphy, sedimentology and ecology. Nevertheless, valuable contributions to carbonate diagenesis have been published on the Upper Miocene reefs of Spain (i.e., Armstrong et al., 1980; Esteban and Calvet, 1983; Franseen, 1989; Goldstein et al., 1990; Oswald, 1992; Pomer et al., this volume), the Middle Miocene reefs of the Red Sea (Aissaoui et al., 1986), the Middle Miocene carbonates of the Paratethys (Dullo, 1983; Pisera 1985) and the Lower and Upper Miocene reefs of Cyprus (Follows, 1992). The 1988 model proposed an increase in cementation, lithification, leaching and dolomitization from early to late phases of the highstand system tract; this was well documented by Oswald et al. (1990) in Mallorca, showing increased cementation in the aggrading part of the 4th-order sequences. Marine cements are commonly present but not abundant (i.e., Franseen, 1989; Follows, 1992), with isopachous fringes, botryoids and bladed mosaics. Skeletal-moldic porosity and karst is mostly developed in the late HST and early LST. Carbonate phosphatization is common at the top of the rhodalgal unit. The 1988 model referred to the transgressive rhodalgal unit as the pink phase because many sections tend to have intense red-brown, ochre and yellow colors, and the coralgal unit is mostly white, light gray, or light yellow (white phase, Fig. 29).

Many of the corals in the late TST and early HST are well preserved, and some are still partially or completely aragonitic (Esteban et al., 1982a, b; Esteban and Prezbindowski, 1985; Santisteban, 1981). Some mollusc shells are also still aragonitic (Santisteban, 1981). The preservation of aragonite in outcrops during 3-4 my is explained by the assumption of a protective evaporite cover, arid environments and encasement in shaly marls. In western Sicily, there are abundant botryoidal aragonite cements, post-dating reef growth, remarkably similar to those from modern reefs (James and Ginsburg, 1979); yet their isotopic compositions show marked oscillations suggesting hypersa-

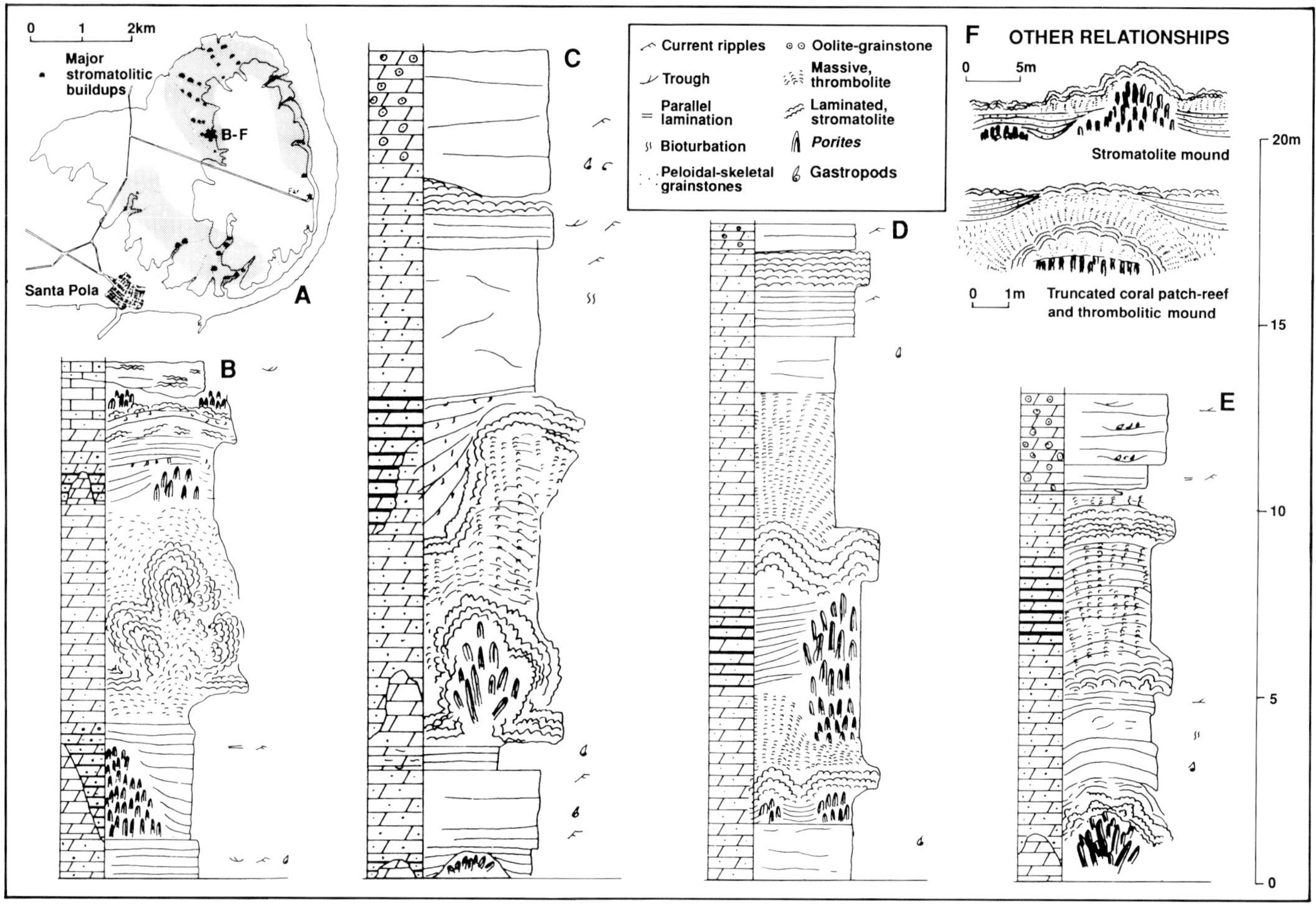


FIG. 32.—Stromatolitic mounds and cyclicity in the Terminal Complex of Santa Pola with shoaling-upwards cycles of stromatolites-coral reef-paleosol. After Vallés-Roca (1985, 1986).

line events periodically intercalated in between normal marine conditions (Esteban and Prezbindowski, 1985).

Many Miocene carbonates preceding basinal evaporitic units (southeastern Spain, Balearic Islands, Italy, Gulf of Suez, Red Sea and Iraq) are pervasively dolomitized and display very similar fabrics. The petrology, geometric relationships and geochemistry of these dolomitized carbonates suggest the involvement of hypersaline seawater during transgressive events (Coniglio et al., 1988; Oswald et al., 1990, 1991a, b; Calvet et al., this volume). Extensive leaching of originally aragonitic constituents is a common feature, commonly interpreted as the result of meteoric freshwater influx (Armstrong et al., 1980; Aissaoui et al., 1986; Coniglio et al., 1988). However, Sun (1992) notes the generalized absence of calcite cements (predating the dolomite) which would be expected in the down-dip freshwater phreatic environment. He suggests that the model of Lazar et al. (1983) and Oswald et al. (1991b) also supports that hypersaline seawater, rather than freshwater, could be involved in the aragonite leaching and dolomitization, without excluding the possibility of mixing with the reduced freshwater lenses in evaporitic environments. A similar interpretation is presented in Franseen and Goldstein (this volume); in any case, most of the blocky calcite cements in the Upper Miocene reefs postdate the dolomite. Finally, poikilotopic calcite cements are common in a zone up to 2 m below the erosional truncation overlain by the Terminal Complex (Franseen and Goldstein, this volume).

The diagenetic patterns of the Upper Miocene reefs in the Mediterranean and the Middle Miocene reefs in the Red Sea-Gulf of Suez are dominated by evaporitic conditions in arid landlocked settings. The Lower and Middle Miocene reefs of the Mediterranean represent more tropical, humid oceanic environments, and diagenetic patterns are expected to be markedly different (Sun and Esteban, 1994). Their diagenesis is poorly known, but there is evidence of less extensive dolomitization (only coastal mixing and burial) and abundant freshwater diagenesis.

Miocene carbonates range from tight, well cemented to porous (moldic, vuggy, intergranular) and chalky (intercrystalline, microvuggy). Most authors interpret these variations in porosity as result of the interplay of the different lithologies and hydrologic regimes in the meteoric diagenetic environment (i.e., Dullo, 1983). However, surface and subsurface occurrences in northeastern Spain, Moroccan Rif and Red Sea suggest important late-diagenetic overprints during Pliocene burial (Esteban,

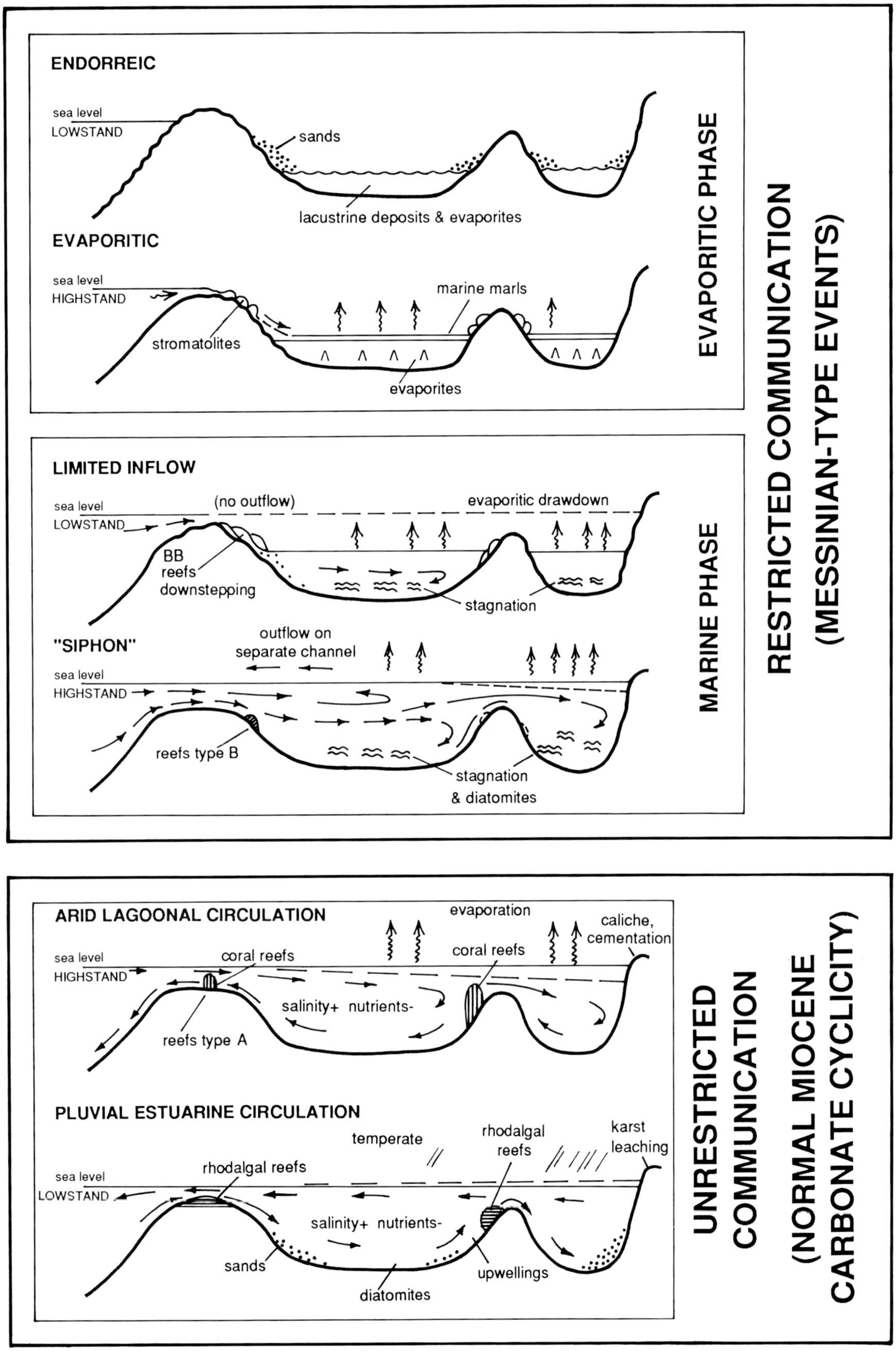

ENDORREIC
sea level
LOWSTAND
sands
lacustrine deposits & evaporites
EVAPORITIC
sea level
HIGHSTAND
marine marls
stromatolites
evaporites
EVAPORITIC PHASE
LIMITED INFLOW
sea level
LOWSTAND
(no outflow)
evaporitic drawdown
BB
reefs
downstepping
stagnation
"SIPHON"
outflow on
separate channel
sea level
HIGHSTAND
reefs type B
stagnation
& diatomites
MARINE PHASE
RESTRICTED COMMUNICATION
(MESSINIAN-TYPE EVENTS)
ARID LAGOONAL CIRCULATION
evaporation
caliche,
cementation
sea level
HIGHSTAND
coral reefs
coral reefs
salinity+ nutrients-
reefs type A
PLUVIAL ESTUARINE CIRCULATION
temperate
rhodalgal
reefs
karst
leaching
sea level
LOWSTAND
rhodalgal reefs
salinity+ nutrients-
sands
upwellings
diatomites
UNRESTRICTED
COMMUNICATION
(NORMAL MIOCENE
CARBONATE CYCLICITY)

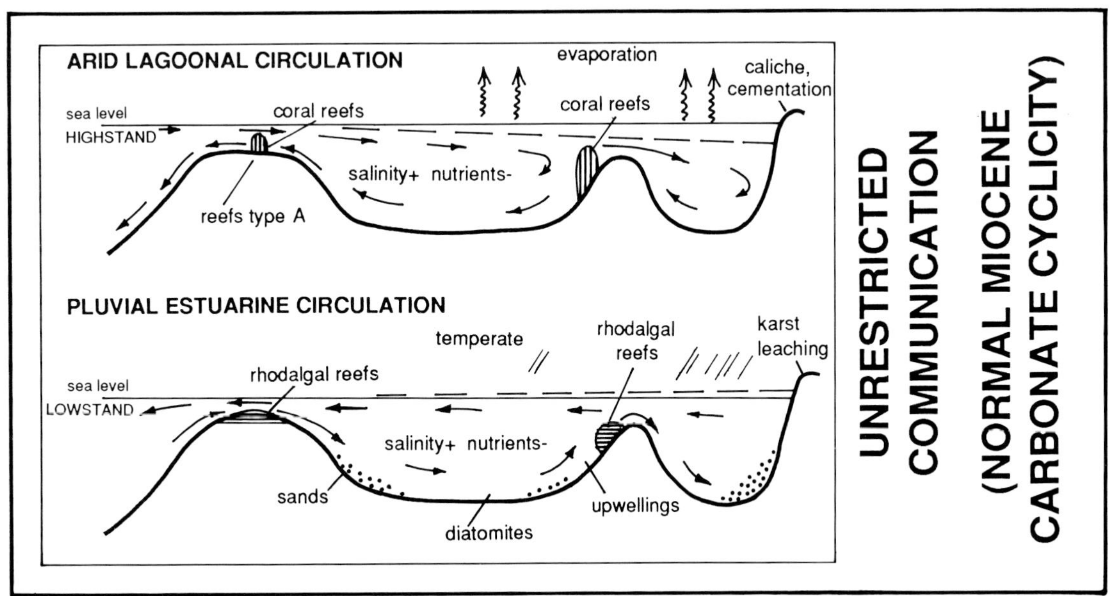

FIG. 33.—Interpretation of the cyclicity of Miocene carbonates in the Mediterranean, based on climatic cycles, sea-level oscillations and water circulation patterns.

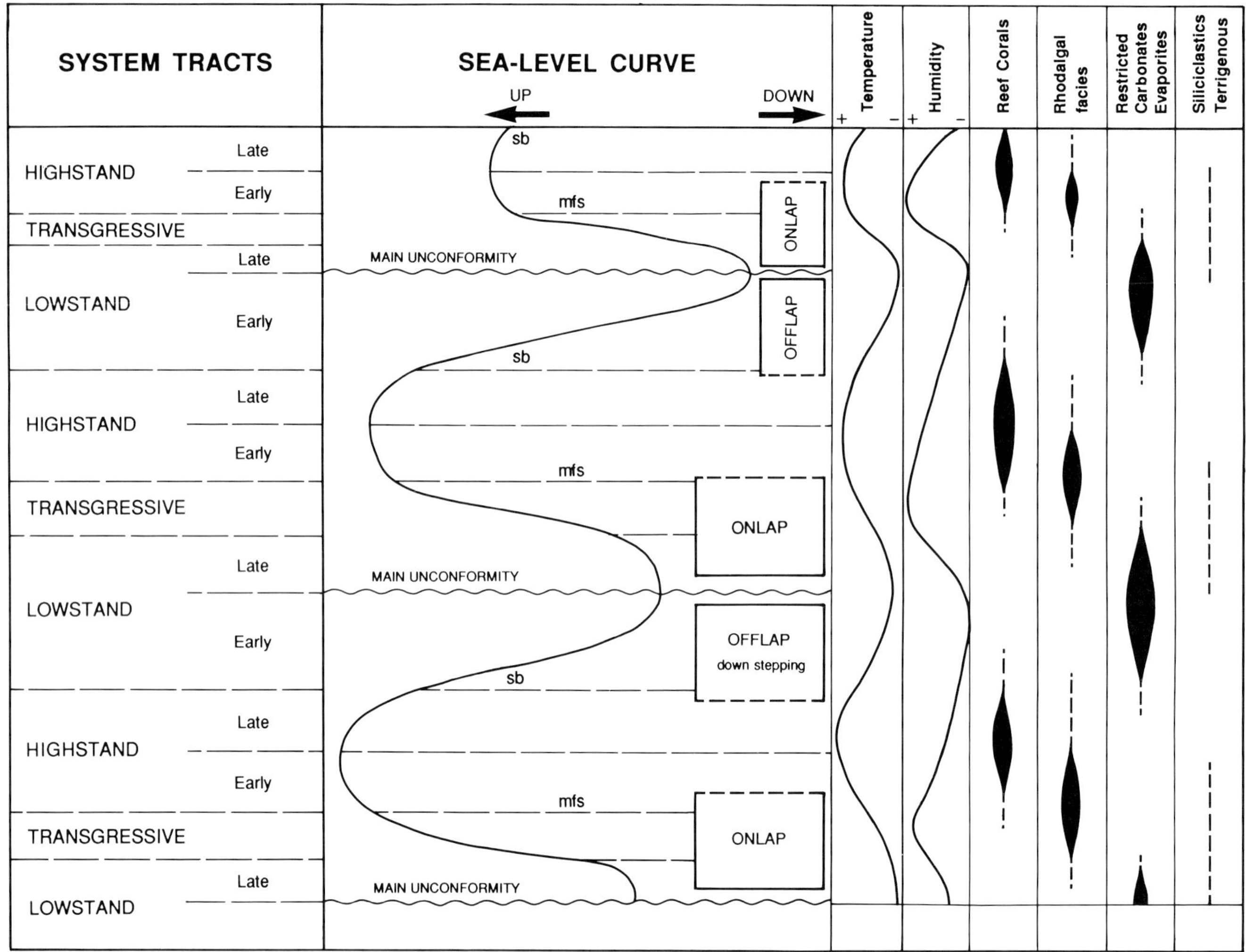


FIG. 34.—System tracts and environmental parameters of Miocene carbonates in the Mediterranean.

unpubl. data), with vuggy corrosion, major dolomitization (zoned ferroan dolomites, baroque dolomite cements) and poikilotopic calcites. It is assumed that reservoir properties of the Miocene carbonates are extensively controlled by their late burial diagenesis.

### *Hypothesis for the Mediterranean Miocene Cyclicity*

The alternating cycles in different frequencies of rhodalgal, coralgal and stromatolitic units is explained in the scheme of Figure 33, which corresponds to a semi-isolated, Mediterranean-type scenario, and it is assumed to be valid for 2nd-, 3rd- and 4th-order cycles (inspired by Sverdrup et al., 1942; Mars, 1963; Sonnenfeld, 1980, 1985; Sonnenfeld and Finetti, 1985; Van Gorsel and Troelstra, 1980; Thunell et al., 1987; Müller and Hsü, 1987; Rouchy, 1982a, b; Benson et al., 1991). This interpretation, already presented in Esteban (1988), has been reproduced by Martín and Braga (1994) with minor modifications. The rhodalgal units are considered to correspond to the transition from climatic minimums or "glacial" periods towards climatic optimums (de-glaciation); this corresponds to pluvial periods in the Mediterranean. This scenario would be characterized by decreased salinity and increased nutrient concentration and can be considered as estuarine-type circulation: discharge (outflow) of surface Mediterranean waters and entry of deep, nutrient-rich Atlantic waters. The interglacial (climatic optimum) periods tend to evolve into semi-arid or arid climates in the Mediterranean-type settings, and would result in a reversal in the water circulation patterns (lagoonal, also called antiestuarine), with entry of Atlantic surface waters, outflow of deep Mediterranean waters. This increases evaporation and salinity and decreases nutrients in the upper water layer, favoring the development of coral reefs. There is good evidence of increased aridity during the transition to the glacial minima (van der Hammen et al., 1971; Street and Groove, 1979; Suc, 1984; Jenkins, 1982; Thunell et al., 1987). The isolation of the Mediterranean basin can be increased by the narrowing (olistostromes, reef growth, sedimentation, uplifts, evaporitic drawdown) of the straits communicating with the Atlantic; this may reduce the volume of discharge of Mediterranean bottom waters into the Atlantic even during a "global" highstand. Dronkert and Pagnier (1977) and Troelstra et al. (1980) emphasize the possible contribution of reef growth in the narrowing of the marine corridors.

As explained by Benson et al. (1991), the "siphon" of Atlantic waters into the Mediterranean during early Messinian time resulted from the reduction in discharge waters from the Mediterranean and requires a discharge channel in a different location (Fig. 33). The obstruction of the discharge channel would result in the rapid evolution to hypersalinity and desiccation in Late Messinian time. To accommodate field observations in the western Mediterranean (see also Esteban et al., this volume), it could be suggested that the "siphon" was a pulsating phenomena interrupted by periods of "estuarine" discharge during the transgressive parts of the 4th-order cycles (Fig. 26). It is likely that the pulsating character of the siphon would be difficult to detect in the marly sections, but it appears well recorded in the expanded marginal carbonate sections.

The interaction between terrigenous clastics, rhodalgal facies and coral reefs in the Mediterranean has been studied in detail by numerous authors. In general, two styles of relationship can be established depending on the role of terrigenous clastics: (1) passive; terrigenous sediments are passive substrates for stabilizing and pioneer communities (oysters, red algae, *Porites*) and (2) active; terrigenous sediments continuously deposited during coral-reef growth. The first case is very common. Coral reefs are observed on inactive channel margins, stream mouth bars, beaches, fan deltas and braided deltas. Reef growth stops or is partly destroyed with each new phase of re-activation of terrigenous sedimentation, which may bury completely the reef unit. After terrigenous sediment deposition, the lack terrigenous clastics allows stabilization and growth of a new coral layer. Major river discharge of freshwater could also trigger nutrient-rich undercurrents unfavorable to coral reefs. It is tempting to see climatic cycles (wet — terrigenous, dry — coral reefs) and/or shifting of distributary channels in this passive style of terrigenous association.

The diagenetic evolution and porosity of the Miocene carbonates also appear affected by climatic cycles (Fig. 33). Pluvial periods coincide with sea-level lowstands and extensive carbonate leaching; groundwater flow is enhanced by increased hydraulic heads and extension of recharge areas. In contrast, during sea-level highstands, carbonate cementation (caliche and speleothems) is predominant, due to the combination of dryer and hotter climates, reduced rainfall and reduced extension of recharge areas and hydraulic heads. This interpretation of diagenetic patterns is likely to have important variations as a result of evaporitic draw-down but could offer a useful working hypothesis.

There seems to be a basic theme of 2nd-order and higher-frequency cyclicity (Fig. 34) controlled by variations in nutrient levels, water temperature, climate and salinity. There are coral cycles, rhodalgal cycles, *Halimeda* cycles, stromatolite cycles and terrigenous cycles; all of different orders of magnitude. In a semi-isolated basin such as the Mediterranean, the normal marine phase involves cycles of evaporation (not evaporite deposition) and salinity increase (not hypersalinity) during global highstands (climatic maximum, arid). When this excess salinity cannot be freely discharged into the Atlantic, there is a fast transition in the evaporitic phase. In the evaporitic phase, the colder Atlantic waters flood the Mediterranean basins with normal-marine waters, but the narrowing of the Rif straits, makes the discharge of excess salts difficult, and the basin rapidly becomes hypersaline after an impressive evaporitic drawdown. Probably there is minor evaporitic drawdown (late highstand, early lowstand, climatic optimum) in each downstepping phase of the normal marine cycle involving "normal" A type reefs. The Messinian evaporitic phase starts by reinforcing a "normal" late highstand-early lowstand downstepping because of the narrowing of the Rif Straits (synchronous tectonic uplift, evaporitic drawdown?). This leads to the local appearance of BB features in some coral reefs in the transition between the "normal" marine phase and the evaporitic phase. All "BB" features in coral reefs are part of the Messinian evaporitic phase (some starting in Fini-Tortonian), but not all Messinian coral reefs show BB features.

## PERSPECTIVES IN HYDROCARBON EXPLORATION

Middle Miocene carbonates are major oil producers in the Gulf of Suez, mostly from rhodalgal units and a few from restricted-marine stromatolitic carbonates; only one oil field has a coral reef as the main reservoir. Major production also occurs in the Lower-Mid Miocene carbonates of the Mesopotamian basin (Asmari and Euphrates limestones). Important shows are known to occur in the Miocene carbonates of the Adana bay in southern Turkey. The Hungarian Panonian basin also has major reserves in Middle Miocene carbonates (most likely of rhodalgal types); similar occurrences are reported in the Vienna basin, Rumania and Moldavia. However, like in most Mediterranean areas, exploration has been traditionally concentrated in Miocene sandstones.

In central and western Mediterranean, Lower and Middle Miocene carbonates produced in small fields in the offshore of southwestern Sicily (i.e., Nilde), northeastern Tunisia, and northeastern Spain. In northern Morocco, the Middle Miocene rhodalgal carbonates present important oil and gas shows under the Nappe complex of the Prerif. Burial diagenesis (dolomitization, leaching and chalky microporosity) is the main contributor to reservoir properties in all these cases; many outcrops of these carbonates in western and central Mediterranean are much less porous than their subsurface equivalents.

From Spain to the Middle East, the Lower Burdigalian marls are considered to be the main source rock. This corresponds to a time of extensive development of rhodalgal ramps, diatomitic marls, phosphatic layers, glauconite and a generalized absence of coral reefs. It is also significant that the Middle Miocene rhodalgal ramps (and locally coral reefs) are the most important oil reservoir of all the Miocene carbonates in the Mediterranean areas. Only subcommercial gas has been found in Upper

Miocene carbonates. In summary, it seems very likely that improved exploration strategies in Miocene carbonates should find more hydrocarbons in the Mediterranean areas, not only in the Gulf of Suez, but also in the Mesopotamian basin, central-south Paratethys, southern Turkey, Malta graben and Gulf of Valencia.

## CONCLUSIONS

There are four major types of "reefs" (organic buildups) in the Miocene Mediterranean: hermatypic coral reefs, ahermatypic mounds, rhodalgal carbonates and stromatolitic mounds. The rhodalgal carbonates (red algae, molluscs, foraminifers and bryozoans) are the most abundant Miocene type.

Despite the variety of tectonic settings, substrate lithologies and morphologies, the general synchronicity in reef events and similarity in facies models suggest an overriding paleoceanographic control. Miocene reefs occur in three major supercycles (2nd-order sequences); most coral reefs occur in the highstands of these 2nd-order cycles: Aquitanian, Langhian, Late Tortonian-Messinian

Upper Miocene reefs have been the most extensively studied, showing a basic facies model with numerous variations depending on the amount of terrigenous mixing and bottom topography. Diversity is relatively low to very low (3-5 coral species to only 1). The zonation in colonial morphology cannot be generalized for the entire basin or province or even for a single reef complex. Fringing reefs predominate; barrier reefs with extensive lagoons are the exception rather than the norm. Different types of morphologies result from the interaction with terrigenous-siliciclastic deposition (patches, ribbons, fringes, etc.). Some outcrops show outstanding preservation of depositional morphologies (reef core, spur-and-grooves, buttresses, slopes, lobes, sand bodies, etc.) and allow a detailed reconstruction of the sea-level fluctuations and the resulting reef geometries: vertical aggradation, progradation and offlapping (downstepping). Backstepping (retrogradation) has been observed, but it is not a common feature.

Late Miocene coral reefs occurred before, during and after the deposition of the basinal evaporite units, following a repeated wedge-on-wedge pattern. To accommodate all the field observations within the biostratigraphic constraints, it is here proposed to consider the 4th-order depositional sequences rather then the 3rd-order sequences. A basic cyclic pattern can be recognized: (1) marine flooding and transgression, and (2) progradation and offlapping, eventually terminating in evaporite deposition. Spectacular offlapping coral reefs present a set of peculiar features (well-developed monogeneric reef with exuberant coral branches coated by cyanobacterial crusts, locally with giant stromatolites) which are considered to reflect the complex interaction of cold, nutrient-rich Atlantic waters and the dense, warmer Mediterranean waters. These features, part of the scenario referred to as the Messinian crises, were associated with basinal diatomites and pre-dated Messinian evaporitic units.

A general model applicable to all the Miocene reefs in the Mediterranean is here proposed for 2nd- to 4th-order sequences. Rhodalgal carbonates predominate during relative rises of sea level and most likely associated estuarine-type circulation in the Mediterranean. Coral reefs occur during the late transgressions, highstands and early (falling) part of the lowstand of sea level during each 2nd-order cycle; this is considered to represent lagoonal-type of circulation in the Mediterranean. However, rhodalgal carbonates could replace the coral reefs in areas of intense vertical mixing, upwelling, strong winds, etc. The efficient discharge of dense Mediterranean waters into the Atlantic is the prerequisite to maintain marine conditions; any obstruction to this discharge could lead to a major evaporite deposition. This basic model of cyclicity (rhodalgal-coral reef) also can be extended to interpret the cyclicity of terrigenous deposition and diagenesis in the Miocene reefs.

The superb outcrops of Mediterranean Miocene reefs offer an unrivalled opportunity for very detailed stratigraphic and sedimentological studies and the elaboration of facies models applicable to other Miocene reefs. There is a common pattern in Miocene coral reefs all over the Mediterranean, more striking where the Miocene reefs are related to basinal evaporites (Paratethys, Red Sea, Gulf of Suez, Mesopotamia). In all cases, cyclicity (3rd-, 4th- and higher order) seems to be the most important common denominator in understanding the complex facies relationships. The detailed stratigraphy and correlation of the Miocene reef events in the Mediterranean is still very problematic. More detailed field work is needed to understand the facies patterns of Lower and Mid Miocene reefs in Mediterranean areas without basinal evaporites, particularly those in the western region and facing the open Atlantic ocean. There is also a need for studies on the peculiar associations of hermatypic and ahermatypic coral mounds that occur in some of the northern limits of the coral reef provinces (e.g., Piemont, Aquitaine). Finally, it has to be stressed that the greatest volume of Miocene carbonates in the Mediterranean is formed by extensive rhodalgal units, which deserve much more attention than the spectacular, but volumetrically limited, coral reefs.

## ACKNOWLEDGMENTS

This review has been improved with the discussions, criticisms and comments on an early manuscript since 1992 by L. Pomar, W. C. Ward, J. M. Rouchy, E. G. Purdy, D. Bosence, J-P. Saint Martin, E. Oswald, S. Q. Sun, C. de Santisteban, J. M. Martin, J. C. Braga, E. K. Franseen, Lucia Simone, G. Carannante, M. Pedley, M. Ziegler, B. Buchbinder, A. Pisera, B. Cahuzac, F. Orszag-Sperber, W. Schlager, F. Sarg, R. H. Benson, C. J. Dabrio and Dana Ulmer-Scholle, although there is no complete agreement with all the data and interpretations here reported. The author apologizes for omissions and misinterpretations; due to the nature of this review, it was impossible to offer a comprehensive and balanced reference to all the valuable publications on the Miocene reefs of the Mediterranean regions. The author acknowledges the support of Petroleum Information (ERICO)

Ltd. in the preparation of an early version of the manuscript and the assistance of numerous colleagues, friends and students that during twenty five years contributed to and shared the excitement in the study of these Miocene carbonates.

## REFERENCES


ADAMS, C. G., 1976, Larger foraminifera and the late Cenozoic history of the Mediterranean region: Palaeogeography, Palaeoclimatology, Palaeoecology, v. 20, p. 47-66.

ADAMS, C. G., 1981, Larger foraminifera and the Paleogene-Neogene boundary: Annales de Géologie des Pays Helléniques, Tome hors serie, v. 4, p. 145-151.

ADAMS, C. G., BENSON, R. H., KIDD, R. B., RYAN, W. B. F., AND WRIGHT, R. C., 1977, The Messinian salinity crisis and evidence of late Miocene eustatic changes in the world ocean: Nature, v. 269, p. 383-386

ADAMS, C. G., GENTRY, A. W., AND WHYBROW, P. J., 1983, Dating the terminal Tethyan event: Utrecht Micropaleontology Bulletin, v. 30, p. 273-298.

ADDICOTT, W. O., SNAVELY, P. D. JR., BUKRY, D., AND POORE, R. Z., 1977, Neogene stratigraphy and paleontology of southern Almeria province, Spain: an overview: Washington, D.C., United States Geological Survey Open-file Report, no.77-716, p. 1-69.

AISSAOUI, D. M., CONIGLIO, M., JAMES, N. P., AND PURSER, B. H., 1986, Diagenesis of a Miocene reef-platform: Jebel Abu Shaar, Gulf of Suez, Egypt, *in* Schroeder, J.H. and Purser, B. H., eds., Reef Diagenesis: Berlin, Springer-Verlag , p. 112-131.

AIR MINISTRY, METEOROLOGICAL OFFICE, 1962, Weather in the Mediterranean: General meteorology. Volume 1 (second edition): London, Her Majesty's Stationery Office, M.O. 391, 150 p

ALVAREZ, G., BUSQUETS, P., PERMANYER, A., AND VILAPLANA, M., 1977, Growth dynamic and stratigraphy of Sant Pau d'Ordal Miocene patch-reef (Prov. of Barcelona, Catalonia): Second International Symposium on Corals and Fossil Coral Reefs, Paris, 1975: Bureau de Recherches Géologiques et Minières, Memoires 89, p. 367-377.

ALVINERIE, J., ANTUNES, M. T., CAHUZAC, B., LAURIAT-RAGE, A., MONTENAT, C., AND PUJOL, C., 1992, Synthetic data on the paleogeographic history of northeastern Atlantic and Betic-Rifian basin, during the Neogene (from Britany, France to Morocco): Palaeogeography, Palaeoclimatology, Palaeoecology, v. 95, p. 263-286.

ARMSTRONG, A. K., SNAVELY, P. D., AND ADDICOTT, W. O., 1977, Porosity in Late Miocene reefs, southern Spain (abs.): American Association of Petroleum Geologists Bulletin, v. 61, p. 761-762.

ARMSTRONG, A. K., SNAVELY, P. D., AND ADDICOTT, W. O., 1980, Porosity evolution of Upper Miocene reefs, Almería Province, southern Spain: American Association of Petroleum Geologists Bulletin, v. 64, p.188-208.

AWRAMIC, S. M. AND RIDING, R., 1988, Role of algal eukaryotes in subtidal columnar stromatolite formation: Proceeding of the National Academy of Sciences of USA, v. 85, p. 1327-1329.

BARNES, D. J., 1973, Growth in colonial scleractinians: Bulletin of Marine Sciences, v. 23(2), p. 280-298.

BARRIER, P., MONTENAT, C., AND ZIBROWIUS, H., 1990, Les Coelenterés, *in* Montenat, C., ed., Les bassins Néogènes du domaine Bétique Oriental (Espagne): Documents et Travaux, Institut Géologique Albert-de-Lapparent, p. 43-44.

BARRON, J. A. AND KELLER, G., 1982, Widespread Miocene deep sea hiatuses, coincidence with periods of global cooling: Geology, v. 10, p. 577-581.

BEN MOUSSA, A., EL HAJJAJI, K., POUYET, S., AND DEMARCQ, G., 1989, Les megafaunes marines du Messinien de Melilla (Nord-Est Maroc): Palaeogeography, Palaeoclimatology, Palaeoecology, v. 73, p. 51-60.

BENSON, R. H., RAKIC-EL BIED, K., AND BANADUCE, G., 1991, An important current reversal (influx) in the Rifian Corridor (Morocco) at the Tortonian-Messinian boundary: the end of Tethys Ocean: Paleoceanography, v. 6, p. 164-192.

BERGER, J. P., ENGESSER, B., BARVIN, V., BOLLIGERT, T., KELLER, B., AND WEIDMANN, M., 1990, Correlative Chart of the European Oligocene and Miocene: Fribourg, Institut de Géologie, Université de Fribourg, 1 table.

BERGGREN, W. A., KENT, D. V., FLYNN, J. J., AND VAN COUVERRING, J. A., 1985a, Cenozoic geochronology: Bulletin of the Geological Society of America, v. 96, p. 1407-1418.

BERGGREN, W. A., KENT, D. V., AND VAN COUVERRING, J. A., 1985b, The Neogene: part 2, Neogene geochronology and chronostratigraphy, *in* Snelling, N. J., ed., The Chronology of the Geological Record: London, Blackwell Sciences Publications, Memoir 10, p. 211-260.

BERNET-ROLLANDE, M., MAURIN, A. F., AND MONTY, C. L., 1980, Porites reef versus stromatolites at Santa Pola, Spain: A Miocene sedimentological puzzle (abs.): Paris, 26th International Geological Congress, Résumés 26, v. 2, p. 435.

BIJU-DUVAL, B., DERCOURT, J., AND LE PICHON, X., 1977, From the Tethys Ocean to the Mediterranean Seas: A plate tectonic model of the evolution of the Western Alpine System, *in* Biju-Duval, B., ed., Structural History of the Mediterranean Basins: Paris, Editions Technip., p. 143-164.

BIJU-DUVAL, B. AND MONTADERT, L., 1976, Histoire structurale des bassins méditerranéens: Split, XXV Congres Assamblé Plénaire de la C.I.E.S.M., Editorial Technip, 445 p.

BORDET, P., MONTENAT, C., OTT D'ESTEVOU, P., AND VACHARD, D., 1982, La "brèche rouge" de Carboneras: Un olistostrome volcano-sédimentaire tortonien (Cordillères bétiques orientales-Espagne): Dijon, Mémoires de Géologie de l'Université de Dijon, v. 7, Livre Jubilaire G. Lucas, p. 285-300.

BOSENCE, D. W. AND PEDLEY, H. M., 1982, Sedimentology and palaeoecology of a Miocene coralline algal biostrome from the Maltese Islands: Palaeogeography, Palaeoclimatology, Palaeoecology, v. 39, p. 9-43.

BOSSIO, A., BRADLEY, F., ESTEBAN, M., GIANNELLI, L., LANDINI, W., MAZZANTI, R., MAZZEI, R., AND SALVATORINI, G., 1981, Alcuni aspetti del Miocene superiore del Bacino del Fine: Pisa, IX Convegno della Società Paleontogica Italiana, v. 3-8, p. 21-54.

BOSSIO, A., ESTEBAN, M., GIANNELLI, L., LONGINELLI, L., A., MAZZANTI, R., MAZZEI, R., RICCI LUCCHI, F., AND SALVATORINI, G., 1978, Some aspects of the upper Miocene in Tuscany: Pisa, Messinian Seminar n. 4, Intenational Geological Correlation Project n. 96 (fasc. spec.), 88 p.

BRAGA, J. C. AND MARTÍN, J. M., 1988, Neogene coralline-algal growth-forms and their palaeoenvironments in the Almanzora River Valley (Almeria, S.E. Spain): Palaeogeography, Palaeoclimotology, Palaeoecology, v. 67, p. 285-303

BRAGA, J. C., MARTÍN, J. M., AND ALCALA, B., 1990, Coral reefs in coarse-terrigenous sedimentary environments (Upper Tortonian, Granada Basin, southern Spain): Sedimentary Geology, v. 66, p. 135-150

BRAITHWAITE, C. J. R., 1971, Seychelles reefs: structure and development: Symposium of the Zoological Society of London, v. 28, p. 39-69.

BUCHBINDER, B., 1975a, Lithogenesis of Miocene reef limestones in Israel with particular reference to the significance of the red algae: Unpublished Ph.D. Thesis, Hebrew University, Jerusalem, 170 p.

BUCHBINDER, B., 1975b, Lithogenesis of Miocene reef limestones in Israel with particular reference to the significance of the red algae: Geological Survey of Israel, Oil Research Division, Report OD/3/75, 173 p.

BUCHBINDER, B., 1979, Facies and environments of Miocene reef limestones in Israel: Journal of Sedimentary Petrology, v. 49, p. 1323-1344.

BUCHBINDER, B., MARTINOTTI, G. M., SIMAN-TOV, R., AND ZILBERMAN, E., 1993, Temporal and spatial relationships in Miocene reef carbonates in Israel: Palaeogeography, Palaeoclimatology, Palaeocology, v, 101, p. 97-116

BURCHETTE, T., 1988, Tectonic control on carbonate platform facies distribution and sequence development: Miocene, Gulf of Suez: Sedimentary Geology, v. 59, p. 179-204.

BUXTON, M. W. N. AND PEDLEY, H. M., 1989, A standardized model for Tethyan Tertiary carbonate ramps: Journal of the Geological Society of London, v. 146, p. 746-748.

CABRERA, L., CALVET, F., GUIMERÁ, J., AND PERMANYER, A., 1991, El registro sedimentario miocénico en los semigrabens del Vallès-Penedès y de el Camp: Organización secuencial y relaciones tectónica sedimentación: Barcelona, Guidebook, I Congreso del Grupo Español del Terciario, 132 p.

Cahuzac, B. and Poignant, A., 1988, Les foraminifères benthiques dans l'Oligocène terminal du Vallon de Poustagnac (Landes, Bassin d'Aquitaine, Sud-Ouest de la France): Benthos'86 (3e Symposium International of Benthic Foraminifera): Genève, Révue de Paléobiologie, v. 2, p. 633-642.

Cahuzac, B. and Poignant, A., 1990, Les foraminifères benthiques intéressant la limite oligo-miocène en Aquitaine (SW de la France). Comparaisons avec la Mésogée occidentale (abs.): Barcelona, IXth Congress Regional Committee on Mediterranean Neogene Stratigraphy, Vol. Abstracts, p. 87.

Carannante, G., Esteban, M., Milliman, J. D., and Simone, L., 1988, Carbonate lithofacies as paleolatitude indicators: problems and limitations: Sedimentary Geology, v. 60, p. 333-346

Catalano, R., 1979, Scogliere ed evaporiti Messiniane in Sicilia: modelli genetici ed emplicazioni strutturali: Lavori dell'Instituto di Geologia dell'Universitata di Palermo, n. 18, p. 1-21.

Catalano, R. and Esteban, M., 1978, Messinian reefs of western and central Sicily (abs.): Rome, Messinian Seminar 4, 1 p.

Chaix, C., Moissette, P., and Saint-Martin, J. P., 1986, Réflexions sur les Biocénoses/Taphocénoses en milieu récifal (Messinien d'Algérie): Bulletin du Muséum National d'Histoire Naturelle de Paris, v. 8C, p. 219-230.

Chamley, H., Meulenkamp, J. E., Zachariasse, W. J., and Van Der Zwaan, G. J., 1986, Middle to Late Miocene marine ecostratigraphy: clay minerals, planktonic foraminifera and stable isotopes from Sicily: Oceanologica Acta, v. 9, p. 227-238.

Charrière, A. and Saint-Martin, J. P., 1989, Relations entre les formations récifales du Miocène supérieure et la dynamique d'ouverture et de fermeture des communications marines à la bordure méridional du sillon sud-rifain (Maroc): Comptes-Rendus de l'Académie des Sciences de Paris, v. 309, serie II, p. 611-614.

Chevalier, J. P., 1961, Recherches sur les Madreporaires et les formations récifales miocènes de la Méditérranée occidentale: Paris, Ph. D. Thesis, Mémoires de la Societé Géologique de France, no. 93, 562 p.

Chevalier, J. P., 1977, Aperçu sur la faune corallienne récifale du Neogène: 2nd Coral Reef Symposium: Paris, Transations of Memoires du Bureau de Recherches Géologiques et Minières, v. 89, p. 359-366.

Cita, M. B., 1982, The Messinian Salinity Crisis in the Mediterranean: A Review: American Geophysical Union, Geodynamics Series, v. 7, p. 113-140.

Cita, M. B. and McKenzie, J., 1986, The Terminal Miocene Event, *in* Hsü, K. J., ed., Mesozoic and Cenozoic Oceans: Boulder, Geological Society of America, Geodynamics Serie, v. 15, p. 123-140.

Coniglio, M., James, N. P., and Aissaoui, D. M., 1988, Dolomitization of Miocene carbonates, Gulf of Suez, Egypt: Journal of Sedimentary Petrology, v. 58, p. 100-119.

Cornée, J. J., Saint-Martin, J. P., Conesa, G., and Müller, J., 1994, Geometry, palaeoenvironments and relative sea-level (accommodation space) changes in the Messinian Murdjadjo carbonate platform (Oran, western Algeria): Consequences: Sedimentary Geololy, v. 89, p. 143-158.

Cunningham, K. J., Farr, M. R., and Kruna, R. B., 1994, Magnetostratigraphic dating of an Upper Miocene shallow-marine and continental sedimentary succession in northeastern Morocco and correlation to regional and global events: Earth and Planetary Science Letters, v. 127, p. 77-93.

Dabrio, C. J., 1974, Los niveles arrecifales del Neógeno de Purchena (SE Cordilleras Béticas): Cuadernos de Geología, v. 5, p. 78-88.

Dabrio, C. J., 1975, La sedimentación arrecifal neógena en la región del rio Almanzora: Estudios Geológicos, Instituto de Investigaciones Geológicas Lucas Mallada, v. 31, p. 285-296.

Dabrio, C. J., 1989, Asociaciones de facies en los fan deltas de las cuencas neógenas y cuaternarias de las Cordilleras Béticas Orientales: Geogaceta, v. 6, p 53-55.

Dabrio, C. J., 1990, Fan-delta facies associations in late Neogene and Quaternary basins of southeastern Spain: Special Publications of the International Association of Sedimentologists, v. 10, p. 91-111.

Dabrio, C. J., Esteban, M., and Martín, J. M.,1981, The coral reef of Nijar, Messinian (uppermost Miocene), Almeria Province, S.E. Spain: Journal of Sedimentary Petrology, v. 51, p. 521-539.

Dabrio, C. J. and Martín, J. M., 1978, Los arrecifes Messinienses de Almería (SE de España): Cuadernos de Geología, v. 8-9, p. 83-104.

Dabrio, C. J., Martín, J. M., and Megías, A. G., 1985, The more tectonosedimentary evolution of Mio-Pliocene reefs in the province of Almería (S. E. Spain), *in* Mila, M. D. and Rosell, J., eds., International Association of Sedimentologists 6th European Regional Meeting: Barcelona, Universitat Autonoma de Barcelona, Institut d'Estudis Ilerdencs, p. 271-305. Dabrio, C. J. and Polo, M. D., 1988, Late Neogene fan deltas and associated coral reefs in the Almanzora Basin, Almería Province, southeastern Spain, *in* Nemec, W. and Steel, R. J., eds., Fan Deltas: Sedimentology and Tectonic Settings: Bergen, Blackie and Sons Publications, p. 354-367.

Decima, A., McKenzie, J. A., and Schreiber, B. C., 1988, The origin of evaporitive limestones: an example from the Messinian of Sicily (Italy): Journal of Sedimentary Petrolology, v. 58, p. 256-272.

Decima, A. and Wezel, F. C., 1971, Osservazioni sulle evaporiti Messiniane della Sicilia centro-meridionale: Rivista di Mineralogia Scienza, v. 22, p. 130-134, p 172-187.

Delfaud, J. and Revert, J., 1974, Observations sur le calcaire à stromatolites d'âge Miocène terminal du Djebel Murdjadjo (Oran, Algérie): Comptes Rendus de l'Académie des Sciences de Paris, v. 279, p. 1979-1982.

Demarcq, G., 1984, Importance des megafaunes marines benthiques dans l'évolution paleothermique de la Méditérranée au Neogène: Annales de Géologie des Pays Hélléniques, v. XXXII, p. 87-95.

Demarcq, G., 1985, Paleothermic evolution during the Neogene in Mediterranea through the marine megafauna (abs.): Budapest, Regional Comittee on Mediterranean Neogene Stratigraphy, 7th Congrés Nóegène Méditerranéen: p. 176-178

Dercourt, J., Zonenshain, L. P., Ricou, L. E., Kazmin, V. G., Le Pichon, X., Knipper, A. L., Grandjacquet, C., Sborshichikov, I. M., Boulin, J., Sorokhtin, O., Geyssant, J., Lepvrier, G., Biju-Duval, B., Subuet, J. C., Savostin, L. A., Westphal, M., and Laner, J. P., 1986, Geological evolution of the Tethys belt from Atlantic to Pamir since Liassic: Tectonophysics, v. 123, p. 241-315.

Dewey, J. F., Helman, M. L., Turco, E., Hutton, D. H. W., and Knott, S. D., 1989, Kinematics of the Western Mediterranean, *in* Coward, M. P., Dietrich, D., and Park, R. D., eds., Alpine Tectonics: London, Blackwell Science Publications, Geological Society of London Special Publication 45, p. 265-283.

Dewey, J. F., Pitman, W. C., Ryan, W. B. F., and Bonnin, J., 1973, Plate tectonics and the evolution of the Alpine system: Bulletin of the Geological Society of America, v. 84, p. 3137-3180.

Dill, R. F., Shinn, E. A., Jones, A. T., Kelly, K., and Steinen, R. P., 1986, Giant subtidal stromatolites forming in normal salinity waters: Nature, v. 324, p. 55-58.

Di Stefano, E. and Catalano, R., 1976, Biostratigraphy, paleoecology, and tectono-stratigraphic evolution of the pre-evaporitic and evaporitic deposits of the Cimina Basin (Sicily): Memoria della Società Geologica Italiana, v. 16, p. 95-110.

Downing, N., 1985, Coral reef communities in an extreme environment: the northwestern Arabian Gulf (abs.): Tahiti, Procedings of the 5th International Coral Reefs Congress, v. 2, p. 112.

Dronkert, H. and Pagnier, H., 1977, Introduction to Mio-Pliocene of the Sorbas basin: Málaga, Field Trip Guidebook 2, International Geologic Correlation Program, Project 96, Messinian Seminar 3, p. 1-18.

Drooger, C. W., 1979, Marine connections of the Neogene Mediterranean, deduced from the evotution and distribution of larger foraminifera: Annales de Géologie des Pays Hélléniques, Tome hors serie, v. 1, p. 361-369.

Dullo, W. C., 1983, Fossildiagenese im miozänen Leitha-Kalk der Paratethys von Österreich: Ein Beispiel für Feunenverschiebungen durch Diageneseunterschiede. Diagenesis of Fossils of the Miocene Leitha Limestone of the Paratethys, Austria: An Example for Faunal Modifications Due to Changing Diagenetic Environments: Facies, v. 8, p. 1-112.

Esteban, M., 1978, Significance of the Upper Miocene reefs in the western Mediterranean (abs.): Rome, Messinian Seminar 4, p. 2.

Esteban, M.,, 1979, Significance of the Upper Miocene coral reefs of the western Mediterranean: Palaeography, Palaeoclimatology, Palaeoecology, v. 29, p. 169-188.

ESTEBAN, M., 1980a, Tertiary reefs of western Mediterranean with emphasis on Miocene reefs: Houston, American Association of Petroleum Geologists Fall Education Conference, p. 1-23.

ESTEBAN, M.,, 1980b, Outer slopes of the Tertiary carbonate platforms: facies and porosity: Ricardo Asseretto Memorial Symposium on Carbonate Platform Margins of the Passive Type: Paris, 26th International Geological Congress, p. 465.

ESTEBAN, M., 1988, Miocene reefs in western Mediterranean (abs.): American Association of Petroleum Geologists Bulletin, v. 72, p. 182.

ESTEBAN, M., CALANDRA, D., CATALANO, R., AND DI STEPHANO, E., 1982a, La Scogliera Messiniana di Mazara Del Vallo, *in* Catalano, R. and D'Argenio, B., eds., Guida alla Geologia Della Sicilia Occidentale: Palermo, Società Geologica Italiana, 1° Centenario della Società Geologica Italiana, p. 148-152.

ESTEBAN, M. AND CALVET, F., 1983, Cementation of Upper Miocene reefs in western Mediterranean (abs.): American Association of Petroleum Geologists Bulletin, v. 67, p. 457.

ESTEBAN, M., CALVET, F., DABRIO, C. J., BARÓN, A., GINER, J., POMAR, L., AND SALAS, R., 1977a, Messinian(Uppermost Miocene) reefs in Spain: morphology, composition, and depositional environments, (abs.): Miami, 3rd International Coral Reef Sympossium Abstracts, p. 172-173

ESTEBAN, M., CALVET, F., DABRIO, C. J., BARON, A., GINER, J., POMAR, L., SALAS, R., AND PERMANYER, A., 1978, Aberrant features of the Messinian coral reefs, Spain: Acta Geológica Hispánica, v. 13, p. 20-22.

ESTEBAN, M., CALVET, F., AND GINER, P., 1977b, Los arrecifes Mesinieneses: El arrecife de Santa Pola, *in* Salas, R., ed., 1er. Seminario Práctico de Asociaciones Arrecifales Evaporíticas: Barcelona-Alacant, Universitat de Barcelona, p. 4.1-4.51.

ESTEBAN, M., CATALANO, R., AND DI STEPHANO, E., 1982b, Scogliere Messiniane a Porites nella Sicilia sud-occidentale: Rendicompti della Società Geologica Italiana, v. 5, p. 61-64.

ESTEBAN, M. AND GINER, J., 1977, Field guide to Santa Pola Reef: Málaga, Field Trip Guidebook, Messinian Seminar 3, p. 23-30.

ESTEBAN, M. AND GINER, J., 1980, Messinian coral reefs and erosion surfaces in Cabo de Gata (Almería, SE Spain): Acta Geológica Hispánica, v. 15, p. 97-104.

ESTEBAN, M. AND PERMANYER, A., 1977, El Mioceno del Penedés: Los arrecifes, *in* Salas, R., ed., lst Seminario Práctico de Asociaciones Arrecifales Evaporíticas: Barcelona-Alacant, Universitat de Barcelona, p. 2.1-2.32

ESTEBAN, M., POMAR, L., AND WARD, W. C., 1992, Peculiarities of Miocene carbonate sequences in the Mediterranean (abs.), *in* Simó, T., ed., Carbonate Stratigraphic Sequences: Sequence boundaries and Associated Facies: La Seu, Society of Economic Paleontologists and Mineralogists and International Association of Sedimentologists Research Conference, p. 36-37.

ESTEBAN, M. AND PREZBINDOWSKI, D. R., 1985, Preserved aragonite cements in Miocene coral reefs: a record of Messinian salinity crises in Mediterranean (abs.): American Association of Petroleum Geologists Bulletin, v. 69, p. 253-254.

EVANS, A. L., 1988, Neogene tectonic and stratigraphic events in the Gulf of Suez rift area, Egypt: Tectonophysics, v. 153, p. 235-247.

FLOS, J., 1985, The Driving Machine, *in* Margalef, R., ed., Western Mediterranean: Oxford, Pergamon Press, p. 60-99.

FOIS, E., 1990, Stratigraphy and palaeogeography of the Capo Milazzo area (NE Sicily, Italy): clues to the evolution of the southern margin of the Tyrrhenian Basin during the Neogene: Palaeogeography, Palaeoclimotology, Palaeoecology, p. 78, v. 87-108.

FOLLOWS, E. J., 1992, Patterns of reef sedimentation and diagenesis in the Miocene of Cyprus: Sedimentary Geology, v. 79, p. 225-253

FORNÓS, J. J. AND POMAR, L., 1983, El Mioceno superior de Mallorca: Unidad de Calizas de Santanyí (Complejo Terminal), *in* Pomar, L., Obrador, A., Fornós, J. J., and Rodríguez-Perea, A., eds., El Terciario de las Baleares (Mallorca-Menorca): Institut d'Estudis Baleàrics, Universitat de Palma de Mallorca, p. 177-206.

FRANSEEN, E. K., 1989, Depositional sequences and correlation of Middle to Upper Miocene carbonate complexes, Las Negras area, southeastern Spain: Unpublished Ph.D., University of Wisconsin, Madison, 374 p.

FRANSEEN, E. K., GOLDSTEIN, R. H., AND WHITESHELL, T. E., 1993, Sequence stratigraphy of Miocene carbonate complexes, Las Negras area, Southeastern Spain: implications for quantification of changes in relative sea level, *in* Loucks, R. G. and Sarg, J. F., eds., Carbonate Sequence Stratigraphy: Tulsa, American Association of Petroleum Geologists Memoir 57, p. 409-434.

FRANSEEN, E. K. AND MANKIEWICZ, C., 1991, Depositional sequences and correlation of middle (?) to late Miocene carbonate complexes, Las Negras and Níjar areas, southeastern Spain: Sedimentology, v. 38, p. 871-898.

FRANSEEN, E. K., MANKIEWICZ, C., AND PRAY, L C., 1988, Depositional sequences and correlation of Middle-Upper Miocene reef complexes, Níjar and Las Negras areas, southeastern Spain (abs.): American Association of Petroleum Geologists Bulletin, v. 72, p. 186-187.

FRIEBE, J. G., 1990, Carbonate sedimentation within a siliciclastic environment: the Leithakalk of the Weissenegg Formation (Middle Miocene, Styrian Basin, Austria): Zentralblatt für Geologie und Paläontologie, v. 11, p. 1671-1687.

GARRIDO-MEGÍAS, A., 1985, Tectosedimentary relationships between Mio-Pliocene reefs and evaporites in Almería and Sorbas basins (S.E. Iberian Peninsula): Lérida, International Association of Sedimentologists 6th European Regional Meeting, p. 292-295

GAUTIER, F., CLAUZON, G., SUC, J. P., CRAVATTE, J., AND VIOLANTI, D., 1994, Age et durée de la crise de salinité messinienne: Comptes Rendus de la Académie des Sciences de Paris, t. 318, serie II, p. 1103-1109.

GILL, D., 1977, The Belle River Mills gas field: Productive Niagaran reefs encased by sabkha deposits, Michigan Basin: Michigan Basin Geological Society Special Paper 2, 188 p.

GILL, D., 1985, Depositional facies of Middle Silurian (Niagaran) pinnacle reefs, Belle River Mills gas field, Michigan basin, southeastern Michigan, *in* Roehl, P. O. and Choquette, P. W., eds., Carbonate Petroleum Reservoirs: New York, Springer-Verlag, p. 121-139.

GOLDSTEIN, R. H., FRANSEEN, E. K., AND MILLS, M. S., 1990, Diagenesis associated with subaerial exposure of Miocene strata, southeastern Spain: implications for sea-level change and preservation of low-temperature fluid inclusions in calcite cement: Geochimica et Cosmoschimica Acta,, v. 54, p. 699-704.

GRASSO, M., LENTINI, F., AND PEDLEY, H. M., 1982, Late Tortonian-Lower Messinian (Miocene) palaeogeography of SE Sicily: information from two new formations of the Sortino Group: Sedimentary Geology, v. 32, p. 279-300.

GRASSO, M. AND PEDLEY, H. M., 1988, The sedimentology and development of Terravecchia Formation carbonates (Upper Miocene) of north central Sicily: possible eustatic influence on facies development: Sedimentary Geology, v. 57, p. 131-149.

GRASSO, M. AND PEDLEY, H. M., 1989, Palaeoenvironment of the Upper Miocene coral build-ups along the northern margins of the Caltanissetta Basin (central Sicily): Atti 3° Simposio di Ecologia e Paleoecologia della Comunità Bentoniche: Società Paleontologica Italiana, Università di Catania, p. 373-389.

GRASSO, M., PEDLEY, H. M., AND ROMEO, M., 1990, The Messinian Tripoli Formation of Central Sicily: palaeoenvironmental and paleoclimatic interpretations based on sedimentological, micropalaeontological and regional tectonic studies: Paléobiologie Contributions, v. 17, p. 189-204.

HALLOCK, P., 1988, The role of nutrient availability in bioerosion: consequences to carbonate buildups: Palaeogeography, Palaeoclimatology, Palaeoecology, v. 63, p. 275-291.

HALLOCK, P. AND SCHLAGER, W., 1986, Nutrient excess and the demise of coral reefs and carbonate platforms: Palaios, v. 1, p. 389-398

HAQ, B. U., HARDENBOL, J., AND VAIL, P. R., 1987, Chronology of fluctuating sea levels since the Triassic: Science, v. 235, p. 1156-1167.

HALLAM, A., 1981, Relative importance of plate movements, eustacy and climate in controlling major biogeographical changes since the early Mesozoic, *in* Nelson, G. and Rosen, B., eds, Vicariance Biogeography: A Critique: New York, Columbia University Press, p 303-340.

HARLAND, W. B., ARMSTRONG, R. L., COX, A. V., GRAIG, L. E., SMITH, A. G., AND SMITH, D. G., 1990, A Geologic Time Scale 1989: Cambridge,

Cambridge Unversity Press, p. 1-263.

HAYWARD, A. B., 1982, Coral reefs in a clastic sedimentary environment: fossil (Miocene, SW Turkey) and modern (Recent, Red Sea) analogues: Coral Reefs, v. 1, p. 109-114.

HECKEL, P. H., 1974, Carbonate buildups in the geologic record: a review: Tulsa, Society of Economic Paleontologists and Mineralogists Special Publication 18, p. 9-154.

HERMITE, H., 1879, Études Géologiques sur les Iles Baléares: Paris, F. Savy Editions, Pinchon Impremerie, 362 p.

HOPKINS, T. S., 1985, Physics of the sea, *in* Margalef, R., ed., Western Mediterranean: Oxford, Pergamon Press, p. 100-125.

HSÜ, K. J., CITA, M. B., AND RYAN, W. B. F., 1973, The origin of the Mediterranean evaporites: International Report Deep Sea Drilling Project 13: Washington, D.C., United States Government Printing Office, p. 1203-1231.

HSÜ, K. J., MONTADERT, L., BERNOULLI, D., CITA, M. B., ERICKSON, A., GARRISON, R. E., KIDD, R. B., MELIERES, F., MÜLLER, C., AND WRIGHT, R., 1977, History of the Mediterranean salinity crisis: Nature, v. 267, p. 399-403.

HUBBARD, J. A. E. B. AND POCOCK, Y. P., 1972, Sediment rejection by scleractinian corals: a key to palaeoenvironmental reconstruction: Geologisches Rundschau, v. 61, p. 598-626.

HUH, J. M., BRIGGS, L. I., AND GILL, D., 1977, Depositional environments of pinnacle reefs, Niagaran and Salina groups, northern shelf, Michigan basin, *in* Fisher, J. H., ed., Reefs and Evaporites- Concepts and Depositional Models: Tulsa, American Association of Petroleum Geologists Studies in Geology 5, p. 1-21.

JAMES, N. P., 1983, Reef, *in* Scholle, P. A., Bebout, D. G., and Moore, C. H., eds., Carbonate Depositional Environments: Tulsa, American Association of Petroleum Geologists Memoir 33, p. 345-440.

JAMES, N. P., CONIGLIO, M., AÏSSOUI, D. M., AND PURSER, B. H., 1988, Facies and geologic history of an exposed Miocene rift-margin carbonate platform: Gulf of Suez, Egypt: Bulletin of the American Association of Petroleum Geologists, v. 72, p. 555-572.

JAMES, N. P. AND GINSBURG, R. N., 1979, The Seaward Margain of Belize Barrier and Atoll Reefs: Oxford, International Association of Sedimentologists, Blackwell Scientific Publications Special Publication 3, 191 p.

JENKINS, J. A., 1982, An isotopic study of Nile Cone sediments and late Pleistocene sapropel formation in the eastern Mediterranean Sea: M.S. Thesis, University of South Carolina, Columbia, 85 p.

JONES, B. AND HUNTER, I. G., 1991, Corals to rhodolites to microbialites- A community replacement sequence indicative of regressive conditions: Palaios, v. 6, p. 54-66.

KENDALL, C. G. ST. G. AND SCHLAGER, W., 1981, Carbonates and relative changes in sea level: Marine Geology, v. 44, p. 181-212.

KINSMAN, D. J. J., 1964, Reef coral tolerance of high temperatures and salinities: Nature, v. 202, p. 1280-1282.

LA CHAPELLE, G., 1988, Le bassin néogène de Níjar-Carboneras (NE Espagne): Les rélations entre la sédimentation et les étapes de la structuration: Ph.D. Thesis Université de Lyon I - Claude Bernard, 253 p.

LAND, L. S. AND GOREAU, T. F., 1970, Submarine lithification of Jamaican reefs: Journal of Sedimentary Petrology, v. 40, p. 457-462.

LAZAR, B., STARINSKY, A., KARTZ, A., AND SASS, E., 1983, The carbonate system in hypersaline solutions: alkalinity and $CaCO_3$ solubitity of evaporated seawater: Limnology and Ocenography, v. 28, p. 978-986.

LEES, A., 1974, Contrasts between recent warm- and cold-water carbonates: significance in the interpretation of ancient limestones: Annales de la Societé Géologique de Belgique, v. 97, p. 159-161.

LEES, A., 1975, Possible influence of salinity and temperature on modern shelf carbonate sedimentation: Marine Geology, v. 19, p. 159-198.

LEES, A. AND BULLER, A. T., 1972, Modern temperate-water and warm-water shelf carbonate sediments contrasted: Marine Geology, v. 13, p. 63-73

LEMOINE, M., 1978, The peri-Tyrrhenian System: Corso-Sardinian Block, Apennines and Padana Basin, Calabro-Sicilian Arc, *in* Lemoine, M., ed., Geological Atlas of Alpine Europe and Adjoining Areas: Amsterdam, Elsevier Science Publications Company, p. 227-311.

LIETZ, J. AND SHWARZBACH, M., 1970, Neue fundpunkte von marinem Tertiär auf der Atlantik-Insel Porto Santo (Madeira-Archipel): Neues Jahrbuch Geologisches undPaläontologisches Monatshefte v. 5, 270-282

MAIKLEN, W. R., 1971, Evaporitic draw-douwn- a mechanism for water-level lowering and diagenesis in the Elk Point basin: Canadian Petroleum Geology Bulletin, v. 19, p. 487-503.

MANKIEWICZ, C., 1987, Sedimentology and calcareous algal paleoecology of Miocene reef complexes, southeastern Spain: Unpublised Ph.D. Dissertation, University of Wisconsin, Madison, 341 p.

MANKIEWITCZ, C., 1988, Occurrence ans paleoecologic significance of *Halimeda* in late Miocene reefs, southeastern Spain: Coral Reefs, v. 6, p. 271-279.

MARGALEF, R., 1968, Perspectives in ecological theory: Chicago, Chicago University Press, 112 p.

MARS, P., 1963, Les faunes et la stratigraphie du Quaternaire Méditerranéen: Recueil des Travaux, Bulletin Station Maritime Endoume, v. 24, p. 61-97.

MARTÍN, J. M. AND BRAGA, J. C., 1990, Arrecifes Messinenses de Almería. Tipologías de crecimiento, posición estratigráfica y relación con las evaporitas: Geogaceta, v. 7, p. 66-68.

MARTÍN, J. M. AND BRAGA, J. C., 1994, Messinian events in the Sorbas Basin in southeastern Spain and their implications in the recent history of the Mediterranean: Sedimentary Geology, v. 90, p. 257-268

MARTÍN, J. M., BRAGA, J. C., AND RIVAS, P., 1989, Coral successions in Upper Tortonian reefs in SE Spain: Lethaia, v. 22, p. 271-286.

MAUFFRET, A., 1976, Etude géodynamique de la marge des Iles Baleares: Unpublished Ph.D. Thesis, Université Pierre et Marie Curie, Paris VI, 141 p.

MAURIN, A. F., MONTY, C. L., AND BERNET-ROLLANDE, M., 1980, The Miocene *Porites* reef of Santa Pola, Spain (abs.): Paris, 26th International Geological Congress, v. 2, p. 515.

MCKENZIE, J. A., JENKYNS, H. C., AND BENNETT, G. G., 1979, Stable isotope study of the cyclic diatomite-claystone from the Tripoli Formation, Sicily: A prelude to the Messinian salinity crisis: Palaeogeography, Palaeoclimatology, Palaeocology , v. 29, p. 125-141

MESOLELLA, K. J., ROBINSON, J. D., MCCORMICK, L. M., AND ORMISTON, A. R., 1974, Cyclic deposition of Silurian carbonates and evaporites in Michigan basin: American Association of Petroleum Geologists Bulletin, v. 58, p. 34-62.

MEULENKAMP, J. E., DRIEVER, B. W. M., JONDERS, H. A., SPAAD, P., ZACHARIASSE, W. J., AND VAN DER ZUANN, G. J., 1979, Late Miocene-Piocene climatic fluctuations and marine "cyclic" sedimentation patterns: Athens, VIIth International Congress on Mediterranean Neogene: Annales de Géologie des Pays Helléniques, Tome hors série, v. II, p. 831-842.

MOISSETTE, P. AND SAINT-MARTIN, J. P., 1990, Upwelling and benthic communities in the Messinian of western Mediterranean (abs.): Barcelona, Published in 1992, IX Congress of the Regional Committee on Mediterranean Neogene Stratigraphy, Barcelona, Paleontologia i Evolució, v. 24-25, p. 245-254.

MONTENAT, C., 1973a, Le Miocène terminal des chaînes Bétiques (Espagne méridonale) Esquisse paleogéographique, *in* Drooger, C. W., ed., Messinian Events in the Mediterranean: Amsterdam, North Holland Publications Company, Geodynamics Science Report 7, p. 180-187.

MONTENAT, C., 1973b, Les formations néogènes et quaternaires du Levant espagnol (Provinces d'Alicante et de Murcia): Unpublished Ph.D. Thesis, Université de Paris-Orsay, Paris, 1170 p.

MONTENAT, C., 1975, Le néogène des Cordillères Bétiques; essai de synthèse stratigraphique et paleogéographique: Paris, Centre National de Recherche, Science Publications, Beicip, 187 p.

MONTENAT, C., 1977, Les bassins neogènes du Levant d'Alicante et de Murcia (Cordillères Bétiques orientale-Espagne): stratigraphie, paleogéographie et évolution dynamique: Documents du Laboratoires de Géologie de la Faculté des Sciences de Lyon, v. 69, p. 1-345.

MONTENAT, C., OTT D'ESTEVOU, P., LAROUZIÈRE, F. D., AND BEDU, P., 1987, Originalité géodynamique des bassins néogènes du domaine bétique oriental (Espagne): Paris, Notes et Mémoires Compagnie Française des Pétroles, v. 21, p. 11-50.

MONTENAT, C., OTT D'ESTEVOU, P., PLAZIAT, J. C., AND CHAPEL, J., 1980, La signification des faunes marines contemporaines des évaporites messiniennes dans le Sud-Est de l'Espagne. Conséquences pour l'interpretation des conditions d'isolement de la Méditérranée occidentale: Géologie Méditérranéenne, v. 7, p. 81-90.

MONTENAT, C., OTT D'ESTEVOU, P., PURSER, B. H., BUROLLET, P. F., JARRIGE, J. J., ORSZAG-SPERBER, F., PHILOBBOS, E., PLAZIEAT, J. C., PRAT, P., ROUSSEL, N., AND THIRIEF, J. P., 1988, Tectonic and sedimentary evolution of the Gulf of Suez and the northwestern Red Sea: Tectonophysics, v. 153, p. 161-177.

MONTY, C. L., 1981, The Upper Miocene in the Benejúzar region, Alicante Province and associated stromatolites, Societé Géologique de Belgique, Annales 104, v. l, p. 109-114.

MONTY, C. L., MAURIN, A., AND BERNET-ROLLANDE, M., 1980, Miocene stromatolitic reefs, Benejúzar, southeastern Spain (abs.): Paris, 26th International Geological Congress, Résumés 26, v. 2, p. 522.

MONTY, C. L., ROUCHY, J. M., MAURIN, A., BERNET-ROLLANDE, M. C., AND PERTHUISOT, J. P., 1987, Reef-stromatolites-evaporites facies relationships from Middle Miocene examples of the Gulf of Suez and the Red Sea, *in* Peryt, T. M., ed., Lecture Notes in Earth Sciences, v. 13, p. 133-188.

MÜLLER, D., 1986, Die Salinitätskrise im Messinian (Späten Miozan) der Becken von Fortuna und Sorbas (Sudöst Spanien): ETH Dissertation 8056, Zurich, p. 1-183.

MÜLLER, D. AND HSÜ, K. J., 1987, Event stratigraphy and paleoceanography in the Fortuna Basin (Southeast Spain): A scenario for the Messinian Salinity Crisis: Paleoceanography, v. 4, p. 75-86.

NEBELSICK, J. H., Temperate water carbonate facies of the Early Miocene Paratethys (Zogelsdorf Formation, Lower Austria): Facies, v. 21, p. 11-40.

NESTEROFF, W. D. AND GLAÇON, G., 1977, Le caractère rythmique des evaporites messiniennes en Mediterranée orientale (coupe d'Eraclea Minoa, Sicile): Bulletin de la Societé Géologique de la France, v. 19, p. 489-500.

OBRADOR, A., POMAR, L., AND TABERNER, C., 1992, Late Miocene breccia of Menorca (Balearic Islands): a basis for the interpretation of a Neogene ramp deposit: Sedimentary Geology, v. 79, p. 203-223.

ORSZAG-SPERBER, F. AND PILOT, M. D., 1976, Grands traits du Néogène de Corse: Bulletin de la Societé Géologique de France, v. 18, p. 1183-1187.

ORSZAG-SPERBER, F. AND ROUCHY, J. M., 1979, Le Miocène supérieure et le Pliocène Inférieur du Sud de Chypre: Field Trip Guidebook: Livret gride de l'excursion du 5e Séminaire International sur le Messinien: Paphos, International Geological Correlation Program, Project 117, 50 p.

ORTÍ-CABO, F. AND SHEARMAN, D. J., 1977, Estructuras y fábricas deposicionales en las evaporitas del Mioceno superior (Messiniense) de San Miguel de Salinas (Alicante-España): Instituto de Investigaciones Geológicas, v. XXXII, p. 5-54.

OSWALD, E. J., 1992, Dolomitization of a Miocene reef complex, Mallorca, Spain: Unpublished Ph.D. Dissertation, State University of New York at Stony Brook, Stony Brook, 437 p.

OSWALD, E. J., FRANSEEN, E. K., AND MEYERS, W. J., 1991a, Similarities in the dolomitization of upper Miocene reef complexes in Mallorca and the Las Negras area, Spain: Possible evidence for a Mediterranean dolomitizing event during the Messinian (abs.): American Association of Petroleum Geologists Bulletin, v. 74, p. 735.

OSWALD, E. J., MEYERS, W. J., AND POMAR, L., 1990, Dolomitization of an Upper Miocene reef complex, Mallorca, Spain: evidence for a Messinian dolomitizing Mediterranean Sea (abs.): American Association of Petroleum Geologists Bulletin, v. 74, p. 735.

OSWALD, E. J., SCHOONEN, M. A. A., AND MEYERS, W. J., 1991b, Dolomitizing sea in evaporitic basins: a model for pervasive dolomitization of upper Miocene reefal carbonates in the western Mediterranean (abs.): American Association of Petroleum Geologists Bulletin, v. 75, p. 649.

OTT D'ESTEVOU, P., 1980, Evolution dynamique du bassin néogène de Sorbas (Cordillères Bétiques orientales, Espagne): Unpublished Ph.D. Thesis, Institut Géologique Albert Lapparent, Paris, 1, 264 p.

PAGNIER, H., 1977, Excursion to Messinian reef deposits in the northern part of Sorbas Basin: an introduction: Field Trip Guidebook, Messinian Seminar 3, International Geological Correlation Program 96, University of Málaga, p. 44-53.

PEDLEY, H. M., 1976, A palaeoecological study of the Upper Coralline Limestone, *Terebratula aphelesia* Bed (Miocene, Malta), based on bryozoan growth-form studies and brachiopod distributions: Palaeogeography, Palaeoclimatology, Palaeoecology, v. 20, p. 209-234.

PEDLEY, H. M., 1979, Miocene bioherms and associated structures in the Upper Coralline Limestone of the Maltese Islands: their lithification and paleoenvironment: Sedimentology, v. 26, p. 577-591.

PEDLEY, H. M., 1981, Sedimentology and palaeoenvironment of the southeast sicilian Tertiary platform carbonates: Sedimentary Geology, v. 28, p. 273-291.

PEDLEY, H. M., 1983, The petrology and palaeoenvironment of the Sortino Group (Miocene) of SE Sicily: evidence for periodic emergence: Journal of the Geological Society of London, v. 140, p. 335-350

PEDLEY, H. M., 1987a, Controls on Cenozoic carbonate deposition in the Maltese Islands: review and reinterpretation: Memoria della Società Geologica Italiana, v. 38, p. 81-94

PEDLEY, H. M., 1987b, The Ghar Lapsi limestones: sedimentology of a Miocene intra-shelf graben: Centro, University of Malta Press, v. l, p. 1-14.

PEDLEY, H. M. AND GRASSO, M., 1993, Controls on faunal and sediment cyclicity within the Tripoli and Calcare di Base basins (Late Miocene) of central Sicily: Palaeogeography, Palaeoclimatology, Palaeoecology, v. 105, p. 337-360.

PERMANYER, A., 1990, Sedimentologia i diagènesi dels esculls miocènics de la Conca del Penedès: Institut d'Estudis Catalans, Arxius de la Secció de Ciències, v. XCII, 320 p.

PERMANYER, A. AND ESTEBAN, M., 1973, El arrecife mioceno de Sant Pau d'Ordal (provincia de Barcelona): Barcelona, Institut d'Investigacions Geològiques, v. 28, p. 45-72.

PIERRE, C., 1982, Teneurs en isotopes stables et conditions de genèse des évaporites marines: applications à quelques milieux actuels et au Messinien de la Mediterranée: Unpublished Ph.D Thesis, Université Paris-Sud, Orsay, 266 p.

PIERRE, C. AND ROUCHY, J. M., 1990, Sedimentary and diagenetic evolution of Messinian evaporites in the Tyrrhenian sea (ODP Leg 107, sites 652, 653 and 654): petrographic, mineralogic and stable isotope records: College Station, Proceedings of the Ocean Drilling Program, Scientific Results, p. 187-210.

PISERA, A., 1985, Paleoecology and lithogenesis of the Middle Miocene (Badenian) algal-vermetid reefs from the Roztocze Hills, southeastern Poland: Acta Geologica Polonica, v. 35, p. 89-155.

POMAR, L., 1991, Reef geometries, erosion surfaces and high-frequency sea-level changes, Upper Miocene Reef Complex, Mallorca, Spain: Sedimentology, v. 38, p. 243-269.

POMAR, L., 1993, High-resolution sequence stratigraphy in prograding Miocene carbonates: Application to seismic interpretation, *in* Loucks, R. G., and Sarg, J. F., eds., Carbonate Sequence Stratigraphy, Recent Developments and Applications: Tulsa, American Association of Petroleum Geology Memoir 57, p. 389-407.

POMAR, L., ESTEBAN, M., CALVET, F., AND BARON, A., 1983, La unidad arrecifal del Mioceno superior de Mallorca, *in* Pomar, L., Obrador, A., Fornós, J., and Rodríguez-Perea, A., eds., El Terciario de las Baleares (Mallorca-Menorca), Guía de las Excursiones del X Congreso Nacional de Sedimentología: Institut d'Estudis Baleàrics and Universidad de Palma de Mallorca, p. 139-175.

POMAR, L., FORNÓS, J. J., AND RODRÍGUEZ-PEREA, A., 1985, Reef and shallow carbonate facies of the upper Miocene of Mallorca: Lleida-Barcelona, Field Trip Guidebook, 6th International Association of Sedimentologists, European Reef Meeting, p. 495-518.

POMAR, L. AND WARD, W. C., 1994, Response of a late Miocene Mediterranean reef platform to high-frequency eustasy: Geology, v. 22, p. 131-134.

PUIGDEFÁBREGAS, C., 1975, La sedimentación molásica en la cuenca de Jaca: Monografía del Instituto de Estudios Pirenaicos, Consejo Superior de Investigaciones Científicas, v. 104, 188 p.

PURSER, B. H., ORSZAG-SPERBER, F., AND PLAZIAT, J. C., 1987, Sédimentation et rifting: les séries néogènes de la marge nord-occidentale de la Mer Rouge (Egypte): Paris, Notes et Mémoires, Compagnie Française des Pétroles, v. 21, p. 111-144.

Purser, B. H., Philobbos, E. R., and Soliman, M., 1990, Sedimentation and rifting in the NW parts of the Red Sea: a review: Bulletin de la Societé Géologique de France, v. 8, p. 371-384

Rehault, J. P., Boillot, G., and Mauffret, A., 1984, The Western Mediterranean Basin geological evolution: Marine Geology, v. 55, p. 447-478.

Riding, R., Braga, J. C., and Martín, J. M., 1991a, Oolite stromatolites and thrombolites, Miocene, Spain: analogues of Recent giant Bahamian examples: Sedimentary Geology, v. 71, p. 121-127.

Riding, R., Martín, J. M., and Braga, J. C., 1991b, Coral-stromatolite reef framework, Upper Miocene, Almería, Spain: Sedimentology, v. 38, p. 799-818.

Rodriguez-Perea, A., 1989, Miocene mixed shelf deposits of Mallorca island (abs.): Budapest, 10th International Association of Sedimentologists Regional Meeting, p. 196-197.

Rögl, F. and Steininger, F. F., 1983, Vom Zerfall der Tethys zu Mediterran und Paratethys. Die Neogene Peleogeographie und painspastik des zirkum-Mediterranen Raumes: Annalen des Naturhistorischen Museums in Wien, v. 85A, p. 135-163.

Rögl, F. and Steininger, F. F., 1984, Neogene Paratethys, Mediterranean and Indo-Pacific Seaways: Implications for the paleobiogeography of marine and terrestrial biotas, *in* Brenchley, P., ed., Fossils and Climate: New York, John Wiley and Sons Ltd., p. 171-179.

Rögl, F., Steininger, F. F., and Müller, C., 1978, Middle Miocene salinity crisis and paleogeography of the Paratethys (middle and eastern Europe): Initial Reports of the Deep Sea Drilling Project, Volume XLII, part 1, p. 985-990

Rouchy, J. M., 1979, La sédimentation évaporitique sur les marges messiniennes: Athens, 7th International Congress on Mediterranean Neogene Annales de Géologie des Pays Hélléniques, v. 3, p. 1051-1060.

Rouchy, J. M., 1980, La genèse des évaporites messiniennes de Méditerranée: un bilan: Bulletin du Centre de Recherche, Exploration et Production Elf-Aquitaine, v. 4, p. 511-545.

Rouchy, J. M., 1982a, La genèse des évaporites messiniennes de Méditerranée: Mémoires du Muséum National d'Histoire Naturelle, v. 50, 267 p.

Rouchy, J. M., 1982b, La crise évaporitique messinienne de Méditerranée: nouvelles propositions pour une interprétation génétique: Paris, Bulletin du Muséum National d'Histoire Naturelle, v. 4, p. 107-136.

Rouchy, J. M., 1986, Les évaporites miocènes de la Méditerranée et de la Mer Rouge et leurs enseignements pour l'interprétation des grandes accumulations évaporitiques d'origine marine: Bulletin de la Societé Géologique de France, v. 8, p. 511-520.

Rouchy, J. M., 1988, Relations evaporites-hydrocarbures: l'association laminites-récifs-evaporites dans le Messinien de Méditérranée et ses ensignements, *in* Busson, J., ed., Evaporites et Hydrocarbures: Paris, Mémoires du Muséum National d'Histoire Naturelle, serie C, v. 55, p. 43-69.

Rouchy, J. M., Bernet-Rollande, M. C., Maurin, A. F., and Monty, C., 1983, Signification sédimentologique et paléogéographique des divers types de carbonates bioconstruits associés aux évaporites du Miocène moyen près du Gebel Esh Mellaha (Egypte): Comptes-Rendus de l'Académie des Sciences de Paris, v. 296, serie II, p. 457-462.

Rouchy, J. M., Chaix, C., and Saint-Martin, J. P., 1982a, Importance et implications de l'existence d'un recif corallien messinien sur le flanc sud du Djebel Murdjadjo (Oranie, Algerie): Comptes-Rendus de l'Académie des Sciences de Paris, serie II, v. 294, p. 813-816.

Rouchy, J. M. and Monty, C. L., 1981, Stromatolites and cryptalgal laminates associated with Messinian gypsum of Cyprus, *in* Monty, C. L., ed., Phanerozoic Stromatolites: Berlin, Springer-Verlag, p. 155-180.

Rouchy, J. M., Monty, C. L., Bernet-Rollande, M., and Maurin, A. F., 1982b, Mid-Miocene stromatolites of Gebel Esh Mellaha (Egypt) and their sedimentological interest (abs.): Hamilton, 11th International Congress on Sedimentology, v. 11, p. 29.

Rouchy, J. M., Noel, D., Wali, A. M. A., and Aref, M. A. M., 1995, Evaporitic and biosiliceous cyclic sedimentation in the Miocene of the Gulf of Suez; depositional and diagenetic aspects: Sedimentary Geology, v. 94, p. 277-297.

Rouchy, J. M. and Pierre, C., 1979, Données sédimentologiques et isotopiques sur les gypses des séries évaporitiques messiniennes d'Espagne méridionale et du Chypre: Revue de Géologie Dynamique et de Géographie Physique, v. 21, p. 267-280

Rouchy, J. M. and Saint-Martin, J. P., 1992, Late Miocene events in the Mediterranean as recorded by carbonate-evaporite relations: Geology, v. 20, p. 269-632.

Rouchy, J. M., Saint-Martin, J. P., Maurin, A., and Bernet-Rollande, M. C., 1986, Evolution et antagonisme des communautés bioconstructrices animales et végétales à la fin du Miocène terminal: biologie et sédimentologie: Bulletin des Centres de Recherche et Exploration Elf-Aquitaine, v. 10, p. 333-348.

Ryan, W. B. F., Hsü, K. J., Cita, M. B., Dumitrica, P., Lort, J. M., Mainc, W., Nesteroff, W. D., Pautot, G., Stradner, H., and Wezel, F. C., 1973, Reports of the Deep Sea Drilling Project 13: Washington, D.C., United States Government Printing Office, 1447 p.

Saint-Martin, J. P., 1984, Le phénomène récifal messinien en Oranie (Algérie): Géobios, Mémoire Spécial, v. 8, p. 159-166.

Saint-Martin, J. P., 1990, Les formations récifales coralliennes du Miocène supérieur d'Algérie et du Maroc: Paris, Muséum National d'Histoire Naturelle, Sciences de la Terre, v. 56, 366 p.

Saint-Martin, J. P. and Chaix, C., 1981, Sur la paléoécologie des formations récifales du Miocène superieur d'Oranie occidentale: Comptes-Rendus de l'Académie des Sciences de Paris, v. 292, p. 1341-1343.

Saint-Martin, J. P. and Chaix, C., 1984, Paléoenvironments des récifs coralliens du Messinien en Méditérranée (abs.): Bordeaux, 10th Reunion Annuelle des Sciences de la Terre, Societé Géologique de France, p. 496.

Saint-Martin, J. P., Chaix, C., and Moissette, P., 1983, Le Messinien récifal d'Oranie (Algérie): une mise au point: Comptes-Rendus de l'Académie des Sciences de Paris, Serie 2, v. 297, p. 545-547.

Saint-Martin, J. P., Cornée, J. J., Conesa, G., Bessedik, M., Delkebir, L., Monsour, B., Moissette, P., and Anglada, R., 1992, Un dispositif particulier de plate-forme carbonatée messinienne: la bordure méridionale du bassin du Chelif (Algérie): Comptes-Rendus de l'Académie des Sciences de Paris, v. 315, p. 1365-1372.

Saint-Martin, J. P., Cornée, J. J., Muller, J., Camoin, G., Andre, J. P., Rouchy, J. M., and Bonmoussa, A., 1991, Controles globaux et locaux dans l'édification d'une plateforme carbonatée messinienne (bassin de Melilla, Maroc): apport de la stratigraphie séquentielle et de l'analyse tectonique: Comptes-Rendus de l'Académie des Sciences de Paris, v. 312, p. 1573-1579.

Saint-Martin, J. P., Moissette, P., and Freneix, S., 1985, Paléoécologie des assemblages de Bivalves dans les récifs messiniens d'Oranie occidentale (Algérie): Géobios, v. 21, p. 280-283.

Saint-Martin, J. P. and Rouchy, J. M., 1986, Intérêt du complexe récifal du Cap des Trois Fourches (Bassin de Nador, Marroc septentrional) pour l'interprétation paléogéographique des évènements messiniens en Méditerranée occidentale: Comptes-Rendus de l'Académie des Sciences de Paris, v. 302, p. 957-962.

Saint-Martin, J. P. and Rouchy, J. M., 1990, The Messinian carbonate platforms in the western Mediterranean: their importance for the reconstruction of the late Miocene sea level variations: Bulletin de la Société Géologique de France, v. 8, p. 83-94.

Saint-Martin, J. P. and Thomassin, B., 1989, Récifs du Miocène superieur de Méditérranée: interpretation selon les modèles actuels (abs.): Marseille, Colloque Biologie et Géologie des Récifs Coraliens, International Society of Reef Studies Meeting, p. 143-144.

Salas, R., 1979, El sistema arrecifal del Eoceno superior de la cuenca de Igualada, Barcelona: Unpublished M.S. Thesis, University of Barcelona, 196 p.

Salas, R. and Esteban, M., 1980, El sistema arrecifal priaboniense de la cuenca de Igualada (abs.): XIII Coloquio Europeo de Micropaleontología, University of Salamanca, p. 81-82.

Santisteban, C., 1980, Composition, morphology, and structure of the fore-reef fan; the Tortonian Desastre Reef, Murcia, Spain (abs.): Bochum, International Association of Sedimentologists, Regional Meeting, p. 212-214.

Santisteban, C., 1981, Petrología y sedimentología de las materiales del Mioceno superior de las cuenca de Fortuna (Murcia) a la luz de la Teoría de la Crisis de Salinidad: Unpublished Ph.D. Thesis,

University of Barcelona, Barcelona, 725 p.

SANTISTEBAN, C. AND DAWANS, J. M., 1985, Essay on the correlation between reefs and evaporites of the Upper Miocene of southeast Spain (abs.): Lleida, International Association of Sedimentologists: 6th Annual Meeting, p. 415-419.

SANTISTEBAN, C. AND TABERNER, C., 1983, Shallow marine and continental conglomerates derived from coral reef complexes after desiccation of a deep marine basin; the Tortonian Messinian deposits of the Fortuna Basin, southeastern Spain: Geological Society of London, v. 140, p. 401-411.

SANTISTEBAN, C. AND TABERNER, C. 1988, Sedimentary models of siliciclastic deposits and coral reefs interrelation, *in* Doyle, L. J. and Roberts, H. H., eds., Carbonate-Clastic Transitions: New York, Elsevier, p. 35-76.

SCOTT, R. W. AND MCGOVEAN, F. M., 1985, Early depositional history of a rift basin: Miocene in western Sinai: Palaeogeography, Palaeoclimatology, Palaeoecology, v. 52, p. 143- 158.

SIMÓ, A. AND GINER, J., 1983, El Neógeno de Ibiza y Formentera (Isla Baleares): Revista de Investigaciones Geológicas, v. 36, p. 67-81.

SMALE, J. L., THUNELL, R. C., AND SCHAMEL, S., 1988, Sedimentological evidence for early Miocene fault reactivation in the Gulf of Suez: Geology, v. 16, p. 113-116.

SONNENFELD, P., 1980, Postulates for massive evaporite formation: Géologie Méditerranéene, v. VII, p. 103-114.

SONNENFELD, P., 1985, Models of Upper Miocene evaporite genesis in the Mediterranean region, *in* Stantey, D. J., and Wezel, F. C., eds., Geological Evolution of the Mediterranean: New York, Springer-Verlag, p. 323-345.

SONNENFELD, P. AND NAD FINELLI, I., 1985, Messinian evaporites in the Mediterranean: a model of continuous inflow and outflow, *in* Stantey, D. J., andWezel, F. C., eds., Geological Evolution of the Mediterranean: New York, Springer-Verlag, p. 347-353

STEININGER, F. F., BERNOR, R. L., AND FAHLBUSCH, V., 1990, European Neogene marine/continental chronologic correlations, *in* Lindsay, E. H., Fahlbusch, V. and Mein, P., eds., European Neogene Mammal Chronology: New York, Plenum Press, p. 155-285.

STEINENGER, F. F., RÖGL, F., AND MARTINI, E., 1976, Current Oligocene-Miocene biostratigraphic concept of the Central Paratethys (Middle Europe): Newsletters in Stratigraphy, v. 4, p. 174-202.

STEININGER, F. F., SENES, J., KLEEMANN, K., AND RÖGL, F., 1985, Neogene of the Mediterranean Tethys and the Paratethys, stratigraphic correlation tables and sediment distribution maps: Institute of Paleontology of Vienna, v. 1, p. 472, and v. 2, p. 504.

STREET, F. A. AND GROVE, A. T., 1979, Global maps of lake-level fluctuations since 30,000 years B.P.: Quarternary Research, v. 12, p. 83-118.

SUC, J. P., 1984, Origin and evolution of the Mediterranean vegetation and climate in Europe: Nature, v. 307, p. 429-432.

SUN, S. Q., 1992, Skeletal aragonite dissolution from hypersaline seawater: a hypothesis: Sedimentary Geology, v. 77, p. 249-257.

SUN, S. Q. AND ESTEBAN, M., 1994, Paleoclimatic controls on sedimentation, diagenesis and reservoir quality: lessons from Miocene carbonates: American Assocition of Petroleum Geologists Bulletin v. 78, p. 519-543.

SVERDRUP, H. V., JONHSON, M. W., AND FLEMMING, R. H., 1942, The Oceans: Prentice-Hall, Englewood Cliffs, 1087 p.

SWAIN, F. M., 1949, Onlap, offlap, overstep and overlap: American Association Petroleum Geologists Bulletin, v. 33, p. 634-636

THUNELL, R. C., WILLIAMS, D. F., AND HOWELL, M., 1987, Atlantic-Mediterranean water exchange during the late Neogene: Paleoceanography, v. 2, p. 661-678.

TROELSTRA, S. R., VAN DE POEL, H. M., HUISMAN, C. H. A., GEERLINGS, L. P. A., AND DRONKERT, H., 1980, Paleontological changes in the latest Miocene of the Sorbas Basin, SE Spain: Géologie Méditérranéenne, v. 7, p. 115-126.

VALLÉS-ROCA, D., 1985, Sedimentologia dels materials carbonatats del Miocè superior a l'àrea de Santa Pola (Alacant): Unpublished M.S. Thesis, University of Barcelona, Barcelona, 222 p.

VALLÉS-ROCA, D., 1986, Carbonate facies and depositional cycles in the upper Miocene of Santa Pola (Alicante, SE Spain): Revista de Investigaciones Geológicas, v. 42/43, p. 45-66.

VAN DE POEL, H. M., 1991, Messinian stratigraphy of the Níjar Basin (S.E. Spain) and the origin of its gypsum-ghost limestones: Geologie en Mijnbouw, v. 70, p. 215-234.

VAN DER HAMMEN, T., WIJMSTRA, T. A., AND ZAGWIJN, W. H., 1971, The floral record of the late Cenozoic of Europe, *in* Turckian, K. K., ed., Late Cenozoic Glacial Age: New Haven, Yale University Press, p. 391-424.

VAN GORSEL, J. T. AND TROELSTRA, S. R., 1980, Late Neogene climate changes and the Messinian salinity crisis: Géologie Méditérranéene, v. 7, p. 127-134.

WELLS, J. W., 1956, Scleractinia, *in* Moore, R. C., ed., Treatise of Invertebrate Paleontology: New York, v. F (Colenterata), p. 239-444.

WERNLI, R., 1987, Micropaléontologie du Néogène post-nappes du Maroc septentrional et description systematique des foraminifères planctoniques: Notes et Mémoires du Service Géologique du Maroc, 331 p.

WIJSMAN-BEST, M. AND BOEKSCHOTEN, S. J., 1982, On the coral fauna in the Miocene reef at Baixo, Porto Santo (eastern Atlantic): Netherlands Journal of Zoology v. 32, p. 412-418

WILSON, J. L., 1975, Carbonate Facies, *in* Carbonate Facies in Geologic History: Heidelberg, Springer-Verlag, 471 p.

ZIBROWNIUS, H., 1980, Les Scléractiniaires de la Méditerranée et de l'Atlantique nord-occidental: Mémoires de l'Institut Océanographique de Monaco, v. 11, p. 1-227.

ZIEGLER, P. A., 1988, Evolution of the Arctic-North Atlantic and the Western Tethys: American Association of Petroleum Geologists Memoir 43, 198 p.

# WESTERN MEDITERRANEAN REEF COMPLEXES

MATEU ESTEBAN
*Carbonates International Ltd., Vilanova 70, E-07190, Esporles, Mallorca, Spain*
JUAN CARLOS BRAGA, JOSÉ MARTÍN, AND
*Departamento de Estratigrafía, Universidad de Granada, Campus de Fuentenueva, 18002 Granada, Spain*
CARLOS DE SANTISTEBAN
*Departamento de Geología, Universidad de Valencia, 46100 Burjasset, Valencia, Spain*

Abstract: The western Mediterranean region contains abundant examples of the different types of Lower, Middle and Upper Miocene reefs (hermatypic coral reefs, ahermatypic mounds, rhodalgal biostromes and stromatolitic reefs). Those corresponding to the Upper Tortonian-Messinian rock units are the ones that have attracted the most attention because of the extraordinary quality of the outcrops and their relation to the polemic Messinian events in the Mediterranean. This section is a general introduction to the region, with a review of the Lower-Middle Miocene rhodalgal biostromes and coral reefs of the Gulf of Valencia-Provençal Basin and the Middle-Upper Miocene reefs of southeastern Spain and northern Morocco. The emphasis in this paper will be on the complex Miocene stratigraphy and paleogeography of southeastern Spain (Betics) and northern Morocco (Rif). This part of the western Mediterranean is important in understanding the paleogeographic evolution of the entire Mediterranean and its connection with the Atlantic Ocean.

## REGIONAL GEOLOGIC SETTING

Figure 1 displays the major structural units of the western Mediterranean with present-day onshore outcrops of marine Miocene basins (this figure is not a paleogeographic map). Most of these basins were affected by the Middle-Late Miocene Alpine orogeny which changed the paleogeography throughout Miocene and Pliocene times. The schematic pattern of the present-day onshore Miocene basins illustrates their inter-connections and complexities in relation to the Alpine foldbelts. Only the Portuguese and Aquitaine basins in the Atlantic passive margins (adjacent to the stable cratonic areas of the Massif Central and the Iberian Meseta) are unrelated to the Alpine foldbelts.

The northern sector of the western Mediterranean region is dominated by the Pyrenees Mountains and the epicratonic foldbelts (Iberian and Catalan Mountains) that were deformed during the Late Cretaceous and Paleogene times. Intense thrust and oblique tectonic styles resulted from the collision of the European plate with the Iberian plate (Durand-Delga and Fontboté, 1980). The Mediterranean margins of these foldbelts were affected by important graben formation during the Neogene times (Julivert et al., 1974; Fontboté et al., 1990; Roca and Guimerà, 1992; Roca and Deselgaux, 1992). This rift system initiated during Mesozoic times and continued towards northwest Europe (Rhine Graben, North Sea). The main Alpine thrustbelt dominated the southern sector of the western Mediterranean with the Betics (Betides) and the Magrebides (Rif and Tell) forming an extensive arch to the west of Gibraltar that terminates abruptly to the east of the Balearic Islands. This termination was related to southeastward drifting of Corsica and Sardinia. Major thrusting occurred in the external zones of the Betic Ranges during Mid-Miocene time as a result of collision of the Iberian and African plates; this was followed by extensive gravitational tectonics during the Late Miocene and Pliocene times. In the internal zones of the Betics, the major thrusting was during Oligocene-Aquitanian times (Comas et al., 1990); during the Middle Miocene times there was major foundering in the Alboran Sea followed by strike-slip faulting (Sanz de Galdeano, 1990). The Gulf of Valencia, Guadalquivir Basin, Gharb Basin and Saïss Basin can be considered as Neogene foredeep basins of the Betic-Magrebides foldbelt. Numerous post-orogenic basins were developed along this foldbelt during the Late Miocene times controlled to a large extent by gravitational re-adjustments and block-faulting associated with strike-slip tectonics that continued through the Pliocene and Quaternary times (Vidal, 1977; Wildi, 1983; Sanz de Galdeano, 1990; Comas et al., 1990; Montenat et al., 1990).

The complexity of the Miocene stratigraphic framework in the western Mediterranean (Fig. 2), with marked variations in facies and thicknesses, reflects the Alpine orogenic evolution. Nevertheless, marine Miocene carbonates were preferentially developed in the Aquitanian, Langhian-Early Serravallian, Tortonian (mostly Late) and Messinian times, during 3rd-order eustatic highstands. Terrigenous deposits are abundant in all Miocene units and are also mixed with the marine carbonates (except in isolated platforms such as the Upper Miocene of Mallorca and Santa Pola). Marine evaporites and carbonates appear intimately associated in the Messinian (and locally in the Uppermost Tortonian) outcrops. Continental lacustrine evaporites are much thinner than the marine evaporites and occur extensively in the Burdigalian, Serravallian and Tortonian rock units. Miocene coral reefs are common in the Upper Tortonian-Messinian (some rare Lower Tortonian-Serravallian?), Langhian and Aquitanian exposures. Rhodalgal, bryozoan-rich carbonate platforms are very extensive on the Atlantic side of the Betic and Rif Basins (Guadalquivir, Gharb) but also occur with or without associated coral reefs in the Mediterranean side of these Miocene basins. Many of these rhodalgal carbonates are mixed with siliciclastic deposits or appear as part of extensively developed ramps (as in the Middle Miocene outcrops of Mallorca).

Models for Carbonate Stratigraphy from Miocene Reef Complexes of Mediterranean Regions,
SEPM Concepts in Sedimentology and Paleontology #5, Copyright © 1996,
SEPM (Society for Sedimentary Geology), ISBN 1-56576-033-6, p. 55-72.

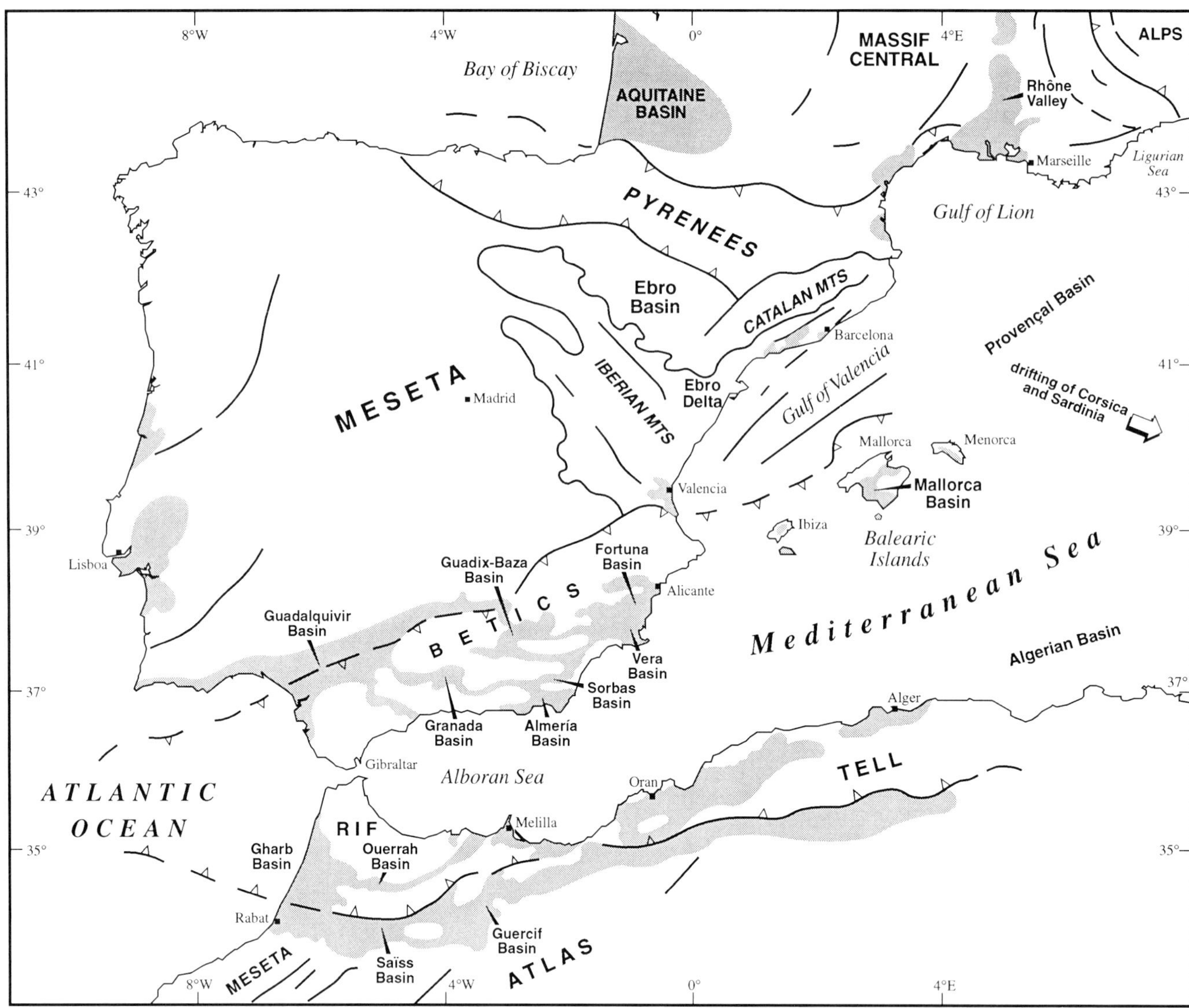


FIG. 1.—Present-day outcrops (hachured areas) of marine Miocene basins in the western Mediterranean in relation to the major structural units.

## LOWER MIOCENE REEFS

Lower Miocene coral reefs in the western Mediterranean have been well documented in a relatively small number of localities (Fig. 3), but there are indications of much more extensive occurrences in association with thick and extensive rhodalgal carbonate platforms and also mixed with terrigenous and carbonate-clastic deposits. Most of these carbonates were intensely deformed during Mid-Miocene orogenic events. Exposures are fragmented, thus making facies studies difficult. There are two major types of settings for these Lower Miocene carbonates and coral reefs: (i) onlapping continental Paleogene deposits and karst breccias on Mesozoic carbonates (La Nerthe, Mallorca), and (ii) in continuity with extensive shallow marine Oligocene platforms (Alacant area).

### *Reefs Onlapping Mesozoic and Continental Paleogene*

The best known Lower Miocene reef occurs in the Aquitanian section of La Nerthe (W. Marseille, Provence, SE France), where Chevalier (1961) described 45 hermatypic coral species, with abundant *Porites, Acropora, Tarbellastraea, Mussismilia* as main frame builders. Catzigras et al. (1972) and Magné et al. (1987) refined the traditional stratigraphy of the 85-m-thick Aquitanian section. Monleau et al. (1989) described four major depositional sequences of which two involved coral reefs. Their description will be summarized in this section. The lower sequence (Cap de Nautes) is 28 m thick and grades downward into the Oligocene continental conglomerates and red marls (Rouet). The sequence contains sands and sandy marls with lenses of reef limestones (branching *Porites*) in the middle and conglomerates and reef corals (several colonial morphologies)

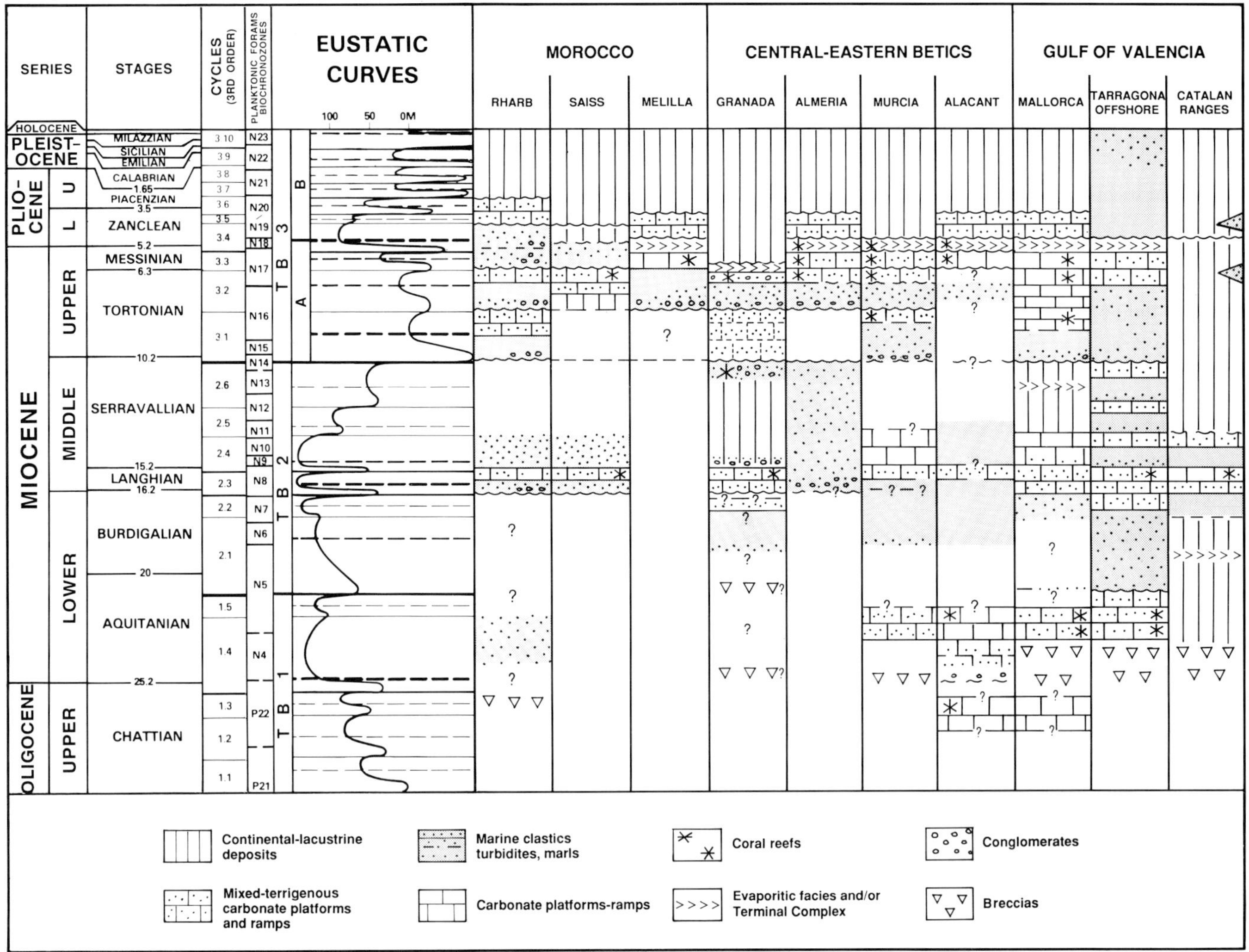


FIG. 2.—General stratigraphic framework proposed for the Miocene units in the western Mediterranean.

at the top. There is a rich community of bryozoans, red algae, molluscs, echinoderms, serpulids, foraminifers (Soritids, Miliolids, arenaceous, *Miogypsinoides, Sphaerogypsina, Nephrolepidina, Amphistegina, Operculina*, encrusting forams), brachiopods and other skeletal material. The thickest coral reef unit is 5-6 m thick, with about three or four coral-rich lenses, each 1-3 m thick. The second sequence (Rousset) is 14 m thick and consists of a basal stromatolite unit and sandy marls with cross-bedded sandstones containing brackish marine mollusks. The third sequence (Carry) is 22 m thick with deeper marine marls at the base, cross-bedded sandstones, and a 6-m-thick skeletal limestone at the top with echinoderms, Lepidocyclines, bryozoans, red algae and corals (*Acropora, Tarbellastraea*, Favidae, *Porites)*. The fourth sequence (Sausset), being 20 m thick, is a shallowing-upwards sequence from marls to sandy carbonates without coral reefs. These four sequences of Monleau et al. (1989) are attributed to the Aquitanian TB 1.4 cycle of Haq et al. (1987), and the TB 1.5 cycle is considered missing (Monleau et al., 1989). Each one of these four, 4th-order sequences is subdivided into 2-4 fith-order sequences, with repeated evidence of subaerial exposure and karstification along many of the sequence boundaries.

Similar Aquitanian reefs are described in Mallorca with up to 11 hermatypic coral species (Rodríguez-Perea, 1984a, b) and are known to occur in other Balearic Islands, in Sardinia, and in the offshore Gulf of Valencia. The Mallorcan reef corals were mentioned by Hermite (1879) and Fallot (1922) as "Burdigalian" in age, but Anglada and Serra-Kiel (1986) demonstrated a Chattian-Aquitanian age (*Miogypsinoides bantamesis, Miogypsina gunteri*). Regional studies (Rodríguez-Perea, 1984a, b; 1989; Rodríguez-Perea et al., 1988) indicate a complex reef paleogeography fringing karstified Mesozoic cliffs and frontal lobes of small delta fans, later intensely deformed by Mid-Miocene thrusting. Red algae are the most important constituents, followed by calcareous lithoclasts, bryozoans, foraminifers, bivalves, quartz grains, ostracods, gastropods and echinoderms. There are also red algae mounds (12 m high, tens of meters long) without corals. Rodríguez-Perea (1984a, b; 1989) distinguished three main types of coral reefs according to the amount and texture of the terrigenous mixing:

1. Reefs in medium-to-coarse terrigenous fans: This is the most common type in Mallorca and presents two cycles of reef development intercalated in delta fan wedges associated with rocky shores of karstified Mesozoic rocks. The lower cycle is only a few meters thick, and it is characterized by branching red algal packstones with bryozoans and planar corals (*Porites* and *Agaricia batalleri*). The upper cycle is up to 15 m thick and is interbedded with up to four coarse-grained terrigenous layers. Coral colony morphologies evolve from planar in the lower part to irregular-elliptical at the top. Coral diversity decreases with increasing terrigenous amounts. *Porites* sp., *Porites collegniana, Thegiostraea aequa, Favia falloti* and *Heliastraea mellanica* occur in the higher diversity intervals; *Porites* is the only coral present where terrigenous deposits are most dominant. Coral reefs in these rocky shore settings, with angular blocks falling-in, survived apparently quite well; the resulting broken corals, still alive, embraced the fallen blocks.

2. Reefs mixed with fine-grained terrigenous deposits: *Stylophora raristella* is the dominant coral, locally exclusive or accompanied with up to 10 coral species (*Agaricia batalleri, Heliastraea* sp., *H. asteroides, H. deformis, H. mellanica, Tarbellastraea aquitaniensis, Thegiostraea burdigaliensis, Th. diversiformis, Th. miocenica, Ellasmoastraea* sp.(?)). Some layers show corals in life position, but mostly they are broken fragments in packstones and wackestones with abundant gastropods, oysters and other bivalves.

3. Terrigenous-free reefs: These types of coral reefs are relatively small (up to 16 m thick, extending laterally for a few tens of meters) and grade laterally into terrigenous deposits. *Porites* sp. is the dominant coral, but others are locally present. Massive and encrusting morphologies of corals are predominant with red algal boundstones in a matrix of wackestones and packstones of intraclasts, coral fragments, serpulids, bryozoans, scallops and oysters.

The Aquitanian reefs in Mallorca show minor karstification at the top and are overlain by Burdigalian carbonate turbidites (Rodríguez-Perea and Pomar, 1983). Monleau et al. (1989) describe a similar situation in southeastern France.

### *Reefs in Continuity with Oligocene Platforms*

The Lower Miocene reefs in the eastern Betics (Alacant region) have a different style (A. Simó, pers. commun., 1982). These reefs are 40-60 m thick, with abundant grainstones and rudstones with red-algae, rhodoliths, corals, large benthic forams (*Lepidocyclina*), bivalves and Miliolids developed on mixed carbonate-terrigenous slopes containing abundant echinoderms and large benthic foraminifera. These carbonates are attributed to the Chattian-Aquitanian sequences and occur on top of extensive Lower Oligocene carbonate platforms (60 m thick, 10 km wide) with similar components and well-developed carbonate slopes and turbidites. Extensive Lower Miocene rhodalgal carbonate platforms with abundant *Lepidocyclina* occur along the northwest margin of the Betics corresponding to the Prebetic zone (Colom, 1975). Further west, coral reefs less than 50 m thick are reported in the Algarve and Lisboa areas of Portugal (Antunes in Steininger et al., 1985). All paleogeographic reconstructions show a wide north Betic (Prebetic) Strait from the Guadalquivir to the Alacant region and the Balearic islands (Fig. 3) with emergent land masses of the Alboran microplate to the south. Along the southern margin of the Rif (Prerif), a wide strait similar to the Prebetic Strait occurred during the Early Miocene connecting the Atlantic (Gharb basin) with the Mediterranean (Guercif basin). However, carbonate platforms are mixed with siliciclastics and appear to be much less developed than in the Betics.

### *Comments*

The Aquitanian reefs of the western Mediterranean are still poorly known. Chevalier (1961) reported 21 common coral species between La Nerthe and the Aquitaine basin; this marked similarity could be explained by a good communication along the Betic region and Portugal. There was no marine connection between the Aquitaine Basin and SE France (Cahuzac, pers. commun., 1991; Alvinerie et al., 1992). In Mallorca (11 coral species), Chevalier recognized 6 coral species in common with the Aquitaine basin and 5 with La Nerthe (17 species). Unfortunately, the absence of more detailed paleontological studies in the western Mediterranean Lower Miocene carbonates prevent more understanding of the paleogeographic trends (Fig. 3). A Lower Miocene coral reef province is defined by the Provençal basin, limited by the Corsica-Sardinian block (drifting to the east) and the Balearic Islands, and connecting with the Atlantic domains across the Betic platforms. To the east of the Provençal province, Aquitanian coral reefs re-appear in southeastern Sicily (Pedley, 1981) and probably in offshore eastern Tunisia (Fournié, 1978). Lower and Upper Oligocene coral reefs occur along the eastern part of Italy (Frost, 1981; Luperto, 1962).

The Burdigalian sequences in the western Mediterranean contains abundant hemipelagic marls, turbiditic deposits (siliciclastics and carbonates) and rhodalgal platforms. In some localities, there are abundant ahermatypic coral mounds in the Burdigalian (Langhian?) marly carbonates (Torino region, northern Italy, with up to 148 species according to Chevalier, 1961). Hermatypic corals (up to 36 genera) may be locally present (possibly in slumped marls). These coral-rich marls (a total of 248 species in Chevalier, 1961) could be correlative with the ahermatypic-hermatypic mounds of Western France (Esteban, this volume). In the western Mediterranean, coral reefs are only known in what is considered the uppermost part of the modern Burdigalian stratotype (Cahuzac and Chaix, this volume), forming part of the Langhian coral reef complexes.

The Lower Miocene carbonates in the Provençal basin show important clues for hydrocarbon exploration in the area. Outcrops are very tight in general, but subsurface occurrences are proven reservoirs, commonly dolomitized and fractured (Dorada, Angula and Salmonete oil fields). Burdigalian hemipelagic

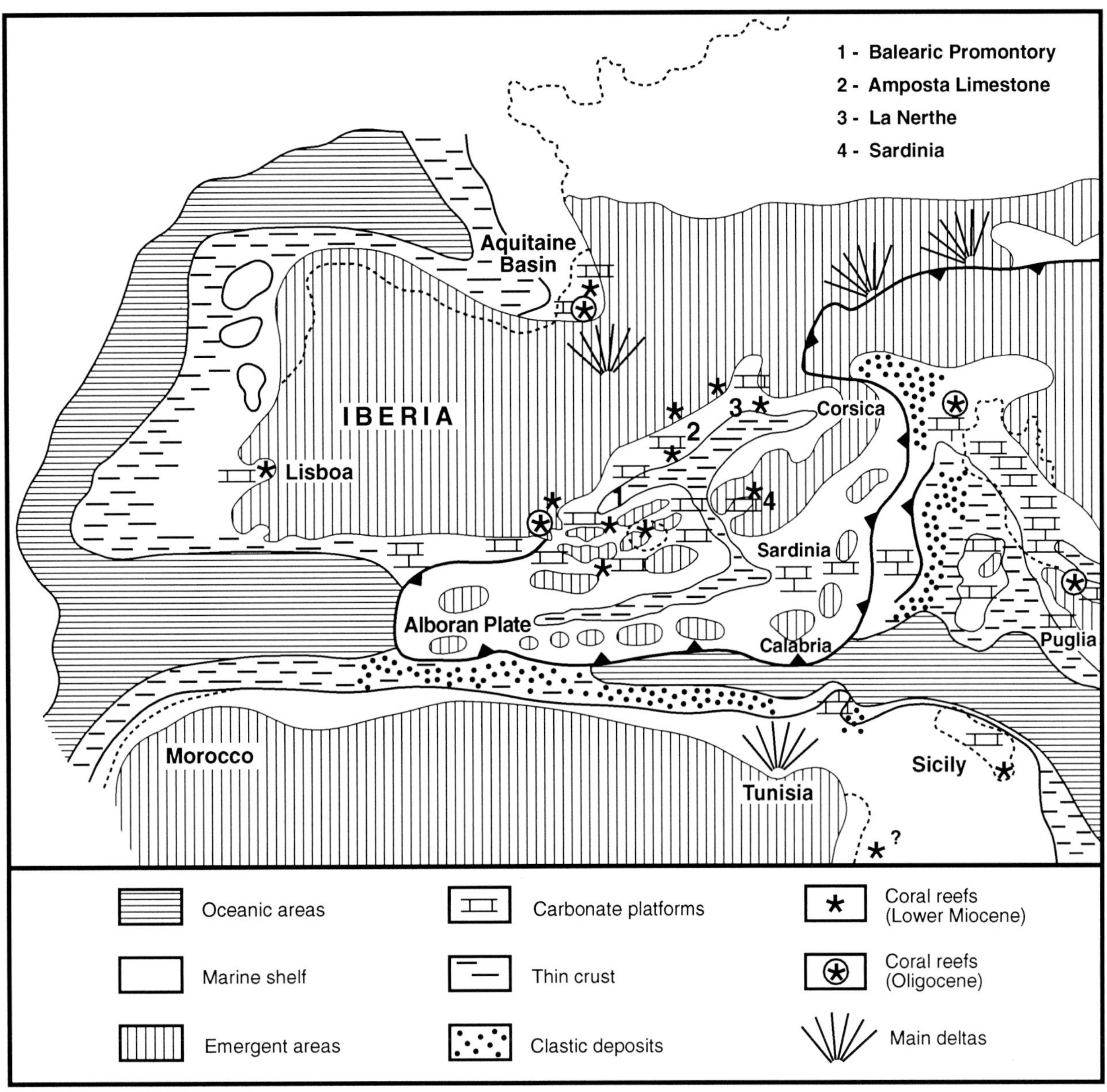


FIG. 3.—Early Miocene paleogeography in the western Mediterranean (modified from Dercourt et al., 1985, 1986) with location of the main reef units.

marly limestones in the Gulf of Valencia are considered as the main source rock for the Casablanca oil field (Demaison and Bourgeois, 1985). Black shales and organic-rich marls are common intercalations in the Aquitanian reef outcrops of La Nerthe as well (Monleau et al., 1989).

## MIDDLE MIOCENE REEFS

The Mid-Miocene carbonates (Fig. 4) are characterized by rhodalgal platforms with localized Langhian (and Upper Burdigalian?) coral reef development known only in a few localities (Languedoc in southeastern France, the Prerif ridges in Morocco, Penedés in northeastern Spain and Granada in south Spain). Many of the Mid-Miocene coral reefs occur directly onlapping a karstified Mesozoic basement (locally with small alluvial fans) and are overlain by lacustrine and continental Serravallian deposits. In other localities, the Langhian reefs appear at the top of thick Burdigalian sequences of marine marls, turbidites, thick alluvial fans, continental red beds and evaporites.

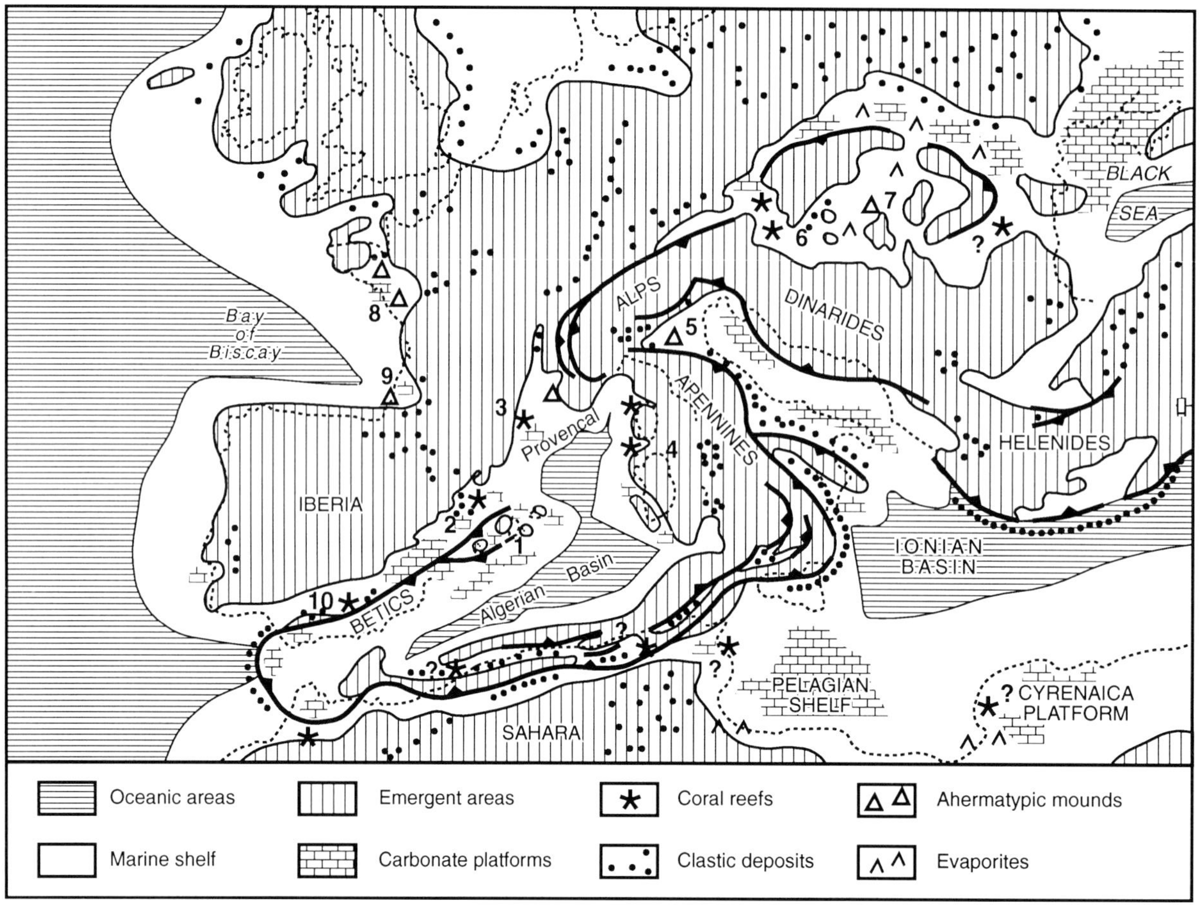


FIG. 4.—Middle Miocene paleogeography in the western Mediterranean (extensively modified from Ziegler, 1988) with location of the main reef units. 1: Balearic Islands (Mallorca), 2: Penedés and offshore Gulf of Valencia, 3: Languedoc, 4: Corsica, 5: Torino-Casale, 6: Viena, Sthyria basin, 7: Pannonian basin, 8: Brittany, Loire, 9: Aquitaine basin, 10: Murchas, Granada.

## *Coral Reefs*

The coral reefs of the Penedés region (southwest of the city of Barcelona) are some of the best known (since Vezian, 1856) in the western Mediterranean. They are very similar to those of the Languedoc in southeastern France. Chevalier (1961) described 24 coral species (4 ahermatypics) in the Penedés, 21 in the Languedoc (no ahermatypics), with 10 species common to both areas. *Porites* in different morphologies is the dominant builder, also with abundant *Heliastraea* and *Tarbellastraea* and local concentrations of Mussidae (*Mussismilia vindobonensis*) and Faviidae. Permanyer and Esteban (1973), Permanyer (1990) and Cabrera et al. (1991) distinguished the following reef morphologies: (i) fringing-reefs, 3-7 km long with well-developed prograding reef slopes (100-200 m of paleodepths with up to 1 km of progradation) and coral frameworks up to 50 m thick, (ii) very patchy fringing-reefs (patch-reefs), less than 1 km long and a few hundred meters wide, with coral framework less than 10 m thick and poorly developed or absent talus slopes (paleodepths of about 20 m), mostly in the northeast shallow corner of the basin, and (iii) coral knobs, 5-10 m in diameter and 1-2 m high, occurring discontinuously on a rhodalgal carbonate ramp (10-to 30-m paleodepths, in the open southern end of the basin. Coral diversity is higher in the larger reefs (12-16 species); 6-8 in the patchy fringing-reefs and 2-3 in coral knobs. Matrix in the reef frameworks contains abundant red-algae, bryozoans, encrusting forams, locally Milleporids and oysters. Planktonic foramiminifera and nannofauna are common in the distal slopes and basin transition; Permanyer (1990) considered a Late Burdigalian (N7) age for all the reef complexes, whereas Magné (1978) determined an early Langhian (N8) age in similar sections.

Middle Miocene carbonates are extensively developed in the offshore of northeastern Spain. Subsurface data indicate that most of them are of the rhodalgal type, but coral reefs appear to be clearly present in some localities (Elvira Alvarez, pers. commun. 1993; Esteban, unpubl. data) and better developed than in the outcrops. Some of these offshore coral reefs could have up to 70-m-thick reef frameworks and 200-m-thick slopes.

The only Mid-Miocene coral reef known in the Betic region is described in Murchas (Granada; Braga et al., this volume). These are small patch-reefs, up to 20 m across and 4 m thick, developed on an open platform with mixed bioclastic and fine-siliciclastic sediments seaward of a shoaling bar. These reefs were built by corals (*Heliastraea, Mussismilia, Tarbellastraea and Porites*) and oysters (*Hyotissa squarrosa*). In Northern Morocco, Lower and Mid-Miocene carbonates are essentially rhodalgal mixed with abundant bryozoans and siliciclastics; coral reefs attributed to the Langhian are only known in the northern margin of the Saïss Basin (Esteban, unpubl. data). In Northern Algeria, Chevalier (1961, p. 105) mentions a large number of Lower and Mid-Miocene hermatypic corals. However, we have no further documentation. In the central Mediterranean (Pedley, this volume), the Mid-Miocene carbonates are dominated by rhodalgal facies; only Orszag-Sperber and Pilot (1976) documented coral reef development in Corsica.

### *Rhodalgal Ramps*

Mid-Miocene rhodalgal carbonate ramps and turbidites are extensively exposed in the Balearic Islands and eastern Betics and show no trace of coral reef development (Rodríguez-Perea, 1984a, b; Pomar et al., 1983b; Alvaro et al., 1984). In central Mallorca, the Randa Limestone (Pomar and Rodríguez-Perea, 1983) is a 150-m-thick rhodalgal carbonate ramp developed contemporaneously with an overthrusting event (and erosion of the Mesozoic substrate). These Langhian-Early Serravallian carbonates (Pomar et al., 1983) are capping a 450-m-thick unit of carbonate turbidites and hemipelagites considered Late Burdigalian-Early Langhian age (González-Donoso et al., 1982). The Randa Limestone is a thickening and coarsening upwards sequence of grain-supported skeletal carbonates (fragments of red algae, bryozoans, *Amphistegina, Heterostegina,* bivalves and lithoclast, locally with rhodoliths and a few planktonic foraminifers; detailed descriptions in Pomar and Rodríguez-Perea, 1983). Most of the lithoclasts are Mesozoic and Tertiary carbonates; clay pebbles and grains of the underlying Langhian-Upper Burdigalian hemipelagic and turbiditic deposits are very conspicuous. The outer ramp (hemipelagic marls and skeletal turbidites with slumps) is overlain by an intermediate ramp unit, 110 m thick, with large sigmoidal lobes of small-volume turbidites with olistoliths, interpreted as deposited in a gentle ramp below storm-wave base (Pomar, pers. commun., 1993). This unit is overlain by 20 meters of massive limestones with large-scale erosion scars, mass-flow convex-up beds and large-medium-scale hummocky and swaley cross stratification (upper intermediate ramp of Pomar, pers. commun., 1993). The depositional slope is considered to be on the order of a few degrees, but the entire unit (and the underlying turbidites) displays a syntectonic fan array geometry (progressive discordance). Pomar interprets the intermediate ramp facies as the consequence of extremely high-energy waves (tsunamis?). The Randa Limestone is an example of a laterally-discrete, skeletal-rich carbonate that could be qualified as a "reef" in areas with limited subsurface data.

### *Comments*

In summary, the available documentation suggests paleogeographic trends somewhat similar to those of the Lower Miocene carbonates. The width of the north Betic (Prebetic) and south Rif (Prerif) straits was reduced by Alpine thrusting but continued to be the major communication between the Atlantic and the western Mediterranean. While most of the Alboran Plate was foundered by Mid-Miocene time, the leading edges of the thrusts included partly emergent areas controlling sediment supply. The best coral reefs occur around the Provençal basin, with possibilities of good developments in the Northern Algerian basins. The other Mid-Miocene carbonates are predominantly rhodalgal, mixed with abundant terrigenous sediments and show poor coral reef development. Along the Atlantic side of Europe, Mid-Miocene carbonates contain up to 36 coral species (12 ahermatypics) distributed in 20 genus (6 ahermatypics), but well developed coral reefs are not reported. Similar associations, but with lower coral diversity, occur in Madeira (Lietz and Schwarzbach, 1970), probably of similar "Helvetian" age. These associations are considered to indicate water temperatures too low to allow coral reef development (Chevalier, 1961); thus they represent the outer limit of the extensive Mid-Miocene coral reef province (Fig. 9 in Esteban, this volume). Within the context of the extensive coral reefs of the eastern Mediterranean-Red Sea regions, it is quite remarkable to see the lack of coral reefs in the central Mediterranean and their concentration in the Provençal Basin of the western Mediterranean. Hypothetically, it could be considered that the Provençal basin was already a coral reef "refuge" in Mid-Miocene time, surrounded by cooler water rhodalgal carbonates and partly isolated from the prolific eastern Mediterranean coral reefs.

Although the age and correlation of the Middle Miocene reefs in the western Mediterranean is not well resolved, there appears to be two distinctive reef events or depositional sequences. Middle Miocene (and uppermost Burdigalian) carbonates in Corsica, Austria and the Red Sea also show evidence of two reef events separated by a subaerial exposure surface. An improved correlation of the Middle Miocene reefs would be a major contribution to the understanding of the complexities of the Alpine regional geology in the Mediterranean.

## UPPER MIOCENE REEFS

Most of the Upper Miocene carbonates in the western Mediterranean occur in the southern sector (Balearic Islands, Betics and Magrebides ranges); in the northern sector, carbonates are poorly known and scarce, mostly occurring as marly rhodalgal facies mixed with terrigenous deposits. The Betic and Balearic provinces offer the highest variety of types and ages of Upper Miocene reefs. During the Late Miocene times, the Neogene basins of the Rif and Betic Ranges occurred as a series of

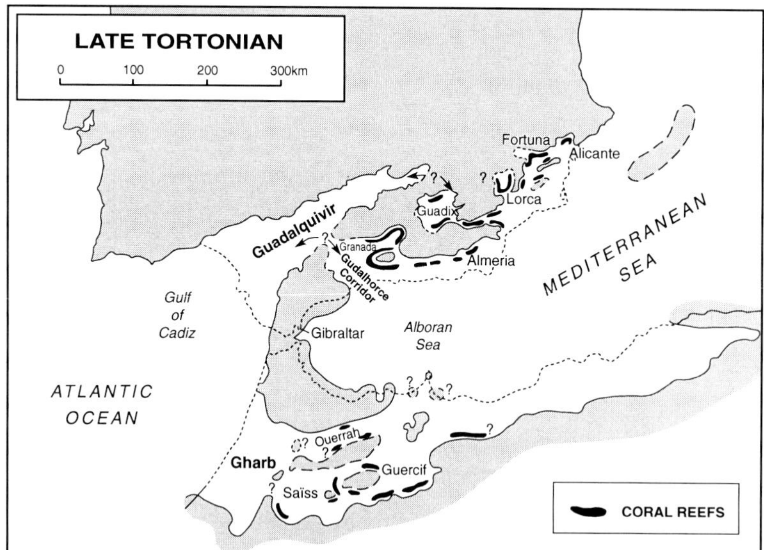
LATE TORTONIAN
0 100 200 300km
Fortuna
Alicante
Lorca
Guadix
Guadalquivir
Granada
Gudalhorce Corridor
Almeria
MEDITERRANEAN SEA
Gulf of Cadiz
Gibraltar
Alboran Sea
ATLANTIC OCEAN
Ouerrah
Gharb
Guercif
Saïss
CORAL REEFS

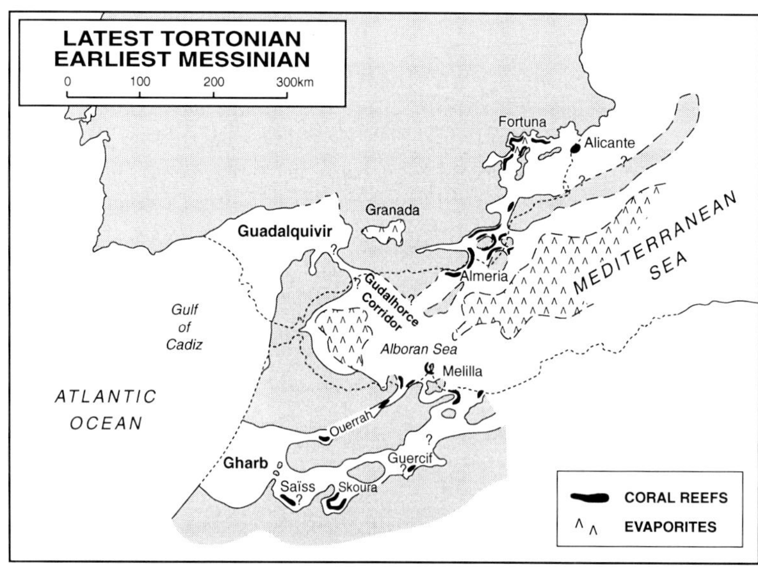
LATEST TORTONIAN EARLIEST MESSINIAN
0 100 200 300km
Fortuna
Alicante
Granada
Guadalquivir
Gudalhorce Corridor
Almeria
MEDITERRANEAN SEA
Gulf of Cadiz
Alboran Sea
Melilla
ATLANTIC OCEAN
Ouerrah
Gharb
Guercif
Saïss
Skoura
CORAL REEFS
EVAPORITES

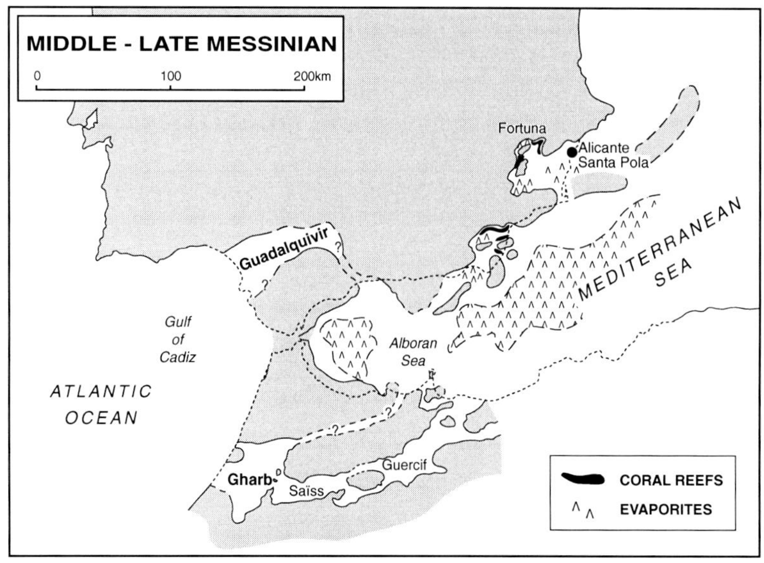
MIDDLE - LATE MESSINIAN
0 100 200km
Fortuna
Alicante
Santa Pola
Guadalquivir
MEDITERRANEAN SEA
Gulf of Cadiz
Alboran Sea
ATLANTIC OCEAN
Gharb
Guercif
Saïss
CORAL REEFS
EVAPORITES

interconnected corridors, passageways and basins, resulting in a complex and changing paleogeography around an emergent archipelago of numerous islands (basin-and-range paleogeography; Fig. 5). Although the main tectonic events occurred from the Oligocene to the Mid-Miocene in different zones of the Betics and Rif, important phases of regional uplift also occurred during the Late Tortonian and Early Messinian times, contributing to the present-day basin-and-range morphology. The progressive severing and restriction of the Betic and Rif Straits during the Late Miocene times was the key factor leading to the isolation of the Mediterranean and the onset of the Messinian salinity crises. The Upper Miocene stratigraphy presents unsuspected complexity and variability in the different Betic basins (Fig. 6). The western part of the Betic basins is occupied by the extensive Guadalquivir basin, the final foredeep of the Betic thrustbelt under predominant or exclusive Atlantic influence; Upper Miocene carbonates are of rhodalgal facies in the Guadalquivir basin. In the central Betics, the Late Miocene basins show clear Mediterranean influence expressed by the presence of coral reefs in the Late Tortonian (Granada basin, Alpujarra corridor, Guadix-Baza basin).

Two different types of Late Miocene basins exist in the eastern Betics. The Almería basins (Níjar, Cabo de Gata, Sorbas, Vera) are characterized by Early Messinian coral reef development with spectacular progradation and a Late Messinian evaporitic unit followed by oolite shoals, small coral patch-reefs and stromatolites (Terminal Complex). The Murcia basins (Fortuna, Mula, Lorca, Elx) contain 3-4 evaporitic units with a wide variety of coral reefs from the Upper Tortonian to the Upper Messinian sequences. The major facies relationships of the different Upper Miocene reefs (Fig. 7) reflect the complexity of the stratigraphic framework of the Betic basins.

The Upper Miocene reefs of the Balearic Islands present many similarities to the Almería-type of reefs of the Betic province, but locally form very extensive reef platforms (Llucmajor, Mallorca; Pomar et al., this volume). In the Magrebides province (Saint Martin, this volume; Saint Martin and Cornée, this volume), the Upper Miocene reefs appear reduced to only one (possibly two) stratigraphic levels: the Lowermost Messinian-Uppermost Tortonian (Saint-Martin and Rouchy, 1990), with rhodalgal accumulations and coral reefs quite similar to those in the Betic province. Major differences in the Upper Miocene sequences of the Betics and the Rif are the abundance and variety of carbonate units in the Betics and the lack of extensive evaporite deposits in the Rif.

FIG. 5.—Evolution of the Late Tortonian and Messinian paleogeography of the Betic and Rif straits (modified from Rouchy, 1982; Santisteban, 1981; Montenat, 1977; Feinberg, 1978; Auzende et al., 1975; Allan and Morelli, 1971; Montenat and Bizon, 1976; Fernex and Szep, 1971; Esteban and Giner, 1980; Dabrio et al., 1981; Rouchy et al., 1986; Martín et al., 1989; Braga et., 1990; Sanz de Galdeano and Vera, 1992; and unpublished data). (A): Late Tortonian, (B): Latest Tortonian-Earliest Messinian, and (C): Middle-Late Messinian.

### *Early Tortonian Reefs*

Most Early Tortonian carbonates in the entire western Mediterranean are formed by rhodalgal facies with variable amounts of associated bryozoans, mollusks and large benthic foraminifera mixed with siliciclastic deposits. These Early Tortonian carbonates are interpreted to typically occur in the highstand portion of a 3rd-order depositional sequence filling-up the deep basinal settings created after the Mid-Miocene orogenic event (i.e. post-Nappe Miocene in the Rif basins). The best examples crop out in the southern part of Menorca, in sections dated as N 16 (Bizon et al., 1973) and interpreted as proximal ramp facies (Obrador and Pomar, 1983; Obrador et al., 1983, 1992). These rhodalgal carbonates include spectacular clinoforms, 40 m high and 2 km-wide, of rhodolithc rudstones and packstones. Individual rhodoliths are up to 8 cm in diameter and contain variable amounts of encrusting foraminifers and bryozoans. There is no evidence of in-situ emerging shoals or buildups, but these prograding clinoforms and associated deposits could be considered as "reefs" in the sense of laterally discrete, biogenic accumulations distinctive from adjacent sediments.

Lower? Tortonian coral reefs are only known to occur in the southern part of the Granada basin, consisting of small (1-10 m in diameter, 1-2 m thick) patchy buildups with abundant siliciclastic mixing. However, there are indications suggesting a Late Serravallian age for these coral mounds. In the Guajares outcrop, the corals (*Porites* and *Tarbellastraea*) colonize a debris-flow breccia and are covered by a deltaic sequence also including small coral patches. In the Albuñuelas outcrops, the corals colonize banks of large oysters in conglomeratic channels of the deltaic sequence.

### *Late Tortonian Reef*

Coral reefs flourished in the Late Tortonian central and eastern Betic basins (Fig. 5A), Rif and Balearic Islands. In the eastern Betics, these basins were moderately deep (about 500-m water depth) and were controlled by important wrench-fault tectonics (Montenat and Ott d'Estevou, 1990). Most of the coral reefs are fringing in character and allow a precise reconstruction of the paleogeography of the region (Fig. 5). There are three major types of Late Tortonian coral reefs in the Betics: (i) small patch-reefs in deltaic systems, (ii) long arcuate fringing reefs in deltaic systems, and (iii) large reef complexes on terraced erosion surfaces without deltaic systems. Individual coral framework units are commonly only 10 m thick and 100 m across when intercalated in coastal terrigenous sections (fan deltas, braid deltas, sandy deltas). However, thicker coral reef frameworks (up to 100 m thick) are the result of the stacking of these individual units in prograding or retrograding patterns without major terrigenous intercalations. Examples of these large coral reef complexes without deltaic associations occur in Lorca (Rouchy et al., 1986), Fortuna basin (Santisteban, 1981), Almanzora corridor (Martín et al., 1989), and Granada basin (Braga et al., 1990); the best developed example is El Desastre

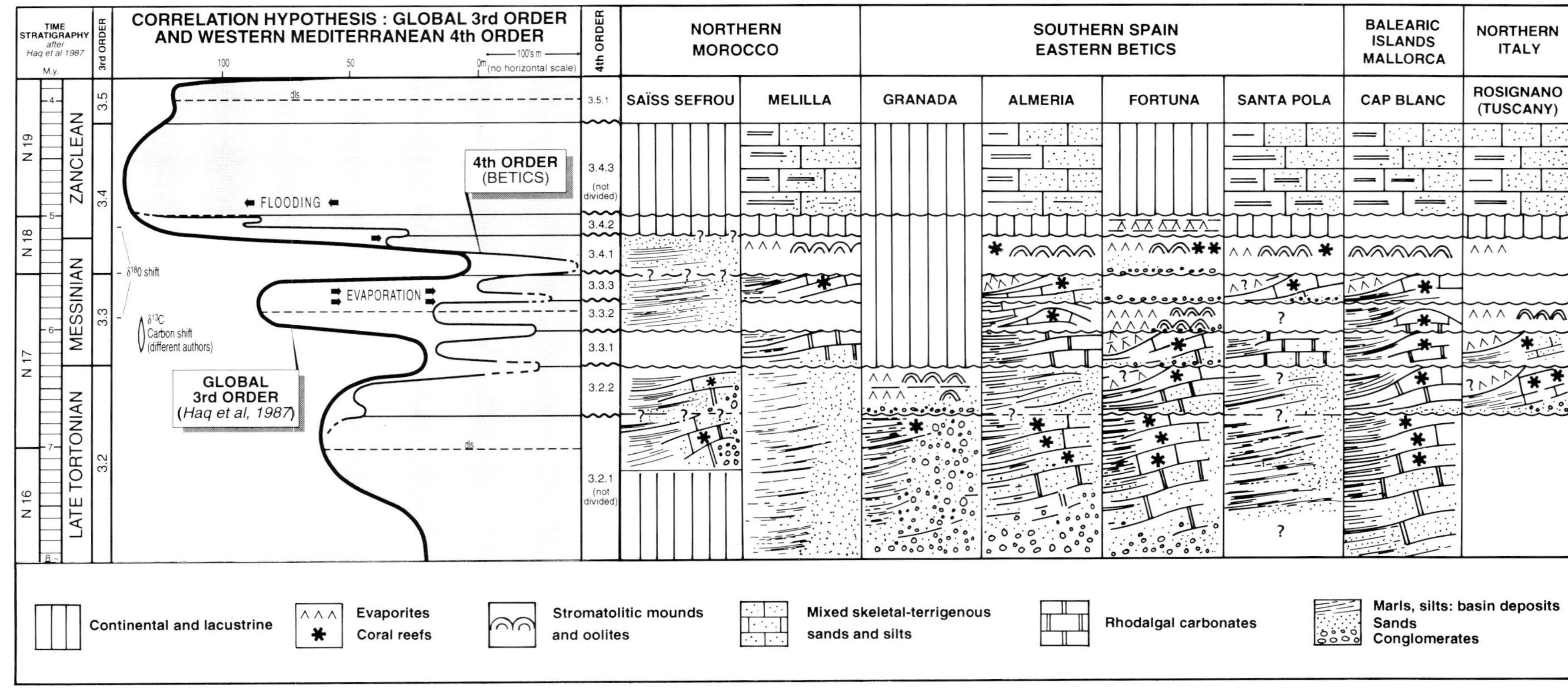


FIG. 6.—Stratigraphic framework of the Upper Tortonian, Messinian and Lower Pliocene sections of the Betic region and comparison with the Rif and Mallorca. Based on and modified from: Saint-Martin (1990) for Morocco; Braga et al. (1990), Martín and Braga (1990), Riding et al. (1991), Dabrio et al. (1981), Franseen and Mankiewicz (1991), Santisteban (1981), Montenat and Ott d'Estevou (1990) for the Betics; Pomar et al. (1983), Pomar (1991) for Mallorca. Note that the proposed correlation in this figure differs slightly from the alternative interpretation of Figure 26 in Esteban (this volume).

## CONCLUDING REMARKS

The luxuriant development of Miocene coral reefs and rhodalgal carbonates in the western Mediterranean was controlled by the Alpine orogeny in the Betic and Rif Straits progressively narrowing and severing the marine communication with the Atlantic. Since Mid-Miocene times, these connections were the only ones known for the entire Mediterranean and particularly narrowed in the Late Tortonian and Messinian times, at the time of the spectacular coral reef development in the region.

Lower Miocene reefs are the least known in the western Mediterranean. There are two major types of coral reefs and carbonates platforms: (i) onlapping continental Paleogene deposits, karst breccias and Neogene fans (La Nerthe, Mallorca) and (ii) in continuity, conformably overlying the Oligocene carbonate platforms (eastern Betics, La Puglia). Most of the coral reefs appear to be Aquitanian age, and are attributed to two 4th-order sequences in the TB 1.4 cycle. However, some could be Chattian age, and there are major differences in the definition and correlation of the Lower Miocene stages. The Provençal basin could define a coral reef province limited by the Corsica-Sardinia block and the Balearic Islands and connecting to the Atlantic domains across the Prebetic basins. The Burdigalian sections present predominant rhodalgal carbonates and marly deposits, some of them with excellent source-rock potential. In northern Italy, the Burdigalian marls contain coral mounds with extremely high diversity.

The Mid-Miocene coral reefs also seem to appear as two episodes corresponding to the 3rd-order sequences TB 2.2 and 2.3 (N7 and 8) but show lower diversity in corals. The Provençal basin is again the area with the best developed coral reefs (Languedoc, Penedés), although that could also reflect the elimination of outcrops by Mid-Miocene thrusting in the southern parts of the western Mediterranean. Nevertheless, it seems that rhodalgal ramps without coral reefs predominate in the tectonically active Betic and Rif thrusts. The Randa Limestone in Mallorca is a good example of a rhodalgal ramp associated with olistostromes, turbidites, slumps and extremely high-energy waves. Similar to other rhodalgal ramps in the Lower and Upper Miocene sequences, in areas of limited subsurface data these skeletal-rich, laterally discrete carbonate bodies could be considered as "reefs".

The Upper Miocene reefs are the best known in the entire Mediterranean because of the superb quality and continuity of most of the outcrops. Coral diversity is very low (up to six corals present, commonly only one or two dominant), but reef platforms can be up to 25 km wide (Mallorca). Different types of fringing and patchy reefs on the substrate or on Late Miocene fan deltas are predominant. Reef progradation is most developed towards the east, southeast and south reflecting that favorable conditions existed in the southern part of western Mediterranean. Only rhodalgal platforms without coral reefs are recorded in the northern part. In Morocco and Algeria, coral reefs occur only as one episode in the Lower Messinian (and also probably in the Upper Tortonian) sections. In contrast, the Betic-Balearic provinces display a large number and variety of coral reef events: Early Tortonian (rare), Late Tortonian (possible two or three 4th-order cycles) and Messinian (four 4th-order cycles). Rhodalgal carbonates show spectacular prograding rhodolithic clinoforms in the Lower Tortonian section of Menorca and the Lower Messinian section of Melilla. Some Messinian reef units show peculiar features (associations with cyanobacterial crusts and stromatolites, large amounts of green algae, large-scale downstepping) related to the concept of Messinian crises (see also Esteban, this volume). Although the detailed stratigraphic correlation is still tentative, it seems that these peculiar features are not everywhere synchronous and that similar lithostratigraphic episodes occurred at different times. This is considered to reflect the complexity of the paleogeographic evolution of the Betic and Rif Straits.

The Betic Straits were probably still recognizable during the Early Tortonian times, but during the Late Tortonian-Early Messinian times the only possible location seems to be reduced to the Guadalhorce valley; the rest of the marine passageways were apparently emerged. There is more evidence of effective marine connection along the Rif Straits during the Late Tortonian-Early Messinian times.

## ACKNOWLEDGMENTS

The authors thank the reviews and comments by E. K. Franseen, D. Bosence, Bill Ward, L. Pomar, Q. Sun, J. Jiménez, E. G. Purdy and Dana Ulmer-Scholle. J. C. Braga and J. Matín acknowledge the support of DGICYT PB90-0854 and PB93-1113.

## REFERENCES

Allan, T. D. and Morelli, C., 1971, A geophysical study of the Mediterranean Sea: Bolletino di Geophisica Teorica e Applicatta, v. 13, p. 99-141.

Alvaro, M., Barnolas, A., Del Olmo, P., Ramírez del Pozo, J., and Simo, A.,1984, El Neógeno de Mallorca: caracterización sedimentológica y biostratigráfica: Boletín Geológico y Minero, v. 95, p. 3-25.

Alvinerie, J., Antunez, M. T., Cahuzac, B., Lauriat-Rage, A., Montenat, C., and Pujol, C., 1992, Synthetic data on the paleographic history of northeastern Atlantic and Betic-Rifian basin, during the Neogene (from Brittany, France to Morocco): Palaeogeography, Palaeoclimatology, Palaeoecology, v. 95, p. 263-286.

Anglada, E. and Serra-Kiel, J., 1986, El Paleógeno y tránsito al Neógeno en el área del Macizo de Randa (Mallorca): Boletín Geológico y Minero, v. 985, p. 580-589.

Auzende, J. M., Rehault, P. J., Pastouret, L., Szep, B., and Olivet, J. L., 1975, Les bassins sédimentaires de la mer d'Alboran: Bulletin de la Societé Géologique de France, v. 7, n. 17/1, p. 98-107.

Benson, R. H., Rakic-El Bied, K., and Banaduce, G., 1987, An important water-mass "reversal" in the rifian portal (Morocco) at the Tortonian-Messinian boundary (abs.):Evolution climatique dans le domain méditerranéen au Néogène: Montpellier-Barcelona, Regional Committee on Mediterranean Neogene Stratigraphy, Interim-Colloquium, p. 10.

Benson, R. H., Rakic-El Bied, K., and Banaduce, G., 1991, An inportant current reversal (influx) in the Rifian Corridor (Morocco) at the Tortonian-Messinian boundary: the end of Tethys Ocean: Paleoceanography, v. 6, p. 164-192.

Bizon, G., Bizon, J. J., Bourrouilh, R., and Massa, D., 1973, Presence aux îles Baléares, (Méditerranée Occidental) de sediments "messiniens" depossé dans une mer ouverte a salinité normale: Comptes Rendues de l'Academie des Sciences de Paris, v. 277, p. 985-988.

BIZON, G., BIZON, J. J., AND MONTENAT, C., 1975, Définition biostratigraphique du Messinien: Comptes Réndues de l'Académie des Sciences de Paris, Serie D, v. 281, p. 359-362.

BRAGA, J. C., MARTÍN, J. M., AND ALCALÁ, B., 1990, Coral reefs in coarse-terrigenous Sedimentary environments (Upper Tortonian, Granada Basin, southern Spain): Sedimentary Geology, v. 66, p. 135-150

CABRERA, L., CALVET, F., GUIMERÁ, J., AND PERMANYER, A., 1991, El registro sedimentario miocénico en los semigrabens del Vallès-Penedès y del Camp: organización secuencial y relaciones tectónica-sedimentación: Barcelona, Guidebook, I Congreso del Grupo spañol del Terciario, 132 p.

CATZIGRAS, F., GLINTZBOECKEL, C., COLOMB, E., MERCIER, H., L'HOMER, A., ANGLADA, R., LORENZ, C., CARBONNEL, G., CHATEAUNEUF, J. J., LEZAUD, L., ANDREIFF, P., LAY, J., PARFENOFF, A., JACOB, C., AND CAVELIER, C., 1972, Contribution à l'étude de l'Aquitanien. La coupe de Carry-le-Rouet (B.d.R.): Bulletin de Recherches Géologiques et Minières, section 2, p. 1-135.

CHARRIÈRE, A., 1984, Evolution néogène de bassins continentaux et marins dans le Moyen Atlas central (Maroc): Bulletin de la Societé Géologique de France, v. 26, p. 1127-1136.

CHEVALIER, J. P., 1961, Recherches sur les Madreporaires et les formations récifales miocènes de la Méditerranée occidentale: Published Ph.D. Thesis, Paris, Mémoires de la Societé Géologique de France, no. 93, 562 p.

CHEVALIER, J. P., 1962, Les Medréporaires miocènes du Maroc: Notes et Mémoires du Service Géologique de France, v. 40, 562 p.

CITA, M. B., 1973, Mediterranean evaporite: paleontological arguments for a deep basin desiccation model, *in* Drooger, C. W., ed., Messinian Event in the Mediterranean: Koninkijke Nederlandse Akademie van Wetenschappen, p. 206-228.

COLALONGO, M. L., DI GRANDE, A., D'ONOFRIO, S., GIANNELLI, L., IACCARINO, S., MAZZEI, R., ROMEO, M., AND SALVATORINI, G., 1979, Stratigraphy of the Late Miocene Italian section stradding the Tortonian-Messinian boundary: Boletino della Società Paleontologica Italiana, v. 18, p. 258-302.

COLOM, G., 1975, Geología de Mallorca: Palma de Mallorca, Gráficas Mallorca, 522 p.

COLOM, G. AND GAMUNDI, J., 1951, Sobre la extensión e importancia de las "moronitas" a lo largo de las formaciones aquitano-burdigalenses del Estrecho Nord-Bético: Estudios Geológicos, v. 14, p. 331-385.

COMAS, M. C., GARCÍA-DUEÑAS, V., MALDONADO, A., AND MEGIÁS, A. G., 1990, The Alboran Basin: Tectonic regime and evolution of the northern Alboran sea: Barcelona, IX Regional Comittee on Mediterranean Neogene Stratigraphy, Congress Abstract, p. 107-108.

DABRIO, C. J., ESTEBAN, C. M., AND MARTÍN, J. M., 1981, The coral reef of Níjar, Messinian (uppermost Miocene), Almería Province, S.E. Spain: Journal of Sedimentary Petrology, v. 51, p. 521-539.

DABRIO, C. J., MARTÍN, J. M., AND MEGIÁS, A. G., 1982, Signification sédimentaire des évaporites de la depression de Grenade (Espagne): Bulletin de la Societé Géologique de France, v. 24, p. 705-710.

DARDER-PERICAS, B., 1928, La Paleogeografia de la Mediterrània Occidental, segons les idees d'Emile Argand: Barcelona, Ciència, v. 21, p. 3-13.

DEMAISON, G. AND BOURGEOIS, F. T., 1985, Environment of deposition of Middle Miocene (Alcanar) Carbonate Source beds, Casablanca field, Tarragona Basin, Offshore Spain, *in* Palacas, J., ed., Petroleum Geochemistry and Source Rock Potential of Carbonate Rocks: Tulsa, American Association of Petroleum Geologists Studies in Geology 18, p. 151-161.

DERCOURT, J., ZONENSHAIN, L. P., RICOU, L. E., KAZMIN, V. G., LE PICHON, X., KNIPPER, A. L., GRANDJACQUET, C., SBORSHICHIKOV, I. M., BOULIN, J., SOROKHTIN, O., GEYSSANT, J., LEPVRIER, G., BIJU-DUVAL, B., SUBUET, J. C., SAVOSTIN, L. A., WESTPHAL, M., AND LANER, J. P., 1985, Présentation de 9 cartes paléogéographiques au 1:20,000,000 s'étendant de l'Atlantique au Pamir pour la période du Lias: état actuel: Bulletin de la Societé Géologique de France, v. 1, p. 637-652.

DERCOURT, J., ZONENSHAIN, L. P., RICOU, L. E., KAZMIN, V. G., LE PICHON, X., KNIPPER, A. L., GRANDJACQUET, C., SBORSHICHIKOV, I. M., BOULIN, J., SOROKHTIN, O., GEYSSANT, J., LEPVRIER, G., BIJU-DUVAL, B., SUBUET, J. C., SAVOSTIN, L. A., WESTPHAL, M., AND LANER, J. P., 1986, Geological evolution of the Tethys belt from Atlantic to Pamir since Liassic: Tectonophysics, v. 123, p. 241-315.

D'ONOFRIO, S., GIANELLI, L., IACCARINO, S., MORLOTTI, E., ROMEO, M., SALVATORINI, G., SAMPO, M., AND SPROVIERI, R., 1975, Planktonic foraminifera from some Italian sections and the problem of the lower boundary of the Messinian: Boletino della Società Paleontologica Italiana, v. 14, p. 177-196.

DURAND-DELGA, M. AND FONTBOTÉ, J. M., 1980, Le cadre structural de la Méditerranée occidentale: Paris, 26e Congrès Géologique International, Colloque 5: Mémoires du Bureau de Recherches Géologiques et Minières, v. 115, p. 67-85.

ESTEBAN, M., 1979, Significance of the Upper Miocene coral reefs of the western Mediterranean: Palaeogeography, Palaeoclimatology, Palaeoecology, v. 29, p. 169-188.

FALLOT, P., 1922, Étude géologique de la Sierra de Majorque: Paris, Librerie Polytéctique Charles Beranger, 418 p.

FEINBERG, H., 1978a, Evolution paléogéographique de l'avant-pays du rif (Maroc) pendant le Miocène supérieur: Bulletin du Muséum National d'Histoire Naturel de Paris, s. 3, 518, Sciences de la Terre, v. 70, p. 149-155.

FEINGERG, H., 1978b, Les séries tertiares du Prérif et des dépendances post-tectoniques du Rif (Maroc): Unpublished Ph.D. Thesis, Travaux du Laboratoire de Géologie Méditerranéen, Toulouse, 211 p.

FERNEX, F. AND SZEP, B., 1971, Le bassin pontien à l'Est de la Province de Murcia (Espagne) et le problème du prolongement du système orogénique bétique vers l'Est en mer: Comptes Rendues Somaires de la Societé Géologique de France, p. 421-426.

FONTBOTÉ, J. M., GUIMERÁ, J., ROCA, E., SABAT, F., SANTANACH, P., AND FERNÁNDEZ-ORTIGOSA, F., 1990, The Cenozoic geodynamic evolution of the Valencia trouch (Western Mediterranean): Revista de la Sociedad Geológica de España, v. 3, p. 31-47.

FOURNIÉ, D., 1978, Nomenclature lithostratigraphique des séries du Cretacé superieur au Tertiare de Tunisie: Bulletin du Centre de Recherche, d'Exploration et de Production Elf-Aquitaine, v. 2, p. 97-148.

FRANSEEN, E. K., 1989, Depositional sequences and correlation of Middle to Upper Miocene carbonate complexes, Las Negras area, southeastern Spain: Unpublished Ph.D. Dissertation, University of Wisconsin, Madison, 374 p.

FRANSEEN, E. K. AND MANKIEWICZ, C., 1991, Deposicional sequences and correlation of middle(?) to late Miocene carbonate complexes, Las Negras and Níjar areas, southeastern Spain: Sedimentology, v. 38, p. 871-898.

FROST, S. H., 1981, Oligocene reef coral biofacies of the Vicentin, northeast Italy, *in* Toomey, D. F., ed., European Fossil Reef Models: Tulsa, Society of Economic Paleontologists and Mineralogists Special Publication 30, p. 483-539.

GENTIL, L., 1916, Sur la trouée de Taza: Comptes Rendues de l'Academie des Sciences de Paris, v. 163, p. 105-708.

GONZÁLEZ-DONOSO, J. M., LINARES, D., PASCUAL, I., AND SERRANO, F., 1982, Datos sobre la edad de las secciones del Mioceno inferior de Port des Canonge y Randa (Mallorca): Bolletí de la Societat d'Història Natural de les Balears, v. 26, p. 229-232.

HAQ, B. U., HARDENBOL, J., AND VAIL, P. R., 1987, Chronology of fluctuating sea levels since the Triassic: Science, v. 235, p. 1156-1167.

HERMITE, H., 1879, Études Géologiques sur les Iles Baléares: Paris, Savy Editions, Pinchon Imprimerie, 362 p.

HSÜ, K. J., 1973, The desiccated deep basin model for the Messinian events, *in* Drooger, C. W., ed., Messinian Event in the Mediterranean: Amsterdam, Koninkijke Nederlandse Akademie van Wetenschappen, p. 60-67.

HSÜ, K. J., CITA, M. B., AND RYAN, W. B. F., 1973, The origin of the Mediterranean evaporites: International Report Deep Sea Drilling Project 13: Washington, United States Government Printing Office, p. 1203-1231.

HSÜ, K. J., MONTADERT, L., BERNOUILLI, L., CITA, M. B., ERICKSON, A., GARRISON, R. E., KIDD, R. B., MELIERES, F., MÜLLER, C., AND WRIGHT, R., 1978, History of the Mediterranean salinity crisis: Initials Reports of the Deep Sea Drilling Project: Washington, United States Government Priting Office, v. 42, p. 1053-1078.

IACCARINO, S., MORLOTTI, E., PAPANI, G., PELOSIO, G., AND RAFFI, S., 1975, Lithostratigrafia e biostratigrafia di alcune serie neogeniche della provincia di Almería (Andalousia orientale-Spagna): Acta Naturale "Ateneo Parmense", v. 11, p. 237-313.

JULIVERT, M., FONBOTÉ, J. M., RIBEIRO, A., AND CONDE, L., 1974, Mapa tectónico de la Península Ibérica y Baleares: Madrid, Memoria del Instituto Geológico y Minero de España, Contribución al mapa tectónico de Europa.

LEBLANC, D., 1979, Etude géologique du Rif extense oriental au nord de Taza (Maroc): Notes et Memoires du Service Géologique du Maroc, v. 281, 159 p.

LIETZ, J. AND SCHWARZBACH, M., 1970, Neue Fundpunkte von marinem Tertiar auf der Atlantik-Insel Porto Santo (Madeira-Archipel), New localities of marine Tertiary on the Atlantic island Porto Santo (Madeira Archipelago): Neues Jahrbuch in Geologie und Paleontologie Monatshefte, no. S, p. 270-282.

LÓPEZ-GARRIDO, A. C. AND SANZ DE GALDEANO, C., 1991, La comunicación en el Tortoniense entre el Atlántico y el Mediterráneo por la Cuenca del Guadalhorce (Málaga): Barcelona, I Congreso del Grupo Español del Terciario, Comunicaciones, p. 190-191.

LUPERTO, E., 1962, L'Oligocene della terra d'Otranto: Memorie della Società Geologica Italiana, v. III, p. 594-621.

MAGNÉ, J., 1978, Etudes microstratigraphiques sur le neogène de la Méditerranée nord-occidental: v. I: Les bassins neogènes catalans: Centre National de Recherches Scientifiques, Centre Regional de Publications de Toulouse, Sciences de la Terre, 259 p.

MAGNÉ, J., GOURINARD, Y., AND WALLEZ, M. J., 1987, Comparaison des étages du Miocène inferieur definis par stratotypes ou par zones paleontologiques: Toulouse, Strata, v. 1, p. 95-107.

MARTÍN, J. M. AND BRAGA, J. C., 1990, Arrecifes Messinenses de Almería. Tipologías de crecimiento, posición estratigráfica y relación con las evaporitas: Geogaceta, v. 7, p. 66-68.

MARTÍN, J. M., BRAGA, J. C., AND RIVAS, P., 1989, Coral successions in Upper Tortonian reefs in SE Spain: Lethaia, v. 22, p. 271-286.

MARTÍN, J. M., ORTEGA-HUERTA, M., AND TORRES-RUIZ, J., 1984, Genesis and evolution of strontium deposits of the Granada Basin (southeastern Spain): evidence of diagenetic replacement of a stromatolite belt: Sedimentary Geology, v. 39, p. 281-298.

MONLEAU, C., ANGLADA, R., ARNAUD, M., MONTAGGIONI, L., THOMASSIN, B. A., AND ROSEN, B., 1989, Constructions a Madreporaires et les dépôts associés de l'Aquitanien de la Nerthe (Provence Occidental, France), Aquitanian coral buildups and associated deposits of the Nerthe coast (Western Provence, France): Marseille, 1989 Annual Meeting of the International Society for Reef Studies: Centre de Stratigraphie et Paléoécologie, Université de Provence, Centre d'Océanologie de Marseille, Université d'Aix-Marseille II, 38 p.

MONTENAT, C., 1973a, Le Miocène terminal des chaines Bétiques (Espagne méridonale) Esquisse paleogéographique, *in* Drooger, C. W., ed., Messinian Events in the Mediterranean: Ansterdam: North Holland Pubications Company, Geodynamics Science Report 7, p. 180-187.

MONTENAT, C., 1973b, Les formations neogènes et quaternaires du Levant espagnol (Provinces d'Alicante et de Murcia): Unpublished Ph.D. Thesis, Université de Paris-Orsay, Paris, 1170 p.

MONTENAT, C., 1977, Les basins neogènes du levant d'Alicante et de Murcia (Cordillères Bétiques orientale-Espagne): stratigraphie, paleogéographie et évolution dynamique: Documents du Laboratoires de Géologie de la Faculté des Sciences de Lyon, v. 69, p. 1-345.

MONTENAT, C. AND BIZON, G., 1976, A propos de l'évolution géodynamique mio-pliocène en méditerranée occidentale. L'example du bassin de Vera. (Cordillères bétiques. Espagne méridionale): Comptes Rendues Sommaires de la Societé Géologique de France, Fascicle 1, p. 15-16.

MONTENAT, C. AND OTT D'ESTEVOU, P., 1990, Eastern Betic Neogene Basins: a review, *in* Montenat, C., Les bassins Néogènes du domain Bétique oriental (Espagne): Paris, Centre National de Recherches Scientifiques, v. 12-13, p. 9-18.

OBRADOR, A. AND POMAR, L., 1983, El Neógeno del sector de Maó, *in* Pomar, L., Obrador, A., Fornós, J., and Rodríguez-Perea, A., eds., El Terciario de las Baleares (Mallorca-Menorca), Guía de las Excursiones del X Congreso Nacional de Sedimentología: Palma de Mallorca, Institut d'Estudis Baleàrics and Universitat de Palma de Mallorca, p. 233-255.

OBRADOR, A., POMAR, L., RODRÍGUEZ, A., AND JURADO, M. J., 1983, Unidades deposicionales del Neógeno Menorquín: Acta Geológica Hispánica, v. 18, p. 87-97.

OBRADOR, A., POMAR, L., AND TABERNER, C., 1992, Late Miocene megabreccia, of Menorca (Balearic Islands): A basis for the interpretation of a megabreccia ramp deposit, *in* Pedley, H. M., ed., Carbonate Ramps: Processes and Diagenesis: Sedimentary Geology, v. 79, p. 203-223.

ORSZAG-SPERBER, F. AND PILOT, M. D., 1976, Grands traits du Neogène de Corse: Bulletin de la Societé Géologique de France, v. 7, p. 1183-1187.

PEDLEY, H. M., 1981, Sedimentology and palaeoenvironment of the southeast sicilian Tertiary platform carbonates: Sedimentary Geology, v. 28, p. 273-291.

PERMANYER, A., 1990, Sedimentologia i diagènesi dels esculls Miocènics de la conca del Penedès: Barcelona, Institut d'Estudis Catalans, 324 p.

PERMANYER, A. AND ESTEBAN, M., 1973, El arrecife mioceno de Sant Pau d'Ordal (provincia de Barcelona) (The Miocene reef of Sant Pau d'Ordal, Barcelona): Barcelona, Institut d'Investigacions Geològiques, Publicació 28, p. 45-72.

POMAR, L., 1991, Reef geometries, erosion surfaces and high-frequency sea-level changes, upper Miocene Reef Complex, Mallorca, Spain: Sedimentology, v. 38, p. 243-269

POMAR, L., ESTEBAN, M., CALVET, F., AND BARON, A., 1983a, La unidad arrecifal del Mioceno superior de Mallorca: Menorca, El Terciario de las Baleares, Guía de las Excursiones del X Congreso Nacional de Sedimentología, p. 139-175

POMAR, L., OBRADOR, A., FORNÓS, J., AND RODRÍGUEZ-PEREA, A., eds., 1983b, Guía de las excursiones, El Terciario de las Baleares (Mallorca-Menorca): Menorca, X Congreso Nacional de Sedimentología, 255 p.

POMAR, L. AND RODRÍGUEZ-PEREA, A., 1983, El Neógeno inferior de Mallorca: Randa, *in* Pomar, L., Obrador, A., Fornós, J., and Rodríguez-Perea, A., eds., El Terciario de las Baleares (Mallorca-Menorca), Guía de las Excursiones del X Congreso Nacional de Sedimentología: Palma de Mallorca, Institut d'Estudis Baleàrics and Universitat de Palma de Mallorca, p. 115-137.

RIDING, R., BRAGA, J. C., AND MARTÍN, J. M., 1991a, Oolite stromatolites and thrombolites, Miocene, Spain: analogues of Recent giant Bahamian examples: Sedimentary Geology, v. 71, p. 121-127.

RIDING, R., MARTÍN, J. M., AND BRAGA, J. C., 1991b, Coral-stromatolite reef framework, Upper Miocene, Almería, Spain: Sedimentology, v. 38, p. 799-818.

ROCA, E. AND DESEGAULX, P., 1992, Analysis of the geological evolution and vertical movements in the València Trough area, western Mediterranean: Marine and Petroleum Geology, v. 9, p. 167-185.

ROCA, E. AND GUIMERÁ, J., 1992, The neogene structure of the eastern Iberian margin: structural constraints on the crustal evolution of the Valencia Trough (western Mediterranean): Tectonophysics, v. 203, p. 203-218.

RODRÍGUEZ-PEREA, A., 1984a, El Mioceno de la Serra de Tramuntana: Estratigrafía, Sedimentología e implicaciones estructurales: Unpublished Ph.D. Thesis, University of Barcelona, Barcelona, 533 p.

RODRÍGUEZ-PEREA, A., 1984b, La formación calcarenítica de Sant Elm: un ejemplo de plataforma mixta terrígeno-carbonatada: Pubicaciones de Geología, Universidad Autónoma de Barcelona, n. 20, p. 399-417.

RODRÍGUEZ-PEREA, A., 1989, Miocene mixed shelf deposits of Mallorca island (abs.): Budapest, 10th International Association of Sedimentologists Regional Meeting on Sedimentology, p. 196-197.

RODRÍGUEZ-PEREA, A. AND POMAR, L., 1983, El Neógeno inferior de Mallorca: Port d'es Canonge-Banyalbufar, *in* Pomar, L., Obrador, A., Fornós, J., and Rodríguez-Perea, A., eds., El Terciario de las Baleares (Mallorca-Menorca), Guia de las Excursiones del X Congreso Nacional de Sedimentología: Palma de Mallorca, Institut d'Estudis Baleàrics and University of Palma de Mallorca, p. 91-114.

RODRÍGUEZ-PEREA, A., POMAR, L., AND FORNOS, J. J., 1988, Towards a model for mixed shelves?: Columbus, Annual Midyear Meeting, v.

5, p. 46.

ROUCHY, J. M., CHAIX, C., AND SAINT-MARTIN, J. P., 1982, Importance et implications de l'existence d'un recif corallien messinien sur le flanc sud du Djebel Murdjadjo (Oranie, Algerie): Comptes Rendues, serie II, v. 294, p. 813- 816.

ROUCHY, J. M., SAINT-MARTIN, J. P., MAURIN, A., AND BERNET-ROLLANDE, M. C., 1986, Evolution et antagonisme des communautés bioconstructrices animales et végétales à la fin du Miocène occidental: biologie et sédimentologie: Bulletin des Centres de Recherche et Exploration Elf-Aquitaine, v. 10, p. 333-348.

SAINT-MARTIN, J. P., 1990, Les formations recifales coraliennes du Miocène superieur d'Algerie et du Maroc: Paris, Muséum National d'Histoire Naturelle, 366 p.

SAINT-MARTIN, J. P. AND ROUCHY, J. M., 1990, The Messinian carbonate platforms in the western Mediterranean: their importance for the reconstruction of the late Miocene sea level variations: Bulletin de Géologie de la Societé de France, v. 8, p. 83-94.

SANTISTEBAN, C, 1981, Petrología y sedimentología de los materiales del Mioceno superior de la cuenca de Fortuna (Murcia) a la luz de la Teoría de la Crisis de Salinidad: Unpublished Ph.D. Thesis, Universitat de Barcelona, Barcelona, 725 p.

SANZ DE GALDEANO, C., 1990, Geologic evolution of the Betic Cordilleras in the Western Mediterranean, Miocene to the present: Tectonophysics, v. 172, p. 107-119.

SANZ DE GALDEANO, C. AND LÓPEZ-GARRIDO, A. C., 1991, Tectonic evolution of the Málaga Basin (Betic Cordillara). Regional implications: Geodinamica Acta (Paris), v. 3, p. 173-186.

SANZ DE GALDEANO, C. AND VERA, J. A., 1992, Stratigraphic record and palaeogeographical context of the Neogene basins in the Betic Cordillera, Spain: Basin Research, v. 4, p. 21-36.

SERRANO-LOZANO, F., 1979, Los foraminíferos planctónicos del Mioceno superior de la Cuenca de Ronda y su comparación con los de otras áreas de las Cordilleras Béticas: Unpublished Ph.D. Thesis, Departamento de Geología, Facultad de Ciencias, Universidad de Málaga, Málaga, 327 p.

SIMÓ, A. AND RAMON, X., 1986, Análisis sedimentológico y descripción de las secuencias deposicionales del Neógeno postorogénico de Mallorca: Boletín Geológico y Minero, v. 97, p. 445-472.

STEININGER, F. F., SENES, J., KLEEMANN, K., AND RÖGL, F., 1985, Neogene of the Mediterranean Tethys and the Paratethys stratigraphic correlation tables and sediment distribution maps: Institute of Paleontology of Vienna, v. 1, p. 472, and v. 2, p. 504.

VEZIAN, A., 1856, Mollusques et zoophytes des terrains nummulitiques et tertiaire marin de la province de Barcelona: Montpellier, Editeur Martel Ainé, Imprimerie de la Faculté de Sciences de Montpellier, 50 p.

VIDAL, J. C. L., 1977, Structure actuelle et évolution depuis le Miocène de la chaine rifaine (partie sud de l'arc de Gibraltar): Bulletin de la Societé Géologique de la France, v. 7, p. 789-796.

WERNLI, R., 1980, Le Messinien à *Globorotalia conomiozea* (foraminifère planctonique) de la côte méditerranéenne marocaine: Eclogae Geologicae Helvetica, v. 73, p. 71-93.

WERNLI, R., 1987, Micropaléontologie du Néogène post-nape du Maroc septentrional et description systématique des Foraminifères planctoniques: Rabat, Notes et Mémoires du Service Géologique du Maroc, 331 p.

WILDI, W., 1983, La chaine tello-rifaine (Algérie, Marroc, Tunisie), structure, stratigraphie, évolution du Trias au Miocène: Revue de Géologie Dynamique et Géographie Physique, v. 24, p. 201-298.

ZIEGLER, P. A., 1988, Evolution of the Arctic-North Atlantic and the Western Tethys: Tulsa, American Association of Petroleum Geologists, Memoir 43, 198 p.

# MIOCENE REEF DISTRIBUTIONS AND THEIR ASSOCIATIONS IN THE CENTRAL MEDITERRANEAN REGION: AN OVERVIEW

MARTYN PEDLEY

*Department of Geology (Leicester) and School of Geography and Earth Resources, University of Hull, HU6 7RX, United Kingdom*

ABSTRACT: A review of the distribution of Miocene coral bioherms and biostromes is presented for Italy, Malta, Libya and Tunisia. These can be grouped into three natural settings related respectively to stable forelands, Alpine fold belts and graben zones. Morphological development within each zone is controlled by such factors as water depth, exposure to water currents, tectonism and siliciclastic sedimentation rates. The more diverse reefs developed in clear seas within stable foreland settings. Three principal reef development episodes are recognised. The earliest reefs (Aquitanian) are modest developments and represent a continuation of Oligocene coral reef growth with a high species diversity. Middle Miocene reefs are dominated by non-coral faunas and coralline algal biostromes; many are related to ramp situations. Corals return in Late Tortonian time though with much lower diversities than their Aquitanian predecessors. Many low diversity coral reefs typically contain up to five species but are always dominated by *Porites* and *Tarbellastraea* together with abundant coralline algae. Quite distinct from these are the reefs containing only *Porites* and *Tarbellastraea* and even more peculiar, the monogeneric *(Porites)* reefs. The latter are particularly typical of the Early Messinian. The slender coral rods are never as long and thin as those from strata of similar age in Spain; however, they are similarly associated with stromatolites and abundant *Halimeda* plates, particularly in the highest beds. It is suggested that locally introduced toxin and nutrient imbalances may contribute significantly to such aberrant reef growth. The driving force for these fluctuations appears to be a combination of tectonic basin deformation and small-scale, intra-Mediterranean eustatic oscillations marking the onset of the principal desiccation event.

## INTRODUCTION

The central Mediterranean region is comprised of mainland Italy, together with the islands of Sardinia and Corsica lying due west on the Tyrrhenian Sea margins. To the south lie the Maltese Islands and Sicily, both bounded to the east by the Ionian Sea. The North African states of Tunisia and Libya also fall into the central Mediterranean region, together with the Pelagian Islands (Italy) which lie within the Pelagian Sea equidistant from Tunisia, Malta and Sicily.

These countries collectively lie on the divide between the Western Mediterranean and eastern Mediterranean basins. At times during the Miocene deposition, this divide severely constricted water exchange. The Upper Miocene reefs of the central Mediterranean region generally have been compared with the western Mediterranean Basin (see Esteban, 1979), though this mainly reflects the lack of descriptive literature on the Cretan, Israeli and Turkish reefs. Saint-Martin and Rouchy (pers. commun.) and Esteban (1979) however suggest that the east Mediterranean Basin reefs show a higher biotal diversity.

Reef types are diverse. Many are dominated by abundant scleractinian coral genera, though there is a marked tendency towards reduction in coral diversity from in excess of fifteen dominant species in Late Oligocene time (Maltese Islands, Pedley, 1975; Colli Berici, north Italy, Geister and Ungaro, 1977; Frost, 1981,1977) to *Porites* and *Tarbellastraea* by Late Tortonian to Early Messinian time (see Chevalier, 1977). Regardless of scleractinian diversity within the region there is a tendency towards three principal forms of bioconstruction with fringing patch reefs being well developed during Aquitanian sedimentation (Libya, east Sicily and Sardinia) and buttress or wall-reefs during Late Tortonian-Early Messinian time (north and west Sicily, Calabria and Tuscany). Rhodalgal ramps are ubiquitous throughout the Miocene in all areas of this study and, as with other regions, provide the only significant reef related associations during the Burdigalian to Late Serravallian interval (Middle Miocene Supercycle). Much of this rhodalgal material is partly reworked, though significant *in situ* coralline algal framework biostromes with bryozoans and occasional corals are common (e.g., Malta). The possible causes of bioconstruction variability are explored later.

## REGIONAL GEOLOGICAL SETTINGS

The central Mediterranean may conveniently be divided into three contrasting depositional domains: stable African foreland, Alpine foldbelts and graben zones.

### *Stable African Foreland*

This domain lies along the southern and eastern margins of the region extending through the Apulian peninsula of south east Italy, south east Sicily and the Pelagian Islands, The Maltese Islands, east Tunisia and north Libya (see Fig. 1). Collectively this part of the African Plate has been referred to as the Pelagian Block (Burollet, 1969) and until Late Oligocene times lay far away to the south-east of the European Plate margin. Only during Miocene deposition did the African Plate (Pelagian Block) commence to impinge upon the European Plate. Initially, plate motion was northwards and the African Plate proceeded to underthrust the European Plate. Since Late Miocene times, the maximum compression has become north-west directed in the Pelagian area and Sicily (Grasso and Pedley, 1985) and both plates have effectively become locked together since Early Quaternary times (Pedley and Grasso, 1991). Figure 2 shows the east-west trending positions of several of the associated superimposed thrust sheets. Similar underthrusting of the Apulian Platform margins has occurred in east Italy with the develop-

Models for Carbonate Stratigraphy from Miocene Reef Complexes of Mediterranean Regions,
SEPM Concepts in Sedimentology and Paleontology #5, Copyright © 1996,
SEPM (Society for Sedimentary Geology), ISBN 1-56576-033-6, p. 73-87.

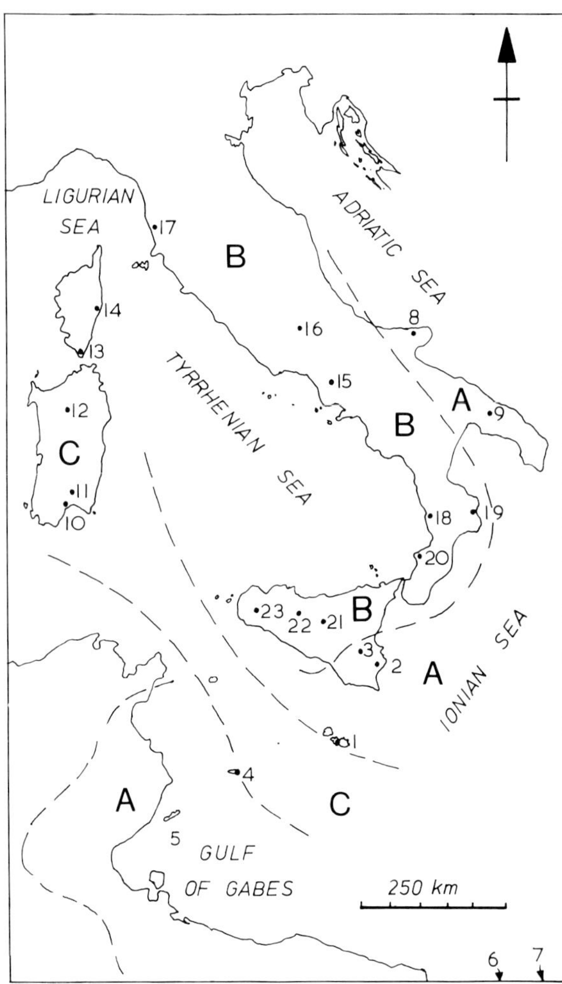


FIG. 1.—Map of the central Mediterranean: A = stable African platform; B = Alpine Foldbelts, and C = Graben zones. Numbers are for Miocene reef localities and are consecutively numbered within each of the above zones; stable African platform- 1, Maltese Islands; 2, Ognina (eastern Hyblea); 3, Carlentini; 4, Lampedusa; 5, Ketatna; 6, Sirte Basin; 7, Cyrenaica; 8, Gargano; 9, Apulia; graben zone- 10, Cagliari and Funtanazza; 11, Marmilla and Villagreca; 12, Loqudoro; 13, Bonifacio, Balistra and Paraguana areas; 14, Plaine Orientale; Alpine fold belts- 15, Matese Mountains; 16, Monte Velino; 17, Rosignano; 18, Amantea; 19, Sila; 20, Vibo Valentia, Palmi and Santa Domenica; 21, Petralia and Rocca Limata; 22, Ciminna; 23, Calatafimi, Grieni and Salemi.

ment of the Bradanic foredeep separating Apulian Platform from Appennine foldbelt. The north-south trending position of the Pliocene external thrust front is marked on Figure 2.

To the south and east of the allochthonous thrusts, the African Plate is relatively untectonised and during Miocene time was the site of extensive carbonate platform development. Pre-Miocene substrates generally consist of long established, continuous carbonate sequences reaching back into the Mesozoic Era. Thicknesses exceeding 3000 m are common (e.g., Malta) though thicknesses decrease progressively southwards as the Saharan shield is approached in south Tunisia and Libya.

### *Alpine Foldbelts*

This Tertiary foldbelt domain extends from the north Italian Alps southward as the Appennines, turning through almost 90 ° (the Calabrian Arc) and westward via the Peloritanian, Nebrodian and Madonie mountains of north Sicily (see Fig. 1). Continuing westward across the Straits of Sicily, the Tertiary foldbelt is developed into the Tunisian Atlas Ranges. These mountain chains developed during Afro-European plate collision with tectonic activity becoming regionally significant from Eocene times onward (for local details see Catalano et al., 1976; Abate et al., 1982; Gissetti et al., 1982). Subsequent neotectonism has significantly modified the Miocene paleogeographies as the result of crustal shortening in both north Sicily and west Apulia, and by extensive collapse of the Tyrrhenian margins (Pedley and Grasso, 1994a).

Reefs associated with the fold belts are principally Late Tortonian to Early Messinian in age, though fragmented reef distributions are sufficient to indicate direction of facing and substrate control. Many reefs developed on submerged thrust generated seafloor topographies (e.g., west, central north Sicily, Grasso and Pedley, 1988; Ruggieri and Torre, 1984; Catalano et al., 1976). The Rosignano reefs of Tuscany are also developed on eroded structural highs (Bossio et al., 1978; Chevalier, 1961). Others (e.g., south Calabria) lie on the shoulders of granitic basement ridges, now up to 350 m above sea level. Those reefs on the southern margins of the Sicilian Nebrodian Mountains often developed at shelf slope breaks, especially on drowned delta top systems (Grasso and Pedley, 1988; Catalano, 1979). Collectively, their development necessitated areas of shallow seafloor, and these became available in great number as the complexity of the alpine fold belts developed.

### *Graben zones*

These structural zones are well developed in the Sardinian area (Campidano Graben) where reef walls are located both on graben shoulder locations and as small barrier reefs associated with volcanics (see Fig. 1). The graben appears to be associated with the development of the so called "external Alpine mollasse trough" (see Alvarez, 1972; Cocozza and Schafer, 1974), though Illies (1980, 1981) relates it to an "intercontinental rift" system which extends southeast to the "Sicily Channel Rift Zone" (Finetti, 1984, 1985; Reuther and Eisbacher, 1985). Similar reef walls are associated with the Sicily Channel graben both on the island of Lampedusa (Grasso and Pedley, 1985) and in south west Malta (Pedley, 1987b). These two islands lie on either side of the Sicily Channel Rift Zone which here showed its first significant movements during Late Tortonian time. This approximately coincides with the timing of Sardinian graben-related reef development.

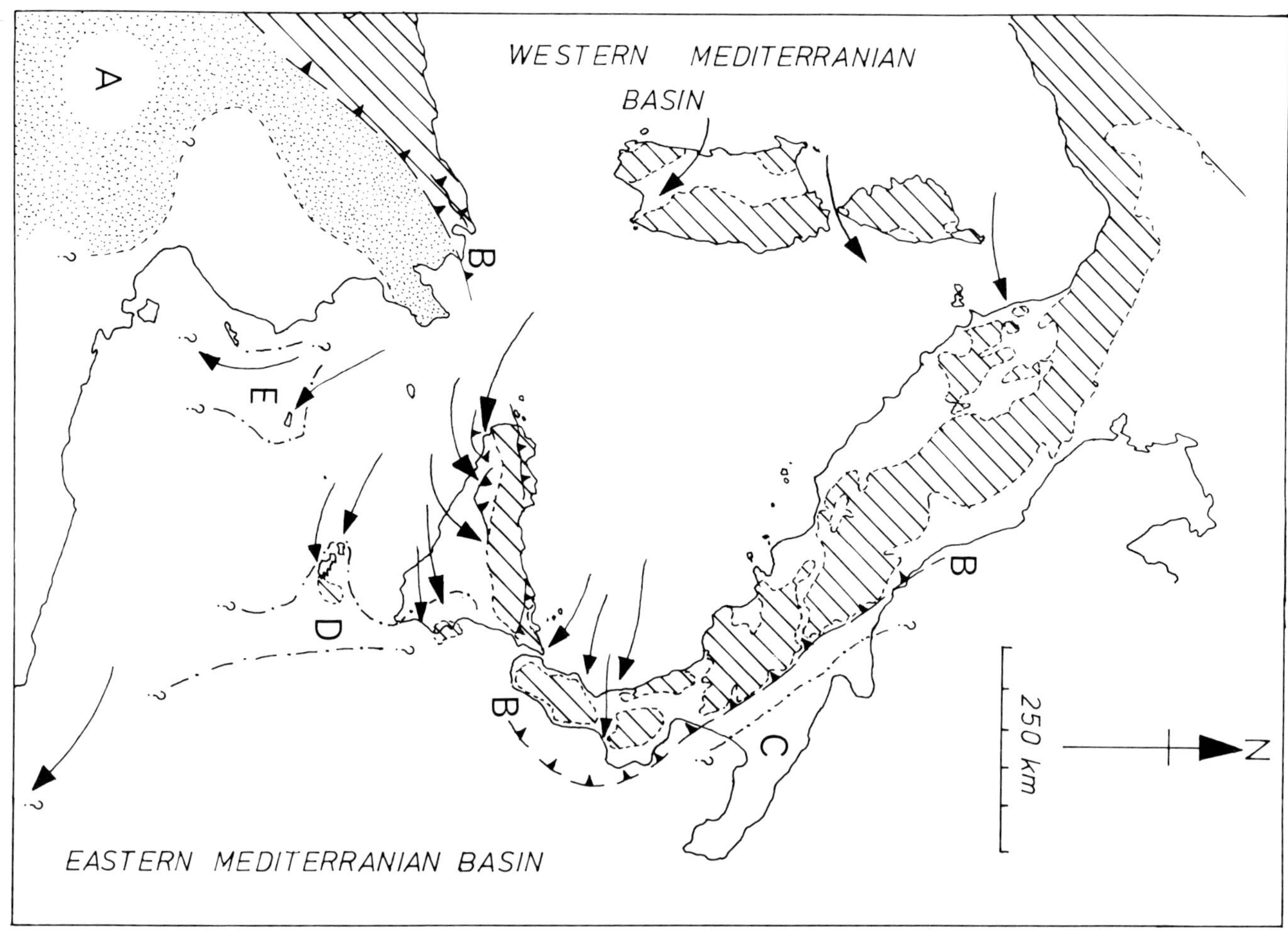


FIG. 2.—Paleogeographic reconstruction of the central Mediterranean region during Late Tortonian-Early Messinian time. Hatched areas are emergent foldbelts. Stippled areas (A) are terrestrial siliciclastics. B represents the front of the Pliocene extensional thrust (effectively here delimiting the western observable margin of the Bradanic Trough). Carbonate platforms of the Pelagian Block are indicated by dot-dash lines thus: C, Apulian Platform; D, Sicily-Maltese Platform; E, Pelagian Platform. Arrows indicate Paleomediterranean water circulation paths as deduced from occurrences and facing directions of Late Miocene coral reefs.

## MIOCENE STRATIGRAPHY

Clearly, each of the three physical reef settings was controlled by distinct tectonic settings which influenced strongly the morphological expression of the reef systems (see later). Equally significant is the stratigraphic control with reefal strata which is principally located in Late Tortonian and Early Messinian strata (see Fig. 3):

### *Sequence Stratigraphy*

The eustatic signal (Haq et al., 1987) is not always clear on account of the regional plate driven tectonism which reached an activity peak during Late Miocene time. In general, the basal Aquitanian transgression (cycle 1.4) caused drowning of several well established reef tracts (e.g., Maltese Islands, Pedley, 1978a; Colli Berici, Geister and Ungaro, 1977). On the other hand, it permitted new colonisation in the Campidano Graben, south east Sicily (Pedley, 1981) and in Tunisia.

The deeper water regime (Cycles 2.2 to 2.6) and cooler climate of Middle Miocene time (Langhian and Serravallian) were less conducive to coral reefs in the central Mediterranean, but favored rhodalgal ramp carbonate developments such as in the south Appennines of Italy (Simone and Carannante, 1988; Barbera et al., 1980). Eustatic sea-level oscillations associated with the proposed 3.2 and 3.3 global eustatic cycles were internally controlled within the Mediterranean due to isolation from the world oceans; these had profound affects on reef development (see later).

### *Lithostratigraphy*

Generally the facies variations of the central Mediterranean region are too variable for anything beyond local correlation purposes. The lateral persistence of coral reef units, related to highstand episodes, is an exception to this (Pedley, 1983; Saint-Martin and Rouchy, 1990). Tectonic instability during Neogene time however, has resulted in the deposition of varied clay

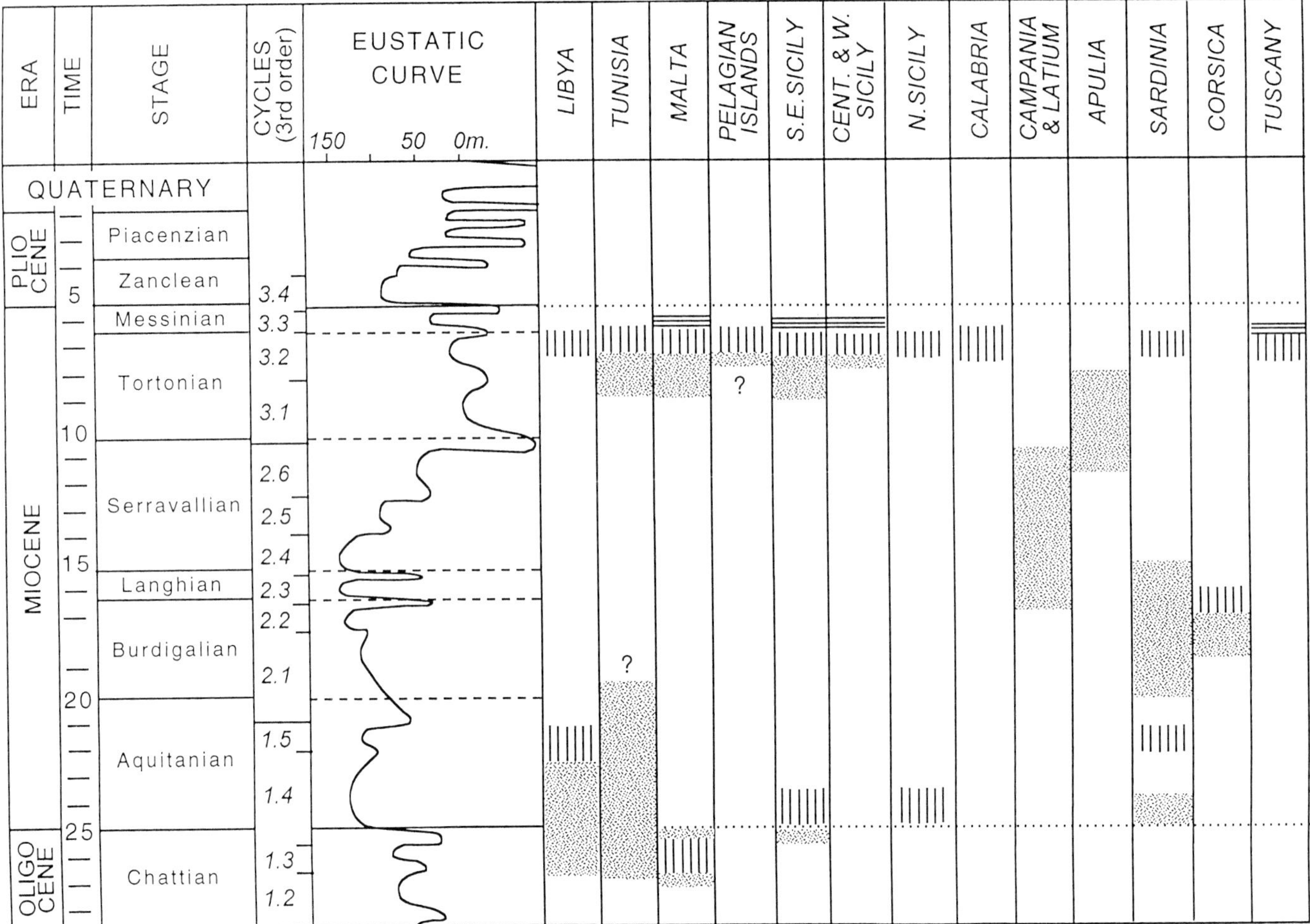


FIG. 3.—Stratigraphic distributions of the Miocene reefs of the central Mediterranean region and their relationship to the Vail curve in Haq et al. (1987). Symbols: stipple represents rhodalgal biostromes; vertical bars represent *Porites* dominated coral reefs; and horizontal bars represent stromatolite associated terminal complex.

mineral suites (see de Visser, 1990) which may in future be of value for correlation purposes.

### *Biostratigraphy*

Reef developments contain only sporadic microfossils of zonal significance, and this has hampered both correlation between areas and correlation between local facies variants. This is nowhere better illustrated that in the Late Miocene sequences. Generalised biostratigraphies have been applied with variable success, especially in the western Mediterranean Basin, using coral species diversity (Esteban, 1979) in order to overcome Messinian correlation difficulties (see discussion in Esteban, this volume).

Considerable recent work using planktonic microfossils has greatly improved the situation and many of these reef bearing units (Esteban, 1979) are now considered to be of Late Tortonian to Early Messinian in age in southern Italy (e.g., Sicily, Fois, 1990; Pedley, 1983; Esteban et al., 1982; Catalano, 1979; Courme-Rault, 1991), the Maltese (Pedley, 1983) and Pelagian Islands (Grasso and Pedley, 1985).

This conclusion is at variance with observed lithological associations in Spain (e.g., Santisteban, 1981) where reefal carbonates also appear to occur between younger Messinian evaporite cycles. The excellent microbiostratigraphic control now available on the Spanish successions (e.g., Serrano, 1990; Martin and Braga, 1990, 1994) confirms the presence of an earliest Messinian coralline algal sequence followed by two coral reef sequences below a later "Messinian Terminal Complex" including coral reefs. In north Italy, the Tuscany reefs (Bossio et al., 1981a, 1981b) show, with the aid of associated microfauna, an Early Messinian "lagoonal interval" capping the earliest reefs but followed by a later Messinian Castelnuovo reef development.

The present consensus is that the southern Italian and Maltese terminal complexes relate to the "lagoonal" events of north Italy, Spain and Algeria and that later Messinian terminal complexes are not represented because of emergence. In west Sicily (Catalano, 1979), central and south east Sicily, however, a restricted marine, Late Messinian "*Congerie*" fauna (Cafici, 1880) is locally present above the Early Messinian evaporites and may correlate with the younger Messinian marine events elsewhere in the Mediterranean or possibly the "Lac Mer" episode.

## CENTRAL MEDITERRANEAN REEF EVENTS

These various approaches to correlation permit the division of the Central Mediterranean Miocene into three episodes of coral reef and rhodalgal biostrome development (see Fig. 3).

### *Early Miocene Coral Reefs*

Aquitanian reef development within the region is best documented from the Ales and Villagreca areas of Sardinia (Fig. 1, location 11). Here, at the south east margins of the Campidano Trough occurs a series of small barrier reefs associated with the basal Miocene transgression. They are underlain by siliciclastic sediments. They lie close above a north-west trending Oligocene ridge and consist of pale grey coral bioherms containing bryozoans, echinoids, bivalves and coralline algae. In the Villagreca area, the reefal sequence is about 25 m thick.

In south east Sicily an Aquitanian patch reef sequence is developed above basic Late Cretaceous lavas in the vicinity of Priolo (Pedley, 1981; Grasso et al., 1979). The pale-grey micritic reef (1.3 m thick) contains several genera of coral, many in life position together with pectinid bivalves, gastropods, echinoids and scattered algal rhodoliths. A foraminiferal microfauna (*Lepidocyclina* and *Miogypsinoides*) underlies the reefs (Grasso et al., 1979). Locally, a pre-Late Serravallian age is suggested by marine faunas above the reefs. Northward the reef gives way to rhodolitic algal pavement facies. The patch reef sequence is truncated by packstones and a dolomitized paleosol with rhizocretion fabrics and succeeding dolomicrites. In all, some 6 m of strata is visible in the road cut.

Fragmentary Chattian-Aquitanian coral reef material associated with *Lepidocyclina* is also seen in a roadside section on the east side of Capo Tindari on the north Sicily coast west of Milazzo. It appears originally to have been developed upon metamorphic basement.

In Libya, the Faidiyah Formation shows development of shallow marine coralline algal carbonates with reefs (probably marking the end of the transgressive cycle) at the top of the formation in the vicinity of Jabal al Akdar (north Libya). The reefs contain abundant branching corals enclosed in a recrystallized micritic groundmass and form prominent hills.

### *Early Miocene Coralline Algal Biostromes*

In Tunisia coralline algal successions are reported to occur in the Ketatna and marginal Fortuna Formations but are only known from well data (Fournie, 1978). They contain abundant coralline algae, molluscs, bryozoans and larger foraminifers.

In the Derna area of Cyrenaica, to the east of the Sirte Basin (see Fig. 1) occur chalky carbonates with coralline algal levels. The lower levels are friable "maerl" facies but these pass up into *Archaeolithothamnion* rhodolitic algal carbonates towards the sequence top (Lower White Limestone of Rose, 1971). Off-reef facies are dominated by miliolid-rich carbonates containing echinoids and molluscs. Rose (1974) presents a case for these strata being Late Oligocene in age.

*Lithothamnion* boundstones occur in Burdigalian to Langhian strata above Oligocene volcanics in the Loqudoro section of northern Sardinia.

### *Middle Miocene Coral Reefs*

The close of Aquitanian times in the central Mediterranean was a time of coral decolonisation. In the Maltese Islands (Pedley, 1978a) and Derna, north Libya (Rose, 1971) deeper water conditions became established in Aquitanian times.

The coral-red algal reefs of Corsica appear unique in Middle Miocene time and accumulated after the arrival of the Corsican alpine nappes (Orszag-Sperber and Pilot, 1976). At the south end of the island, the Bonifacio Basin contains extensive developments of Burdigalian bioherms with corals, associated rhodolitic coralline algae and a diverse echinoid and mollusc fauna. Smaller coral developments lie within the basal transgressive conglomerate sequence (Arnaud et al., 1989a). Smaller coral reef development of Langhian age occurs at Balistra and Paraguano (Arnaud et al., 1989a). Further east, the Plaine Orientale succession (Aleria region) contains coral-red algal bioherms with pectinids and echinoids within the upper part of the Langhian sequence associated with local limestones and sands.

### *Middle Miocene Coralline Algal Biostromes*

In south east Sicily, some shallowing in Early Burdigalian times (base of cycle 2.1) around Ragusa is indicated by coralline algal fragments in small percentages in an otherwise pelagic succession. True reefal coral growth in Sicily is unknown though widely scattered outcrops of algal rhodolith and bryozoan biostromes of Late Burdigalian to Early Langhian age (Floresta Formation) occur in north Sicily (e.g., Patti, Carbone et al., 1993), with smaller outliers around the towns of Gangi and Sperlinga to the east of Petralia (see Fig. 1, locality 21) on the south side of the Madonie Mountains (Fravega et al., 1993). More widespread coralline algal colonization (rhodolitic algal biostromes) occurs throughout Latium, Abruzzi, Molise and Campania (Barbera et al., 1980) being well displayed in the eastern Matese Mountains of Campania (Barbera et al., 1978; Simone and Carannante, 1985; Carannante and Simone, this volume). This Cusano Limestone succession spans a Burdigalian to Langhian age and is dominated by bryozoans, macroforaminifers, echinoids, serpulid worms and rhodolitic coralline algae. The association developed upon a step-like faulted submarine topography in an open shelf location down to 80 m water depth (Barbera et al., 1978). Carannante et al. (1981) conclude that a warm temperate to sub-tropical environment is indicated by the biota.

The only other developments of coralline algae within the region during the Middle Miocene are in Sardinia and Corsica. The Loqudoro sections contain biostromal *Lithothamnion*-rich beds of Burdigalian age overlying conglomerates. The Triso valley and Funtanazza sections of north Sardinia and the Sunai sections of Sardinia also appear to be Burdigalian to lower Langhian in age.

Burdigalian rhodolitic algal carbonates are developed in the Saint-Florent Basin in the north of Corsica in association with bryozoans, pectinids, oysters and echinoids (Arnaud et al., 1989b).

### *Late Miocene Coral Reefs*

The Tortonian succession (especially the Late Tortonian) witnessed a return of reefal carbonate deposition within the central Mediterranean. These were dominated by coralline algae, however, *Mesophyllum* and *Lithophyllum* became dominant in these younger reefs (cf. *Archaeolithothamnium* in the Late Oligocene reefs). As with other Mediterranean areas there was a marked reduction in coral diversity with *Porites* and *Tarbellastraea* dominating the scleractinian genera. Codiacean algae, such as *Halimeda,* characteristically dominate the off-reef sediments, and microbial micrites are increasingly important reef contributors towards the top of the marine sequences. Reef development continued into Early Messinian time but appears to have been extinguished by the onset of evaporitic conditions (1st cycle), except in Tuscany. The Mediterranean eustatic curve departs from the "Vail curve" towards the close of Late Tortonian times due to isolation from the world oceans (see Haq, 1991 for World Cenozoic trends). This manifests itself, particularly in southeast Sicilian and Maltese strata, by the development of two or three coral levels, each truncated by erosion before a final shallow marine "Terminal Complex" (often ooidal to restricted lagoonal). Evaporites were never deposited on these Early Messinian carbonates, being restricted to basinal areas .

Extensive coral-algal patch reef developments occur in south east Sicily (Grasso et al., 1982), Lampedusa (Grasso and Pedley, 1985) and Malta (Pedley, 1979, 1981; Saint-Martin and Andre, 1992). These micrite dominated mounds accumulated about a sparse framework of *Porites* and *Tarbellastraea* but contain an abundant burrowing bivalve infauna. *Halimeda* often dominates the coarser sediments between the patch reefs. These small scattered patch reefs appear to have developed at shallow water depths in either mid-ramp or shelf edge settings.

These lenticular patch reefs are associated with biostromal spreads of coral and crustose coralline algal-rich carbonates with the same biotal composition. The truncated tops of the biostromes testify to having been planed off periodically by temporary sea-level draw-down events peculiar to the Mediterranean Basin (Pedley, 1983). The surfaces can be correlated not only within the areas described but also with the two reefal levels within the upper part of the Terravecchia Formation (Schmidt di Friedburg, 1965) of north central Sicily (Grasso and Pedley, 1988 and later in this chapter).

Northern Sicilian Late Miocene environments, like many other synorogenic marginal shelf settings in the Mediterranean at this time (e.g., Hayward, 1982; Dabrio, 1975), were dominated by siliciclastic deposition. Reefs were limited to outer shelf coral patches which developed on spreads of gravel generally at the mouths of abandoned fluvial channels and delta lobes. Eustatic rises and associated highstands effectively drowned the delta systems and permitted short intervals with more continuous reef development across the tops of the inactive fans and at the mouths of the abandoned distributary channels. The north central Sicilian reefs of Portella di Landro, Petralia and Cacchiamo are dominated by *Palaeoplesiastraea, Favites, Tarbellastraea* and *Coeloria*, with the commonest genus (*Porites*) adopting a characteristic "organ-pipe" or rod-like growth form where it is developed along steep slope breaks (see Pedley, this volume). This habit is also seen in similar positioned reefs in Malta (Pedley, 1987a) and Lampedusa (Grasso and Pedley, 1985).

Similar coral reef developments in western Sicily are associated with the Baucina Formation (Aruta and Buccheri, 1971) which replaces the upper part of the Terravecchia Formation in that area. Esteban et al. (1982) Catalano (1979) and Catalano et al., (1976) drew attention to rhodalgal strata containing coral barrier reefs and associated patch reefs at Calatafimi, Ciminna, Salemi and Gibellina. These are thought, on the basis of coral diversity, to be, in part, Late Tortonian in age (reef type A, Esteban, 1979) despite containing *Porites* rods in the reef front location. The diverse coral associations, in addition, include *Tarbellastraea, Palaeoplesiastraea, Heliastraea, Coeloria,* and *Siderastraea.* At Ciminna, coral and bryozoan mounds dominate and are succeeded by rhodolitic coralline algal carbonates (Catalano et al., 1976). Four stacked reef episodes are developed in the Grieni sections of west Sicily, each conforming to the model proposed by Esteban (1979). Basically, they show a *Porites*-rod reef wall, a reef flat with domed coral heads at the seaward margin and smaller coral heads and patch reef development in the lagoon. The more typical patch reefs at Salemi and Calatafimi contain only a few *Porites* but additionally contain *Siderastraea* and developed on sand shoals within broad lagoons.

It is difficult to correlate accurately the western Sicilian reefs with the north central Sicilian reefs or with minor "draw-down" events in the other Sicilian areas. Nevertheless, both the Baucina and north central Sicilian Terravecchia reef successions lie between Late Tortonian marine marls and clays with pelagic faunas in the basinward direction and coastal Terravecchia developments immediately to the north. Calcare di Base Formation and evaporites belonging to the first cycle (Early Messinian) are closely associated with the overlying sequences which often lie with sharp basal contacts on the reefs (Ruggieri and Torre, 1984; Catalano et al., 1976; Pedley and Grasso, 1988). In Sicily the field evidence (Rouchy, 1988; Pedley and Grasso, 1993) suggests that the diatomaceous Messinian unit (Tripoli Formation) is, in part, time synchronous with the *Porites* reefs. The youngest Tripoli levels intercalate with the base of the Calcare di Base (often represented by laminar carbonate beds but sometimes dominantly evaporitic in aspect).

The Capo Milazzo area of north east Sicily also provides reefal Late Miocene localities (Fois, 1989, 1990). The successions commence, above basal conglomerates and sands, with encrusting red algae, bryozoans and subordinate *Porites* which all lie directly on crystalline basement. This initial algal-rich unit passes upwards into *Porites*-dominated reef. The clastic-dominated off-reef successions and abundance *Porites* within the reefs is taken by Fois (1990) to indicate a correlation with the Baucina carbonates of western Sicily and hence a Late Tortonian-Early Messinian age. Much of the succession has subsequently been dismantled by later Messinian tectonism and several subsequent episodes of erosion. Other contemporaneous reefs occur to the west along the northern Sicilian coast at Santo Stefano di Camastra (Pedley et al., 1994).

In Calabria, the Upper Tortonian has much in common with north Sicilian reefal strata with terrigenous sands dominating the succession. Tortonian corals were first recorded from the area by Seguenza (1880). Generally, the deposits rest on crystalline (often granitic) basement above which upper Tortonian conglomerates with coral fragments occur (e.g., Amanthea section, Ortolani and Esteban, pers. commun.). In some areas (e.g., Tropea and Santa Domenica) upper Tortonian sections may be represented by under 5 m of terrigenous strata with coral fragment bearing conglomerates at the base (*Porites* and *Tarbellastraea*) and locally with up to 5 m of rhodolitic algal biostrome with bryozoan intervals above this (Pedley and Grasso, 1994a).

Chevalier (1961) listed the principal coral genera as *Tarbellastraea, Porites, Heliastraea, Solenastraea, Favites* and *Goniastraea.* They appear to have been developed in a chain of westerly facing fringing reefs on the Tyrrhenian Sea side of basement ridges (Chevalier, 1961). Within the part dismantled structures (present day elevations of between sea level and 350 m), there is evidence of a coral head zone, in the (?lagoonal) patch reefs in the Vibo Valentia area; a branching coral zone, mainly of *Porites* and *Tarbellastraea*, locally seen around Palmi, Calabria (external reef wall); a coral breccia zone, also seen in the vicinity of Palmi (fore reef); and a rhodalgal and *Heterostegina* pavement external to the reefs which is dominated by siliciclastic sediments (e.g., the Tropea sections). Further details are in Pedley, and Grasso (1994a).

The Langhian to Serravallian coral reef developments of Sardinia continue intermittently into Tortonian age. For example, Pomesano Cherchi (1974) and Tilla Zuccari (1969) record conglomerates with coralline algae, bryozoans and coral debris sandwiched within a shale sequence in the San Michele sections near Cagliari ("Pietra Forte" and "Tramezzario"). Arnaud et al. (1989c) record limited occurrences of corals associated with coralline algae and operculine foraminifers from the Tortonian units of the Aleria district (Plaines Orientale).

In the northern Italian Tuscany region, the Calcare di Rosignano succession contains well developed reefs of Late Tortonian-Early Messinian age in Monti Livornesi (Rosignano). These strata have been extensively described by Bartoletti, et al. (1985); Bossio et al. (1978); Bossio et al. (1981c); Giannini (1962); Chevalier (1961); Esteban et al. (this volume). Reefs associated with the Castelnuovo della Misericordia, Parrana and the Popogna river sections are probably "patch reefs" (Bossio et al., 1978). It is difficult to demonstrate contemporaneity of all these reefs as a major post depositional erosion surface now separates the outcrops (Esteban, et al., this volume). The reefs at Volterra and Casaglia may also belong here.

The reefal levels are terminated by erosion surfaces during which time much of the complex was removed prior to deposition of the "lagoonal terminal complex."

The "lagoonal complex", exposed at Monte Livornese and Casaglia presents a wide range of shallow water carbonate facies. Above the "Argille a *Pycnodonta*" lies the Castelnuovo Formation which also contains important coral developments.

### *Late Miocene Coralline Algal Biostromes*

In Tunisia, a Late Tortonian transgressive sequence (Melquart Formation) occurs in association with the Beglia Formation. It contains carbonate-rich beds, with bryozoans, calcareous algae and corals, interbedded with clayey sandstones and evaporites. In the Libyan north Sirte Basin, continued subsidence permitted deeper water limestone and shale development only.

Further north, on the Pelagian Block, shallow shelf sea conditions became established late in Tortonian time after a widespread deeper water marl episode. Initially, they were dominated by coralline algal biostromes (e.g., Bosence and Pedley, 1982; Pedley, 1981) similar in many ways to the Middle Miocene developments of Campania. These coarse bioclastic wackestones and packstones, containing abundant frameworks of crustose and rhodolitic coralline algae and rare coral heads, cover large areas of Malta (basal Upper Coralline Limestone Formation) and south east Sicily (Siracusa Limestone Member of the Monti Climiti Formation, Grasso et al., 1982).

Coralline algal strata were locally developed in Apulia during Tortonian deposition. In the Salento area in the east where Serravallian-Tortonian carbonates are developed above a post-Oligocene unconformity, no coral development is seen. The Apricena limestones generally unconformably overlie Mesozoic strata to the north in the Gargano peninsula. Often there is a basal "terra rossa" containing terrestrial vertebrates. Tortonian proximal facies contain, at their base, fragments of *Porites* and *Tarbellastraea* associated with solitary corals, oysters and other molluscs, barnacles, and some bryozoa and coralline alga but never show true bioherm development. These grade into distal, fine grained calcarenites containing pelagic microfaunas. Finally, in excess of 3000 m of syntectonic sediments accumulated in the Bradanic Trough.

Collectively these Tortonian coralline algal biostromal associations represent the earliest Late Miocene reestablishment of the reefal environment.

### *Late Miocene Microbial Developments*

The youngest reefal developments in the central Mediterranean are associated with microbial build-ups (stromatolites). Colonies are generally low-relief laminar encrusting mounds (LL-H growth form, Logan et al., 1964). They can be extensive laterally and may locally build up into large domal masses. Although not always developed, they are seen in the capping sequences of west-central Tuscany (Castellina Formation) and western Sicily (e.g., Baucina Formation, Grieni, Esteban et al., 1982). In south east and north central Sicily and Malta, stromatolitic laminites are present as coating around corals but entire stromatolite mounds are rare. However, small (1-2 m diameter) cushion-like serpulid or sebellariid worm and vermetid gastropod patch reefs are developed in the youngest carbonate levels. (Pedley, 1978; Grasso et al., 1982; Esteban et al., 1982). Occasional serpulid-stromatolite colonisation is also found in Tuscany (Bossio et al., 1978).

Stromatolite-bearing sequences are principally developed as a capping sequence to the coral reef building episodes and usually are associated with terminal complexes (Esteban, 1979). Generally in the central Mediterranean, these Terminal Complexes are associated with ooidal and peloidal beds, occasionally with pectinid bivalves intercalated with tabular beds of lime mudstone containing *Limnocardium* and *Abra* species (Monte Carrubba Formation of Sicily, Grasso et al., 1982; Ghar Lapsi limestones of Malta, Pedley, 1987b). Stromatolitic levels also occur in the Calcare di Base Formation of southern Calabria (M. Ortolani, pers. commun.) and central Sicily. These low diversity faunas associated with emergent cycles record hyposaline conditions and have much in common with the "Sarmatian" faunas of Hungary and the Sea of Aral (Von Fuchs, 1874).

In Spain identical lithological associations typify the "Terminal Complex" (Esteban, 1979) which generally is of later Messinian age (i.e., post-evaporitic). In Calabria, Sicily (perhaps with the exception of the Salemi section, Esteban, pers commun.) and the Maltese Islands, these associations can be dated as Early Messinian age (immediately pre-evaporitic Messinian) on stratigraphic and microfaunal grounds, although they have also been referred to as "Terminal Complex" (e.g., Catalano, 1979).

Magnetostratigraphic studies on the volcanics associated with the youngest coral debris (associated with the Monte Carrubba Formation) in south east Sicily confirm an Early Messinian age for the strata (Grasso et al., 1983). Field studies also confirm their lateral contemporaneity with the Tripoli Formation on the western Hyblean margins of south east Sicily. In Hyblean areas, sea-level never returned to the subaerially exposed reefal tracts until Pliocene times.

## DISCUSSION

### *Tectonism*

The central Mediterranean region lies at the meeting point of eastern and western Mediterranean basins. As such, it provides a valuable control point at which to correlate basin wide events. The Appennine Mountains were already well established by the Early Miocene and in many areas show an orogenic history extending back into the Eocene (see Boccaletti and Manetti, 1978). Many areas suffered an accelerating deformational regime from Burdigalian times (e.g., central and southern Appennines, D'Argenio et al., 1973), with Late Miocene to Pliocene deformation climax events in most regions of Italy and along the North-South Axis of Tunisia, (Burollet et al., 1978). The Appennine-related movements resulted in the production of vast gravity slides and the production of extensive siliciclastic sediment prisms in areas flanking the emergent orogenic belts. Extremely inhospitable environments for reef colonization resulted; some dominated by exceptionally high sedimentation rates, except for brief highstand events which permitted coral colonisation (e.g., Late Tortonian-Early Mesinian interval, see Fig. 4).

The stable African foreland areas (e.g., SE Sicily and Malta) were far removed from this orogeny but were subjected to Miocene rifting processes in the Sicily Channel (Finetti, 1985; Pedley, 1987a). Seafloor topographies were controlled by an interaction between this newly initiated rifting and reactivation of Mesozoic tensional structures (Pedley, 1990).

### *Water Circulation*

Throughout the Oligocene to Aquitanian record Paleomediterranean water flow was towards the west (Pedley, 1987a). With the closure of the Indian Ocean connection after Aquitanian time this flow suffered a reversal (perhaps by the end of the proposed TB1.5 cycle of Haq et al., 1987). Furthermore, the developing Appennine foldbelts presented a new physical barrier to water circulation patterns. This profound change in circulation must have caused fundamental changes in nutrient supply, velocity and direction of local water flow. Many former areas of coral colonisation certainly become untenable under the new regime. It is probably no coincidence that middle Miocene time was a time of Mediterranean coral reef retrenchment within the central Mediterranean, with marked reductions in generic diversity. Nevertheless, major middle Miocene reefs do occur in other Mediterranean areas.

The environmental dominance by coralline algae during Middle Miocene deposition may reflect the availability of open ramp and shelf sites temporarily freed from the restrictive effects on water circulation which are normally provided by coral reef development (see other theories in Martin and Braga, 1994). Certainly, conditions appear still to have been subtropical, though perhaps a little cooler.

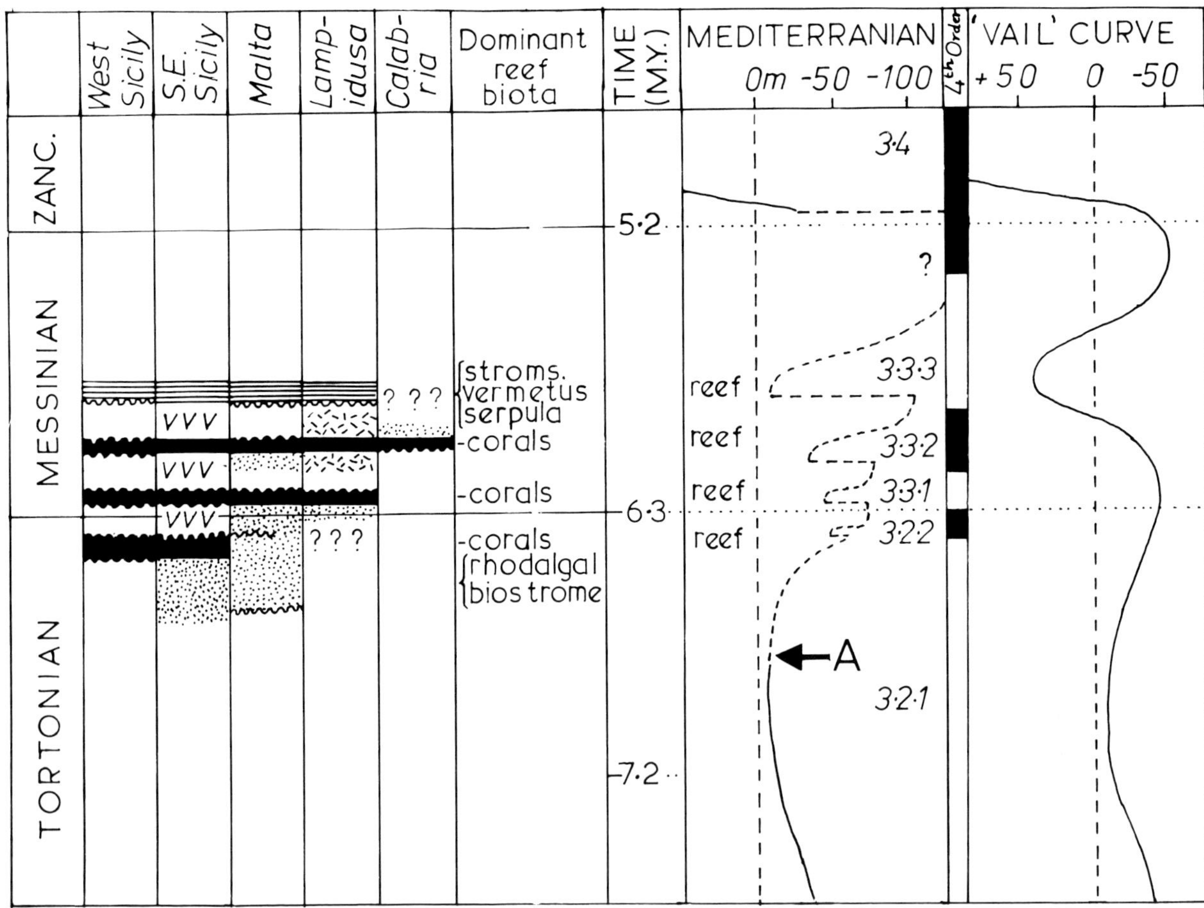


FIG. 4.—Detailed eustatic curve for Late Tortonian and Messinian deposition illustrating the position and number of reefs within southern Italy and the Maltese Islands. The fourth-order curve for the region is produced from field data on reef and erosion surface occurrences (cf. western Mediterranean curve of Esteban, this volume). A indicates the position in time when isolation from Atlantic influences started. The dashed line traces the local intra-basinal Central Mediterranean curve.

Only across a relatively narrow sill, covered by shallow shelf seas, extending from southern Calabria to the Cap Bon Peninsula was there any water exchange between the western and eastern Mediterranean basins. From Burdigalian time onwards, Atlantic waters were drawn over this narrow threshold to compensate for water loss by evaporation in the eastern basin (anti-estuarine circulation pattern). Inevitably this led to Late Miocene coral reef development becoming focussed on the southern part of the central Mediterranean region and in Sardinia, though somewhat surprisingly also in the Livorno region of northern Italy.

The virtual absence of observable coral reefs of Tortonian age in central Italy is more likely to reflect adverse water circulation patterns resulting from the new circulatory regime. This is best seen in the area between the Italian mainland and the island of Corsica (see Fig. 2). All coral reefs on the eastern margin of the western Mediterranean Basin face westward or southwesterly (Fig. 2). Only in the Rosignano area, Tuscany; Calabria; the Campidano Graben and Sardinia do coral reefs develop, albeit principally during Late Tortonian-Messinian time. Significantly, these are the only areas north of Sicily likely to be exposed to the new anti-estuarine flow regime. A possible exception, seen only on seismic in offshore eastern Corsica (Thomas, 1987), may have been encouraged by water flowing eastward through the Bonifacio Straits between Corsica and Sardinia. The Adriatic margins of the Appennines and Apulia must have faced a blind embayment with stagnated circulation by late middle Miocene time and consequently little or no coral reef development was possible from this date.

The presence of typical Sarmatian faunas associated with the Early Messinian "terminal complexes" in eastern Sicily and Malta suggests a direct link eastwards to Paratethys during post-reefal Early Messinian deposition. It is difficult to confirm this without corroborative data: however, the fauna is an oligohaline to hypersaline one and is very similar, even at the species level, to that of southeastern Europe. If this interpretation is correct then it lends further support to the concept of a progressively closing Adriatic arm to the Mediterranean at this time.

### *Eustatic Curves*

Late Tortonian-Early Messinian reef episodes are cyclic and well documented for the southern part of the central Mediterranean. The proposed new fourth-order eustatic curves for the western Mediterranean (see Esteban, this volume) fit well the observed data of the central region and fully vindicate the speculations of Pedley (1983). Figure 4 illustrates the interrelationships between the local fourth-order cyclicity and the reef development and is detailed as follows: continued world regression from Late Burdigalian until Tortonian time not only brought drowned platforms back within reach of reefal communities generally but served to isolate further the Mediterranean basins from the world oceans. Firstly, during Tortonian time the deeper water coralline algal communities regained a foothold in the shallowing water depths of (c. 65 m in Malta, Pedley, 1976). Shortly afterwards, associated with the first minor draw down (3.2.2, see Esteban, this volume) in later Tortonian, times those coral survivors from the reversed circulation event recolonized suitable current-facing sites. Eustatic events from Late Tortonian times (later part of cycle 3.2 of Haq et al., 1987) were probably controlled by major evaporation in the eastern Mediterranean Basin (for eastern basin details see Buchbinder et al., 1989) at a time when the Atlantic connection had effectively been severed.

By correlating the sequence stratigraphic events of the region, it is possible to recognise three pulses of coastal onlap built onto an overall regressive regime prior to the final Messinian "draw down" associated with deposition of the first evaporite cycle (Early Messinian). Each fourth-order cycle (3.3.1; 3.3.2; 3.3.3 in Fig. 4) represents a basinwide eustatic rise possibly bringing the Mediterranean water level back closely in line with world sea level. The timing of these cycles fit well the local geomagnetic data (Grasso et al., 1983), biostratigraphic data (Grasso et al., 1979) and positioning of the 3.2 transgressive peak of the Vail curve (Haq et al., 1987) but does not follow precisely the conclusions of Saint-Martin and Rouchy (1990) regarding the 3.2 peak position. Estimations, based on sediment thickness and palaeoenvironments in the most tectonically stable areas of south east Sicily and Malta suggest minimum draw downs of between 20 to 60 m. On each of the temporary draw downs, faunal diversity was generally maintained, thus excluding the possibility of a marked salinity change. Water replenishment may have been from the Atlantic, possibly resulting from temporary breaching of the Betic and Rif Straits by the rising world ocean sea level during Early Messinian time (see Vail curve in Haq et al., 1987).

Each transgressive pulse produced a new reef development which lies upon or is overlain by an erosional surface (cf. with the stepped later Messinian reefs of south west Spain, Esteban, 1979). The lowest level is generally associated with coralline algal biostrome development; the highest usually is stromatolitic or contains vermetid-surpulid reefs rather than corals (e.g., Malta and south east Sicily) as it is anticipating the major draw down event. This leaves the two central events as the most likely to produce coral reef development in any Late Tortonian-Early Messinian tectonic regime.

### *Biota Diversity*

Much has been made of the low coral diversity of the Late Miocene reefs (e.g., Esteban, 1979; Esteban et al., 1978). Valuable data on biotal diversity has also been presented for many Mediterranean areas permitting both reefs and lagoons to be characterised (e.g., Esteban, 1979; Pedley, 1979; Bossio et al., 1978; Dabrio et al., 1981; Santisteban, 1981; Saint-Martin et al., 1983), and it is now clear that the biota overall is a fully marine one throughout all the coral and coralline algal biostrome episodes. In the Sicilian and Maltese reefs there is a clear distinction between lagoonal and mid-ramp patch reefs and shelf margin "reef-wall" or buttress reef despite their contemporaneous occurrence. Patch reefs are always micrite dominated; this sediment frequently being the home of a diverse infaunal bivalve community. The four or five coral genera present are linked into a framework in some patches by the abundant crustose coralline algae. This linked framework is further reinforced by an abundant encrusting community of serpulid worms, homotrematid foraminifera and bryozoa with a substantial contribution from the microbial community in the form of structureless micritic coatings (compact crust, Pedley, 1983; coatings of Riding et al., 1991). Extensive, pervasive borings are usually present, generally being created by clionid sponges, endoproct bryozoa, filamentous cyanobacteria and bivalves. Off-reef sediments are bioclastic, sometimes oolitic, but rarely contain significant quantities of *Halimeda.*

Shelf edge reef walls contain similar limited coral assemblages but with a dominance of the rod-like poritid growth form. There is, in fact, a tendency for the limited species to adopt multiple growth form strategies in order to maximise reef colonization. This is a clear indication of niche availability. Micrite (often peloidal) fills inter-coral areas but an infaunal bivalve community is lacking. Microbial micrite encrustation (Pedley, 1979, 1983) is much more extensive than in the seaward, lagoonal margin patch reefs, but the boring community is at times poorly developed or even lacking in the central Sicilian and Maltese examples. The reduced frequency of borings has also been observed in the deeper parts of the Spanish and Tuscan reefs (Esteban, pers. commun., Pomar et al., 1983). Coralline algae and invertebrate encrusters are present in subordinate amounts in the shelf edge reef walls but are never so extensive as in the lagoonal bioconstructions. *Halimeda* plates abound in the off-reef sediments. The reef wall foreslope breccias pass basinwards into poorly exposed fine siliciclastics and marls containing *Globorotalia mediterranea.*

According to Rouchy (1988), the Late Tortonian coralline algal biostromes pass laterally into diatomaceous laminites. This is also seen in Malta where fish-bearing laminites lie downslope from the "Coralline Algal Biostrome" (Pedley,

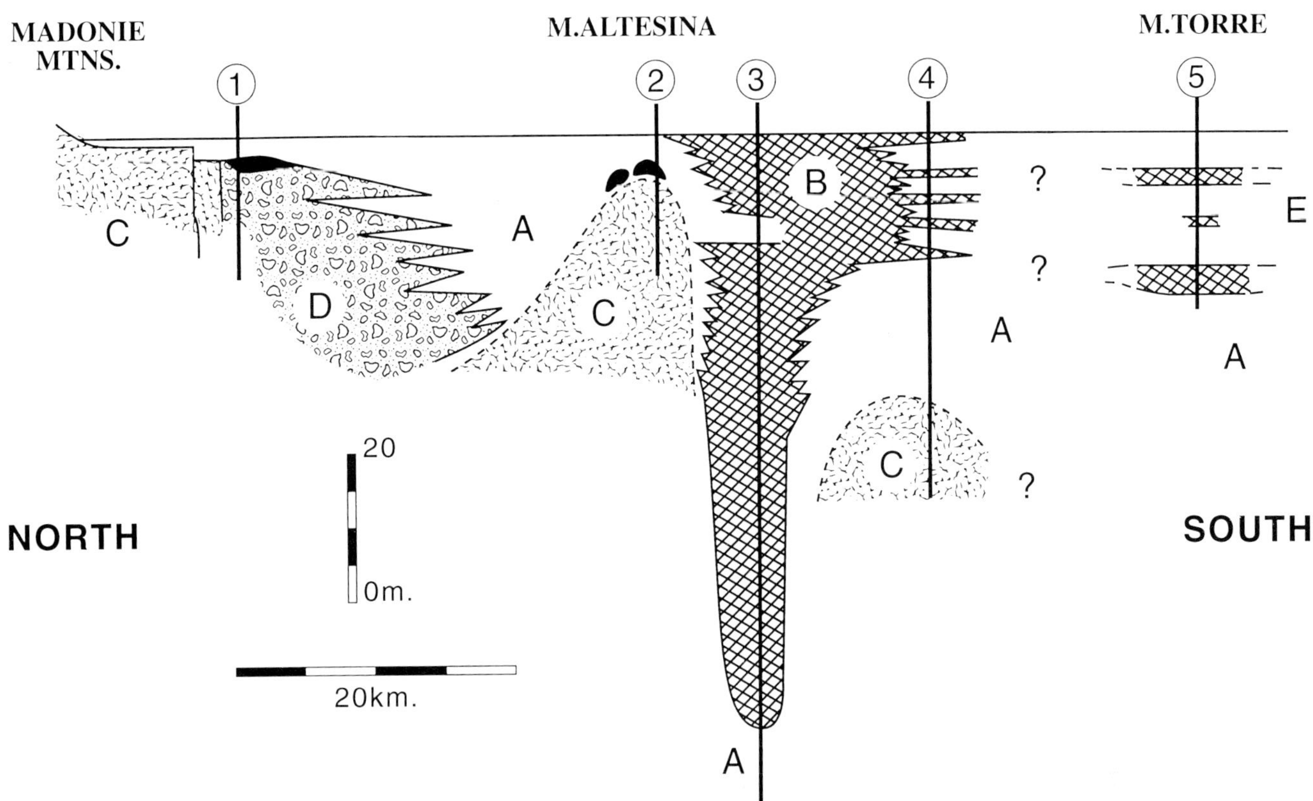


FIG. 5.—North-south Profile through central Sicily showing the interrelationships between the fully marine northern basin, the Altesina Ridge and the localised Tripoli basin to the south. Symbols: A, normal marine Late Tortonian-Early Messinian marls and clays; B, diatomite laminites; C, tectonised basement; D, siliciclastic Terrevecchia Formation; and E, alternations of lime mudstones marls and "tripoli" beds. Note how the *Porites* reefs (solid black) face the normal marine basin but could be adversly affected by elevated salinity spill-over from the "tripoli" basin. Localities: 1, Petralia towns; 2, Cacchiamo, near Villadoro; 3, Contrada Gaspa, near Villapriolo; 4, Monte Pasquasia; 5, Bessima (Monte Torre section) east of Barrafranca.

1978b). Similar fish-bearing laminites are also seen in the western margins of the Hyblean region of south east Sicily intercalated between Late Tortonian Tellaro Formation and Early Messinian Monte Carrubba Formation (Pedley and Grasso, unpubl. data).

Field evidence in central northern Sicily, however, shows a close lateral juxtapositioning of Early Messinian reefs with tripolaceous diatomite laminites (Fig. 5) containing fish and associated dinoflagellate-bearing torba levels (Tripoli Formation, Pedley and Grasso, 1993; Suc et al., 1995). Unfortunately contacts have not been observed directly (see Fig. 5). The Contrada Gaspa Tripoli section (Figs. 3, 5) in central Sicily shows six cycles in a tight syntectonic east-west orientated basin, with a seventh involving the Calcare di Base (Grasso et al., 1990). Planktonic foraminifera, belonging to the *Globorotalia conomiozea* Zone (Messinian), indicate a progressive deterioration in conditions with progressively elevating nutrient levels early within each cycle. This developed into cyclical foraminiferal impoverishment and finally to extinction associated with hypersalinity in the topmost cycle. Elsewhere in southern Italy, the same details pertain, with eutrophic *G. conomiozea* attaining high frequencies in the early levels whereas eurytopic *G. bulloides* and *O. universa* dominate later cycles. *T. quinqueloba* and *T. multiloba* dominate the nutrient depleted and elevated salinity levels (Zachariasse and Spaak, 1983). A five-fold cyclicity also occurs at the Bessima section (Fig. 5, column 5).

It is not clear yet how these cycles relate to the reef episodes, however, the central Sicilian Tripoli Formation is clearly too young to be a lateral development of the same strata as contain the Tortonian rhodalgal biostromes (cf. Rouchy, 1988; Pedley, 1978b). The cyclicity and high organic content, sometimes with distinct thin sapropel beds, in each Sicilian diatomite cycle suggests the development of stratified water bodies in multiple local syntectonic basins not directly linked to reef development. Significantly, in north-central Sicily, the diatomaceous laminites are virtually undeveloped in the normally marine basin, and coral reefs alone occur north of the Altesina Ridge (see Fig. 5 in Grasso and Pedley 1988). The evidence to date suggests that the various Tripoli basins of Sicily developed progressively later in a southern direction (for mechanism see Butler et al., 1995)

All the indications in the well-exposed Central Sicilian sections are of a complexly linked association of intermittently stagnating Early Messinian diatomite basins. These cyclically restricted basins were later to be the focus for halite and gypsum precipitation (Decima et al., 1989). They are associated with Calcare di Base developments, characteristically autobrecciated and in parts intercalated with the upper parts of the Tripoli Formation laminites.

Clearly the situation in Sicily (and possibly many other Italian areas) is one of laminite development commencing locally in Late Tortonian time in association with coralline algal biostromes but with the main diatomaceous laminite phase being of Early Messinian age. The localised syntectonic deformation of the substrate into several basins precludes the possibility of upwelling. Normal salinity but nutrient-enriched waters could, however, spill over from these, stagnant diatomite basins into adjacent reef areas. The resulting elevated nutrient levels around the reefs could then effectively encourage bacterial biofilm colonisation (now stromatolite coatings) which would readily choke out or poison all but the most tolerant elements of the coral community. Additionally, the general reduction in water turbulence consequent to closing off of the Mediterranean basin from the world oceans must have significantly affected coral feeding habits and may further have encouraged intraspecific selection towards those growth forms able to reach the more turbulent, and fresher, surface water zone (see discussion in Pedley and Grasso, 1994b). Conversely, spillover of normal marine waters was responsible for the cyclic replenishment of planktonic foraminifers in the diatomite basins.

Rapid-growth strategies such as found in the rod-like poritid growth form could aid in avoiding burial by microbial coatings. Variable growth form strategies within a species would also maximize survival chances by increasing the span of the ecological niche. Similarly, the high rates of standing crop turnover in the codiacean algae would also be likely of success in an opportunistic setting. Vagrant benthos (e.g., echinoids), less likely to be affected by microbial blooms, are present throughout the coral reefs and testify to continuing normal salinity conditions.

Against this hypothesis are the recorded occurrences of coral reefs and diatomites within the same basins (e.g., Tuscany, Morocco and Algeria (Saint-Martin and Rouchy, 1990) and south east Spain. Much more data are required (see Pedley and Grasso, 1993 and discussion in Esteban, this volume).

## DIAGENESIS

Diagenetic trends are complex. Many reefs contain well developed early submarine isopachous fringe cements (Pedley, 1983). Aragonite dissolution is the most common early diagenetic fabric modifier and occurred shortly after submarine cementation. Generally, the resulting coral and molluscan cavities remain unfilled imparting a high moldic porosity to the reef strata. Aragonite survives generally only where bioclasts were entombed in clays. A notable exception is the thick, fibrous botryoidal aragonite cement present in the Mazzara reef, west Sicily (Esteban and Prezbindowski, 1985). This developed after the reef episodes and is either Late Messinian or Pliocene in age. Many of the youngest reefs are undolomitized. This possibly results from their relatively high elevations which precluded them from evaporite burial during the latest Miocene desiccation (cf. the Rosignano reefs and the Spanish reefs which, in both cases, are sandwiched between evaporites). Where Messinian evaporites were juxtaposed to reefs or closely overlay them (e.g., Ghar Lapsi, Malta) fabric retentive dolomite replacement results, unless primary lithologies were aragonitic, in which case frequently they were removed prior to the dolomitization.

## ACKNOWLEDGMENTS

This review benefits from lengthy discussions with Mateu Esteban to whom I am most grateful. The background research in Italy was supported by the Royal Society of London and National Environment Research Council Grant GR3/8610.

## REFERENCES

ABATE, B., CATALANO, R., D'ARGENIO, B., DISTEFANO, E., DISTEFANO, P., CICERO, G., MONTANARI, L., PECORARO, C., AND RENDA, P., 1982, Evoluzione delle zone di ceriera tra piattaforme carboniche e bacin nel Mesozoico e nel Paleogene della Sicilia occidentale, *in* Catalano, R. and D'Argenio, B., eds., Guida alla Geologia Della Sicilia Occidentale: Palermo, Societa Geologica Italiana, p. 53-81.

ALVAREZ, W., 1972, Rotation of the Corsica-Sardinia microplate: Nature Physical Sciences, v. 235, p. 553-559.

ARNAUD, M., MONLEAU, C., AND NEGRETTI, B., 1989a, II-1 La regione de Bonifacio, *in* Le Neogene Corse, Groupe Francais d'etude du Neogene: Marseilles, Universite de Provence, Excursion, p. 8-19.

ARNAUD, M., MONLEAU, C., AND NEGRETTI, B., 1989b, II-3 La region deSaint Florent: Quatrieme journee, *in* Le Neogene Corse, Groupe Francais d'etude du Neogene: Marseilles, Universite de Provence, Excursion, p. 23-26.

ARNAUD, M., MONLEAU, C., NEGRETTI, B., LOYE-PILOT, M. D., AND MAGNE, J., 1989c, II-2 La regione D'Aleria: Troisieme Journee, *in* Le Neogene Corse, Groupe Francais d'etude du Neogene: Marseilles, Universite de Provence, Excursion, p. 20-22.

ARUTA, L. AND BUCCHIERI, G., 1971, Il Miocene preevaporitico in facies carbonatico-detritico dei dintorni di Baucina, Ciminna, Ventimiglia di Sicilia, Calatafimi (Sicilia): Rivista Mineraria Siciliana, v. 130-132, p. 188-194.

BARBERA, C., CARANNANTE, G. D'ARGENIO, B., AND SIMONE, L., 1980, Il Miocene calcareo dell'Appennino meridionale: contributo della paleoecologia alla costruzione di un modello ambientale: Annali Universita Ferrara (new series) Section 9, v. 6, suppliment, p. 281-299.

BARBERA, C., SIMONE, L., AND CARANNANTE, G., 1978, Depositi circalittorale di piattaforma aperta nel Miocene Campano analisi sedimentologica paleoecologica: Bollettino della Societa Geolgica Italiana, v. 97, p. 821 834.

BARTOLETTI, E., BOSSIO, A., ESTEBAN, M., MAZZANTI, R., MAZZEI, R.,SALVATORINI, G., SANESI, G., AND SQUARCI, P., 1985, Studio geologico del territorio comunale di Rosignano Marittimo in relazione alla carte geologica alla scala 1:25000: Quaderni del Museo di Storia Naturale di Livorno, Supplimento 1, v. 6, p. 33-170.

BOCCALETTI, M. AND MANSETTI, P., 1978, The Tyrrhenian Sea and adjoining areas, *in* Nairn, A. E. M., Kanes, W. H., and Stehli, F.G., eds., The Ocean Basins and Margins, The Western Mediterranean: New York, Plenum Press, p.149-200.

BOSENCE, D. AND PEDLEY, H. M., 1982, Sedimentology and palaeoecology of a Miocene coralline algal biostrome from the Maltese Islands: Palaeogeography, Palaeoclimatology, Palaeoecology, v. 38, p. 9-43.

BOSSIO, A., BRADLEY, F., ESTEBAN, M., GIANELLI, L., LANDINI, W.,MAZZANTI, R., MAZZEI, R., AND SALVATORINI, G., 1981, Alcune aspetti del Miocene superiore del Bacino del Fine: Pisa, IX Convegno della Societa Paleontologica Italiana, p. 21-53.

BOSSIO, A., ESTEBAN, M., GIANELLI, L., LONGINELLI, A., MAZZANTI,

R.,Mazzie, R., Ricci Lucchi, F., and Salvatorini, G., 1978, Some aspects of the upper Miocene of Tuscany: Pisa, International Geological Correlation Program Project 96, Messinian Correlation, Seminar 4, 88 p.

Bossio, A., Gianelli, L., Mazzanti, R., and Salvatorini, G., 1981a, Il passaggio dalla facies lacustre alla evaporitica e le <<Argille a Pycnodonta>> presso Radicondoli (Siena): Pisa, IX Convegno della Societa Paleontologica Italiana, p. 161-194.

Bossio, A., Gianelli, L., Mazzanti, R., and Salvatorini, G., 1981b, Gli strati alti del Messiniano, il passaggio Miocene-Pliocene e la sezione plio pleistocenica di Nugola nelle colline a NE del Monti Livornesi: Pisa, IX Convegno della Societa Paleontologica Italiana, p. 55-90.

Buchbinder, B., Martinotti, G. M. Siman-Tov, R., and Zilberman, E., 1989, Temporal and spatial relationships in Miocene reef carbonates in Israel: Palaeogrography, Palaeoclimatology, Palaeoecology, v. 101, p. 97-116.

Burollet, P. F., 1969, Petroleum geology of the western Mediterranean Basin: Brighton, Joint Conference of the Institute of Petroleum and the Ammerican Association of Petroleum Geologists, p. 19-30.

Burollet, P. F., Mugniot, J. M., and Sweeney, P., 1978, The geology of the Pelagian Block: the margins and basins off southern Tunisia and Tripolitania, *in* Nairn, A. E. M., Kanes, W. H., and Stehli, F. G., eds., The Ocean Basins and Margins, The Western Mediterranean: New York, Plenum Press, p. 331-359.

Butler, R. H. W., Lichorish, W. H., Grasso, M., Pedley, H. M., and Ramberti, L., 1995, Tectonics and sequence stratigraphy in Messinian basins, Sicily: constraints on the initiation and termination of the Mediterranean "salinity crisis": Bulletin of the Geological Society of America, v. 107, p. 425-439.

Cafici, I., 1880, La formazione gessosa del Vizzinese e del Licodiano (provincia di Catania): Bollettino del Royale Comitato Geologica d'Italia, v. 11, p. 37-54.

Carannante, G., Simone, L., and Barbera, C., 1981, Calcare a briozoie litotamni of southern Appennines. Miocenic anologous of Recent Mediterranean rhodolitic sediments (abs.): Bologna, International Association of Sedimentologists 2nd European Meeting, p. 17-20.

Carannante, G., Esteban, M., Milliman, J. D., and Simone, L., 1988, Carbonate lithofacies as paleolatitude indicators: problems and limitations: Sedimentary Geology, v. 60, p. 333-346.

Carbone, S., Pedley, H. M., Grasso, M., and Lentini, F., 1993, Origin of the <<Calcareniti di Floresta>> of NE Sicily: late orogenic sedimentation associated with a middle Miocene sea-level high stand: Giornale di Geologia, v. 55, p. 105-116.

Catalano, R., 1979, Scogliere ed evaporiti messiniane in Sicilia. Modelli genetici ed implicazioni strutturali: Lavoro Istituto Geologia della Universita di Palermo, v. 18, 21 p.

Catalano, R., Channell, E. T., D'Argenio, B., and Napoleone, G., 1976, Mesozoic paleogeography of the southern Appennines and Sicily: Memorie Societa Geologica Italiana, v. 15, p. 95-118.

Chevalier, J-P., 1961, Recherches surles Madreporaires et les formations recifales Miocenes de la Mediterranee occidentale: Memoire Societe Geologique de France, Paris, n. s. 40, v. 93, 562 p.

Chevalier, J-P., 1977, Apercu sur la faune corallienne recifale du Neogene: Second Fossil Coral Reefs Symposium, Paris: Transactions Memoires Bureau Recherches Geologiques et Mineres, v. 89, p. 359-366.

Cocozza, T. and Schafer, K., 1974, Cenozoic graben tectonics in Sardinia: Rendiconti Seminario della Facolta di Scienze Universita di Cagliari, (Atti Convegno su "Paleogeografico dell Terziario Sardo nell' ambito del Mediterraneo occidentale", Cagliari, 1973), p. 145-162.

Courme-Rault, M. D. 1991, Precisions sur le Neogene de l'avant-pays ibleen (bordure nord-Sicile S.E.): implications paleogeographiques: Geologie Mediterranene, v. 18, p. 171-187.

Dabrio, C. J., 1975, La sedimentacion arrecifal Neogena en la region del vio Almanzora Estudios: Geologicos, v. 41, p. 285-296.

Dabrio, C. J., Esteban, M., and Martin, J. M., 1981, The coral reef of Nijar, Messinian (Uppermost Miocene), Almera Province, S. E. Spain: Journal of Sedimentary Petrology, v. 51, p. 521-539.

D'Argenio, B., Pescatore, T., and Scandone, P., 1973, Schemageologico dell'Appennino meridionale (Campania Lucania): Rome, Atti dell Conv. sul tema "Moderne vedute sulla geologia dell'Appennino," Quad no. 183, Accademia Nazionale dei Lincei, Atti class di Scienze, Fisiche, Matematiche Naturali, p. 49-81.

Decima, A., McKenzie, J. A., and Schreiber, B. C., 1989, The origin of "evaporative" limestones: an example from the Messinian of Sicily (Italy): Journal of Sedimentary Petrology, v. 58, p. 256-272.

Distefano, E. and Catalano, R., 1976, Biostratigraphy palaeoecology and tectonosedimentary evolution of the preevaporitic and evaporitic deposits of the Ciminna Basin (Sicily), *in* Catalano, R., Ruggieri,G., and Sprovieri, R., eds., Messinian Evaporites in the Mediterranean: Erice Seminar, Memorie Societa Geologica Italiana, v. 16, p. 95-110.

Esteban, M., 1979/80, Significance of the upper Miocene coral reefs of the Western Mediterranean, *in* Cita, M. B. and Wright, R., eds., Geodynamic and Biodynamic Effects of the Messinian Salinity Crisis in the Mediterranean, 4th Messinian Seminar: Palaeogeography, Palaeoclimatology, Palaeoecology, v. 29, p.169-188.

Esteban, M., Calvet, F., Dabrio, C., Baron, A., Giner, J., Pomar, L., Salas, R., and Permanyer, A., 1978, Aberrent features of the Messinian coral reefs, Spain: Acta Geologica Hispanica, tome 13, no. 1, p. 20-22.

Esteban, M., Catalano, R., and Distefano, E., 1982, Scogliere Messiniane a Porites nella Sicilia sud-occidentale: Rendiconti di Societa Geologica Italiana, v. 5, p. 61-64.

Esteban, M. and Prezbindowski, D. R., 1985, Preserved aragonite in Miocene coral reefs: a record of the Messinian salinity crisis in the Mediterranean (abs.): American Association of Petroleum Geologists Bulletin, v. 62, p. 253-254.

Finetti, I., 1984, Geophysical study of the Sicily Channel Zone: Bollettino di Geofisica Teorica ed Applicata, v. 26, p. 3-28.

Finetti, I., 1985, Structure and geological evolution of the Central Mediterranean, *in* Stanley, D. J. and Wezel, F. C., eds., Geological Evolution of the Mediterranean Basins: New York, Springer-Verlag, p. 215-230.

Fois, E., 1989, La successione neogenica di Capo Milazzo (Sicilia NE): Rivista Italiana di Paleontolologia e Stratigrafia, v. 95, p. 397-440.

Fois, E., 1990, Stratigraphy and palaeoecology of the Capo Milazzoarea (NE Sicily, Italy): clues to the evolution of the southern margin of the Tyrrhenian Basin during the Neogene: Palaeogeography, Palaeoclimatology, Palaeoecology, v. 78, p. 87-108.

Fournie, D., 1978, Nomenclature lithostratigraphique des series du Cretace superieur au Tertiare de Tunisie: Bulletin des Centre Recherches Exploration Producton Elf-Aquitain, v. 2, p. 97-148.

Fravega, P., Grasso, M., and Pedley, H. M., 1993, Sedimentology, palaeoenvironment, age and tectonic setting of the Sperlinga bioclastic carbonate deposits, Central-North Sicily: Bollettino della Societa Geologica Italiana, v. 112, p. 191-200.

Frost, S., 1977, Ecologic controls of Carribean and Mediterranean Oligocene reef coral communities: Proceedings Third International Coral Reef Symposium, Rosenstiel School of Marine and Atmospheric Science, University of Miami, p. 367-373.

Frost, S. H., 1981, Oligocene reef coral biofacies of the Vicentin, northeast Italy, *in* Toomey, D. F., ed., European Fossil Reef Models: Tulsa, Society of Economic Paleontologists and Mineralogists Special Publication 30, p. 483-539.

Geister, J. and Ungaro, S., 1977, The Oligocene coral formations of the Colli Berici (Vicenza, northern Italy): Eclogae Geologicae Helvetiae, v. 70, p. 811-823.

Giannini, E., 1962, Geologia del bacino della Fine (province di Pisae di Livorno): Bollettino della Societa Geoleologica Italiana, v. 81, p. 101-224.

Gisetti, F., Scarpa, R., and Vezzani, L., 1982, Seismic activity, deep structures and deformation processes in the Calabrian Arc, southern Italy: Earth Evolution Sciences, v. 3, p. 248-260.

Grasso, M., Lentini, F., Lombardo, G., and Scamarda, G., 1979, Distribuzione delle facies Cretaceo-Mioceniche lungo L'alliniamento Augusta-M. Lauro (Sicilia sud-orientale): Bollettino della Societa Geologica Italiana, v. 98, p.175-188.

Grasso, M., Lentini, F., Nairn, A. E. M., and Vigliotti, L., 1983, A geological and paleomagnetic study of the Hyblean volcanic rocks of Sicily: Tectonophysics, v. 98, p. 271-295.

Grasso, M., Lentini, F., and Pedley, H. M., 1982, LateTortonian-Lower Messinian (Miocene) palaeogeography of SE Sicily: information from two new formations of the Sortino Group: Sedimentary Geology, v. 32, p. 279 300.

Grasso, M. and Pedley, H. M., 1985, The Pelagian Islands: a new geological interpretation from sedimentological and tectonic studies and its bearing on the evolution of the central Mediterranean Sea (Pelagian Block): Geologica Romana, v. 24, p. 13-34.

GRASSO, M. AND PEDLEY, H. M., 1988, The sedimentology and development of Terravecchia Formation carbonates (Upper Miocene) of north Sicily: possible eustatic influence on facies development: Sedimentary Geology, v. 57, p. 131-149.

GRASSO, M., PEDLEY, H. M., AND ROMEO, M., 1990, The Messinian Tripoli Formation of Central Sicily: palaeoenvironmental interpretations based on sedimentological, micropalaeontological and regional tectonic studies: Paleobiologie Continentale, Montpelier, v. 17, p.189-204.

HAYWARD, A. B., 1982, Coral reefs in a clastic sedimentary environment: Fossil (Miocene, S. W. Turkey) and modern (Recent, Red Sea) analogues: Coral Reefs, v. 1, p. 109-114.

HAQ, B. U., 1991, Sequence stratigraphy, sea-level change, andsignificance for the deep sea, *in* Macdonald, D. I. M., ed., Sedimentation, Tectonics and Eustasy, Sea level Change at Active Plate Margins: London, International Association ofSedimentologists Special Publication 12, p. 3-40.

HAQ, B. U., HARDENBOL, J., AND VAIL, P., 1987, Chronology of fluctuating sea levels since the Triassic: Science, v. 235, p. 1156-1167.

ILLIES, J. H., 1980, Form and function of graben structures: The Maltese Islands, *in* Cloos, H., Von Gehlen, K., Illies, J. H., Kuntz, E., Neumann, J., and Seibold, E., eds., Mobile Earth: Berlin, Harald Boldt Verlag Boppard, p. 161-184.

ILLIES, J. H., 1981, Graben formation- the Maltese Islands-a case study, *in* Illies, J. H., ed., Mechanisms of Graben Formation: Tectonophysics, v. 73, p. 151-168.

LOGAN, B. W., REZAK, R., AND GINSBURG, R. N., 1964, Classification and environmental significance of algal stromatolites: Journal of Geology, v. 72, p. 62-83.

MARTIN, J. M. AND BRAGA, J. C., 1990, Arrecifes Messinienses de Almeria Tipologias de crecimiento, posicion estratigrafica y relacioncon las evaporitas: Geogaceta, v. 7, p. 66-68.

MARTIN, J. M. AND BRAGA, J. C., 1994, Messinian events in the Sorbas Basin in southeastern Spain and their implications in the Recent history of the Mediterranean: Sedimentary Geology, v. 90, p. 257-268.

ORSZAG-SPERBER, F. AND PILOT, M-D., 1976, Grands traits du Neogene de Corse: Bulletin Societe Geologique de France 7, v. 18, p. 1183-1187.

PEDLEY, H. M., 1975, The Oligo-Miocene sediments of the Maltese Islands: Unpublished Ph.D. Dissertation, University of Hull, Hull, 205 p.

PEDLEY, H. M., 1976, A palaeoecological study of the Upper Coralline Limestone Terebratula-Aphelesia Bed (Miocene, Malta) based on bryozoan growth forms and brachiopod distributions: Palaeogeography, Palaeoclimatology, Palaeoecology, v. 20, p. 209-234.

PEDLEY, H. M., 1978a, A new lithostratigraphical and palaeoenvironmental interpretation for the coralline limestone formations (Miocene) of the Maltese Islands: Institute of Geological Sciences, Overseas Geology and Mineral Resources, v. 54, 17 p.

PEDLEY, H. M., 1978b, A new fish horizon from the Maltese Islands and its palaeoecological significance: Palaeogeography, Palaeoclimatology, Palaeoecology, v. 24, p. 73-83.

PEDLEY, H. M., 1979, Miocene bioherms and associated structures in the Upper Coralline Limestone of the Maltese Islands; their lithification and palaeoenvironment: Sedimentology, v. 26, p. 577-591.

PEDLEY, H. M., 1981, Sedimentology and palaeoenvironment of the southeast sicilian Tertiary platform carbonates: Sedimentary Geology, v. 28, p. 273 291.

PEDLEY, H. M., 1983, The petrology and palaeoenvironment of the Sortino Group (Miocene) of SE Sicily: evidence for periodic emergence: Journal Geological Society of London, v. 140, p. 335-350.

PEDLEY, H. M., 1987a, Controls on Cenozoic carbonate deposition in the Maltese Islands: review and reinterpretation: Memorie della Societa Geologica Italiana, v. 38, p. 81-94.

PEDLEY, H. M., 1987b, The Ghar Lapsi limestones: sedimentology of a Miocene intra-shelf graben: Centro, v. 1, p. 1-14.

PEDLEY, H. M., 1990, Syndepositional tectonics affecting Cenozoic and Mesozoic deposition in the Malta and SE Sicily areas (central Mediterranean) and their bearing on Mesozoic reservoir development in the N Malta offshore region: Marine and Petroleum Geology, v. 7, p. 171-180.

PEDLEY, H. AND GRASSO, M., 1988, The sedimentology and development of the Terravecchia Formation carbonates (Upper Miocene) in north central Sicily: possible eustatic influences on facies development: Sedimentary Geology, v. 57, p. 131-149.

PEDLEY, H. M. AND GRASSO, M., 1991, Sealevel change around the margins of the Catania-Gela Trough and Hyblean Plateau, southeast Sicily (African European plate convergence zone): a problem of Plio-Quaternary plate buoyancy, *in* MacDonald, D., ed., Sea Level Change at Active Plate Margins: London, International Association of Sedimentologists Special Publication 12, p. 451-464.

PEDLEY, H. M. AND GRASSO, M., 1993, Controls on faunal and sediment cyclicity within the Tripoli and Calcare di Base basins (Later Miocene) of Central Sicily: Palaeogeography, Palaeoclimatology, Palaeoecology, v. 105, p. 337-360.

PEDLEY, H. M. AND GRASSO, M., 1994a, Upper Miocene peri-Tyrrhenian reefs in the Calabrian Arc: sedimentological, tectonic and palaeogeographic implications: Geologie Mediterranene, v. 21, p. 123-136.

PEDLEY, H. M. AND GRASSO, M., 1994b, Tectonic-palaeoenvironmental model for the Late Miocene reef-tripolaceous associations of Sicily and its relevance to aberrant growth-forms and reduced biological diversity within the Palaeomediterranean: Geologie Mediterranene, v. 21, p. 109-121.

PEDLEY, H. M., LAMANNA, F., AND GRASSO, M., 1994, A new record of upper Miocene carbonates from Santo Stefano di Camastre-Caronia (northern Sicily) and its regional significance: Bollettino della Societa Geologica Italiana, v. 113, p. 435-444.

POMAR, L., OBRADOR, J., FORNOS, J., AND RODRIGUEZ-PEREA, A., 1983, El Terciario de las Baleares (Mallorca-Menorca), Guia de las excursiones del X Congresso Nacional de Sedimentologia: Mallorca, Institut d'Estudis Balearics and Universidad de Palma de Mallorca, 175 p.

POMESANO CHERCHI, A., 1974, Appunto biostratigrafici sul Miocene della Sardegne: Lyon, Fifth Congress du Neogene Mediterraneen, Memiore Bureau de Recherches Geologiques et Minieres, tome 1, no. 78, p. 433-445.

REUTHER, C-D. AND EISBACHER, G. H., 1985, Pantelleria Rift-crustal extension in a convergent intraplate setting: Geologica Rundshau, v. 73/74, p. 585-597.

RIDING, R., MARTIN, J. M., AND BRAGA, J. C., 1991, Coral-stromatolite reef framework, Upper Miocene, Almera, Spain: Sedimentology, v. 38, p. 799-818.

ROSE, E. P. F., 1971, Stratigraphical and facies distribution of irregular echinoids in Miocene limestones of Gozo, Malta and Cyrenaica, Libya: Lyon, 5th Congres du Neogene Mediterraneen, Memoire Bureau de Recherches Geologiques et Minieres Tome 1, no. 78, p. 349-355.

ROUCHY, J-M., 1988, Relations evaporites-hydrocarbures: l'association laminites-recifs-evaporites dans le Messinien de Mediterranee et ses ensignements, *in* Busson, G., ed., Evaporites et Hydrocarbures: Paris, Memoires Museum National d'Histoire Naturelle, series C, 55, p. 43-69.

RUGGIERI, G. AND TORRI, G., 1984, Il Miocene superiore di Terravecchia (Sicilia centrale): Giornale Geologia, serie 3, v. 46, p. 33-43.

SAINT MARTIN, J-P. AND ANDRE, J-P., 1992, Les constructions coralliennes de la plateforme carnonatee messinienne de Malte (Mediterranee centrale): Geologie Mediterraneenne, tome XIX, no. 3, p. 145-163.

SAINT-MARTIN J. P., CHAIX, C., AND MOISSETTE, J., 1983, Le Messinien recifal d'Oranie (Algerie): une mise au point: Comptes Rendus Seances Academie des Sciences Paris, v. 297, p. 545-547.

SAINT-MARTIN, J-P. AND ROUCHY, J-M., 1990, Les plates-forms carbonatees messiniennes en Mediterranee occidentale: leur importance pour la reconstitution des variations du niveau marin au Miocene terminal: Bulletin Societe Geologique de France. v. 8, p. 83-94.

SANTISTEBAN, C., 1981, Petrologia y sedimentologia de los materiales del Mioceno superior de las cuenca de Fortuna (Murcia) a la luz de la"Teoria de la Crisis de Salinidad": Unpublished Ph.D. Dissertation, University of Barcelona, Barcelona, 725 p.

SANTISTEBAN, C. AND TABERNA, C., 1983, Shallow marine and continental conglomerates derived from coral reef complexes after dessication of a deep marine basin: the Tortonian-Messinian deposits of the Fortuna Basin, SE Spain: Journal of the Geological Society of London, v. 140, p. 401-411.

SCHMIDT DI FRIEDBERG, P., 1965, Litostratigrafia petrolifera della Sicilia: Rivista Mineraria Siciliana, 88-90 (1964), 91-93 (1965), 43 p.

SEGUENZA, G., 1880, Le formazione nella provincia di Reggio (Calabria): Accademia Nazionale dei Lincei, Class di Scienze, Fisica, Matematiche, Naturali, Memorie, serie 3, v. 6, p. 3-446.

SERRANO, F., 1990, El Mioceno medio en el area de Nijar (Almeria Espana): Rev. Soc. Geol. Espana, v. 3, p. 66-77.

SIMONE, L. AND CARANNANTE, G., 1985, Evolution of a carbonate open shelf up to its drowning: Rendiconti dell' Accademia delle Scienze Fisiche e Matematiche Napoli, serie 4, v. 53, p. 1-43.

SIMONE, L. AND CARANNANTE, G., 1988, The fate of foramol ("temperate type") carbonate platforms: Sedimentary Geology, v. 60, p. 347-354.

SUC, J-P., VIOLANTI, D., LONDEIX, L., POUMOT, C., ROBERT, C., CLOUZON, G., GAUTIER, F., TURON, J-L., FERRIER, J., CHIKHI, H., AND CAMBON, G., 1995, Evolution of the Messinian Mediterranean environments: the Tripoli Formation at Capodarso (Sicily, Italy): Review of Palaeobotany and Palynology, v. 87, p. 51-79.

TILLA ZUCCARI, A., 1969, Relazione stratigrafica sul "Pozzo Oristano 2" (Riola Sardo, Sardegna occidentale): Bollettino della Societa Geologica Italiana, v. 88, p. 183-215.

THOMAS, B., 1987, Marges continentales sardes: geologie, geodynamique: Unpublished Ph.D. Dissertation, University of Paris, Paris, 200 p.

VISSER, J. P. DE., 1990, Clay mineral stratigraphy of Miocene to Recent marine sediments in the Central Mediterranean: Utrecht, Istituut voor Aardwetenschappen der Rijksuniversiteit te Utrecht, Geologica Ultraiectina, 243 p.

VON FUCHS, T., 1874, Intorno alla esistenza presso Siracusa di strati Miocenici che presentano i caratteri del Piano Sarmatico: Bollettino del Royale Comitato Geologica d'Italia, v. 5, p. 373-377.

ZACHARIASSE, W. J. AND SPAAK, P., 1983, Middle Miocene to Pliocene paleoenvironmental reconstruction of the Mediterranean and adjacent Atlantic Ocean: planktonic foraminiferal record of southern Italy, *in* Meulenkamp, J. E., ed., Reconstruction of Marine Paleoenvironments: Utrecht Micropaleontological Bulletins, v. 30, p. 91-110.

# MIOCENE CARBONATES OF THE EASTERN MEDITERRANEAN, THE RED SEA AND THE MESOPOTAMIAN BASIN: GEODYNAMIC AND EUSTATIC CONTROLS

BINYAMIN BUCHBINDER
*Geological Survey of Israel, 30 Malkhei Yisrael Street, 95501 Jerusalem, Israel*

ABSTRACT: Miocene carbonates of the eastern Mediterranean and Middle East areas are characterized by subtropical-temperate rhodalgal or foramol facies. The geological setting of most carbonate occurrences is of platforms or low-energy ramps, whereas reef buildups are usually subordinate. The convergence of Euro-Asian and the African-Arabian plates in Late Oligocene and Miocene times resulted in the narrowing of the Tethys seaway and eventual separation of the Mesopotamian basin from the eastern Mediterranean basin. In Early Miocene (Aquitanian) to early Middle Miocene (Langhian) times, the Mesopotamian basin was occupied by a shallow low-energy ramp. Carbonate sedimentation of locally dolomitized wackestones and packstones, with red algae, mollusks and benthic foraminifers, prevailed during sea-level highstands and evaporites prevailed during lowstands. Three carbonate evaporite cycles are distinguished: (1) Middle Asmari-Kalhur; (2) Euphrates-Dhiban, of Aquitanian Burdigalian age; and (3) Jeribe-Lower Fars of Langhian Serravalian age. Early Miocene carbonates are poorly developed along the southeastern Mediterranean coasts, probably due to terrigenous influx by the pre-Nile river system draining the northern part of the African continent. Low sea levels at the beginning of cycle TB2 may have enhanced land erosion and subsequently siliciclastic deposition in the southeastern Mediterranean. Early Miocene carbonate deposition was limited to Cyprus, southern Turkey and to the Suez-Red Sea basin. Ubiquitous and uniform carbonate deposition throughout the entire Middle East took place in early Middle Miocene (Langhian) times when highstand seas spilled over the barrier separating the eastern Mediterranean from the Mesopotamian basin, depositing the Jeribe Formation in the Mesopotamian basin, the Ziqlag and Terbol Formations in Israel, Lebanon and northwest Syria and the Marmarica Formation, west of the Nile River, in Egypt. Lowstand seas in the Serravalian, coupled by continuous plate convergence, resulted in evaporite deposition in the Mesopotamian basin (Lower Fars) and the Red Sea (Belayim Formation). Carbonate deposits were absent from the Mediterranean during most of the Serravalian because of a surface water salinity decrease and a cooling trend. Late Miocene (Tortonian early Messinian) carbonates are scarce in the Middle East because of widespread tectonic emergence. Continental sedimentation prevailed in the Mesopotamian basin, and evaporitic deposition prevailed in the Red Sea. Patchy development of coral reefs and rhodalgal carbonates took place along the coasts of Israel, northern Sinai, Cyprus and Crete until their demise during the Messinian salinity crisis.

## INTRODUCTION

From the Jurassic throughout the Eocene time, a broad seaway, the ancient Tethys, linked the area occupied by the present-day Mediterranean to the Indian Ocean. The convergence of the Euroasian and the African-Arabian plates, during the Late Oligocene and Miocene times, resulted in the narrowing and severing the Tethys in the Middle East area (Figs. 1, 2, 3), eventually leading to the closure of Tethys seaway and separation of the Mediterranean Tethys from the Mesopotamian Basin (Iraq and northeastern Syria) in the east (Buchbinder and Gvirtzman, 1976; Adams et al., 1983). The interplay of a continuous plate convergence process and a cyclic eustatic process resulted in a succession of desiccation events which began to affect the Mesopotamian basin in Early Miocene times, the Suez-Red Sea basin in Middle Miocene (Serravalian) times and finally the Mediterranean Sea in Late Miocene times. Carbonates usually developed during highstands on elevated platforms along the peripheries of the basins and evaporites usually developed during lowstands. Exceptions are the Early Miocene carbonates of the Mesopotamian basin which occupy a previously deep basin which became shallower in Oligocene times due to tectonic uplift of the basin floor.

The first distinct barrier separating the Indian Ocean from the Mediterranean occurred in middle Burdigalian times (Adams et al., 1983). This barrier was temporarily lifted during the Langhian, when highstand seas spilled over it, resuming the connection between the Mediterranean and the Mesopotamian basins. By Late Miocene times, the Mediterranean was permanently disconnected from the Indian Ocean (Fig. 3), and continental deposition prevailed in the Mesopotamian basin. Unlike the western Mediterranean, most Miocene carbonates in the eastern Mediterranean are of Early and Middle Miocene age. Late Miocene carbonates are more scarce, as this part of the Mediterranean was structurally emerged at that time.

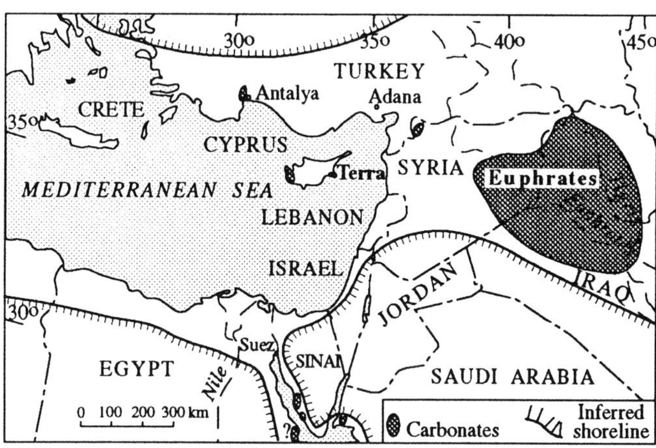


FIG. 1.—Distribution of Early Miocene carbonates in the Middle East.

Most of the studies on Miocene carbonates in the eastern Mediterranean and Middle East areas deal with stratigraphic or paleontological problems with only minor reference to facies

Models for Carbonate Stratigraphy from Miocene Reef Complexes of Mediterranean Regions,
SEPM Concepts in Sedimentology and Paleontology #5, Copyright © 1996,
SEPM (Society for Sedimentary Geology), ISBN 1-56576-033-6, p. 89-96.

FIG. 2.—Distribution of Middle Miocene (Langhian early Serravalian) carbonates in the Middle East.

FIG. 3.—Distribution of Late Miocene (Tortonian Messinian) carbonates in the Middle East.

analysis or environmental distribution. Detailed facies models have only been presented for the Abu Shaar reef in the Suez rift, for the Ziqlag reef in Israel and for the Miocene reefs in Cyprus.

Climatic deterioration which has been affecting carbonate deposition since Oligocene times resulted in the deposition of bioclastic rhodalgal or foramol carbonates which indicate subtropical-temperate climate (Carannate et al., 1988, Sun and Esteban, 1994). These were deposited on platforms or ramp settings, while reef buildups were rare. The distribution of coral reefs and platform carbonates along the southeast coasts of the eastern Mediterranean was highly influenced by the availability of flooded shelf areas. The terrigenous influx of the ancient Nile River system and the dispersion of its clastics by the counterclockwise circum-Mediterranean current (Salem, 1976; Buchbinder, this volume) affected reefal distribution in Egypt and Israel. In Lebanon-Syria, Cyprus, Turkey and in the Suez-Red Sea rift, reefal distribution was more a function of tectonic movements which formed basins and elevated platforms.

## EARLY MIOCENE CARBONATES

In the Early Miocene, carbonate development along the eastern Mediterranean coasts was quite restricted. In contrast, significant carbonate deposition took place in the Mesopotamian basin and in faulted basins around the eastern Mediterranean (i.e., the Suez-Red Sea rift, which started to form at that time, the Antalya basin (southern Turkey) and in Cyprus, Figs. 1, 4). Aquitanian reefs in the Mesopotamian basin are reported from Lurestan (Iran) only (Middle Asmari, Adams et al., 1983). The first Miocene evaporites were also deposited at that time (Kalhur gypsum, Prazak, 1978; Adams et al., 1983) pointing to the first evaporation phase of the Mesopotamian basin.

The first development of Aquitanian coral reefs around that time in the Suez rift is represented by the Basal Carbonates of the Nukhul Formation, with *Montastrea schweinfurthei*, *M. elliptica*, *Diplostrea lyonsi* and *Porites pusilla* (Scott and Govean, 1985). The lower part of the Mutaysh Formation from the northeast Red Sea coast of Saudi Arabia includes reefs of similar age (Purser and Hotzl, 1988). Early Miocene reefs of Burdigalian age are also reported from the northwest margins of the Red Sea (Purser et al., 1987, 1990). The review of their stratigraphic data (M. Esteban, pers. commun. 1992), however, suggests transforming the Burdigalian age of these reefs into the late Burdigalian-Langhian of Haq et al. (1987). Other Burdigalian reefs were reported from the Red Sea coast of Saudi Arabia (Purser and Hoztl, 1988; Jado et al., 1990).

Burdigalian carbonates with faviid corals were reported from the Mesopotamian basin in northeast Syria and Iraq (the Euphrates Limestone, Fig. 1, Van Bellen, 1959; Adams et al., 1983; Prazak, 1978); they include the large benthic foraminifera *Miogypsina globulina* and *Borelis melo* (Ctyroky and Karim, 1975). The Euphrates Limestone either interfingers with the Dhiban Anhydrite Formation (Adams et al., 1983) or is overlain by it (Fig. 4). It was deposited in a large marine embayment that was in connection with the Mediterranean Tethys, forming a low-energy ramp (Sun and Esteban, 1994). The sediments are dominated by mud-rich skeletal wackestones and packstones with subordinate oolitic packstones and grainstones and are locally dolomitized. Red algae, mollusks and benthic foraminifera (especially miliolids and peneroplids) are the predominant skeletal components (Banat and Al-Dyani, 1981; Sun and Esteban, 1994).

Dubertret (1955) and Daniel (1963) reported Burdigalian reefal limestones with *Miogypsina* and *Lepidocyclina* from the northwest Syrian Aafrine Valley (the Parsa Limestone). These are located within the seaway that connected the Mediterranean with the Mesopotamian Basin.

In Cyprus, coral reefs and carbonate platforms of the Terra Member (of the pelagic Pakhna Formation) are regarded as Aquitanian to middle Burdigalian in age on the basis of their *Miogypsina-Lepidocyclina* assemblage (Follows and Robertson, 1990; Robertson et al., 1991; Follows, 1992). Baroz and Bizon (1985) place the Terra Member in the early Burdigalian stage, whereas Orzag-Sperber et al. (1989) ascribe it to the Burdigalian-Langhian stage. The coral community of the Terra Member is diverse and includes mostly faviid, poritid and *Acropora* species

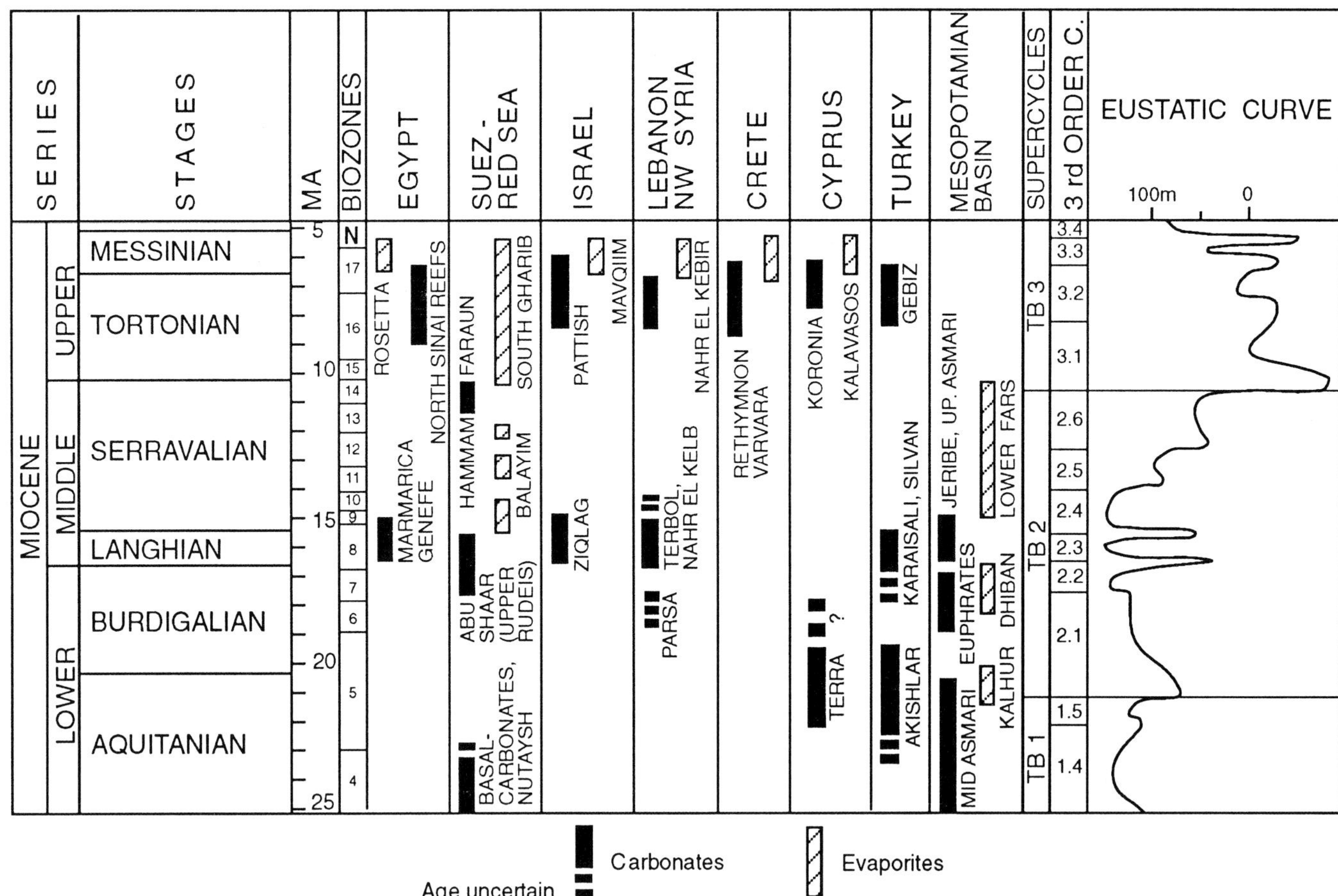


FIG. 4.—Stratigraphic correlation of Miocene carbonates and evaporite formations in the Middle East. See text for references.

(Follows, 1992).

Miocene reefal limestones overlying pre-Miocene buried hilltops are found along the margins of the Adana and Antalya basins in southern Turkey (Karaisali and Akishlar Formations, Erentöz and Öztemür, 1964; Rigassi, 1971; Yetis, 1988). Their geological setting resembles that of reefs along margins of the Suez and the Red Sea rift systems. More recent biostratigraphic studies (Bizon et al., 1974; Öztemür et al., 1985) indicate that the Lower Miocene reefs in the Antalya basin (probably the Akishlar Formation) are of Aquitanian to Burdigalian age only, while in the Adana basin the Karaisali Formation is of Burdigalian to Langhian age (Yetis, 1988). The Karaisali Formation includes *Borelis melo-Miogypsina-Lepidocyclina* assemblage and should therefore be correlated with the early Middle Miocene carbonates.

## MIDDLE MIOCENE CARBONATES

The most extensive carbonate development in the Middle East occurred during early Middle Miocene times (Langhian, Fig. 2). This is obviously connected with the high sea-level stands of cycle TB2 (Haq et al., 1988). High sea level may have caused a reduction in water turbidity in the eastern Mediterranean, as clastic influx by the Nile River system was shifted landwards. Marine circulation was resumed between the eastern Mediterranean and the Mesopotamian basin, and carbonate deposition was restored after the Dhiban evaporation stage. The free circulation between the Mediterranean and the Mesopotamian basin existed for a relatively short duration (about 2 my). The continuous convergence of the Arabian-African plates and the Eurasian plates in Serravalian times caused the severance of the Mesopotamian basin and the Suez graben from the Mediterranean Tethys, leading to the accumulation of evaporites (the Belayim Formation in the Suez Rift and the Lower Fars in the Mesopotamian basin). Lower sea levels during the later part of the Serravalian age together with the freshening of the Mediterranean surface waters were possibly responsible for the interrupted evolution of the *Globorotalia fohsi* lineage zone in the Mediterranean at that time (Buchbinder, this volume).

Early Middle Miocene reefs usually show transgressive onlap over earlier Miocene clastic sediments. The reef limestones are characterized by diversified coralline fauna and rhodalgal platform deposits. The alveolinid foraminifer *Borelis melo curdica* is the main diagnostic microfossil of Middle Miocene reef limestones (Adams et al., 1983).

In Egypt, platform carbonates developed in the Western Desert and were named the Marmarica Formation (Said, 1962; Gindy, 1971; Gindy and El Askary, 1969). They continue westward into Libya as the Al Jaghbub Formation (Bellini, 1968). Coral reefs were reported, in Egypt, only from Salum area, on the Mediterranean Coast, close to the Libyan border (Gindy and El Askary, 1969). Salem (1976) compared the Marmarica platform in northern Egypt to the Recent carbonate platform west of Florida, where east to west currents move away from the carbonate bank of Florida towards the Mississippi delta. In a similar way the Miocene circum-Mediterranean current in Egypt moved away from the Marmarica carbonate platform, transporting deltaic sediments of the Nile towards the east. Limited carbonate reefal sediments developed, however, east of the delta close to the Suez isthmus, in Gebel Gharra (Souaya, 1963) and Genefe (El-Heiny, 1982; Said, 1990). El-Heiny (1982) assigned the Marmarica and the Genefe Formations to the Langhian-Serravalian. However, Said, (1990) attributed a Langhian age only to these formations.

In Israel, similar carbonates, named the Ziqlag Formation (Buchbinder, 1979; Buchbinder et al. 1993; Buchbinder, this volume) form an extensive platform along the Judean foothills area. The formation onlaps older Tertiary and Cretaceous rocks; it is of early Middle Miocene (N8-N9) age (Buchbinder, this volume) and not of Tortonian age, as was previously assumed (Buchbinder, 1979). Coral reefs are patchy and are mostly developed on the platform edge. They contain faviid and stylophorid corals and encrusting poritids. The platform carbonates consist of rhodalgal facies with the large benthic foraminifera, *Heterostigina*, *Operculina*, *Amphistegina* and *Borelis*, and with rare coral colonies.

Similar carbonates are found in Lebanon: the Nahr el Kelb Limestone and Jebel Terbol Limestone (Dubertret, 1963; Edgell and Basson, 1975). Krashenennikov (1985a) assigned a Langhian to Serravalian age to these carbonates. In Syria, platform skeletal limestones with coral reefs of similar age and similar onlap distribution are developed in the Haleb (Aleppo) Plateau and in the Latakia, Aafrin, Nahar el Kabir and El-Ghab areas (Ponikarov, 1966; Dubertret, 1963; Daniel, 1963; Krashenennikov, 1985b).

In the Mesopotamian basin, Middle Miocene carbonates of the Jeribe Formation (Figs. 2, 4) and the upper Asmari Formation (Van Bellen, 1959; Ponikarov, 1966; Prazak, 1978; Adams et al., 1983) also show transgressive onlap relationships. The Jeribe Formation postdates the Dhiban evaporatic phase. It consists mostly of locally dolomitized rhodalgal carbonates and is overlain by the Lower Fars evaporites. The Jeribe Formation is characterized by small-scale cycles of open marine carbonates intercalated with restricted marine carbonates and evaporites (Al-Murani, 1986). There is a continuous geographical connection between the Jeribe Formation in the Mesopotamian Basin in north Syria and the Middle Miocene limestones of the eastern Mediterranean basin (Fig. 2).

In the fold and foothill belt of southeast Turkey, reefal carbonates of the Silvan Formation cap structural highs (Rigo De Righi and Cortesini, 1964; Beydoun, 1988). The Karaisali reef limestone in the Adana basin is probably also part of the early Middle Miocene phase.

In the Suez and the Red Sea basins, carbonate platforms fringed by coral reefs developed on elevated blocks along the southwest coast of the Gulf of Suez and the northwest coast of the Red Sea. Siliciclastics and evaporites were confined to the deeper parts of the basin. The carbonates include the well known Abu Shaar reef and are incoporated in Group B deposits (Montenant et al., 1988). Their age has been widely debated. Recent studies favor a late Burdigalian-Langhian (N7/N8) age (Scott and Govean, 1985; Evans, 1988; James et al. 1988a, b; Burchette, 1988; Esteban, pers. commun. 1992) According to Monty et al. (1987) and Purser et al. (1990) the Abu Shaar reef together with similar reefs along the northwest coast of the Red Sea terminate the Middle Miocene marine sequence (Group B, upper Rudeis Formation) and precede the Middle to Upper Miocene evaporites (Group C). Its coral community includes *Stylophora* and hemispherical faviids dominated by *Montastrea* and *Acanthastrea* (James et al., 1988b).

Other carbonate deposits in the Suez rift are connected with the Middle Miocene evaporation phase (the evaporites of the Belayim Formation, Fig. 4). Marine carbonates developed occasionally when sea level rose above the sill, separating the Mediterranean from the Gulf of Suez. The carbonates where either of normal, or restricted marine origin. Stromatolites of the Esh el Mellaha Member of the Abu Shaar reef complex may have been the precursor of the Belayim Evaporation phase (Rouchy et al., 1983; M. Esteban, pers. commun. 1992). In this regard, they should be considered as an independent unit which postdates the coral buildup (Monty et al., 1987). Notably, reef-stromatolite associations are common to both the Middle Miocene of the Red Sea and the Messinian reefs of the western Mediterranean, reflecting a similar carbonate-evaporite evolution (Rouchy, 1986; Rouchy and Saint Martin, 1992).

The Hammam Faraun Member of the Belayim Formation consists of rhodalgal platforms and coral reefs with *Heterostigina* and *Borelis*. They characterize the elevated periphery of the basin where they either onlap older pre-Miocene formations or overlie older Miocene formations, while the center of the Suez basin is filled with the evaporites and shales of the Belayim Formation. The Hammam Faraun Member is overlain by the evaporites of the South Gharib Formation. The age of the Hammam Faraun Member is not clearly defined. It ranges between N9 and N13 (Wasfi, 1968; El Heiny and Martini, 1981; El Heiny, 1982; Scott and Govean, 1985). On the other hand, Evans (1988) assigned the Hammam Faraun Member to the N14 Zone. Esteban (pers. commun., 1992) differentiates three carbonate buildup events above the Rudeis Formation: the Shagar Member of the Kareem Formation, the Sidri Member and the Hammam Faraun Member of the Belayim Formation. These carbonates span the N10 to N14 zones.

## LATE MIOCENE CARBONATES

Unlike in the western Mediterranean area, Late Miocene reefs are less common in the eastern Mediterranean, and most are found only in the subsurface or in tectonically elevated areas, such as Cyprus. Continental to shallow marine clastic sedimentation (Upper Fars) prevailed in the Mesopotamian basin, which was mostly emerged during that time. Along the coasts of the eastern Mediterranean, coral debris mixed with hemi-pelagic mudstones underlie the Late Miocene evaporites of the Mavqi'im Formation (Buchbinder et al., 1993; Buchbinder, this volume). These debris sheets belong to the Late Miocene N16-N17 zones, and are considered slope sediments of the Late Miocene Pattish Reef belt which is distributed along the shelf-edge (Fig. 3). The carbonates of the Pattish Formation consist of poritid and faviid corals and rhodalgal limestones. In northern Sinai, Late Miocene carbonates form a continuous belt and are mostly dominated by rhodalgal facies (see also Jenkins, 1990). There are no reports of Late Miocene carbonates west of Sinai or from the Gulf of Suez area, where thick evaporates of the South Gharib Formation were deposited.

Ponikarov (1966) reported an Upper Miocene serpulid reef overlain by gypsum in northern Syria (Nahr el Kebir).

In Cyprus, Late Miocene reefs of the Koronia Limestone are aligned parallel to graben-bounding faults. They include the benthic foraminifera assemblage *Operculina-Heterostegina-Borelis* and are ascribed to the Tortonian, to early Messinian (Follows and Robertson, 1990; Follows, 1992; Robertson et al., 1991). Although dominated by poritid corals, *Tarbellastraea* and *Montastraea* are also present. The overlying evaporitic facies of the Messinian Kalavasos Formation is located between the reefs in a topographically lower position (Follows and Robertson, 1990), and locally it onlaps pre-evaporites carbonates (Rouchy, 1982). In the central part of the depositional basins, the evaporites overlie a section of laminated diatomites and stromatolites (Orzag-Sperber et al., 1989, Rouchy, 1988). Laminated diatomites underlie the evaporites in many basins around the Mediterranean; they are precursory to the salinity crisis, reflecting increase in the Mediterranean water residence time, during the initial restriction stage (Rouchy, 1982; Rouchy and Saint Martin, 1992).

In Turkey, Late Miocene coral algal reefs (Gebiz Formation) are reported from the Antalya basin (Rigassi, 1971) and Tortonian patch-reefs fringe alluvial-fan deposits along the ancient sea shore in southwest Turkey (Kasaba Formation, Hayward, 1982). These reefs are dominated by large branching colonies of *Tarbellastraea* and large hemispheral colonies of *Montastraea.*

Late Miocene (Tortonian to lower Messinian) carbonates developed in Crete in the basins of Heraklion and Rethymnon (the Rethymnon and Varvara Formations, Meulenkamp, 1969; Meulenkamp et al., 1977). Small coral patches and larger carbonate platforms are located along the peripheries of the Heraklion basin, while evaporites occupy lower topographic positions as in Cyprus. Locally, the evaporites onlap reefal sediments (Rouchy, 1982). Tortonian reefs in the Heraklion basin show a transgressive (retrogradational) stacking pattern and their coralline genera include *Tarbellastraea, Palaeoplesiastraea, Acanthastraea, Caryophyllia, Flabellum, Siderastraea* and *Porites* (Chaix and Delrieu, 1994).

## DISCUSSION AND CONCLUSIONS

The biostratigraphic zonation of the Miocene reefs in the eastern Mediterranean is not always clear, especially when the age determination is based on benthic assemblages only. This makes the correlation between different basins difficult. The dating of the reefs in the Suez-Red Sea area, Cyprus and southern Turkey is especially controversial. Some of the "Burdigalian" reefs should, possibly, be included in the Middle Miocene phase. In addition, the time span of reefal deposition is probably much shorter than often assumed, and where dating is precise, does not exceed 1-2 my.

Lower Miocene carbonates are poorly developed along the Levant coast where fine siliciclastic deposition prevailed. In contrast, they are extensively developed in the Mesopotamian Basin (Fig. 1) and moderately developed in the tectonically affected basins of the Suez-Red Sea region, Cyprus and southern Turkey. Their poor development in the southeastern margins of the Mediterranean could be due to terrigenous influx by the river system draining the African continent. Low sea level at the beginning of cycle TB2 could have contributed to an increase in erosion and transportation of its products into the eastern Mediterranean. More favorable conditions for carbonate deposition existed eastwards toward the Mesopotamian basin, probably because of reduced siliciclastic shedding. Higher sea levels in Middle Miocene time resulted in reduced clastic influx into the Mediterranean. This together with the confinement of the dispersed Nile-related rivers system into a single river valley (Salem, 1976) enhanced carbonate production along the eastern Mediterranean shores. Reefal deposition was inhibited within the delta region and eastward in the direction of the counter-clockwise circum-Mediterranean current which carried Nile-derived sediments. Away from the Nile Delta, platform and reefal carbonates flourished in the Western Desert (of Egypt), in Israel, Lebanon and in northern Syria (Fig. 2).

In the Suez and Red Sea rifts, reefal carbonates dominated elevated horsts, especially along the basin margins, whereas clastics were trapped in depressions. In the Mesopotamian basin, skeletal production proliferated from Oligocene through Middle Miocene times with brief interruptions of desiccation, emergence and denudation.

Ubiquitous and uniform carbonate deposition in the entire Middle East area took place in early Middle Miocene times only (Fig. 2). The biostratigraphic evidence published on the exact age of Middle Miocene reefs is in many cases not satisfactory. Ages which span most of the Middle Miocene (6 my) should be regarded with reservation. There is accumulating evidence that most of the Serravalian (N10-N13, *G. fohsi fohsi* lineage zone, Buchbinder, this volume) is characterized by a lowering of surface water salinities and a cooling trend in the Mediterranean (Buchbinder et al., 1993), while hypersaline conditions existed in the Mesopotamian basin and in the Gulf of Suez. Reef development should therefore have been very limited at that time. Normal marine conditions were resumed in the Mediterranean in late Serravalian time (N14 Zone). It is highly probable that the Middle Miocene reef phase spans N8-N9 zones only (i.e., from the uppermost Burdigalian to lowermost Serravalian). This period is marked by high sea-level stands of Haq et al's. (1988) third-order cycles 2.2, 2.3 and 2.4. On the other hand, Burdigalian coral reefs from Cyprus failed to keep pace with the rapid sea-level rise during Langhian times and were succeeded by pelagic facies (Robertson et al., 1991). The widespread development of carbonate deposits during early Middle Miocene times marks the final inter-connected carbonate production in the Tethys seaway. During the following Serravalian stage the connection with the Mesopotamian basin was permanently closed and the connection with the Red Sea basin was highly severed. Serravalian (N10-N14) carbonates characterized the Suez Rift only, when intermittent marine flooding resulted in carbonate production on rift shoulders while stromatolites developed during periods of evaporitic-drawdowns.

Studies of Miocene coral populations in the Middle East area are scarce and the existing knowledge is very general. Middle Miocene reefs are characterized by hemispheric faviids, while poritids are less common. Branching stylophorids were found both in the Ziqlag Formation (N8/N9) in Israel and in the Abu Shaar reef (N7/N8) in the Gulf of Suez rift.

Distribution of Late Miocene carbonates is relatively limited (Fig. 3) due to lower sea-levels (compared with the Middle Miocene phase) and a continuous uplift following plate convergence. Carbonate deposits appear to be limited to basin-margin areas and are found mostly in the subsurface and therefore are less known. They are composed mostly of poritid corals with subordinate faviids.

Sun and Esteban (1994) divided the facies patterns of worldwide Miocene carbonates into two end members: (1) humid oceanic, tropical-subtropical carbonates and (2) arid, land-locked, subtropical-temperate carbonates. The Mediterranean and Middle East carbonates were included in the second group. These carbonates are characterized by rhodalgal platforms or ramps, with limited coralline buildups. Lower Miocene carbonates in the Mesopotamian basin (in Iraq and Iran) are particularly poor in coral reefs. The geological setting of the Mesopotamian basin is characterized by an areally extensive low-energy ramp with mud-rich skeletal wackestones and packstones. In Iraq and northeast Syria, the Lower to Middle Miocene carbonates are made up of two third-order depositional sequences. The first sequence consists of the lowstand evaporites of the Kalhur Formation and the Euphrates Limestone which was deposited during the trangression and highstand stages. The second sequence consists of the lowstand evaporites of the Dhiban Formation, followed by the transgressive and highstand carbonates of the Jeribe Formation. Lowstand deposits of the Lower Fars Evaporites overlie the Jeribe Formation. The latter terminates the marine deposition in the Mesopotamian basin, except for the eastern Iranian side which was occasionally flooded from the direction of the Indian Ocean. Late Miocene marine deposition was confined to the peripheries of the eastern Mediterranean only.

## ACKNOWLEDGMENTS

Thanks are due to M. Esteban for his comments and suggestions and to J. M. Rouchy and M. Coniglio for their constructive reviews.

## REFERENCES

AL-MURANI, G. S. G., 1986, Sedimentology and petrophysical aspects of the Middle Miocene Jeribe Formation, East Baghdad Field, Iraq: Unpublished Ph.D. Dissertation, University of Oxford, Oxford, 256 p.

ADAMS, C. G., GENTRY, A. W., and WHYBROW, P. J., 1983, Dating the terminal Tethyan event: Utrecht Micropaleontological Bulletins, v. 30, p. 273-298.

BANAT, K. M. and AL-DYANI, 1981, Petrology and mineralogy of selected sections from the Euphrates and Jeribe formations (Miocene), Iraq: Journal of the Geological Society of Iraq, v. 14, p. 1-9.

BAROZ, F. and BIZON, G., 1985, Polemi basin, west Cyprus, *in* Steininger, F. F., Senes, J., Kleemann, K., and Rogl, F., eds., 1985, Neogene of the Mediterranean Tethys and Paratethys: Vienna, Institute of Paleontology, University of Vienna, v. 2, p. 125.

BELLINI, E., 1968, Biostratigraphy of the "Al Jaghbub (Giarabub) Formation" in Eastern Cyrenaica (Libya): Cairo, Proceedings of the Third African Micropaleontological Colloquium, p. 165-183.

BEYDOUN, Z. B., 1988, The Middle East: Regional Geology and Petroleum Resources: Beaconsfield, Scientific Press, 292 p.

BIZON, G., BIJU-DUVAL, B., LETOUZEY, J., MONOD, O., POISSON, A., ÖZER, B., and ÖZTEMÜR, E., 1974, Nouvelles precisions stratigraphiques concernant les bassins tertiaires du sud de la Turquie (Antalya, Mut, Adana): Revue de l'Institut Français du Pétrole, v. 29, p. 305-320.

BUCHBINDER, B., 1979, Facies and environments of Miocene reefs limestones in Israel: Journal of Sedimentary Petrology, v. 49, p. 1323-1344.

BUCHBINDER, B. and GVIRTZMAN, G., 1976, The breakup of the Tethys Ocean into the Mediterranean Sea, the Red Sea, and the Mesopotamian Basin during the Miocene: a sequence of fault movements and desiccation events (abs.): Tokyo, First International Congress on Pacific Neogene Stratigraphy, p. 32-35

BUCHBINDER, B., MARTINOTTI, G. M., SIMAN-TOV, R., and ZILBERMAN, E., 1993, Temporal and spatial relationships in Miocene reef carbonates in Israel: Palaeogeography, Palaeoclimatogy, Palaeoecology, v. 101, p. 97-116.

BURCHETTE, T. P., 1988, Tectonic control on carbonate platform facies distribution and sequence development: Miocene, Gulf of Suez: Sedimentary Geology, v. 59, p. 179-204.

CARANNATE, G., ESTEBAN, M., MILLIMAN, J. D., and SIMONE, L., 1988, Carbonate lithofacies as paleolatitude indicators: problems and indications: Sedimentary Geology, v. 60, p. 333-346.

CHAIX, C. and DELRIEU, B., 1994, Les récif coralliens du Miocene Supérieur en Crete centrale (Grèce): Mineraux et Fossiles, no. 214, p. 7-16.

CTYROKY, P. and KARIM, S. A., 1975, *Miogypsina* and *Borelis* in the Euphrates Limestone Formation in the Western Desert of Iraq: Neus Jahrbuch fuer Geologie und Palaontologie, Abhandlungen, v. 148, p. 33-49.

DANIEL, E. J., 1963, Lexique stratigraphique International, v. 3, Syrie Interieure: Paris, Centre National de la Recherch Scientifique, p. 159-289.

DUBERTRET, L., 1955, Géologie des roche vertes du nord ouest de la Syrie et du Hatay (Turquie): Notes et Mémoires sur le Moyen Orient, v. 7, p. 193-220.

DUBERTRET, L., 1963, Lexique stratigraphique International v. III, Liban, Syrie: Paris, Centre National de la Recherch Scientifique, p. 1-153.

EDGELL, H. S. and BASSON, P. W., 1975, Calcareous algae from the Miocene of Lebanon: Micropaleontology, v. 21, p. 165-184.

EL-HEINY, I., 1982, Neogene Stratigraphy of Egypt: Newsletters on Stratigraphy, v. 11, p. 41-54.

EL-HEINY, I. and MARTINI, E., 1981, Miocene foraminiferal and calcareous nannoplankton assemblages from the Gulf of Suez region and correlations: Géologie Méditérranéenne, v. 7, p. 101-108.

ERENTÖZ, L. and ÖZTEMÜR, C., 1964, Aperçu general sur la stratigraphie du Néogène de la Turquie et observations sur ses limites inférieure et supérieure: Instituto "Lucas Mallada", C.S.I.C. (Espana), Cursillos y Conferencias, IX, p. 259-266.

EVANS, A. L., 1988, Neogene tectonic and stratigraphic events in the Gulf of Suez rift area, Egypt: Tectonophysics, v. 153, p. 235-247.

FOLLOWS, E. J., 1992, Patterns of reef sedimentation and diagenesis in the Miocene of Cyprus: Sedimentary Geology, v. 79, p. 225-253.

FOLLOWS, E. J. and ROBERTSON, A.H.F., 1990, Sedimentology and structural settings of Miocene reefal limestones in Cyprus, *in* Malpas, J., et al., eds., Ophiolites: Oceanic Crust Analogues: Nicosia, Proceedings, Symposium "Troodos 1987", Cyprus Geological Survey, p. 207-215.

GINDY, A. R., 1971, The stratigraphic succession of the exposed coastal Miocene sediments north of Salum town, its diagenesis and correlation with that of Siwa Depression, Western Desert of Egypt: Cairo, Proceedings of the Egyptian Academy of Science, no. 23, p. 9-28.

GINDY, A. R. and EL-ASKARY, M. A., 1969, Stratigraphy, structure and origin of Siwa depression, Western Desert of Egypt: American Association of Petroleum Geologists Bulletin, v. 53, p. 603-625.

HAQ, B. V., HARDENBOL, J., and VAIL, P. R., 1987, Chronology of fluctuating sea levels since the Triassic: Science, v. 253, p. 1156-1167.

HAQ, B. V., HARDENBOL, J., and VAIL, P. R., 1988, Mesozoic and Cenozoic chronostratigraphy and cycles of sea-level changes, *in* Wilgus, C. K., Ross, C. A., Posamentier, H., and Kendall, C. G. St. C., eds., Sea-level Changes: An Integrated Approach: Tulsa, Society of Economic Paleontologists and Mineralogists Special Publication 42, p. 71-108

HAYWARD, A. B., 1982, Coral reefs in a clastic sedimentary environment: fossil (Miocene, SW Turkey) and Modern (Recent, Red Sea) analogues: Coral Reefs, v. 1, p. 109-114.

JADO, A. R., HOTZL, H., and ROSCHER, B., 1990, Development of sedimentation along the Saudi Arabian Red Sea coast. Journal of the King Abdulaziz University, Faculty of Earth Science, v. 3, p. 47-62.

JAMES, N. P., CONIGLIO, M., AISSAOUI, D. M., and PURSER, B. H., 1988a, Facies and geologic history of an exposed Miocene rift-margin carbonate platform: Gulf of Suez, Egypt: American Association of Petroleum Geologists Bulletin, v. 72, p. 555-572.

JAMES, N. P., ROSEN, B., and CONIGLIO, 1988b, Miocene platform-margin reef, Gulf of Suez, Egypt: American Association of Petroleum Geologists Bulletin, v. 72, p. 200-201.

JENKINS, D. A., 1990, North and central Sinai, *in* Said, R., ed., The Geology of Egypt (Chapter 19): Rotterdam, A. A. Balkema, p. 361-380.

KRASHENENNIKOV, V. A., 1985a, Beirut - Tripoli Basin, Lebanon, *in* Steininger, F. F., Senes, J., Kleemann, K., and Rogl, F., eds., 1985, Neogene of the Mediterranean Tethys and Paratethys: Vienna, Institute of Paleontology, University of Vienna, v. 2, p. 196

KRASHENENNIKOV, V. A.,1985b, Aleppo plateau, north-west Syria, *in* Steininger, F. F., Senes, J., Kleemann, K., and Rogl, F., eds., 1985, Neogene of the Mediterranean Tethys and Paratethys: Vienna, Institute of Paleontology, University of Vienna, v. 2, p. 199.

MEULENKAMP, J. E., 1969, Stratigraphy of Neogene deposits in the Rethymnon province, Crete, with special reference to the phylogeny of universal *Uvigirina* from the Mediterranean region: Utrecht Micropaleontological Bulletins, no. 2, 168 p.

MEULENKAMP, J. E., JONKERS, A., and SPAAK, P., 1977, Late Miocene to early Pliocene development of Crete: Athens, Proceedings, VI Colloquium on the Geology of the Aegean Region, p. 137-149.

MONTY, C. L. V., ROUCHY, J. M., Maurin, A., BERNET-ROLLANDE, M. C., and PERTHUISOT, J. P., 1987, Reef-stromatolites-evaporites facies relationships from Middle Miocene examples of Suez and the Red Sea, *in* Peryt, T. M., ed., Evaporite Basins: Berlin, Springer, p. 134-188.

MONTENANT, C., OTT D'ESTEVOU, P., PURSER, B., BUROLLET, P. F., JARRIGE, J. J., ORSAG-SPERBER, F., PHILOBBOS, E., PLAZIAT, J. C., PRAT, P., RICHERT, J. P., ROUSSEL, N., and THIRIET, J. P., 1988, Tectonic and sedimentary evolution of the Gulf of Suez and the northwestern Red Sea: Tectonophysics, v. 153, p.161-177.

ORZAG-SPERBER, F., ROUCHY, J. M., and ELION, P., 1989. The sedimentary expression of regional tectonic events during the Miocene-Pliocene transition in the southern Cyprus basins: Geological Magazine, v. 126, p. 291-299.

ÖZTEMÜR, E G., BIZON, J. J., and FEINBERG, H., 1985, Antalya basin, Kizilagac, Turkey, *in* Steininger, F. F., Senes, J., Kleemann, K., and Rogl, F., eds., Neogene of the Mediterranean Tethys and Paratethys: Vienna, Institute of Paleontology, University of Vienna, v. 2, p. 178.

PONIKAROV, V. P., ed., 1966, The geological map of Syria: Scale 1:1000 000, Explanatory Notes: Damascus, Syrian Arab Republic, Ministry of Industry.

PRAZAK, J., 1978, The Development of the Mesopotamian Basin during the Miocene: Journal of the Geological Society of Iraq, v. 11, p. 170-189.

PURSER, B. H., ORZAG-SPERBER, F., and PLAZIAT, J. C., 1987, Sédimentation et rifting: les series neogenes de la marge nord-occidentale de la mer Rouge: Notes et Mèmoires Total, no. 21, p. 111-144.

PURSER, B. H. and HOTZL, H. ,1988, The sedimentary evolution of the Red Sea rift: a comparison of the northwest (Egyptian) and the northeast (Saudi Arabian) margins: Tectonophysics, v. 153, p. 193-208.

PURSER, B.H., PHILOBBOS, E. R. and SOLIMAN, M., 1990, Sedimentation and rifting in the parts of the Red Sea: a review: Bulletin de la Société Géologique de France, v. 6, p. 371-384.

RIGASSI, D., 1971, Petroleum geology of Turkey, *in* Campbell, A. S, ed., Geology and History of Turkey: Tripoli, Petroleum Exploration Society of Libya, 13th Annual Field Conference, p. 453-482.

RIGO DE RIGHI, M. and CORTESINI, A., 1964, Gravity tectonics in foothills structure belt of southeast Turkey: American Association of Petroleum Geologists, v. 48, p. 1911-1937.

ROBERTSON, A. H. F., EATON, S., FOLLOWS, E. J., and MCCALLUM, J. E., 1991, The role of local tectonics versus global sea-level change in the Neogne evolution of the Cyprus active margin, *in* McDonald, D. I. M. ed., Sedimentation, Tectonics and Eustasy: London, International Association of Sedimentologists Special Publications 12, p. 331-369.

ROUCHY, J. M., 1982, La genèse des évaporites messiniennes de Méditerrannee: Mémoires du Museum National D'Histoire Naturelle (C), v. 30, 267 p.

ROUCHY, J. M., 1986, Les évaporites miocene de la Méditerranée et de la Mer Rouge et leur enseignements pour l'interpretation des grandes accumulations evaporitiques d'origine marine: Bulletin de la Société Géologique de France, v. 2, p. 511-520.

ROUCHY, J. M., 1988, Relation evaporites-hydrocarbures: l'association laminites-recif-évaporites dans le Messinian de Méditerranée et ses enseignements, *in* Busson, G., ed., Evaporites et Hydrocarbures: Mémoires Du Muséum National D'Histoire Naturelle (C), v. 55, p. 43-69.

ROUCHY, J. M., BERNET-ROLLANDE, M. C., MAURIN, A. F., and MONTY, C., 1983, Signification sédimentologique et paléogeographique des divers types de carbonates bioconstruits associes aux evaporites du Miocene moyen pres du Gebel Esh Mellaha (Egypte): Paris, Comptes Rendus Hebdomadaires de Seances, Academie de Sciences, v. 296, Serie II, p. 457-462.

ROUCHY, J. M. and SAINT MARTIN, J. P., 1992, Late Miocene events in the Mediterranean as recorded by carbonate evaporite relations: Geology, v. 20, p. 629-632.

SAID, R., 1962, Das Miozan in der Westlichen Wuste, Agyptens: Geologie Jahrbuch, v. 80, p. 349-366.

SAID, R., 1990, Cenozoic, *in* Said, R., ed., The Geology of Egypt: Rotterdam, A. A. Balkema, p. 451-486.

SALEM, R., 1976, Evolution of Eocene-Miocene sedimentation patterns in parts of northern Egypt: American Association of Petroleum Geologists, v. 60, p. 34-64.

SCOTT, R. W. and GOVEAN, F. M., 1985, Early depositional history of a rift basin: Miocene in western Sinai: Palaeogeography, Palaeoclimatology, Palaeoecology, v. 52, p. 143-158.

SOUAYA, F. J., 1963, On the foraminifera of Gebel-Gharra (Cairo-Suez Road) and some other Miocene samples: Journal of Paleontology, v. 37, p. 433-457.

SUN, S. Q. and ESTEBAN, M., 1994, Paleoclimatic control on sedimentation, diagenesis, and reservoir quality: lessons from Miocene carbonates: American Association of Petroleum Geologist, v. 78, p. 519-543.

YETIS, C., 1988, Reorganization of the Tertiary stratigraphy in the Adana basin, southern Turkey: Newsletters on Stratigraphy, v. 20, p. 43-58.

VAN BELLEN, R. C., 1959, Lexique Stratigraphique International, v. III, Iraq: Paris, Centre National de la Recherch Scientifique, 333 p.

WASFI, S. M., 1968, Miocene planktonic foraminiferal zones from the Gulf of Suez, Egypt: Cairo, Proceedings of the Third African Micropaleontological Colloquium, p. 461-474.

# MIOCENE REEFS OF THE PARATETHYS: A REVIEW

ANDRZEJ PISERA
*Institute of Paleobiology, Polish Academy of Sciences, Al. Żwirki i Wigury 93, 02-089 Warsaw, Poland*

ABSTRACT: Following successive transgressive events, two main reef episodes occurred during Miocene deposition in the Paratethys: the Badenian and Sarmatian. Small coral patch reefs and coral carpets are known from Austria, Hungary and Bulgaria from the Badenian. Algal-vermetid reefs form the main Badenian reef chain extending from Poland through the Ukraine to Moldavia. The Sarmatian reefs are composed of peloidal (thrombolitic) limestones with large amounts of early fibrous cement and varying admixture of serpulid tubes, sessile forams *Nubecularia*, monostromatic red algae and encrusting cheilostomatous bryozoans. The main Sarmatian reef chain occurs in the same paleogeographic position as the Badenian algal-vermetid reefs and follows the former shallow-water platform margin from Poland through the Ukraine, Moldavia to Bulgaria; these reefs occur also on the Crimea and in the Forecaucasian region. The development of the Badenian reefs had been controlled by sedimentation patterns (clastics supply) and climatic factors. The Sarmatian reefs composition had been controlled by the chemistry of the Sarmatian basin water.

## INTRODUCTION

During Miocene time, as a result of alpine orogenic movements, the Northern branch of the Tethys had been divided into a complex system of basins called Paratethys (Fig. 1). The extra-Carpathian and intra-Carpathian parts are included into the Central Paratethys (Fig. 1), whereas the larger part consisting of the Black Sea region, Caucasian region, and Caspian and Aral Seas represents the Eastern Paratethys (Rögl and Steininger, 1984). There are several well defined small basins with quite different histories (i.e., intramontane Pannonian basin, Alpine-Carpathians foredeep, Black Sea and basins of the Ponto-Caspian region; Rögl and Müller, 1978; Steininger and Rögl, 1979; Rögl and Steininger, 1983, 1984; Steininger et al., 1985). The separated small basins caused a complex mosaic facies pattern making correlation very difficult.

Lower Miocene units are still represented by flysh deposits in the Carpathian orogene, while in other regions they are represented by molasse deposits. Middle-Upper Miocene postorogenic deposits are much better developed and widespread. Central parts of the basins are infilled with fine clastic sediments often reaching several thousand meters in thickness (Ney et al., 1974; Kazmer, 1990). Shallow-water, mostly carbonate rhodalgal, sedimentation, prevailed during particular intervals especially along the northern and eastern shores (which are usually developed on stable cratonic settings) and on swells in the Pannonian Basin (Kolesnikov, 1940; Korolyuk, 1952; Radwański, 1965; Ney et al., 1974; Steininger et al., 1985; Studencki, 1988; Kazmer, 1990; Lelkes and Studencki, 1990). There are several evaporitic horizons (Fig. 2) known mainly in the Carpathian foredeep (Ney et al., 1974; Rögl and Müller, 1978; Rögl and Steininger, 1984; Steininger et al., 1985), the most important being the Middle Badenian horizon (=Lower Serravallian). During this time, in the entire Carpathian foredeep and part of the Pannonian basin, evaporites (gypsum and halite) were deposited.

In the Carpathian foredeep, folding and overthrusting were active during the Middle-Upper Miocene time as proved by the presence, for example, in Poland of shallow-water Badenian (also rhodalgal) deposits folded together with Carpathian flysh rocks and next overthrusted onto the Lower Sarmatian deposits infilling the Carpathian foredeep (Ney et al., 1974).

The Miocene Paratethys Sea at different times, especially during Badenian time, had connections with the Indian Ocean or the Mediterranean and thus normal marine conditions prevailed (except during periods of evaporite deposition) or was a closed or semiclosed basin, as during the Sarmatian time, with conditions strongly differing from normal marine (Iljina et al., 1976; Rögl and Steininger, 1983, 1984; Steininger et al., 1985). The Sarmatian Paratethys has been described as "brackish-water", this being, however an oversimplification, as areas and periods with increased salinity also occurred (Belokrys, 1967).

This paper deals with the reefs (i.e., non-bedded bio-constructed carbonates displaying the presence of framework made by builders as diversified as corals, red-algae, sessile vermetid gastropods, bryozoans, serpulids, sessile forams and microbial communities). All other carbonates (especially bioclastic rhodalgal limestones), which are common in many parts of the Miocene Paratethys, are out of the scope of this review.

There were two main periods of extensive reef development in the Paratethys during the Middle-Upper Miocene time (Fig. 2), and no Lower Miocene reefs are known from the whole area. The first period occurred during Badenian time, when marine conditions prevailed over the area (Rögl and Steininger, 1983, 1984) thus allowing the development of normal-marine reefs (i.e., coral and red-algal reefs). The second period occurred during Sarmatian time, when the Paratethys was a closed or semiclosed basin causing conditions to strongly deviate from normal marine and when the so-called cryptalgal, serpulid, bryozoan and *Nubecularia* reefs developed. In effect, the Badenian and Sarmatian reefs structures are completely different in composition, despite the fact that they occur in the same paleogeographical situation within the basin and are separated by a narrow span of time.

## BADENIAN REEFS

The main reef building organisms during Badenian deposition in the Paratethys were red algae and not corals, as a result (most

Models for Carbonate Stratigraphy from Miocene Reef Complexes of Mediterranean Regions,
SEPM Concepts in Sedimentology and Paleontology #5, Copyright © 1996,
SEPM (Society for Sedimentary Geology), ISBN 1-56576-033-6, p. 97-104.

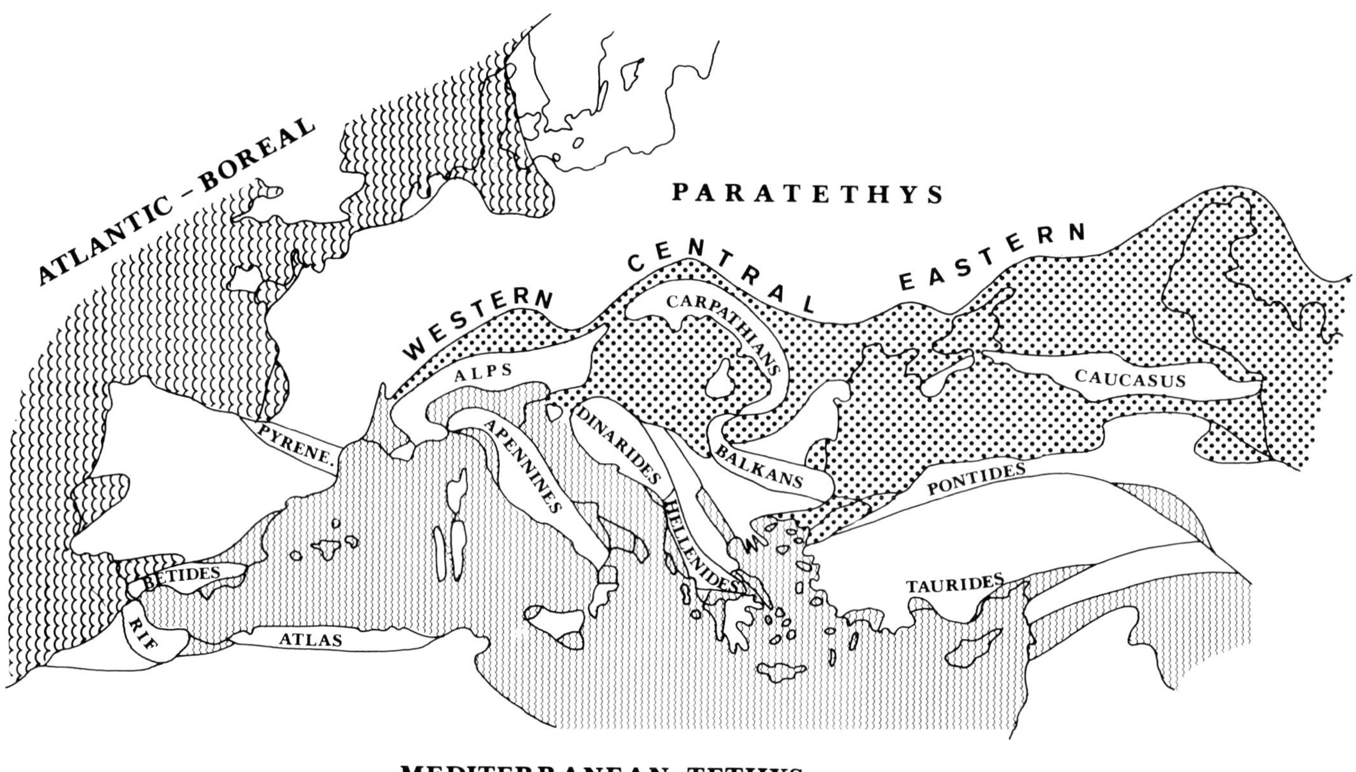


FIG. 1.—The European Neogene paleogeography and bioprovinces (after Rögl and Steininger, 1984).

probably) of the fact that most of the region was then outside of the tropical zone. However, more southern parts of the basin (i.e., the Vienna and Styrian basins, as well as Hungarian and Bulgarian basins) have some Badenian coral reef constructions (Figs. 3, 4).

In the Vienna Basin, only small patch reefs not exceeding about 4 m in thickness occur and are dated as Upper Badenian deposits (Steininger and Papp, 1978; Dullo, 1983, and literature herein). According to Dullo (1983), they are composed of *Porites* and *Tarbellastraea*, while Piller et al. (1991) list also *Caulastrea*. From the Styrian Basin (Southern Austria), similar small patch reefs and coral carpets are composed of *Porites*, *Tarbellastrea* and *Montastrea*, are dated as Lower Badenian and are described by Piller et al. (1991) and Friebe (1990, 1993). Small patch reefs of a similar type are also reported from the Budapest area (Kókay et al., 1984; Müller, 1984; Lelkes and Studencki, 1990). There are also reports of Lower Badenian coral reefs (only coral rubble in fact) from Northern Hungary composed mainly of *Heliastraea*, *Porites*, *Tarbellastraea*, *Stylophora*, *Favia*, *Cyphastraea* (Hegedüs, 1970; Scholz, 1970; Müller, 1984; Oosterbaan, 1990). Corals are reported in some rhodalgal reefs also in Rumania (Ghiurca, 1974).

Much more extensive coral reef development can be observed in north-western Bulgaria (i.e., along the southern shore of the Paratethys; Figs. 3, 4). Most of the coral reefs in Bulgaria are Lower Badenian in age, more rarely they occur in the Upper Badenian (see Kojumdgieva and Strashimov, 1960; Kojumdgieva, 1976, 1978). The Lower Badenian reefs are known only as coral-algal rubble and no *in situ* reef framework is observed, but some displaced coral colonies are often thickly encrusted with red algae. As particular coral colonies often reach over 50 cm in diameter, these deposits may represent in fact the reef core itself. The Bivolare reef (NW Bulgaria), which is relatively well exposed, was no less than 15 meters thick and over 50 m long. Corals are relatively diversified in these reefs and represented mostly by *Stylophora*, *Favites*, *Favia*, *Heliastraea*, *Tarbellastraea*, *Porites*, *Turbinaria*, *Sideastrea* and *Solenastrea* (Kojumdgieva and Strashimov, 1960; Kojumdgieva, 1978). On the other hand, the Upper Badenian reef known from Okhrid is practically built of one species (i.e., *Tarbellastraea conoidea*) and consists of large (up to 2 m) branching coral colonies. The reef front has an algal ridge developed similar to those in Recent reefs. The reef is about 100 m long and about 15 m thick; its upper surface represents a rather sharp erosional surface (suggesting emersion). The reef is covered with well-stratified bioclastic limestones. Lagoonal sediments are developed as intermingly burrowed bioclastic foraminiferal limestones.

The far more important main Badenian (exactly Upper Badenian in age) reef system is composed of red algae with varying amount of vermetid gastropods and locally bryozoans, while stratified deposits contemporaneous with the reefs are developed as various red-algal limestones (see Korolyuk, 1952; Janakievitsch, 1977; Pisera, 1978, 1985; Krach, 1981, and literature herein). This represents the long reef tract (Fig. 4)

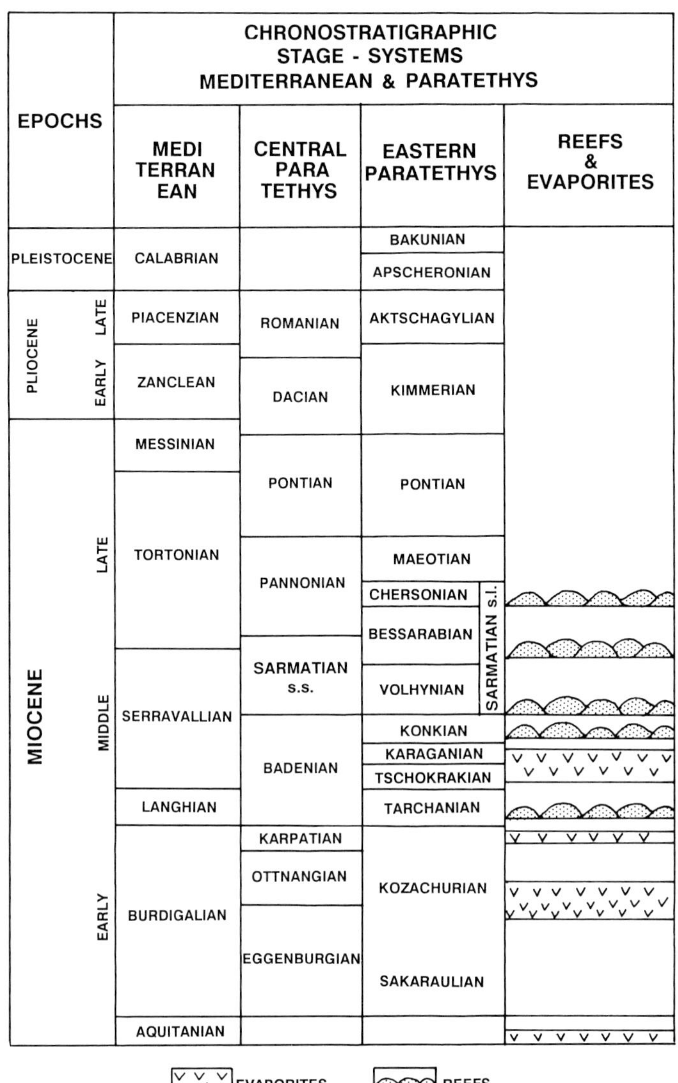


FIG. 2.—Correlation of Mediterranean and Paratethys stages and distribution of reefs and evaporites in the Paratethys (stages correlation after Rögl and Steininger, 1984).

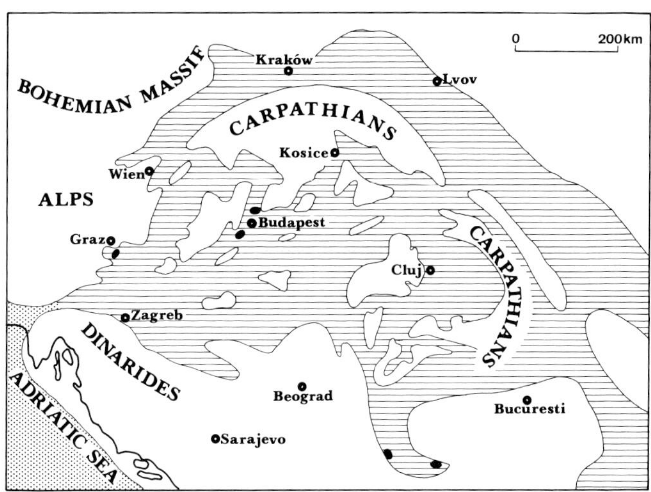


FIG. 3.—Paleogeography and reef distribution in the Central Paratethys during the Early Badenian time (paleogeography modified after Rögl and Steininger, 1984; reefs marked as black oval dots).

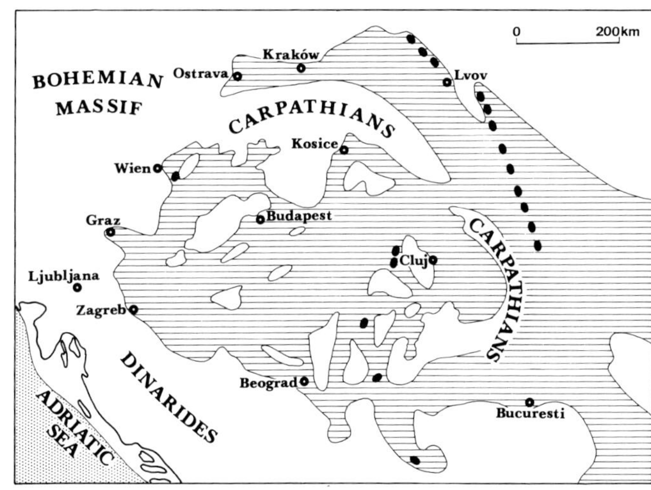


FIG. 4.—Paleogeography and reef distribution in the Central Paratethys during the Late Badenian time (paleogeography modified after Rögl and Steininger, 1984; reefs marked as black oval dots).

which extends along the northern and eastern shores of the sea which infilled the Fore-Carpathian Depression. It begins in eastern Poland (Pisera, 1985) extends through the Ukraine (Teisseyre, 1895; Korolyuk, 1952; where the hills formed by the reef rocks are called Miodobory, in opposition to Toltry which is the local name for hills formed by the Sarmatian reefs developed in the same area) and ends in Moldavia (Roshka and Sajanov, 1966; Janakievitsch, 1977), representing a distance of several hundred kilometers.

The algal vermetid reefs of Badenian age are usually associated with topographic highs, erosional and/or tectonic in origin both in Poland (Fig. 5) and in the Ukraine and Moldavia (Korolyuk, 1952; Janakievitsch, 1977; Pisera, 1985).

The size of the individual reefs in Poland, which are irregularly oval in shape, is up to 15 m in thickness, 500 m long and about 100 m wide. In the Ukraine and Moldavia, the reefs reach even 40 to 80 m in thickness, 800-1500 m in length and 80-400 m in width (Korolyuk, 1952; Janakievitsch, 1977). Janakievitsch (1977, Figs. 8-11) described details of the large reef massifs which are composed of smaller oval or irregular biohermal bodies with sizes up to 4 m x 3 m x 10 m.

Crustose red algae, which composes up to 70% of the reef rock, are represented by 21 species, but only *Lithophyllum albanense*, *Lithothamnion ishigakiensis*, *L. praefruticulosum*, *L. lacroixi*, *Mesophyllum* cf. *schenki* are real frame builders (Pisera, 1985). Most of the algae developed thin lamellar thalli, while thick mamillate crusts are less common (Pisera, 1985). Secondary encrusters are represented by bryozoans, serpulids, sessile foraminifers *Miniacina* and *Nubecularia*. Similar composition is known from the Ukrainian and Moldavian reefs (Korolyuk, 1952; Janakievitsch, 1977).

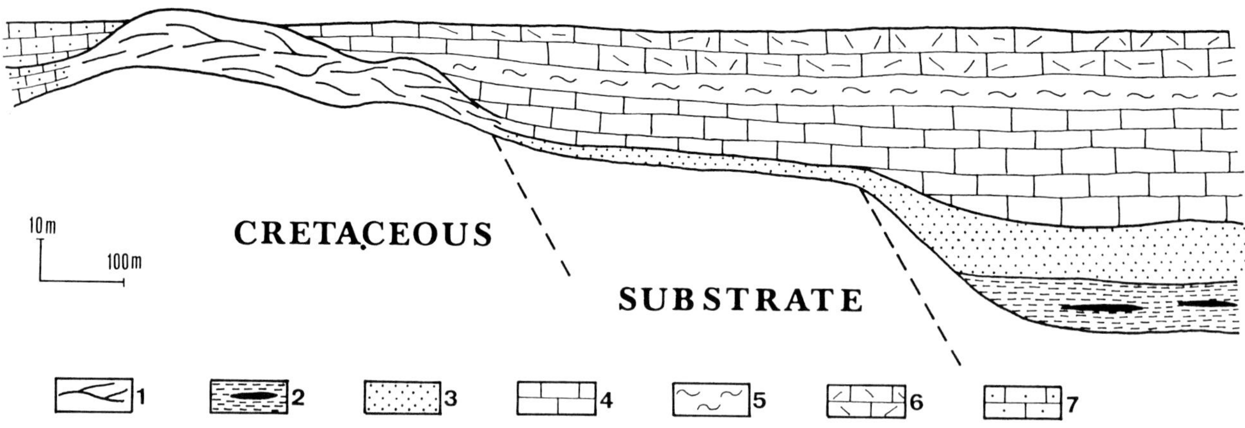


FIG. 5.—Schematic facies relationship of the Upper Badenian algal-vermetid reef and stratified facies from the Roztocze Hills, SE Poland. 1: Red-algal-vermetid reef, 2: clays with brown-coal, 3: quartz sands, 4: red-algal limestones, 5: marls with red-algae, 6: detrital red-algal limestones, and 7: detrital red-algal limestones with quartz.

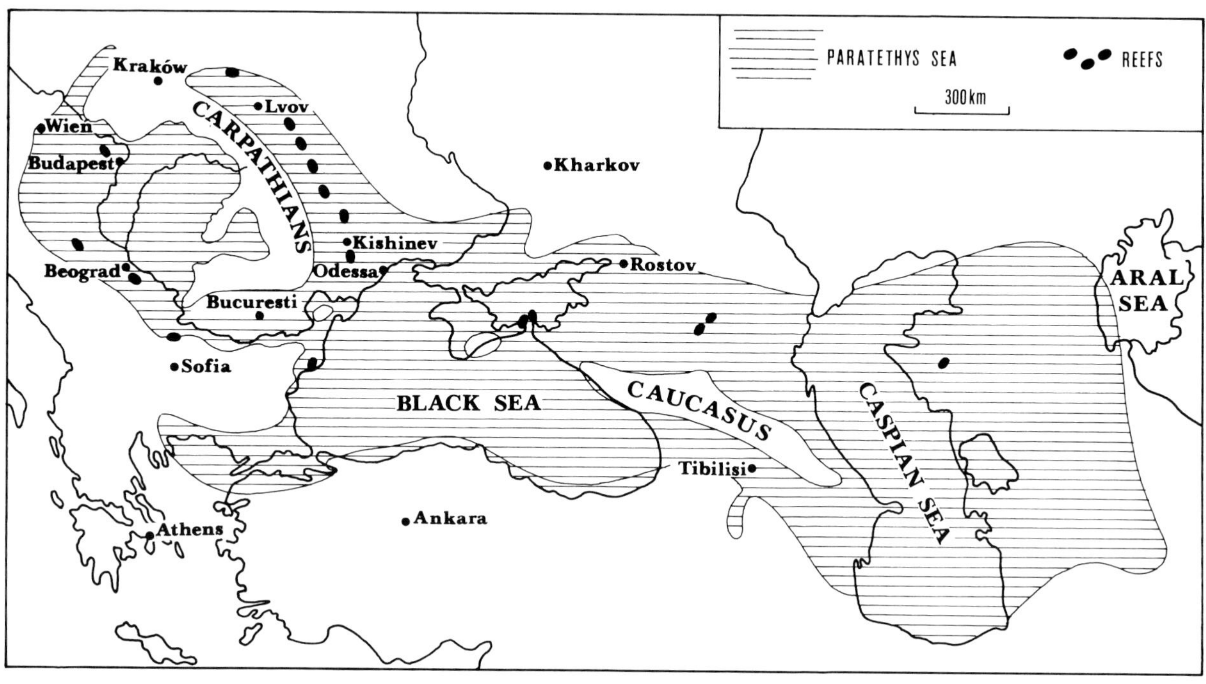


FIG. 6.—Paleogeography and reef distribution during the Lower-Middle Sarmatian (Volhynian to Lower Bessarabian) time (modified after Iljina et al., 1976). Paleogeography shown for Lower Sarmatian (Volhynian) units.

## SARMATIAN REEFS

This reef system (Figs. 6, 7) extends also from Poland to Moldavia and Rumania, often parallel to the Upper Badenian red-algal-vermetid reefs, but is additionally known from north-western and north-eastern Bulgaria (Kojumdgieva, 1969, 1978), Crimea (Kertsch and Taman Peninsulas, Andrusov, 1961), Caucasian Region (Kolesnikov, 1940; Iljina et al., 1976). Biohermal limestones of similar character are also reported from Yugoslavia and Hungary (Pikija et al., 1989). Thick reef complexes, consisting of numerous superimposed smaller biohermal bodies, may attain 15 m in thickness, about 100 m in width, and several hundred meters in length in Poland (Liszkowski and Muchowski, 1969); up to 20 m in thickness in the Ukraine (Korolyuk, 1952) and up to 60 m in thickness in Moldavia (Sajanov, 1968). In Moldavia, there are two roughly parallel belts of reef complexes: a western belt (slightly thinner) of Lower Sarmatian age (Volhynian) and an eastern belt of Middle Sarmatian age (Bessarabian). Thick reef complexes occur in Poland along the former platform edge (Fig. 8) and on its slope

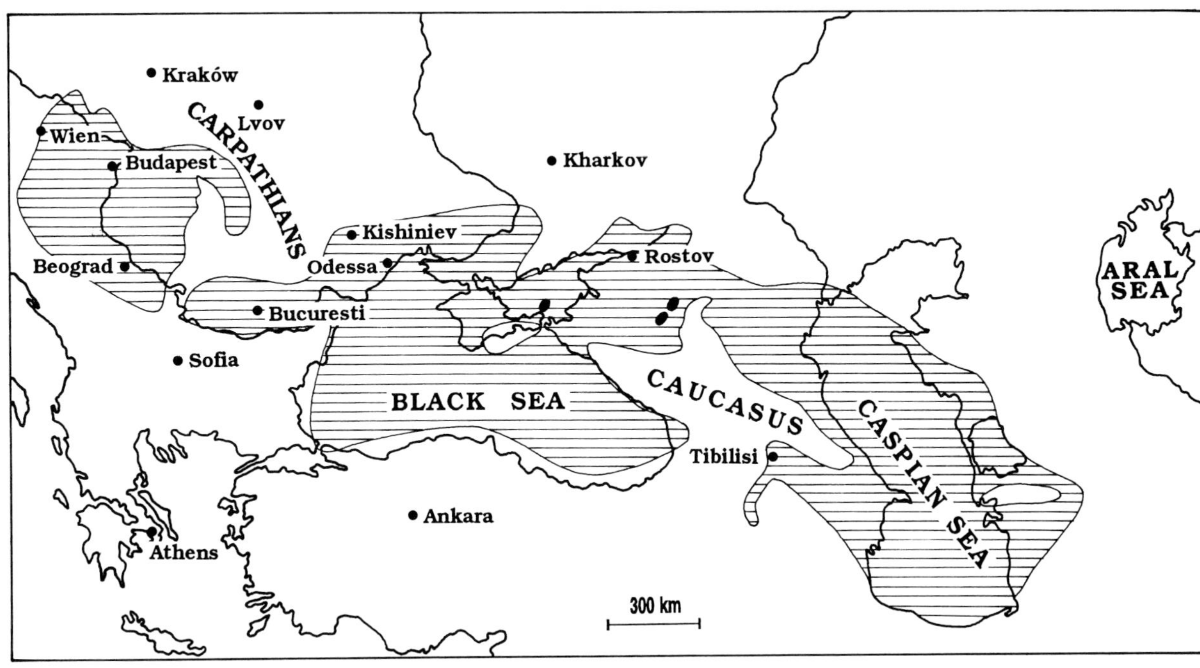


FIG. 7.—Paleogeography and reef distribution during the Upper Sarmatian (Upper Bessarabian to Chersonian) time (modified after Iljina et al., 1976, explanations as for Fig. 6).

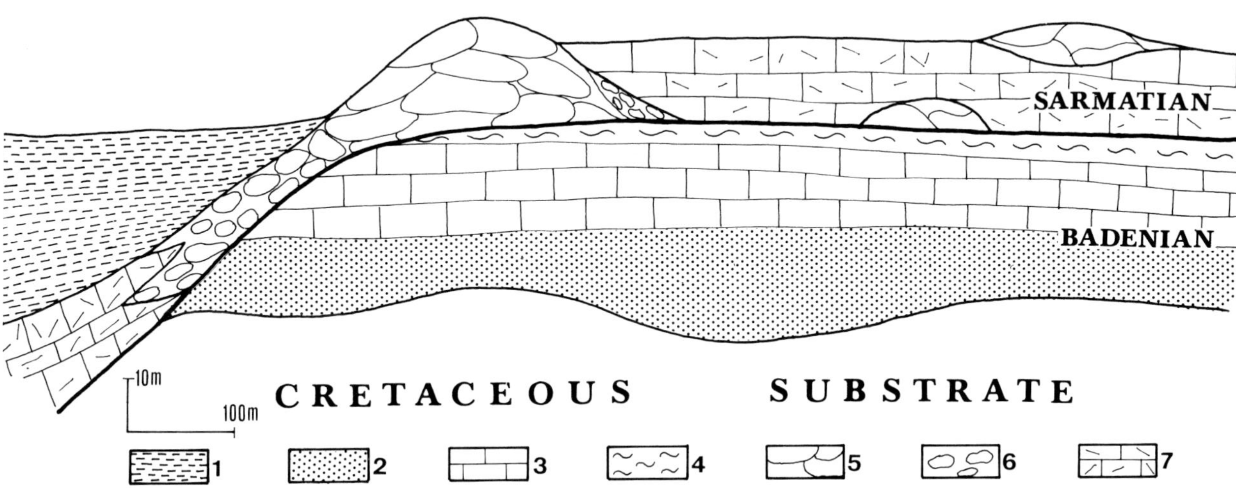


FIG. 8.—Schematic facies relationships of the Lower Sarmatian reefs and stratified Badenian and Lower Sarmatian deposits from the Roztocze Hills, SE Poland. 1: fine clastic deposits of the Carpathian foredeep, 2: quartz sands, 3: red-algal limestones, 4: marls with red-algae, 5: cryptalgal-serpulid reef, 6: reef tallus composed of reef rock conglomerates, and 7: detrital red-algal limestones.

(Liszkowski and Muchowski, 1969; Pisera, 1978). Behind it, there are individual smaller (several meters in size usually) biohermal bodies dispersed on the wide shallow-water platform (usually bioclastic carbonate sedimentation). A similar distribution pattern exists along the entire chain of these reefs (i.e., in the Ukraine and Moldavia as well; Teisseyre, 1884; Korolyuk, 1952; Roshka and Sajanov, 1966; Sajanov, 1968). A somewhat different picture is observed on the Crimean Peninsula where reefs sometimes formed atoll-like structures (Fig. 9) or irregular masses enclosed within clay sediments (Figs. 10, 11); they are associated there with local syncline margins, rather than fault lines (Andrusov, 1961; Krylov and Krasnov, 1966), as is the case for the main chain of the Sarmatian reefs. All these reefs were usually named after the most obvious constructing organisms (i.e., serpulid reefs, Liszkowski and Muchowski, 1969; bryozoan reefs, Andrusov, 1961; *Nubecularia* reefs, Gillet and

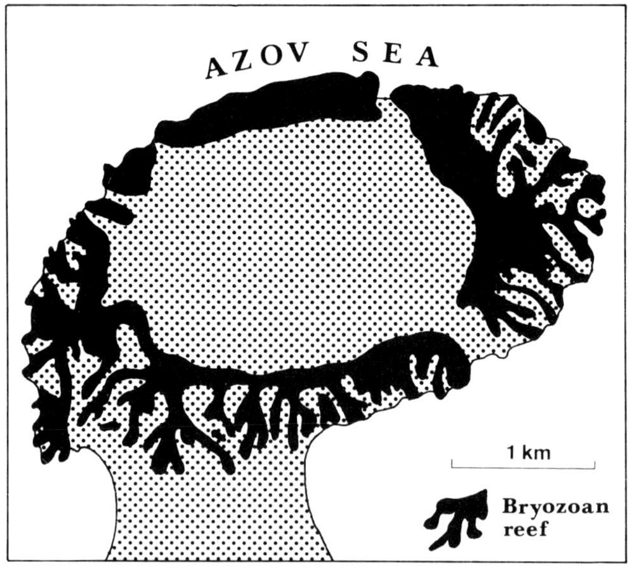


FIG. 9.—Upper Sarmatian (Chersonian) atoll-like bryozoan reef from the Crimean (Kazantip) Peninsula (after Andrusov, 1961). Stippled are bedded deposits.

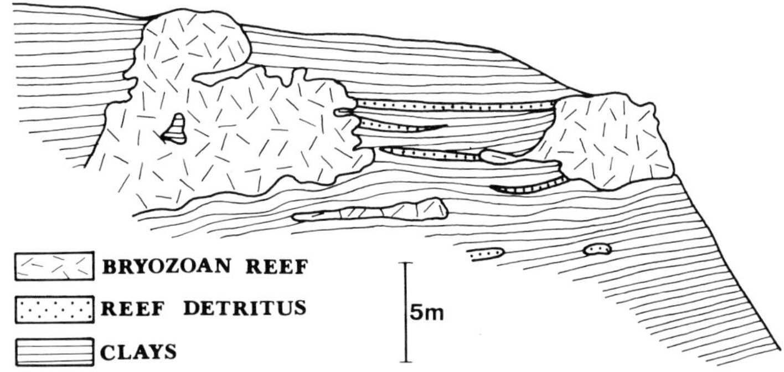


FIG. 10.—Relationships of the Upper Sarmatian (Chersonian) bryozoan reefs and stratified deposits from the Crimean (Takyl-Burun; from Kolesnikov, 1940).

Derville, 1931). Mixtures of organisms, however, are present as well. In many cases, various stromatolitic and thrombolitic structures (composed of peloidal sediment) and early fibrous carbonate cements are often very important, if not dominating, elements. The reef rock is extremely porous and usually contains numerous internal sediments. There is a trend in the composition of reefs through time. For example, Lower Sarmatian (Volhynian) reefs show a dominance of cryptalgal fabrics, while skeletal components (bryozoans, *Nubecularia* and monostromatic red algae) are of subordinate importance and usually form only 2- to 3-cm-thick encrustations on individual biohermal bodies (Pisera, 1978, 1990). On the other hand, the Middle Sarmatian (Bessarabian) reefs often contain more skeletal than cryptalgal structures (Andrusov, 1961; Sajanov, 1968). Early submarine cement is always of considerable importance and its role was simply not recognized earlier.

Among skeletal components, serpulids (most probably

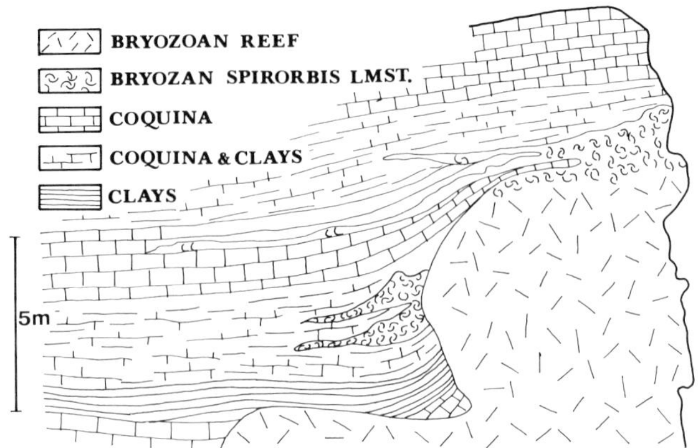


FIG. 11.—Details of the contact between the Upper Sarmatian (Chersonian) bryozoan reef and surrounding stratified deposits from the Crimean (Nasyr; from Kolesnikov, 1940).

*Hydroides*) are the most important in the Lower Sarmatian (Volhynian) reefs in Poland and the Ukraine. Bryozoans (both cheilostomates and cyclostomates) are second in importance. They are represented mainly by *Schizoporella* and *Cryptosula* and usually encrust only cryptalgal bioherms. Sometimes they construct their own small biohermal bodies in Poland and the Ukraine. The only reef forming cyclostomatous bryozoan in Poland is *Fascicularia* (Malecki, 1980). The same bryozoan genera participated in the construction of other reefs (Ghiurca, 1974; Ghiurca and Stancu, 1974; Bagdasarian and Ponomarieva, 1982; Veis, 1988), but in the case of the Kertsch reefs, *Membranipora* is the most important reef builder (Veis, 1988).

There are clearly several discrete episodes of reef formation: Lower Sarmatian (Volhynian), Middle Sarmatian (Bessarabian) and Upper Sarmatian (Chersonian) (see Fig. 2). Each reef episode starts from an erosional (transgressive) surface separating it from older deposits. In Poland and the Ukraine, only Lower Sarmatian (Volhynian) reefs occur; whereas more to the southeast, in Moldavia, Bulgaria and Crimea, there are also Middle Sarmatian (Bessarabian) reefs; and on the Kertsch and Taman Peninsulas (Crimea), Middle (Bessarabian) and Upper Sarmatian (Chersonian) reefs are known (Kojumdgieva, 1969; Iljina et al., 1976; Veis, 1988). No Upper Sarmatian (Chersonian) reefs are known from the Pannonian Basin, as during that time it was a nearly fresh-water lake (Kazmer, 1990) separated from the rest of the Paratethys (see Fig. 7).

## DISCUSSION

There were many factors controlling reef development during Miocene deposition in the Paratethys realm, but most important for all the reef types was the development of a vast and shallow (usually carbonate) nearshore platform and the presence of positive denivelation (erosional or tectonic) of the sea bottom on which the reef growth had started. Such situations originated following the Badenian and Sarmatian transgressions (see Rögl and Steininger, 1983). All other factors responsible for the reef growth were specific for particular reef types. The presence of coral reefs had been limited by climatic factors and salinity. The northern part of the basin was, during Badenian time, outside the tropical zone and thus devoid of coral reefs. Instead extensive of algal-vermetid reefs were developed. In more southern parts of the basin, the more tropical climate allowed for coral reef growth, but these reefs still were probably at their northern limit as the corals are poorly diversified. One cannot exclude also some increase in salinity as a limiting factor, keeping in mind how resticted the connections were with the open ocean at that time. Another local factor controlling reef development was the supply of clastics. Reefs are absent along the Carpathian shore of the foredeep sea, because growing Carpathians were an important source of clastics.

The origin of the Sarmatian reefs depended mostly on the chemistry of waters in a closed or semiclosed basin. The Sarmatian and younger Paratethys sea was previously described as a freshening to brackish water basin. However, such hypothesis seems to be an oversimplification as indicated by the importance of early submarine fibrous cement (now calcite but most probably formerly aragonite as suggested by crystal morphology and high (>2000 ppm) Sr content) and the common occurrence of ooids (Pisera, 1990). Rather one would expect a sea with a water chemistry that is not simply an effect of dilution of the sea water but rather is highly supersaturated in respect to calcium carbonate and probably also with high alkalinity (flourishing cyanobacterial communities responsible for thrombolitic and stromatolitic fabric). Development of a local hypersaline environment without precipitation of evaporitic minerals cannot be ruled out. Such specific water chemistry variations seem to be responsible for the development of the peculiar and complex reef constructions.

## REFERENCES


ANDRUSOV, N. I., 1961, Iskopajemyje mshankovyje rify Kertschenskovo i Tamanskovo poluostrovov, *in* Andrusov, N. I., Izbrannyje Trudy (1909-1912): Moskva, Akademia Nauk SSSR, p. 395-540.

BAGDASARIAN, K. G. AND PONOMARIEVA, L. D., 1982, Polozenije mshanok v ekosistiemakh sarmatskovo moria: Paleontologitscheskij Sbornik, v. 19, p. 76-81.

BELOKRYS, L. S., 1967, Opresnialos li juzhnoukrainskoje morie: Sovietskaja Gieologija, v. 7, p. 97-110.

DULLO, Ch.-W., 1983, Fossildiagenese im miozänen Leitha-Kalk der Paratethys von Österreich: Ein Beispiel fur Faunenverschreibungen durch Digeneseunterschiede: Facies, v. 8, p. 1-112.

FRIEBE, J. G., 1990, Lithostratigraphische Neugliederung und Sedimentologie der Ablagerungen des Badenium (Miozän) um die Mittelsteirische Schwelle (Steirischen Becken, Österreich): Jahrbuch der Geologischen Bundesanstalt, v. 132, p. 223-257.

FRIEBE, J. G., 1993, Sequence stratigraphy in a mixed carbonate-siliciclastic depositional system (Middle Miocene: Styrian Basin, Austria): Geologische Rundschau, v. 82, p. 281-294.

GHIURCA, V., 1974, Les caractères stratigraphiques de litho- et biofacies du Tortonien récifal de la République Populaire Roumaine et des pays voisins: Bulletin du VI Congrès de l'Association Géologique Carpatho-Balkanique, Stratigraphie, v. 3, p. 285-291.

GHIURCA, V. AND STANCU, J., 1974, Les Bryozoaires sarmatiens du Paratethys Central, *in* Papp, A., Marinescu, F., and Seneš, J., eds., Chronostratigraphie und Neostratotypen. Miozän der zentralen Paratethys. M5. Sarmatien, v. 4: Bratislava, VEDA, p. 298-310.

GILLET, S. AND DERVILLE, H., 1931, Nouveau gisement d'un récif à *Nubecularia* à Cricov près de Chisinau (Bessarabie): Bulletin de la Société géologique de France, 5 série, v. 1, p. 721-738.

HEGEDÜS, G., 1970, Tortonai korallok Herendöl: Földtani Közlöny, v. 100, p. 185-191.

ILJINA, L. B., NEVESSKAJA, L. A. AND PARAMONOVA, N. O., 1976, Regularities of molluscs development in the Neogene semimarine and brackishwater basins of Eurasia (late Miocene-early Pliocene): Moskva, Trudy Paleontologitscheskovo Instituta, v. 155, p. 1-287.

JANAKIEVITSCH, A. N., 1977, Srednemiocenovye rify Moldavii: Kishiniev, Shtintza, 115 p.

KAZMER, M., 1990, Birth, life and death of the Pannonian Lake: Palaeogeography, Palaeoclimatology, Palaeoecology, v. 79, p. 171-188.

KOJUMDGIEVA, E., 1969, Sarmat, *in* Fosilite na Balgaria, v. 8: Sofia, Académie Bulgare des Sciences, 135 p.

KOJUMDGIEVA, E., 1976, Paléoécologie des communautés des Mollusques du Miocène en Bulgarie du Nord-Ouest. II. Communautés des Mollusques du Badenien (Miocène moyen) en Bulgarie du Nord-Ouest: Geologica Balcanica, v. 6, p. 63-94.

KOJUMDGIEVA, E., ed., 1978, Guide de l'excursion du IX Symposium du groupe "Paratethys". Néogène en Bulgarie du Nord-Ouest: Sofia, 42 p.

KOJUMDGIEVA, E. AND STRASHIMOV, B., 1960, Tortonien, *in* Fosilite na Balgaria, v.7: Sofia, Académie Bulgare des Sciences, 317 p.

KÓKAY, J., MIHÁLY, S., AND MÜLLER, P., 1984, Bádeni korú rétegek a budapesti Örs vezér tere környékén: Földtani Közlöny, v. 114, p. 285-295.

KOLESNIKOV, V. P., 1940, Vierkhnij miocen, *in* Arkhangielskij, A. D., ed., Stratigrafija SSSR: Moskva, Akademia Nauk SSSR, p. 229-406.

KOROLYUK, I. K., 1952, Podolskiye toltry i uslovya ikh obrazovanya: Trudy Instituta Geologitscheskikh Nauk SSSR, v. 110, p. 1-115.

KRACH, W., 1981, The Baden reef formation in Roztocze Lubelskie: Prace Geologiczne PAN, v. 121, p. 5-115.

KRYLOV, I. N. AND KRASNOV, E. V., 1966, Neogenovyje rifogennyje obrazovanija Kertschenskovo poluostrova, *in* Heckker, R. F. and Niegadajev-Nikonov, K. N., eds., Putievoditiel Ekskursij Tschetviertoj Paleoekologo-Litologitshceskoj Sessi: Kishiniev, p. 3-13.

LELKES, G. AND STUDENCKI, W., 1990, Badenian (Middle Miocene) carbonates of Central Paratethys: Tropical or temperate? (abs.): Barcelona, Abstracts of IXth Congress of Regional Committee on Mediterranean Neogene Stratigraphy, p. 211.

LISZKOWSKI, J. AND MUCHOWSKI, J., 1969, Morfologia, budowa wewnętrzna oraz geneza masywów wapieni biogenicznych dolnego sarmatu strefy progów zewnętrznych poludniowej krawędzi wyżyny lubelskiej: Biuletyn Geologiczny, v. 11, p. 5-29.

MAŁECKI, J., 1980, A new reef-building bryozoan species from the Miocene of Roztocze: Acta Palaeontologica Polonica, v. 25, p. 91-99.

MÜLLER, P., 1984, Decapod Crustacea of the Badenian: Geologica Hungarica, Series Palaeontologia, v. 42, p. 1-317.

NEY, R., BURZEWSKI, W., BACHLEDA, T., GÓRECKI, W., JAKÓBCZYK, K., AND SŁUPCZYŃSKI, K., 1974, Zarys paleogeografii i rozwoju litologiczno-facjalnego utworów miocenu Zapadliska Przedkarpackiego: Prace Geologiczne, v. 82, p. 1-65.

OOSTERBAAN, A. F. F., 1990, Notes on a collection of Badenian (Middle Miocene) corals from Hungary in the National Museum of Natural History at Leiden (The Netherlands): Contribution to Tertiary and Quaternary Geology, v. 27, p. 3-15.

PIKIJA, M., ŠIKIC, K., TIŠLJAR, J., AND ŠIKIC, L., 1989, Biolititni i prateci karbonati facjesi sarmata u području Krašic-Ozallj (središnja Hrvatška): Geoloski Vjesnik, v. 42, p. 15-28.

PILLER, W., KLEEMANN, K., AND FRIEBE, J. G., 1991, Middle Miocene reefs and related facies in eastern Austria, *in* Excursion B4 Guidebook of the VI International Symposium on Fossil Cnidaria including Archaeocyatha and Porifera, Münster 1991: Münster, International Association for the Study of Fossil Cnidaria and Porifera, 47 p.

PISERA, A., 1978, Miocene reef deposits of the western Roztocze: Przegląd Geologiczny, v. 3, p. 159-162.

PISERA, A., 1985, Paleoecology and lithogenesis of the Middle Miocene (Badenian) algal-vermetid reefs from the Roztocze Hills, south-eastern Poland: Acta Geologica Polonica, v. 35, p. 89-155.

PISERA, A., 1990, Upper Miocene cryptalgal-serpulid reefs from Poland: interplay of biological and inorganic factors (abs.): Barcelona, Abstracts of IXth Congress of Regional Committee on Mediterranean Neogene Stratigraphy, p. 269.

RADWAŃSKI, A., 1965, Transgresja dolnego tortonu na południowych stokach Gór Świętokrzyskich (strefa zatok i ich przedpola): Acta Geologica Polonica, v. 19, p. 1-160.

RÖGL, F. AND MÜLLER, C., 1978, Middle Miocene salinity crisis and paleogeography of the Paratethys (Middle and Eastern Europe): Washington, D.C., Initial Reports of the Deep Sea Drilling Project, v. 42, p. 985-990.

RÖGL, F. AND STEININGER, F. F., 1983, Vom Zerfall der Tethys zu Mediterran und Paratethys: Annalen des Naturhistorischen Museum Wien, v. 85/A, p. 135-163.

RÖGL, F. AND STEININGER, F. F., 1984, Neogene Paratethys, Mediterranean and Indo-Pacific Seaways, *in* Brenchley, P. J., ed., Fossils and Climate: London, Wiley, p. 171-200.

ROSHKA, V. H. AND SAJANOV, V. S., 1966, Iskopajemyje rify Moldavskoj SSR, *in* Hecker, R. F. and Niegadajev-Nikonov, K. N., eds., Putievoditiel Ekskursij Tschetviertoj Paleoekologo-Litologitscheskoj Sessi: Kishiniev, p. 46-52.

SAJANOV, V. S., 1968, Sostav, strojenije i proiskhozdienije sredniesarmatskikh biogermov Moldavskoj SSR, *in* Iskopajemyje rify i metodika ikh izutschenija: Sverdlovsk, Uralskij Filial Akademi Nauk SSSR, Institut Gieologi i Gieofiziki, p. 210-225.

SCHOLZ, G., 1970, A visegrádi Fekte-hegy tortonai korall faunája: Földtani Közlöny, v. 100, p. 192-206.

STEININGER, F. AND PAPP, A., 1978, Faziostratotypus: Gross Hofleins NNW, Steinbruch "FENK", Burgenland, Osterrreich, *in* Seneš, J., ed., Chronstratigraphie und Stratotypen. Miozän M4. Badenien: Bratislava, VEDA, v. 6, p. 194-199.

STEININGER, F. F. AND RÖGL, F., 1979, The Paratethys history- A contribution towards the Neogene geodynamics of the Alpine orogene (An abstract): Annales Géologiques des Pays Helleniques, Hors serie, v. 3, p. 1153-1165.

STEININGER, F. F., SENEŠ, J., KLEEMANN, K., AND RÖGL, F., 1985, Neogene of the Mediterranean Tethys and Paratethys. Stratigraphic correlation tables and sediment distribution maps. Wien: Institute of Paleontology, University of Vienna; v. 1 and 2, 189 and 536 p.

STUDENCKI, W., 1988, Facies and sedimentary environment of the Pinczów Limestones (Middle Miocene; Holy Cross Mountains, central Poland): Facies, v. 18, p. 1-26.

TEISSEYRE, L., 1884, Der Podolische Hagelzug der Midoboren als ein sarmatisches Bryozoenriff: Jahrbuch der Geologischen Reichsanstalt, v. 34, p. 299-312.

TEISSEYRE, W., 1895, O charakterze fauny kopalnej Miodoborów: Rozprawy Wydziału Matematyczno-Przyrodniczego Akademii Umiejętności, v. 30, p. 1-11.

VEIS, O. B., 1988, Miocenovyje mshanki seviernovo Kavkaza i Kryma: Trudy Paleontologitscheskovo Instituta, v. 232, p. 1-101.

# STRUCTURAL AND FAUNAL EVOLUTION OF CHATTIAN — MIOCENE REEFS AND CORALS IN WESTERN FRANCE AND THE NORTHEASTERN ATLANTIC OCEAN

BRUNO CAHUZAC

*Laboratoire de Recherches et d'Applications Géologiques (LARAG), Université Bordeaux-I, 351 Cours de la Libération, F-33405 Talence, France*

AND

CHRISTIAN CHAIX

*Laboratoire de Paléontologie, Muséum National d'Histoire Naturelle, 8 Rue Buffon, F-75005 Paris, France*

ABSTRACT: Chattian and Miocene coral reefs and faunas of the western France basins are reviewed within a paleogeographic context. Several new outcrops have been discovered, and extensive new and historic collections have been studied. Coral diversity was very high in the Aquitaine basin during the Chattian(1) (ca. 150 species) and a little less so during the Early Miocene (110 species); during these times, relatively small reefal buildups formed in a tropical climate. The Mid-Miocene coral faunas show a marked decrease in diversity (some 75 species in all), with "subreefal" facies in the Langhian of southwestern and northwestern France. The Upper Miocene fauna is even poorer (just about 20 species) and only known in northwestern France. Throughout the Miocene, the proportion of hermatypic taxa also decreased notably; in the coral assemblages, these species were strongly predominant from Chattian to Burdigalian. Afterwards, the ahermatypic taxa became progressively predominant. Other northeastern Atlantic areas (Portugal, Morocco) are also investigated. Some biogeographic data sketch the evolutionary trend of these coral communities. During the Chattian, an (eastern and western) Atlantic-Mediterranean bioprovince was differentiated. During the Early Miocene, this bioprovince was restricted to eastern Atlantic and Mediterranean. From the Mid-Miocene, the coral faunas were disconnected from the Mediterranean, and an impoverished eastern Atlantic bioprovince became established without real renewal. A comparison with Mediterranean reefs shows that maximum coral building took place within the Mid-Miocene in the Mediterranean realm (with continuation of reefs in the Late Miocene), instead of Chattian (and Early Burdigalian) as in the Atlantic areas.

## INTRODUCTION

Abundant corals are present throughout the Chattian and the Miocene along the western coast of France (Figs. 1, 2) and in the Lower Miocene of Portugal. Coral reefs occur in the Aquitaine Basin (Burdigalian, Aquitanian and Upper Oligocene) and in Low Tage Gulf near Lisbon (Aquitanian, with only some ten species). Important ahermatypic mounds and blankets associated with hermatypic corals formed in the Mid-Late Miocene of western France but did not develop any real coral reef buildup. The Miocene corals in Atlantic France have been well known since the early 19th Century (e.g., Lamarck, Defrance, Michelin, Milne-Edwards and Haime, and others: ref. in Chevalier, 1961; Oosterbaan, 1988). Chevalier (1961) presented an extensive monograph on this abundant and well-preserved coral fauna, currently under revision by Chaix (unpubl. data).

First, the geologic and paleogeographic setting will be summarized. Then, this work will review the key stratigraphic and paleoecologic features of the Miocene reefs cropping out in western France, stressing that in contrast with this rich coral fauna, the actually observable reef buildups are very thin and small, generally less than few meters thick (unlike, for instance, some large bioherms and fringing reefs known in the Aquitaine Rupelian (="Stampian"): Tuc du Saumon, Fig. 3, cf. Cahuzac, 1980). On the other hand, Mediterranean Chattian and Miocene reefs are thick and well developed, commonly with comparatively few reef genera. Moreover, we present the basic overall trends in western France reef development and attempt to relate them to larger-scale patterns of climate change and/or marine current circulation. Several other more regional environmental factors are discussed, too. In addition, a comparison is made with other eastern Atlantic areas (Portugal or Morocco basins with corals), as well as with Mediterranean.

The main marine Miocene basins cropping out in western France are the Aquitaine Basin in the southwest and the basins around Brittany in northwestern France (Fig. 1). These Atlantic basins had no direct marine connection with the Mediterranean Miocene basin of southern France (Rhône Valley, Languedoc, Basse-Provence), but they are included in this volume because of their stratigraphic, paleogeographic and paleontological importance; moreover, there are striking relationships with the Mediterranean coral faunas. The Aquitanian and Burdigalian stratotypes were defined in the Aquitaine Basin (Alvinerie et al., 1977, 1992); and biostratigraphic revisions (Poignant and Pujol, 1976, 1978; Müller and Pujol, 1979; Cahuzac et al., 1994, 1995), the study of benthic forams (Drooger et al., 1955; Cahuzac, 1980, 1984; Cahuzac and Poignant, 1988, 1992, 1993) and the "grade-datings" (Magné et al., 1987) allow a precise correlation of the standard Miocene stages with the local stratigraphic units (Tables 1, 2). The Miocene carbonates of western France are typically poorly consolidated and friable, and contain abundant gastropod and bivalve coquina, with variable amounts of terrigenous siliciclastics: the traditional French term "falun" (nearly synonymous with the English term "crag") refers to this characteristic lithology (Moyes, 1966). The excellent preservation of fossils (forams, ostracods, otoliths, mollusks, corals, echinoids, bryozoans, etc.) in these rocks explains the extensive paleontological literature in the region. A remarkable feature of all the Miocene carbonates of western France is the absence or rarity of calcareous algae.

---

[1] In this text, when we use a term of epoch or geologic stage on its own, as "the Chattian", we refer to Chattian time.

Models for Carbonate Stratigraphy from Miocene Reef Complexes of Mediterranean Regions,
SEPM Concepts in Sedimentology and Paleontology #5, Copyright © 1996,
SEPM (Society for Sedimentary Geology), ISBN 1-56576-033-6, p. 105-127.

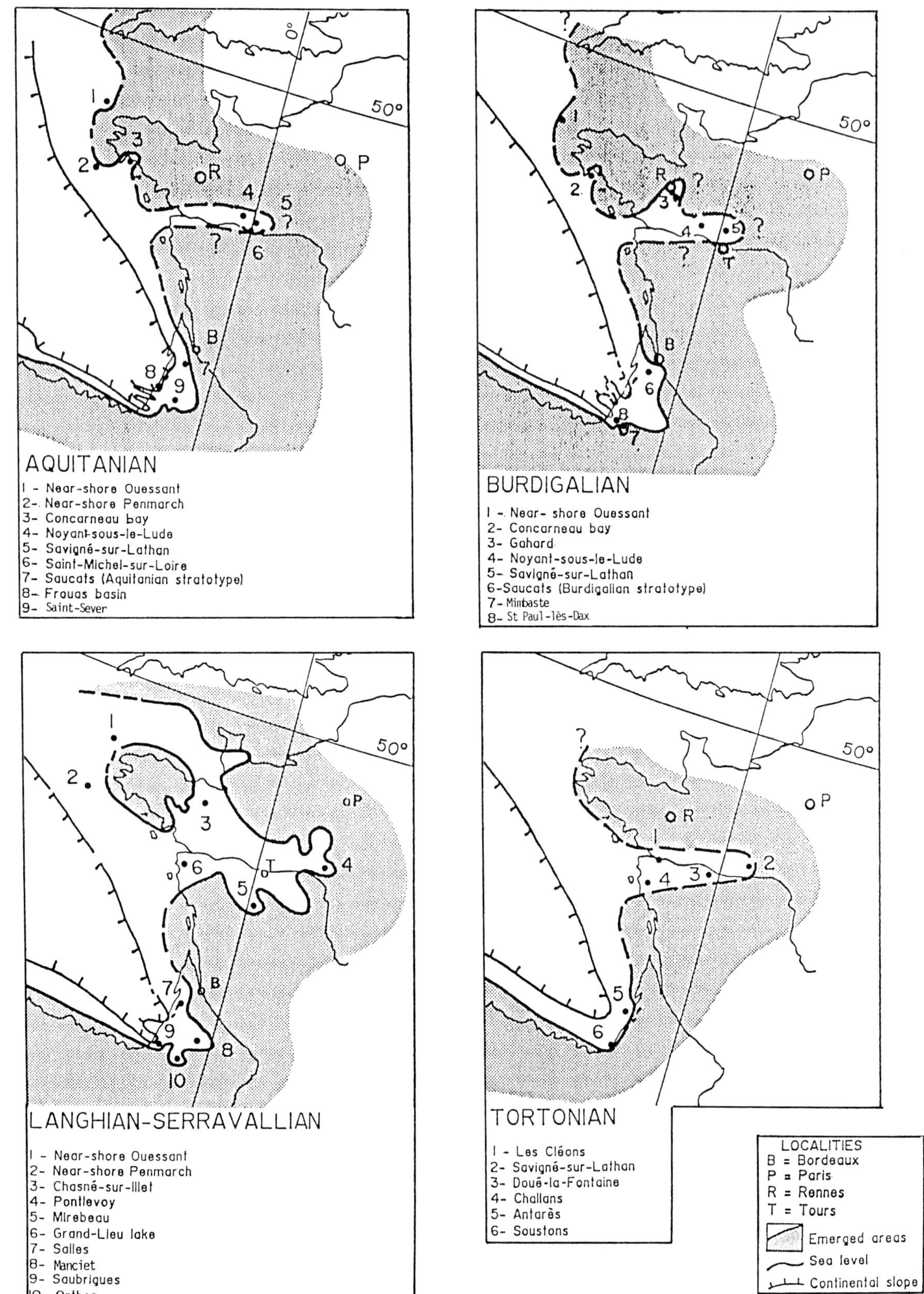


FIG. 1.—Miocene paleogeographic maps of western France (after Alvinerie et al., 1992).

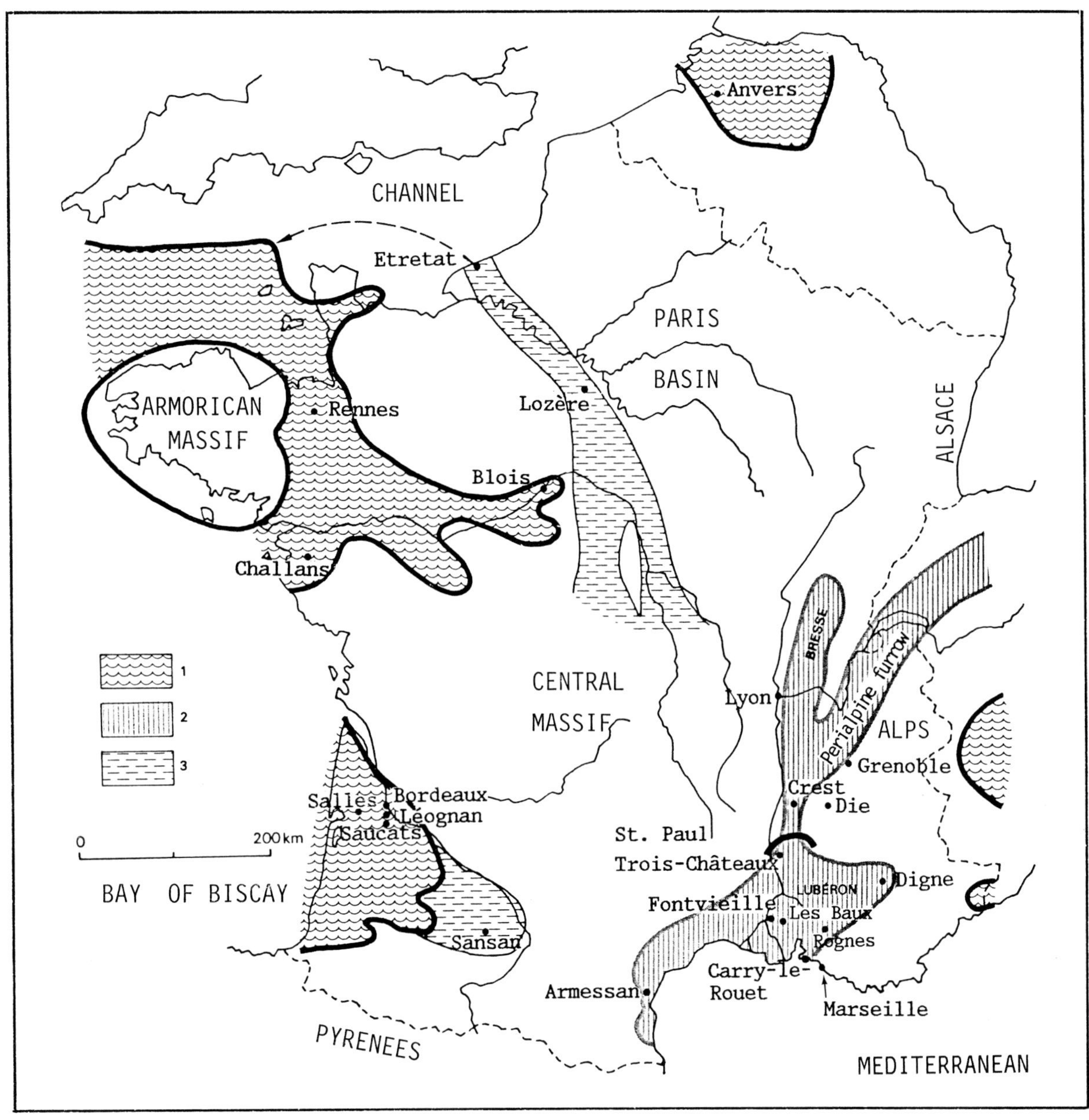


FIG. 2.—The Miocene deposits in France (after Pomerol, 1973); 1- Epeiric seas; 2- Perialpine "molassic" trench; 3- Continental deposits.

## AQUITAINE BASIN AND COMPARISON WITH OTHER EASTERN ATLANTIC AREAS DURING CHATTIAN AND EARLY-MID MIOCENE

During the Miocene, the Aquitaine Basin was a large embayment (Fig. 2-4) opening westward into the Bay of Biscay (Atlantic Ocean). The northern part was a stable shallow shelf with gentle anticlines. The southern part contains thick Cenozoic sections controlled by west-northwest/east-southeast-trending diapiric ridges (Fig. 3). The Alpine Pyrenean Range is the southern limit of the Aquitaine Basin, where major deformation occurred at the end of the Late Lutetian (Mid Eocene), and thrust movement finished at the end of the Oligocene (Cahuzac, 1980). The Miocene was a calm period in the Aquitaine Basin, only interrupted by repeated diapiric movements of Triassic evaporites and minor re-activation of pre-Neogene structures at the end of the Mid-Miocene (Alvinerie, 1969) and before the Pliocene (Durand, 1974). Miocene strata are essentially subhorizontal or gently dipping, progressively filling up the Aquitaine embayment.

The marine Miocene carbonates of the Aquitaine Basin interfinger with and grade laterally into sandy carbonates, marls and clastics of the margino-littoral domain (swamps, lagoons and fluvio-lacustrine environments). During this period, several

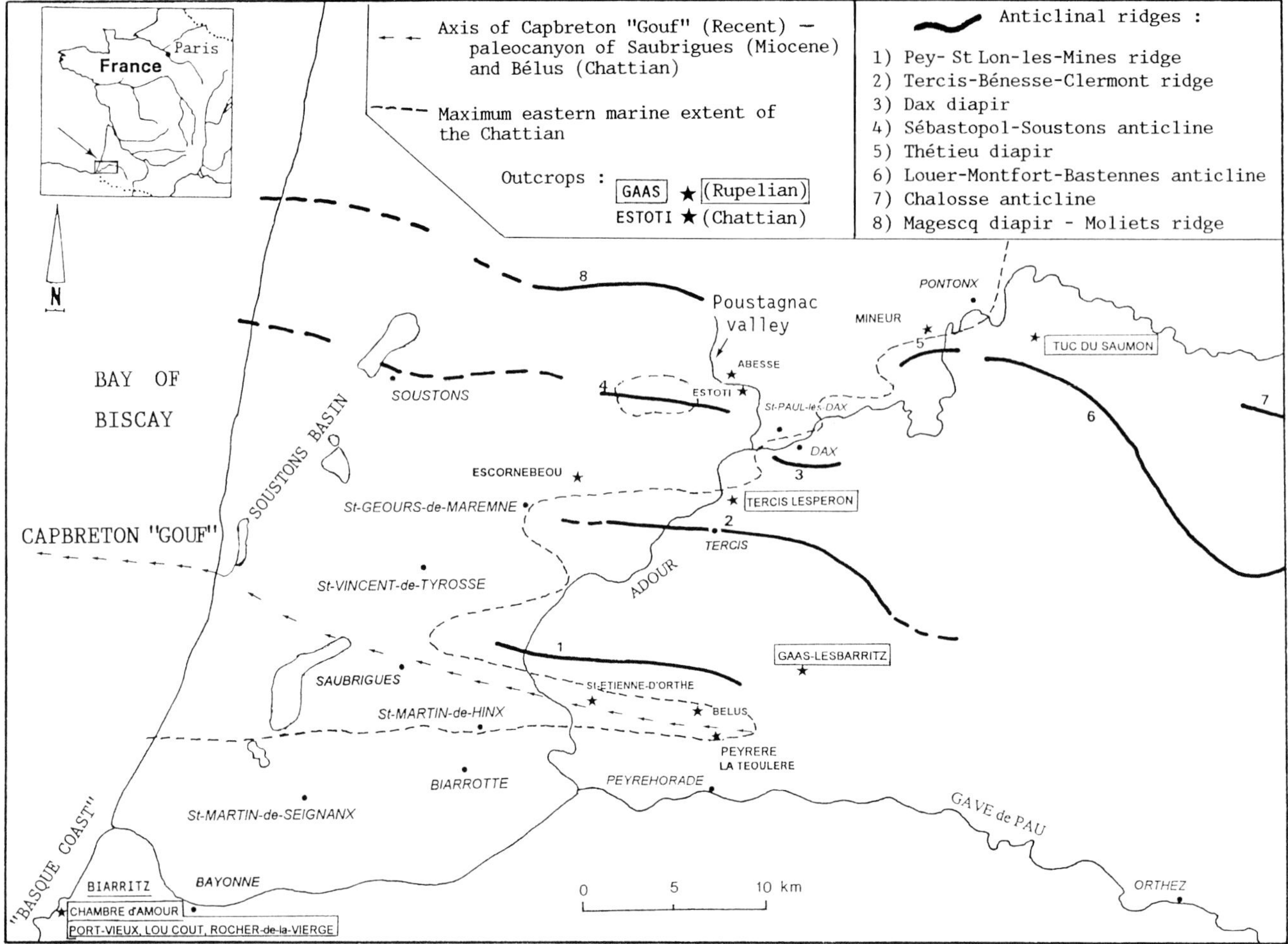


FIG. 3.—Stuctural framework and paleogeography of the Chattian southwestern Aquitaine Basin.

transgressions extended more or less eastward from the Atlantic. The depths of these epeiric seas were generally shallow in the Aquitaine Basin, except in the western areas where pelagic marls were deposited (for instance Soustons and Frouas basins, Saubrigues Gulf; Figs. 1, 3). These transgressive overlaps approximately reflect the eustatic cycles of the Miocene epoch (Table 1). The main difference with Haq et al. (1987) is in the absolute age of the Oligocene-Miocene boundary, considered to be at about 23 Ma rather than 25.2 Ma; this difference is due to more refined datings of the Aquitanian and Burdigalian stratotypes that are now available (Magné et al., 1987; Alvinerie et al., 1992) and also of many other Miocene series (Odin, 1994). The location of most of the Miocene reef outcrops is controlled by the trend of the emergent diapiric ridges (Figs. 3, 4), notably in southern Aquitaine. There is no surface or subsurface evidence of a marginal or frontal Miocene barrier reef protecting the Aquitaine embayment from the open Atlantic Ocean.

Some of the well exposed sections of the Upper Oligocene (Chattian) series of southern Aquitaine were previously believed to be Lower Miocene age (e.g., Chevalier, 1961, 1963) particularly on the basis of mollusk studies. Several new Chattian outcrops rich in corals have been discovered recently. Their faunal content has been revised in detail, and a Chattian age has been established notably thanks to the microfauna assemblages which generally contain the marker species of classic biozonations: numerous large forams (e.g., Nummulitidae, *Miogypsinoides*), planktonic forams and nannoplankton indicating the Zones P22 Blow (1969) and NP25 Martini (1971) (e.g., G.F.E.N., 1974; Cahuzac, 1980, 1984; Cahuzac and Poignant, 1988, 1992; Cahuzac et al., 1995). Also, there are Sr isotopes age dates (Cahuzac, in prep.). In this southern Aquitaine area, there is a practically continuous section from the Lower Chattian to the Langhian. The Serravallian series seems incomplete and less well-exposed and corresponds in part to the traditional "Helvetian" and "Sallomacian" local units. The marine Upper Miocene and Pliocene deposits are only known from wells on the Atlantic coast (ref. in Alvinerie et al., 1992).

TABLE 1.—STRATIGRAPHIC FRAMEWORK, WITH THE REGIONAL STAGES AND FACIES IN THE WESTERN FRANCE (PRO PARTE AFTER HAQ ET AL., 1987; ALVINERIE ET AL., 1992, MODIFIED).

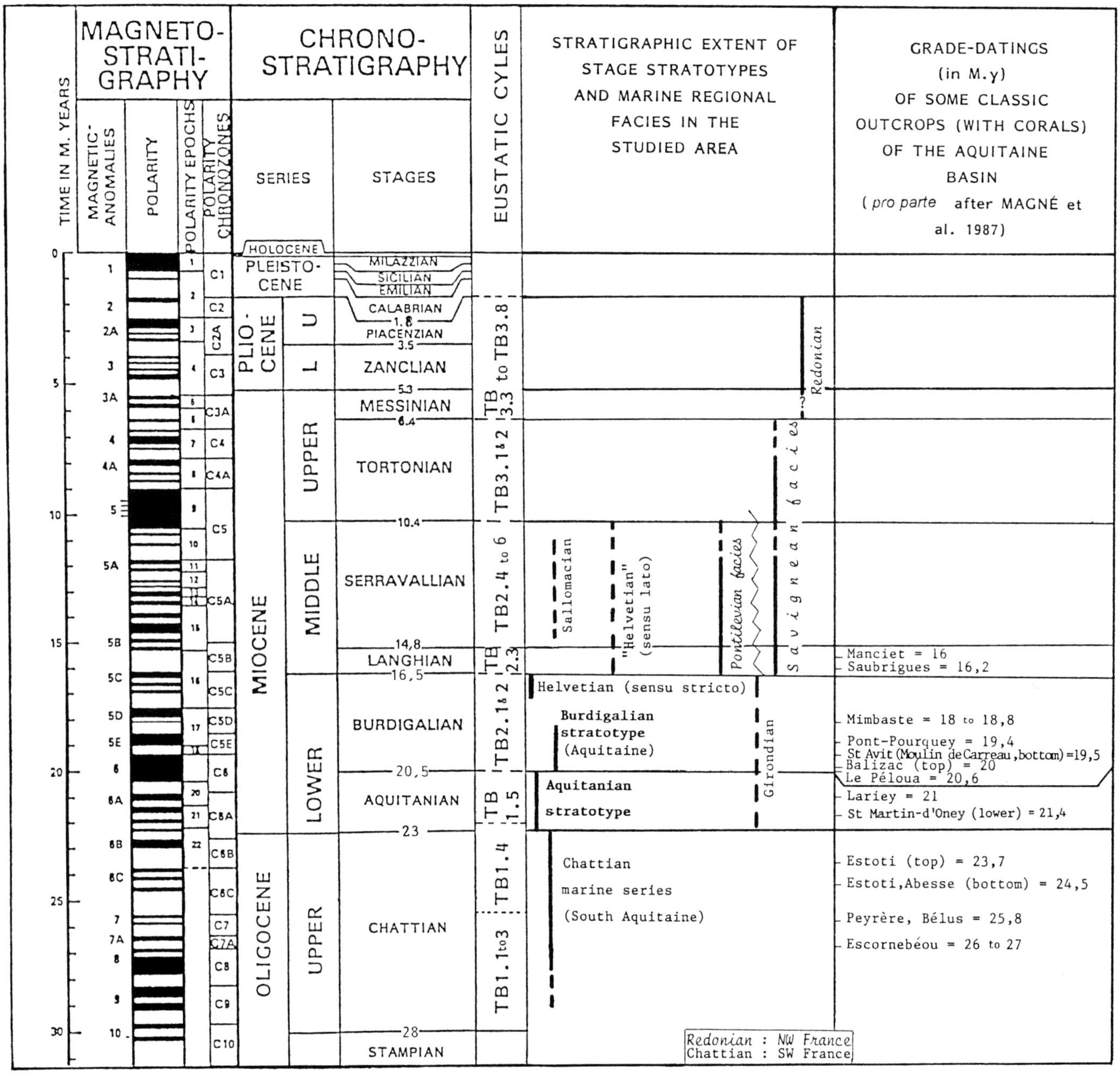

| TIME IN M. YEARS | MAGNETO-STRATIGRAPHY: MAGNETIC ANOMALIES | POLARITY CHRONOZONES | CHRONO-STRATIGRAPHY: SERIES | | STAGES | EUSTATIC CYCLES | STRATIGRAPHIC EXTENT OF STAGE STRATOTYPES AND MARINE REGIONAL FACIES IN THE STUDIED AREA | GRADE-DATINGS (in M.y) OF SOME CLASSIC OUTCROPS (WITH CORALS) OF THE AQUITAINE BASIN (*pro parte* after MAGNÉ et al. 1987) |
|---|---|---|---|---|---|---|---|---|
| 0 | | | HOLOCENE | | | | | |
| | 1 | C1 | PLEISTOCENE | | MILAZZIAN, SICILIAN, EMILIAN | | | |
| | 2 | C2 | | | CALABRIAN — 1.8 | | | |
| | 2A | C2A | PLIOCENE | U | PIACENZIAN — 3.5 | TB3.3 to TB3.8 | *Redonian* | |
| 5 | 3 | C3 | | L | ZANCLIAN — 5.3 | | *Redonian* ? | |
| | 3A | C3A | MIOCENE | UPPER | MESSINIAN — 6.4 | TB3.3 | | |
| | 4, 4A, 5 | C4, C4A, C5 | | | TORTONIAN — 10.4 | TB3.1&2 | *Savignean facies* | |
| 10–15 | 5A, 5B | C5A, C5B | | MIDDLE | SERRAVALLIAN — 14,8 | TB2.4 to 6 | Sallomacian; "Helvetian" (sensu lato); *Pontilevian facies*; *Savignean facies* | |
| | 5B | C5B | | | LANGHIAN — 16,5 | TB2.3 | | Manciet = 16; Saubrigues = 16,2 |
| | 5C, 5D, 5E | C5C, C5D, C5E | | LOWER | BURDIGALIAN — 20,5 | TB2.1&2 | Helvetian (sensu stricto); Burdigalian stratotype (Aquitaine); Girondian | Mimbaste = 18 to 18,8; Pont-Pourquey = 19,4; St Avit (Moulin de Carreau, bottom) = 19,5; Balizac (top) = 20 |
| 20 | 6, 6A | C6, C6A | | | AQUITANIAN — 23 | TB1.5 | Aquitanian stratotype; Girondian | Le Péloua = 20,6; Lariey = 21; St Martin-d'Oney (lower) = 21,4 |
| 25 | 6B, 6C, 7, 7A, 8, 9 | C6B, C6C, C7, C7A, C8, C9 | OLIGOCENE | UPPER | CHATTIAN — 28 | TB1.4; TB1.1to3 | Chattian marine series (South Aquitaine) | Estoti (top) = 23,7; Estoti, Abesse (bottom) = 24,5; Peyrère, Bélus = 25,8; Escornebéou = 26 to 27 |
| 30 | 10 | C10 | | | STAMPIAN | | | |

*Redonian* : NW France
Chattian : SW France

## Upper Oligocene Reefs

*Overview.—*

During the Chattian, a shallow-water platform with coral reefs occurred in the southwestern part of the Aquitaine Basin (Cahuzac, 1980); this facies is well represented in the area of Adour valley (Fig. 3). The climate of this neritic realm was clearly tropical of the western-Tethyan province type as shown by numerous faunal groups: microfauna, bryozoans, malacofauna, scleractinians and echinoids (Lauriat-Rage et al., 1993). Thermophile large benthic forams (e.g., *Nummulites, Operculina, Spiroclypeus, Heterostegina, Cycloclypeus, Grzybowskia, Lepidocyclina* and *Miogypsinoides*) are abundant and diverse in association with the reefs and allow the precise dating of these formations (Table 2; cf. Cahuzac, 1984; Cahuzac and Poignant, 1988, 1993). The mollusk fauna is particularly rich and abundant, with up to 1,200 species in total, dominated by tropical gastropods including Strombidae, Volutidae, 40 species of Cypraeidae, Olividae, Cassidae, Conidae, Terebridae, Nassariidae, Turridae, etc. (Cahuzac, 1980, 1983; Lozouet, 1986). This type of high diversity is only present today in the southeastern Asia and Caribbean regions. Endemic features are

TABLE 2.—DISTRIBUTION OF SOME BENTHIC FORAMINIFERS DURING THE OLIGO-MIOCENE, IN THE AQUITAINE BASIN (AFTER CAHUZAC AND POIGNANT, 1992, SLIGHTLY MODIFIED).

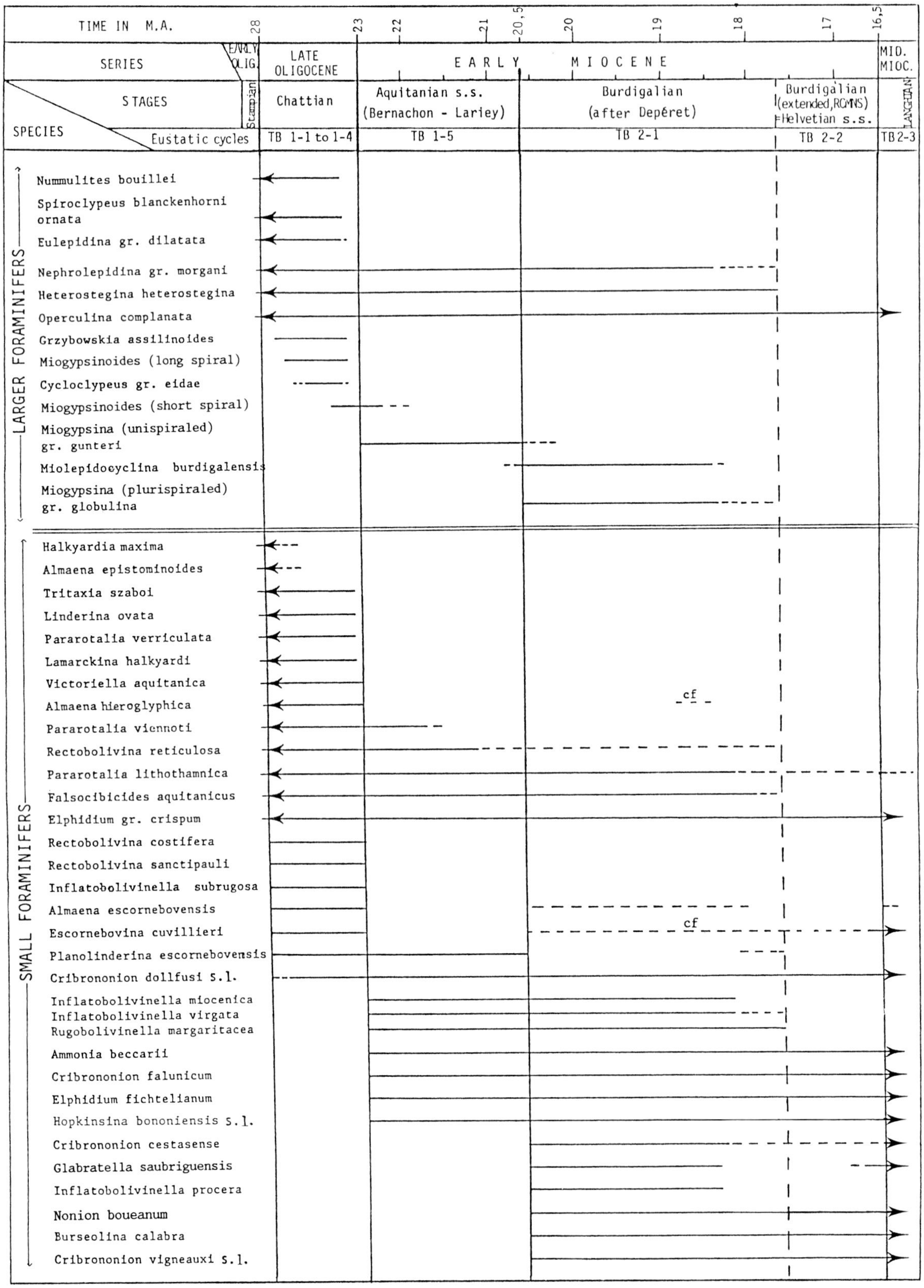

TIME IN M.A. 28 23 22 21 20,5 20 19 18 17 16,5

SERIES: EARLY OLIG. | LATE OLIGOCENE | EARLY MIOCENE | MID. MIOC.

STAGES: Stampian | Chattian | Aquitanian s.s. (Bernachon - Lariey) | Burdigalian (after Depéret) | Burdigalian (extended, RCMNS) = Helvetian s.s. | LANGHIAN

SPECIES / Eustatic cycles: TB 1-1 to 1-4 | TB 1-5 | TB 2-1 | TB 2-2 | TB 2-3

LARGER FORAMINIFERS

Nummulites bouillei
Spiroclypeus blanckenhorni ornata
Eulepidina gr. dilatata
Nephrolepidina gr. morgani
Heterostegina heterostegina
Operculina complanata
Grzybowskia assilinoides
Miogypsinoides (long spiral)
Cycloclypeus gr. eidae
Miogypsinoides (short spiral)
Miogypsina (unispiraled) gr. gunteri
Miolepidocyclina burdigalensis
Miogypsina (plurispiraled) gr. globulina

SMALL FORAMINIFERS

Halkyardia maxima
Almaena epistominoides
Tritaxia szaboi
Linderina ovata
Pararotalia verriculata
Lamarckina halkyardi
Victoriella aquitanica
Almaena hieroglyphica
Pararotalia viennoti
Rectobolivina reticulosa
Pararotalia lithothamnica
Falsocibicides aquitanicus
Elphidium gr. crispum
Rectobolivina costifera
Rectobolivina sanctipauli
Inflatobolivinella subrugosa
Almaena escornebovensis
Escornebovina cuvillieri
Planolinderina escornebovensis
Cribrononion dollfusi s.l.
Inflatobolivinella miocenica
Inflatobolivinella virgata
Rugobolivinella margaritacea
Ammonia beccarii
Cribrononion falunicum
Elphidium fichtelianum
Hopkinsina bononiensis s.l.
Cribrononion cestasense
Glabratella saubriguensis
Inflatobolivinella procera
Nonion boueanum
Burseolina calabra
Cribrononion vigneauxi s.l.

cf
cf

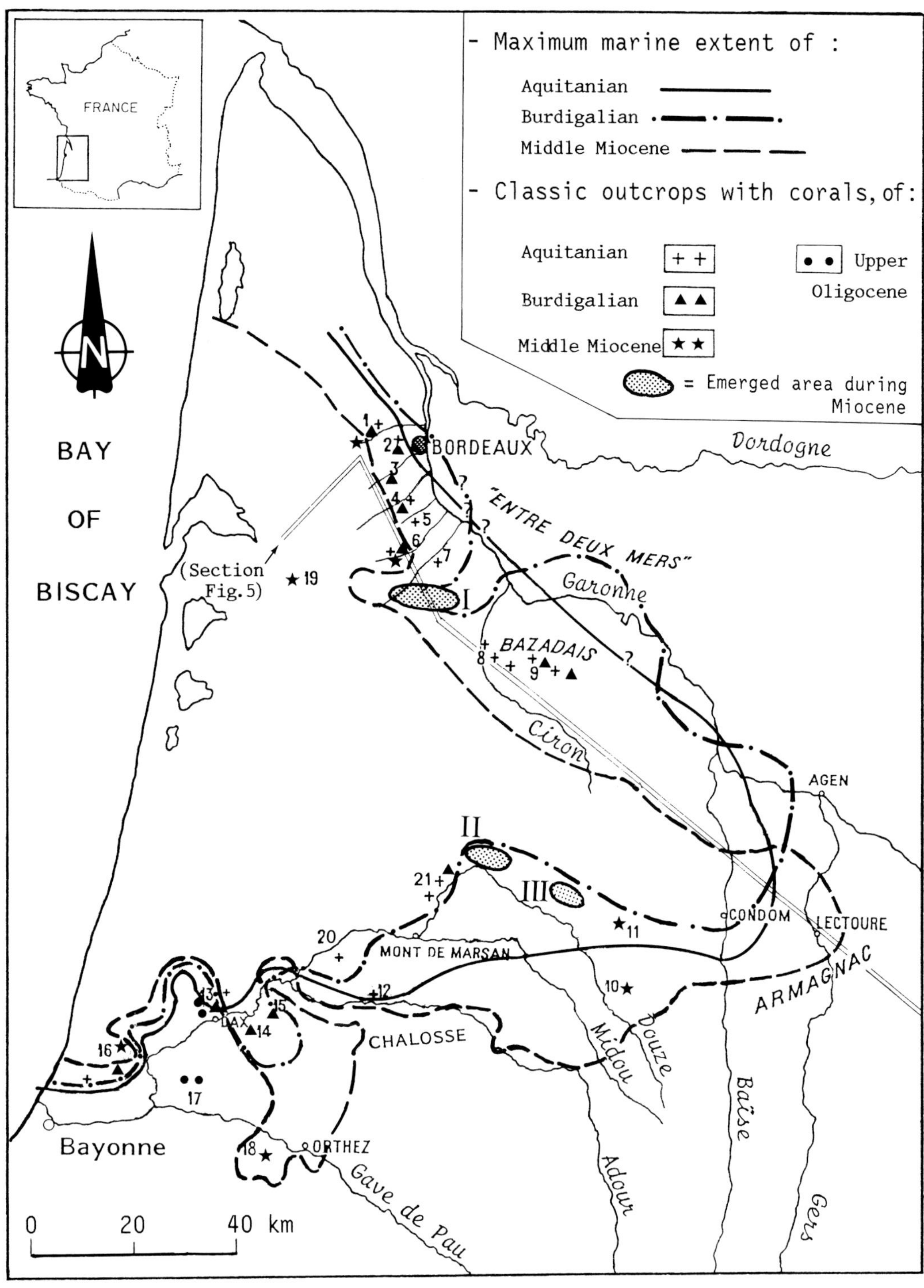


FIG. 4.—The coral formations of the Miocene of Aquitaine, in their paleogeographic setting (after Chevalier, 1961; Cahuzac, 1980, modified). Main outcrops : 1- "La Jalle de St Médard" valley; 2- Mérignac; 3- "L'Eau Bourde" valley; 4- "L'Eau Blanche" valley (Léognan); 5- Breyra; 6- Saucats; 7- "Gât-Mort" valley; 8- Villandraut (Gamachot); 9- Bazas; 10- Manciet; 11- Parleboscq; 12- St Sever; 13- St Paul-lès-Dax; 14- Mimbaste; 15- Ozourt; 16- Saubrigues area; 17- Peyrère, La Téoulère (Peyrehorade); 18- Salies-de-Béarn; 19- Salles; 20- St Martin-d'Oney (lower); 21- St Avit. The emerged areas ("isles") during the whole Miocene are the Cretaceous anticlines of: I- Villagrains-Landiras, II- Roquefort, and III- Créon-d'Armagnac.

noticeable in this rich assemblage (not less than 100 new species of mollusks) and in the associated coral fauna (several endemic species described *in litteris* and about 10 new taxa, under revision, cf. Cahuzac and Chaix, 1993, Table 3[2]). The examina-

TABLE 3.—LIST OF THE UPPER OLIGOCENE CORALS IN THE SOUTHWESTERN FRANCE (THE FEW INDICATED SYNONYMYS CONCERN ONLY THE TAXA PREVIOUSLY QUOTED IN AQUITAINE BASIN).

## UPPER OLIGOCENE (SOUTHERN AQUITAINE CHATTIAN)

### Family Stylophoridae Milne-Edwards and Haime, 1857

*Stylophora pistillata* (Esper, 1797) = *S. affinis* Duncan, 1863
*Stylophora rugosa* (d'Archiac, 1848)
*Stylophora mutata* (Michelotti *in* Sismonda, 1871)
*Stylophora tenuissima* Gerth, 1923
*Stylophora* sp. 2 Barta-Calmus, 1973
*Seriatopora regulata* (Chevalier, 1961)
*Madracis myriaster* (Milne-Edwards and Haime, 1850) = *M. asperula* (Milne-Edwards and Haime, 1850)
*Madracis decactis* (Lyman, 1859) = *M. pulchella* (Matheron *in* Chevalier, 1961)
*Madracis decaphylla* (Matheron *in* Chevalier, 1961)

### Family Pocilloporidae Gray, 1842

*Pocillopora madreporacea* (Lamarck, 1816)

### Family Astrocoeniidae Koby, 1890

*Astrocoenia bistellata* (Catullo, 1856)
*Astrocoenia pyrenaica* (Matheron *in* Chevalier, 1961) = *Actinastraea(?) pyrenaica* Matheron *in* Chevalier, 1961
*Stephanocoenia intersepta* (Lamarck, 1816)
*Platycoenia palmata* (Catullo, 1856)
*Glyphastraea* nov.sp. 1 (= *Platycoenia palmata sensu* Chevalier, 1956, *non* Catullo, 1856)

### Family Stylinidae d'Orbigny, 1851

#### - Subfamily Cyathophorinae Vaughan and Wells, 1943

*Miophora peyrerensis* Chevalier, 1963

### Family Montlivaltiidae Dietrich, 1926 emend. Chevalier, 1961

*Nerthastraea* nov.sp. 1

### Family Faviidae Gregory, 1900

*Caulastraea tournoueri* (Matheron *in* Chevalier, 1963) = "?*Desmocladia* sp." *in* Cahuzac and Chaix, 1993
*Aphrastraea* sp. (*in* Chevalier, 1963)
*Ellasmoastraea multilateralis* (Michelin, 1842)
*Ellasmoastraea intermedia* (d'Achiardi, 1867)
*Ellasmoastraea parvistella* Chevalier, 1961
*Favia favus* (Forskal, 1775)
cf. *Favia* sp.
*Favites aranea* (Defrance, 1826)
*Favites virens* (Dana, 1846)
*Favites neglecta* (d'Achiardi, 1868)
*Favites detecta* (Michelotti *in* Sismonda, 1871)
*Heliastraea (Heliastraea) columnaris* Reuss, 1868
*Heliastraea (Athecastraea) stellata* Defrance, 1826
*Heliastraea (Athecastraea) fragilis* Chevalier, 1961
*Heliastraea (Athecastraea)* nov. sp. 1
*Heliastraea (Heliastraeopsis)* nov. sp.1
*Heliastraea (Aquitanastraea) incrustans* (Osasco, 1897)
*Heliastraea (Aquitanastraea) pruvosti* Chevalier, 1954
*Antiguastraea alveolaris* (Catullo, 1856)
*Astroria (=Defrancia) irregularis* (Defrance, 1826)
*Astroria (=Defrancia) peyrerensis* (Chevalier, 1963)
*Colpophyllia multisepta* Chevalier, 1961
*Diploastraea heliopora* (Lamarck, 1816)
*Thegioastraea diversiformis* (Michelin, 1842)
*Thegioastraea planulata* (d'Achiardi, 1868)
*Thegioastraea multisepta* (Sismonda, 1871)
*Thegioastraea roasendai* Michelotti *in* Sismonda, 1871
*Cladocora granulosa* (Goldfuss, 1833) = *C. vermiculata* Matheron *in* Chevalier, 1961
*Cladocora manipulata* (Michelin, 1842)
*Cladocora gamachotensis* Chevalier, 1961
*Antillophyllia alloiteaui* (Grange, 1956)
*Plesiastraea (Palaeoplesiastraea) desmoulinsi* (Milne-Edwards and Haime, 1851)

### Family Astrangiidae Verrill, 1870

*Astrangia manthelanensis* Chevalier, 1961
*Cladangia semispherica* (Defrance, 1826)
*Culicia parasitica* (Michelin, 1847)
*Cryptangia reptans* Chevalier, 1961 = *Diplothecangia minima* Chev., 1961 = *"Endopsammia minima" in* Cahuzac and Chaix, 1993
*Oulangia speyeri* (Reuss, 1865)

### Family Oculinidae Gray, 1847

*Amphihelia (= Madrepora)* nov. sp. 1
*Diplohelia conferta* (Milne-Edwards and Haime, 1850) = *Psammocyathus pyrenaicum* Matheron *in* Chevalier, 1963
*Diplohelia moravica* (Quenstedt, 1881)

### Family Meandrinidae Gray, 1847

*Meandrina subcircularis* Catullo, 1856

### Family Mussidae Ortmann, 1890

*Lithophyllia michelottii* (Michelin, 1841)
*Lithophyllia detrita* (Michelin, 1842)
*Lithophyllia plana* (Duncan, 1864)
*Lithophyllia patula* (Sismonda, 1871)
*Syzygophyllia (Syzygophyllia) elongata* (Sismonda, 1871)
*Syzygophyllia (Syzygophyllia) asymmetrica* (Gregory, 1900)
*Syzygophyllia (Syzygophyllia) grandis* Chevalier, 1961
*Syzygophyllia (Syzygophyllia)* nov. sp.1
*Syzygophyllia (Aquitanophyllia) grandistellae* Chevalier, 1961
*Mussismilia (Protomussa)* nov. sp. 1
*Lobophyllia aff. aspera* Milne-Edwards and Haime, 1849
*Lobophyllia pasottii* (Zuffardi-Commerci, 1932)
*Leptomussa brauni* (Michelin, 1847)
*Leptomussa variabilis* d'Achiardi, 1867
*Leptomussa lehmani* Vinson, 1956
*Leptomussa falloti* Chevalier, 1961
*Scolymia aff. lacera lacera* (Pallas, 1766)
*Scolymia* nov. sp. 1

### Family Pectiniidae Vaughan and Wells, 1943

*Echinophyllia* sp. (aff. *rugosa* Chevalier, 1975)
*Oxypora* sp.

### Family Caryophylliidae Gray, 1847

#### - Subfamily Caryophylliinae Gray, 1847

cf. *Caryophyllia (Acanthocyathus)* sp.
*Ceratotrochus (Edwardsotrochus) subcristatus* (Milne-Edwards and Haime, 1848)
*Ceratotrochus (Edwardsotrochus) pentaradiata* Chevalier, 1961
*Deltocyathus* nov. sp. 1
*Cyathoceras dertonensis* (Michelotti *in* Sismonda, 1871)
*Trochocyathus (Trochocyathus) plicatus crassus* (Milne-Edwards and Haime, 1848)
*Trochocyathus (Aplocyathus) armatus* (Michelotti, 1838)
*Trochocyathus (Platycyathus) crassicostatus* (Seguenza, 1864)

#### - Subfamily Turbinoliinae Milne-Edwards and Haime, 1848

*Conocyathus* sp.
*Sphenotrochus intermedius* (Goldfuss, 1826)

#### - Subfamily Eusmiliinae Milne-Edwards and Haime, 1857

*Eusmilia* aff. *fastigiata* (Pallas, 1766)
*Clonosmilia tournoueri* Matheron *in* Chevalier, 1963
*Plerogyra* sp. (= "? Genus A. sp." *in* Cahuzac and Chaix, 1993)

#### - Subfamily Desmophyllinae Vaughan and Wells, 1943

*Desmophyllum pyrenaicum* Matheron *in* Chevalier, 1961
*Desmophyllum subcostatum* Matheron *in* Chevalier, 1961

#### - Subfamily Parasmiliinae Vaughan and Wells, 1943

*Coelocyathus macneili* (Weisbord, 1971)
*Smilotrochus* sp.
*Dendrosmilia bainbridgensis* Durham, 1942
*Parasmilia minimaxis* (Chevalier, 1963)

**Family Flabellidae Bourne, 1905**
*Flabellum avicula* (Michelotti, 1838)
*Flabellum trapezoidale semiovoidale* Osasco, 1895
*Flabellum* sp.

**Family Agariciidae Gray, 1847**
*Agaricia okeni* (Matheron *in* Chevalier, 1961)
*Agaricia (?) granulata* (Matheron *in* Chevalier, 1961)
*Pavona (Pavona) profunda* (Reuss, 1868)
*Pavona (Pavona) panamensis* Vaughan, 1919
*Pavona (Pavona) ruvida* (Prever, 1921)
*Pavona (Pavona) paronai* (Prever, 1921)
*Pavona (Polyastra) matheroni* Chevalier, 1961
*Cyathoseris infundibuliformis* (de Blainville, 1830)
*Pachyseris (Pavonaraea)* sp.

**Family Siderastraeidae Vaughan and Wells, 1943**
*Siderastraea radians* (Pallas, 1766)
*Siderastraea siderea* (Ellis and Solander, 1786)
*Siderastraea crenulata crenulata* (Goldfuss, 1826)
*Siderastraea crenulata gamachotensis* Chevalier, 1961
*Siderastraea miocenica miocenica* Osasco, 1897
*Siderastraea miocenica regularis* d'Orbigny, 1852
*Rhizangia brevissima* (Deshayes, 1834)

**Family Micrabaciidae Vaughan, 1905**
*Stephanophyllia* sp.

**Family Poritidae Gray, 1842**
*Goniopora raulini* (Milne-Edwards and Haime, 1857)
*Goniopora granosa* Matheron *in* Chevalier, 1961
*Goniopora globulosa* Chevalier, 1961 = *G. daxitertia sensu* Chevalier, 1961, *non* Bernard, 1903
*Goniopora tampaensis* Weisbord, 1973
*Alveopora elegans* (Leymerie, 1836)
*Alveopora meridionalis* Chevalier, 1961
*Alveopora discors* De Angelis, 1894 = *A. matheroni* Chevalier, 1963
*Alveopora intricata* (Matheron *in* Chevalier, 1963)
*Porites collegniana collegniana* Michelin, 1842 = *P. collegniana martinii* (d'Orbigny, 1852) = *P. collegniana girundiensis prima* (Bernard, 1903)
*Porites nummulitica* Reuss, 1864
*Porites pusilla* Felix, 1884
*Porites matheroni* Chevalier, 1956

**Family Dendrophylliidae Gray, 1847**
*Balanophyllia (Balanophyllia) caulifera* (Conrad, 1847)
*Balanophyllia (Balanophyllia) varians* Reuss, 1854
*Balanophyllia (Balanophyllia) concinna* Reuss, 1871
*Balanophyllia (Balanophyllia)* nov. sp. 1
*Balanophyllia (Eupsammia) striata* (Defrance, 1826)
*Balanophyllia (Eupsammia) cylindrica* (Michelotti, 1838)
*Dendrophyllia ramea* (Linné, 1758)
*Dendrophyllia cornigera* (Lamarck, 1816)
*Dendrophyllia amica* (Michelotti, 1838)
*Paleoastroides theotvoldensis* (Michelin, 1847)
*Paleoastroides provincialis* Chevalier, 1961 (="? *Duncanopsammia* nov. sp." *in* Cahuzac and Chaix, 1993)
Nov. gen., nov. sp. 1
*Reussopsammia granulosa* (Reuss, 1864) (= "cf. *Bathypsammia* sp." *in* Cahuzac and Chaix, 1993)

**Family Acroporidae Verrill, 1902**
*Acropora ornata* (Defrance, 1823)
*Acropora solanderi* (Defrance, 1828) = *A. exarata* (Michelotti, 1838)
*Acropora lavandulina* (Michelin, 1842)
*Acropora haidingeri* (Reuss, 1864)
*Acropora pachymorpha* Chevalier, 1956
*Astreopora subcylindrica* Matheron *in* Chevalier, 1961
*Astreopora densata* Chevalier, 1961 (= *A. microcalix* Chevalier, 1963)
*Astreopora sp.* [cf. *A. auvertiaca* (Michelin, 1845)]

**Family Turbinariidae Milne-Edwards and Haime, 1857**
*Turbinaria cyathiformis* (de Blainville, 1830)

The genus *Anomopora* (*A. ramosa* Chevalier, 1961) is like an *Astreopora* worn in a particular way, which masks the calice structures.

tion of this Chattian coral assemblage is particularly interesting, because it is richer than all other synchronous European-African assemblages. Moreover, it includes many components of the following Lower Miocene faunas, which it foreshadows.

*Reef development.—*

Corals are particularly abundant in two areas. One is the Bélus-Peyrehorade and St.Etienne-d'Orthe area (Fig. 3) where Chattian marly and calcareous deposits partially filled up the head of a paleo-canyon with abundant malacofauna (400 species, with Dentaliidae, *Corbula*, Volutidae, Ringiculidae, Rissoidae, and others) and foraminifers. Up to 60 species of corals (more than half are hermatypic) have been recorded in this area (mostly in Peyrehorade: Peyrère and La Téoulère sections). Morphologically, the colonies are sometimes locally delicate and either globular (*Alveopora*) or lamellar (*Agaricia, Pavona, Cyathoseris, Turbinaria*), but they are much more frequently big and massive (solid and resisting the action of the waves), either branched (*Acropora, Stylophora*) or more commonly globular (*Astreopora, Astroria, Favites, Colpophyllia, Heliastraea, Antiguastraea, Porites*). Among this reefal fauna, there are ahermatypic corals (e.g., genera *Antillophyllia, Cladocora, Diplohelia, Flabellum*).

Westward (St Etienne-d'Orthe), the marly deposits indicate a greater depth (about 200 m or more, i.e., the circalittoral to epibathyal zones); the coral assemblage is ahermatypic with *Deltocyathus, Stephanophyllia, Conocyathus, Trochocyathus, Clonosmilia*, and *Flabellum*, for instance.

Eastward (Peyrère, Peyrehorade), the coral fauna (commonly *Acropora* and Poritidae) developed on often calm and muddy bottoms at about 30 to 50 m in depth or slightly less (in contrast to the conclusions of Chevalier, 1963); this depth is indicated by the occurrence of such genera as *Goniopora, Porites, Seriatopora, Stylophora, Acropora, Eusmilia*, and *Favites* (after Kühlmann, 1983). In some localities (La Téoulère: about 40 scleractinian species), coral colonies are larger (and mainly hermatypic), and the faunal associations are indicative of more littoral biotopes with bioclastic calcareous deposits (Cahuzac, 1980; Cahuzac and Chaix, 1994; Chaix and Cahuzac, in prep.). Alcyonarian corals and calcareous algae are also present.

The other area of abundant Upper Oligocene corals is St. Paul-lès-Dax (Poustagnac valley; Fig. 3), where deposits consist of sandy, shelly limestones with abundant bryozoans, mollusks (amongst them the tropical Bivalvia *Trisidos grateloupi*, cf. Cahuzac et al., 1992a) and microfauna (Cahuzac, 1983, 1984). Echinoids are more or less common (e.g., Cidaridae, *Echinolampas, Maretia* and the sand dollars urchins *Amphiope* and *Parascutella*, cf. Cahuzac and Roman, 1994). Corals (up to

2 For this table and the others (Tables 4-7), it is impossible to detail the numerous references of coral taxa; the reader should consult principally Wells (1956) and Chevalier (1961); see also Chevalier (1954, 1963, 1972b, 1975, 1977), Weisbord (1971, 1973), Barta-Calmus (1973), Frost (1977), Cairns (1979), Oosterbaan (1988) and Cahuzac and Chaix (1993). This list was drawn up on October 1, 1995 and might be modified/revised according to newer discoveries.

100 species in 2 primary sections at Estoti and Abesse, Fig. 3) are exceptionally well preserved, and the taxonomic composition of this fauna indicates western Atlantic and chiefly Mediterranean influences. For instance, about 16 species come from the Mediterranean Rupelian, 5 from the Mediterranean Eocene, and also 12 are known in the Cenozoic of the Indian Ocean province. The corals are also mostly hermatypic (e.g., *Stylophora, Astreopora, Antiguastraea, Astroria, Acropora, Porites, Favia, Heliastraea (Heliastraeopsis),* and *Nerthastraea*). *Pocillopora madreporacea*, a Miocene species, appears only in the uppermost Chattian from the "Abesse cliff" outcrop. Among colonial species, almost 50 are massive and globular, only 15 are branched and 10 phaceloid; just 6 have a lamellar morphology. Therefore, a strong wave exposure was probable for these reefs. Big colonies of *Millepora* (hydrozoans) are associated with the scleractinians. Calcareous algae are rare, as in most the reefs in the Aquitaine Basin. The exposed reef framework is surprisingly quite small, rarely exceeding 3 m in thickness (traceable laterally for 200 m); most of the reefs are considered as blankets and patches on a carbonate ramp on the eastern flank of an emergent anticlinal dome (Sébastopol, Fig. 3) without development of barrier reef trends (Cahuzac, 1980). The "peri-reefal" facies are predominant with channels, bars, and riddens; the deposits are generally coarse-grained and contain accumulations of larger benthic forams. In terms of the paleogeography and depositional setting, the Poustagnac reef has been compared with the modern Yucatan shelf (Cahuzac, 1980).

Corals are also present, but less abundant, in other geographic sectors of the Adour Basin. In Pontonx/Mineur and in St-Geours-de-Maremne/Escornebeou areas (Fig. 3; cf. sections in G.F.E.N., 1974; Cahuzac, 1980; Cahuzac and Roman, 1994), new discoveries have revealed small faunas with some twenty species per locality. The Escornebeou sections are famous (for the discussions about their stratigraphic dating, cf. G.F.E.N., 1974), but corals were not noted till now. In infralittoral deposits (silty and sandy marls and bioclastic limestones, more or less lentoid, rich in Pectinids), we found corals including *Pavona, Trochocyathus, Cyathoceras, Antillophyllia, Favites, Astrocoenia, Lithophyllia,* and *Astreopora*. In the Pontonx/Mineur area, a shallow sequence consists of marine shelly, marly calcareous deposits (interfingering with brackish clays); on these soft bottoms, some corals lived such as *Seriatopora, Stylophora, Acropora, Dendrosmilia, Clonosmilia, Syzygophyllia, Turbinaria,* and *Porites*.

*Conclusions.—*

The Chattian coral fauna (part of it previously considered as Aquitanian) is much richer than the Aquitanian one and also richer than the Burdigalian one and shows a relatively strong Paleogene ancestry. About 150 coral species (more than 90 hermatypics) and more than 70 genera are now recognized in this area; the complete list, previously given (Cahuzac and Chaix, 1993), has been slightly corrected and modified according to new discoveries (Table 3). Many genera are present in modern tropical seas, particularly either in West Indies or in the Indo-Pacific region. The richness of this coral fauna was unsuspected until now. In the Aquitaine Basin deposits, hermatypic corals and also larger benthic forams attained their maximum generic diversity during the Late Oligocene. These two groups are extremely consistent paleotemperature indicators in the Cenozoic (if we suppose that these fossil taxa had similar biotas to their living counterparts, e.g., like the modern zooxanthellate scleractinians: cf. Adams et al., 1990). So, in the Aquitaine Chattian series, such a diversity and abundance in these groups and in the rich thermophile-associated fauna show that this basin was a very warm tropical region, where sea-surface temperatures were 25°C or more.

A comparison with Mediterranean coral reefs is interesting; Chattian reefs seem well developed and locally thick all over this area, but often they are poorly documented or of imprecise or mistaken age (previously "Aquitanian"). For instance, they are known from southeastern France (Nerthe, Carry-le-Rouet), Sicily and southern Italy (Salento peninsula), Malta, Baleares, southeastern Spain (near Alicante) and Catalogne (M. Esteban, pers. commun.; Esteban et al., this volume). The Nerthe reef (near Marseille, Fig. 2) is known for a long time, but its dating has been debated. The lower sequences (previously as "Aquitanian" age) are now considered as uppermost Chattian age, from nannoplankton (NP25 zone) and forams studies (ref. in Martini, 1988; Monleau et al., 1988); they correspond to TB1-4 eustatic cycle that is mainly Late Oligocene age (Table 1). The Nerthe coral communities are fairly different from those of Atlantic Chattian; there is a part of common species (about 14), but few genera are there predominant (*Tarbellastraea, Acropora, Porites*), with diverse associated faunal groups (cf. Nury and Thomassin, 1994; Esteban et al., this volume). To our knowledge, in the Mediterranean Chattian deposits of which the age has been confirmed, the total number of coral species is generally considerably reduced in comparison with the eastern Atlantic (e.g., some 40 species in the Nerthe Chattian, some 30 in Malta, under study). Owing to this fairly low diversity, the Mediterranean does not seem to be, at that time, a center of scleractinians expansion. The important proportion of taxa in common with the Aquitaine Basin associations (16 genera out of 18 in Malta, 2/3 of the Nerthe assemblage) is explained by good marine communication between the Atlantic and western Mediterranean across the Betic platforms during the Chattian (Cahuzac et al., 1992b).

During the Late Oligocene, the coral fauna in the Aquitaine Basin still contained western Atlantic elements (presence of *Antillophyllia, Eusmilia, Scolymia, Colpophyllia, Madracis, Meandrina*, and others). Furthermore, several taxa may have also migrated from the Mediterranean Rupelian (e.g., Castelgomberto fauna, Italy) to the Aquitaine where they are present in the Chattian. Generally speaking, the Upper Oligocene coral associations suggest the existence of a continuous Atlantic (eastern and western) and Mediterranean bioprovince. After Frost (1977), the Oligocene (Rupelian concerned) marked a "time of world-wide maximum abundance and diversity of Tertiary reefs with their widespread development in both the

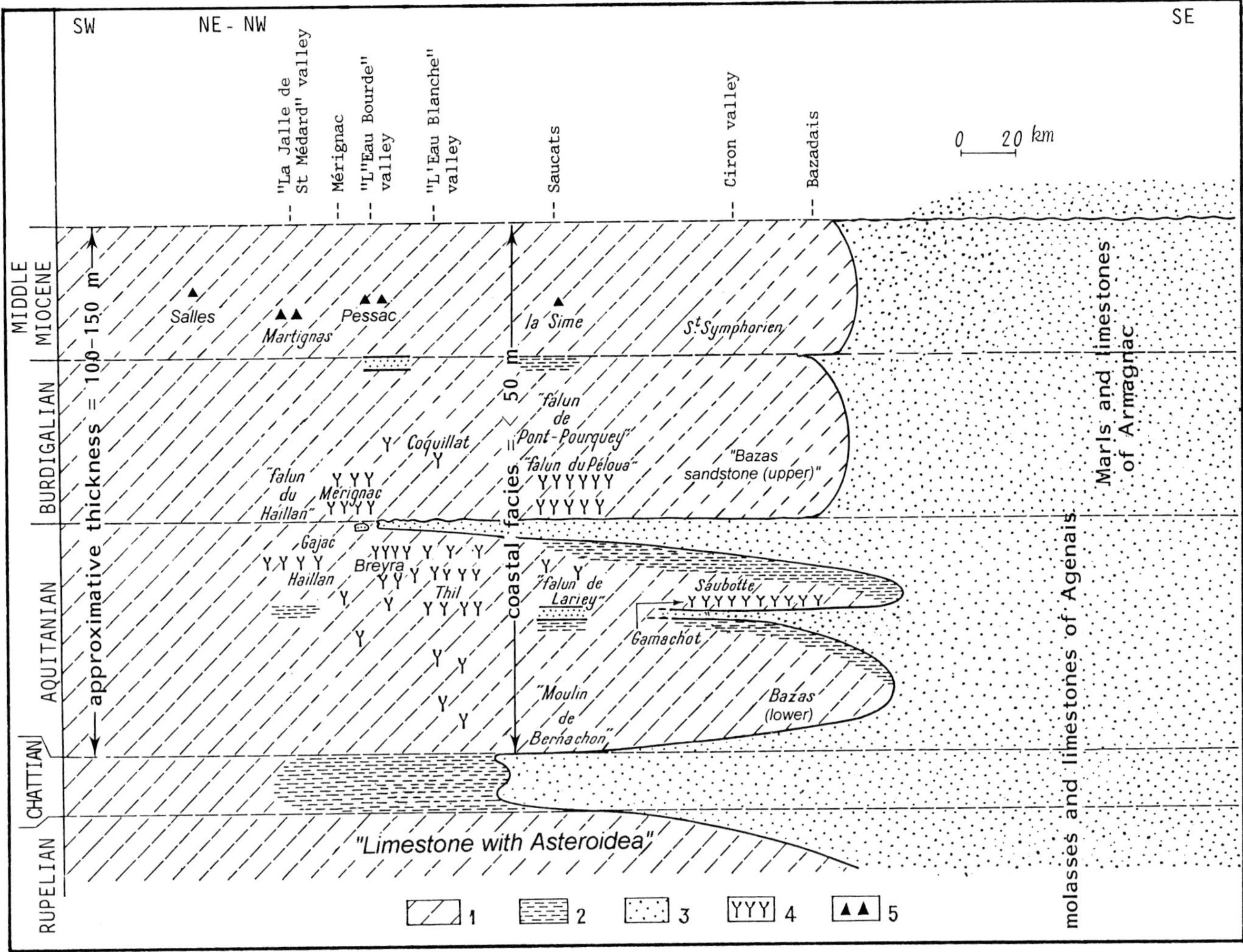


FIG. 5.—Distribution of some Miocene coral facies in north Aquitaine (after Chevalier, 1961, modified). Marine facies (1), brackish (2) and lacustrine (3) facies; hermatypic (4) and ahermatypic (5) corals (Section line, see Fig. 4).

Caribbean and Tethys regions." Our data seem to indicate that this assumption also is true for the Late Oligocene; the tropical westward-flowing surface currents (evidenced through western Tethys and Atlantic in the Rupelian; Frost, 1977) may had been still circumnavigating the low latitudes of northern hemisphere during the Late Oligocene. So, at that time, the Aquitaine Basin seems to be a center of dispersal of corals in eastern Atlantic. Nearly 60 species seem to have originated during the Chattian in the Aquitaine realm.

## Lower Miocene Reefs

*Overview.—*

During the Aquitanian and Burdigalian, the Aquitaine embayment contained marine deposits interfingering with brackish and lacustrine facies (Figs. 4, 5). Chevalier (1961) recognized patch reefs and small fringing reefs intimately associated with terrigenous deposits in shallow-water, often protected, coastal environments. These reefs are also characterized by the absence or rarity of calcareous algae and the good preservation of coral specimens in the poorly consolidated sandy, locally marly, limestones. This is in contrast with similar-age reefs in the Mediterranean or with older Lower Oligocene limestones of the Aquitaine Basin. Broadly speaking, the marine paleoenvironments were more open to oceanic effects during the Burdigalian than during the Aquitanian; Burdigalian faunal assemblages are more stenohaline, and reef buildups are more widely developed.

*Aquitanian reefs.—*

The Oligocene-Miocene boundary (at about 23 Ma) is characterized by extinctions, either in neritic or in oceanic domains. In the Aquitaine Basin, a significant phase of replacement among several faunal groups occurred at the outset of the Aquitanian, and numerous thermophile Chattian taxa disappeared (for instance, such larger forams as *Nummulites bouillei, Spiroclypeus, Cycloclypeus, Grzybowskia assilinoides, Miogypsinoides complanatus, Eulepidina dilatata*); as well as, the scleractinians

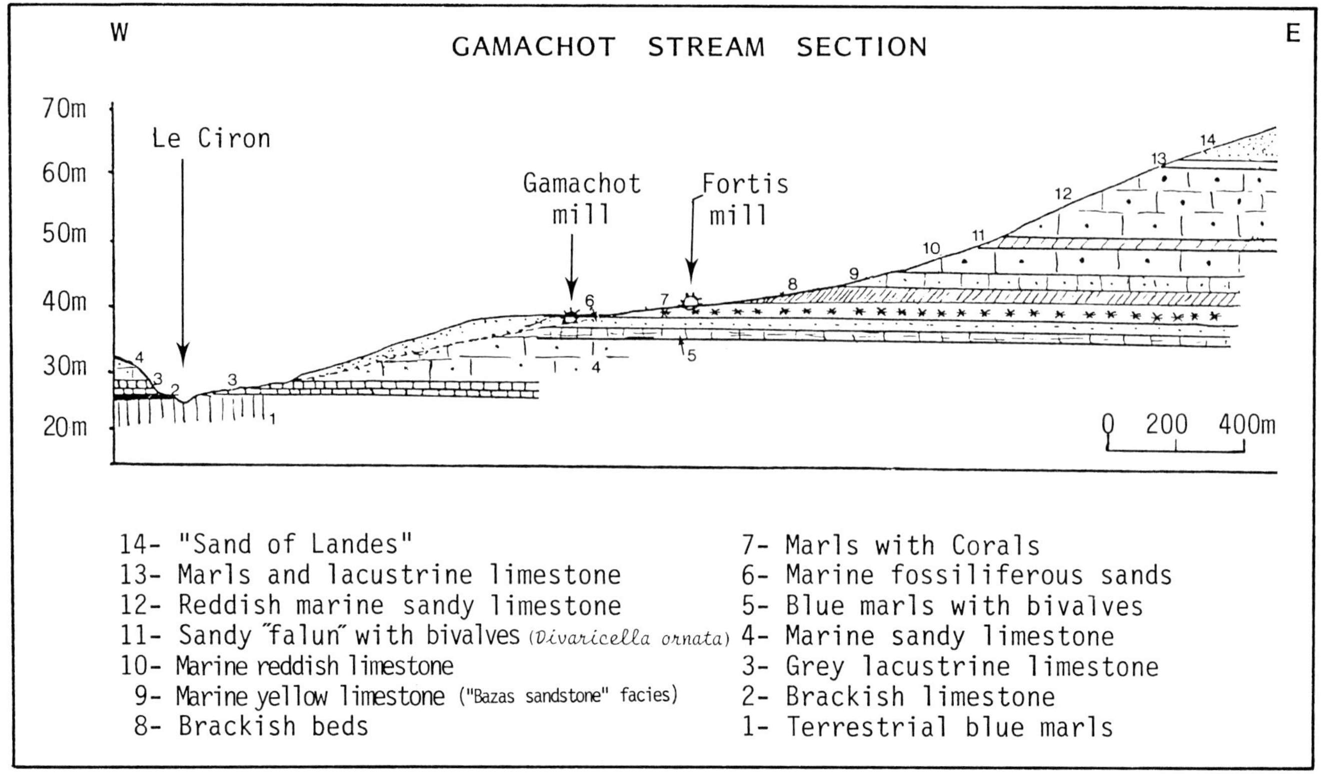


FIG. 6.—Example of Gamachot section in the Lower Miocene (Aquitanian) of Bazadais (Gironde, northern Aquitaine), showing marly layers with hermatypic corals accumulations.

species decreased in number (such genera as *Madracis, Astrocoenia, Stephanocoenia, Miophora, Colpophyllia, Meandrina, Mussismilia, Scolymia, Echinophyllia, Stephanophyllia*, etc., disappeared from the Aquitaine province only). That could indicate a slight drop in temperature. Such an earliest Miocene cooling, probably short, is confirmed by oxygen isotope analysis and by the planktonic microfossils distribution; in the nannoplankton assemblages (NN1 zone), "a very low diversity and a remarkable decrease in size of the species" can be observed at that time (Müller, 1984) related to a decrease of surface-water temperature. Then, the Aquitaine Basin climate became tropical again during the Mid-Late Aquitanian.

In the Aquitanian, reef localities are more abundant in the northern part of the Aquitaine Basin but also are present in the south (Chevalier, 1961; Cahuzac, 1980; Cahuzac and Gautret, 1993). For instance, Gamachot (Gironde, northern Aquitaine, Fig. 6) is a typical locality with in-place *Porites* progressively colonizing marls (in a calm environment) and forming a bank up to 1 m thick interrupted by restricted brackish deposits. This was a "coral carpet" on soft muddy substrate. In the Lariey outcrop (which is the upper part of the Aquitanian stratotype, Saucats), the transgressive shallow sub-reef facies contains about 20 coral species, mainly hermatypic, with isolated colonies in bioclastic coastal deposits produced under turbulent conditions.

In southern Aquitaine, a newly discovered outcrop (St Martin-d'Oney lower section, Fig.4), Upper Aquitanian in age (Cahuzac and Gautret, 1993), contains an abundant coral reef fauna, now under study, which allowed us to add several species to our previous Aquitanian list (Cahuzac and Chaix, 1993). These deposits contain rich microfauna, malacofauna and also hydrozoans, Milleporidae. The coral association is hermatypic; the rolled aspect of the fossils and the presence of numerous pebbles indicate that the reef formed in a high-energy hydrodynamic regime.

A total of about 85 coral species (about 35 genera) are now recorded in the (especially Upper) Aquitanian of the Aquitaine Basin with only some 10 ahermatypic corals (Table 4). Of this assemblage, 24 genera are still present in the modern Indo-Pacific province. The Aquitaine Basin seems to have been, still during the Aquitanian, the "climax" area of eastern Atlantic for scleractinians faunas. In Portugal, only a hermatypic coral faunule of 11 species (all present in Aquitaine) is known in the Low Tage Gulf near Lisbon. There, small coral bioherms (with *Siderastraea, Stylophora, Acropora, Porites,* and *Palaeoplesiastraea*; Antunes and Chevalier, 1971) developed during the later transgressive oscillation of the Aquitanian.

*Burdigalian reefs.—*

During the Burdigalian, there were many thin banks rich in corals and mollusks, chiefly in the Early Burdigalian (i.e. from about 20.5 to 18 Ma, *sensu* stratotype of Gironde, northern Aquitaine, which corresponds to the N5-6 Blow (1969) and NN2

TABLE 4.—LIST OF THE AQUITANIAN CORALS IN THE SOUTHWESTERN FRANCE.

## AQUITANIAN (SOUTHWESTERN FRANCE)

**Family Stylophoridae Milne-Edwards and Haime, 1857**
*Stylophora pistillata* (Esper, 1797) = *S.affinis* Duncan, 1863
*Stylophora raristella* (Defrance, 1826)
*Stylophora thirsiformis* (Michelotti, 1847)
*Stylophora rugosa* (d'Archiac, 1848)
*Stylophora mutata* (Michelotti *in* Sismonda, 1871)
*Stylophora goethalsi* Vaughan, 1919
*Stylophora tenuissima* Gerth, 1923
*Stylophora parvistella* Chevalier, 1961

**Family Pocilloporidae Gray, 1842**
*Pocillopora madreporacea* (Lamarck, 1816)

**Family Astrocoeniidae Koby, 1890**
*Platycoenia turonensis* (Michelin, 1847)
*Platycoenia tarbellensis* Chevalier, 1961
*Glyphastraea bejaensis* (Chevalier, 1972)

**Family Montlivaltiidae Dietrich, 1926 emend. Chevalier, 1961**
*Nerthastraea* nov. sp. 1

**Family Faviidae Gregory, 1900**
*Aphrastraea deformis* (Lamarck, 1816)
*Favia corollaris* Reuss, 1871
*Favia aquitaniensis* Chevalier, 1961
*Favites virens* (Dana, 1846)
*Favites neglecta neglecta* (d'Achiardi, 1868)
*Favites neugeboreni burdigalensis* Chevalier, 1961
*Favites neuvillei* Chevalier, 1961
*Favites* nov. sp. 1
*Ellasmoastraea multilateralis* (Michelin, 1842)
*Ellasmoastraea intermedia* (d'Achiardi, 1867)
*Goniastraea aff. pectinata* (Ehrenberg, 1834)
*Astroria (= Defrancia) irregularis* (Defrance, 1826)
*Caulastraea matheroni* Chevalier, 1961
*Heliastraea (Heliastraea) inaequalis* (Gümbel, 1861)
*Heliastraea (Heliastraea) saucatsensis* (Chevalier, 1954)
*Heliastraea (Heliastraea) pelouaensis* (Chevalier, 1954)
*Heliastraea (Heliastraeopsis) dallagoi* (Osasco, 1902)
*Heliastraea (Aquitanastraea) piveteaui* Chevalier, 1954
*Heliastraea (Aquitanastraea) pruvosti* Chevalier, 1954
*Montastrea oligophylla* (Kopek, 1954)
*Montastrea alloiteaui daxensis* (Chevalier, 1961)
*Montastrea* sp. (aff. *curta* (Dana, 1846))
*Tarbellastraea ellisiana* (Defrance, 1826)
*Tarbellastraea reussiana* (Milne-Edwards and Haime, 1850)
*Tarbellastraea* cf. *eggenburgensis* (Kühn, 1925)
*Antiguastraea alveolaris* (Catullo, 1856)
*Thegioastraea diversiformis* (Michelin, 1842)
*Thegioastraea miocenica* (Michelotti *in* Sismonda, 1871)
*Thegioastraea multisepta* (Sismonda, 1871)
*Thegioastraea roasendai* Michelotti *in* Sismonda, 1871
*Thegioastraea superficialis* (Sismonda, 1871)
*Cladocora gamachotensis* Chevalier, 1961
*Plesiastraea (Palaeoplesiastraea) desmoulinsi* (Milne-Edwards and Haime, 1851)
*Plesiastraea (Palaeoplesiastraea) corrugata* (Michelotti *in* Sismonda, 1871)

**Family Astrangiidae Verril, 1870**
*Cryptangia woodi michelini* Chevalier, 1961
*Culicia parasitica* (Michelin, 1847)
*Cladangia semisphaerica tubiformis* (Michelin, 1847)
*Cladangia carryensis galasciformis* (Matheron *in* Repelin, 1900)
*Cladangia aquitaniensis* Chevalier, 1961
*Astrangia vasconiensis* (Milne-Edwards and Haime, 1850)
*Astrangia manthelanensis* Chevalier, 1961

**Family Mussidae Ortmann, 1890**
*Lithophyllia michelottii* (Michelin, 1841)
*Lithophyllia patula* (Sismonda, 1871)
*Syzygophyllia (Syzygophyllia) crenaticosta* (Reuss, 1868)
*Syzygophyllia (Syzygophyllia) elongata* (Sismonda, 1871)
*Syzygophyllia (Syzygophyllia) grandis* Chevalier, 1961
*Syzygophyllia (Aquitanophyllia) grandistellae* Chevalier, 1961

**Family Caryophylliidae Gray, 1847**
*Parasmilia montaldoensis* Chevalier, 1961
*Phyllangia thilensis* Chevalier, 1961
*Phyllangia(?) conferta saubottensis* Chevalier, 1961

**Family Flabellidae Bourne, 1905**
*Flabellum* sp.

**Family Siderastraeidae Vaughan and Wells, 1943**
*Siderastraea radians* (Pallas, 1766)
*Siderastraea bertrandiana* (Michelin, 1847)
*Siderastraea miocenica miocenica* Osasco, 1897
*Siderastraea crenulata crenulata* (Goldfuss, 1826)
*Siderastraea crenulata gamachotensis* Chevalier, 1961
*Rhizangia brevissima* (Deshayes, 1834) = *R. martini* (Milne-Edwards and Haime, 1849)

**Family Poritidae Gray, 1842**
*Alveopora discors* De Angelis, 1894
*Porites porites* (Pallas, 1766)
*Porites solida* (Forskal, 1775)
*Porites arenosa* (Esper, 1797)
*Porites collegniana collegniana* Michelin, 1842 = *P. collegniana girundiensis prima* (Bernard, 1903)
*Porites diversiformis* (Michelotti *in* Sismonda, 1871)
*Porites leptoclada* Reuss, 1871
*Porites maigensis maigensis* Kühn, 1925
*Porites maigensis gamachotensis* Chevalier, 1961
*Porites pachysepta* Chevalier, 1961

**Family Dendrophylliidae Gray, 1847**
*Dendrophyllia ramea* (Linné, 1758)
*Paleoastroides subirregularis* (Osasco, 1897)

**Family Acroporidae Verrill, 1902**
*Acropora ornata* (Defrance, 1823)
*Acropora solanderi* (Defrance, 1828) = *A. exarata* (Michelotti, 1838)
*Acropora lavandulina* (Michelin, 1842)
*Astreopora densata* Chevalier, 1961

**Family Turbinariidae Milne-Edwards and Haime, 1857**
*Turbinaria cyathiformis cyathiformis* (de Blainville, 1830)

---

Martini (1971) planktonic zones: Müller and Pujol, 1979). A noticeable regressive trend can be observed in the Upper Burdigalian of the Aquitaine Basin. The only outcropping deposits attributed to the upper part of the Burdigalian stage, and linked up to NN3-4 Martini (1971) nannoplankton zones (Cahuzac et al., 1994, 1995), and to N6-8 *pars* Blow (1969) zones (Pujol, 1970) are in the eastern area of Saubrigues Gulf (in southwestern Aquitaine, Fig. 3, 4). There, the facies are very marly with a few corals.

The reefal bioclastic sandy limestones of Peloua (one of the outcrops of the stratotype in Saucats, northern Aquitaine, Fig. 7) contain more than 40 coral species in a variety of colonial morphologies (mainly massive, more rarely branching or lamellar). The hydrodynamics of this reef must have been turbulent.

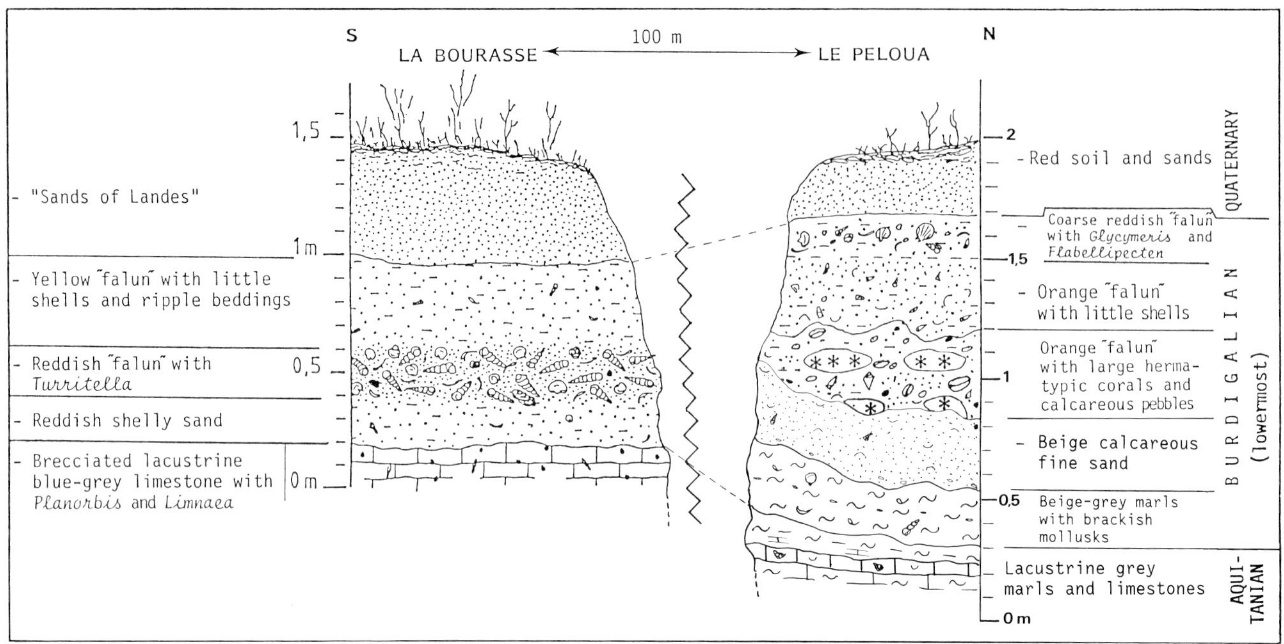


FIG. 7.—Schematic geological section at the La Bourasse (after Londeix, 1991) and Le Peloua outcrops (Saucats), the latter with hermatypic coral reefs.

In northern Aquitaine, many of the coral reefs occur directly onlapping a lacustrine grey limestone (which corresponds to a short regressive episode at the Aquitanian-Burdigalian boundary, Fig. 5). The most famous Burdigalian coral localities (e.g., St.Paul-lès-Dax: Cabanes, Mandillot; Mimbaste, etc.) occur in the southern part of the Aquitaine Basin in the area of Adour. Corals are associated with a rich littoral malacofauna (*Tudicla rusticula, Turritella terebralis, Euthriofusus burdigalensis, Athletaficulina, Arca, Glycymeris cor, G. bimaculata, Cardium*) and a rich microfauna including tropical larger foraminifers (*Miogypsina globulina, Miolepidocyclina burdigalensis, Nephrolepidina morgani, Operculina, Heterostegina, Amphistegina*). In the Cabanes outcrop, there are up to 50 coral species, mostly hermatypic (e.g., *Acropora, Heliastraea* (s.s.), *Heliastraeopsis, Aquitanastraea, Pocillopora*, and *Favites*); the massive colonies are predominant, but a few are branching, phaceloid or lamellar. Some of these colonies reach 1 m in diameter and are considered part of small fringing reefs on a platform flanking an emergent peninsula (Magescq, Sébastopol, Dax; cf. Cahuzac, 1980). To the southeast, the transgressive sea rose above the shoal between the Dax and Thétieu diapiric ridges and formed a small gulf in the western "Chalosse" (Fig. 4). There, the conditions were favorable for small reef buildups (in Mimbaste, Ozourt, Saugnac, with *Favites, Tarbellastraea, Hexastraea,* and *Acropora*).

A total of about 40 genera and nearly 100 species are recorded in the Burdigalian, only 11 species of which are strictly ahermatypic (Table 5). Certain families strongly diversified, as for example the Faviidae. In relation to the Aquitanian, the Burdigalian association in the Aquitaine Basin is characterized by the first occurrence (in Aquitaine) of *Hexastraea, Dyctioastraea* and *Astrohelia*, and the reintroduction (genera already present in the Chattian) of *Heliastraea (Athecastraea), Goniopora, Sphenotrochus, Oulangia, Pachyseris, Pavona, Diplohelia* and *Cyathoceras*. At the same time, we observe the likely absence of *Cladocora, Phyllangia* and *Rhizangia*. All these facts, notably the proliferation of reef genera and the diversity of the whole fauna, may be attributed to a general warming trend in the Burdigalian (e.g., Müller, 1984). Indeed, a thermic optimum (for all the Miocene) has been evidenced during the Burdigalian (reminiscent of that, more raised, of the Late Oligocene) in the marine neritic realm of the northeastern Atlantic frontage (Lauriat-Rage et al., 1993).

In the Portugal Burdigalian deposits, the coral assemblage is very poor; 5 species are recorded (all hermatypic) without reefal buildup.

*Conclusions.—*

The Lower Miocene in the Aquitaine Basin contains a total of about 110 coral species. The richest assemblages occur in the Lower Burdigalian. Upper Burdigalian hermatypic corals tend to decrease in number. In contrast with this richness in corals, the coral reefs are thin (rarely over a few meters), commonly with abundant marls and/or terrigenous sediments. Erosional truncation and burial could limit the observations, but there are no traces of larger reef development in the abundant coral-rich outcrops of the Aquitaine Basin.

As early as the beginning of the Aquitanian, the coral fauna of the Aquitaine Basin was disconnected from the western Atlantic (just 3 species in common). During the Early Miocene, the coral

TABLE 5.—LIST OF THE BURDIGALIAN CORALS IN THE SOUTHWESTERN FRANCE.

## BURDIGALIAN (SOUTHWESTERN FRANCE)

### Family Stylophoridae Milne-Edwards and Haime, 1857
*Stylophora raristella* (Defrance, 1826)
*Stylophora rugosa* (d'Archiac, 1848)
*Stylophora pistillata* (Esper, 1797) = *S. affinis* Duncan, 1863
*Stylophora granulata* Duncan, 1864
*Stylophora mutata* (Michelotti *in* Sismonda, 1871)
*Stylophora goethalsi* Vaughan, 1919

### Family Pocilloporidae Gray, 1842
*Pocillopora madreporacea* (Lamarck,1816)

### Family Astrocoeniidae Koby, 1890
*Platycoenia turonensis* (Michelin, 1847)
*Platycoenia tarbellensis* Chevalier, 1961
*Hexastraea fromenteli* Bellardi *in* Sismonda, 1871

### Family Montlivaltiidae Dietrich, 1926 emend. Chevalier, 1961
*Nerthastraea* nov. sp. 1

### Family Faviidae Gregory, 1900
*Aphrastraea deformis* (Lamarck, 1816)
*Dyctioastraea profunda* (d'Achiardi, 1868)
*Favia corollaris* Reuss, 1871
*Favites aranea* (Defrance, 1826)
*Favites neglecta neglecta* (d'Achiardi, 1868)
*Favites neglecta taurinensis* Chevalier, 1961
*Favites neugeboreni burdigalensis* Chevalier, 1961
*Favites mimbastensis* Chevalier, 1961
*Ellasmoastraea multilateralis* (Michelin, 1842)
*Goniastraea aff. pectinata* (Ehrenberg, 1834)
*Goniastraea* sp. (aff. *speciosa* (Felix, 1913))
*Astroria irregularis irregularis* (Defrance, 1826)
*Astroria irregularis granulata* Chevalier, 1961
*Caulastraea matheroni* Chevalier, 1961
*Heliastraea (Heliastraea) delicata* Osasco, 1897
*Heliastraea (Heliastraea) pelouaensis* (Chevalier, 1954)
*Heliastraea (Heliastraea) saucatsensis* (Chevalier, 1954)
*Heliastraea (Heliastraea) solenastroides* (Chevalier, 1954)
*Heliastraea (Heliastraea) laticosta* Chevalier, 1961
*Heliastraea (Heliastraea) nerthensis* Chevalier, 1961
*Heliastraea (Athecastraea) vesiculosa* (Milne-Edwards and Haime, 1850)
*Heliastraea (Heliastraeopsis) dallagoi* (Osasco, 1902)
*Heliastraea (Heliastraeopsis) alloiteaui alloiteaui* Chevalier, 1954
*Heliastraea (Heliastraeopsis) alloiteaui parva* Chevalier, 1961
*Heliastraea (Aquitanastraea) incrustans* (Osasco, 1897)
*Heliastraea (Aquitanastraea) piveteaui* Chevalier, 1954
*Heliastraea (Aquitanastraea) pruvosti* Chevalier, 1954
*Heliastraea (Aquitanastraea) pachyformis* Chevalier, 1954
*Heliastraea (Aquitanastraea) tenuitabulata* Chevalier, 1961
*Montastrea forbesi* (Duncan, 1865)
*Montastrea oligophylla* (Kopek, 1954, *non* Reuss, 1871)
*Montastrea alloiteaui daxensis* (Chevalier, 1961)
*Antiguastraea alveolaris* (Catullo, 1856)
*Antiguastraea* nov. sp. 1
*Tarbellastraea ellisiana* (Defrance, 1826)
*Tarbellastraea reussiana reussiana* (Milne-Edwards and Haime, 1850) = *T. raulini* (Milne-Edwards and Haime, 1850)
*Tarbellastraea reussiana echinulata* Chevalier, 1961
*Tarbellastraea aquitaniensis* Chevalier, 1961
*Tarbellastraea mimbastensis* Chevalier, 1961
*Thegioastraea diversiformis* (Michelin, 1842)
*Thegioastraea taurinensis* (d'Achiardi, 1868)
*Thegioastraea superficialis* (Sismonda, 1871)
*Thegioastraea multisepta multisepta* (Sismonda, 1871)
*Thegioastraea multisepta zuffardii* Chevalier, 1961
*Thegioastraea roasendai* Michelotti *in* Sismonda, 1871
*Thegioastraea variabilis* (Sismonda, 1871)
*Thegioastraea miocenica* (Michelotti *in* Sismonda, 1871)
*Thegioastraea aequalicostata* (Osasco, 1897)
*Thegioastraea asymmetrica* (Gregory, 1898)
*Thegioastraea* cf. *rosacea* (Zuffardi-Comerci, 1932)
*Thegioastraea alternaticosta* Chevalier, 1961
*Thegioastraea burdigalensis* Chevalier, 1961
*Plesiastraea (Palaeoplesiastraea) desmoulinsi* (Milne-Edwards and Haime, 1851)
*Plesiastraea (Palaeoplesiastraea) corrugata* (Michelotti *in* Sismonda, 1871)

### Family Astrangiidae Verrill, 1870
*Cladangia carryensis galasciformis* (Matheron *in* Repelin, 1900)
*Cladangia aquitaniensis* Chevalier, 1961
*Astrangia vasconiensis* (Milne-Edwards and Haime, 1850)
*Oulangia speyeri* (Reuss, 1865)
*Culicia parasitica* (Michelin, 1847)

### Family Oculinidae Gray, 1847
*Astrohelia palmata* (Goldfuss, 1826)
*Diplohelia reflexa* (Michelotti, 1847)

### Family Mussidae Ortmann, 1890
*Lithophyllia michelottii* (Michelin, 1841)
*Lithophyllia patula* (Sismonda, 1871)
*Syzygophyllia (Syzygophyllia) crenaticosta* (Reuss, 1868)
*Syzygophyllia (Syzygophyllia) elongata* (Sismonda, 1871)
*Syzygophyllia (Syzygophyllia) grandis* Chevalier, 1961
*Syzygophyllia (Aquitanophyllia) grandistellae* Chevalier, 1961

### Family Caryophylliidae Gray, 1847
*Sphenotrochus cestasensis* Chevalier, 1961
*Ceratotrochus (Edwardsotrochus) duodecimcostatus magnei* Chevalier, 1961
*Ceratotrochus (Conotrochus)* aff. *subrectus* (De Angelis, 1894)
*Cyathoceras* nov. sp. 1
*Cylindrophyllia duncani* (Reuss, 1871)

### Family Agariciidae Gray, 1847
*Pachyseris crassatheca* Chevalier, 1961
*Pavona burdigalensis* Chevalier, 1961

### Family Siderastraeidae Vaughan and Wells, 1943
*Siderastraea crenulata crenulata* (Goldfuss, 1826)
*Siderastraea bertrandiana* (Michelin, 1847)
*Siderastraea miocenica miocenica* Osasco, 1897
*Siderastraea miocenica italica* (Defrance, 1826)
*Siderastraea lecointrei* Chevalier, 1961

### Family Micrabaciidae Vaughan, 1905
*Stephanophyllia* sp.

### Family Poritidae Gray, 1842
*Alveopora discors* De Angelis, 1894
*Alveopora daxensis* Chevalier, 1961
*Goniopora raulini* (Milne-Edwards and Haime, 1857)
*Goniopora globulosa* Chevalier, 1961
*Goniopora chevalieri* Oosterbaan, 1988
*Porites arenosa* (Esper, 1797)
*Porites collegniana collegniana* Michelin, 1842 = *Porites collegniana girundiensis prima* (Bernard, 1903)
*Porites leptoclada* Reuss, 1871
*Porites diversiformis* (Michelotti *in* Sismonda, 1871)
*Porites pusilla* Felix, 1884
*Porites maigensis maigensis* Kühn, 1925

### Family Dendrophylliidae Gray, 1847
*Dendrophyllia ramea* (Linné, 1758)
*Paleoastroides subirregularis* (Osasco, 1897)

### Family Acroporidae Verrill, 1902
*Acropora ornata* (Defrance, 1823)
*Acropora solanderi* (Defrance, 1828) = *A. exarata* (Michelotti, 1838)
*Acropora lavandulina* (Michelin, 1842)
*Acropora duncani* (Reuss, 1867)
*Acropora pachymorpha* Chevalier, 1956
*Astreopora densata* Chevalier, 1961

### Family Turbinariidae Milne-Edwards and Haime, 1857
*Turbinaria cyathiformis cyathiformis* (de Blainville, 1830) = *T. cyathiformis lamelliformis* Chevalier, 1961
*Turbinaria grandis* Chevalier, 1961

bioprovince was restricted to the eastern Atlantic and Mediterranean. At that time, the western Mediterranean continued to maintain open connections with the Atlantic (Berggren and Hollister, 1974; Alvinerie et al., 1992). As in the Aquitaine Basin, the Mediterranean Early Miocene series is often rich in reefal buildups (and also in rhodalgal deposits), the reefs thickness being sometimes great (150 m in Egypt Burdigalian; Chevalier, 1977), but their dating still is locally under discussion. In northern Italy, Burdigalian coral mounds occurred with very high diversity (Esteban et al., this volume).

### *Middle Miocene Coral Faunas*

Hermatypic coral reefs are absent in the Mid-Miocene series of the Aquitaine Basin; only subreefal facies locally persist in the Langhian. Recent studies of extensive material from new exposures indicate that the coral fauna contains in all 60 species, belonging to 29 genera; about half of these genera are hermatypic (*Porites, Siderastraea, Tarbellastraea, Montastrea, Heliastraea, Platycoenia*, etc.; cf. Table 6). This hermatypic fauna, not abundant anywhere, is principally restricted to the Langhian. Among the complete fauna, 30 species have a massive morphology, 6 are branching, 4 are phaceloid, 3 are reptoid and 17 were solitary. The marked decrease in diversity and abundance in comparison with the Early Miocene faunas is considered notably the result of an important climatic change from the Burdigalian. The eastern Atlantic climate, previously tropical, underwent a general progressive cooling. Particularly along the northeastern Atlantic neritic realm, a latitudinal thermic gradient has been evidenced on the basis of the evolution of faunas and floras, with cooling from north to south through the Neogene period, this being more marked from the Mid-Miocene (Lauriat-Rage et al., 1993). This deterioration is marked among several faunal groups and seems more pronounced during the Serravallian (probably linked to an increase in Antarctic glaciation at that time and perhaps also to the incursion of waters from the north). In the northern Atlantic especially, the surface- and bottom-water temperature deterioration is quite obvious (ca. 13 to 11 Ma ago; Vergnaud-Grazzini et al., 1979; Müller, 1984). Besides this, other regional environmental factors may have been responsible for this coral fauna impoverishment. For instance, one factor may have been the development of marly or sandy and gravelly siliciclastic facies in the studied areas.

The Langhian series in the southern part of the Aquitaine Basin corresponds to the N8 *pars*-N9 *pars* Blow (1969) and NN5 Martini (1971) planktonic zones. It contains small reefal coral colonies and locally numerous solitary corals; this fauna, including 28 hermatypic species out of 40 in all, occurs mainly in two areas. On the one hand, in the southwestern area ("Bas-Adour": Saubrigues, Fig. 3), thick marly deposits fill up one canyon resulting from the erosional unconformity at the Rupelian-Chattian boundary (Kieken, 1973; Cahuzac et al., 1992b, 1994). Most of these marls are considered to be deposited at water depths up to 100-200 m. On the other hand, towards the east and in central Aquitaine, isolated colonies of hermatypic corals also are known in coeval "falun" deposits (e.g., Le Houga, Manciet, Parleboscq, Baudignan). There, the transgressive sea developed a small shallow gulf, the "Gulf of Manciet" (to the east of Roquefort and Créon domes, Fig. 4, cf. Cahuzac et al., 1995), with sub-reefal facies rich in littoral malacofauna and coarse clastics. In this area, the marine series onlapped an Upper Burdigalian continental sequence ("marls and limestones of Armagnac").

The Serravallian corresponds to the "Helvetian" regional denomination and to the former local stage "Sallomacian" in northern Aquitaine with bioclastic sandy deposits in the Salles area (Fig. 4; Folliot et al., 1993). At that time, the transgression spread extensively in the Aquitaine Basin and formed gulfs, notably farther than during the previous periods, to the south (Chalosse and Orthez/Salies-de-Béarn areas) and to the east ("Armagnac Gulf"). These deposits may be attributed to the NN6-7 (even early NN8 locally) nannoplankton zones (Cahuzac et al., 1995); a few Sr isotopes datings from bivalve coquinas indicate an age of about 12 to 13 Ma. During the Serravallian, corals notably decreased and at the end of the stage disappeared locally from the Aquitaine Basin. Only very rare hermatypic colonies of *Cladangia, Antiguastraea* and *Paleoastroides* mixed with ahermatypic genera (*Culicia, Astrangia, Sphenotrochus*, etc.) are present, with a total of about 25 species, of which 5 hermatypics (Table 6). To the only 6 species previously known (Cahuzac and Chaix, 1993), we recently added several taxa (particularly from upper part of the Pessac-Magonty new exposure near Bordeaux), indicating a few affinities with the Netherlands Mid-Miocene coral fauna (e.g., *Sphenotrochus, Ceratotrochus*, and *Flabellum*). The disappearance of other tropical forms among benthic forams, mollusks, and bryozoans corroborates a general cooling trend during the Mid-Miocene epoch. Besides, Serravallian deposits formed in a dominantly terrigenous clastic regime under often turbulent hydrodynamic conditions. So, in the whole Aquitaine Basin, the Serravallian upper marine series generally ends with widespread ferrugineous coarse sandstones containing rare macrofauna (bivalves *Crassostrea crassissima/gryphoides*, Pectinids and *Megacardita jouanneti*, and locally sand dollar urchins); this was certainly unfavourable to reefal conditions.

During the Mid-Miocene, few Mediterranean influences were still present in the Aquitaine Basin, but an exclusive eastern Atlantic coral bioprovince began to differentiate (about 15 species common with the Loire Basin). In Portugal, few sub-reefal facies are locally known in the "Upper Helvetian" (i.e., probably the Serravallian) series of Low Tage Gulf. They contain a faunule of less than 10 taxa, with mixed solitary and very rare hermatypic (one genus only, *Hexastraea*) species, as in the western France Langhian. There also, the proportion of reefal genera strongly decreased (Chevalier and Nascimento, 1975; Chevalier, 1977). In contrast, the reefal phenomenon developed then very plainly in the western Mediterranean (e.g., in Provençal basin, Catalogne, Languedoc, with a fauna still well

TABLE 6.—LIST OF THE MIDDLE MIOCENE CORALS IN THE SOUTHWESTERN FRANCE.

## MIDDLE MIOCENE (SOUTHWESTERN FRANCE)

### LANGHIAN

**Family Stylophoridae Milne-Edwards and Haime, 1857**
*Stylophora rugosa* (d'Archiac, 1848)

**Family Astrocoeniidae Koby, 1890**
?*Stylocoeniella* sp.
*Platycoenia turonensis* (Michelin, 1847)
*Glyphastraea* nov. sp. 1 ( = *Platycoenia palmata sensu* Chevalier, 1956, *non* Catullo, 1856)

**Family Faviidae Gregory, 1900**
*Favites neugeboreni* (Reuss, 1871)
*Heliastraea (Heliastraea) saubriguensis* Chevalier, 1961
*Heliastraea (Aquitanastraea) incrustans* (Osasco, 1897)
*Heliastraea (Athecastraea) fragilis* Chevalier, 1961
*Montastrea forbesi* (Duncan, 1865)
*Montastrea parva* (Chevalier, 1961)
*Montastrea* nov. sp. 1
*Antiguastraea alveolaris* (Catullo, 1856)
*Antiguastraea* nov. sp. 1
*Tarbellastraea ellisiana* (Defrance, 1826)
*Tarbellastraea reussiana* (Milne-Edwards and Haime, 1850)
*Tarbellastraea eggenburgensis formosa* (Kühn, 1925)
*Tarbellastraea eggenburgensis andalousianensis* Chevalier, 1961
*Thegioastraea miocenica* (Michelotti *in* Sismonda, 1871)
*Cladocora multicaulis* (Michelin, 1842)
*Cladocora prevostiana* Milne-Edwards and Haime, 1849

**Family Astrangiidae Verrill, 1870**
*Cladangia semispherica tubiformis* (Michelin, 1847)
*Cladangia crassoramosa* (Michelin, 1847)
*Cladangia pachyphylla* (Reuss, 1847)
*Astrangia manthelanensis* Chevalier, 1961
*Oulangia speyeri* (Reuss, 1865)
*Culicia parasitica* (Michelin, 1847)

**Family Oculinidae Gray, 1847**
*Astrohelia meneghiniana* (d'Achiardi,1868)
*Astrohelia* nov. sp. 1 ( = *A. neglecta sensu* Chevalier, 1961, *non* Osasco, 1897)

**Family Caryophylliidae Gray, 1847**
*Sphenotrochus intermedius* (Goldfuss, 1826) (= *S. milletianus* Defrance, 1828)
*Paracyathus turonensis* Milne-Edwards and Haime, 1848
*Paracyathus pedemontanus alternaticosta* Osasco, 1895
*Cylindrophyllia duncani* (Reuss, 1871)

**Family Flabellidae Bourne, 1905**
*Flabellum basteroti* Milne-Edwards and Haime, 1848

**Family Siderastraeidae Vaughan and Wells, 1943**
*Siderastraea radians* (Pallas, 1766)
*Siderastraea miocenica italica* (Defrance, 1826)
*Siderastraea felixi* Roszkowska, 1932

**Family Poritidae Gray, 1842**
*Porites collegniana collegniana* Michelin, 1842
*Porites mancietensis* Chevalier, 1961
*Goniopora decaturensis* Vaughan, 1919

**Family Dendrophylliidae Gray, 1847**
*Balanophyllia (Balanophyllia) concinna* Reuss, 1871
*Dendrophyllia amica* (Michelotti, 1838)

### SERRAVALLIAN

**Family Faviidae Gregory, 1900**
*Antiguastraea* nov. sp. 1
*Cladocora michelottii popognae* Chevalier, 1961

**Family Astrangiidae Verrill, 1870**
*Cladangia semispherica semispherica* (Defrance, 1826)
*Cladangia crassoramosa* (Michelin, 1847)
*Cladangia carryensis galasciformis* (Matheron *in* Repelin, 1900)
*Astrangia vasconiensis* (Milne-Edwards and Haime, 1850)
*Astrangia manthelanensis* Chevalier, 1961
*Cryptangia woodi* Milne-Edwards and Haime, 1848
*Cryptangia reptans* Chevalier, 1961
*Culicia parasitica* (Michelin, 1847)

**Family Mussidae Ortmann, 1890**
*Lithophyllia detrita* (Michelin, 1842)

**Family Caryophylliidae Gray, 1847**
*Ceratotrochus (Edwardsotrochus) bellingherianus* (Michelin, 1841)
*Ceratotrochus (Edwardsotrochus) kefersteini* (Krejci, 1926)
*Sphenotrochus intermedius* (Goldfuss, 1826)

**Family Flabellidae Bourne, 1905**
*Flabellum intermedium* Milne-Edwards and Haime, 1848
*Flabellum waeli waeli* Nyst, 1861

**Family Dendrophylliidae Gray, 1847**
*Balanophyllia (Balanophyllia) italica* (Michelin, 1841)
*Balanophyllia (Balanophyllia) calyculus* Wood, 1844
*Balanophyllia (Balanophyllia) varians* Reuss, 1854
*Balanophyllia (Balanophyllia)* sp. (aff. *wellsi* Cairns, 1977)
*Balanophyllia (Eupsammia) sismondiana* (Michelin, 1841)
*Balanophyllia (Eupsammia) bossolensis* (Chevalier, 1961)
*Dendrophyllia longaeva* Michelotti *in* Sismonda, 1871
*Dendrophyllia trifurcata* Michelotti *in* Sismonda, 1871
*Dendrophyllia crassa* Osasco, 1895
*Paleoastroides subirregularis* (Osasco, 1897)

diversified in number of genera). During the Mid-Miocene, the biogeographic evolutionary trends of coral communities may have been related to several significant events. For one thing, until the Burdigalian, the Mediterranean was a seaway connecting the Indo-Pacific and the Atlantic (the open communication through this extensive Tethys sea helps explain the continuation of a tropical climate in low latitudes). In the Mid-Upper Burdigalian (about 18 Ma ago; Berggren and Hollister, 1974), a closure occurred separating the eastern and western Tethys. The communication was disconnected between the Mediterranean and Indo-Pacific. Consequently, coral faunas became globally impoverished in diversity from that time in Atlantic-Mediterranean realm, because of restricted renewal from the east. Another result of this is the strong modification of current circulation through the Atlantic which was increased by the following second event.

Within the Mid-Miocene (from ca. 15 Ma ago), a junction occurred between Europe and North Africa with generation of the Gibraltar Sill (Berggren and Hollister, 1974). So, the marine domain between Iberia and Africa became narrower in consequence of submeridian compressive strains (Alvinerie et al., 1992). The western Mediterranean became further restricted, which explains the persistance of reef development in this "protected" area fairly apart from cold currents at least during the Langhian; very few faunal exchanges with Atlantic occurred until the end of the Mid-Miocene. There ensued a modified and enhanced marine circulation in the northern Atlantic (Berggren and Hollister, 1974) caused by exclusion of the Gulf Stream from its eastward former Mediterranean route, initiation of a North Atlantic drift northwards, extrusion of Arctic waters, and development of vigorous bottom currents. These paleoceanographic events, concomitant with the general cooling and the related differentiation of a distinctive latitudinal zonation and provincialization of faunas in Atlantic marine domain, may explain the characteristics of coral assemblages at that time.

The Miocene coral fauna from the Madeira archipelago sets a great problem of dating. This association of hermatypic/ahermatypic mixed species was labelled as "Vindobonian" age (Chevalier, 1972a; Boekschoten and Wijsman Best, 1981), and it is close to the Mediterranean faunas. It might date, at least partly, from Mid (?) and perhaps Upper Miocene.

## MIOCENE OF NORTHWESTERN FRANCE AND COMPARISON WITH OTHER EASTERN ATLANTIC AREAS DURING UPPER MIOCENE

Two major gulfs formed in northwestern France during the Neogene, the Channel Gulf and the Ligerian (Loire) Gulf (Fig. 1). The fossiliferous coastal-marine Lower Miocene deposits of northwestern France contain no corals (Charrier et al., 1980; Alvinerie et al., 1992); but until now, these deposits are known essentially from few borings, which allows only to outline hypothetically a narrow gulf eastward.

The Mid-Miocene epoch represents the most extensive marine transgression in the region (Fig. 1) with coarse-grained shelly sands and poorly consolidated skeletal carbonates ("faluns") (Charrier and Palbras, 1978; Charrier et al., 1980; Cavelier et al., 1980; Cavelier, 1989; Margerel and Cousin, 1989; Anonymous, 1989). Ripple cross-lamination, channels and bars are common sedimentary features. The macrofaunas (corals, bryozoans, bivalves, gastropods, echinoids) are abundant and diversified with the mollusks forming most of the biomass (more than 500 species). Although there are a large number of endemic species (bryozoans, gastropods and echinoids), there is a marked affinity with the Aquitaine Basin. Important endofaunas of bivalves and echinoids characterize the predominant soft bottoms. These Mid-Miocene faunal assemblages (notably for scleractinians) suggest a warm climate of a subtropical type with western Africa-Caribbean characteristics notable for gastropods (Brébion, 1988). Hermatypic corals are mixed with a variety of ahermatypic corals (Table 7, modified in comparison with the one in Cahuzac and Chaix, 1993). These all occur as isolated colonies without developing reef buildups. A part of this Mid-Miocene series has been dated as probable Langhian age (locally NN5 nannoplankton zone in Rennes area at Chasné-sur-Illet; Margerel and Bréhéret, 1984), but Serravallian deposits are also present in the southern part of the region (Mirebeau) and in the Noyant/Savigné-sur-Lathan basin (Fig. 1). Two primary facies have been distinguished, as follows:

The "Pontilevian" facies contains 37 coral species (22 strictly ahermatypic) and 22 genera (12 ahermatypic). The presence of *Cladocora, Paracyathus, Balanophyllia, Phyllangia, Cryptangia, Culicia, Dendrophyllia, Oulangia*, and of the hermatypic "Mediterranean" genus *Acanthastraea* (which seems unknown earlier in the Aquitaine Basin but exists in Portugal), together with the disappearance or the absence of a large number of hermatypic corals (all of them known in the Burdigalian of Aquitaine: *Stylophora, Pocillopora, Hexastraea, Favia, Favites, Ellasmoastraea, Astroria, Caulastraea, Montastrea, Athecastraea, Heliastraeopsis, Aquitanastraea, Thegioastraea, Paleoplesiastraea, Syzygophyllia, Aquitanophyllia, Pachyseris, Pavona, Alveopora, Goniopora* and *Astreopora*) is characteristic of these Mid-Miocene facies. This is consistent with the general cooling trend. Chevalier (1961) already noted the abundance of Dendrophylliidae and Astrangiidae, commonly inhabiting bottoms at water depths of 20-150 m, and also the rarity of Caryophylliidae, typical of greater water depths. Nevertheless, the occurrence of *Oulangia* (not cited by Chevalier, 1961) indicates shallow-water conditions, probably between 15 and 20 m (*Sphenotrochus* is not known before 20 m in Recent seas). The hermatypic corals present in the "faluns" of Touraine and Anjou (*Acropora, Turbinaria, Porites, Siderastraea, Acanthastraea*) could occur at similar water depths (*Acanthastraea* seems to live now only between 20 and 40 m according to Kühlmann, 1983, but it is very rare in these deposits). The morphology of the colonies (among the 15 identified hermatypic species, 13 are massive, just one is lamellar and one branched), their small size and the absence or rarity of corals abundant so far in the Mediterranean reefs during the same time period or in the Aquitaine reefs until the Burdigalian (Faviidae, Mussidae, Poritidae, etc.) further support the idea of unfavorable ecological conditions (temperature, depth, turbulence of sedimentary regime) for the development of coral reefs. The most important point seems to be the hydrodynamics: the kind of sedimentation of these "faluns" or "crags" (interbeddings, cross-beddings, coarse gravel deposits containing rolled and broken corals) indicate a great wave-exposure with probably numerous tempests in this gulf. Chevalier (1961) considered these accumulations of corals as "sub-reefal" and stressed the similarities with modern coral assemblages in the sea of Japan and the western coast of California. They look also like the western African coast assemblages (Gulf of Guinea).

The "Savignean" facies of the Mid-Miocene is characterized by very abundant bryozoan colonies and occurs extensively in northwestern France; it only contains 4 species of ahermatypic corals. The most frequent are *Sphenotrochus intermedius* and *Culicia parasitica*, this last one is an exclusive symbiont of the

TABLE 7.—LIST OF THE MIDDLE AND UPPER MIOCENE CORALS OF THE NORTHWESTERN FRANCE.

## MIOCENE OF NORTHWESTERN FRANCE

### "PONTILEVIAN" (MID MIOCENE)

**Family Astrocoeniidae Koby, 1890**
*Platycoenia turonensis* (Michelin, 1847)

**Family Faviidae Gregory, 1900**
*Heliastraea (Heliastraea) saucatsensis* (Chevalier, 1954)
*Heliastraea (Aquitanastraea) pruvosti* Chevalier, 1954
*Tarbellastraea ellisiana manthelanensis* Chevalier, 1961
*Cladocora multicaulis* (Michelin, 1842)

**Family Astrangiidae Verrill, 1870**
*Cladangia semispherica semispherica* (Defrance, 1826)
*Cladangia semispherica tubiformis* (Michelin, 1847)
*Cladangia crassoramosa* (Michelin, 1847)
*Cladangia aquitaniensis* Chevalier, 1961
*Astrangia manthelanensis* Chevalier, 1961
*Cryptangia woodi woodi* Milne-Edwards and Haime, 1848
*Cryptangia woodi michelini* Chevalier, 1961
*Cryptangia reptans* Chevalier, 1961
*Oulangia gracilis* (Zuffardi-Commerci, 1932)
*Culicia parasitica* (Michelin, 1847)

**Family Mussidae Ortmann, 1890**
*Acanthastraea turonensis* Chevalier, 1961

**Family Caryophylliidae Gray, 1847**
*Bathycyathus* nov. sp. 1
*Paracyathus turonensis* Milne-Edwards and Haime, 1848
*Paracyathus incrustans* Zuffardi-Commerci, 1932
*Phyllangia (?) conferta* (Milne-Edwards and Haime, 1849)
*Sphenotrochus intermedius* (Goldfuss, 1826)

**Family Flabellidae Bourne, 1905**
*Flabellum* sp.

**Family Siderastraeidae Vaughan and Wells, 1943**
*Siderastraea miocenica miocenica* Osasco, 1897
*Siderastraea miocenica italica* (Defrance, 1826)
*Siderastraea miocenica regularis* d'Orbigny, 1852
*Siderastraea bertrandiana* (Michelin, 1847)

**Family Poritidae Gray, 1842**
*Porites maigensis maigensis* Kühn, 1925
*Porites turonensis* (Chevalier, 1961)

**Family Dendrophylliidae Gray, 1847**
*Balanophyllia (Balanophyllia) italica* (Michelin, 1841)
*Balanophyllia (Balanophyllia) varians* Reuss, 1854
*Balanophyllia (Balanophyllia) concinna* Reuss, 1871
*Balanophyllia (Balanophyllia) angusticalix* Chevalier, 1961
*Dendrophyllia cornigera* (Lamarck, 1816)
*Dendrophyllia amica* (Michelotti, 1838)
*Dendrophyllia digitalis* Michelin, 1842
*Dendrophyllia taurinensis* Milne-Edwards and Haime, 1848
*Dendrophyllia trifurcata* Michelotti *in* Sismonda, 1871
*Dendrophyllia alternaticosta* Chevalier, 1961
*Paleoastroides michelini* Chevalier, 1961

**Family Acroporidae Verrill, 1902**
*Acropora solanderi* (Defrance, 1828) = *A. exarata* (Michelotti, 1838)

**Family Turbinariidae Milne-Edwards and Haime, 1857**
*Turbinaria cyathiformis turonensis* (d'Orbigny, 1852)

### "SAVIGNEAN" (MID-UPPER MIOCENE)

**Family Astrangiidae Verrill, 1870**
*Culicia parasitica* (Michelin, 1847)

**Family Caryophylliidae Gray, 1847**
*Sphenotrochus intermedius* (Goldfuss, 1826)

**Family Dendrophylliidae Gray, 1847**
*Balanophyllia (Balanophyllia) varians* Reuss, 1854
*Dendrophyllia digitalis* Michelin, 1842

### UPPER MIOCENE
([T] "TORTONIAN" of Doué-la-Fontaine and [M] hypothetic "MESSINIAN" of Beugnon and Renauleau)

**Family Faviidae Gregory, 1900**
*Cladocora multicaulis* (Michelin, 1842) [T, M]
*Cladocora prevostiana* Milne-Edwards and Haime, 1849 [M]

**Family Astrangiidae Verrill, 1870**
*Astrangia manthelanensis* Chevalier, 1961 [T, M]
*Astrangia talquinensis* Weisbord, 1971 [M]
*Cryptangia reptans* Chevalier, 1961 [T, M]
*Cladangia semispherica semispherica* (Defrance, 1826) [T]
*Cladangia crassoramosa* (Michelin, 1847) [T]
*Cladangia aquitaniensis* Chevalier, 1961 [T]
*Culicia parasitica* (Michelin, 1847) [T, M]

**Family Oculinidae Gray, 1847**
*Diplohelia parvistella* (Reuss, 1871) [M]
*Diplohelia quenstedti* Chevalier, 1961 [M]

**Family Caryophylliidae Gray, 1847**
*Cyathoceras* sp. [M]
*Peponocyathus* nov. sp. 1 [M]
*Sphenotrochus intermedius* (Goldfuss, 1826) [T, M]

**Family Dendrophylliidae Gray, 1847**
*Balanophyllia (Balanophyllia) varians* Reuss, 1854 [T, M]
*Dendrophyllia cornigera* (Lamarck, 1816) [T]
*Dendrophyllia amica* (Michelotti, 1838) [T, M]
*Dendrophyllia taurinensis* Milne-Edwards and Haime, 1848 [T, M]
*Dendrophyllia longaeva* Michelotti *in* Sismonda, 1871 [T]

bryozoan *Celleporaria palmata* in this facies.

Another particular facies is the "*Arca* sands," generally very bioclastic (or even lumachellic) and more or less indurated, which often occur above, or as lateral facies of, the "Savignean" deposits. They may be dated as the upper part of Mid-Miocene: an age of about 12,5 Ma has been obtained with few Sr datings from both *Arca* sands and some Savignean deposits (Cahuzac et al., in prep.). So, this series seems more or less coeval with the Serravallian transgressive one in the Aquitaine Basin. No corals were found there until now, but in some outcrops (Lublé, Milvraut, Savigné), we recently have found a few species, with *Cladangia* associated with ahermatypic taxa (e.g., *Dendrophyllia,*

*Culicia, Balanophyllia*). All genera were already known from the Miocene Loire Basin.

The Upper Miocene deposits are present in Anjou (Doué-la-Fontaine), Touraine (Savigné-sur-Lathan area) and probably in Loire-Atlantique (Les Cléons) and Vendée (Challans), with "savignean" type of facies, locally very sandy (Fig. 1). There are only few localities with Upper Miocene corals in northwestern France, which are the only ones of this age in all the western coast of France. The Tortonian deposits with corals (dated as the lower part of this stage) occur solely in Doué-la-Fontaine with 13 species and 8 genera; only 3 species are strictly hermatypic (of the genus *Cladangia*), but they are rare (Table 7). All these corals are survivors from the regional Mid-Miocene; only one species is new (ahermatypic *Dendrophyllia longaeva*), it is present in the Aquitaine Serravallian. Another one, the hermatypic *Paleoastroides theotvoldensis* (Michelin, 1847), has been mentioned only once *in litteris* by the author of the taxon; its presence is very doubtful, because it has never been found here again (in contrast, this coral is most common in the Chattian deposits of the Aquitaine Basin).

The "Messinian" deposits are probably present in southeastern Angers (Renauleau and Beugnon; Cavelier et al., 1980; Alvinerie et al., 1992) with a total of 14 coral species and 10 genera. None of these corals is truly hermatypic; 6 species (and 3 genera: *Peponocyathus, Cyathoceras, Diplohelia*) are new in the region. This fauna, of a dubious Messinian (or perhaps Upper Tortonian at Beugnon) age, shows a clear trend towards the totally ahermatypic coral fauna of the Pliocene when there were cooler and deeper waters than in the Late Miocene (Chaix, 1989; Anonymous, 1989). In the "Redonian" local facies of the Pliocene (Table 1), of which the deposits show two transgressive phases of large extent (Lauriat-Rage, 1981; Alvinerie et al., 1992), the coral association is somewhat diversified (37 species), corresponding to a chiefly temperate climate.

In Portugal, the Upper Miocene corals are very rare (only 2 taxa, ahermatypic, in the Tortonian; Antunes and Chevalier, 1971); in the Azores, a faunule of 8 ahermatypic species has been recorded in the "Vindobonian." On the western Morocco Atlantic coast (Rabat area), the Upper Miocene coral fauna is more diversified than the European Atlantic one (more than 20 species; Chevalier, 1962); all the corals are ahermatypic isolated colonies. There, a widespread transgression occurred; the series began with littoral deposits (Tortonian) rapidly followed by a pelagic marl sedimentation which persisted through the Messinian (ref. in Alvinerie et al., 1992). During the Late Miocene, this absence of any Atlantic reefal phenomenon made a very strong contrast with the Mediterranean domain which contained spectacular large coral reefs (stressing that there also, the number of genera decreased to less than 10 in Tortonian, and only a few in Messinian). Nevertheless, at least for the Tortonian transgression, the Atlantic-Mediterranean communications seem more open than during the Mid-Miocene, through Betic and Rifian basins and through Alboran archipelago (Alvinerie et al., 1992; Esteban et al., this volume). Then, the communications became very restricted during part of the Messinian stage. The continuing cooling of Atlantic waters, the hydrodynamics (e.g., the Messinian cold deep-water current circulation through South-Rifian basin, Benson et al., 1991) and the Messinian relative isolation of the (climatically protected) Mediterranean basin, linked to the eustatic drop(s) in the sea level and to the tectonic events which occurred in the Betic-Rifian complex orogenic area, can account for the significant differences in reefal development between the Atlantic and Mediterranean.

## CONCLUSIONS

The Atlantic basins of western France are then interesting for outlining the Miocene evolution of coral communities and reef buildups. Thanks to an extensive revision of the scleractinians faunas and reefs and to an accurate stratigraphy, the main lines of this evolution have been specified and linked to paleoclimatic changes. Particularly, southwestern France is of a previously unsuspected interest for the knowledge of the Chattian series (formerly often considered as Aquitanian). During the Chattian, this was a "climax" tropical area for scleractinians diversification (about 150 species). Based on all the faunal groups, the Late Oligocene was the warmest period (of the time interval considered here) in the neritic domain of northeastern Atlantic. After an important renewal of coral species at the Chattian-Aquitanian boundary, the reefs developed again in the Early Miocene of Aquitaine, characterized by a persistent tropical climate. For these periods, the comparison with Mediterranean reefs still often lacks accurate dating of Mediterranean reefal series. Nevertheless, in terms of biogeography, it looks as if the Aquitaine Basin was a center of dispersal of corals during the Late Oligocene and Early Miocene epochs, in the eastern Atlantic realm. An Atlantic-Mediterranean coral bioprovince was differentiated, related to a widespread Tethys, open either westward or eastward. Surprisingly, the Atlantic reef buildups seem fairly thin in the studied areas, perhaps owing to local causes. In Aquitaine for example, the carbonate platforms were widely open westward to the ocean, thus little sheltered from the often turbulent seas.

The fossil hermatypic corals, as modern species with symbiotic zooxanthellae, are reliable paleothermic indicators distributed within the shallow tropics. Therefore, the evolutionary trends in the coral reefs and faunas of Atlantic Europe seem to be closely linked to general climatic variations and chiefly to progressive sea-water cooling, observed since the Eocene in the marine domain. Since the Langhian, the number of hermatypic coral taxa, previously highly diversified, and the total number of species have strongly decreased in the whole eastern Atlantic. Reefs disappeared and the coral bioprovince was restricted to the eastern Atlantic. A latitudinal north-south thermic gradient was developed along the Atlantic frontage. At that time, in contrast, the Mediterranean domain shows a maximum of coral building with a high faunal diversity. The same contrast recurred during the Late Miocene between the two domains. In addition to the climatic variations, different factors are related to these observed evolutionary trends, such as locally unfavourable sedimentary environments, changes in the larger-scale patterns of the marine hydrodynamics, the tectonic events in the westernmost Mediterranean, or the stop in the faunal renewal from the east. These factors need to be supported by further research. Moreover, a study of the areas south of Morocco and north of Brittany could improve our data about the Miocene latitudinal faunal zonation and about the Atlantic reef development.

## REFERENCES


ADAMS, C. G., LEE, D.E., AND ROSEN, B. R., 1990, Conflicting isotopic and biotic evidence for tropical sea-surface temperatures during the Tertiary: Palaeogeography, Palaeoclimatology, Palaeoecology, v. 77, p. 289-313.

ALVINERIE, J., 1969, Contribution sédimentologique à la connaissance du Miocène aquitain. Interprétation stratigraphique et paléogéographique: Unpublished Thèse de Doctorat, University of Bordeaux, Bordeaux, 475 p.

ALVINERIE, J., ANGLADA, R., CARALP, M., AND CATZIGRAS, F., 1977, Stratotype et parastratotype de l'Aquitanien: Paris, Centre National de la Recherche Scientifique, Stratotypes, v. 4, 105 p.

ALVINERIE, J., ANTUNES, M. T., CAHUZAC, B., LAURIAT-RAGE, A., MONTENAT, C., AND PUJOL, C., 1992, Synthetic data on the paleogeographic history of Northeastern Atlantic and Betic-Rifian basin, during the Neogene (from Brittany, France to Morocco): Palaeogeography, Palaeoclimatology, Palaeoecology, v. 95, p. 263-286.

ANONYMOUS, 1989, Le Tertiaire du Massif armoricain: Orléans, Géologie de la France, Bureau de Recherches Géologiques et Minières, fasc. 1-2, 311 p.

ANTUNES, M. T. AND CHEVALIER, J. P., 1971, Notes sur la géologie et la paléontologie du Miocène de Lisbonne. VII- Observations complémentaires sur les madréporaires et les faciès récifaux: Lisboa, Revista da Faculdade de Ciências, série 2, v. XVI, p. 291-306.

BARTA-CALMUS, S., 1973, Révision de collections de Madréporaires provenant du Nummulitique du Sud-Est de la France, de l'Italie et de la Yougoslavie septentrionale: Unpublished Thèse de Doctorat, University of Paris, Paris, 693 p.

BENSON, R. H., RAKIC-EL-BIED, K., AND BONADUCE, G., 1991, An important current reversal (influx) in the Rifian Corridor (Morocco) at the Tortonian-Messinian boundary: the end of Tethys Ocean: Paleoceanography, v. 6, p. 164-192.

BERGGREN, W. A. AND HOLLISTER, C. D., 1974, Paleogeography, paleobiogeography and the History of Circulation in the Atlantic Ocean, *in* Hay, W. W., ed., Studies in Paleo-oceanography: Tulsa, Society of Economic Paleontologists and Mineralogists Special Publication 20, p. 126-186.

BLOW, W. H., 1969, Late Middle Eocene to Recent planktonic foraminiferal biostratigraphy: Geneva, Proceedings of the First International Conference on Planktonic Microfossils (1967), v. 1, p. 199-422.

BOEKSCHOTEN, G. J. AND WIJSMAN BEST, M., 1981, *Pocillopora* in the Miocene reef at Baixo, Porto Santo (Eastern Atlantic): Amsterdam, Proceedings of the Koninklijke Nederlands Akademie van Wetenschappen, series B, v. 84, p. 13-20.

BRÉBION, P., 1988, Evolution dans le temps et dans l'espace des Gastéropodes marins dans la province nordique depuis le Miocène: Paris, Bulletin du Muséum national d'Histoire naturelle, 4e série, section C, v. 10, p. 163-173.

CAHUZAC, B., 1980, Stratigraphie et paléogéographie, de l'Oligocène au Miocène moyen, en Aquitaine Sud-Occidentale: Unpublished Thèse de Doctorat, University of Bordeaux, Bordeaux, 586 p.

CAHUZAC, B., 1983, Un exemple de site d'âge oligocène terminal dans le Bassin de l'Adour: le gisement d'Estoti (Vallon de Poustagnac, Saint-Paul-lès-Dax, Landes): Dax, Bulletin de la Société de Borda, v. 108, no. 390, p. 297-314.

CAHUZAC, B., 1984, Les faunes de Miogypsinidae d'Aquitaine méridionale (France), *in* Oertli, H. J., ed., Benthos '83: Pau, 2nd International Symposium on Benthic Foraminifera (1983), p. 117-129.

CAHUZAC, B., ALVINERIE, J., CLUZAUD, A., AND LESPORT, J. F., 1992a, Les *Trisidos* (Bivalvia, Arcidae) du Chattien du Bassin de l'Adour (Aquitaine, France). Systématique, intérêt paléoécologique et paléobiogéographique: Lyon, Géobios, Mémoire spécial 14, p. 87-96.

CAHUZAC, B., ALVINERIE, J., LAURIAT-RAGE, A., MONTENAT, C., AND PUJOL, C., 1992b, Palaeogeographic maps of the Northeastern Atlantic Neogene and relation with the Mediterranean Sea: Barcelona, IXth Congress of Regional Committee of the Mediterranean Neogene Stratigraphy (1990), Sabadell, "Paleontologia i Evolucio", Tomes 24-25, p. 279-293.

CAHUZAC, B. AND CHAIX, C., 1993, Les faunes de coraux (Anthozoaires Scléractiniaires) de la façade atlantique française au Chattien et au Miocène: Lisbon, 1st Congress of Regional Committee of the Atlantic Neogene Stratigraphy (1992), "Ciências da Terra" (Universidade Nova Lisboa), no. 12, p. 57-69.

CAHUZAC, B. AND CHAIX, C., 1994, La faune de coraux de l'Oligocène supérieur de La Téoulère (Peyrehorade, Landes): Dax, Bulletin de la Société de Borda, v. 119, no. 436, p. 463-484.

CAHUZAC, B. AND GAUTRET, P., 1993, Découverte, dans le Miocène inférieur des Landes (Bassin Aquitain, France) de constructions squelettiques flottantes attribuées aux Hydrozoaires et signalées pour la première fois dans le Cénozoïque français: Paris, Comptes-Rendus Académie des Sciences, v. 316, Série II, p. 853-860.

CAHUZAC, B., JANIN, M. C., AND STEURBAUT, E., 1994, Biostratigraphie dans l'Oligo-Miocène du Bassin d'Aquitaine (S.O. France) grâce au nannoplancton calcaire. Le remplissage du paléogolfe de Saubrigues: Toulouse, 1er Congrès Français de Stratigraphie (1994), Strata, série 1, v. 6, p. 174-175.

CAHUZAC, B., JANIN, M. C., AND STEURBAUT, E., 1995, Biostratigraphie de l'Oligo-Miocène du Bassin d'Aquitaine fondée sur les nannofossiles calcaires. Implications paléogéographiques: Orléans, Géologie de la France, Bureau de Recherches Géologiques et Minières and Société Géologique de France, no. 2, p. 57-82.

CAHUZAC, B. AND POIGNANT, A., 1988, Les foraminifères benthiques dans l'Oligocène terminal du Vallon de Poustagnac (Landes, Bassin d'Aquitaine, Sud-Ouest de la France): Genève, 3rd International Symposium on Benthic Foraminifera (1986), Revue de Paléobiologie, Benthos '86, volume spécial 2, p. 633-642.

CAHUZAC, B. AND POIGNANT, A., 1992, Les foraminifères benthiques intéressant la limite oligo-miocène en Aquitaine (SW de la France). Comparaisons avec la Mésogée occidentale: Barcelona, IXth Congress of Regional Committee of the Mediterranean Neogene Stratigraphy (1990), Sabadell, "Paleontologia i Evolucio", Tomes 24-25, p. 15-28.

CAHUZAC, B. AND POIGNANT, A., 1993, Répartition des foraminifères benthiques dans les gisements de surface du Miocène d'Aquitaine (Sud-Ouest de la France): Lisbonne, 1st Congress of Regional Committee of the Atlantic Neogene Stratigraphy (1992), "Ciências da Terra" (Universidade Nova Lisboa), no.12, p. 71-81.

CAHUZAC, B. AND ROMAN, J., 1994, Les Echinoïdes de l'Oligocène supérieur (Chattien) des Landes (Sud-Aquitaine, France): Genève, Revue de Paléobiologie, v. 13, p. 349-371.

CAIRNS, S. D., 1979, The deep-water scleractinia of the Caribbean sea and adjacent waters, *in* Hummelinck, P. W. and Van der Steen, L. J., eds., Studies on the Fauna of Curaçao and other Caribbean islands: Utrecht, Natuurwetenschappelijke studiekring voor Suriname en Curaçao, v. 57, 250 p.

CAVELIER, C., 1989, Le Bassin Parisien au Néogène. Progrès récents: Paris, 114ème Congrès National des Sociétés Savantes, Section des Sciences, Géologie du Bassin de Paris, Comité des Travaux Historiques et Scientifiques, p. 41-54.

CAVELIER, C., KUNTZ, G., LAUTRIDOU, J. P., MANIVIT, J., PAREYN, C., RASPLUS, L., AND TOURENQ, J., 1980, Miocène et Pliocène, *in* Mégnien, C., ed., Synthèse géologique du Bassin de Paris, vol. 1: Orléans, Mémoires du Bureau de Recherches Géologiques et Minières, no. 101, p. 415-436.

CHAIX, C., 1989, Les Scléractiniaires du Pliocène de Normandie: Paris, Bulletin du Muséum national d'Histoire naturelle, 4e série, section C, v. 11, p. 3-13.

CHARRIER, P., CARBONNEL, G., CHATEAUNEUF, J. J., GARDETTE, D., MARGEREL, J. P., RIVELINE, J., AND ROUX, M., 1980, Découverte dans le bassin de Savigné-sur-Lathan (Indre-et-Loire) d'une microfaune et d'une microflore du Miocène inférieur correspondant aux premiers nivaux transgressifs de la mer des Faluns de Touraine: Paris, Comptes-Rendus Académie des Sciences, v. 290, D, p. 1325-1328.

CHARRIER, P. AND PALBRAS, N., 1978, Mise en évidence, dans le bassin de Savigné-sur-Lathan, du passage latéral entre faciès savignéen et pontilévien au sein des faluns miocènes de Touraine: Paris, Comptes-Rendus Académie des Sciences, v. 287, D, p. 915-918.

CHEVALIER, J. P., 1954, Contribution à la révision des Polypiers du genre *Heliastraea*: Paris, Annales Hébert et Haug, v. 8, p. 105-190.

CHEVALIER, J. P., 1961, Recherches sur les Madréporaires et les formations récifales miocènes de la Méditerranée occidentale: Paris, Mémoires de la Société Géologique de France, no. 93, 562 p.

CHEVALIER, J. P., 1962, Les Madréporaires miocènes du Maroc: Rabat, Notes et Mémoires du Service Géologique, no.173, 75 p.

CHEVALIER, J. P., 1963, Les Madréporaires de l'Aquitanien inférieur de Peyrère près de Peyrehorade (Landes): Reims, Annales de l'Association Régionale pour l'Etude et la Recherche Scientifique, v. I, fasc. 2, p. 3-15.

CHEVALIER, J. P., 1972a, Les Scléractiniaires du Miocène de Porto Santo (Archipel de Madère). Etude paléontologique: Paris, Annales de Paléontologie, v. 58, p. 141-160.

CHEVALIER, J. P., 1972b, Etude paléontologique, *in* Biély, A. and Chevalier, J. P., eds., Présence de Scléractiniaires dans le Miocène inférieur de la Tunisie Septentrionale: Tunis, Notes du Service Géologique, no. 40, Travaux de Géologie Tunisienne 8, p. 55-69.

CHEVALIER, J. P., 1975, Les Scléractiniaires de la Mélanésie française (Nouvelle-Calédonie, Iles Chesterfield, Iles Loyauté, Nouvelles Hébrides): Paris, Fondation Singer-Polignac, 2e partie, 407 p.

CHEVALIER, J. P., 1977, Aperçu sur la faune corallienne récifale du Néogène: Paris, 2e Symposium international sur les coraux et récifs coralliens fossiles (1975), Orléans, Mémoires du Bureau de Recherches Géologiques et Minières, no. 89, p. 359-366.

CHEVALIER, J. P. AND NASCIMENTO, A., 1975, Notes sur la géologie et la paléontologie du Miocène de Lisbonne. XVI- Contribution à la connaissance des madréporaires et des faciès récifaux du Miocène inférieur: Lisboa, Boletim da Sociedade Geologica de Portugal, v. XIX, p. 247-281.

DROOGER, C. W., KAASSCHIETER, J. P. H., AND KEY, A. J., 1955, The Microfauna of the Aquitanian-Burdigalian of Southwestern France: Amsterdam, Verhandelingen der Koninklijke Nederlandse Akademie van Wettenschappen, v. 21, 136 p.

DURAND, A., 1974, Stratigraphie des terrains d'âge paléogène supérieur et néogène du plateau continental basque et asturien, d'après l'étude des foraminifères planctoniques: Unpublished Thèse de Doctorat, University Rennes, Rennes, 118 p.

FOLLIOT, M., PUJOL, C., CAHUZAC, B., AND ALVINERIE, J., 1993, Nouvelles données sur le Miocène moyen marin ("Sallomacien") de Gironde (Bassin d'Aquitaine-France). Approche des paléoenvironnements: Lisbonne, 1st Congress of Regional Committee of the Atlantic Neogene Stratigraphy (1992), "Ciências da Terra," Universidade Nova Lisboa, no. 12, p. 117-131.

FROST, S. H., 1977, Oligocene reef coral biogeography Caribbean and western Tethys: Paris, 2e Symposium International sur les Coraux et récifs Coralliens Fossiles (1975),Orléans, Mémoires du Bureau de Recherches Géologiques et Minières, no. 89, p. 342-352.

G.F.E.N. (GROUPE FRANÇAIS D'ETUDE DU NEOGENE), 1974, Etude biostratigraphique des gisements d'Escornebeou (Aquitaine méridionale, France): Lyon, Documents des Laboratoires de Géologie de la Faculté des Sciences de Lyon, no. 59, 86 p.

HAQ, B. L., HARDENBOL, J., AND VAIL, P. R., 1987, The new chronostratigraphic basis of Cenozoic and Mesozoic sea level cycles, *in* Ross, C. A. and Haman, D., eds., Timing and Depositional History of Eustatic Sequences: Constraints on Seismic Stratigraphy: Washington, D. C., Cushman Foundation Foraminiferal Research Special Publication 24, p. 7-13.

KIEKEN, M., 1973, Evolution de l'Aquitaine au cours du Tertiaire: Paris, Bulletin de la Société Géologique de France, 7e série, Tome XV, p. 40-50.

KÜHLMANN, D. H. H., 1983, Composition and ecology of deep-water coral associations: Hambourg, Helgoländer Meeresuntersuchungen, v. 36, p. 183-204.

LAURIAT-RAGE, A., 1981, Les Bivalves du Redonien (Pliocène atlantique de France); signification stratigraphique et paléobiogéographique: Paris, Mémoires du Muséum National d'Histoire Naturelle, série C, Tome XLV, 174 p.

LAURIAT-RAGE, A., BRÉBION, P., CAHUZAC, B., CHAIX, C., DUCASSE, O., GINSBURG, L., JANIN, M. C., LOZOUET, P., MARGEREL, J. P., NASCIMENTO, A., PAIS, J., POIGNANT, A., POUYET, S., AND ROMAN, J., 1993, Paleontological data about the climatic trends from Chattian to Present along the Northeastern Atlantic frontage: Lisbonne, 1st Congress of Regional Committee of the Atlantic Neogene Stratigraphy (1992), "Ciências da Terra," Universidade Nova Lisboa, no. 12, p. 167-179.

LONDEIX, L., 1991, Actualisation de quelques coupes classiques du Miocène inférieur et moyen bordelais (France): Bordeaux, Bulletin de la Société Linnéenne, Tome 19, p. 59-74.

LOZOUET, P., 1986, Les gastéropodes prosobranches de l'Oligocène supérieur du Bassin de l'Adour (systématique, paléoenvironnements, paléoclimatologie, paléobiogéographie): Paris, Unpublished Mémoire de l'Ecole Pratique des Hautes Etudes, 465 p.

MAGNÉ, J., GOURINARD, Y., AND WALLEZ, M. J., 1987, Comparaison des étages du Miocène inférieur définis par stratotypes ou par zones paléontologiques: Toulouse, Strata, série 1, v. 3, p. 95-107.

MARGEREL, J. P. AND BRÉHÉRET, J. G., 1984, Révision de l'attribution stratigraphique du gisement de Chasné-sur-Illet (Ille-et-Vilaine) à l'aide de la faune de foraminifères et de la nannoflore calcaire: Paris, Cahiers de Micropaléontologie, no.1, p. 1-20.

MARGEREL, J. P. AND COUSIN, M., 1989, Les faluns de Doué-la-Fontaine (Maine-et-Loire): Angers, Bulletin de la Société d'Etudes Scientifiques d'Anjou, v. 13, p. 27-33.

MARTINI, E., 1971, Standard Tertiary and Quaternary calcareous nannoplankton zonation, *in* Farinacci, A., ed.: Proceedings of the Second Planktonic Conference, Roma (1970), v. 2, p. 739-785.

MARTINI, E., 1988, Late Oligocene and Early Miocene calcareous nannoplankton. (Remarks on French and Moroccan sections): Berlin, Newsletters on Stratigraphy, v. 18, p. 75-80.

MICHELIN, H., 1840-1847, Iconographie zoophytologique. Description par localités et terrains des polypiers fossiles de France et pays environnants: Paris, Bertrand, 348 p.

MONLEAU, C., ARNAUD, M. AND CATZIGRAS, F., 1988, L'Oligocène supérieur marin de la Nerthe (Bouches-du-Rhône): nouvelles données sédimentologiques et paléogéographiques, dans le cadre de la géodynamique de la Méditerranée occidentale: Paris, Comptes-Rendus Académie des Sciences, v. 306, Série II, p. 487-491.

MOYES, J., 1966, Les faluns néogènes du Bordelais: Bordeaux, Bulletin de l'Institut de Géologie du Bassin d'Aquitaine, no.1, p. 85-111.

Müller, C., 1984, Climatic evolution during the Neogene and Quaternary evidenced by marine microfossil assemblages: Montpellier, Paléobiologie continentale, v. XIV, p. 359-369.

MÜLLER, C. AND PUJOL, C., 1979, Etude du nannoplancton calcaire et des foraminifères planctoniques dans l'Oligocène et le Miocène en Aquitaine (France): Marseille, Géologie Méditerranéenne, Tome 6, p. 357-368.

NURY, D. AND THOMASSIN, B. A., 1994, Paléoenvironnements tropicaux, marins et lagunaires d'un littoral abrité (fonds meubles à bancs coralliens, lagune évaporitique) à l'Oligocène terminal en Basse-Provence (région d'Aix-en-Provence—Marseille, France): Marseille, Géologie Méditerranéenne, Tome 21, p. 95-108.

ODIN, G. S., 1994, Geological Time Scale (1994): Paris, Comptes-Rendus Académie des Sciences, v. 318, Série II, p. 59-71.

OOSTERBAAN, A. F., 1988, Early Miocene corals from the Aquitaine Basin (SW France): Leiden, Mededelingen Werkgroep Tertiaire Kwartaire Geologie, v. 25, p. 247-284.

POIGNANT, A. AND PUJOL, C., 1976, Nouvelles données micropaléontologiques (foraminifères planctoniques et petits foraminifères benthiques) sur le stratotype de l'Aquitanien: Lyon, Géobios, no.9, p. 607-663.

POIGNANT, A. AND PUJOL, C., 1978, Nouvelles données micropaléontologiques (foraminifères planctoniques et petits foraminifères benthiques) sur le stratotype bordelais du Burdigalien: Lyon, Géobios, no.11, p. 655-712.

POMEROL, C., 1973, Stratigraphie et Paléogéographie. Ere Cénozoique: Paris, Doin, 269 p.

PUJOL, C., 1970, Contribution à l'étude des foraminifères planctoniques dans le Bassin d'Aquitaine: Bordeaux, Bulletin de l'Institut de Géologie du Bassin d'Aquitaine, no.9, p. 201-219.

Vergnaud-Grazzini, C., Müller, C., Pierre, C., Létolle, R., and Peypouquet, J. P., 1979, Stable isotopes and Tertiary paleontological paleoceanography in the Northeast Atlantic, *in* Montadert, L. and Roberts, D. G., eds., Initial Reports of the Deep Sea Drilling Project: Washington, D. C., United States Government Printing Office, v. XLVIII, p. 475-491.

Weisbord, N. E., 1971, Corals from the Chipola and Jackson Bluff formations of Florida: Tallahassee, State of Florida, Department of Natural Resources, Bureau of Geology, Geological Bulletin 53, 100 p.

Weisbord, N. E., 1973, New and little known corals from the Tampa formation of Florida: Tallahassee, State of Florida, Department of Natural Resources, Bureau of Geology, Geological Bulletin 56, 146 p.

Wells, J. W., 1956, Scleractinia, *in* Moore, R. C., ed., Treatise on Invertebrate Paleontology: Lawrence, Geological Society of America and University of Kansas Press, part F, p. 328-444.

# Part II:
# DETAILED STUDIES

# MIDDLE MIOCENE CORAL-OYSTER REEFS, MURCHAS, GRANADA, SOUTHERN SPAIN

JUAN C. BRAGA, ANTONIO P. JIMENEZ, JOSE M. MARTIN, AND PASCUAL RIVAS

*Departamento de Estratigrafía y Paleontología, Facultad de Ciencias, Universidad de Granada, Campus de Fuentenueva s.n., 18002 Granada, Spain*

ABSTRACT: Middle Miocene coral-oyster patch reefs crop out at Murchas, south of the city of Granada in southern Spain. They are irregularly shaped masses of coral-oyster boundstone, up to 18 m wide and 3-4 m high, that developed on the outer part of a homoclinal ramp, seaward of some sand shoals, in a mixed carbonate-terrigenous enviroment. In these patch reefs, oysters and hermatypic corals are the main frame-builders, their association being entirely fortuitous. *Heliastrea* is the predominant coral. *Porites*, *Tarbellastraea* and the phaceloid coral *Mussismilia* are also important components. These corals show no clear pattern in their distribution and appear embedded in a silty (bioclastic) matrix. Oysters in the reef community belong to the species *Hyotissa squarrosa*. They grew vertically one upon another, anchored directly to coral skeletons or, more commonly, attached to other oysters. *Hyotissa* is irregularly distributed but in places accounts for up to 70% of the reef. Encrusting organisms are restricted to sediments between individual coral colonies or between reefs.

## INTRODUCTION

This paper analyzes the paleoenvironmental context and internal structure of Middle Miocene coral-oyster reefs which crop out at Murchas, south of the city of Granada in southern Spain (Fig. 1). These are small patch reefs that developed seaward of sand shoals in an open-platform environment with mixed carbonate and terrigenous sedimentation. These are the only Middle Miocene reefs known in southern Spain, although reefs of this age have been described in other areas of the western Mediterranean, such as Catalonia (Chevalier, 1961; Permanyer and Esteban, 1973; and Esteban et al., this volume) and Languedoc (Chevalier, 1961).

Neogene and Recent oyster reefs have been described by Norris (1953), Stenzel (1971), Edwards (1982), Herb (1984), and Jimenez et al. (1991), among others. In these reefs, a single species of oyster (usually belonging to the genus *Crassostrea*) builds thin blankets of closely aggregated individuals, growing one upon the other. Oysters are also common accessory fossils in shallow-water, coral reef sediments of Cenozoic age, where they appear as secondary components scattered among scleractinians. In the Murchas reefs, however, oysters are associated with corals, forming a reef structure of coral-oyster boundstone. Such a reef framework has neither been described in detail nor referred to previously, except for some small buildups mentioned by Kidwell (1988) in the Pliocene deposits of the northern Gulf of California.

## REGIONAL SETTING

Middle Miocene sedimentary rocks containing coral-oyster reefs occur at the southern margin of the Granada Basin. The outcrop studied is 1 km west of the town of Murchas (Fig. 1).

Neogene-Quaternary deposits in this area (Fig. 2) lie on a basement of metamorphic rocks belonging to the Alpujárride Complex, one of the major domains of the Internal Zones of the Betic Cordillera. Middle Miocene rocks containing the Murchas reefs are preserved in a small half-graben, where they have been protected from various erosional episodes that have occurred

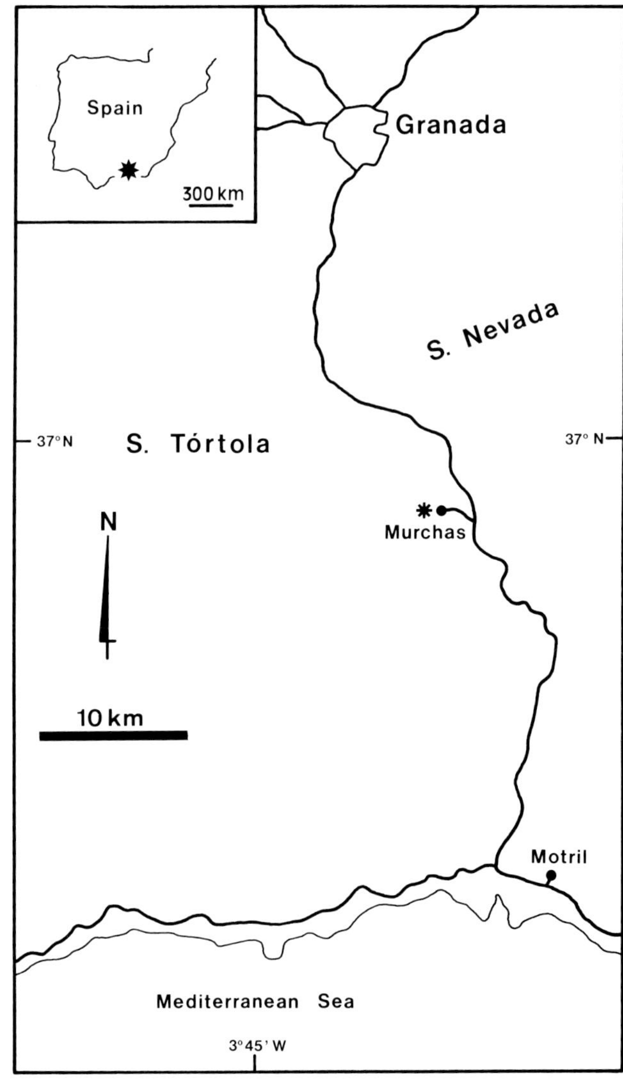


FIG. 1.—Geographical location and main access roads to the Murchas outcrop (asterisk).

Models for Carbonate Stratigraphy from Miocene Reef Complexes of Mediterranean Regions,
SEPM Concepts in Sedimentology and Paleontology #5, Copyright © 1996,
SEPM (Society for Sedimentary Geology), ISBN 1-56576-033-6, p. 131-139.

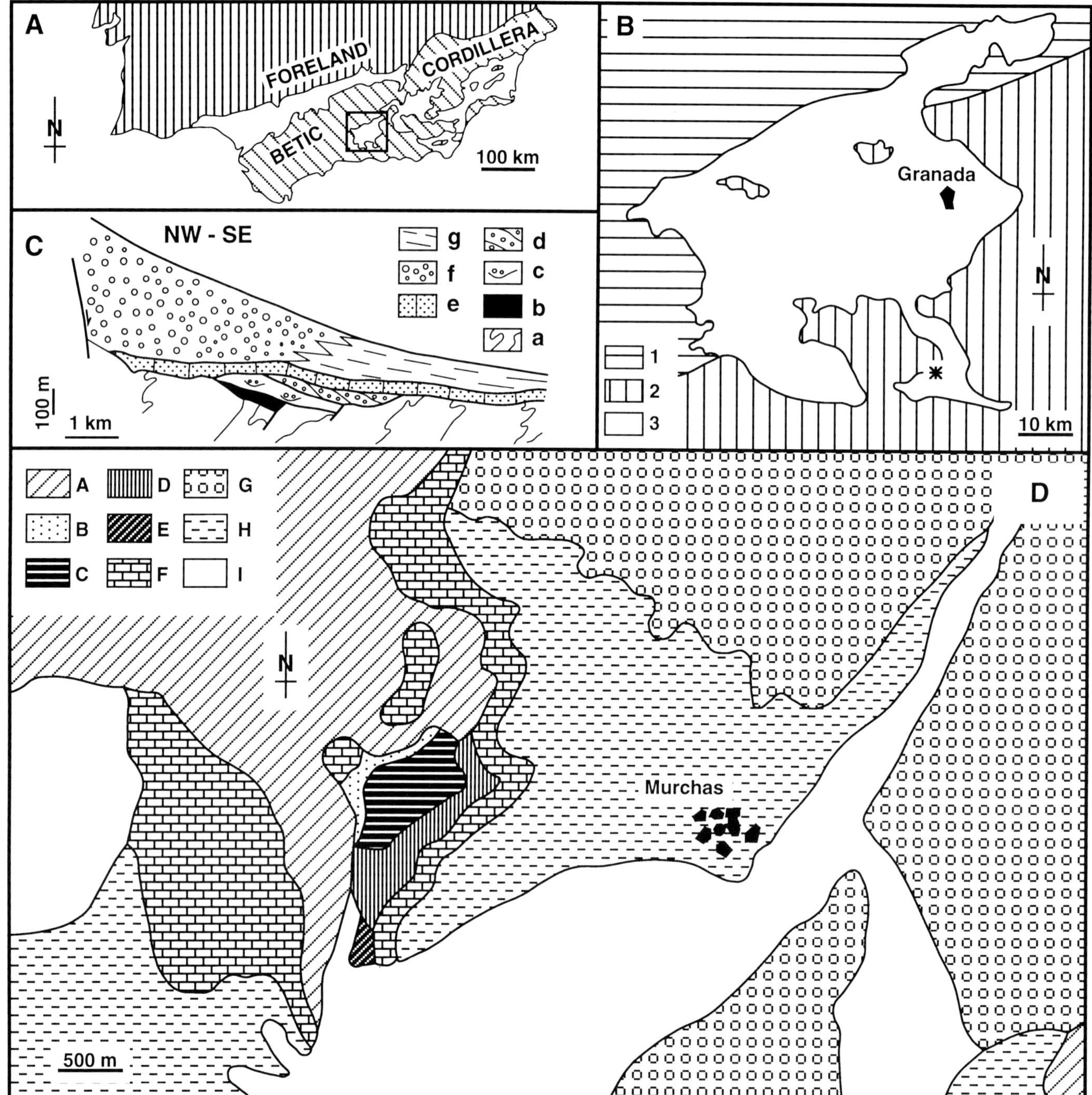


FIG. 2.—(A) Neogene basins of the Betic Cordillera (southern Spain); inset shows the position of the Granada Basin (non-hachured area). (B) Location of the Murchas reef outcrops (asterisk) in the Granada Basin. 1: External Zones of the Betic Cordillera; 2: Internal Zones; 3: Neogene to Quaternary deposits. (C) Lithofacies cross section of Miocene strata in the southern part of the Granada Basin. a: basement (Paleozoic and Triassic); b: undifferentiated breccias, marls, laminated planktonic-foraminiferal grainstones, and conglomerates/sandstones/cross-bedded skeletal grainstones-rudstones/coral-oyster boundstones (Aquitanian to Langhian); c: siltstones, sandstones and conglomerates (Lower Serravallian); d: conglomerates and sandstones (Serravallian-Lower Tortonian?); e: bioclastic sandstones/conglomerates and skeletal packstones/grainstones (Lower Tortonian); f: conglomerates and sandstones (Upper Tortonian); and g: silts and silty marls (Upper Tortonian). (D) Detailed geological map of the Murchas area. A: basement; B: undifferentiated breccias, marls and laminated planktonic-foraminiferal grainstones (Aquitanian to Burdigalian); C: reef unit consisting of conglomerates, sandstones, cross-bedded skeletal grainstones/rudstones and reef limestones of coral-oyster boundstone (Upper Langhian); D: siltstones, sandstones and conglomerates (Lower Serravallian); E: conglomerates and sandstones (Serravallian-Lower Tortonian?); F: bioclastic sandstones/conglomerates and skeletal packstones/grainstones (Lower Tortonian); G: conglomerates and sandstones (Upper Tortonian); H: silts and silty marls (Upper Tortonian); I: undifferentiated Pliocene to Quaternary deposits.

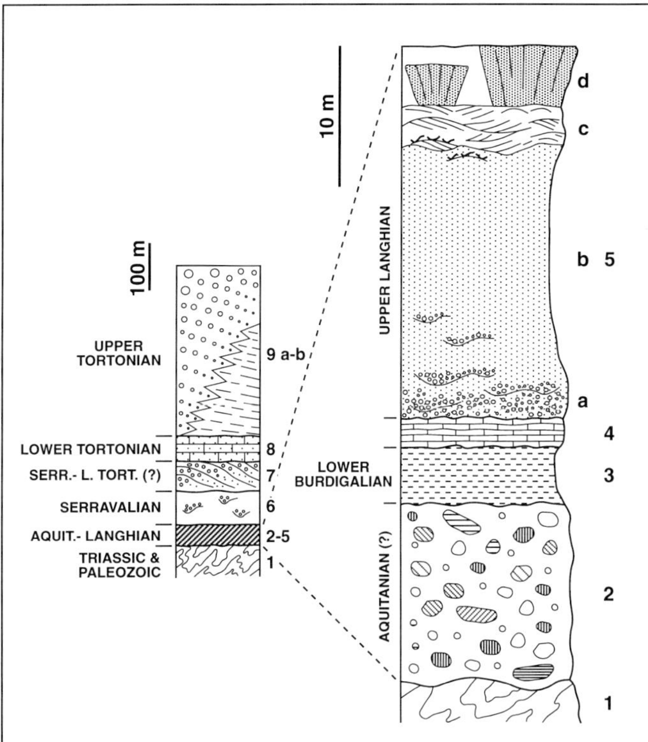


FIG. 3.—Miocene stratigraphy of the Murchas area and detailed stratigraphic column of the lowermost units of the sequence. 1: basement; 2: polymictic breccia; 3: marls; 4: laminated planktonic-foraminiferal grainstones; 5a: clast-supported polymictic conglomerates; 5b: litharenithic sandstone; 5c: cross-bedded skeletal grainstones/rudstones including lenticular banks of large *Crassostrea* at the base; 5d: reef limestones of coral-oyster boundstone; 6: siltstones, sandstones and conglomerates; 7: conglomerates and sandstones; 8: bioclastic sandstones and conglomerates and skeletal packstones/grainstones; 9a: conglomerates and sandstones; 9b: silts and silty marls.

from the Middle Miocene to the present. They form part of a monoclinal structure dipping 40°S, which is cut by a system of normal faults (trending N60E and dipping 50°N) and vertical faults (trending N170E).

Miocene stratigraphy of the southern margin of the Granada basin can be summarized as follows (Fig. 3). A 12-m-thick polymictic breccia, probably Aquitanian in age, lies unconformably on a basement composed of metamorphic rocks belonging to the Alpujárride Complex. This breccia is in turn overlain unconformably by marls (4 m thick), which, according to planktonic foraminifera (Gonzalez-Donoso, 1978) and calcareous nannoplankton (Martín-Pérez et al., 1989), are Early Burdigalian in age. This is followed unconformably by 2 m of finely laminated, turbiditic grainstones containing abundant reworked planktonic foraminifera. The unit that includes the Murchas reefs lies unconformably on top of these. Calcareous nannoplankton, including *Sphenolithus heteromorphus*, *S. moriformis*, and *Calcidiscus macintyrei*, date this unit as Late Langhian in age (Martín-Pérez and Aguado, 1990). The overlying sequence consists of 60 m of siltstones, sandstones and conglomerates of Lower Serravallian age (Martín-Suarez et al., 1993); 50 m of conglomerates with well rounded clasts in a sandy matrix, and sandstones of probable Serravallian-Lower Tortonian (?) age; up to 45 m of Lower Tortonian bioclastic sandstones/conglomerates and skeletal packstones/grainstones; and more than 300 m of polymictic, matrix-supported, Upper-Tortonian conglomerates (Braga et al., 1990), which grade laterally to sandstones and to basinal silts and silty marls.

## DEPOSITIONAL FACIES

The Upper Langhian reef unit consists of the following sequence, from bottom to top (Fig. 3):

1. 2 m of poorly rounded, cobble-sized, clast-supported conglomerates.

2. 14 m of sandstones that interfinger with conglomerates to the north and northeast and incorporate meter-wide conglomerate channels to the south. They are slightly silty on top (uppermost 2 m). Small bivalves (*Amusium*) and irregular echinoids (*Spatangus* sp.) are the most common fossil remains.

3. 5 m of cross-bedded skeletal grainstones/rudstones, with some dispersed terrigenous clasts, exhibiting poorly preserved bar morphologies. They can be subdivided into two units:

a. A basal bioclastic interval, 1-2 m thick, with abundant and well-fragmented remains of bivalves (mainly pectinids and oysters), echinoids, bryozoans, coralline algae, serpulid worms, barnacles, gastropods (vermetids and conids), and, locally, corals. Vertical burrowing is evident in places. The tops of bars are often colonized by thin, lenticular banks of large *Crassostrea*.

b. An upper bioclastic interval, 3 m thick, similar in composition to the lower one but containing abundant coral fragments. Infaunal bivalves, *Lutraria* and *Panopea*, together with specimens cemented to bioclasts of the pelecypod *Chama*, are also common.

4. Small (up to 4 m high and 18 m across), closely spaced, coral-oyster patch reefs surrounded by a silty (bioclastic) sediment. The vertical transition from the underlying grainstones/rudstones is marked by a horizon (approximately 0.5 m thick) extremely rich in *Chlamys seniensis* (Lamarck) and crowned in places by thin banks of *Hyotissa squarrosa* (De Serres).

### *Depositional Model*

The sedimentary model that is inferred, both from the vertical evolution and lateral transition of lithofacies described above, corresponds to a platform with predominantly terrigenous inner-shelf sedimentation and carbonate outer-shelf sedimentation. In a generalized north-to-south cross-section, the following lithofacies belts can be distinguished (Fig. 4):

a. a coastal belt with conglomerates,

b. a lagoonal facies belt with sands, locally cut by conglomerate channels,

c. a shoal zone composed mainly of cross-bedded skeletal grainstones/rudstones, and

d. an outer-shelf area where coral-oyster boundstone reefs

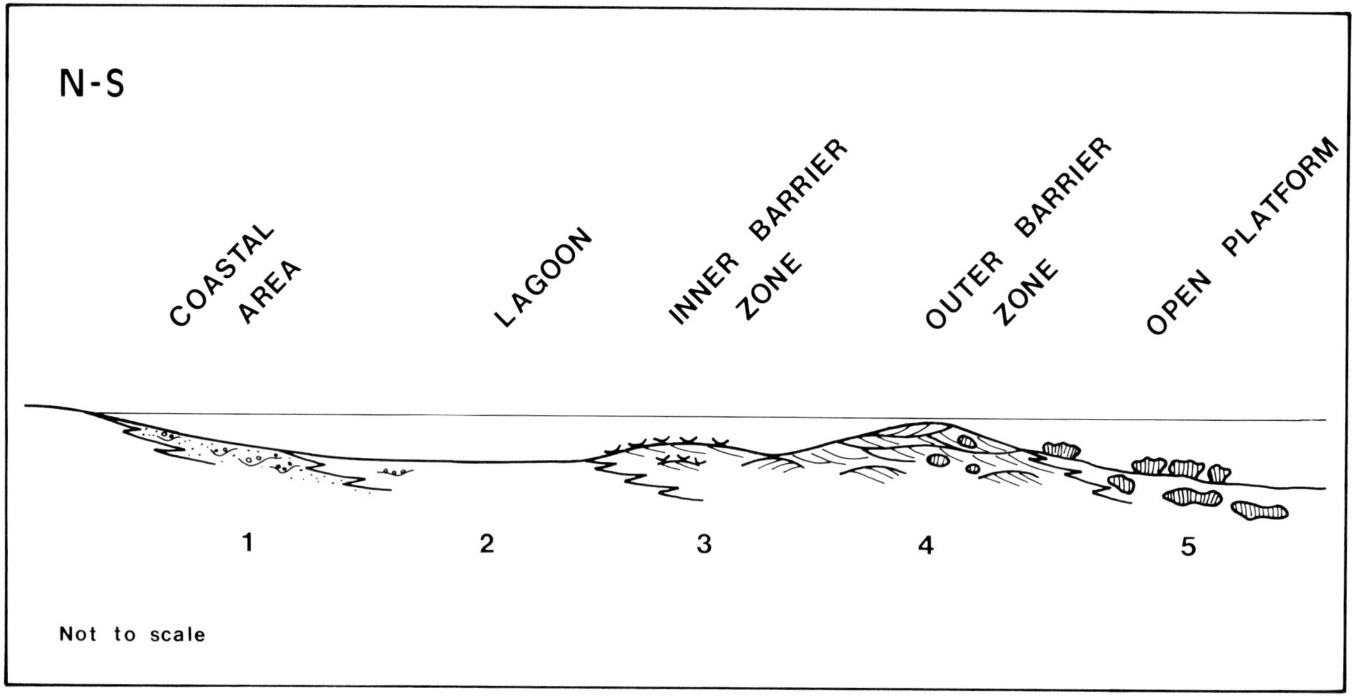


FIG. 4. —Facies distribution and sedimentary model for the Murchas reef unit. 1: clast-supported conglomerates; 2: sandstones with some local conglomerate channels; 3: cross-bedded skeletal grainstone/rudstone shoals (with *Crassostrea* banks appearing on top of some bars); 4: outer shoals of cross-bedded skeletal grainstones/rudstones with fragments of coral colonies increasing in seaward direction; 5: coral-oyster patch reefs growing among bioclastic silts.

developed.

The vertical sequence is clearly trangressive, and open-marine facies are consistently found overlying more restricted coastal ones.

## CORAL-OYSTER REEFS

The reefs of Murchas have not been studied before, although their existence has been referred to in regional works dealing with Neogene deposits of the Granada Basin (Gonzalez-Donoso, 1967, 1978; Rodriguez-Fernández, 1982; Braga et al., 1990). Our study of the Murchas reefs is based on field observations and data obtained from eight measured sections, separated by distances of 20-100 m. Good exposures in these sections allow the observation of three-dimensional reef geometries and the recognition of the distribution of reef facies and inter-reef sediments. About 50 thin sections were used for petrographic analysis and for the identification of certain biotic constituents.

The predominant coral found in the Murchas patch reefs (Fig. 5) is *Heliastrea*. It grew in the shape of clubs, fused together into flowerpot-like colonies. Coral heads of the same species are also found. Individual coral colonies may be up to 2 m in height and 1.5 m wide. Small heads of *Porites* and *Tarbellastraea*, up to 0.5 m high, and the phaceloid coral *Mussismilia* are also important components; colonies of the latter may be up to 1.5 m in height. *Siderastrea*, *Thegioastraea*, *Stylophora* and *Cladangia* appear in minor quantities as well. There is no clear pattern to the distribution of the various coral genera within the reefs, although *Porites* tends to be concentrated at the bottom of the reefs and also appears sporadically in isolation away from coral patches. The coral colonies are embedded in a silty (bioclastic) matrix. Most of them remain in their original life position, and even the few that have fallen over are not fragmented.

Oysters in the reef community belong to the species *Hyotissa squarrosa*. They grow vertically one upon another, and their valves are perfectly preserved. They may be anchored directly to coral skeletons (Figs. 6, 7) but more commonly are attached to other oysters near corals (Fig. 8). As the reefs grew, new coral colonies settled on oyster valves (Fig. 9), as well as on coral colonies.

*Hyotissa* is a common dweller among the corals in modern Pacific reefs (Stenzel, 1971, Morton, 1983). It also occurs, as does *Ostrea*, as a secondary component in Upper Miocene reefs in southern Spain (Jimenez et al., 1991), where it either grows upon coral fragments or provides the substrate for later coral colonies. It is only in Middle Miocene reefs, however, that *Hyotissa* plays an important role as a reef frame builder. In patch reefs near Murchas, numerous individuals of *H. squarrosa* grew vertically, either one upon another or attached to coral skeletons of several genera, and thus contributed to the framework of the reefs (Figs. 5, 6, 7, 8). *Hyotissa* is irregularly distributed but in places accounts for up to 70% of the reef.

*H. squarrosa* is not confined to coral-oyster reefs but also forms small oyster banks landward of coral-oyster patch reefs.

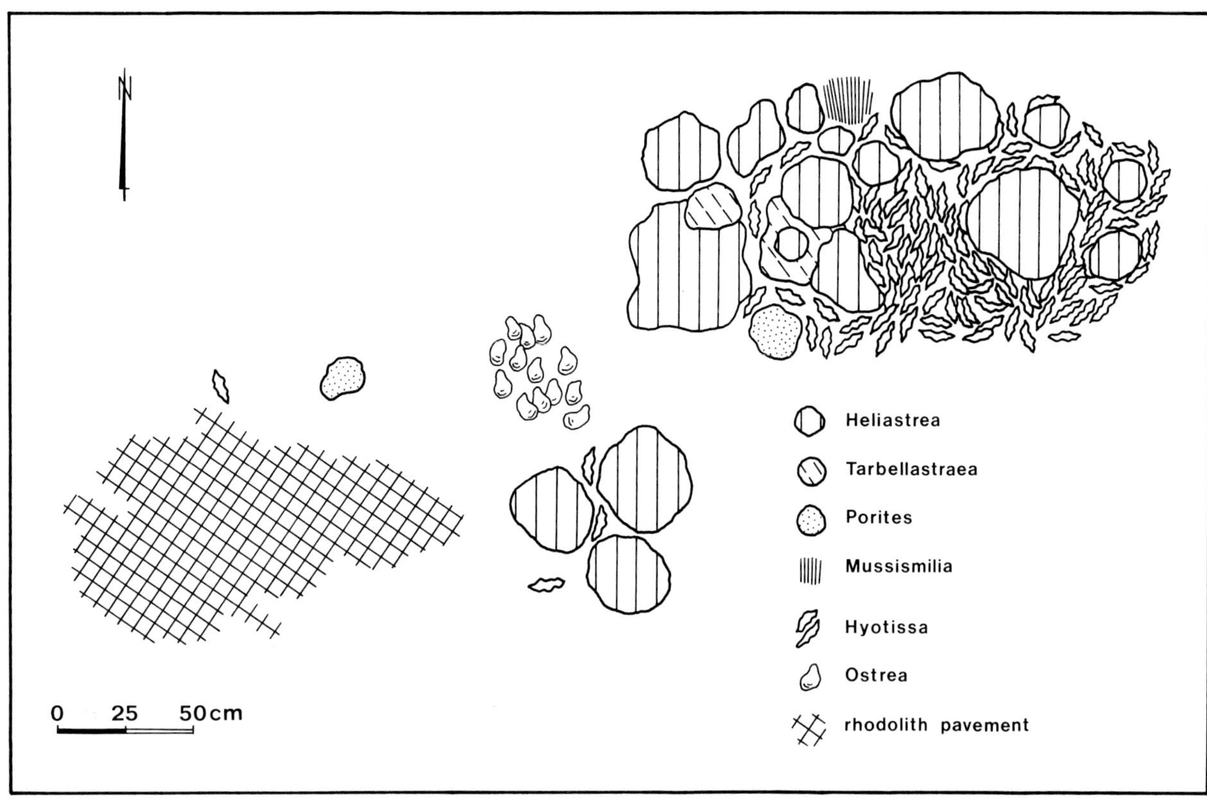


FIG. 5.—Schematic distribution of framework elements in a small coral-oyster patch reef taken from a field sketch. A rhodolith pavement and some *Ostrea* aggregations developed in coral-free areas to the left of the patch.

In these banks, reclining oyster specimens are more common than vertical ones. Rare examples of this species also appear in *Crassostrea* banks that developed on top of the innermost bars of the shoals.

It seems that the consortium of corals and *Hyotissa* in the structure of these reefs was entirely fortuitous, even within the confines of this platform, as both were perfectly able to develop independently of each other and neither was obliged to form part of the consortium.

Small aggregations of reclining *Ostrea edulis* var. *lamellosa* (Brocchi) are found on top of the reefs and also in sediments between the patches of reef framework (Fig. 5). Solitary corals, echinoids, gastropods (mainly vermetids and conids), serpulid worms and bivalves (mainly *Cardita* and small forms of *Chlamys*) are embedded in sediments within the framework of the reefs and between the reefs themselves.

Centimeter-sized rhodoliths, composed of *Neogoniolithon*, *Mesophyllum*, *Lithophyllum* and *Spongites* (listed in order of abundance), are dispersed throughout silty (bioclastic) sediment between patches of reef framework (Braga et al., 1989). They tend to be concentrated in small rhodolith pavements in coral-free areas between reefs (Fig. 5). Their nuclei consist of small fragments of corals or oysters. Red-algal coatings are relatively thin, tending to be thicker on the upper side of rhodoliths (Fig. 10). Individual crusts are predominantly laminar and sprout into branching growths on the upper side of the rhodoliths. The algae themselves are intergrown with vermetid gastropods, foraminifera and serpulid worms (Fig. 10).

Coral growth was not limited to the outer platform; it also occurred to a limited extent within conglomerate channels that cut through lagoonal deposits and also on top of some bars. Reefs in these environments were not well developed, presumably because coral growth was disturbed as soon as channels and bars became active. Similarly, patch reefs situated close to shoal areas on the outer platform are not well preserved and contain more fragmented and tilted colonies than those growing in more seaward positions.

## BIVALVE DISTRIBUTION

Bivalve remains have distinct distribution patterns throughout the various environments on the small platform where patch reefs developed (Fig. 11). Fine-grained lagoonal sediments are characterized by rare occurrences of infaunal bivalves, preserved in the form of internal molds, together with some specimens of *Amusium*. Globular, boulder-sized individuals of

FIG. 6.—Upright-growing *Hyotissa* (h) attached to a *Mussismilia* (m) colony; scale bar equals 5 cm.

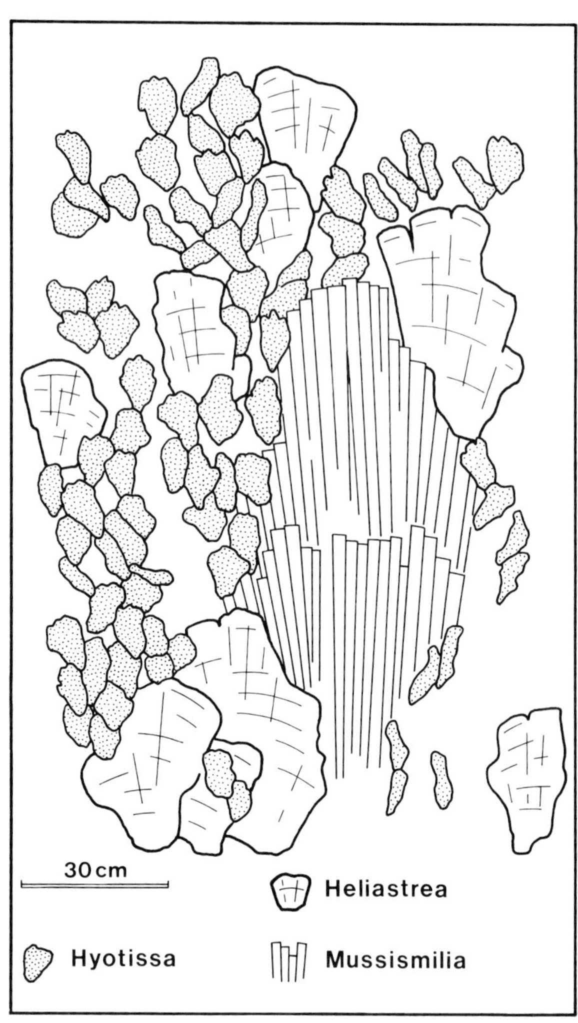


FIG. 8.—Field sketch of a vertical section of a patch reef of coral-oyster boundstone.

FIG. 7.—Plan view of *Hyotissa* (h) individuals growing attached to and around a *Heliastrea* (he) colony (upper right); the hammer head is 17.8 cm across.

*Crassostrea* are concentrated landward on the shoals that enclose lagoonal environments. In these zones, the first colonizers lie horizontally on skeletal grainstone/rudstone substrates, and successive individuals grew upon them, adapting themselves to the irregular bottom created by earlier ones and eventually forming thin lenticular banks. The shoals have a distinct association of *in situ* bivalves dominated by infaunal *Lutraria* and *Panopea* and common occurrences of *Chama* cemented to bioclasts. Seaward of the shoals are accumulations of well-preserved, single valves of *Chlamys seniensis*. *Hyotissa* also forms thin banks, irregularly distributed throughout this subenvironment. Besides *Hyotissa*, *Ostrea edulis* var. *lamellosa* is common on the reef, cemented on top of corals or forming small aggregations irregularly dispersed among reefs. Isolated

FIG. 9.—A *Heliastrea* (he) colony on top of a *Hyotissa* (h) framework, the latter consisting of vertically stacked individuals growing one on top of the other; the hammer lenght is 33 cm.

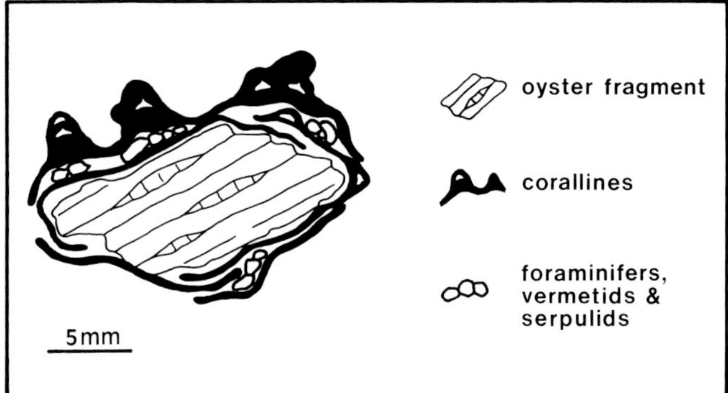


FIG. 10.—Scheme of a rhodolith from an interpatch pavement, showing the asymmetrical distribution of coralline algal crusts.

individuals of *Cardita* and *Chlamys* are also common in sediments around coral colonies. Fragments of all the bivalves mentioned above are incorporated into skeletal grainstones/rudstones of the shoal environments.

## DIAGENETIC OVERPRINT

Skeletal grainstones/rudstones underlying the reefs and coral-oyster boundstones both share a common diagenetic history. Skeletal grainstones/rudstones show an early (eogenetic) dissolution phase, in which aragonite components such as gastropod and bivalve shells (excluding oysters and pectinids) were leached, forming molds. This was followed by a compactional phase which crushed their micrite envelopes somewhat. During a later (mesogenetic) cementation phase, a sparry mosaic calcite cement precipitated and occluded most of both the primary voids (interparticle voids, intraparticle voids, and shelter voids beneath single bivalve shells) and secondary voids (bivalve and gastropod molds). In the reefs, the internal structure of corals has also been profoundly transformed by these dissolution-precipitation processes and, consequently, they are now highly recrystallized. The absence of binding organisms, together with the abundance of silty sediment between frame-builders, is probably the reason for the scarcity of extra-skeletal cements inside reef structures.

In all the above-mentioned lithofacies, only residual porosity remains within the center of primary and secondary voids that have not been completely filled in by sparry cement. It constitutes up to 10% of the total volume of the rock. The open voids (mesopores to megapores) are isolated and unconnected to each other, resulting in an almost complete lack of permeability. In reef facies, this is emphasized by the presence of silt in the intersticial sediment.

## CONCLUSIONS

The patch reefs near Murchas are irregularly shaped masses of coral-oyster boundstone, up to 18 m wide and 3-4 m high. They developed on the outer shelf of a homoclinal ramp with mixed carbonate and terrigenous sedimentation. Both inter- and intra-reef sediments are bioclastic silts.

Hermatypic corals, although represented by only eight genera, attain their highest generic diversity in the Neogene reefs of southern Spain during this Middle Miocene episode. *Heliastrea* and *Mussismilia* are the main reef builders, together with *Tarbellastraea* and *Porites*.

The oyster *Hyotissa*, either growing vertically on coral skeletons or one upon another, contributes in varying degrees to the framework of the reef themselves, sometimes forming up to seventy percent of the structure.

*Hyotissa* and hermatypic corals are the main framebuilders in the Murchas reefs. Encrusting organisms, such as coralline algae, encrusting foraminifera, bryozoan and vermetid gastropods, grew neither upon *in situ* corals nor upon oysters but were restricted to sediments between individual coral colonies or between reefs, being developed only on terrigenous grains or skeletal fragments.

The diagenetic history of the Murchas patch reefs consists of an early dissolution of aragonite skeletons, and a later cementation phase in which sparry calcite cement almost completely

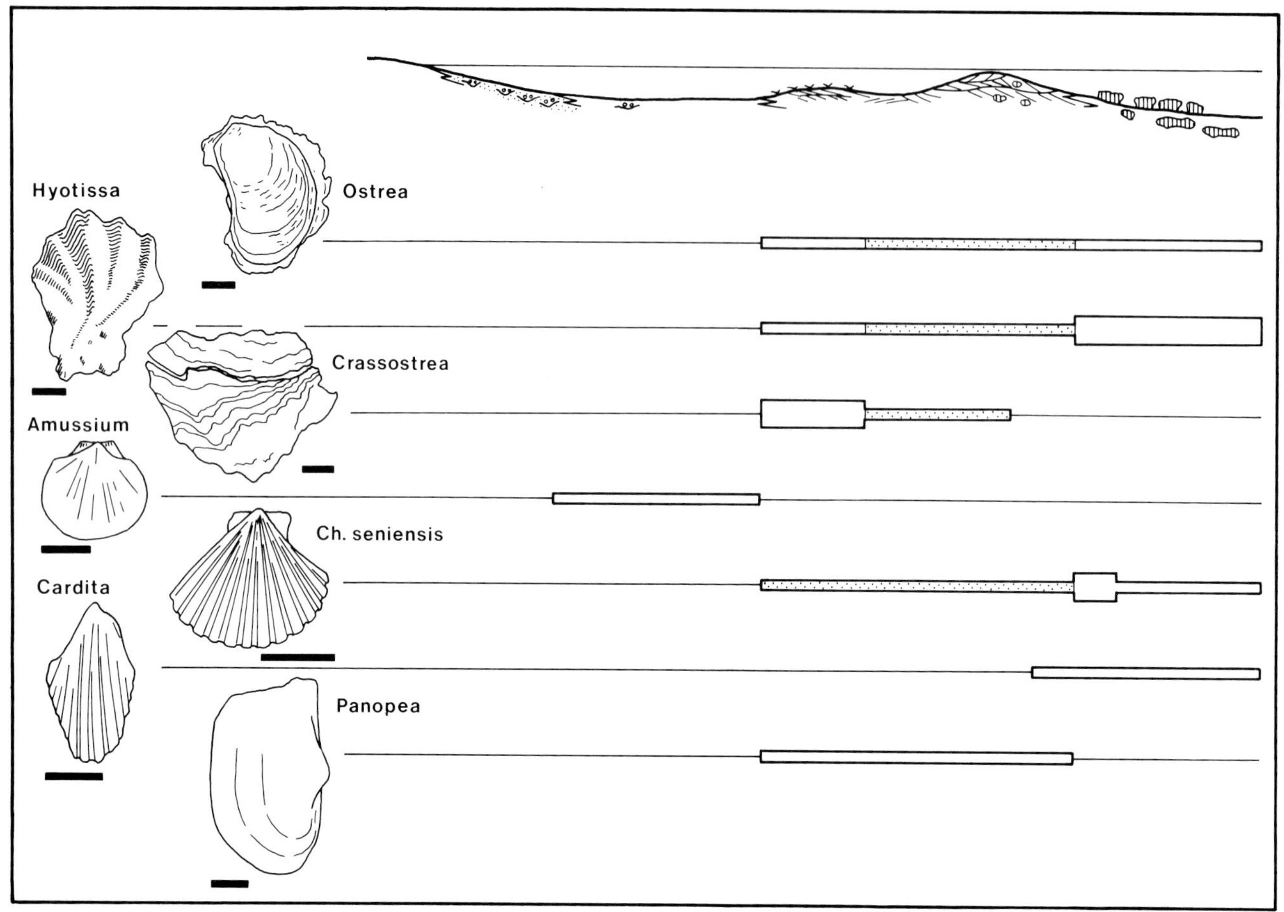


FIG. 11.—Paleoenvironmental distribution of the most important bivalves of the reef unit. "Dotted bars" indicate the extent of reworked remains. The width of the bars refers to the relative abundance of bivalves. Sedimentary model appears as in Figure 4. All scale bars equal 2.5 cm.

filled both primary and secondary pores. The resulting porosity is restricted to the center of voids that have not been completely occluded and represents less than 10% of the rock volume.

## REFERENCES

BRAGA, J. C., JIMENEZ, A. P., MARTÍN, J. M., AND RIVAS, P., 1989, Coralline growths in coral-oyster, Middle-Miocene reefs in southern Spain (abs): Granada, Algae in Reefs, p. 21.

BRAGA, J. C., MARTÍN, J. M., AND ALCALÁ, B., 1990, Coral reefs in coarse-terrigenous sedimentary environments (Upper Tortonian, Granada Basin, southern Spain): Sedimentary Geology, v. 66, p. 135-150.

CHEVALIER, J. P., 1961, Recherches sur les madréporaires et les formations récifales miocènes de la Méditerranée occidentale: Mémoir Societé Géologique France, v. 93, p. 1-562.

EDWARDS, L., 1982, Oyster reefs, valuable to more than oysters: Sea Frontiers, v. 28, p. 23-25.

GONZALEZ-DONOSO, J. M., 1967, Estudio geológico de la Depresión de Granada: Unpublished Ph.D. Dissertation, University of Granada, Granada, 149 p.

GONZALEZ-DONOSO, J. M., 1978, Los materiales miocénicos de la Depresión de Granada: Cuadernos de Geología Universidad de Granada, v. 8-9, p. 191-203.

HERB, R., 1984, Récifs a huitres recents et miocènes, *in* Geister, J. and Herb, R., eds., Géologie et paleoécologie des récifs: Berne, Institute Géologie Université Berne, p. 22.1-22.12.

JIMENEZ, A. P., BRAGA, J. C., AND MARTÍN, J. M., 1991, Oyster distribution in the Upper Tortonian of the Almanzora Corridor (Almería, S.E. Spain): Geobios, v. 24, p. 725-734.

KIDWELL, S. M., 1988, Taphonomic comparison of passive and active continental margins: Neogene shell beds of the Atlantic Coastal Plain and Northern Gulf of California: Palaeogeography, Palaeoclimatology, Palaeoecology, v. 63, p. 201-223.

MARTÍN-PÉREZ, J. A. AND AGUADO, R., 1990, Nannoplancton calcáreo de los sedimentos marinos del Mioceno inferior y medio en el valle de Lecrín (zonas internas de la Cordillera Bética): Revista Sociedad Geológica de España, v. 3, p. 335-344.

MARTÍN-PÉREZ, J. A., MARTÍNEZ-GALLEGO, J., BRAGA, J. C., AND MARTÍN, J. M., 1989, Preliminary data on the distribution of calcareous nannoplankton in the marine units of the Granada Basin: Tetouan (Maroc), Premier Colloque du Néogene Atlantico-Méditerranéen, Resumés des Communications, p. 69-73.

MARTÍN-SUAREZ, E., FREUDENTHAL, M., AND AGUSTÍ, J., 1993, Micromammals from the Middle Miocene of the Granada Basin (Spain): Geobios, v. 26, p. 377-387.

MORTON, B., 1983, Coral-associated bivalves of the Indo-Pacific, *in* Russell-Hunter, W. D., ed., The Mollusca, Ecology: Orlando, Academic Press, p. 140-224.

NORRIS, R. M., 1953, Buried oyster reefs in some Texas bays: Journal of Paleontology, v. 27, p. 569-576.

PERMANYER, A. AND ESTEBAN, M., 1973, El arrecife mioceno de Sant Pau d'Ordal (provincia de Barcelona): Revista del Instituto de Investigaciones Geológicas Diputación Provincial-Universidad de

Barcelona, v. 28, p. 45-72.

Rodríguez-Fernández, J., 1982, El Mioceno del sector central de las Cordilleras Béticas: Ph.D. Dissertation, University of Granada, Granada, 224 p.

Stenzel, H. B., 1971, Oysters, *in* Moore, R. C., ed., Treatise on Invertebrate Paleontology, Mollusca 6: Lawrence, Geological Society of America and Kansas University Press, p. 953-1224.

# THE MIDDLE TO UPPER MIOCENE CARBONATE COMPLEX OF NÍJAR, ALMERÍA PROVINCE, SOUTHEASTERN SPAIN

CAROL MANKIEWICZ

*Departments of Biology and Geology, Beloit College, 700 College Street, Beloit, WI 53511 USA*

ABSTRACT: Field and petrographic studies of middle? to upper Miocene carbonate rocks near the town of Níjar (Almería Province, southeastern Spain) document the effects of relative sea-level change on sedimentation and biotic composition. Erosion, prograding and down-stepping beds (producing down-lap surfaces), or alteration due to subaerial exposure signify relative sea-level fall. Onlapping geometries, extensive bioturbation, channeling, or crossbedding characterize periods of relative rise.

Geometries of and within seven depositional units, variation in grain size and fossil constituents, sedimentary structures, and slight variation in color help differentiate the carbonate rocks in the field. The seven recognizable depositional units can be grouped into three broad associations: pre-reef, reef, and post-reef. Reefs did not develop during deposition of the older half of the section at Níjar; reef complexes (fringing types) are evident only in the younger half of the section. Reef-slope sediments are well preserved and exposed, but erosion truncated part of the reef core. The coral *Porites* and associated micritic coatings (probably microbial) constructed the reef framework. A variety of mollusks, coralline algae, serpulid worms, and bryozoans lived in association with *Porites*. The green calcareous alga *Halimeda* and the coralline alga *Mesophyllum* seem to have colonized the reef slope.

Syndepositional alteration (micro- and macro-boring) and cementation, fabric-selective dissolution of predominantly aragonitic components, dolomitization, and calcite cementation modified original mineralogy, fabric, and porosity.

## INTRODUCTION

A Miocene carbonate succession is beautifully exposed near the town of Níjar, Almería Province. Because tectonic activity has not altered the carbonate strata appreciably, the present-day topography mimics Miocene topography (Dabrio et al., 1981). In addition, present-day dip of strata reflects depositional dip. These features and the accessibility of the area greatly facilitate study of the Níjar Miocene carbonate rocks.

Dabrio et al. (1981) suggested that the reef near Níjar might be representative of all uppermost Miocene reefs of the western Mediterranean. Reef deposits, however, only are evident in the upper half of the section at Níjar. Unlike correlative carbonate successions throughout southeastern Spain, carbonate shelf deposits are more important at Níjar. From a distance, the layering of the rocks within the section looks simple: stacked, shallow-dipping strata. Detailed studies, however, reveal many stratigraphic complexities that seem to be a result of numerous fluctuations in relative sea level.

In this paper, I will summarize earlier descriptions of the reef complex and post-reef deposits, present new data on the reef complex, and describe the strata underlying the reef complex, which had previously been included as reef deposits. A description of the entire sequence, pre-reef as well as reef, will further the interpretation of regional correlation.

### *Previous Work*

Since the mid 1960s, geologists from The Netherlands, Spain, France, the United States, and Switzerland have completed detailed stratigraphic studies of carbonate strata in the Neogene basins of southeastern, mainland Spain (Esteban et al., this volume). Several have concentrated on the Miocene strata exposed in Almería Province. Dabrio and Martín (1978) summarized joint work with Esteban, Calvet, and Gíner showing characteristics and distribution of reef complexes throughout the province. Dabrio et al. (1981) focused on an area (4 km x 2 km) within the central part of the province to the east of the town of Níjar. On the basis of field and petrographic studies, these geologists proposed a generalized reef model for late Miocene reef sedimentation throughout the western Mediterranean. Other stratigraphic studies of the area include Addicott et al. (1978), Mankiewicz (1987, 1988), and Megías (1985).

### *Location*

Miocene carbonate strata crop out east of the town of Níjar, Almería Province (Fig. 1). This study concentrates on the succession exposed on the first prominent ridge one mile (1.5 km) east of Níjar—the "Níjar section" of Dabrio et al. (1981). Miocene strata are well exposed and easily accessible along the ridge that trends north to south over a distance of one km. An extension of the reef strata occurs to the east of the ridge for about 4 km.

### *Stratigraphic Sequence*

Geologists from The Netherlands first undertook detailed stratigraphic studies of the Neogene basins in southeastern mainland Spain. Völk (1967) studied the stratigraphy in the Vera Basin and laid out the stratigraphic framework that has been adopted by most geologists. Earlier workers had provided some of the groundwork for Völk's study, notably Ruegg (1964, in Dabrio and Martín, 1978) in the Sorbas Basin, and Völk and Rondeel (1964, in Dabrio and Martín, 1978) in the Vera Basin. Others have refined Völk's work, including Dronkert and Pagnier (1977) in the Sorbas Basin, Dabrio and Martín (1978) in the Almería area, Dabrio et al. (1981) in the Níjar Basin, and Veeken

Models for Carbonate Stratigraphy from Miocene Reef Complexes of Mediterranean Regions,
SEPM Concepts in Sedimentology and Paleontology #5, Copyright © 1996,
SEPM (Society for Sedimentary Geology), ISBN 1-56576-033-6, p. 141-157.

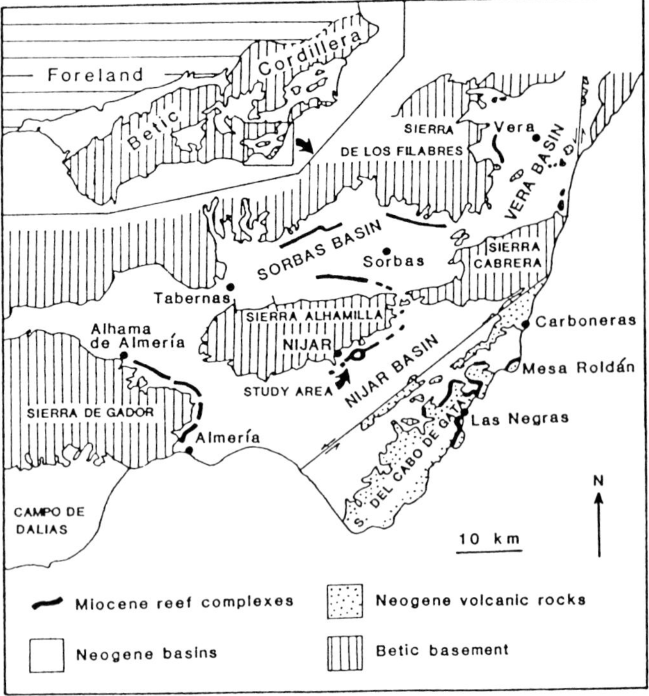


FIG. 1.—Index map of Níjar area showing major Neogene basins and late Miocene reef exposures (after Dabrio et al., 1981).

(1983) in the Pulpi Basin. The resulting interpretations of the stratigraphy are schematically shown in Figures 2A and 2C.

The ridge to the east of Níjar is composed of the Turre Formation that comprises the Azagador, Abad, and Cantera Members. The Azagador Member is characterized by a basal conglomerate that contains clasts derived from the basement rocks of the Betic Mountains. Most of the sediments are grainstones and packstones that were deposited on an open-marine platform. The Abad Member contains marls rich in planktic foraminifers that accumulated in a basin-margin to basin environment. The Cantera Member consists of coral framework and associated grainstones and packstones that formed in reef and reef-slope environments. The Azagador and Cantera Members may represent shallow-marine facies of the deeper marine Abad Member (Dabrio et al., 1981; Fig. 2C).

Dabrio et al. (1981) noted that boundaries between members of the Turre Formation do not correspond to depositional events. For example, the Cantera Formation comprises reef and reef-talus deposits. The proximal reef sediments, however, would be included in the Azagador Member and the distal reef sediments would be categorized as Abad Member strata. The discrepancy between lithostratigraphic terms and depositional units prompted these workers to abandon the lithostratigraphic framework in favor of informal terms (Fig. 2D). Mankiewicz (1987) followed their lead and identified an additional, pre-reef depositional unit that Dabrio et al. (1981) had designated as reef deposits. A modified version of the facies used by Mankiewicz (1987) is adopted for this report (Fig. 2F).

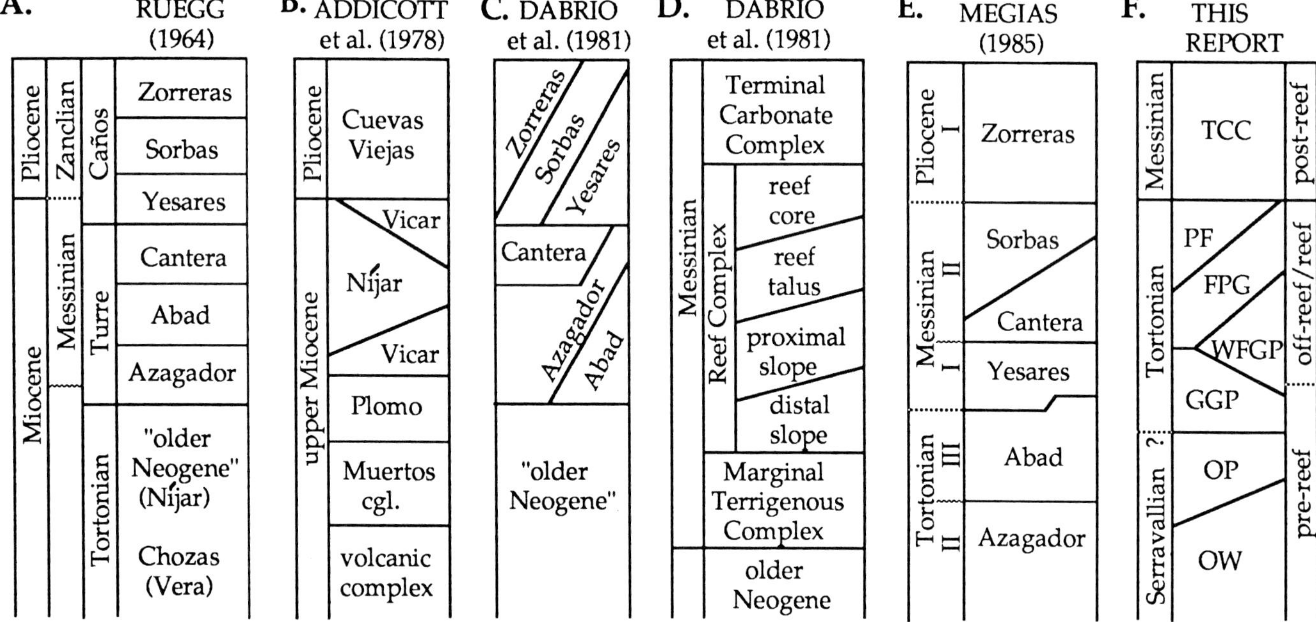


FIG. 2—Various interpretations of stratigraphy in the Níjar area showing (A) lithostratigraphic framework of Ruegg (1964, in Dabrio et al., 1978) and later workers; (B) interpretation of Addicott et al. (1978); (C) possible facies relationships suggested by Dabrio et al. (1981); (D) informal terminology of Dabrio et al. (1981); (E) interpretation of Megías (1985); and (F) units used in this report. Abbreviations used in (F) are as follows: OW (Orange Wackestones), OP (Orange Packstones), GGP (Gray Grainstones/Packstones), WFGP (White Fine-Grained Packstones), FPG (Fossiliferous Packstones/Grainstones), PF (Porites Framestones), and TCC (Terminal Carbonate Complex). OW and OP make up Depositional Sequence (DS) 1; GGP is equivalent to DS2; and WFGP, FPG, and PF make up DS3 of Franseen and Mankiewicz (1991).

To the east of the Níjar ridge, the flat-lying beds of the "Terminal Carbonate Complex" (Esteban, 1979) unconformably overlie the reef facies. Major erosional contacts characterize both the upper and lower boundaries of the Terminal Carbonate Complex; the contact is demonstrably subaerial in some outcrops in the Almería area (Esteban and Gíner, 1980; Dabrio et al., 1981). The Terminal Carbonate Complex contains a mix of lithologies including dolomitized *Porites* framestones, stromatolitic boundstones, and oölitic and gastropod grainstones. The oölitic grainstones can be grouped into packages that typically show fining-upward trends. Dabrio et al. (1981, p. 534) interpreted the fining trends "as the result of shoaling storm waves of progressively decreasing energy" and suggested that the Terminal Carbonate Complex of the Níjar area is equivalent to the Sorbas Member (a shallow-marine, mixed carbonate-siliciclastic unit) of the Caños Formation (Figs. 2A, C, E).

Other stratigraphic terminology has been presented. For example, Megías (1985) has offered an alternative interpretation (Fig. 2E) of the lithostratigraphy discussed above; he based his interpretation on "tectonosedimentary analysis" that included regional mapping of disconformities. Addicott et al. (1978) used different lithologic terminology (Fig. 2B) that is not easily applied to the Níjar ridge, particularly the reef deposits.

### *Geologic Age of the Reef*

Most previous workers (e.g., Martin and Braga, 1993) have assigned the Níjar and equivalent reef complexes to the Messinian Stage (uppermost Miocene); they attribute the pre-reef carbonate deposits to the upper part of the Tortonian Stage (Fig. 2). Dating, however, was tenuous due to the paucity of well-preserved biostratigraphic markers.

Mankiewicz (1987) noted the apparent correlation between (1) inferred sea-level changes at Níjar and (2) third-order eustatic sea-level curves of Haq et al. (1987). On the basis of matching eustatic sea-level curves, Mankiewicz (1987) suggested a Serravallian (middle Miocene) through Messinian age for the Níjar succession; the same age range is suggested by correlation to a similar succession in Las Negras where dating of microfossils help to correlate the inferred curves to eustatic curves (Franseen and Mankiewicz, 1991). Serrano (1990) recently reported the occurrence of Langhian to Serravallian planktic foraminifers from lowermost Níjar strata and demonstrated a Messinian age for the reef section. Esteban et al. (this volume) offer a tentative correlation of the Níjar reef with the rest of the western Mediterranean units.

### *Regional Geologic Framework*

The Betic Cordillera represents the westernmost Alpine chain. Extensive compressional deformation occurred from Late Cretaceous into middle Miocene (Rehault et al., 1985). Since that time, shear deformation has prevailed and has resulted in NW-SE regional shortening and concomitant NE-SW extension (Montenat et al., 1987; Sanz de Galdeano, 1990). The so-called Trans-Alboran shear zone extends over 250 km from the Almería area northeastward to the Alicante area and is characterized by numerous NE-SW trending strike-slip faults (Hernandez et al., 1987; Larouzière et al., 1988). Miocene (largely late Miocene) calcalkaline volcanism typically occurred alongside strike-slip faults; partial melting of the crust yielded dacitic lavas in the Níjar area (Hernandez et al., 1987; Sanz de Galdeano, 1990).

Numerous Neogene basins developed within the Betic Cordillera as a consequence of shearing (Hernandez et al., 1987). During late (and middle?) Miocene time, fringing reefs such as the Níjar complex formed along the rims of the basins (Dabrio and Martin, 1978; Esteban, 1979). Around the Níjar Basin, several other upper Miocene reef complexes crop out including those south of the coastal village of Las Negras (see Franseen and Goldstein, this volume) and at Mesa Roldán (Fig. 1).

Although the region is tectonically active, there is no evidence of syndepositional tectonic activity within the Miocene carbonate succession near Níjar. Geopetal fabrics throughout the succession at Níjar suggest that dips of the beds are depositional. Postdepositional, ENE-WSW trending faults do cut the section, but show a maximum displacement of 3 m.

## LITHOFACIES ASSOCIATED WITH THE REEF

### *Data Base*

The following interpretation of the depositional facies relies heavily on the work of Dabrio et al. (1981). My detailed field and petrographic studies corroborate many of their ideas, but also provide the basis for recognizing a more extensive pre-reef depositional system. I measured and described twelve stratigraphic sections along the 1-km ridge (Fig. 3), mapped facies on composite photographs of the hillsides, collected and examined

FIG. 3.—Cross-sectional view of geology exposed southeast of the town of Níjar. Large arrow points to quarry shown in Figure 4. *Porites* Framestones occur on the four hilltops designated with the small arrows.

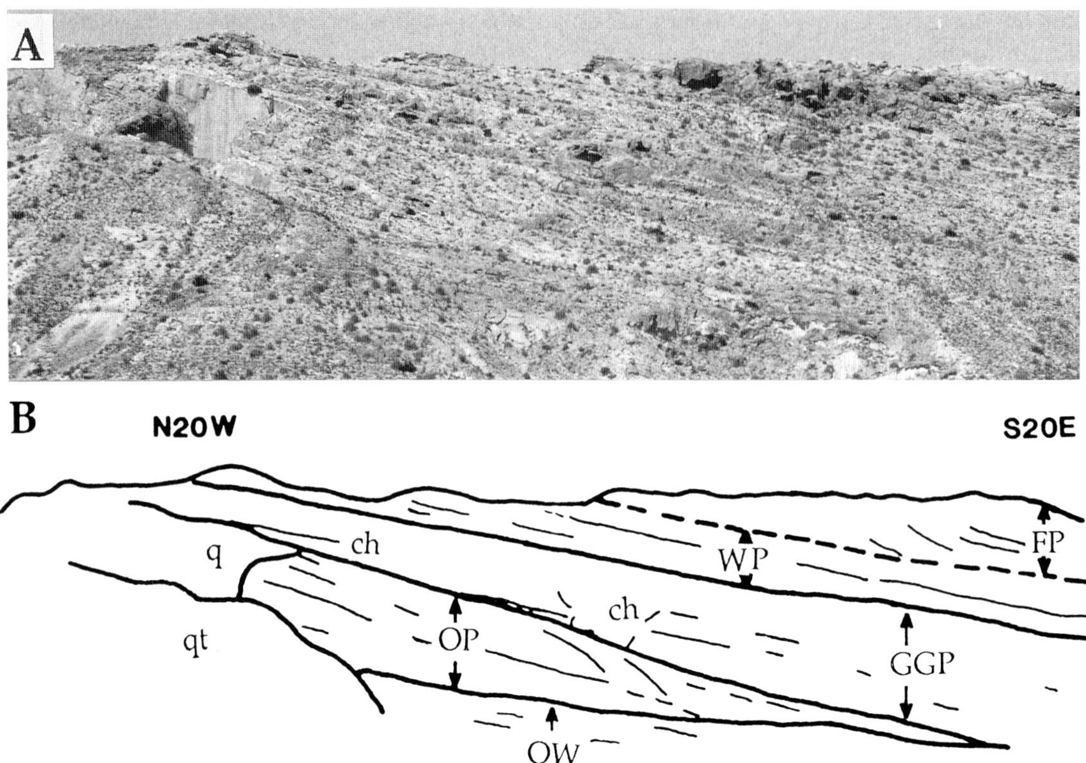


FIG. 4.—View of hill with quarry (A) and interpretive sketch (B). Diagram shows shallow-dipping Orange Wackestones (OW), wedge-shaped, steeper dipping Orange Packstones (OP), onlapping sedimentary packages of the Gray Grainstones/Packstones (GGP), and disconformable Fossiliferous Packstones/Grainstones (FP). Beds dip slightly into the hill to the southeast. "ch" designates two channels (the southern channel is illustrated in Figures 6C and 6D); "q" indicates a quarry; and "qt" shows quarry-talus cover.

over 300 hand specimens, and studied about 150 thin sections.

### *Depositional Facies*

Most of the sediments at Níjar consist of packstones and grainstones that display differences (commonly subtle) in color, grain size, and composition that can be identified in the field. Color (orange to gray to white) may be a function of concentration and grain size of the terrigenous-derived component (much of which are iron-rich altered metamorphic rock fragments). The color variations help differentiate seven depositional "units" (Fig. 4) perhaps best defined by geometric relationships with overlying and underlying units. In this paper, I restrict the term "unit" to large-scale sedimentary successions that are on the order of tens of meters thick. Many of the units are composed of smaller scale "packages" that are typically 2 to 8 m thick.

The units, characteristics of which are summarized in Table 1, are the pre-reef Orange Wackestones, Orange Packstones, and Gray Grainstones/Packstones and the reef/off-reef *Porites* Framestones, Fossiliferous Packstones/Grainstones, and White Fine-Grained Packstones. A modern caliche crust covers parts of the geology, particularly in the topographically higher parts. This crust obscures details making study of some of the section difficult or even impossible.

*Orange Wackestone Unit.—*

Only the upper part of the Orange Wackestone Unit has been examined; it is composed predominantly of dolomitized fine-sand- to silt-size wackestones and fine-grained packstones. Extensive bioturbation produced a mottled texture destroying most evidence of bedding that consistently dips 10° or less to the southeast. Irregular patches on the scale of 1 to 4 cm contain concentrations of very coarse sand (quartz and metamorphic rock fragments) derived from the basement; the patchy distribution may result from bioturbation. Macro- and microfossils are rare and include marine mollusk fragments and glauconite-filled planktic foraminifers (Fig. 5A).

The paucity of biotic constituents led Addicott et al. (1978) to suggest that this unit, which I believe is equivalent to their informal "upper Níjar formation," might have been deposited in a brackish water environment. The occurrence of glauconite-filled planktic foraminifers, however, suggests a low-energy marine environment (assuming that the foraminifers are not reworked).

*Orange Packstone Unit.—*

The Orange Packstones are mud rich; part of the mud is noncarbonate. HCl-insoluble residues from 13 samples of this unit range from 7 to 22 weight percent; mud makes up much of

TABLE 1.—CHARACTERISTICS OF STRATIGRAPHIC UNITS

| UNIT* | UNIT GEOMETRY | INTERNAL GEOMETRY | DIP | HCl INSOLUBLE RESIDUES (wt. %) | GRAIN SIZES | INVERTEBRATE FOSSILS | CALCAREOUS ALGAE GENERA | CORALLINE ALGAE MORPHOLOGIES | POROSITY (visual estimates; %) | CEMENTS** |
|---|---|---|---|---|---|---|---|---|---|---|
| *Porites* Framestones | spurs? | massive | 0°?, 20°-30° | 4 (n = 5) | mud to boulder | *Porites* dominant (in situ) | *Halimeda* (rare), *Lithoporella Lithophyllum?* | crusts, fragments | trace to 20 | dolomite, void-filling calcite, pendant calcite |
| Fossiliferous Packstones/ Grainstones | prograding wedges | rhythmic pattern of biotic constituents & sedimentary structures | 10°-25° | 7 (n = 88) | mud to pebble | mollusks, serpulids, *Porites,* bryozoans, echinoids (all transported) | *Halimeda, Lithothamnion, Archaeolithothamnium, Mesophyllum* | crusts, branching thickets, rhodoliths, fragments | trace to 40 | isopachous calcite, dolomite, void-filling calcite, pendant calcite |
| White Fine-Grained Packstones | sheet | parallel bedding that thins basinward | <12° | 12 (n = 18) | mud to fine sand | planktic & small benthic foraminifers | unidentifiable fragments | fragments | trace to 5 | isopachous calcite, dolomite |
| Gray Grainstones/ Packstones | lense | fining & thinning upward packages | ≈10° | 16 (n = 30) | mud to pebble | mollusks, serpulids, benthic foraminifers, echinoids (all transported) | *Cymopolia* (rare) *Lithophyllum, Lithothamnion, Archaeolithothamnium, Mesophyllum, Corallina, Amphiroa* | rhodoliths, fragments | 2 to 40 | isopachous calcite, dolomite, poikilotopic calcite, pendant calcite |
| Orange Packstones | wedge | oblique to parallel pattern of bedding | 10°-20° | 13 (n = 12) | mud to pebble | mollusks, serpulids, benthic foraminifers, bryozoans, echinoids (all transported) | *Lithothamnion,* unidentifiable fragments | rhodoliths | 2 to 40 | dolomite, poikilotopic calcite, pendant calcite |
| Orange Wackestones | sheet | parallel, poorly defined bedding | <10° | 31 (n = 6) | mud to fine sand | planktic foraminifers (glauconitized) | ------ | ------ | 2 to 5 | dolomite |

* The six units can be grouped into three facies. The pre-reef facies comprises the Orange Wackestones and Packstones, the Gray Grainstones/Packstones, and possibly the stratigraphically lower parts of the White Fine-Grained Packstones. The off-reef facies comprises the Fossiliferous Packstones/Grainstones and most of the White Fine-Grained Packstones. The reef facies is equivalent to the *Porites* Framestones.

** Cements are listed in order of occurrence (earliest to latest).

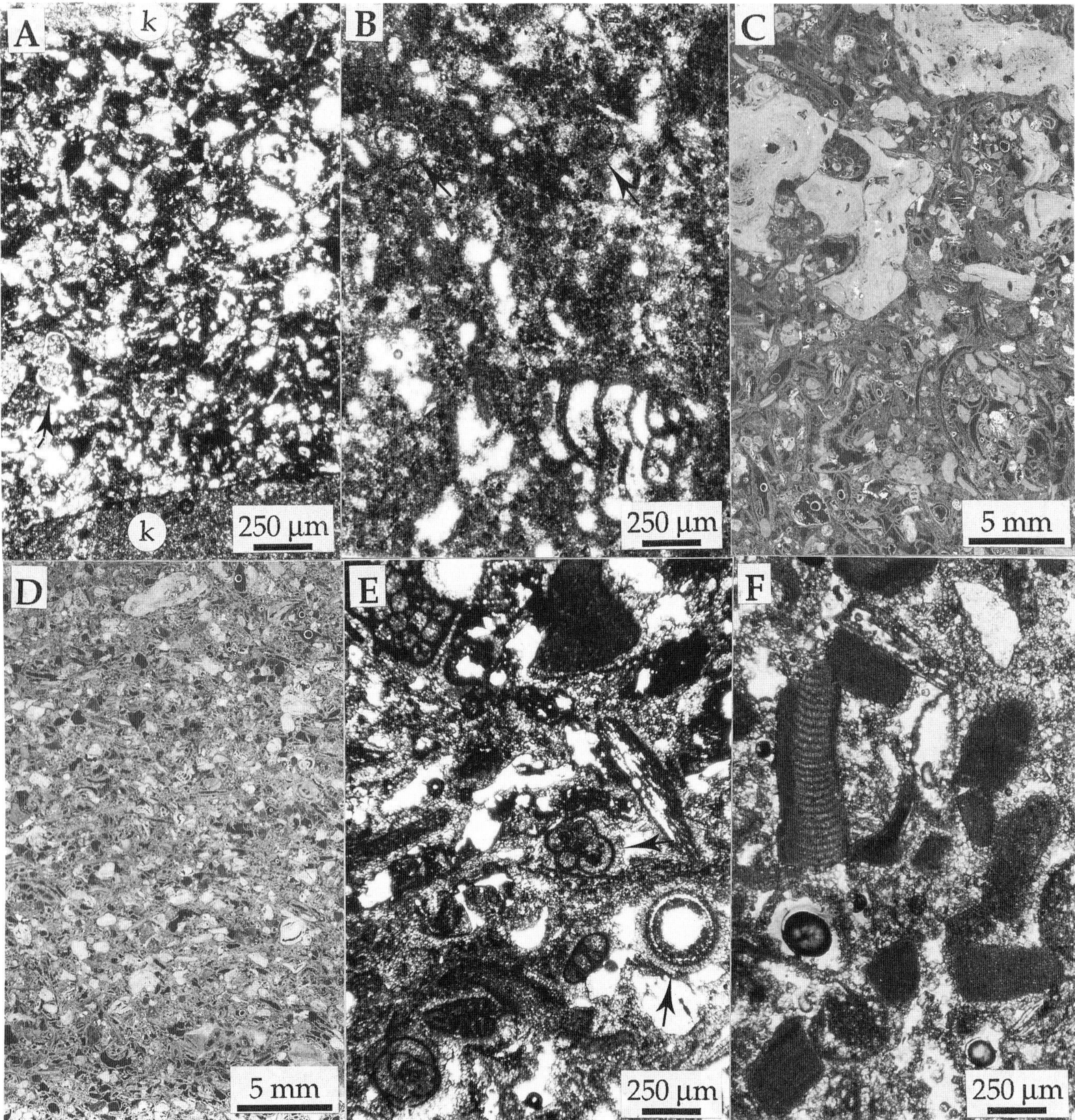


FIG. 5.—Photomicrographs in plane-polarized light (PPL) (A, B, E, F) and negative prints (C, D) of thin sections of the pre-reef facies. (A) Photomicrograph of siliciclastic-rich Orange Wackestones showing single glauconite-filled foraminifer (arrow). The large clasts (k) are basement derived. Most of the white subangular areas represent quartz grains. The matrix is a combination of turbid dolomite rhombs up to 50 µm in diameter and siliciclastic muds. (B) Photomicrograph of mud-rich beds of the Orange Packstones showing a large benthic foraminifer in lower right and two planktonic foraminifers (arrows). Note the dirty dolomite matrix. White represents pore space (moldic and intraparticle). (C) Negative print of bioclastic-rich channel fill of the Gray Grainstones/Packstones. Fossils include serpulid worm tubes, coralline algae (dense white), large benthic foraminifers, and mollusks (crescent shapes defined by micritic rims and calcite cement). (D) Negative print of medium-grained sediment of the Gray Grainstones/Packstones. Note preferred orientation of grains parallel to bedding. (E) Photomicrograph of D showing foraminifers and bryozoan clast (upper left). Bladed to fibrous cement (arrows) partly occludes inter- and intraparticle pore space. (F) Photomicrograph of Gray Grainstones/Packstones illustrating predominance of coralline algae grains. Many of the dense micritic grains display cellular structure characteristic of coralline algae; some are almost peloidal in appearance showing only a faint relict structure. Dolomite rhombs with darkened centers make up matrix.

this residue along with variable amounts of quartz and metamorphic rock fragments derived from the basement.

Skeletal grains of the Orange Packstones include mollusks (some pectinid bivalves and oysters), coralline algae (*Lithothamnion*), echinoids, benthic foraminifers, and bryozoans (Fig. 5B). Most fossil constituents are fragmented. In general, however, within a vertical section the abundance of unfragmented skeletal grains (predominantly disarticulated but unbroken bivalve valves and 1- to 7cm rhodoliths) increases upward, possibly indicating a change from more offshore to onshore components. No fossils appear to display in-place growth. Development of micritic envelopes has enhanced the preservation of fossil form, but in general, internal details of the skeletal grains have been obliterated.

*Gray Grainstone/Packstone Unit.*—

The Gray Grainstone/Packstone Unit is composed of lenticular packages. Each package fines and thins upward and attains a maximum thickness of about 7 m (Fig. 6A). Medium to coarse sand typifies grain size of the matrix in the lower sediments (predominantly grainstones and some packstones) of each package (Figs. 5C-5F) and fine sand to silt characterizes the uppermost layer (muddy packstones). The bedding also changes from the base to top of the package from channelized or crossbedded to parallel layered to burrow-mottled beds. Beds dip 10° or less to the southeast.

Channels commonly occur in the lower part of each fining- and thinning-upward package and trend oblique or parallel to the inferred NE-SW-oriented shoreline, which is assumed to have paralleled strike direction. Erosion carved into coarse-grained parallel-layered strata and commonly cut down into the underlying package of Gray Grainstones/Packstones (or into the Orange Packstones; Figs. 6C, D). Channels attain dimensions up to about 6 m across by 1.7 m in depth. Some channels appear to be amalgamated (Fig. 6C); each subsequent channel cuts less deeply into and more northward (shoreward) of the previous cut-and-filled channel. Variation in the abundance or type of macrofossils helps delineate each channel.

Channels are filled with massive and very coarse-grained sediments, the composition of which is similar to that through which the channels cut: rounded coralline algae fragments dominate matrix grains (Figs. 5C). Other components include quartz and metamorphic rock fragments derived from the basement, mollusk fragments, large benthic foraminifers, echinoids, and articulated terebratulid brachiopods. Some channels contain boulders (up to 0.5 m across) of the finer bioturbated sediments of the underlying fining- and thinning-upward package (Figs. 6D, E); the presence of the boulders suggests at least semi-lithification of the underlying bed occurred. In addition, rhodoliths, which range in size from 2 to 10 cm in diameter, are concentrated within some channel fills (Fig. 6E).

As with channeling, crossbedding occurs in the basal part of some fining- and thinning-upward packages but is not common (Fig. 6B). It typically occurs as single sets up to 1.3 m thick that can be traced laterally for up to about 10 m. Crossbeds are tabular and dip 3° to 6° to the northwest, towards the inferred shoreline. Coarse to very coarse sand-size grains make up the crossbeds; variations in the concentration of pebble-size fossil fragments delineate crossbedding.

Bioturbation is evident throughout the Gray Grainstones/Packstones except within channels and crossbeds (Figs. 6B, E). In the coarser sediments, *Ophiomorpha nodosa*, a bumpy pellet-lined burrow, dominates. Simple vertical and horizontal networks characterize the *Ophiomorpha* at Níjar (Fig. 6E). Shafts of different burrows range from 1 to 2.5 cm in diameter and burrow linings are up to 1 cm thick. The occurrence of this trace fossil probably indicates rapidly shifting, relatively high-energy marine environments, but not necessarily shallow water (Ekdale et al., 1984). Although *Ophiomorpha* is obvious, it is not abundant. Bioturbation does not disrupt bedding appreciably in the coarser grained facies (Figs. 6B, E). Either rapid sedimentation or too few burrowers promoted preservation of the sedimentary layers.

Burrowers do mottle the finer capping bed of each package (Fig. 6D). Individual burrows are commonly not distinct although back-filling of larger (2 cm in diameter) burrows and infill of burrows with overlying coarse-grained sediments define some burrow outlines in the finer sediments.

Coralline algae are the most abundant fossils in the Gray Grainstones/Packstones and show greater diversity than in any underlying or overlying unit; some genera are characteristic of only this unit. Genera include *Lithophyllum* (dominant), *Lithothamnion*, *Mesophyllum*, *Archaeolithothamnium*, and *Lithoporella* as well as segments of articulated coralline algae (*Corallina* and *Amphiroa*). Fragments of coralline algae make up about 20 to 30% by volume in the coarse-grained sediments (Figs. 5C, F). In addition, rhodoliths are very common. Concentrations of rhodoliths occur in channels, in layers of crossbeds, and in layers within the parallel-layered coarse sediments. Most rhodoliths are ovoid in shape, 2 to 10 cm in diameter, and show moderately-dense branching. The tips of branches or columns (stubby branches) typically broaden and commonly merge with their neighbors to form a crust, a morphology that suggests frequent turning (Bosence, 1983).

Chlorophycean (green) algae are rare: no *Halimeda* and only two dasycladacean (*Cymopolia*?) fragments were found.

*Porites Framestone Unit.*—

The coral *Porites* is the main biotic component that constructs the framestones in the Níjar area. Two morphologies of *Porites* dominate: a laminar form occurs predominantly near the base of the framestones and a stick-like form (Fig. 7C) occurs throughout most of the framestones. Bryozoans, coralline algae (*Lithoporella* and *Lithophyllum*?), foraminifers, and serpulid worms encrust *Porites* sticks and laminar forms but typically contribute little to framebuilding.

Most of the rigidity of the framework may have resulted from the thick micritic crusts that coat *Porites* sticks (Figs. 8A, C). Volumetrically, the micritic crusts can be more important than the *Porites*. The laminated form and peloidal texture of some of

A
B
C
b
b
D
b
E
5 cm
F
3
2
1
G
1cm

these crusts led Dabrio et al. (1981) to believe that they are of cement origin. Alternatively, the crusts could be stromatolitic (microbial) in origin (Riding et al., 1991). Similar crusts recently have been described from more recent reefs. For example, Montaggioni and Camoin (1993) described laminar to columnar stromatolitic crusts on coralgal reefs in Tahiti; they interpreted these growths to have formed in shallow-water, high-energy, and presumably "healthy" conditions for growth. Similarly, Jones and Hunter (1991) studied Pleistocene reef rocks from Grand Cayman and San Salvador Island and reported that filamentous cyanobacteria colonized in-place coral and fragments of coral; the cyanobacteria trapped micrite and thus produced microbialites. They interpreted the change from coral-dominated to cyanobacteriadominated communities to reflect an environmental change effected by Pleistocene sea-level fall. The crusts on the *Porites* of Níjar are similar structurally to the Pleistocene examples (laminated, peloidal form that can surround thin coralline algae crusts); the paleoenvironment, one in which sea level fell, was probably similar also. In Níjar, however, dolomitization has destroyed evidence of calcified filaments within the micritic crusts.

Sediments that fill in around the framework consist of packstones and to a lesser extent, grainstones and rudstones. Recognizable biotic components include fragments of mollusks, *Porites* (with and without crusts), serpulid worms, echinoids, coralline algae, bryozoans, *Halimeda*, and very rarely, non-poritid corals. Most of these biotic constituents were originally aragonitic and are now moldic, making identification difficult. Extensive boring (predominantly by clionid sponges) of corals, mollusks, and coralline algae also hinders identification. Some fragments of mollusks, however, can be identified and include *Conus, Arca,* and *Glycymeris*; Dabrio et al. (1981) also reported *Spondylus.*

*Fossiliferous Packstone/Grainstone Unit.—*

The Fossiliferous Packstones/Grainstones can be divided into about eight wedge-shaped packages of sediment. Within each wedge, dips of beds decrease basinward from about 25° to 12°. Individual beds thin and fine basinward. Thus, the beds of each package are arranged as prograding clinoforms. In the more proximal parts, the contact between each successive wedge represents a surface of downlap. Dabrio et al. (1981, Fig. 6A) depicted some of these downlap surfaces, but more are recognized here.

The basal surface of each wedge as traced laterally typically displays about 50 to 60 m of vertical relief. Each wedge of sediments has an erosional base that shows up to 50 cm of erosional relief. A poorly sorted mixed-fossil rudstone to floatstone commonly occurs at the base (Fig. 8A). Beds rich in the green alga *Halimeda* also occur in the lower parts of some of the wedges (Figs. 7D, F). In some wedges, a branching network of the coralline alga *Mesophyllum* occurs in the more distal parts of the *Halimeda* beds (Fig. 7F). No evidence that branches of this alga were fragmented, or concentrated or aligned by currents has been seen. I therefore interpret the branching alga to have grown in place as a thicket that baffled and trapped fine sediment on the reef slope.

Sediments above the *Halimeda*-rich beds are varied in nature. Coarser-grained sediments (coarse sand to pebble particles with a minor siliciclastic component) commonly display parallel lamination and some beds up to 75 cm thick are weakly graded. A few of the parallel-layered beds show 2- to 4-cm-thick layers characterized by parallel layering at the base that grades upward to a fine-sand rippled or laminated cap (Fig. 7B). I interpret the small-scale fining-upward sediments to have been deposited during a single event—perhaps a Bouma b-c/d sequence. Subsequent deposition typically eroded the finer grained capping layer yielding the characteristic monotonous parallel layering composed solely of coarser grained particles. Bivalves (*Glycymeris, Arca,* and others), gastropods, serpulid worms, and coralline algae (as fragments and rhodoliths) dominate the biotic constituents in these beds (Figs. 8B, C, D, F); *Porites* fragments are not common and *Halimeda* plates are rare or absent. Most skeletal grains are highly fragmented; only in the more proximal parts are skeletal grains whole or relatively mildly abraded. No fossils appear to be in situ. Many skeletal grains display

FIG. 6.—Field photographs of Gray Grainstones/Packstones. (A) Two fining- and thinning-upward packages that are relatively resistant, massive, and/or channelized at the base. Geologist in center of photograph (arrow) is standing on the boundary between the two packages. (B) Crossbedding from the lower part of a fining- and thinning-upward package. Crossbeds in the lower part of the photograph dip about 6° northwestward (landward; to the left in the photograph). The 13-cm-thick layer (double arrow) at the top of the crossbed set contains crossbedding with more steeply inclined layers. Parallel-layered strata that dip about 10° to the southeast (to right of photograph) down the paleoslope overlie the cross-sets; in this example, the parallel-layered strata are in fault contact with the crossbedded strata. Note the lack of burrows in the lower crossbedded unit. In the upper part, *Ophiomorpha* burrows (arrows) are common but do not produce mottled bedding. Scale is 15 cm long; smallest divisions on scale equal 1 cm. (C) Amalgamated channels (approximate position of bases indicated by dashed lines) defined by slight variations in texture of the fill. Note that base of channel follows top of fine-grained white bioturbated capping beds (b) of fining- and thinning-upward sequences, suggesting that the capping beds were partly lithified at the time of channeling. Area within square (about 1.5 m on each side) is enlarged in D. (D) Close-up of boxed area of C showing truncation of bioturbated capping (b) and underlying parallel-layered beds. The channel fill contains some clasts (5 to 25 cm in diameter) of the white bioturbated bed. Smaller white nodules in the channel fill are rhodoliths. Staff is marked in 10-cm intervals. (E) Simple-branching network of *Ophiomorpha* (arrows) in parallel-layered beds. Other smaller burrow traces are also evident. Burrowing does not appreciably disrupt bedding. (F) Vertical face of zone of possible subaerial exposure. Lower massive part (1) is well cemented and in sharp contact with the overlying discontinuous layer (few cm thick) (2); the thin layer (2) is friable, fine grained, contains 2- to 3-cm nodules, and is in sharp contact with the overlying well-cemented marine fine-grained massive layer (3). Area in lower right (boxed) is enlarged in G. Arrow points to change in color between solution altered rock above and unaltered rock below. (G) Enlargement of lower right corner of F showing the burrow trace *Ophiomorpha nodosa*; the bumpy burrow lining has dissolved. Note also the laminated caliche? in upper left (arrow) that cuts bedding obliquely. Bedding is more obvious in F; it dips about 10° to the right and slightly away from observer.

FIG. 7.—Field photographs (A-D) and photograph of cut slabs (E, F) of the Fossiliferous Packstones/Grainstones and *Porites* Framestones. (A) Distant view of south part of Níjar ridge showing about 75 m of vertical relief and, in the upper half of the photograph, illustrating several wedge-shaped units of Fossiliferous Packstones/Grainstones. (B) Vertical face showing weakly defined parallel layering in the Fossiliferous Packstones/Grainstones. The 5-mm-thick fine-grained layer (arrows) is discontinuous laterally and defines direction of dip (about 14° to the right and into the photograph). Such fine sand to silt layers are rarely preserved. (C) Contact (arrows) between steeply dipping (about 25°) Fossiliferous Packstones and overlying *Porites* Framestones. Vertical fabric evident in the framestones results from vertically oriented stick-like *Porites*. Hammer at center right (circled) is 33 cm long. (D) Vertical face showing *Halimeda*-rich rock with encrusting *Porites* (arrow). Geopetal fill (circled) indicates up; weakly defined bedding dips about 25° to the right. (E) Cut slab of *Halimeda*-rich bed; up is at top. Note paucity of other biotic constituents (some small moldic gastropods circled), lack of alignment of *Halimeda* segments, and moldic porosity. (F) Cut slab of *Mesophyllum*- and *Halimeda*-rich beds in a muddy matrix. D, E and F represent a trend from more proximal to distal reef, respectively.

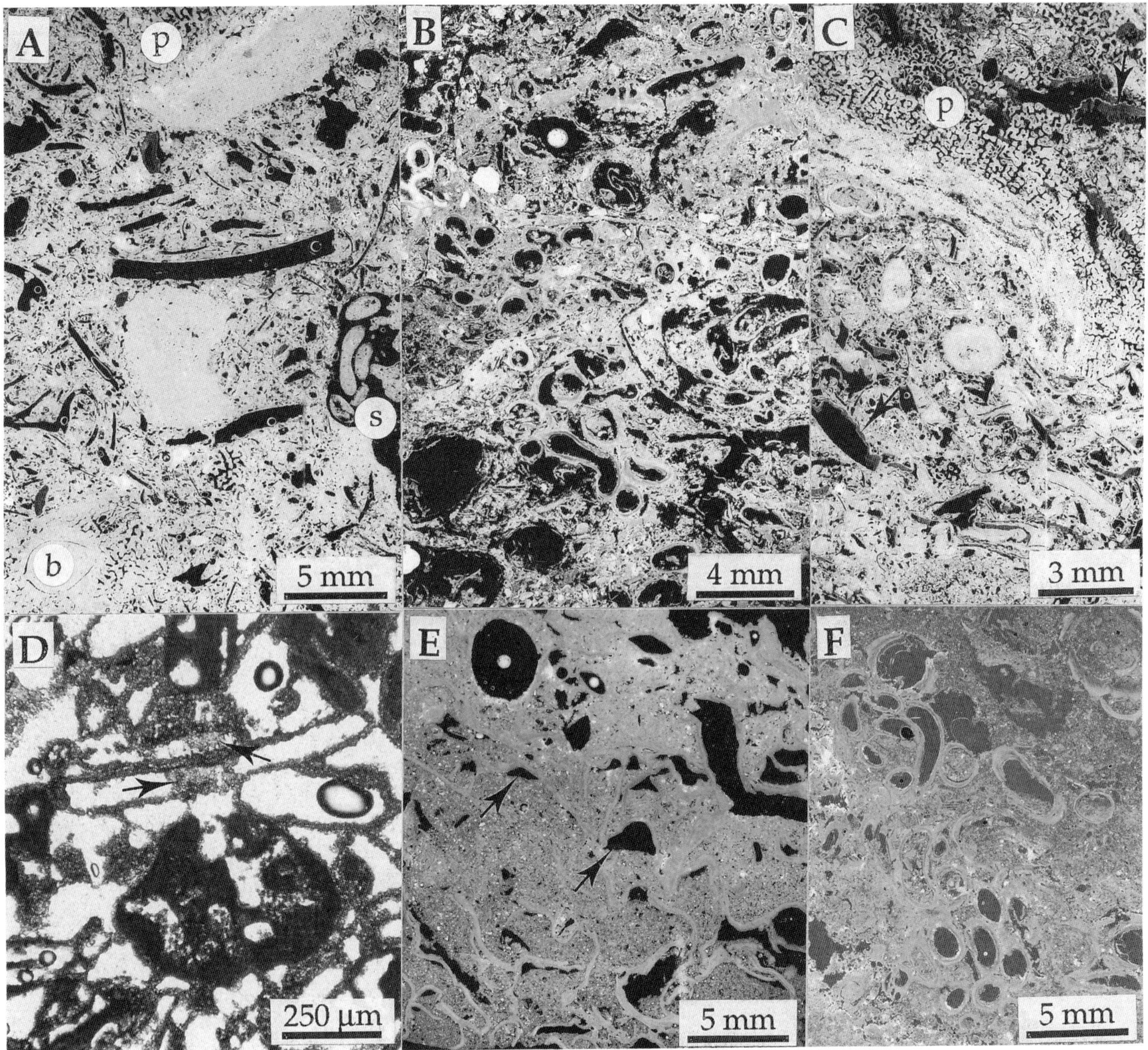


FIG. 8.—Photographs of thin sections from the Fossiliferous Packstones/Grainstones and *Porites* Framestones. (A) Negative print of mixed-fossil hash at base of wedges of the Fossiliferous Packstones/Grainstones. Note lack of orientation of bioclasts, mud-rich matrix, bivalve boring (b) in *Porites* in lower left, clionid sponge borings (s) in unidentified moldic host (possibly a mass of vermetid gastropods) at center right, and crescent and circular shapes that represent moldic mollusk fragments. The abundance and type of borings suggest that clasts originated in shallow water. *Porites* fragment (p) at top center has part of a micritic/stromatolitic crust attached. (B) Negative print of parallel-layered Fossiliferous Packstones/ Grainstones. Serpulid worms and bivalves (moldic) dominate. (C) Negative print of thin section near reef framestones that is characterized by horizontal fabric produced by bored encrusting *Porites* (p), serpulid worm tubes, coralline algae, and pendant cement (arrows). (D) Photomicrograph (PPL) illustrating high moldic porosity. Dolomitized micritic rims define clast shape. Late-stage pendant cement (arrows) reduces the amount of pore space. (E) Fine-grained packstones that are bound together by a lacy framework of coralline algae (*Lithothamnion*?). Note shelter porosity (arrows) under thin, arched crusts of algae. (F) Negative print of thin section from sample associated with reef talus. Serpulid worm tubes dominate.

evidence of boring; clionid sponges produced the chambered borings that are commonly preserved as casts, and bivalves made the clavate-shaped forms (Fig. 8A; Ekdale et al., 1984).

Finer, relatively macrofossil-free beds are interbedded with the parallel-layered beds. The finer beds are composed of coarse-sand to mud-size unidentifiable carbonate particles, and commonly show evidence of bioturbation. The fine bioturbated beds typically form the uppermost bed of each sedimentary wedge. In the more proximal parts, some fine-grained beds are not obviously bioturbated and are bound by a thin crust of the coralline alga *Lithothamnion* (Fig. 8E).

*White Fine-Grained Packstone Unit.—*

The White Fine-Grained Packstone Unit consists of beds of

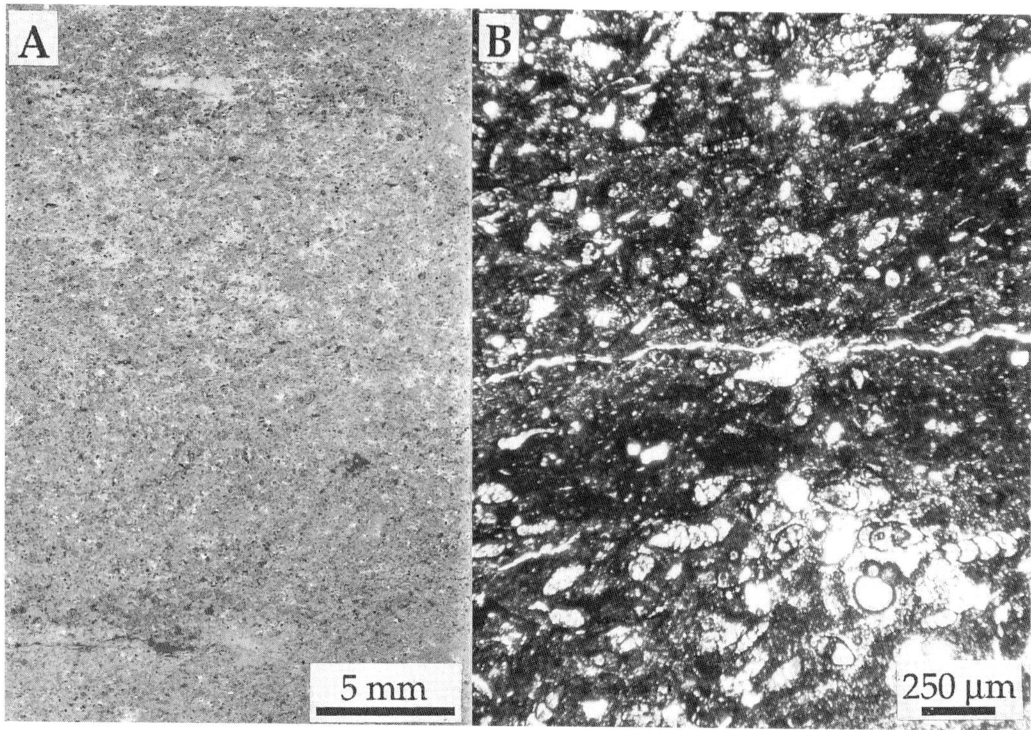


FIG. 9.—Photographs of thin sections from White Fine-Grained Packstones. (A) Negative print of relatively microfossil-rich sample. (B) Photomicrograph (PPL) of thin section in A showing many small, unidentified benthic foraminifers that are concentrated in layers and aligned parallel or subparallel to bedding.

fine-sand to mud carbonate particles. Low angles of dip (less than 12°), fine grain size, and burrowing characterize this unit. Laminated and thin-bedded beds (about 10 to 25 cm thick) are interlayered with massive, bioturbated beds (25 to over 100 cm thick). Macrofossils are rare as are microfossils, which include planktic and small benthic foraminifers (Figs. 9A, B).

### *Depositional Facies: Geometry and Distribution*

Pre-reef sediments dominate the succession at Níjar and are best studied along the northcentral part of the ridge (Figs. 3, 4). Reef facies only occur along the topographically highest south and central parts of the ridge. Off-reef Fossiliferous Packstones dominate the reef sediments; volumetrically speaking, *Porites* Framestones compose but a small percentage (less than 5%) of the preserved reef deposits.

*Pre-reef facies: Orange Wackestone and Orange Packstone Units.—*

Orange Wackestones crop out all along the lower part of the Níjar ridge. Orange Packstones occur only in the northern part of the area where they overlie the Orange Wackestones. The orange units are equivalent to the Marginal Terrigenous Complex of Dabrio et al. (1981) and Depositional Sequence I of Franseen and Mankiewicz (1991). The boundary between these two units represents a surface of downlap (Fig. 4).

The geometry of the Orange Packstone Unit is a wedge that represents about 50 m of depositional relief and attains a maximum thickness of about 35 m (Fig. 4). In general, the unit is a prograding clinoform (Mitchum et al., 1977).

Bedding within the clinoform is poorly defined in part due to bioturbation. Thickness of beds ranges from about 25 to 50 cm and dips vary from about 14° to 20°. Patterns of bedding (Mitchum et al., 1977) range from predominantly shingled to parallel oblique in which beds retain their thickness down dip and bluntly terminate against the underlying more shallow-dipping (less than 10°) Orange Wackestones.

The change from wackestone to packstone deposition probably reflects a shallowing upward trend. Thus, the Orange Packstone Unit is interpreted as a wedge-shaped unit that prograded out over the deeper water Orange Wackestones.

*Pre-reef facies: Gray Grainstone/Packstone Unit.—*

The Gray Grainstone/Packstone Unit occurs in the northern and central parts of the study area where it onlaps the wedge formed by the Orange Packstones (Fig. 4). The unit is lenticular in shape, attaining a maximum thickness of about 20 m in the north central part and thinning northward to 0 m and southward to about 7 m where talus covers the unit.

Dabrio et al. (1981) included the Gray Grainstones/Packstones within the proximal and distal reef-slope sediments. In the field and in hand sample, rocks from the gray unit are difficult to differentiate from those of the reef slope, though *Porites* and *Halimeda* do not occur in the Gray Grainstones/Packstones. Rhodoliths occur in both units but are more abundant in the Gray Grainstones/Packstones where they are mainly constructed by

*Lithophyllum*. In addition, the onlapping geometry, internal fining- and thinning-upward packages, channeling, and crossbedding characterize only the Gray Grainstones/Packstones.

The Gray Grainstones/Packstones represent a marked lithologic change from the underlying orange units. The gray unit is coarser grained and relatively mud free. Coarse grain size, channeling, crossbedding, morphology of rhodoliths, and presence of *Ophiomorpha* all indicate relatively shallow-marine or at least energetic conditions.

The fining- and thinning-upward packages characteristic of the Gray Grainstones/Packstones suggest that conditions fluctuated several times from high to relatively low energy. The onlapping nature of the contact between the Gray Grainstones/Packstones and the Orange Packstones suggests a relative rise in sea level during deposition of the grainstones (Vail et al., 1977); the fining-upward, onlapping sequence logically could have accumulate during a relative rise in sea level. I infer that channeling and deposition of the coarser-grained sediments occurred during minor low stands and that bioturbated finer sediments accumulated during relative high stands. If the Níjar crossbedding is analogous to the Canadian example of Davidson-Arnott and Greenwood (1976), depth of the water over the inferred bars at Níjar can be estimated as about 5 m. On the basis of this depth estimate, the topographically lowest occurrence of cross-sets, and the topographically highest occurrence of Orange Packstones, sea level fell a minimum of about 50 m prior to deposition of the Gray Grainstones/Packstones. The stacked and irregular arrangement of the fining- and thinning-upward packages suggests a net relative sea-level rise of about 50 m punctuated by pulses of sea-level fall. Another drop in sea level occurred following deposition of the Gray Grainstones/Packstones as evidenced by alteration attributed to subaerial exposure of the uppermost Gray Grainstones/Packstones (Fig. 6F). For a few meters below the contact in both the north and central parts of the field area, burrow linings of *Ophiomorpha* dissolved (Fig. 6G), which may be related to the exposure surface (linings are preserved in the lower parts of the Gray Grainstones/Packstones; Fig. 6E).

*Off-reef and pre-reef? facies: White Fine-Grained Packstones.—*

The White Fine-Grained Packstones represent the distal equivalents of the Gray Grainstones/Packstones as well as of the *Porites* Framestones and Fossiliferous Packstones. Perhaps more detailed microfacies analysis could differentiate the two distal facies.

The unit overlies the Orange Packstones in the south but the contact is not well exposed. In the central and northern parts of the field area, these whiter beds overlie the coarser grained sediments of the Gray Grainstones/Packstones (Fig. 4). The contact between the two units shows evidence of some erosion probably due to the lowered sea level (as evidenced from subaerial exposure as discussed above); a coarser mixed-fossil hash occurs sporadically along the contact.

*Reef and off-reef facies: Porites Framestone and Fossiliferous Packstone/Grainstone Units.—*

*Porites* Framestones crop out on the tops of the southern hills of the ridge (Fig. 3). Each successive hill to the south, and therefore each framestone outcrop, is topographically lower than the one to the north. This down-stepping distribution of reef framestones led Dabrio et al. (1981) to suggest that reefs developed and prograded basinward during sea-level fall. Previously, Esteban et al. (1978) and Esteban (1979) had recognized this pattern of reef distribution in many Miocene reef complexes of southeastern Spain, which supported the idea of regional sea-level fall.

At Níjar, the *Porites* Framestones mainly occur as promontories (pinnacles) that are less than 10 to 20 m across and less than 10 m high; they protrude down slope as salient features. No backreef facies was found; if ever present, it has been eroded. Mollusk, serpulid worm, coralline algae, and bryozoan packstones, rudstones, and floatstones fill in between and over each framestone salient.

Each pinnacle of *Porites* Framestone is laterally equivalent to a wedge of Fossiliferous Packstone/Grainstone Unit (Fig. 10) that represents the proximal reef-slope facies as described by Dabrio et al. (1981). The Fossiliferous Packstones locally interfinger with the White Fine-Grained Packstones or distal reef-slope facies. In part, however, the contact between the Fossiliferous Packstone/Grainstone and White Fine-Grained Packstone Units is not transitional; the sharp boundary between the units can appear conformable or erosional, and locally represents a surface of downlap.

Any interpretation of the *Porites* Framestones and Fossiliferous Packstones/Grainstones must address several points: (1) the geometric relationship, in which overlying sediments are both of shallower water origin and topographically lower than preceding packages of similar sediment; (2) the seemingly repetitive stratigraphy within each sedimentary wedge of the Fossiliferous Packstone/Grainstone Unit; (3) the sharp contact between the wedges; and (4) the stratigraphic "isolation" of *Halimeda* from most other biotic constituents within each wedge.

I agree with Dabrio et al. (1981) that the Níjar reef complex prograded basinward during a relative fall of sea level. The geometry of the *Porites* Framestones as well as that of successive wedges of the Fossiliferous Packstone/Grainstone Unit supports this interpretation. Dabrio et al. (1981, p. 535-536) only recognized three major episodes of progradation and suggested eustatic control. They stated, however, that this down-stepping progradation was not associated with evidence for abrupt drops of sea level.

The presence of eight identifiable sedimentary wedges in the Fossiliferous Packstones, the characteristic succession of strata within the wedges, and the sharp contact between these packstones and the underlying White Fine-Grained Packstones are consistent with high-frequency oscillations during relative sea-level fall. The sharp erosional contact between wedges is not absolute evidence of subaerial exposure or erosion, however.

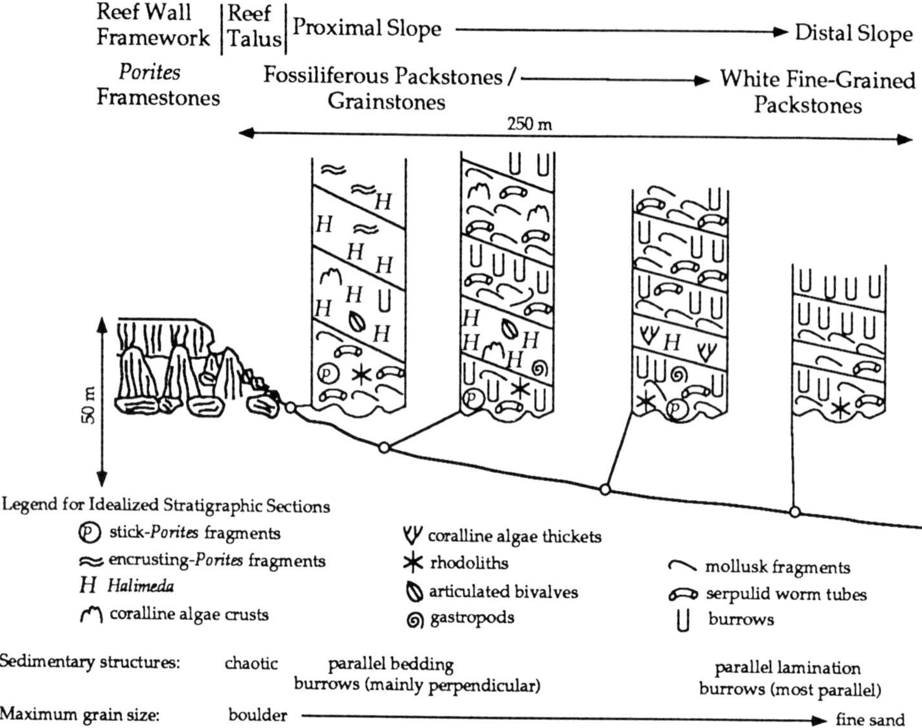


FIG. 10.—Diagram showing trends in composition, dip, sedimentary structures, and grain size along a transect from reef core to distal reef slope. Bedding shown in schematic stratigraphic sections depicts true dip (after Dabrio et al., 1981).

On the basis of the types of algae present and the vertical relief represented by the depositional surface of the Fossiliferous Packstone/Grainstone Unit, *Halimeda* may have grown or accumulated in water depths of up to 50 m, and *Mesophyllum* may have flourished at depths of 60 to 70 m (Mankiewicz, 1988). The laminar and stick-like morphotypes of *Porites* also suggest not-too-shallow water depths, or at least low-energy conditions.

With time, perhaps due to decreased nutrient supply or changes in light intensity, temperature, or grazing pressure, conditions became more favorable for growth of a more diverse assemblage of organisms including *Porites*, mollusks, a greater variety of coralline algae, serpulid worms, and bryozoans.

Episodic sea-level fall followed by a rise is the only scenario I can propose to account for all of the major characteristics of the reef facies. Because the geometries suggest a net fall in relative sea level of at least 60 to 70 m, each rise following a fall flooded the reef complex to a lower sea level than represented by the previous maximum high. Thus there was a net relative fall of sea level (of 60 to 70 m) punctuated by episodic falls and rises of less than 10 m.

## DIAGENETIC OVERPRINT

The following section presents a generalized summary of diagenesis; several inferred sea-level changes probably produced more complex, local diagenesis.

Syndepositional, subaerial, and/or shallow-burial diagenesis altered the section exposed near Níjar. Evidence of deep-burial diagenesis such as pressure solution is lacking. Study of primary Miocene fluid inclusions contained in correlative rocks at nearby Mesa Roldán (Fig. 1) corroborates a shallow-burial interpretation because the fluid inclusions have not re-equilibrated as a result of significant heating that would be expected after deep burial (Goldstein et al., 1990).

### *Petrologic Information*

Thin-section examination documents the following paragenetic sequence of diagenesis.

1. Syndepositional alteration of clasts. Abundant micritic envelopes (Figs. 5C, 8D) suggest syndepositional boring of carbonate particles by microorganisms. Macro-borings produced by bivalves and sponges aided in the breakdown of larger particles (Figs. 8A, C). Evidence of boring is most obvious in coarse-grained packstones and grainstones.

2. Syndepositional cementation. Microcrystalline carbonate is ubiquitous throughout the section (Figs. 5, 8, 9). Some appears to have a void-filling form and thus seems to be cement. Much, however, is difficult to impossible to differentiate from micrite (depositional mud). Less commonly, fibrous or bladed cement coats some of the particles (Fig. 5E). Fibrous and bladed cements are most common in the coarser, pre-reef deposits.

3. Fabric-selective dissolution. All originally aragonitic components (*Halimeda*, most mollusks, and *Porites*) have been dissolved, but micritic rims betray their occurrence (Figs. 8A, B, D). Originally high-magnesian-calcite components (foramini-

fers, coralline algae, and serpulid worm tubes) and low-magnesian-calcite fragments (pectinid bivalves, oysters, bryozoans, and brachiopods) typically display good preservation of fabric (Figs. 5F, 8C, F, 9A).

4. Dolomitization and dolomite cementation. Dolomitization may have been concurrent with fabric-selective dissolution. Because dolomite cement partly occludes moldic porosity, however, dolomitization continued after dissolution.

The entire succession shows evidence of dolomitization (Figs. 5A, B, F, 8D), but it is most prevalent in the post-reef and reef facies and in the earliest pre-reef deposits; the middle part of the section (the part deposited during net rise in sea level) retains more calcitic components, particularly in the topographically highest areas. Most dolomite is finely crystalline (less than 10 μm in diameter) and rarely attains sizes over 100 μm. Larger crystals display a clear rim surrounding a cloudy center. Crystal morphologies are euhedral to spheroidal. Some fabric-selective dolomitization is evident. For example, staining that differentiates calcite from dolomite shows that alternate layers (seasonal banding) in some coralline algae fragments are dolomitized; others are not.

5. Blocky calcite cementation. Blocky calcite cement partly occludes many pores throughout the succession. The form is suggestive of cementation by meteoric waters (James and Choquette, 1984).

6. Poikilotopic calcite cementation/replacement. Large, interlocking calcite cement crystals locally obliterate original depositional fabric and greatly diminish porosity and permeability. The poikilotopic calcite is common but patchy throughout the uppermost part of the section and may be related to subaerial exposure that halted extensive reef development in the Níjar area.

7. Pendant calcite cementation. Pendant calcite is ubiquitous but volumetrically unimportant in all coarse-grained deposits throughout the section (Figs. 8C, D). The pendant cement fluoresces bright blue-white, standing out against other depositional and diagenetic components, which produce no or pale blue fluorescence.

Selected thin sections (12) were studied with cathodoluminescence. Blocky, poikilotopic, and pendant calcite cements are nonluminescent. This finding is consistent with cementation and alteration in a shallow-marine or shallow-meteoric environment. Many calcitic skeletal and nonskeletal grains (e.g., coralline algae fragments and pellets, respectively), however, show pale orange luminescence suggesting post-depositional alteration of the original mineralogy. Dolomitized grains and dolomite cements show no to dull to moderate orange luminescence.

## POROSITY ASSOCIATED WITH THE REEF

### *Porosity Types*

Four types of porosity dominate in the Níjar succession: moldic, interparticle, intraparticle, and shelter (Figs. 5, 8, 9). Solutions have enlarged some pores; cements have partly or wholly occluded many pores. The type of porosity present is not so much dependent on the facies (pre-reef, reef, or post-reef), but rather on grain size, which diminishes with increasing distance from shore, and original skeletal mineralogy. Moldic porosity is the most important pore type and occurs throughout the section; it is most common, however, in the coarser, coralline-algae-poor deposits. Interparticle porosity ranks second in importance. Grainstones and coarser, mud-poor packstones of the pre-reef, reef-talus, and proximal reef deposits retain much interparticle porosity. Intraparticle porosity is most common in subfacies rich in coralline algae, serpulid worm tubes, bryozoans, and foraminifers. Shelter porosity is the least important and occurs in some coralline-algae-rich deposits (Fig. 8E).

### *Porosity Distribution*

Porosity ranges from a trace to about 40% (visual estimates). In general, porosity decreases with increasing distance from the reef (or from the paleoshore in the case of the pre-reef deposits). This trend relates to the grain-size distribution. The nearshore grainstones typically have moldic porosity and, due to the paucity of mud, interparticle porosity. Interparticle porosity tends to be less important in the packstones where moldic porosity dominates. The offshore finer packstones and wackestones typically have less than 5% porosity in the form of moldic and intraparticle porosity. Because cementation is more common in the reef facies compared to pre-reef facies, the pre-reef facies generally retains the greatest porosity.

Porosity in the reef core is quite variable and mainly occurs as moldic porosity. Microcrystalline cement characteristically occludes most to all of the interframework space. Because the originally aragonitic coral has dissolved and because the coral was extensively bored by clionid sponges, the stick-like *Porites* framework represents most of the nonoccluded pore space and commonly provides porous and permeable conduits.

## SUMMARY AND CONCLUSIONS

Large- and small-scale geometry, grain size, and fossil composition help differentiate seven depositional units in an upper Miocene carbonate sedimentary succession southeast of the town of Níjar, southeastern Spain. Types and abundance of calcareous algae in particular distinguish several of the units. Reefs apparently did not develop in the Níjar area during deposition of the three oldest units. Calcareous algae (*Halimeda* and *Mesophyllum*) seemed to colonize the reef slopes (Fossiliferous Packstone/Grainstone Unit); a relatively diverse biota, though depauperate compared to most reefs, succeeded these two genera of algae.

Sea-level changes likely produced many of the depositional and diagenetic characteristics of each unit. Down-lap surfaces and subaerial exposure indicate times of relative sea-level fall whereas onlapping geometries and channeling suggest periods of relative transgression. Perhaps episodic physical disturbance effected by sea-level fluctuations contributed to the characteristic low coral diversity (virtually monospecific) of the Níjar reef complex and others of similar age (Franseen and Goldstein, this volume; Esteban et al., this volume). The fluctuating sea level presumably also influenced or was responsible for stages of dolomitization and meteoric or subaerial diagenesis. Ongoing research (e.g., Lu and Meyers, 1994) that focuses on the details of diagenetic change will better refine and constrain field data summarized in this paper.

## ACKNOWLEDGMENTS

This research represents part of my Ph.D. work at the University of Wisconsin-Madison. L. C. Pray advised me throughout the project and helped in obtaining financial aid. C. V. Mendelson provided field, technical, and moral support throughout the study and reviewed the manuscript. The Comisión Nacional de Geología of Spain granted permission to undertake this research.

I thank reviewers M. Esteban, J. M. Rouchy, and C. Taberner for their insightful comments. The following institutions provided financial and/or material support: American Association of Petroleum Geologists; Geological Society of America; Tinker Foundation; Department of Geology and Geophysics, UW-Madison; Departments of Biology and Geology, Beloit College; Marathon Oil Company, Denver (particularly J. L. Wray); and ERICO International Research, London (particularly M. Esteban).

## REFERENCES

ADDICOTT, W. O., SNAVELY, P. D., JR., BUKRY, D., AND POORE, R. Z., 1978, Neogene stratigraphy and paleontology of southern Almería Province, Spain: an overview: Washington, D.C., United States Geological Survey Bulletin 1454, p. 1-49.

BOSENCE, D. W. J., 1983, The occurrence and ecology of recent rhodoliths—a review, *in* Peryt, T. M., ed., Coated Grains: Berlin, Springer-Verlag, p. 225-242.

DABRIO, C. J., ESTEBAN, M., AND MARTÍN, J. M., 1981, The coral reef of Níjar, Messinian (uppermost Miocene), Almería Province, S.E. Spain: Journal of Sedimentary Petrology, v. 51, p. 521-539.

DABRIO, C. J. AND MARTÍN, J. M., 1978, Los arrecifes Messinienses de Almería (SE de España): Cuadernos de Geología, v. 8-9, p. 85-100.

DAVIDSON-ARNOTT, R. G. D. AND GREENWOOD, B., 1976, Facies relationship on a barred coast, Kouchibouguac Bay, New Brunswick, Canada, *in* Davis, R. A., Jr. and Ethington, R. L., eds., Beach and Nearshore Sedimentation: Tulsa, Society of Economic Paleontologists and Mineralogists Special Publication 24, p. 149-168.

DRONKERT, H. AND PAGNIER, H., 1977, Introduction to Mio-Pliocene of the Sorbas basin: Málaga, Messinian Seminar 3, Field Trip 2, p. 1-21.

EKDALE, A. A., BROMLEY, R. G., AND PEMBERTON, S. G., 1984, Ichnology: Tulsa, Society of Economic Paleontologists and Mineralogists Short Course 15, 315 p.

ESTEBAN, M., 1979, Significance of the upper Miocene coral reefs in the western Mediterranean: Palaeogeography, Palaeoclimatology, Palaeoecology, v. 29, p. 169-182.

ESTEBAN, M., CALVET, F., DABRIO, C., BARÓN, A., GÍNER, J., POMAR, L., SALAS, R., AND PERMANYER, A., 1978, Aberrant features of the Messinian coral reefs, Spain: Acta Geológica Hispánica, v. 13, p. 20-22.

ESTEBAN, M. AND GÍNER, J., 1980, Messinian coral reefs and erosion surfaces in Cabo de Gata (Almería, SE Spain): Acta Geológica Hispánica, v. 15, p. 97-104.

FRANSEEN, E. K. AND MANKIEWICZ, C., 1991, Depositional sequences and correlation of middle(?) to late Miocene carbonate complexes, Las Negras and Níjar areas, southeastern Spain: Sedimentology, v. 38, p. 871-898.

GOLDSTEIN, R. H., FRANSEEN, E. K., AND MILLS, M. S., 1990, Diagenesis associated with subaerial exposure of Miocene strata, southeastern Spain: implications for sea-level change and preservation of low-temperature fluid inclusions in calcite cement: Geochimica et Cosmochimica Acta, v. 54, p. 699-704.

HAQ, B. U., HARDENBOL, J., AND VAIL, P. R., 1987, Chronology of fluctuating sea levels since the Triassic: Science, v. 235, p. 1156-1166.

HERNANDEZ, J., LAROUZIÈRE, F. D. DE, BOLZE, J., AND BORDET, P., 1987, Le magmatisme néogène bético-rifain et le couloir de décrochement trans-Alboran: Bulletin de la Société Géologique de France, Series 8, v. 3, p. 257-267.

JAMES, N. P. AND CHOQUETTE, P. W., 1984, Limestones—the meteoric diagenetic environment: Geoscience Canada, v. 11, p. 161-194.

JONES, B. AND HUNTER, I. G., 1991, Corals to rhodolites to microbialites—a community replacement sequence indicative of regressive conditions: Palaios, v. 6, p. 54-66.

LAROUZIÈRE, F. D. DE, BOLZE, J., BORDET, P., HERNANDEZ, J., MONTENAT, C., AND OTT D'ESTEVOU, P., 1988, The Betic segment of a lithospheric Trans-Alboran shear zone during Upper Miocene: Tectonophysics, v. 152, p. 41-52.

LU, F.H. AND MEYERS, W.J., 1994, Dolomitization of a late Miocene carbonate platform by brines mixing with freshwater in Nijar, Spain: Geological Society of America Abstracts with Programs, v. 26, p. A67.

MANKIEWICZ, C., 1987, Sedimentology and calcareous algal paleoecology of middle and upper Miocene reef complexes near Fortuna (Murcia Province) and Níjar (Almería Province), southeastern Spain: Unpublished Ph.D. Dissertation, Univiversity of Wisconsin, Madison, 341 p.

MANKIEWICZ, C., 1988, Occurrence and paleoecologic significance of *Halimeda* in late Miocene reefs, southeastern Spain: Coral Reefs, v. 6, p. 271-279.

MARTIN, J.M. AND BRAGA, J.C., 1993, Discussion (of Franseen and Mankiewicz, 1991, Depositional sequences and correlation of middle(?) to late Miocene carbonate complexes, Las Negras and Níjar areas, southeastern Spain): Sedimentology, v. 40, p. 351-356.

MEGÍAS, A. G., 1985, Tectosedimentary relationships between Mio-Pliocene reefs and evaporites in Almería and Sorbas basins (SE Iberian Peninsula), *in* Milá, M. D. and Rosell, J., eds., International Association of Sedimentologists, 6th European Regional Meeting, Abstracts: Lleida, Institut d'Estudis Ilerdencs, p. 292-295.

MITCHUM, R. M., JR., VAIL, P. R., AND SANGREE, J. B., 1977, Seismic stratigraphy and global changes of sea level, Part 6: Stratigraphic interpretation of seismic reflection patterns in depositional sequences, *in* Payton, C. E., ed., Seismic Stratigraphy—Aplications to Hydrocarbon Exploration: Tulsa, American Association of Petroleum Geologists Memoir 26, p. 117-133.

MONTAGGIONI, L.F. AND CAMOIN, G.F., 1993, Stromatolites associated with coralgal communities in Holocene high-energy reefs: Geology, v. 21, p. 149-152.

MONTENAT, C., OTT D'ESTEVOU, P., AND MASSE, P., 1987, Tectonic-sedimentary characters of the Betic Neogene basins evolving in a crustal transcurrent shear zone (SE Spain): Bulletin des Centres de Recherches Exploration-Production Elf-Aquitaine, v. 11, p. 1-22.

REHAULT, J.-P., BOILLOT, G., AND MAUFFRET, A., 1985, The western Mediterranean basin, *in* Stanley, D. J. and Wezel, F.-C., eds., Geological Evolution of the Mediterranean Basin: New York, Springer-Verlag, p. 101-129.

RIDING, R., MARTIN, J. M., AND BRAGA, J. C., 1991, Coral-stromatolite reef framework, Upper Miocene, Almería, Spain: Sedimentology, v. 38, p. 799-818.

SANZ DE GALDEANO, C., 1990, Geologic evolution of the Betic Cordilleras in the Western Mediterranean, Miocene to the present: Tectonophysics, v. 172, p. 107-119.

SERRANO, F., 1990, Presencia de Serravalliense marino en la cuenca de Níjar (Cordillera Bética, España): Geogaceta, v. 7, p. 95-97.

VAIL, P. R., MITCHUM, R. M., JR., AND THOMPSON, S., 1977, Seismic stratigraphy and global sea level Part 4: Global cycles of relative change of sea level, *in* Payton, C. E., ed., Seismic Stratigraphy—Applications to Hydrocarbon Exploration: Tulsa, American

Association of Petroleum Geologists Memoir 26, p. 83-98.

VEEKEN, P. C. H., 1983, Stratigraphy of the Neogene-Quaternary Pulpi Basin, provinces Murcia and Almería (SE Spain): Geologie en Mijnbouw, v. 62, p. 255-265.

VÖLK, H. R., 1967, Zur Geologie und Stratigraphie des Neogenbecken von Vera, Südost-Spanien: Unpublished Ph.D. Thesis, University of Amsterdam, Amsterdam, 160 p.

# PALEOSLOPE, SEA-LEVEL AND CLIMATE CONTROLS ON UPPER MIOCENE PLATFORM EVOLUTION, LAS NEGRAS AREA, SOUTHEASTERN SPAIN

EVAN K. FRANSEEN
*Kansas Geological Survey, University of Kansas, 1930 Constant Ave., Lawrence, KS 66047, USA*
AND
ROBERT H. GOLDSTEIN
*University of Kansas, Department of Geology, 120 Lindley Hall, Lawrence, KS 66045, USA*

ABSTRACT: Carbonate platforms in the Las Negras area (southeastern Spain) evolved from onlapping ramps to fringing reef complexes later draped by cyclic shallow marine strata. Although sea-level history and paleoclimate had an effect on platform evolution, substrate topography played a dominant role. The strata are divided into five depositional sequences of Tortonian and Messinian age. From base to top these depositional sequences are labeled as DS1A (newly dated at 8.5±0.1 Ma) mostly consisting of coarse-grained carbonates and volcanic detritus; DS1B consisting of fining upward carbonate cycles; DS2 mostly consisting of reef megabreccia clasts and fine-grained carbonates; DS3 mostly consisting of reef and forereef carbonates and some volcanic detritus; and TCC (Terminal Carbonate Complex) dominated by oolitic cyclic carbonates.

Ramp strata of DS1A and DS1B mostly onlap against volcanic basement. DS1A and lower DS1B strata have characteristics that are consistent with deposition in cool water, probably resulting from a temperate climate and possibly coinciding with upwelling. DS1B strata consist of fining-upward cycles of red algal packstone-grainstone and fine grained wackestone-packstone that suggest repeated variations of water depth from 40-100 m to greater than 100 m. Beds of both facies lap out against basement without any indication of facies change at or approaching the point of onlap. Although sediments likely were generated upslope above the point of onlap in shallow marine water, paleoslope was too steep for those sediments to accumulate permanently. Therefore, deposits of DS1B represent a type of carbonate ramp in which sediment production was in water depths of approximately 40-100 m or more and any sediments generated in shallower water upslope positions were bypassed downslope until areas of low basement paleoslope were reached. Importantly, the onlap in this type of ramp is not necessarily tracking baselevel. The fining-upward character of DS1B cycles may result from an overall high or rising sea level punctuated by higher frequency fluctuations, lower sedimentation rates related to a more temperate climate, or bypass into the distal toe-of-slope setting. The sediments that accumulated during DS1B filled in significant basement topography and created a more gently sloping substrate on which later deposits could accumulate.

Unlike DS1A and most of DS1B, the last phase of DS1B, DS2, and DS3 contain significant amounts of chlorozoan facies indicating a shift toward a more subtropical to tropical climate. Normal marine platform deposits of DS2 essentially draped the gentle paleotopography created by DS1B deposition, but megabreccias composed mostly of reef clasts indicate instability in upslope areas due to steep paleoslopes and relative sea-level falls. During deposition of the *Porites*-dominated fringing reef strata of DS3, sea level was at a position near the tops of relatively flat hills created by earlier deposition. Reef aggradation, progradation and downstepping in DS3 created steep forereef slope topography. These reefal platforms were preserved only on substrates of relatively low paleoslope, where earlier deposits had filled in the steep substrate topography. In contrast, steep-sided volcanic hills in the area were sites where DS3 and DS2-equivalent reefs may have been formed but were bypassed downslope because of the steep paleoslope. Finally, during deposition of latest Miocene strata (TCC), sea level was at a position that allowed for accumulation of shallow-water carbonate cycles, but these cycles were only preserved on areas where the substrate, created either by earlier deposition or erosion, was relatively flat.

The model of platform evolution provided by the Las Negras area exposures likely has applications for other Miocene carbonate complexes in the Mediterranean and could apply to other carbonate complexes in the rock record.

## INTRODUCTION

The Upper Miocene carbonate and siliciclastic strata exposed along the Mediterranean coast near Las Negras, southeastern Spain (Fig. 1) were deposited prior to, during, and after the well-known Messinian "salinity crisis" events in the Mediterranean. These exposures can be used to illustrate paleoslope, sea-level, and climate controls on platform morphology. Tortonian and Messinian carbonate deposition on the earlier Neogene calc-alkaline volcanic substrate in this area started as generally onlapping, non-rimmed foramalgal platform (ramp) strata that evolved upward into a chlorozoan-dominated reef-rimmed platform characterized by *Porites* fringing reefs. Final Miocene (Messinian) deposition consisted of predominantly cyclic stromatolite-oolite grainstone platform strata. The carbonate complex can be divided into five depositional sequences (DS1A, DS1B, DS2, DS3, and TCC) that developed in response to quantifiable relative sea-level fluctuations (Fig. 2; Franseen, 1989; Goldstein et al., 1990; Franseen and Mankiewicz, 1991; Franseen and Goldstein, 1992; Goldstein and Franseen, 1993; Franseen et al., 1993; Goldstein and Franseen, 1995). These previous studies focused on developing the sequence-stratigraphic framework for the carbonate complex, documenting relative sea-level fluctuations and constructing quantitative relative sea-level curves for the sequences.

This paper focuses on the controls on evolution of the carbonate complex and presents a new model that incorporates sea level, paleoslope, and climate as controlling factors on type and location of sediments and depositional sequences throughout platform development. Most studies of Upper Miocene strata in southeastern Spain have focused on details of the fringing reef strata alone. The excellent exposures in the Las Negras area provide one of the few locations in which the entire platform evolution can be documented. In this paper we show that the

Models for Carbonate Stratigraphy from Miocene Reef Complexes of Mediterranean Regions,
SEPM Concepts in Sedimentology and Paleontology #5, 
SEPM (Society for Sedimentary Geology), ISBN 1-56576-033-6, p. 159-176.

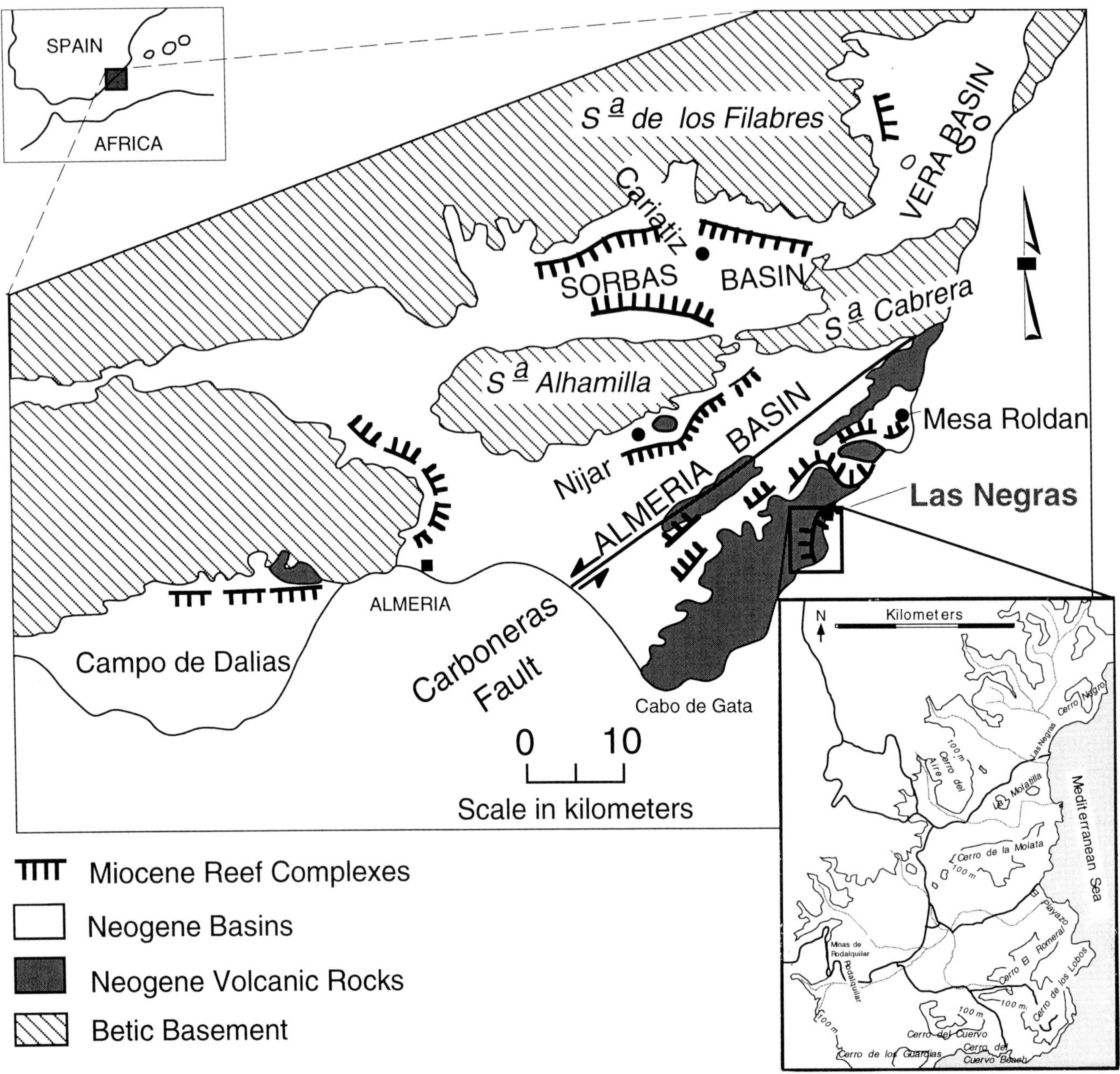


FIG. 1.—Location and general geologic map showing distribution of Upper Miocene reefs in southeastern Spain. Lower inset topographic map shows the location of the hills in the Las Negras area (La Molatilla, La Molata, El Romeral, Cerro del Cuervo). Modified from Dabrio et al. (1985) and Franseen et al. (1993).

earlier-deposited ramp strata in that area represent an important part of the carbonate complex, that their location was controlled largely by volcanic basement paleoslope, and that they were important in development and location of the later fringing reef strata.

## GEOLOGIC SETTING

The study area is near the small coastal village of Las Negras in the Cabo de Gata region on the southern margin of the Almeria basin (Fig. 1). This area is within a shear zone and is bordered on the northwest by the major sinistral strike-slip Carboneras Fault (Fig. 1) that was variously active throughout Tortonian-Messinian and Pliocene time (Montenat et al., 1987). This fault separates the Cabo de Gata region of Neogene volcanic basement to the southeast from Mesozoic-Paleozoic basement of the Betic Range to the northwest. In the Cabo de Gata region, calc-alkaline volcanism occurred from about 17 Ma to 6 Ma (Lopez-Ruiz and Rodriguez-Badiola, 1980; Serrano, 1992). Except for an occurrence of interbedded carbonate and volcanic strata in the

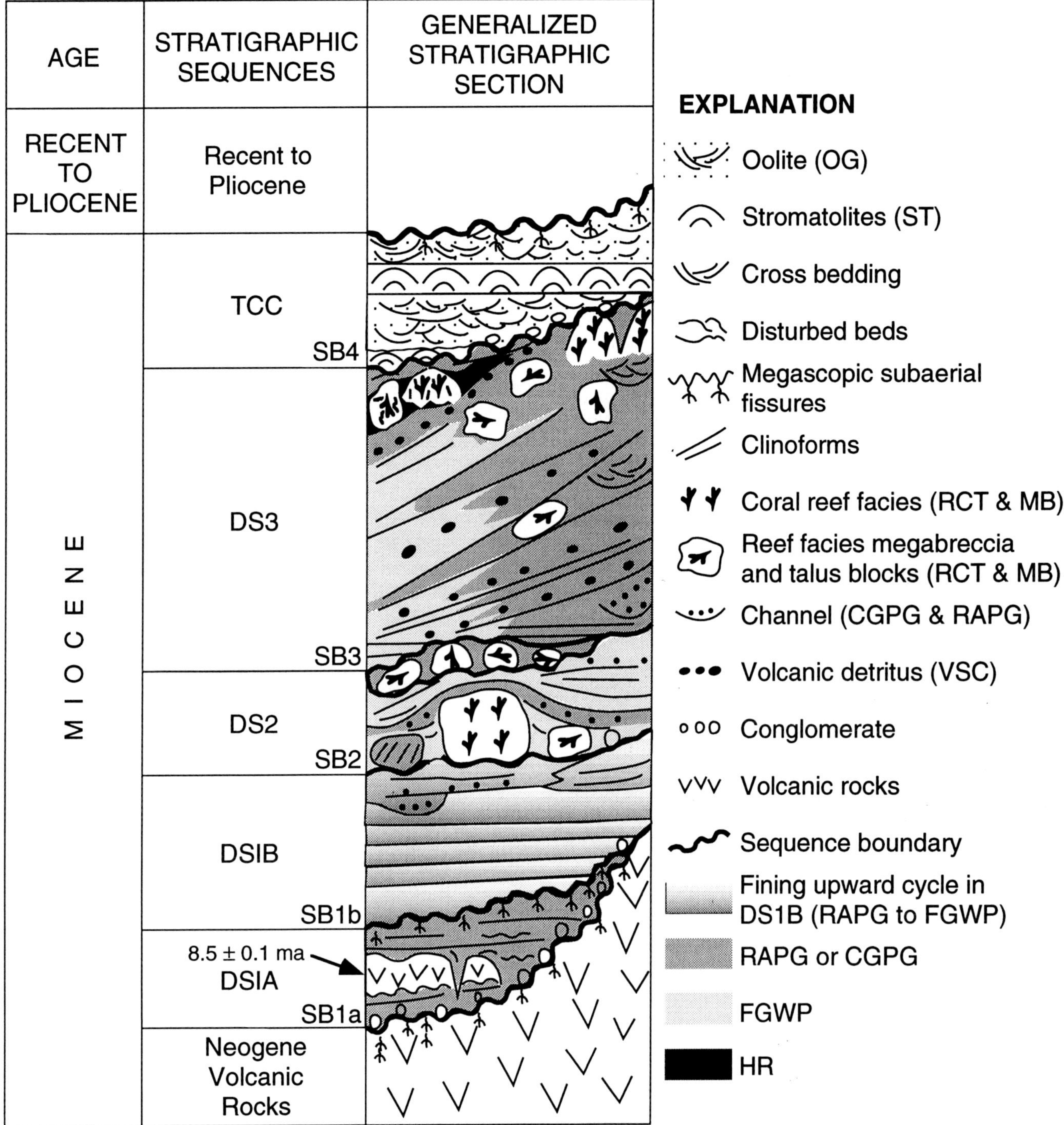


FIG. 2.—Generalized stratigraphic section of the depositional sequences in the Las Negras area. Note the interbedded volcanic unit in DS1A that yielded $^{40}Ar/^{39}Ar$ dates of $8.7 \pm 0.1$ Ma, $8.7 \pm 0.3$ Ma, and $8.5 \pm 0.1$ Ma (see Fig. 6). Symbols and patterns illustrate generalized facies relationships. See text for facies descriptions. OG=Oolitic Grainstone, ST=Stromatolite Boundstone, RCT=Reef Core/Talus, MB=Megabreccia, CGPG=Coarse-Grained Packstone/Grainstone, RAPG=Red Algal Packstone/Grainstone, VSC=Volcanic Sandstone/Conglomerate, FGWP=Fine-Grained Wackestone/Packstone, HR=*Halimeda* rich.

lowest part of the section (Franseen and Goldstein, 1992; Franseen et al., 1993), volcanism mostly predated deposition of the marine carbonates in the Las Negras study area. The volcanic highs created an archipelago (Esteban and Giner, 1980; Esteban et al., this volume) and formed the substrate for the later carbonate strata. The present-day exposures in the area result from regional uplift and Mediterranean sea-level drop since Pliocene deposition. They preserve Miocene topography and have not been altered significantly by tectonic tilting (Franseen and Mankiewicz, 1991; Franseen et al., 1993) as evidenced by regional mapping, numerous geopetal fabrics, stratal geometries, and lack of significant displacement on faults in the carbonate strata. The exposures on several hills near Las Negras (La Molata, El Romeral, and Cerro del Cuervo) represent some of the most complete carbonate sections in the Cabo de Gata area and are the focus of this paper (Fig. 1).

## LITHOFACIES

The data used in this paper were derived from detailed field studies that included mapping and measurement of stratigraphic sections (Franseen, 1989). Additional lithologic and diagenetic details were obtained from petrographic studies using standard petrographic and cold cathodoluminescence techniques.

Ten general lithofacies have been distinguished for the carbonates near Las Negras based on grain type, grain size, presence or absence of framework building organisms, bedding geometries and stratigraphic position. Figure 2 illustrates the generalized distribution of lithofacies in the depositional sequences. More detailed documentation of lithofacies and their distribution within the sequences can be found in Franseen et al. (1993).

### *Fine Grained Wackestone-Packstone (FGWP)*

FGWP occurs throughout the Las Negras carbonate complex, but it is much more common in DS1B and DS2. FGWP has a characteristic very pale orange (10YR8/2) color in outcrop and typically weathers recessively. Primary dips are generally less than 10°. This facies is characteristically massive with original bedding destroyed by intense burrowing.

FGWP is composed predominantly of carbonate mud and silt to very-fine sand size grains. Identifiable components of this facies include planktonic foraminifera (apparently the dominant coarser grains), peloids, red algal fragments (mostly crustose), small bivalves (whole shells and broken fragments), echinoderm fragments, gastropods, minor volcaniclastic grains and some fish fragments as well as unidentified skeletal fragments.

### *Coarse Grained Packstone-Grainstone (CGPG)*

Although this facies occurs throughout the entire section, it is most abundant in DS3. CGPG generally is resistant to weathering and forms ledges and cliffs; it typically has a light brown (5YR6/4) color. Dominant grain sizes for CGPG are medium-coarse sand, granule-pebble and some cobble sized grains. Thin to very-thick beds (and laminations) are generally well-developed. Primary dips range from less than 10° to more than 35°; the higher-angle dips occur in DS3. CGPG also occurs as channel-fill deposits.

The predominant grain types are mollusc and red algal fragments. Biotic grain types include bivalves (whole shell and broken fragments), red algae (including articulated and crustose fragments and rhodoliths), echinoderm fragments, bryozoan fragments, gastropods, serpulid worm tubes, coral fragments, planktonic and benthic foraminifera, and brachiopods. Significant amounts of *Halimeda* plates occur locally only in youngest DS3 strata. Non-skeletal carbonate grains include peloids, coated grains, intraclasts, and volcaniclastic grains.

### *Red Algal-Rich Packstone-Grainstone (RAPG)*

RAPG is differentiated from CGPG because of the abundance of red algae in DS1A and DS1B. RAPG commonly forms resistant ledges and cliffs and has a light brown color (5YR6/4). Primary dips are generally less than 12°. Grains are coarse sand to granule size and locally are in a finer-grained (silt to medium sand) carbonate "matrix". Beds are either massive, possibly indicating extensive reworking of the sediment by burrowing organisms, or laminated to cross stratified indicating reworking and transport by currents. Other features indicating transport include normally graded beds, horizontal planar lamination, modified wave or current ripple lamination, flame structures, and volcanic or carbonate clasts. Red algal fragments (mostly crustose) are the predominant grain type; rhodoliths are common in channels and in layers within beds. Minor grain types include fragments of solitary coral, bryozoans, bivalves (some large whole pectinids), gastropods, benthic and planktonic foraminifera, echinoderms, serpulid worm tubes, peloids, intraclasts and volcaniclastic grains.

### *Volcaniclastic-Carbonate Conglomerate (VCC)*

VCC occurs directly overlying the volcanic basement and is typically poorly exposed. Volcanic clasts, similar to the underlying volcanic basement, are the most characteristic feature of this facies. Clasts are angular to well-rounded and sizes range from sand to boulder. Clasts are chaotically oriented and have a clast-support texture or locally are supported in carbonate matrix. In some localities oxidized rinds rim clasts and are associated with concentric cracks, fissures, and autoclastic breccia. These features provide evidence for subaerial alteration of the volcanic rocks before deposition of overlying marine carbonates.

Carbonate "matrix" consists of carbonate mud, and grains that include bryozoans, bivalves (including some whole pectins), solitary corals, crustose red algae fragments, rhodoliths, gastropods, serpulid worm tubes, planktonic and benthic (e.g., *Heterostegina*) foraminifera and echinoderms.

### *Volcaniclastic Sandstone-Conglomerate (VSC)*

VSC occurs in DS2 and DS3 but is most abundant in DS3. VSC generally forms recessive intervals and occurs in strata with primary dips that are essentially flat to approximately 30°. Bedding commonly is obscured, but thin to thick beds defined by alternating finer and coarser grained layers are identifiable in some places.

The most characteristic feature of the VSC is the abundance of volcanic clasts mixed with minor amounts of carbonate material.

Most volcanic clasts are coarse sand to cobble size. The larger clasts are typically round to sub-angular, and the finer clasts are sub-angular to angular. Many of the clasts have been encrusted by marine organisms. Carbonate "matrix" predominantly is composed of coral (*Porites*) fragments, serpulid worm tubes, bryozoan fragments, crustose and articulated red algae fragments, rhodoliths, bivalve fragments, and gastropods. Micrite and peloids occur in minor amounts.

### *Megabreccia*

*Megabreccia 1 (MB1).—*

MB1 is a distinctive carbonate megabreccia that marks the base of DS2. It is characterized by isolated or groups of two or more allochthonous reef-facies clasts and debris in channelized or tabular bodies. Volcanic, grainstone and wackestone clasts are admixed within MB1. The clasts show signs of transport (tilted geopetal fabrics, scoured bases), and encasing and underlying lithologies show soft sediment deformation structures and incorporation into the megabreccia. MB1 clasts are encased in FGWP or CGPG facies. Locally, the MB1 unit is capped by a normally graded, fine-grained carbonate layer. Elsewhere individual clasts or groups of several clasts have steep-angled, coarse-grained carbonate beds on their flanks.

Clasts range in size from decimeters to a few 10's of meters; most are meters to several meters in diameter. Most clasts are well rounded and only rarely exhibit clast-support textures. A distinctive and important characteristic of the reef clasts is the abundance of both *Tarbellastraea* and *Porites* corals (tabular, horizontal platy, stick, and head morphologies) as the dominant framework organisms. Corals within the clasts are commonly encrusted by micrite, foraminifera, red algae, serpulid worms and bryozoans. Micritic and fibrous-to-bladed cements (interpreted as marine) also are important in some clasts. Borings by algae (micritic envelopes), vermitid gastropods, pholad bivalves, and clionid sponges are common in the framework.

Matrix in the clasts of coral framework is volumetrically more abundant than coral framework and consists of micrite, non-skeletal grains, and skeletal material. The skeletal constituents include planktonic and benthic foraminifera, serpulid worm tubes, crustose and articulated red algae fragments, rhodoliths, bivalves, gastropods, echinoderms, fish fragments and minor occurrences of dasycladacean algae(?) and barnacles. Non-skeletal grains are peloids, composite grains and intraclasts. Volcaniclastic grains are locally admixed in the matrix.

*Megabreccia 2 (MB2).—*

MB2 is a carbonate megabreccia that marks the base of DS3 in distal slope locations. Clast size ranges from coarse sand to boulder (up to several meters). The most conspicuous feature is the local occurrence of reef-framework clasts and coarse grained, reef-derived fossils. *Porites* coral (horizontal platy and stick morphologies) is the exclusive framebuilding coral in MB2 clasts. The coarser grained facies containing reef derived detritus also contain abundant coral specimens of only *Porites*.

The *Porites* framework in the clasts in places is encrusted by micrite, red algae, serpulid worms and foraminifera. Micritic and fibrous-to-bladed cements (interpreted as marine) locally are important in some clasts. Thick micrite coatings on *Porites* are interpreted by Riding et al. (1991) as cyanobacterial stromatolites. These micritic coatings, and matrix within coral framework clasts, are volumetrically more abundant than coral framework. Matrix consists of red algae (fragments and some rhodoliths), serpulid worm tubes, bryozoans, foraminifera, gastropods, bivalves, echinoderms, micrite, peloids, intraclasts, composite grains and some volcaniclastic grains. Similar to MB1, MB2 occurs as channelized deposits, as individual clasts or groups of several isolated clasts or as tabular layers. The MB2 clasts are encased in FGWP or CGPG facies, locally incorporating clasts of the underlying facies. In some localities the MB2 unit is capped by a normally graded, fine-grained carbonate layer. Elsewhere individual clasts or groups of several clasts developed steep-angled, coarse-grained carbonate beds on their flanks.

### *Reef Core or Talus (RCT)*

The uppermost carbonate deposits of DS3 contain in-place reef core and reef talus clasts. RCT is mostly massive to vaguely bedded. Some horizontal fabrics (bedding planes?) may indicate different stages of reef growth. *Porites*, almost exclusively, is the framebuilding organism. Minor amounts of *Tarbellastraea* and *Siderastraea* corals occur but do not appear to be major framebuilders. Locally, serpulid worm tubes are clustered together and likely were forming a rigid framework. Some red algal boundstone "crusts" apparently were binding sediment. Encrusters such as red algae, serpulid worms, foraminifera and bryozoans enhanced framework rigidity of *Porites*. Micritic crusts interpreted as cyanobacterial by Riding et al. (1991) also added to framework stability. The framework shows extensive borings by algae (micritic envelopes), clionid sponges, pholad bivalves and vermitid gastropods. Matrix that infills primary interframework porosity includes micrite, peloids and skeletal material. Biotic grains include serpulid worm tubes, red algae (fragments and rhodoliths), bivalves, foraminifera (mostly benthic), gastropods, bryozoans, echinoderms, and more minor occurrences of green algae (dasycladaceans, *Halimeda*) and fish fragments. Non-skeletal carbonate grains include coated grains, intraclasts and composite grains. Volcaniclastic grains are variously admixed in the matrix.

### *Halimeda-Rich (HR) Facies*

HR facies typically consists of packstone and minor wackestone in which *Halimeda* plates are the dominant grain type. This facies occurs in only the youngest strata of DS3 in reef core, reef talus, and proximal-to-distal foreslope settings. In reef core and reef talus, *Halimeda* plates occur as "matrix" in inter-framework areas. *Halimeda* also occurs as wackestone, packstone and grainstone of the foreslope, and in some beds, is the only

FIG. 3.—Aerial photograph of the Las Negras area looking to the east. The darker hills in the foreground consist of volcanic rocks lacking any carbonate cover. The lighter hills in the background (arrows) consist of volcanic rocks covered by late Miocene carbonate rocks. From left to right the arrows point to the hills of La Molatilla, La Molata, and El Romeral respectively. Photograph courtesy of A. Arribas Jr.

recognizable grain in the rock.

HR typically is massive to thick bedded, some obscure very thin to thin bedding, and lamination (some defined by oriented *Halimeda* plates) occurs as well. In some locations intense bioturbation is apparent. Predominant grain size is coarse sand to granule. Associated grains include crustose and articulated red algae (both as fragments and encrusting layers), serpulid worm tubes, bivalves, solitary corals, bryozoans, gastropods, *Porites* fragments, and volcaniclastic grains. *Halimeda* plates either are aligned parallel to bedding, randomly oriented, or concentrated in "pockets". Some normally graded beds of *Halimeda* plates occur. Most of the *Halimeda* plates (originally aragonitic) are now molds with their shape preserved by micritic envelopes.

### *Oolite Grainstone (OG)*

Oolite grainstone is the most common facies in the Terminal Carbonate Complex and occurs in each of the four TCC cycles. This facies is characterized by thick trough-crossbed sets. Ooids predominantly have skeletal and peloidal nuclei and are variably mixed with fragmented skeletal grains, peloids and volcanic detritus. Some peloids and skeletal grains show only superficial coatings. Fenestrae occur in the uppermost part of oolitic units, at the tops of cycles in the TCC.

### *Stromatolite Boundstone (ST)*

Stromatolite boundstone (up to about 1.5 m thick) occurs at the bases of three of the four TCC cycles and immediately overlies sequence or cycle boundaries. Several of the stromatolite horizons are laterally continuous along the base of the cycles on La Molata. The stromatolites typically occur as wavy laminated beds to laterally-linked hemispheroids. Local digitate and domal stromatolite morphologies are also developed. The stromatolite boundstones typically are overlain by packstones or oolitic grainstones.

## DEPOSITIONAL HISTORY

### *Introduction*

The Las Negras carbonate complex can be divided into five depositional sequences (DS) that we label from bottom (oldest) to top (youngest) DS1A, DS1B, DS2, DS3 and TCC (Terminal Carbonate Complex after Esteban, 1979, Esteban and Giner, 1980, and Esteban et al., this volume; Fig. 2). Each of these five depositional sequences and the sequence boundaries that separate them are interpreted to have formed from several different magnitudes of relative sea-level fluctuation (Franseen, 1989; Goldstein et al., 1990; Franseen and Mankiewicz, 1991; Franseen

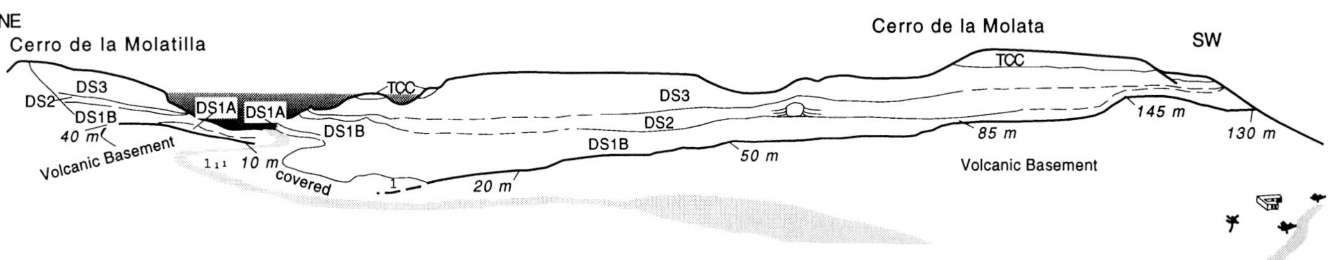


FIG. 4.—Photograph and line drawing showing details of the carbonate complex as exposed on La Molata and La Molatilla. DS1A, DS1B, DS2, DS3, and TCC are the five depositional sequences identified in the carbonate complex. Although orientation of the exposures vary relative to the true strike of the strata and some of the geometries may be only apparent, the elevations listed along the volcanic-carbonate contact demonstrate the flattening of the volcanic basement slope in distal directions, especially as traced from the high point on La Molata (SW end) across the rambla (stippled pattern) and back up on La Molatilla. This relationship is also demonstrated by exposures on El Romeral (see Fig. 8). Note one-story building for scale.

and Goldstein, 1992; Goldstein and Franseen, 1993; Franseen et al., 1993; Goldstein and Franseen, 1995).

### *Basement*

Basement in the area consists of isolated volcanic highs that formed in association with development of the Rodalquilar caldera that initially developed at approximately 11 Ma (Rytuba et al., 1990). The caldera is oval shaped and about 8 km in widest dimension and 4 km at its most narrow dimension. Volcanic rocks associated with the caldera range between andesitic and rhyolitic compositions (Rytuba et al., 1990). Radiometric dating indicates the youngest known volcanic rocks of the local sequence may be as young as 7.5 my (Di Battistini et al., 1987) which generally is corroborated by biostratigraphic data from underlying and overlying sedimentary beds (Serrano, 1992). Deposition of marine carbonates and sandstones occurred in the area after deformation of the caldera as evidenced by marine sediments within the caldera that are largely undeformed (Franseen, 1989; Rytuba et al. 1990; Franseen and Mankiewicz, 1991).

In the Las Negras area, there are volcanic hills that have a carbonate cover and bare volcanic hills that lack a carbonate cover (Fig. 3). Many of the bare volcanic hills are the highest in the area and preserve the steepest slopes (typically higher than 15°). For some of the bare volcanic hills, Miocene carbonates (that dip away from the crests of the hills) are present at or near the base of the hills, arguing against strong erosional steepening of the slopes since the Miocene. For these hills (Cerro del Aire, Cerro de los Lobos, Cerro de las Guardias; Fig. 1) slopes calculated between the outcrop of Miocene carbonate (at the bases of the hills) and the crests of the volcanic hills record maximum slopes of 13° to 17°. The overall steep slope preserved today must therefore approximate the minimum Miocene basement paleoslope. For hills containing a carbonate cover, there is consistent evidence that the volcanic basement surface had flat areas, perhaps resulting from terracing by wave planation (Esteban and Giner, 1980; Franseen, 1989; Franseen and Mankiewicz, 1991). Several of these hills have basement surfaces with ~100-130 m of relief over a lateral distance of ~1 - 0.5 km, yielding overall basement paleoslope angles of 6° to 15°, with the slopes generally becoming flatter as traced downslope (Fig. 4). Thus, it is likely that the location of carbonate cover is related to slope angle on volcanic basement. Steep-sided volcanic highs preserved no carbonate cover because produced carbonate was bypassed downslope, but volcanic highs that had areas with lower slope accumulated a carbonate cover.

### *DS1A*

DS1A is the lowest depositional sequence and ranges in thickness from 0 to approximately 22 m. The lower sequence boundary (SB1a) occurs at the contact between volcanic basement and marine carbonate. In some places, this contact is marked by a polymict volcanic conglomerate (VCC). The top of the volcanic basement rocks contain subvertical fissures several centimeters wide, vugs 10-20 cm wide, autoclastic breccia and concentric cracks that wrap around large volcanic clasts (Franseen et al., 1993). In some areas, this evidence for subaerial exposure along the basal volcanic contact can be traced from 180 m above sea level down to present-day sea level.

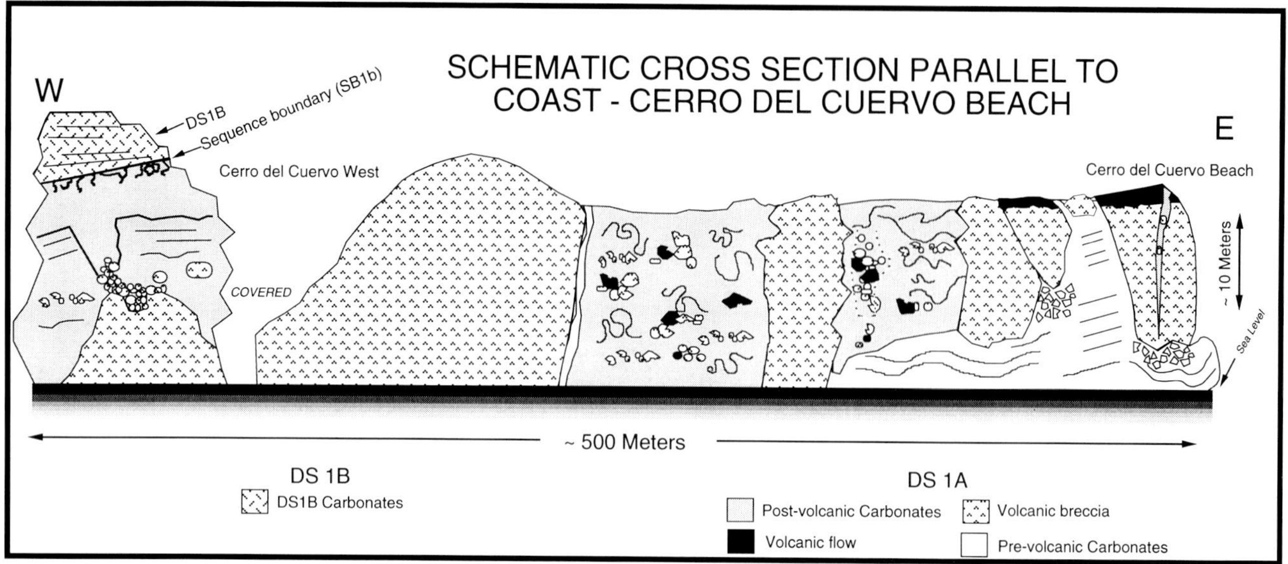


FIG. 5.—Schematic cross section of DS1A, SB1b, and DS1B relationships as exposed along the coast in the Cerro del Cuervo Beach and Cerro del Cuervo Beach West locations (see inset topographic map in Fig. 1 for location).

DS1A consists mostly of shallow-dipping (< 10°), normal marine strata that were deposited on a shallow-water ramp. For the most part, DS1A strata onlap volcanic basement up to approximately 30 m on the paleoslope and are massive to planar bedded with some low-angle crossbedding developed. Basal DS1A beds are predominantly CGPG facies consisting mostly of fragments of bivalves, gastropods, bryozoans, benthic foraminifera, echinoderms and various amounts of admixed volcaniclastic grains. A 10- to 15-m-thick volcanic breccia and overlying volcanic flow is interstratified with DS1A marine carbonates, apparently only at the Cerro del Cuervo Beach locality (Fig. 5). This volcanic unit was extensively eroded into coastal headlands shortly after deposition, likely by wave erosion in a very shallow marine environment. This erosion formed deep reentrants in the volcanic deposit that were subsequently filled by later DS1A carbonates and clasts eroded off of the volcanic unit itself (Fig. 5). Upper DS1A deposits consist of CGPG, RAPG and VSC facies.

The upper sequence boundary (SB1b) can be traced along 25-30 m of relief over a lateral distance of 150-700 m, from present day sea level upslope to where DS1A onlaps volcanic basement and SB1b merges with SB1a.

The volcanic unit interbedded within DS1A carbonates provides the first opportunity for an absolute date for the carbonate sequences in this area, specifically for DS1A strata. $^{40}Ar/^{39}Ar$ dating was accomplished by Larry Snee (USGS) on plagioclase phenocrysts and groundmass of the volcanic unit (Fig. 6 ). Two of the three groundmass samples produced excellent step heating plateaus. One sample, from a clast in the volcanic breccia at the base of the interbedded volcanic unit (Fig. 5), yielded an age of 8.7±0.1 Ma (Fig. 6A). Another groundmass sample from the volcanic breccia had problems with excess argon. The low age intermediate temperature step yields a maximum estimate for the extrusion age of 8.7 ± 0.3 Ma (Fig. 6B). A sample of groundmass from the volcanic flow at the top of the unit (Fig. 5) yielded an age of 8.5 ±0.1 Ma (Fig. 6C). Plagioclase phenocrysts were run from all samples but did not produce valid dates because of their low potassium content. Because clasts in the volcanic breccia may have been reworked from slightly older deposits, we interpret the age of the volcanic unit as 8.5±0.1 Ma. These new dates are the first reliable age control obtained for the five carbonate depositional sequences in the Las Negras area, which has remained controversial up until now (Franseen and Mankiewicz, 1991, 1993; Martin and Braga, 1993). The dates on the interbedded volcanic unit now indicate clearly that DS1A deposition started well within the Tortonian, as opposed to latest Tortonian or Messinian time (Martin and Braga, 1993), which is generally consistent with biostratigraphic data in the area reported by Serrano (1992).

The facies of DS1A closely match the rhodalgal facies of Carannante et al. (1988) or bryomol and foramol facies of Lees and Buller (1972), indicating deposition likely in temperate conditions (Franseen et al., in press). Importantly, significant amounts of chlorozoan grains were not identified. Similar, time equivalent strata occur in surrounding local areas and regionally supporting the interpretation that the Mediterranean was experiencing more temperate conditions (Franseen et al., in press). Sedimentologic features and biota in DS1A indicate deposition in alternating inner and outer ramp settings (Franseen et al., in press). The interbedded volcanic mass flow deposit that was eroded into coastal headlands indicates erosion in relatively shallow water.

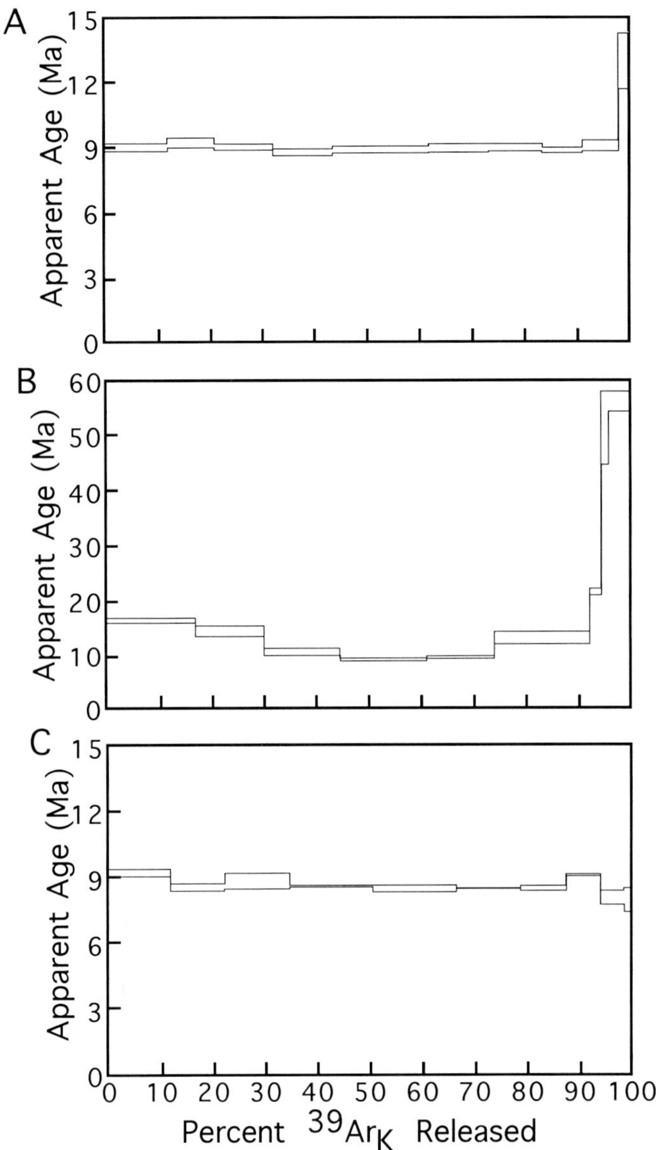


FIG. 6.—$^{40}Ar/^{39}Ar$ step heating runs for samples of volcanic groundmass from which phenocrysts of plagioclase and pyroxene had been removed. (A) Interior of clast in volcanic breccia from lower part of volcanic unit. Despite minor excess $^{40}Ar$, an age of 8.7±0.1 Ma is indicated. (B) Groundmass from volcanic breccia, lower part of volcanic unit exhibiting significant excess $^{40}Ar$. The low-age intermediate temperature step suggests a maximum extrusion age of 8.7±0.3 Ma. (C) Groundmass from volcanic flow at top of volcanic unit. Despite minor excess $^{40}Ar$ or minor $^{39}Ar$ recoil, the age indicated is 8.5±0.1 Ma.

FIG. 7.—Outcrop photograph of portion of typical fining upward cycle in DS1B. White line is drawn along the base of a cycle. The base of the cycle is underlain by fine-grained wackestone-packstone (FGWP) of the underlying cycle. It is overlain by cross-bedded and burrowed red algal packstone-grainstone (RAPG) which fines upward to FGWP. Pen in center of photograph is 14 cm.

FIG. 8.—Photograph of the Big Foot location of El Romeral. Arrows point to the contact between volcanic basement and the carbonate complex containing the depositional sequences (DS1A, DS1B, DS2, DS3, TCC). Note the flattening of the volcanic basement paleoslope in a distal direction, similar to the relationship shown in Figure 4. These exceptional exposures reveal onlap of DS1A strata and the six cycles within DS1B at the toe of the previously subaerially exposed volcanic basement paleoslope. Deposition of these lower two sequences filled in much of the pre-existing relief, creating a gentler paleoslope in this location during deposition of DS2 and DS3 strata.

### *DS1B*

DS1B strata range from 20 to more than 100 m thick and can be traced from present-day sea level up to a present-day elevation of about 150 m. Lower strata of DS1B beds were deposited on a surface of subaerial exposure at the top of DS1A and volcanic basement. Most of DS1B represents deposition on an open marine ramp.

Typically, beds of red algal packstone-grainstone (RAPG) fine upward and grade to fine grained wackestone-packstone (FGWP) on a scale of 2 to 15 m (Fig. 7). Six of these fining upward cycles were identified at El Romeral. Similar cycles are apparent on La Molata and Cerro del Cuervo. Both RAPG and FGWP facies of the cycles lap out upslope against volcanic basement (Figs. 2, 8, 9) with no evidence of significant facies change, including no evidence of a transition to shoreline or shallower water facies. This significant relationship suggests that the onlap may be controlled by a mechanism unrelated to baselevel shift (see later interpretation).

RAPG portions of cycles are generally laminated to cross stratified indicating current reworking. Other features indicat-

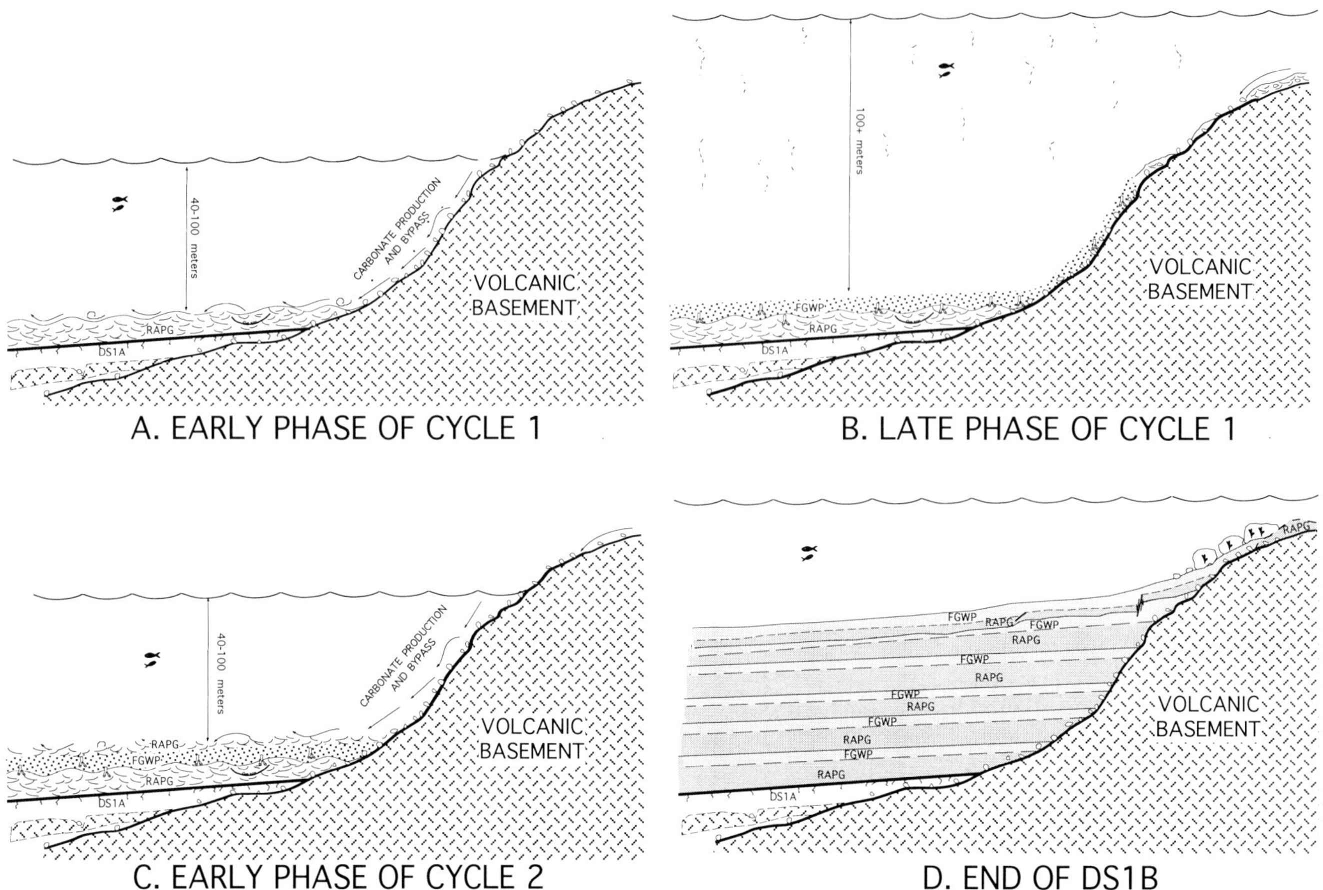


FIG. 9.—Schematic illustration of DS1B cycle deposition demonstrating proposed model for bypass ramp development. (A) Early phase of Cycle 1. During the initial phases of cycle deposition, sea level is interpreted to have been relatively low with water depths between approximately 40 and 100 m for red algal packstone and grainstone (RAPG) deposition. Carbonate production took place upslope but could not accumulate permanently. These deposits were bypassed downslope until they reached an area of low paleoslope where they were also reworked by currents forming crossbeds and some channels. (B) Late phase of Cycle 1. Cycle 1 deposition continued as the water deepened to 100 m or more. This deeper water was dominated by fine-grained wackestone and packstone (FGWP) deposition containing a significant component of pelagic sediment and characterized by abundant burrows (branched symbol on diagram). Red algal sediment likely was produced in shallower water but bypassed downslope. Some FGWP probably was deposited on the steep slope, only to be reworked into deeper water later during the start of the next cycle. (C) Early phase of Cycle 2. As sea level fell, sediments previously deposited on steep basement paleoslope were reworked and bypassed into deeper water areas of low paleoslope. During times of relatively low sea level RAPG deposition began on areas of low and high paleoslope but was bypassed across the areas of high paleoslope. (D) End of DS1B. Six complete fining upward cycles (RAPG to FGWP) were deposited during DS1B (a total of 20 to 30 m in thickness). Note minor synsedimentary faulting occurred in cycle 6. The cycles onlap against basement without facies change and reflect the bypass of sediments onto areas of low paleoslope. The record of onlap is not necessarily directly related to base level. Note that after initial accumulation in upslope locations, the upper DS1B reef deposits were eroded and reworked downslope during an interpreted relative sea level fall (Franseen et al., 1993). The relatively flatter paleoslope created by deposition of DS1A, DS1B, and DS2, and an interpreted shift to a more tropical climate allowed for accumulation of chlorozoan reefs of DS3.

ing transport include normally graded beds, horizontal planar lamination, modified wave or current ripple lamination, soft sediment deformation, volcanic or carbonate clasts, and evidence of mass movement such as detachment, slumping, conglomerates and breccias. Carbonate clasts composed of FGWP and RAPG facies occur locally throughout much of DS1B, whereas clasts of reef facies composed of *Tarbellastraea* and *Porites* corals are confined to uppermost DS1B strata. Many DS1B beds have evidence of burrowing that modified original depositional textures.

FGWP portions of cycles are composed predominantly of mud to silt-sized particles which are mostly unidentifiable. It appears that at least in some beds, the predominant coarser carbonate grains are planktonic foraminifera. The strata are typically intensively burrowed. Some FGWP beds contain interbedded RAPG and CGPG facies and conglomerates. Some of these RAPG and CGPG beds are characterized by eroded, scoured bases, rip-up clasts, Bouma ABC (D?) units, normal grading, and soft sediment deformation. The conglomerates are characterized by scoured bases, carbonate and volcanic clasts admixed with carbonate grains, fragments and matrix, and by both matrix and clast support textures.

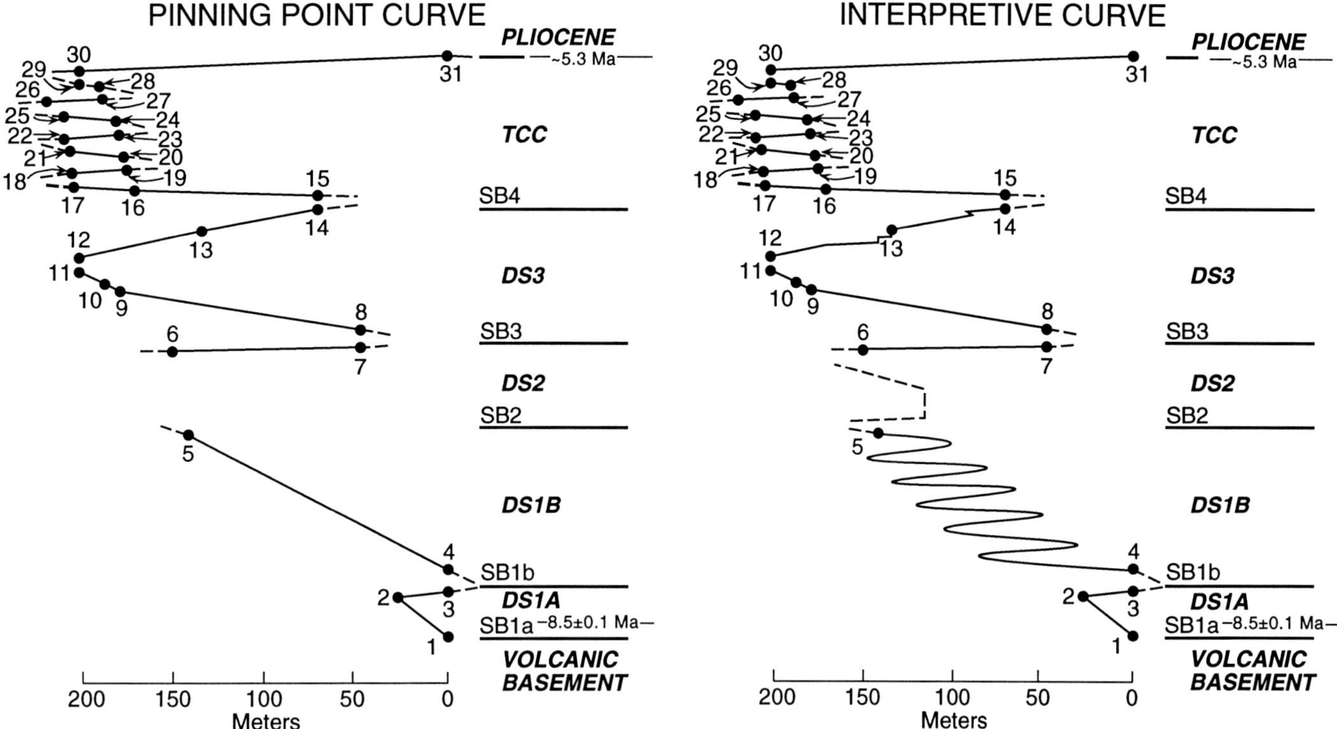


FIG. 10.—Pinning point and interpretive relative sea-level curves for the Las Negras area. Horizontal axis is relative sea-level position in meters; zero is present-day sea level. Vertical axis is time, here depicted in only a general sense but reflecting the new $^{40}Ar/^{39}Ar$ dates derived for the interbedded volcanic unit in DS1A and the date for the Miocene-Pliocene boundary as reported in Shackleton et al. (in press). See Franseen et al. (1993) and Goldstein and Franseen (1995) for details of construction of the pinning point and interpretive relative sea-level curves.

*Interpretation of DS1A and DS1B.—*

Both DS1A and DS1B were deposited as normal marine ramps in which units lapped out upslope against volcanic basement (Fig. 8). The basal sequence boundary (SB1a) was subaerially exposed down to present-day Mediterranean sea level. DS1A strata (all marine deposits), as traced to their most proximal locations, represent an overall rise in relative sea level of at least 25 m (Fig. 10; Franseen et al., 1993). DS1A strata were then subaerially exposed down to at least present-day sea level indicating a relative drop in sea level of at least 25 m (Franseen et al., 1993). DS1B strata (all marine deposits), as traced to their most proximal locations, represent an overall rise in relative sea level of at least 150 m (Fig. 10; Franseen et al., 1993).

The cyclic coarse-grained and fine-grained carbonate strata that compose DS1B are very similar to rhodalgal facies described by Carannante et al. (1988) for Lower Miocene carbonates that they interpret to have been deposited on circalittoral bottoms greater than 50 m. Present-day Mediterranean sediments (as described in Carannante et al., 1988) also provide an excellent analog for the DS1B strata. In the Mediterranean, facies similar to the coarse-grained, red algal-rich facies of DS1B are found at depths of 40-100 m, affected by currents and locally colonized by communities adapted to low-light conditions. Facies similar to the fine-grained, planktonic foraminiferal wackestones of DS1B are found at depths greater than 100 m. Franseen et al. (1993) tentatively interpreted the DS1B fining-upward cycles to be the result of higher frequency relative sea-level fluctuations during an overall relative sea-level rise. If the similar facies described by Carannante et al. (1988) from the modern Mediterranean are analogs, then depths of deposition for the red algal packstones (40-100 m) and for the planktonic foraminiferal wackestones (>100 m) may indicate significant relative changes in sea level (Figs. 9, 10 ).

*Climate.*—DS1A and DS1B strata (predominantly the coarse-grained packstone facies) fall into the rhodalgal lithofacies of Carannante et al. (1988) or foramol and bryomol lithofacies of Lees and Buller (1972) and suggest temperate to subtropical conditions during deposition. However, Carannante et al. (1988) warned that ancient rhodalgal sediments from the tropics are not easily distinguishable from those in temperate zones. They suggested that a possible diagnostic element to differentiate these environments could be the presence or absence of chlorozoan facies in, or resedimented from, shallow-water areas. In light of this, DS1A and DS1B ramp strata provide a valuable example for examining the nature of resedimented shallower-water components in helping to determine climate.

DS1A and lower DS1B strata lack evidence of significant chlorozoan facies even though shallow-water environments existed above the point of onlap. This and other evidence described in Franseen et al. (in press) supports an interpretation for temperate conditions during DS1A and lower DS1B deposition. Conversely, uppermost rhodalgal strata of DS1B (and the

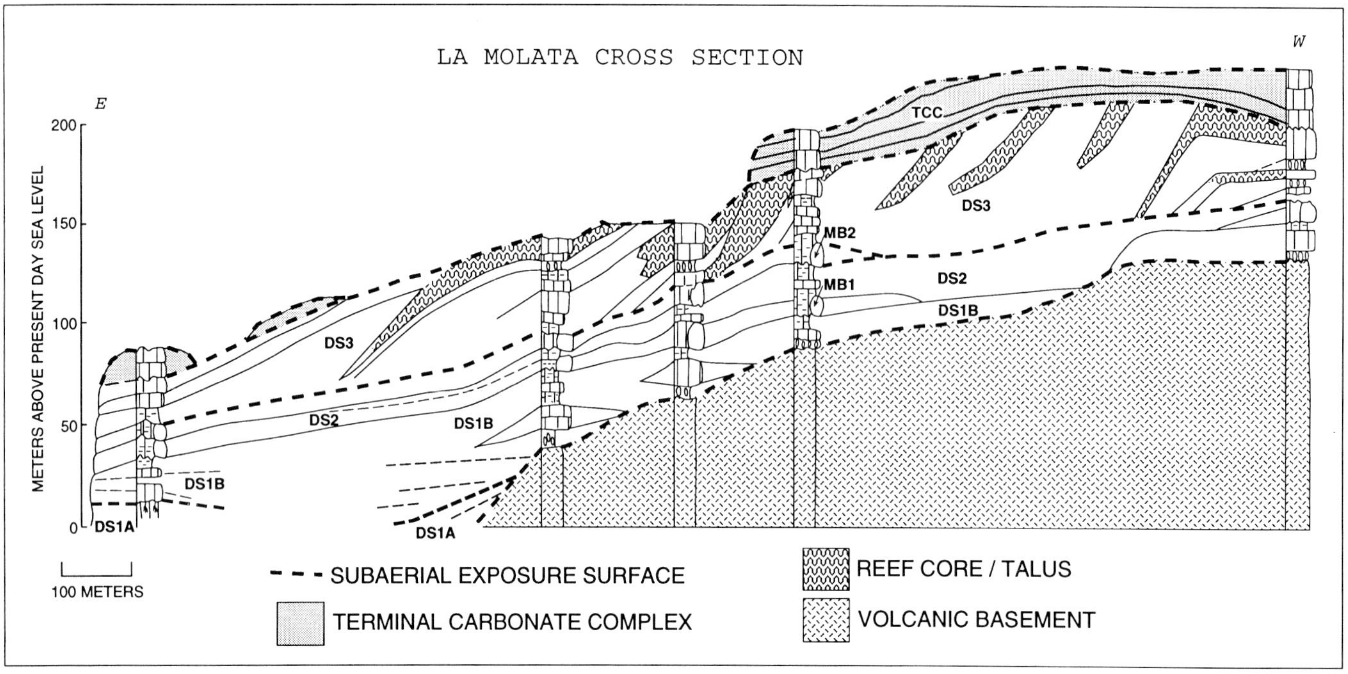


FIG. 11.—Composite cross section of depositional sequences, modified from actual cross-section and outcrop sketches of La Molata in Franseen et al. (1993).

immediately overlying sequence, DS2) contain significant amounts of resedimented chlorozoan facies consisting of pebble to boulder-sized coral clasts of *Tarbellastraea* and *Porites* framework. This suggests that some patch reefs were present in proximal, shallow-marine locations during latest DS1B time (Franseen and Mankiewicz, 1991; Franseen et al., 1993) and indicates a climatic shift towards more tropical conditions.

*Platform Geometry.*—The onlapping relationship and overall geometry of the DS1B ramp deposits may have been controlled by a mechanism quite different from onlap that traces base-level change. The six fining upward cycles lap out against volcanic basement without any significant facies changes that would indicate there was an approach to a shoreline environment at the point of onlap. Even the fine-grained foraminiferal facies lap out against the basement without facies changes. If these presumed deeper water deposits were forming at and below the point of lap out, then it is apparent that shallower water environments existed upslope during their deposition and that any sediment generated upslope must have been bypassed down the slope, thereby keeping the volcanic basement above the point of onlap relatively clean of carbonate deposition. The lack of any shoaling upward evidence and the fining upward character of the cycles suggest deepening upward during cycle deposition as opposed to shallowing upward that is so commonly observed in carbonate platform cycles. Likely important factors on cycle development were the overall lower accumulation rates (Franseen et al., in press) typical for more temperate climate carbonates (Boreen and James, 1993), asymmetry in sea-level fluctuations during an overall large relative rise in sea level, and the lack of a direct base-level control.

The controls for location of onlap appear to be the presence of accommodation space, after an initial or continuing relative sea-level rise, combined with the location of a low substrate slope angle. Kenter (1990) showed the importance of slope angles and grain sizes for controlling the location of carbonate deposition. Our study emphasizes the influence of slope angle, largely independent of grain size, on carbonate deposition. In areas where basement slope was steep, basal DS1B sediment did not permanently accumulate, and instead, these sediments accumulated only in toe-of-slope areas where basement slopes flattened out (Fig. 9). Additionally, in our study area, relative sea-level lowstands during deposition of the coarse-grained portion of cycles probably encouraged bypass of previously deposited sediments across steep slopes. Downslope transport and reworking in DS1B is evidenced by calciturbidites, conglomerates, mass wasting, soft sediment deformation, crossbedding, and scoured surfaces. Abundant volcaniclastic grains, even in some of the fine-grained wackestones, attest to erosion and repeated transport of volcanic sediment from the topographically higher areas. Sediment also was reworked by currents in the deeper water toe-of-slope areas as evidenced by the abundance of lamination and crossbeds in the coarse-grained facies, by the lenticular shape of some red algal-rich beds (channels and possible sand waves), and by the sharp, relatively planar contacts of coarse-grained packstones with underlying wackestones that may have resulted from "sweeping" and abrasion by normal or storm waves, a process that has been described as important in subtidal cycles in other studies (e.g., James and Bone, 1991;

Osleger, 1991). Also, strikes of onlapping beds appear to be roughly parallel to strikes of volcanic basement surfaces that are onlapped and mass flow and detachment structures in some onlapping strata indicate downslope transport roughly parallel to adjacent basement slope direction. These observations argue against distal onlap related to any slope other than the basement slopes observed, and instead, strongly indicate the influence of adjacent basement slope for the onlap relationships (Fig. 9).

In summary, DS1B strata reflect distal onlap caused by production of carbonate predominantly in 40-100 m and more water depths and by currents that swept the steep portions of basement slopes clean of sediments and piled them up at the toes of slopes, thereby partially filling in accommodation during an overall relative sea-level rise. DS1B deposits effectively filled in much of the topography created by the original volcanic slopes which locally created a substrate of low depositional slope on which reefs could later develop.

### *DS2*

DS2 ranges in thickness from 1-30 m. The lower DS2 sequence boundary (SB2) is placed at the base of MB1 which consists of a megabreccia (Figs. 4, 11). Correlation of the sequence boundary and MB1 indicate a minimum shelf-to-basin relief of 160 m over a lateral distance of 1.5-2.0 km during initial DS2 deposition. MB1 consists mostly of allochthonous reef facies clasts (composed of *Tarbellastraea* and *Porites*) with more minor amounts of volcaniclastic clasts, coarse-grained and red algal-rich packstone-grainstone clasts and fine-grained wackestone-packstone clasts. The basal sequence boundary of DS2 and the erosion and deposition of MB1 clasts are interpreted to have resulted from a lowering of relative sea level (evidence for a relative sea-level drop is described in detail in Franseen et al., 1993). No evidence of subaerial exposure was identified in DS2 strata. However, vugs in some transported red algal-rich and *Porites*-rich MB1 clasts are filled with green volcaniclastic debris indicating the vugs likely formed prior to transport, possibly from subaerial exposure upslope (Franseen et al., 1993).

Except for some areas where CGPG facies immediately overlies some of the MB1 clasts, the rest of DS2 is characterized mostly by burrowed FGWP facies with onlapping and draping geometries.

*Interpretation.—*

The reef clasts in MB1 are the first significant evidence of reef development in the Las Negras area and they represent the initial transition in platform type from foramol ramp to fringing reef. The allochthonous MB1 clasts likely originated as upslope patch(?) reefs that were developing on the flanks of RAPG wedges and distal portions of eroding volcanic fan delta material marginal to volcanic islands at the end of DS1B time (Franseen and Mankiewicz, 1991; Franseen et al., 1993). This setting likely is similar to patch reefs that developed on fan deltas in other southeast Spain basins (Santisteban, 1981; Braga and Martin, 1988). A relative sea-level fall (Franseen, *et al.*, 1993) likely triggered transport of MB1 to downslope positions via several mechanisms including submarine debris flows, turbidity currents, slides, and by individual clasts rolling down steep slopes. The predominant grainstone texture of strata flanking some MB1 clasts (composed predominantly of serpulid worm tubes, red algae, and echinoderms), the occurrence of some whole oyster coquinas, and deposition of some coarse-grained, granule and pebble-sized volcaniclastic beds immediately overlying MB1 in downslope positions are suggestive of shallower water conditions compared to those interpreted for the FGWP facies that immediately underlie the MB1 clasts (upper DS1B strata).

The strata that overlie MB1 (uppermost DS2) are dominated by burrowed FGWP facies indicating a return to deeper water conditions (as indicated on the relative sea-level curve in Fig. 10) on an open marine platform dominated by pelagic and hemipelagic deposition (Franseen and Mankiewicz, 1991; Franseen et al., 1993).

Prior to deposition of DS2, the sediments of DS1B had filled in some of the steep substrate paleotopography to produce a gently sloping carbonate ramp surrounded by bare volcanic highs. Most of the DS2 deposits drape that carbonate ramp. Although reefs likely formed on topographically higher volcanic substrates, they apparently could not accumulate permanently in those locations. The steep slopes and a relative fall in sea level at the end of DS2 deposition (Fig. 10) resulted in erosion and bypass of reef facies (MB2), other carbonate facies and volcanic rocks onto the relatively gentle slopes created by earlier deposition.

The occurrence of both rhodalgal (Carannante et al., 1988) or foramol (Lees and Buller, 1972; Lees, 1975) lithofacies, and significant accumulations of chlorozoan lithofacies (MB1 and MB2 reef facies clasts) reflects the continuance of more subtropical/tropical conditions that started near the end of DS1B deposition.

### *DS3*

DS3 has a preserved thickness ranging from 20-70 m and contains the first evidence of in-place fringing reef framework in the Las Negras area (Figs. 4, 11). Reef core, reef talus (both RCT facies) and proximal to distal forereef slopes characterize this depositional sequence.

The base of DS3 is a sharp and locally erosional surface that can be traced throughout the study area. Tracing this surface indicates minimum proximal to distal relief of 120-130 m. The surface is overlain either by a second megabreccia unit (MB2) composed predominantly of transported reef lithofacies debris, coarse-grained packstones or grainstones (CGPG), or volcaniclastic sandstones and conglomerates (VSC). In places, the top of MB2 is stained reddish, autoclastically brecciated and cross cut by possible rhizoliths suggesting subaerial exposure prior to deposition of the rest of DS3 (see Franseen et al., 1993 for further documentation). Thus, MB2 was deposited during a period of falling relative sea level and the sequence boundary was defined to encompass MB2 and the surfaces below and above (Franseen et al., 1993; Fig. 2). MB2 is significantly different from MB1 in that the reef clasts and detritus are composed primarily of *Porites*. This change to *Porites* as the predominant reef building organism is characteristic of many of the latest Miocene reefs in the Western Mediterranean prior to the salinity crisis (Esteban, 1979).

Except in some locations where basal CGPG or FGWP beds onlap the sequence boundary, deposits immediately overlying the MB2 unit or other basal beds are characterized by low-angle to high-angle prograding and downlapping fringing reef complex strata containing either VSC, CGPG, or FGWP facies. The in-place fringing reef core strata are a relatively minor proportion of the DS3 reef complex, which volumetrically is dominated by foreslope strata (CGPG and FGWP facies). Locally, in-place massive to flat bedded reef-core strata (RCT facies) are present. The early part of reef development began with combined aggradation and progradation. This phase was followed by a period of progradation in which sub-TCC erosion has precluded the possibility of identifying aggradation. During the latest stages of reef development, the reefs prograded in a downstepping style in response to a falling relative sea level, with in-place reef deposits forming on the forereef slopes of previous reef cycles (Franseen et al., 1993; Figs. 10, 11). Stratal geometries of all five depositional sequences, preservation of geopetal fabrics, amount of documented downstepping, and lack of evidence for significant faulting or regional tilting indicate that the downstepping cannot merely be an artifact of tectonic tilting as has been suggested by Cornee et al. (1994) for similar downstepping patterns elsewhere.

The only abundant *Halimeda* is in the latest exposed reef, proximal foreslope and distal foreslope. Although transport of *Halimeda* to distal locations is evident (aligned *Halimeda* plates and erosional bases to some beds), *Halimeda* apparently grew all along the 100 m of relief. Evidence for autochthonous growth in distal positions includes: (1) lack of other carbonate grains, a variety of other carbonate grains occur upslope and if *Halimeda* had all been transported, then other grains should have been transported as well; (2) accumulation of *Halimeda* in pockets that may represent in-place clusters deposited after death; and (3) a vertical trend from base of bed upwards shows a general gradational increase in *Halimeda* abundance possibly reflecting start-up and dominance stages of growth.

The fringing reefs developed in close association with volcaniclastic material shed from the adjacent volcanic highs. Locally, volcaniclastic sand is abundant in the interframework matrix of the *Porites* reefs. There are about eight separate reef cycles preserved (Fig. 11), and several of these cycles developed on thick wedges of volcaniclastic debris. After deposition, the volcaniclastic strata typically were encrusted (and stabilized?) by red algae and serpulid worms, and then by *Porites*. Reef-building *Porites* coral generally show a vertical transition from laminar-dish morphology dominant at the base of the reef to more massive and stick morphologies during the last stages of reef growth. *Porites* framework was relatively minor compared to interframework matrix. Most of the *Porites* are molds in which shape is preserved by early encrusters (red algae, foraminifera), fibrous (marine) cements and micritic coatings (interpreted by Riding et al., 1991 as cyanobacterial stromatolites). *Porites* framework apparently had local synoptic relief of a few cm to possibly 30 cm. In some areas, reef crest facies (after Riding, et al., 1991) and flat bedded morphologies (terraces?) indicate growth to sea level.

Other features characteristic of DS3 strata are abundant reef talus clasts and reef material that was shed onto forereef slopes, turbidites in foreslope and distal slope positions, abundant burrowing, channels filled with CGPG in foreslope positions, and some channels developed within and just behind in-place reef core strata.

*Interpretations.—*

*Sea level and paleoslope.*—Deposition of MB2 and formation of the surfaces above and below it are all interpreted to have resulted from a relative drop in sea level for reasons discussed in detail in Franseen et al. (1993; Fig. 10). Reef clasts in MB2 were reworked from reefs that had apparently formed upslope at the end of DS2. After MB2 deposition, a large relative rise in sea level is interpreted to account for the subsequent upslope shift in marine facies and development of the upper DS3 *Porites* fringing reef complex strata (Franseen et al., 1993).

The early phases of reef development are characterized by massive to faintly bedded reef core (1-5 m thick) that grades laterally to talus and forereef slopes with steep (25-35°) clinoforms. The early reefs appear to have prograded (200-300 m) with minor aggradation (about 5 m; Fig. 11). Although no lagoons have been identified, the interpreted aggradation and progradation may have caused some minor restriction behind the reefs as evidenced by some channeling and crossbedded strata landward of reef core, by the occurrence of dasycladacean algae in reef core matrix, and by coated grains (peloids and ooids) and composite grains (grapestones) that are in places closely associated with the reef core facies. This slightly aggradational and largely progradational geometry existed through much of DS3 deposition.

Later stage reef deposition is characterized by downstepping progradation with successive reef strata formed in topographically lower positions, on the forereef slopes of previous reef cycles, as a result of falling relative sea level (Franseen et al., 1993; Figs. 10, 11). The only downstepped reef cycle (last reef cycle, Fig. 11) that has preserved reef crest facies (after Riding et al., 1991) occurs 65 m lower than the previous reef cycle with preserved reef crest facies, indicating a minimum relative sea-level fall of the same magnitude (Fig. 10; Franseen et al., 1993). The latest stage of reef development is characterized by clinoforms with steep proximal dips (25-30°) that flatten abruptly basinward. This feature is a result of basinward thinning and draping of the flatter topography in basinward locations. These later clinoforms have reef core to distal slope relief of 50-90 m over a distance of 300-700 m. All stages of reef development reveal abundant reefal material that was transported to foreslopes by mechanisms not necessarily related to sea-level falls. These include turbidity currents, debris flows, and rock falls. It is likely that the apparent rapid production and progradation of the reef itself was a major contributor (self erosion) of reef talus clasts and reefal debris to the foreslopes.

The importance of mixed siliciclastic/carbonate systems has been underscored by some workers (Hubbard, 1982; Mount, 1984; Doyle and Roberts, 1988; Budd and Harris, 1990; Lomando and Harris, 1991). The DS3 fringing reef complex strata in the Las Negras area superbly display the close association of carbonate fringing reef development and volcaniclastic deposition. The volcaniclastic sandstone and conglomerate wedges exposed in the area are interpreted as marine portions of fan delta lobes that developed marginal to the eroding volcanic islands. The setting is comparable to that described for other areas in southeastern Spain by Dabrio and Martin (1978), Santisteban (1981), Santisteban and Taberner (1988), and Esteban (this volume). The marine interpretation for the volcaniclastic wedges is supported by marine encrustations on the clasts and by admixed marine skeletal material in the matrix. Volcaniclastic influx periodically interrupted reef development and progradation. However, these volcanic clasts eventually formed the substrate for subsequent reef development. It is currently unknown if relative sea-level changes were responsible for alternation between reef and volcaniclastic facies as suggested by Franseen (1990) and Franseen and Mankiewicz (1991), but because evidence for significant relative changes in sea level are lacking between these alternations, any relative sea-level changes must have been minor, and an autogenic mechanism for facies alternation remains a possibility.

After deposition of DS3, relative sea level continued to drop and resulted in subaerial exposure of all DS3 strata in the area (Fig. 11; Franseen et al., 1993). This major drop of over 130 m may coincide with the major drawdown of the Mediterranean Sea during the Messinian salinity crisis (Fig. 10; Franseen and Mankiewicz, 1991; Franseen et al., 1993).

As in the previous depositional sequences, paleoslope was a significant factor in determining the location of DS3 carbonate deposition and the geometries that were preserved. However, significant differences from the earlier depositional sequences involve the close interplay between sea-level position and carbonate accumulation, availability of significant areas of gentle paleoslope exposed to shallow water, importance of siliciclastic debris wedges, and a predominance of chlorozoans able to construct rigid frameworks resulting in the initiation of steep constructional paleotopography. DS3 contains reef and forereef strata displaying aggradational to progradational to downstepping geometries, that closely tracked the history of relative sea-level change. Autochthonous accumulation of reef strata and development of forereef slopes created steep constructional paleotopography. The early formation of this topography created the slopes on which the youngest reef cycles could step downward during the later period of falling relative sea level.

The in-place reef and foreslope, and volcaniclastic fan-delta deposits produced geometries that are not preserved in earlier deposits. Their preservation is mainly attributed to the gentle substrate slope that was created by deposition of earlier deposits. Shallow water conditions just above this gentle slope created a suitable environment for growth and preservation of reefs. The controls on deposition of DS3 in the Las Negras area are strongly supported by nearby DS3-equivalent deposits on Mesa Roldán located approximately 15 km to the northeast (Fig. 1). In that area, DS3 reefs formed and are preserved on a relatively flat surface on the top of the mesa, at approximately the same elevation as DS3 reefs at Las Negras. During that same time and at the same elevation, other volcanic highs that had steep slopes probably were the sites of at least minor reef growth, but reefs were not preserved in place because a flat substrate had not been created either by erosion or by deposition of previous units. Also at this time, in upslope positions, volcanic highs were actively being eroded to deposit fan delta conglomerates and sandstones on the gentle slopes. Thus, without deposition on gentle slopes, reef morphologies and siliciclastic wedges would not likely have been preserved and instead would have been eroded and bypassed downslope similar to the reef framework deposits and other facies of earlier units (e.g., MB1, MB2).

*Climate and environmental controls.*—Overall, DS3 carbonate strata indicate clear, normal marine water for most of DS3 time (despite the abundance of volcaniclastic material) as indicated by the abundance of normal marine faunas and reef growth. The isolated occurrence of coated grains (ooids, grapestones) and dasycladacean algae suggest some local restriction. The predominance of chlorazoan assemblages indicates a more subtropical to tropical climate (Lees, 1975; Lees and Buller, 1972; Carannante et al., 1988) for DS3 deposition as compared to DS1A, DS1B, and DS2 deposition. The importance of chlorozoan components could also be explained partially by the admixture of terrigenous debris in the environment. Mount (1974) suggested that carbonate systems punctuated by mixing of terrigenous material show the highest proportion of chlorozoan components.

The reasons for the predominance of *Porites* as fringing reef framebuilder, and its peculiar stick morphology are problematic. These features have been tied into unusual stresses related to isolation of the basin, proximity to colder waters, and circulation changes in the Mediterranean prior to and during initial stages of the salinity crisis (Esteban, 1979; Rouchy, 1982; Saint Martin and Rouchy, 1990; Esteban, this volume).

The last stage of reef development is characterized by an abundance of *Halimeda*. The occurrence of *Halimeda* in youngest Miocene reefs is not unique to the Las Negras area. Other occurrences in similar stratigraphic positions have been reported throughout southeast Spain (e.g., Esteban, 1979; Dabrio et al., 1985; Mankiewicz, 1988; Braga and Martin, 1992; Mankiewicz, this volume). Mankiewicz (1988) described and interpreted similar *Halimeda* beds in Níjar as event strata that may warrant their use as time-correlative units. She postulated that episodic upwelling conditions of a regional nature may be responsible for the cycles of *Halimeda*-rich and non-*Halimeda* beds, with more nutrient-rich upwelling water favoring *Halimeda* growth.

### *TCC*

Terminal Carbonate Complex (TCC) strata, up to 30 m thick, form the uppermost Miocene depositional sequence in the Las Negras area (Fig. 11). These marine strata overlie the upper DS3 sequence boundary (SB4), that is marked by erosional truncation, caliche laminated crust, and rhizoliths, and represent a minimum relative sea-level rise of 130 m (Fig. 10; Franseen et al., 1993).

TCC strata are preserved predominantly on relatively flat erosional or constructional paleoslopes, with apparently little preserved on steep paleoslopes. In the Las Negras area, TCC strata were deposited at the top of the highest hills covered by previous carbonate strata (e.g., top of La Molata hill at ~ 200 m; Fig. 11) and locally in more distal locations where latest DS3 foreslopes were becoming flat basinward (e.g., La Molata hill at ~75-85 m; Fig. 11). This relationship holds for other hills in the study area (e.g., El Romeral and Cerro del Cuervo) and regionally as well. Just to the north of Las Negras, in the La Rellena, El Plomo, Agua Amarga, and Mesa Roldán areas, the significant accumulations of TCC strata are largely confined to the areas of relatively flat paleoslopes (see Fig. 5 of Esteban and Giner, 1980).

In some locations, the basal deposit of the TCC consists of volcaniclastic conglomerate with some admixture of carbonate clasts and oolites. TCC strata consist predominantly of topography-draping strata containing three to four cycles (each several meters thick) consisting of stromatolitic carbonates at the base and passing upward to fossiliferous packstone and cross-bedded oolite that may become fenestral towards the top. The cycles are capped by a surface of subaerial exposure as evidenced by meniscus cements (Franseen et al., 1993). Each of these cycle components tend to drape at least 25 m of subtle paleotopography above DS3.

The subaerial exposure surface (sequence boundary) at the top of the TCC is represented by a well-developed caliche breccia similar to and probably the same as what Esteban and Giner (1980) termed the "Pre-Pliocene breccia". This exposure surface is probably equivalent to the karstic "Pre-Pliocene" surface that can be traced down to present day sea level at Mesa Roldan (Goldstein et al., 1990; Franseen et al., 1993).

*Interpretation.—*

TCC strata were deposited initially upon the major transgression onto the shelf after exposure of the entire pre-TCC carbonate complex. As interpreted in Franseen et al. (1993), each of the 3-4 TCC cycles that were deposited on the highest portions of hills indicate relative rises and falls of sea level on the order of 25-30 m (Fig. 10). The stromatolite, coated-grain (ooid) grainstone and red algal(?) packstone facies within the cycles indicate both restricted and normal marine conditions for cycle deposition, and the fenestral fabrics and meniscus cements indicate subaerial exposure between cycles. Similar types and numbers of cycles have been described elsewhere for the TCC in southeast Spain (e.g., Esteban and Giner, 1980; Roca, 1986; Braga and Martin, 1992; Calvet et al., this volume; Esteban et al., this volume) lending support to an allochthonous control on cycle development. The style of cyclicity developed in TCC strata is in sharp contrast to that developed in DS1B. The predominant shoaling upward character within TCC cycles reflects the direct control of fluctuating relative sea level on carbonates deposited in shallow marine water. Although sea level may have been fluctuating similarly during DS1B cycle deposition (Fig. 10), DS1B fining upward cycles reflect deeper water depositional controls that are distinctly different from those of the TCC (Fig. 9).

Bypass and accumulation of TCC strata was directly related to substrate paleoslope, with steeper slopes unable to support permanent accumulation. Only substrates with relatively flat areas preserved TCC deposits (Fig. 11). The flat areas on La Molata resulted from erosional planation and progradation of DS3 strata on the higher areas and distal constructional paleotopography where DS3 foreslopes flattened out towards the basin.

In other nearby areas (e.g., Mesa Roldan), Pliocene strata overlie Miocene strata near present-day sea level. The contact is marked by karstic fissures that cut into Miocene strata indicating the pre-Pliocene exposure surface represents a relative drop in sea level of at least 200 m (Franseen and Goldstein, 1992; Goldstein and Franseen, 1993; Franseen et al., 1993; Goldstein and Franseen, 1995).

## SUMMARY

This paper illustrates the various controlling factors on evolution of platform geometry and lithofacies in Upper Miocene, Tortonian and Messinian, strata from near Las Negras, southeastern Spain. Whereas many studies emphasize the importance of relative sea-level change as a dominant controlling factor, this paper emphasizes that the interaction between sea level, paleoslope, and paleoclimate were all important for platform evolution. A dominant control on accumulation of Miocene carbonates of the Las Negras area was the changing availability and distribution of gently sloping substrates in shallow-marine water versus distribution of steeply sloping substrates.

The carbonate platforms in the Las Negras area evolved from a ramp to a fringing reef complex to a topography-draping open-to-restricted carbonate platform. The change from a more temperate to subtropical climate, as evidenced by the predominant rhodalgal or foramol assemblages in ramp strata, to more subtropical or tropical, as evidenced by the predominance of a chlorozoan assemblage in the later fringing reef complex, may reflect a marginal biogeographical setting as interpreted by Esteban (this volume) for Upper Miocene carbonate platforms in Spain (see Franseen et al., in press for more discussion).

Many studies have presented ramp models and documented controls on ramp development (e.g., Read, 1982; Read, 1985; Burchette and Wright, 1992). Some workers have proposed ramp models for the Tertiary of the Tethyan and Mediterranean realms (e.g., Buxton and Pedley, 1989) and others have shown the importance of sea level and climate in controlling the development of those ramps (e.g., Sun and Esteban, 1994). However, the entire DS1B ramp of our study fits within the outer ramp and basin facies belts described for other types of ramps (e.g., Buxton and Pedley, 1989; Burchette and Wright, 1992). The features which set the ramp described herein apart from other carbonate platforms is the lack of significant relatively flat areas in proximal positions (platform tops) that were exposed to shallow water. Instead, relatively deep water ramp strata onlap against volcanic basement and show no significant facies changes approaching the point of onlap. This illustrates that bypass across steep substrate paleoslope was a dominant control on ramp development and that this ramp may not have been greatly controlled by sea-level changes.

The character of DS1B was significantly controlled by basement paleoslope and its accumulation in turn significantly changed substrate morphology for subsequent deposition. Unlike the typical pattern of shoaling upward for carbonate platform cycles, the six ramp cycles in DS1B are characterized by fining (deepening) upward likely as a result of the overall rising or high sea level punctuated by higher frequency fluctuations, lower accumulation rates (Franseen et al., in press) that are typical of temperate climates (Boreen and James, 1993) and distal toe of slope setting. At least some of the carbonate production occurred in water depths of 40-100 m or more. Carbonate sediments that were generated in upslope shallow-water positions were bypassed downslope until areas of low basement paleoslope were reached. Sediments continued piling up and onlapping at the toe of slope of some of the hills forming a sloping platform (ramp) that filled in much of the basement paleotopography, thereby creating a substrate on which later deposits could accumulate. Importantly, no sediment accumulated on other hills that lacked areas of low basement paleoslope.

Similar processes are reflected in later strata of DS2 including associated megabreccias. For these units, some reefs were forming in upslope positions but the relatively high angled slopes and relative falls in sea level resulted in transport of that material downslope until it reached areas of low paleoslope that were formed by deposition of earlier deposits.

During deposition of DS3, sea level was at a position near the tops of relatively flat hills created by earlier deposition to allow for accumulation of reefs. Because the substrate was relatively flat, constructional topography was preserved without significant erosion, and steep forereef slope topography was created through reef aggradation, progradation and downstepping. On other hills in the area, that still had steep paleoslopes, DS3 equivalent reefs may have formed but were bypassed downslope. Finally, during deposition of TCC strata, sea level was at a position that allowed for accumulation of shallow-marine cycles (with subaerial exposure caps) on the tops of flat hills capped by DS3 strata. As in earlier deposits, TCC strata could only accumulate permanently on areas of low paleoslope. Therefore, TCC strata are found on areas of flat erosional topography and on areas atop the toes of DS3 foreslopes in which dip decreases to relatively flat slopes. TCC strata typically are not found draping steeply dipping forereef slopes of DS3.

The model for platform evolution in the Las Negras area shows that sea-level position allowed for the accumulation of carbonate sediments, but substrate paleoslope was a dominant control on determining the location of deposition and depositional sequence geometry. Similarity in style of platform evolution in other Mediterranean areas may indicate similar controls were important in those locations. The platform development model from the Las Negras area may apply to other carbonate complexes throughout the rock record.

## ACKNOWLEDGMENTS

Exxon Production Research Company, and Texaco Research Company provided financial support for our research in southeastern Spain. $^{40}Ar/^{39}Ar$ dates were done at the USGS in Denver, Colorado under the direction of Larry Snee. A. Arribas, Jr. provided the photograph of Figure 3. L. C. Pray contributed insightful observations in the field. The research presented in this paper has benefited from discussions with A. Arribas, Jr., S. Dorobek, M. Esteban, E. J. Oswald, L. C. Pray, J. F. Sarg, W. Schlager, and A. Simo. Critical reviews by M. Esteban, J. Gimenez, and A. Simo were helpful in clarifying ideas and improving the overall content of the paper. M. Esteban and D. Ulmer-Scholle are thanked for handling editorial duties for the paper.

## REFERENCES

Boreen, T. D. and James, N. P., 1993, Holocene sediment dynamics on a cool-water carbonate shelf: Otway, Southeastern Australia: Journal of Sedimentary Petrology, v. 63, p. 574-588.

Braga, J. C. and Martin, J. M., 1988, Neogene coralline-algal growth-forms and their palaeoenvironments in the Almanzora river valley (Almería, S.E. Spain): Palaeogeography, Palaeoclimatology, Palaeoecology, v. 67, p. 285-303.

Braga, J. C. and Martin, J. M., 1992, Messinian carbonates of the Sorbas basin: Sequence stratigraphy, cyclicity, and facies, *in* Franseen, E. K., Goldstein, R. H., Braga, J. C., and Martin, J. M., eds., A Guidebook for the Las Negras and Sorbas Areas: La Seu, SEPM/IAS Research Conference on Carbonate Stratigraphic Sequences: Sequence Boundaries and Associated Facies, p. 78-108.

Budd, D. A. and Harris, P. M., compilers, 1990, Carbonate-Siliciclastic Mixtures: Tulsa, SEPM Reprint Series Number 14, 272 p.

Burchette, T. P. and Wright, V. P., 1992, Carbonate ramp depositional systems: Sedimentary Geology, v. 79, p. 3-57.

Buxton, M. W. N. and Pedley, H. M., 1989, A standardized model for Tethyan Tertiary carbonate ramps: Journal of the Geological Society, London, v. 146, p. 746-748.

Carannante, G., Esteban, M., Milliman, J. D., and Simone, L., 1988, Carbonate lithofacies as paleolatitude indicators: problems and limitations: Sedimentary Geology, v. 60, p. 333-346.

Cornee, J.-J., Saint Martin, J.-P., Conesa, G., and Muller, J., 1994, Geometry, palaeoenvironments and relative sea-level (accommodation space) changes in the Messinian Murdjadjo carbonate platform (Oran, western Algeria): consequences: Sedimentary Geology, v. 89, p. 143-158.

Dabrio, C. J. and Martin, J. M., 1978, Las Arrecifes Messiniense de Almería (SE de España): Cuadernos Geologia, v. 8-9, p. 85-100.

Dabrio, C. J., Martin, J. M., and Megias, A. G., 1985, The Tectosedimentary Evolution of Mio-Pliocene Reefs in the Province of Almería (SE Spain): Lleida, International Association of

Sedimentologists 6th European Regional Meeting, Excursion Guidebook, p. 271-305.

DI BATTISTINI, G., TOSCANI, L., IACCARINOI, S., AND VILLA, J. M., 1987, K/Ar ages and the geological setting of calc-alkaline volcanic rocks from Sierra de Gata, SE Spain: Neues Jahrbuch fuer Mineralogie Monatshefte, v. H8, p. 369-383.

DOYLE, L. J. AND ROBERTS, H. H., eds., 1988, Carbonate-Clastic Transitions: Amsterdam, Developments in Sedimentology 42, Elsevier, 304 p.

ESTEBAN, M., 1979, Significance of the Upper Miocene Coral Reefs of the Western Mediterranean: Palaeogeography, Palaeoclimatology, Palaeoecology, v. 29, p. 169-188.

ESTEBAN, M. AND GINER, J., 1980, Messinian Coral Reefs and Erosion Surfaces in Cabo de Gata (Almería, Southeastern Spain): Acta Geologica Hispanica, v. 15, p. 97-104.

FRANSEEN, E. K., 1989, Depositional Sequences and Correlation of Middle to Upper Miocene Carbonate Complexes, Las Negras Area, Southeastern Spain: Unpublished Ph.D. Dissertation, University of Wisconsin-Madison, Madison, 374 p.

FRANSEEN, E. K., 1990, Middle to Upper Miocene mixed carbonate and volcaniclastic slope deposits, Las Negras and Rodalquilar area, southeastern Spain: Nottingham, Abstracts of Posters, 13th International Sedimentological Congress, p. 80-81.

FRANSEEN, E. K. AND GOLDSTEIN, R. H., 1992, Sequence stratigraphy of Miocene strata, Las Negras area, southeastern Spain: Implications for quantification of relative change in sea level, *in* Franseen, E. K., Goldstein, R. H., Braga, J. C., and Martin, J. M., eds., A Guidebook for the Las Negras and Sorbas Areas: La Seu, SEPM/IAS Research Conference on Carbonate Stratigraphic Sequences: Sequence Boundaries and Associated Facies, p. 1-77.

FRANSEEN, E. K., GOLDSTEIN, R. H., AND FARR, M. R., in press, Substrate slope and temperature controls on carbonate ramps: Revelations from Upper Miocene outcrops, SE Spain, *in* Clarke, J. and James, N. P., eds., Cool-water Carbonates: Tulsa, SEPM Special Publication.

FRANSEEN, E. K., GOLDSTEIN, R. H., AND WHITESELL, T. E., 1993, Sequence stratigraphy of Miocene carbonate complexes, Las Negras area, southeastern Spain: implications for quantification of changes in relative sea level, *in* Loucks, R. G. and Sarg, J. F., eds., Carbonate Sequence Stratigraphy: Recent Developments and Applications: Tulsa, American Association of Petroleum Geologists Memoir 57, p. 409-434.

FRANSEEN, E. K. AND MANKIEWICZ, C., 1991, Depositional Sequences and Correlation of middle(?) to late Miocene carbonate complexes, Las Negras and Nijar areas, southeastern Spain: Sedimentology, v. 38, p. 871-898.

FRANSEEN, E. K. AND MANKIEWICZ, C., 1993, Depositional Sequences and Correlation of Middle(?) to Late Miocene Carbonate Complexes, Las Negras and Nijar Areas, Southeastern Spain: Reply: Sedimentology, v. 40, p. 353-356.

GOLDSTEIN, R. H., FRANSEEN, E. K., AND MILLS, M. S., 1990, Diagenesis associated with subaerial exposure of Miocene strata, southeastern Spain: Implications for sea-level change and preservation of low-temperature fluid inclusions in calcite cement: Geochimica et Cosmochimica Acta, v. 54, p. 699-704.

GOLDSTEIN, R. H. AND FRANSEEN, E. K., 1993, Pinning points: A method that provides quantitative constraints on relative sea-level history: American Association of Petroleum Geologists Abstracts with Programs, p. 109.

GOLDSTEIN, R. H. AND FRANSEEN, E. K., 1995, Pinning points: a method providing quantitative constraints on relative sea-level history: Sedimentary Geology, v. 95, p. 1-10.

HUBBARD, J. A. E. B., 1982, Siliciclastics in reefs and carbonate sequences: The conflict between theory and fact: Utrecht, Abstracts of Papers, Eleventh International Congress on Sedimentology, International Association of Sedimentologists, p. 109.

JAMES N. P. AND BONE, Y., 1991, Origin of a cool water Oligo-Miocene deep shelf limestone, Eucla Platform, southern Australia: Sedimentology, v. 38, p. 323-341.

KENTER, J. A. M., 1990, Carbonate platform flanks: slope angle and sediment fabric: Sedimentology, v. 37, p. 777-794.

LEES, A., 1975, Possible influence of salinity and temperature on Modern shelf carbonate sedimentation: Marine Geology, v. 19, p. 159-198.

LEES, A. AND BULLER, A. T., 1972, Modern temperate-water and warm-water shelf carbonate sediments contrasted: Marine Geology, v. 13, p. M67-M73.

LOMANDO, A. J. AND HARRIS, P. M., eds., 1991, Mixed Carbonate-Siliciclastic Sequences: Tulsa, SEPM Core Workshop No. 15, 569 p.

LÓPEZ-RUIZ, J. AND RODRÍGUEZ-BADIOLA, E., 1980, La Región Volcánica Neógena del Sureste de España: Estudios Geológicos, v. 36, p. 5-63.

MANKIEWICZ, C., 1988, Occurrence and paleoecologic significance of *Halimeda* in Late Miocene reefs, southeastern Spain: Coral Reefs, v. 6, p. 271-279.

MARTIN, J. M. AND BRAGA, J. C., 1993, Depositional Sequences and Correlation of Middle(?) to Late Miocene Carbonate Complexes, Las Negras and Níjar Areas, Southeastern Spain: Discussion: Sedimentology, v. 40, p. 351-353.

MONTENAT, C., OTT D'ESTEVOU, P., AND MASSE, P., 1987, Tectonic-Sedimentary Characters of the Betic Neogene Basins Evolving in a Crustal Transcurrent Shear Zone (SE Spain): Bulletin des Centres de Recherches Exploration Production Elf-Aquitaine, v. 11, p. 1-22.

MOUNT, J. F., 1984, Mixing of siliciclastic and carbonate sediments in shallow shelf environments: Geology, v. 12, p. 432-435.

OSLEGER, D., 1991, Subtidal carbonate cycles: Implications for allocyclic vs. autocyclic controls: Geology, v. 19, p. 917-920.

READ, J. F., 1982, Carbonate platforms of passive (extensional) continental margins: types, characteristics and evolution: Tectonophysics, v. 81, p. 195-212.

READ, J. F., 1985, Carbonate platform facies models: American Association of Petroleum Geologists Bulletin, v. 69, p. 1-21.

RIDING, R., MARTIN, J. M. AND BRAGA, J. C., 1991, Coral-stromatolite reef framework, Upper Miocene, Almeria, Spain: Sedimentology, v. 38, p. 799-819.

ROCA, D. V., 1986, Carbonate facies and depositional cycles in the upper Miocene of Santa Pola (Alicante, SE Spain): Revista d'Investigacions Geologiques, v. 42/43, p. 45-66.

ROUCHY, J. M., 1982, La crise evaporitique messinienne de Mediterranee: nouvelles propositions pour une interpretation genetique: Bulletin de Museum National d'Histoire Naturelle, Paris, 4e ser., 4, p. 107-136.

RYTUBA, J. J., ARRIBAS A., JR., CUNNINGHAM, C. G., MCKEE, E. H., PODWYSOCKI, M. H., SMITH, J. G., KELLY, W. C., AND ARRIBAS, A., 1990, Mineralized and unmineralized calderas in Spain; Part II, evolution of the Rodalquilar caldera complex and associated gold-alunite deposits: Mineralium Deposita, v. 25 (Suppl), p. S29-S35.

SAINT-MARTIN, J. P., AND ROUCHY, J. M., 1990, Les plates-formes carbonatees messiniennes en Mediterranee occidentale: leur importance pour la reconstitution des variations du niveau marin au Miocene terminal: Bulletin de la Societe Geologique de France, v. 6, p. 83-94.

SANTISTEBAN, C., 1981, Petrología y sedimentología de los materiales del Mioceno superior de la cuenca de Fortuna (Murcia), a la luz de la "teoría de la crisis de salinidad": Unpublished Ph.D. Thesis, Universidad de Barcelona, Barcelona, 725 p.

SANTISTEBAN, C. AND TABERNER, C., 1988, Sedimentary models of siliciclastic deposits and coral reef interrelation, *in* Doyle, L. J. and Roberts, H. H., eds., Carbonate-Clastic Transitions: Amsterdam, Developments in Sedimentology 42, Elsevier, p. 35-76.

SERRANO, F., 1992, Biostratigraphic control of Neogene volcanism in Sierra de Gata (south-east Spain): Geologie en Mijnbouw, v. 71, p. 3-14.

SHACKLETON, N. J., CROWHURST, S., HAGELBERG, T., PISIAS, N., AND SCHNEIDER, D. A., in press, A new late Neogene timescale: application to ODP leg 138 sites: Proceedings of the Ocean Drilling Program, Scientific Results, v. 138 (1995).

SUN, S. Q. AND ESTEBAN, M., 1994, Paleoclimatic controls on sedimentation, diagenesis, and reservoir quality: Lessons from Miocene carbonates: American Association of Petroleum Geologists, v. 78, p. 519-543.

# LATE MIOCENE REEFS OF THE ALICANTE-ELCHE BASIN, SOUTHEAST SPAIN

FRANCESC CALVET
*Facultat de Geologia, Universitat de Barcelona, 08028 Barcelona, Spain*
ISABEL ZAMARREÑO
*Institut J. Almera, C.S.I.C., c. Marti Franques s.n., 08028 Barcelona, Spain*
AND
DOLORS VALLÈS
*Facultat de Geologia, Universitat de Barcelona, 08028 Barcelona, Spain*

ABSTRACT: The Late Miocene Alicante-Elche Basin, located in southeastern Spain, is filled with marls and evaporites in the depocenters and with a variety of carbonate facies (including reefs) and proximal deposits in the shallower areas. The Late Miocene deposits are composed of five lithostratigraphic units, which from base to top are Tabarca Unit, Torremendo Marls Unit, Reef Complex Unit, Terminal Complex and Gypsum and Marly Unit.

The Reef Complex Unit presents two morphologies: fringing reefs and atoll-like reefs. The fringing reefs trend E-W, between Alicante and Elche, and contain a variety of coral taxa. An asymmetrical atoll-like reef forms the Santa Pola hill and is composed of *Porites*.

The Terminal Complex Unit is well exposed at Santa Pola. It is composed of two distinct lithological units described here as sub-units: a basal calcarenitic sub-unit (transgressive deposits) and a cyclic stromatolitic sub-unit. The cyclic sub-unit consists of four main outcropping metric-scale shallowing-upward cycles bounded by erosion surfaces. The stromatolites occur at the base of each cycle and are interpreted as subtidal deposits. They grade upward into different facies (oolites, *Porites* patch reefs and sandy deposits).

The Late Miocene Reef System (Reef Complex Unit and Terminal Complex Unit) in the Santa Pola hill exhibits pervasive nondestructive dolomitization. The dolomite occurs as microcrystalline to subhedral-euhedral (7-45 μm) replacement fabrics and euhedral to rounded-anhedral crystal (7-30 μm) cements. The heavy stable isotope values ($\delta^{18}O$ = +3.3 to 4.7 ‰; $\delta^{13}C$ = +0.9 to 2.5 ‰) of the dolomites suggest a hypersaline influx, which may be related to the late Messinian evaporite event.

## INTRODUCTION

Upper Miocene deposits of the Alicante-Elche Basin, southeastern Spain, crop out to the north of the village of Santa Pola, 14 km south of Alicante, and between Alicante and Elche, bordering the pre-Tortonian and Prebetic basement (Fig. 1).

In the Betic Cordillera, extensional tectonics during Neogene time led to several Neogene basins being superimposed on older structures of the Cordillera. The Alicante-Elche Basin, located in the easternmost part of the Cordillera, has been interpreted as having formed in a transtensional shear zone trending northeast (Montenat et al., 1990). The resultant basins and horsts were filled with marls and evaporites in the depocenters and with a variety of carbonate facies (including reefs) and proximal deposits in the shallower areas.

The Alicante-Elche Basin contains numerous Upper Miocene reefs, but it is the round platform of Santa Pola (5 km in diameter) that has received the most attention (Fig. 1). The Santa Pola sequence has been interpreted either as: (1) two reefs (the "récif à Madrépores" and the "récif à stromatolithes") separated by a major discontinuity considered to be an erosion surface (Montenat, 1973, 1977; Bernet-Rollande et al., 1980; Rouchy et al., 1986) or (2) as several coral-stromatolite interdigitations (Esteban and Giner, 1977, Esteban, 1977, 1979; Esteban and Pray, 1981). According to Esteban (1979), the eastern margin of the Santa Pola reef is dominated by *Porites* sticks and laminar colonies although the entire platform is predominantly composed of well-developed stromatolites, repeatedly encrusting *Porites* reefs and oolite shoals.

Vallès (1985, 1986) and Calvet et al. (1991, 1994) divided the Messinian deposits of the Santa Pola area into the Lower and Upper Depositional Units, separated by an erosional unconformity surface displaying a terraced morphology. The Upper Depositional Unit is made up of stromatolites, *Porites* patch reef, oolite and sandy facies. Feldmann and McKenzie (1993) analysed the stromatolites from Santa Pola.

## LITHOSTRATIGRAPHIC UNITS

Five lithostratigraphic units are recognized (Fig. 2) in the Upper Miocene deposits of the Alicante-Elche Basin which onlap folded Mesozoic (Prebetic basement) and Middle to Upper Miocene rocks. These units, from base to top, are as follows:

### *Tabarca Unit*

The lower boundary of this unit forms an angular unconformity with respect to the underlying deposits (Prebetic basement and Middle Miocene to Middle Tortonian deposits). A paleokarst surface is developed where the unit onlaps carbonate Prebetic substrates, as occurs on the island of Tabarca. The Tabarca unit (defined in this paper) reaches up to 200 m in thickness. The lowest part of the unit consists of conglomeratic and calcarenitic (branching coralline algae, rhodoliths, bryozoans) facies which grade upwards into calcisiltitic facies. This unit has been interpreted as an open temperate carbonate platform grading upward into more distal deposits. An Upper Tortonian age can

Models for Carbonate Stratigraphy from Miocene Reef Complexes of Mediterranean Regions,
SEPM Concepts in Sedimentology and Paleontology #5, Copyright © 1996,
SEPM (Society for Sedimentary Geology), ISBN 1-56576-033-6, p. 177-190.

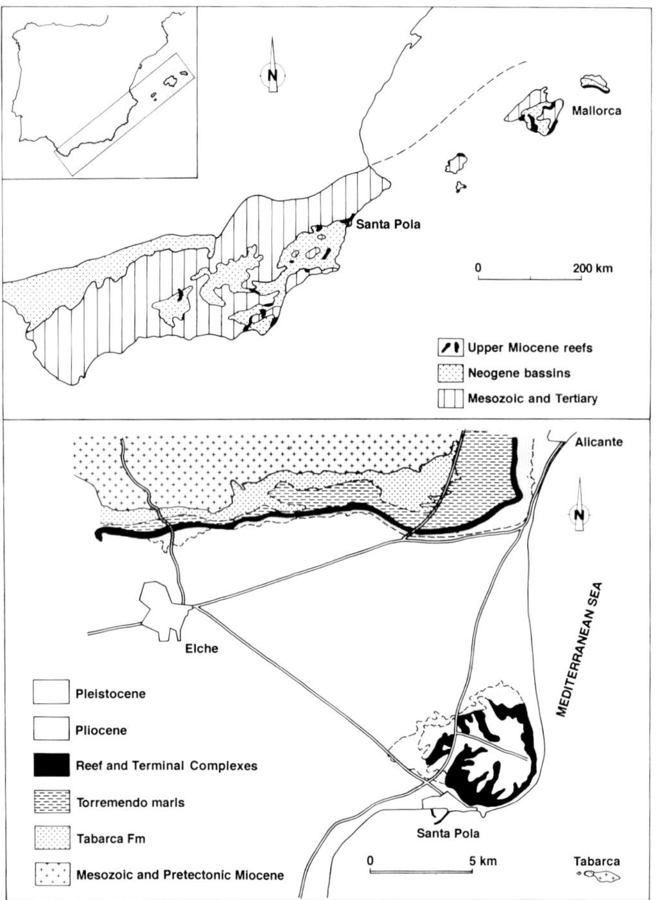


FIG. 1.—(A) Late Miocene sedimentary basins in the eastern Betic Cordillera. (B) Simplified geological map of the Alicante-Elche basin.

be assigned to this unit due to the presence of *Globorotalia pseudomiocenica* (Kampschuur and Simon, 1969; Montenat, 1977). This unit is the equivalent of the Tortonian II deposits described by Montenat (1973, 1977).

### *Torremendo Marls Unit*

This unit (Montenat, 1973) consists mainly of basinal blue-grey marls with sandstone intercalations. It attains a thickness of 400 m in the center of the basin. The Torremendo Marls Unit is Messinian in age (Bizon et al., 1972), but the lowermost part could be Upper Tortonian age. The Torremendo marls are interpreted as basinal deposits.

### *Reef Complex Unit*

The Reef Complex Unit (Esteban,1979, modified) probably overlies the Tabarca Unit deposits in the Santa Pola area, but the contact between them is not visible. The Reef Complex Unit overlies the Torremendo marls in the Alicante-Elche area. It consists of a basal coralline algae pavement, followed by a calcarenite slope, coral reef-front and lagoon deposits. This unit is up to 100 m thick at Santa Pola and ranges from 8 to 20 m in thickness in the Alicante-Elche area. The age of the Reef Complex is unresolved. Several authors date it as Messinian (Montenat,1977; Esteban, 1979).

### *Terminal Complex Unit*

This unit (Esteban 1979, modified) onlaps an erosional unconformity surface with evidence of subaerial exposure located at the top of the Reef Complex in the Santa Pola area. It

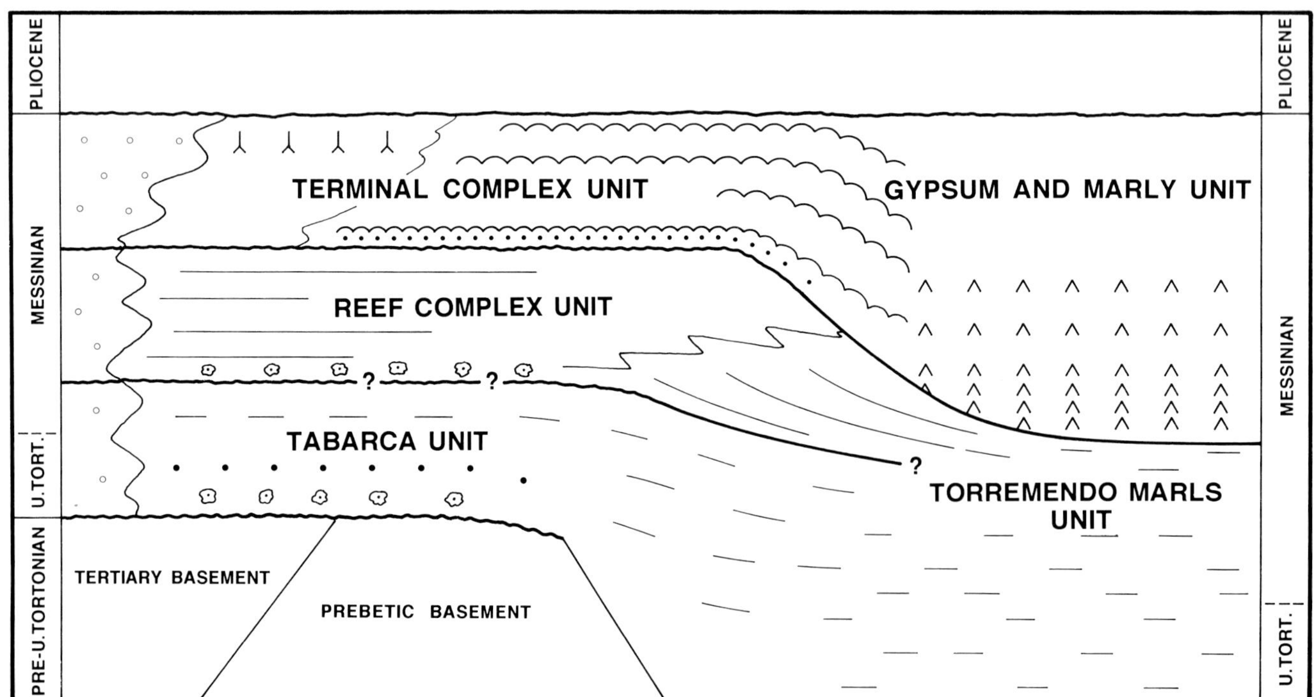


FIG. 2.—Stratigraphic cross section of the Late Miocene units in the Alicante-Elche basin.

is composed of two sub-units: a basal calcarenitic sub-unit and a cyclic stromatolitic sub-unit. The Terminal Complex Unit is Messinian in age. In the central and western parts of the Alicante-Elche area, the unit is composed of marls and calcisiltites whereas in the eastern part near Alicante the facies are similar to those of Santa Pola. This unit, 40 m thick, progrades and passes laterally into the overlying Gypsum and Marly Unit towards the center of the basin. Together the Reef Complex and the Terminal Complex Units form the Late Miocene Reef System described in this paper.

### *Gypsum and Marly Unit*

This unit (Montenat, 1973, modified) consists of alternating gypsum and marl deposits, located in the basin zone overlying the pre-evaporitic Torremendo marls. The metric-scale gypsum beds exhibit selenitic fabrics (Shearman and Ortí, 1976; Ortí and Shearman, 1977). The marls are mainly composed of planktonic and benthonic foraminifera (Martínez and Perconig, 1977; Garcín, 1989). The gypsum and marl interbeds grade upwards into shallow siliciclastic and carbonate facies (Monty, 1981; Garcín, 1989). This unit passes laterally into the Terminal Complex Unit and is Messinian in age on the basis of planktonic fauna (Garcín, 1989).

## THE REEF COMPLEX UNIT

Two reef morphologies can be recognized in the Alicante-Elche Basin: fringing belts and atoll-like reefs. The fringing reef trends from east to west over 20 km between Alicante and Elche (Fig. 1). The asymmetric atoll-like platform forms the Santa Pola hill with excellent exposures, and as it is undeformed, the present topography is close to its original depositional morphology. The reef development in Santa Pola was conditioned by the existence of an isolated Prebetic basement horst located some 15 km off the paleo-coastline (Fig. 1).

The fringing reefs contain a large variety of coral taxa and morphologies (*Porites*, *Tarbellastraea*, *Solenastraea*, *Palaeoplesiastraea*), compared to the atoll-like reefs, which is predominantly composed of *Porites* displaying only stick and dish morphologies.

### *The Reef Complex Unit in the Alicante-Elche Area*

The fringing reefs border the paleomainland with thicknesses varying from 8 m to 30 m. The reefs developed on calcarenitic beds locally containing siliciclastic grains. The main components of the calcarenitic beds are rhodoliths and branching coralline algae (*Neogoniolithon*). Secondary components are bivalves, echinoderms and fragmented corals.

The reef front comprises different lithologies from base to top. A clear framebuilder zonation is lacking in the smaller reefs. Colonies of laminar *Porites*, locally with branching sticks more than 0.5 m high, are the main framebuilders. The internal sediment is calcisiltite or marly limestone.

The most common distribution in the zoned reefs consists of dish to planar *Porites* or *Tarbellastraea* in the lowermost section, followed by branching stick colonies of *Porites* and *Tarbellastraea* up to 4 m long. Locally, calcarenite or rudstone sediments, interpreted as groove deposits, occur among the branching colonies. This zone passes laterally and upwards into a zone of massive corals (columnar morphology of Montenat, 1977; head morphology of Santisteban and Taberner, 1983). The massive corals belong to the genus *Palaeoplesiastraea* according to Montenat (1977).

The reefs are typically cut by an erosion surface or are onlapped by siliciclastic deposits, which locally contain either scattered coral colonies or patch reefs (Fig. 3A). The siliciclastic deposits (proximal fan-deltas, deltas) suggest an important siliclastic input related to a sea-level fall.

### *The Reef Complex Unit in the Santa Pola Area*

The Reef Complex in Santa Pola hill shows an asymmetric atoll-like isolated platform morphology, 5 km in diameter, with a maximum thickness, of 100 m. The reef front and reef slope facies are well exposed in the eastern margin of the hill (Fig. 3B). Towards both the inner and the western hill area, it decreases in thickness and is overlain by the Terminal Complex and Pliocene deposits. The back-reef facies crop out in some gullies of the inner hill area. The following facies associations can be recognized (Fig. 4): reef-slope, reef-front and back-reef facies.

The reef-slope facies only crops out in the lowest section of the eastern cliffs of Santa Pola hill (Esteban and Giner, 1977). It shows steep clinoforms and truncational surfaces. Two subfacies can be distinguished: (a) distal slope deposits, consisting of white to light tan calcisiltites and *Halimeda* packstones, and (b) proximal slope deposits with coral breccias, blocks of *Porites* colonies and *Halimeda* lenses. The breccia deposits exhibit fibrous marine cement locally.

The reef-front facies is up to 40 m thick, and its most noteworthy feature in Santa Pola hill is the presence of spur and groove morphologies (Fig. 3C) similar to those of Recent reefs (Land and Moore, 1977; Shinn et al., 1981). The spurs, built by coral framestones, are up to 250 m wide. The spurs are separated by grooves up to 30 m in width primarily consisting of white *Halimeda* packstones. The boundary between the lower section of the reef-front and the reef-slope facies is generally sharp, except when the grooves intersect the reef front (Esteban and Giner, 1977).

The coral framestones consist solely of *Porites* genera showing a morphological zonation of laminar, branching and massive zones. Similar morphological zonations have been described in Upper Miocene reefs from other areas (Esteban, 1979; Pomar et al., 1985, 1990; Pomar, 1991; Riding et al., 1991b) and in Recent reefs (Goreau and Goreau, 1973). The laminar-coral zone in the lowermost section of the reef front consists of *Porites* with platy morphologies up to 30 cm in diameter grading into platy-stick fabrics. The branching-coral zone constitutes the main part of

FIG. 3.—(A) General view of the Reef Complex Unit in the Santa Pola eastern cliff section. (B) Detail of spurs (s) and grooves (g). Santa Pola lighthouse (arrow). (C) Thickets of *Porites* with stick morphology showing moldic coral porosity. (D) Dense micritic crusts around vertical *Porites* sticks (P).

the reef-front facies and exhibits *Porites* buttresses up to 4 m high. The thickets consist of vertical *Porites* sticks up to 2.5 cm in diameter. Thin, dense grey crusts up to 2 cm thick of micritic dolomite surround most of the sticks (Fig. 3D). Similar micritic crusts have been interpreted as microbial in origin in the Upper Miocene reefs from Almeria (Riding et al., 1991b). The massive-coral zone is locally exposed in the uppermost section of the reef front and consists of decimeter-scale hemispheric *Porites* knobs.

The back-reef facies in Santa Pola only crops out at the bottom of some gullies (Fig. 5) and consists of meter-scale patch reefs and calcarenites with *Halimeda,* bivalves, and meter-scale horizontal beds with essentially planar stromatolites at the base which grade upwards into patch reefs.

### *The Reef Complex Unit: Discussion*

The fringing reefs with *Tarbellastraea*, *Porites*, *Palaeoplesiatraea*, and others are similar to the type A reefs of Esteban (1979). The asymmetric atoll-like reef of Santa Pola, composed solely of *Porites*, is similar to type B reefs of Esteban (1979) which are thought to be Messinian in age. A similar coral evolution has been found in Las Negras, Almeria (Franseen, 1984, written commun.). The coral *Tarbellastraea* is an important component in the Depositional Sequence 2 of Franseen and Mankiewicz (1991), whereas the Depositional Sequence 3 of the same authors consists of in-place fringing-reef framework composed almost exclusively of *Porites*. A major erosion surface with subaerial exposure evidence separates both depositional sequences.

We postulate that the fringing reefs in the study area, with its large diversity of coral genera, are older (Upper Tortonian?-Messinian in age) than the Santa Pola reef (Messinian in age) which is predominantly composed of *Porites*.

## THE TERMINAL COMPLEX UNIT

Two major facies associations can be recognized in the Alicante-Elche Basin: marly facies and stromatolitic facies. The marly facies crops out in the western and central parts of the

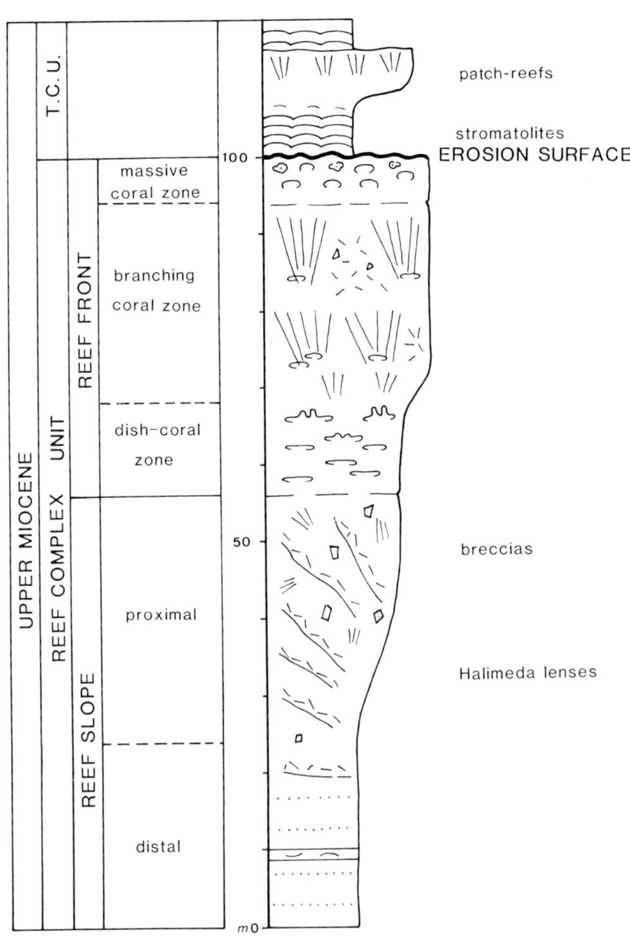


FIG. 4.—Vertical zonation of the Reef Complex Unit in the eastern cliff of the Santa Pola area.

FIG. 5.—Erosion surface (solid line) between the Reef Complex Unit (RCU) and the Terminal Complex Unit. The basal calcarenitic sub-unit (T.S.T.) onlaps (arrows) the erosion surface, and the cyclic stromatolitic sub-unit (H.S.T.). Intermediate part of the Fondo gully, south of the Santa Pola hill.

Alicante-Elche area. The stromatolitic facies are well exposed at the Santa Pola hill and at several outcrops near Alicante.

The Terminal Complex Unit in the western and central parts of the Alicante-Elche area, where it reaches 40 m in thickness, is formed by grey-green marls and calcisiltites with root casts interpreted as mangrove facies, passing upward into grey-pink marls with hydromorphic soil horizons.

### *The Terminal Complex Unit in the Santa Pola Area*

The Terminal Complex is well exposed at Santa Pola hill and reaches 40 m in thickness. It is dolomitic and is composed of two distinct lithological sub-units: a basal calcarenitic sub-unit and a cyclic stromatolitic sub-unit (Figs. 6, 7).

*Basal calcarenitic sub-unit.—*

This sub-unit, a few meters thick, onlaps the Reef Complex Unit. The basal calcarenitic sub-unit consists of either one or two meter-scale shallowing-upward cycles. Each cycle is composed of the following lithofacies (Fig. 8): (1) bioturbated wackestones with scattered pelletoidal packstone beds grading upwards into (2) bioclastic packstones, whose main components are foraminifera (Miliolidae, small Rotaliida, etc.), bivalves, and thin foliose coralline algae, locally abundant tubes (serpulids ?) also occur; (3) coquina-bioclastic packstones with gastropods and bivalves exhibiting cross stratification and current ripples; (4) marly-lime mudstones with vertical root-bioturbation; and (5) millimeter-scale pseudolaminated calcrete crust.

These shallowing upwards cycles are interpreted as the progradation of low-energy subtidal deposits (lithofacies 1) and medium-to-high energy shallow deposits (lithofacies 2 and 3), followed by a regressive trend and subaerial exposure (lithofacies 4 and 5).

*Cyclic stromatolitic sub-unit.—*

This sub-unit, up to 30 m thick, consists of four main stromatolite intervals, termed here intervals 0, 1, 2 and 3, interspersed with coral patch reefs, oolite and bioclastic facies. These intervals form four main meter-scale cycles. The stromatolites occur at the base of each cycle, grading upwards into oolite, sand or coral facies (Fig. 7). The stromatolitic intervals are good "marker beds" which can be traced over lateral distances of as much as 1 km. Four lithofacies can be distinguished: stromatolite, *Porites* patch reef, oolite, and peloidal and bioclastic facies.

*Stromatolite facies.*—Stromatolites are prominent within the Terminal Complex, with some forming spectacular 5-m-high domes. These stromatolites consist of one discontinuous interval (0) and three laterally continuous (Fig. 6) intervals (1, 2, and 3). Each interval is characterized by diverse stromatolite morphologies, sizes and microstructures. In gross morphology, they consist primarily of domal, laterally-linked stromatolite biostromes, domal bioherms or stratiform stromatolites. First-order domes are of meter and decameter scale (Fig. 9A); second-order hemispheroids (Fig. 9B), decimeter scale, are present in some stromatolite domes whereas others are relatively smoothly

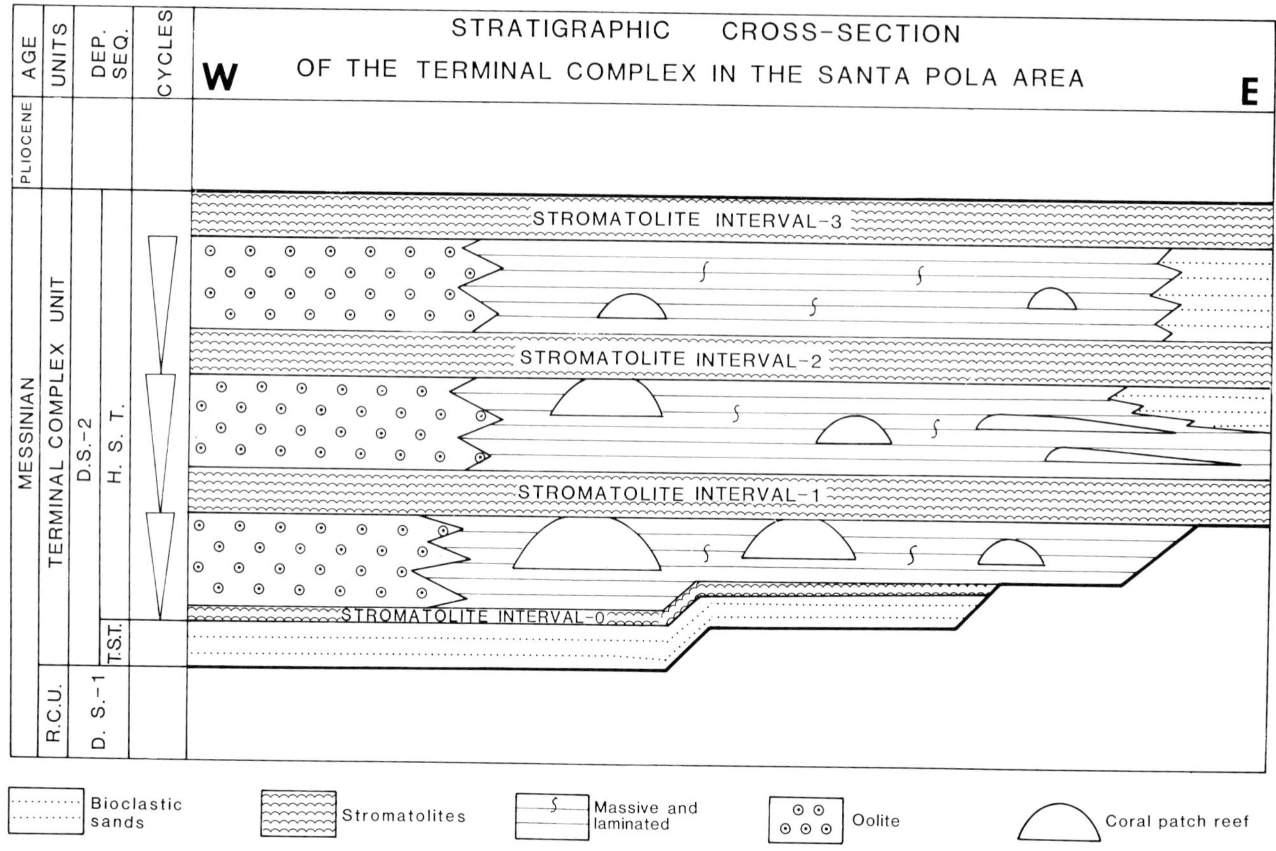


FIG. 6.—Stratigraphic cross section of the Terminal Complex Unit in the Santa Pola area.

laminated (Fig. 9C). Calcareous sands, coral patch reefs, and less frequently biohermal thrombolites, constitute the core of domal stromatolites.

Three basic varieties of even lamination occur in the stromatolites: smooth, wavy and wrinkled. The smooth lamination commonly builds broad domal structures. A composite alternating lamination also occurs, where fine micritic laminae alternate with thicker clotted laminae with abundant peloids (thrombolitic fabric). This composite lamination is peculiar to stratiform stromatolites, and some laminations display laminoid fenestrae.

Thrombolitic fabrics are particularly prominent in the stromatolitic interval 2 in the southern margin of the platform. They occur as discrete domical bioherms covered by laminated stromatolites and/or as extensive biostromes overlying laminated stromatolitic domes. Both types of thrombolites exhibit a clotted texture which consists of micrite peloids and lumps associated with small irregular fenestral pores. Elongated to tubular fenestrae varying in size occur occasionally. Locally, irregular patches of homogeneous micrite are also common.

Although the origin of peloids is still the subject of much debate, the peloids in the Santa Pola thrombolites are considered to be microbial in origin given their textural characteristics and close association with laminated stromatolites.

The lack of similar stromatolite cycles in modern environments makes it difficult to interpret this cyclic stromatolitic subunit. The widespread occurrence and high degree of inheritance of laminae, coupled with the absence of features indicative of subaerial exposure, provides evidence that stromatolites from intervals 0, 1 and 2 developed in a subtidal environment as has been postulated by Montenat (1977), Vallès (1985, 1986) and Riding et al. (1991a). The predominant stratiform morphologies and the presence of mud cracks and abundant intraformational breccias point to a shallower (submergent to emergent) environment of deposition of stromatolites from interval 3.

The striking features of Santa Pola stromatolites are: coarse-grained fabric (ooids, peloids, and locally skeletal grains), scarcity of micrite, crude lamination, coarse fenestrae, large size (up to 5 m high) and close association with mobile ooid and bioclastic sands indicative of high-energy environments. Petrographic analysis indicates that the dominant process involved in their formation was by mechanical trapping and binding of particles displaying extensive grain micritization. Complex processes of diagenesis and dolomitization are conspicuous. According to their microstructure and dimensions, they are remarkably similar to recent giant subtidal open marine stromatolites from the channels between the Exuma Islands on the eastern Bahama Bank (Dill et al., 1986; Riding et al., 1991c). These giant stromatolites are built by a diverse microscopic

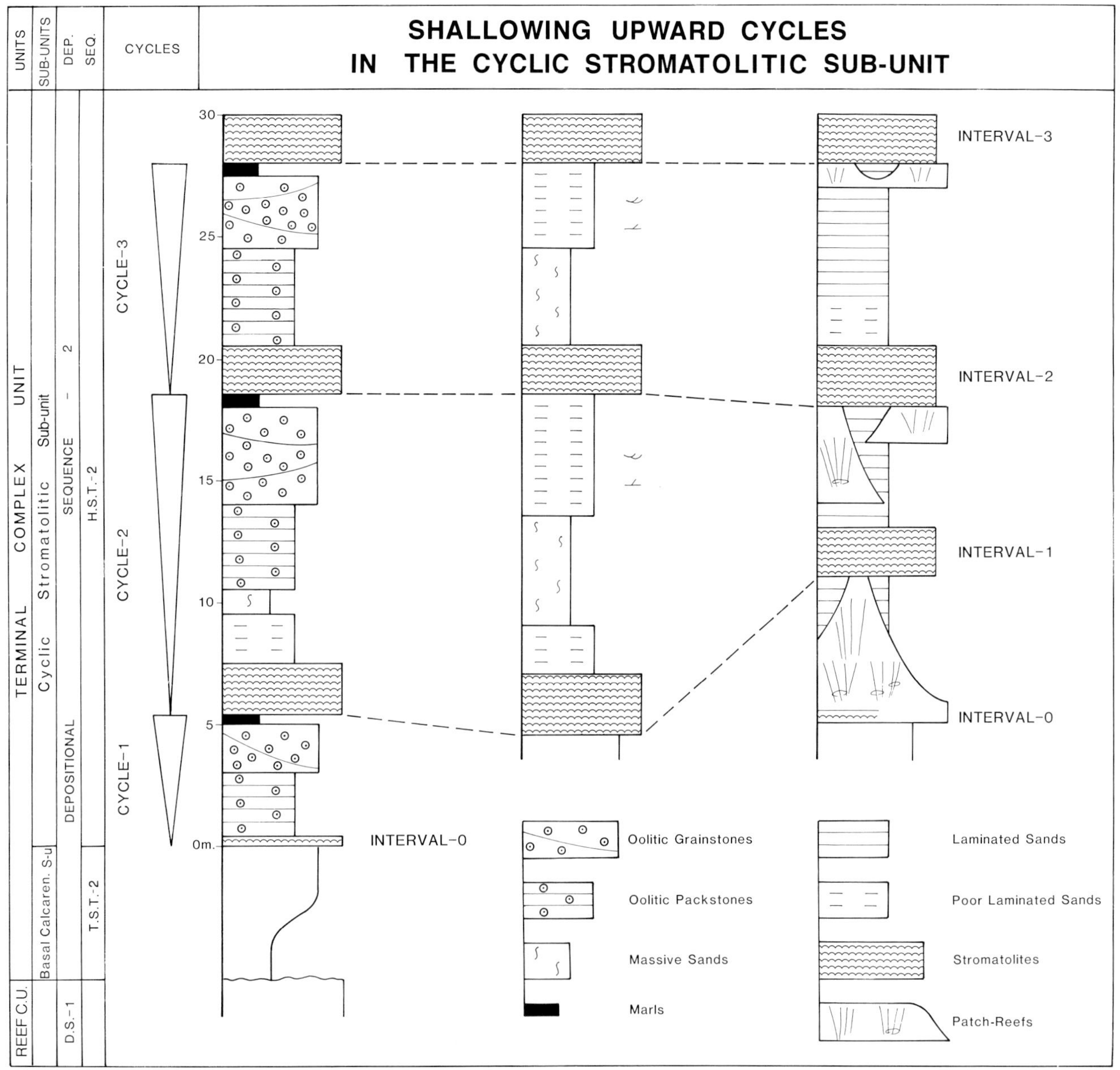


FIG. 7.—Shallowing-upward cycles of the Terminal Complex Unit in the Santa Pola area. The stromatolite facies occur at the base of each cycle and grade upwards to oolite, sand or coral facies. The left column is located in the western margin, the central column in the southern margin, and the right column in the northern margin of the platform.

community of algal eukaryotes (particularly diatoms and chlorophytes) and cyanobacteria producing large amounts of gel capable of trapping and binding coarse sediment. Although such microbiotas have not been identified in the Santa Pola stromatolites, their similarities with the modern example from the Bahamas support an analogous origin. Riding et al. (1991a) reported Miocene subtidal stromatolites and thrombolites from southern Almería, which formed in oolitic shoals probably under normal salinity conditions and compared them with giant Bahamian examples. Normal marine conditions may also be postulated for the Santa Pola stromatolites, although hypersaline conditions, particularly during accretion of stromatolites from interval 3 are not excluded.

*Porites patch-reef facies.*—The *Porites* patch reefs are sandwiched between the main stromatolite intervals. The number and size of coral patch reefs decrease towards the top of the cyclic stromatolitic sub-unit. The size of patch reefs increases towards the margin of Santa Pola platform inside each package (Fig. 6).

The *Porites* patch reefs display hemispheric (Fig. 9D) and occasionally tabular morphologies. The former vary from 1 to 15 m in height and from 1 to 12 m in width, and the latter vary

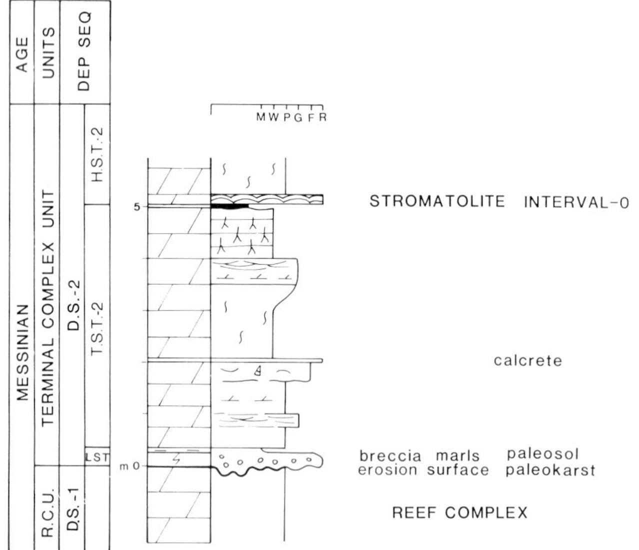


FIG. 8.—Columnar section of the basal calcarenitic sub-unit of the Terminal Complex Unit in the intermediate part of the Fondo gully. South of the Santa Pola hill.

from 1 to 3 m in height, and 10's of meters in width. The lower boundary is gradational with respect to the underlying deposits; whereas, the upper boundary is an erosion surface overlain by a coquina lag deposit.

The patch reefs are exclusively constructed of *Porites* sp. displaying a variety of morphologies: stick, dish or both (Fig. 9E). The vertical sticks vary from 0.5 m to 1 m in height. The dishes are 0.1 to 0.3 m in diameter, occurring in the lowermost parts of the patch reefs.

The matrix consists of bioclastic packstones where the main components are bivalves (*Chama* sp., *Arca* sp., *Pecten* sp.), gastropods, foraminifera and rhodoliths. The corals show little evidence of boring (*Clyona*). An important part of the reef framework is formed by a white-grey laminated micrite crust, coating the *Porites* sticks. Locally, the micrite crusts include crustose coralline algae, encrusting foraminifera and serpulids (Fig. 9F).

*Oolitic facies.*—Oolite sands occur between the different stromatolite intervals (Fig. 6) directly overlying the stromatolites either with a gradational boundary or locally lying on top of an erosion surface.

The oolitic facies show meter-scale coarsening- and thickening-upward sequences. Each sequence consists, from base to top, of the following lithologies: (1) massive lime-mudstone-wackestone-packstones, intensely bioturbated. The components are peloids, gastropods, and ooids. Centimeter-scale beds, exhibiting cryptalgal laminations are present. This facies is interpreted as restricted platform deposits; (2) oolite packstone-grainstones that gradually overlie the massive deposits. Cross lamination and current ripples are present and increase towards the top. Locally the ripples are covered by a band of fine-laminated stromatolites. This facies represents the transitional deposits between those of the restricted interior platform and those corresponding to the oolite shoal; (3) oolite grainstones with cross- and planar-bedding. The ooids, 1 mm in diameter, are the main components whereas grapestones are minor constituents. This facies is considered to be the active zone of the oolite complex. Its upper boundary is an erosion surface. (4) The oolitic shallowing sequence is capped by green marls interpreted as paleosol deposits. Similar coarsening and thickening-upward sequences have been described by Hine (1977) in Bahamas, and Smona (1984) in the Silurian of West Virginia.

*Peloidal and bioclastic facies.*—Three main sand facies (Fig. 6) have been differentiated on the basis of composition and sedimentary structures: (1) massive sands, ranging from 3 to 6 m in thickness, with a smooth wedge morphology passing laterally and vertically to faintly laminated sands. They consist of packstones composed of peloids, gastropods and bivalves. Local thin stromatolite horizons are also present. Meter-scale channel structures occur in the upper part of the massive sands. The massive sands may grade upward into sands with traction-current structures that are interpreted as subtidal deposits; (2) laminated sands occur as filling meter-scale interdomal stromatolitic depressions that display channel morphologies. The lamination consists of an alternation of packstone-grainstone laminae and thinner lime-mudstone laminae. This facies, located in the northeastern margin, is interpreted as high-energy subtidal deposits; and (3) bioclastic sands have a small wedge morphology and range in thickness from 0.5 to 3 m. This facies displays traction-current structures and is composed of bioclastic grainstones with *Porites* fragments, rhodoliths and gastropods.

### *The Terminal Complex Unit: Environments and Cyclicity*

The Terminal Complex Unit in the studied area shows a variety of facies and environments. The deposits of the Terminal Complex in the western and central parts of the Alicante-Elche area are interpreted as restricted and transitional facies. Continental influxes (distal mud flats) increase towards the west and north. These deposits are related to the paleomainland and the mangrove facies may approximately represent the paleo-coastline.

In the Santa Pola area, these deposits are related to the previous atoll-like morphology which was influenced by the existence of an isolated Prebetic horst basement. The variety and distribution of facies indicates shallow marine platform deposits.

The subsiding area located between the mainland (Alicante-Elche) and the isolated horst of Santa Pola area consists of marine deposits (Gypsum and Marly Unit) grading into the Terminal Complex deposits.

*Paleogeography of the Terminal Complex in the Santa Pola area.*—

The basal calcarenitic sub-unit is composed of one or two meter-scale shallowing-upward cycles which are similar to those reported by Belperio et al. (1988) from Holocene units of southern Australia.

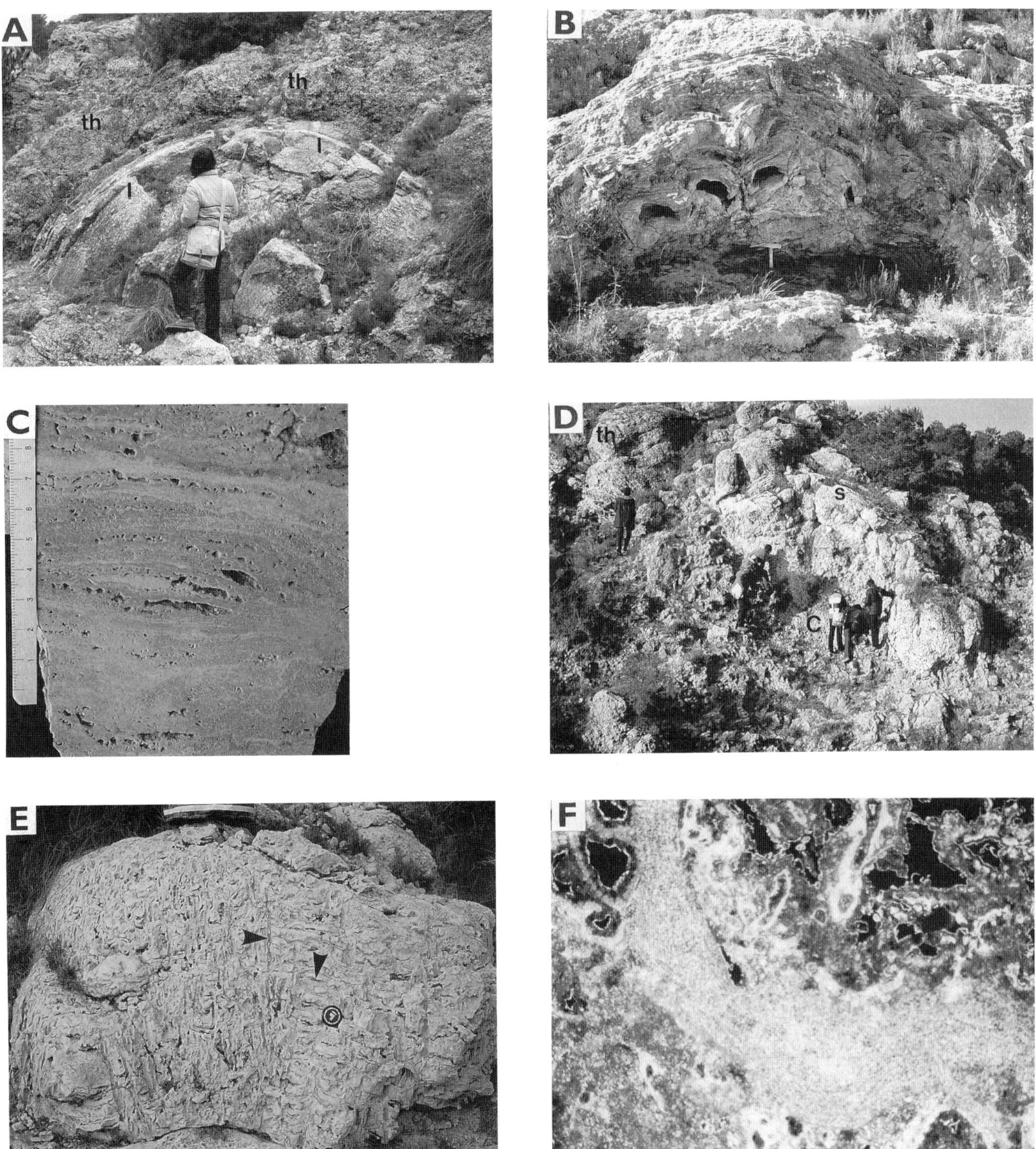


FIG. 9.—(A) Hemispheric domal stromatolite showing laminated structure (l), passing upwards to thrombolitic facies (th). Alejos gully, N-E of the Santa Pola hill. (B) Meter-scale hemispheric domal structure showing decimeter size second-order domes. Lower part of the Paco el Maño gully, S-W of the Santa Pola hill. (C) Detail of laminated stromatolite with predominantly smooth laminations alternating with coarse-grained laminae. Note coarse fenestral fabric and abundant large cavities. Polished slab. (D) Meter size coral patch reef (C) constructed of *Porites* and overlain by a stromatolitic interval (l) and thrombolitic facies (th). Alejos gully, N-E of the Santa Pola hill. (E) Detail of a patch reef showing leached branching and laminar (arrows) *Porites* morphologies. (F) Cathodoluminescence photomicrograph of dense micritic crust showing the cell structure of coralline algae.

The cyclic stromatolitic sub-unit shows a geometric progradation towards the basin center (south, west and northern directions). It consists of four main meter-scale shallowing-upward cycles, and stromatolites occur at the base of each cycle. The stromatolites grade upwards into oolite, sand or coral facies.

In contrast, the upper part of the cycles shows a variation of facies (oolite, sand and coral facies) and paleogeographic distribution. The lateral and vertical facies variation can be determined since the regressive phase deposits are sandwiched between the continuous stromatolite intervals ("marker beds") and the erosion surface.

The oolite shoals are located in the southwestern margin of the Santa Pola platform where high-energy conditions (wind, ocean, tidal currents ?) probably prevailed. The mobile oolitic shoals grade towards the interior of the Santa Pola platform to a stabilized sand flat and finally to restricted deposits. The oolite shoals display a marine sand belt disposition.

Concerning the patch reef distribution, small buildups developed in the inner area of the platform whereas the larger ones grew close to the margin of the platform, especially in the northwestern margin. This margin probably faced the open sea and was an area of moderate energy conditions.

The massive sandy facies grading upwards into sands with traction-current structures (sandy shoals, sandy wedges) are located in the southern margin of the platform where high-to-moderate energy conditions were likely to have prevailed. This facies grades towards the interior of the platform to massive sands, interpreted as restricted subtidal deposits laid down under low-energy conditions. By contrast, the laminated sands, located mainly along the northwestern margin, partially filled the inner patch reef depressions.

In brief, the Santa Pola platform is characterized by restricted deposits in the interior of the platform whereas high-to-moderate energy deposits are located along the margins albeit in different palaeogeographic settings. This facies distribution could have been controlled by the previous paleotopography and by physical conditions (wind, ocean, tidal ? currents).

*Cyclicity of the Terminal Complex Unit in the Santa Pola area.—*
The cyclic stromatolitic sub-unit consists of three complete meter-scale shallowing-upward cycles bounded by erosion surfaces. Each cycle is composed of a transgressive and a regressive phase.

The stromatolitic intervals, interpreted as subtidal deposits, represent the transgressive phase of each cycle. The thickness of the stromatolitic intervals gradually decreases towards the inner area of the platform. Similar thickness reduction in the transgressive phase of shallowing-upward cycles has been described by Gray (1981) in the Dinantian limestones from North Wales. The stromatolitic intervals show a vertical trend of shallowing and increasing salinity, intervals 0 and 1 being the deepest and interval 3 the shallowest.

The oolite, sand and coral facies represent the regressive phase in each cycle, showing shallowing-upward facies distribution in accordance with their paleogeographic position. The vertical trend of regressive phases is as follows: (1) the number and size of patch reefs decrease from Cycle-1 to Cycle-3, indicating a shallowing trend; (2) the oolitic shoals migrate basinward according to the vertical trend and display an offlap geometry; and (3) the sandy facies varies from high-energy deposits in Cycle-1 to low-to-moderate energy deposits in the remaining cycles, indicating a shallowing trend.

The cycles are capped by an erosion surface which is interpreted as a paleokarst surface. Marl layers occur locally and may be interpreted as paleosol deposits.

## DEPOSITIONAL SEQUENCES

The Tabarca Unit deposits, which are related to paleohigh areas, could be related to Depositional Sequence 1 (DS1) of Franseen and Mankiewicz (1991). The fringing reefs of the Reef Complex Unit in the Alicante-Elche area, which contain a variety of corals, could be related to the Depositional Sequence 2 of Franseen and Mankiewicz (1991) and Franseen and Goldstein (this volume). The Sequence Boundary corresponds to the erosional surface between the Torremendo Marls Unit and the basal coralline pavement of the Reef Complex Unit. The basal coralline pavement is interpreted as the transgressive systems tract whereas the reef facies corresponds to the highstand systems tract. However, the outcrops do not permit the observation of the geometric relationships between these units.

Santa Pola hill allows a detailed depositional sequence analysis. In this area, the Late Miocene deposits can be divided into two depositional sequences (Figs. 10, 5), here termed depositional sequence 1 (= Reef Complex Unit in Santa Pola area) and depositional sequence 2 (= Terminal Complex Unit).

### *Depositional Sequence 1*

The lower boundary of this depositional sequence and the transgressive deposits do not crop out in the Santa Pola area. The reef facies with progradation geometry corresponds to the highstand systems tract, here called HST-1. This depositional sequence could be related to the depositional sequence 3 defined by Franseen and Mankiewicz (1991) and Franseen and Goldstein (this volume) in Las Negras, Almería.

### *Depositional Sequence 2*

The lower boundary is an unconformable erosional surface between the Terminal Complex and the Reef Complex deposits (Montenat, 1977; Monty et al., 1980; Vallès, 1986; Rouchy et al., 1986). Similar erosion surfaces have been reported in other Messinian basins (Esteban and Giner, 1980; Dabrio et al., 1981; Pomar et al., 1985, 1990; Saint-Martin, 1990; Saint-Martin and Rouchy, 1990; Franseen and Mankiewicz, 1991; Riding et al., 1991b; Martin et al., 1993; Franseen and Goldstein, this volume). The regional morphology of the unconformity in the Santa Pola area shows a series of platform surface areas located

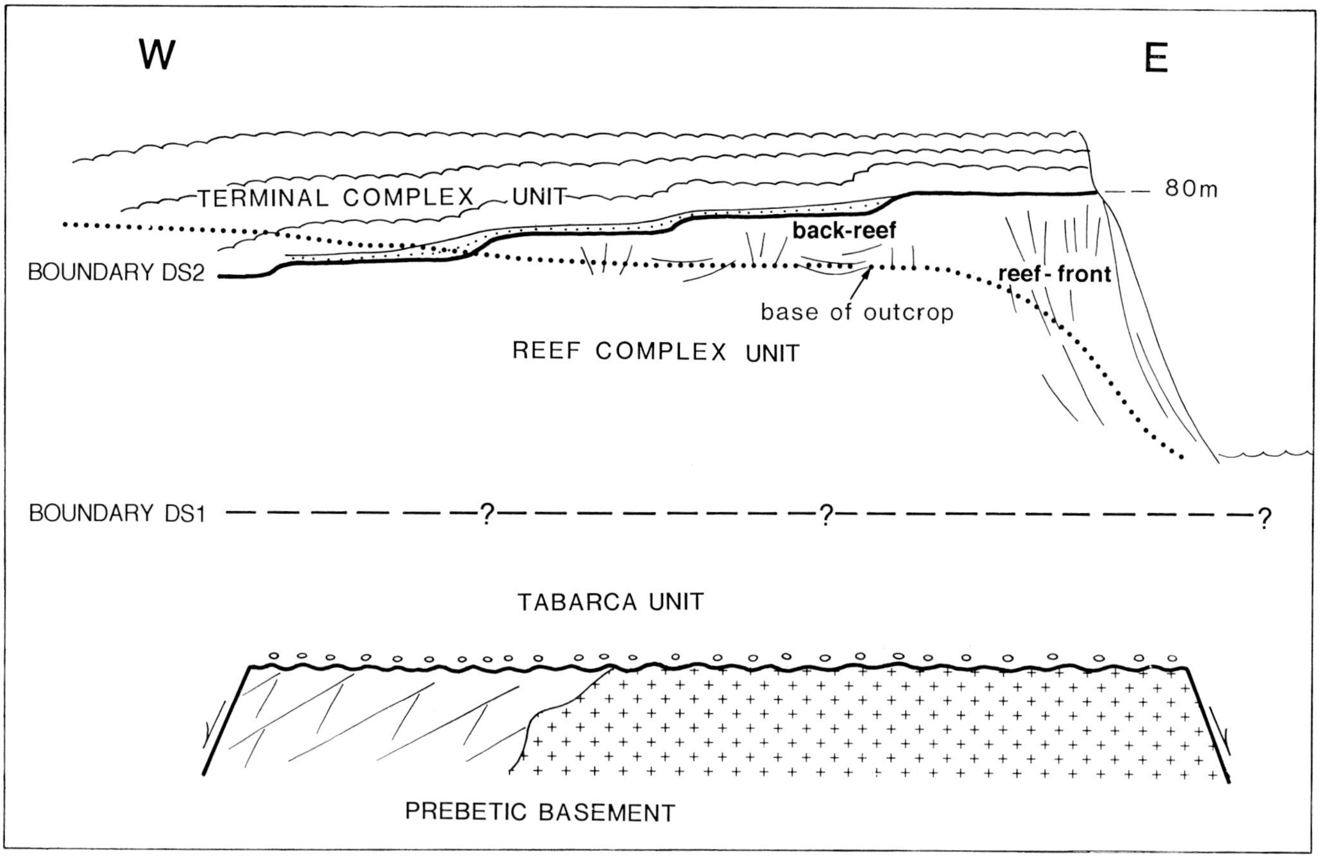

FIG. 10.—General cross section of the Santa Pola area.

at different levels (Figs. 10, 11). The main platform surfaces are placed approximately at 100 m, 80 m, 60 m, and 45 m respectively. The platform surfaces are 100's meters in scale, dip gently and exhibit an irregular and hummocky morphology being cut by meter-scale steps. They are interpreted as coastal marine planation terraces, and the steps correspond to paleocliffs. The platform surfaces were subsequently affected by karstification and calichification processes after planation with continuing sea-level fall. The upper boundary of depositional sequence 2 is an erosion surface with evidence of subaerial exposure which is overlain by Pliocene marine deposits.

The lowstand systems tract of depositional sequence 2 (LST-2) is represented by an erosional surface together with the continental deposits related to this surface. The basal calcarenitic sub-unit onlaps onto the Reef Complex Unit and corresponds to the transgressive systems tract (TST-2). The cyclic stromatolitic sub-unit displays a geometric progradation towards the basin center and represents the highstand systems tract (HST-2).

## DIAGENESIS

### *Dolomitization*

The Reef Complex Unit and the Terminal Complex in the Santa Pola area show a pervasive nondestructive dolomitization, varying in intensity according to the stratigraphic unit and

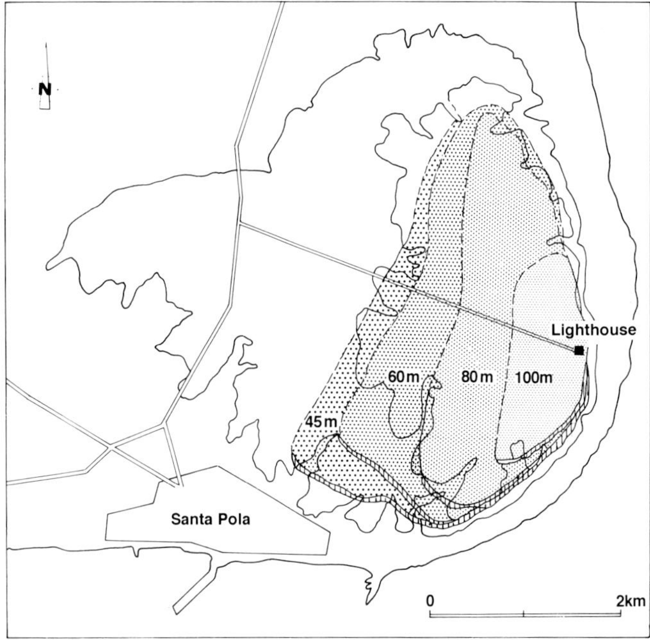


FIG. 11.—Map illustrating the boundary between the Reef Complex Unit and the Terminal Complex Unit. Stippled patterns show the different platform surface elevations and morphologies.

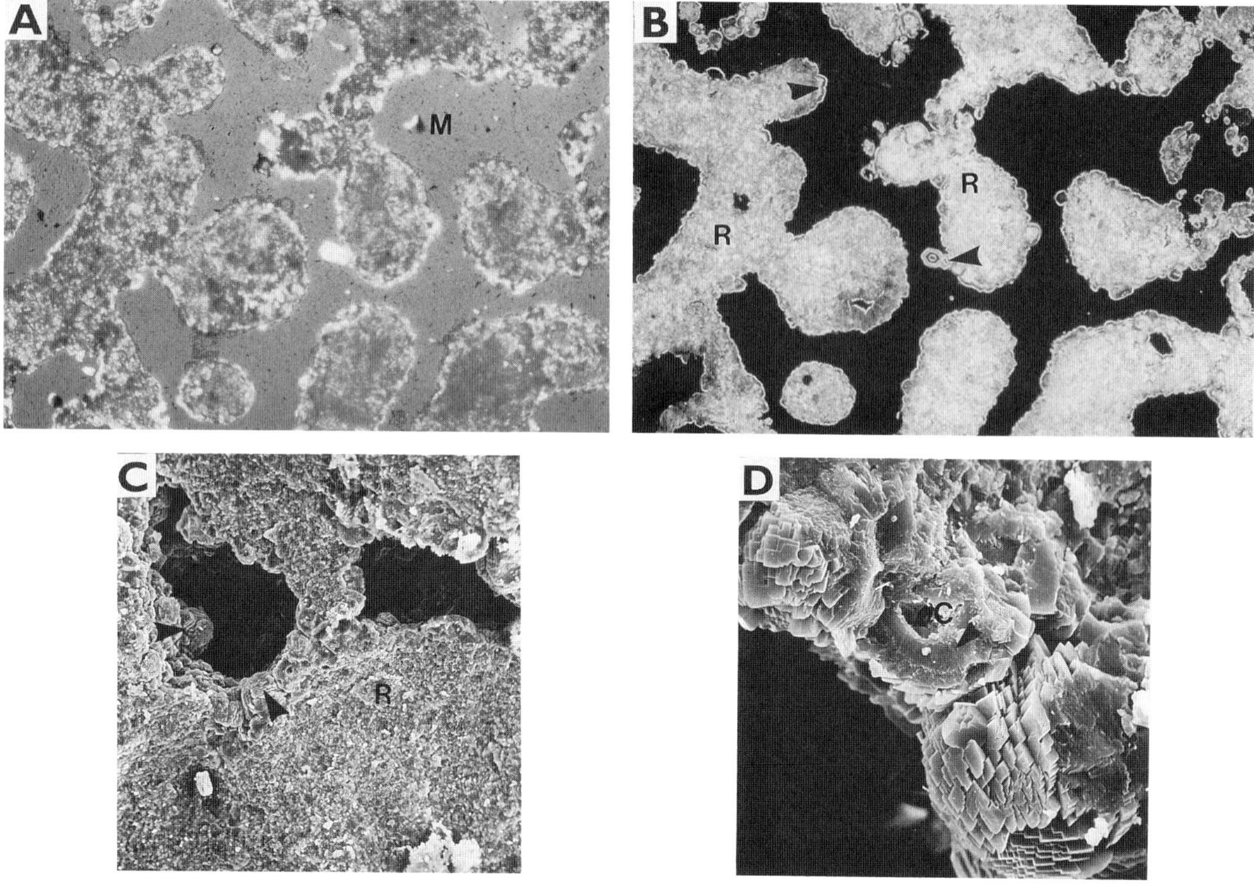


FIG. 12.—(A) Thin-section photomicrograph of *Porites* structure showing moldic porosity (m) and dolomitized matrix. (B) Cathodoluminescence photomicrograph of replacive dolomitic fabric (R), and zoned luminescent and nonluminescent dolomitic cements (arrows). (C) SEM photomicrograph of fine-grained replacive dolomitic fabric (R) and dolomitic cement (arrow) infilling vug porosity. (D) Detail of dolomitic cement showing crystals with dissolved core (C) and zoned rim (arrow).

depositional facies. The Terminal Complex deposits are more dolomitized than those of the Reef Complex.

Two main dolomite fabrics may be recognized (Fig. 12A-D): (1) microcrystalline to subhedral-euhedral (7-45 μm) replacement fabric. This dolomite preserves original Mg-calcite skeletons (foraminifera and coralline algae) which appear better preserved than the aragonite skeletons (corals). This dolomite is non-luminescent to slightly dull and (2) euhedral to rounded-anhedral crystals (7-30 μm) constitute cements and rims lining vugs, interparticle and moldic porosity. These crystals exhibit cloudy or partially dissolved cores and clear, zoned rims. The latter may present a luminescent-nonluminescent-luminescent zonation (Fig. 12A, B).

These dolomite fabrics are broadly comparable with those described by Coniglio et al. (1988) in Middle Miocene units from the Red Sea and with those described by Oswald et al. (1990) and Meyers and Lu (1994) in Upper Miocene strata from Mallorca and Almeria, respectively.

### *Stable Isotopes*

Oxygen isotope values from bulk dolomitic samples of the Terminal Complex Unit in the Santa Pola area have an average of +4.1 ‰ (PDB) varying from +3.3 to +4.7 ‰ and the carbon isotope data are on average of +1.7 ‰ (PDB) ranging from +0.9 to +2.5 ‰ (Fig. 13). These values are similar to those reported by Oswald et al. (1990) from the dolomitized Late Miocene Reef Complex in Mallorca.

The heavy stable isotope values of the Santa Pola dolomites suggest hypersaline influx related to the Messinian evaporitic deposition. Similar values have been reported from sabkha dolomites in Abu Dhabi (McKenzie, 1981).

Locally, the Terminal Complex deposits contain poikilotopic calcite cement which postdates dolomitization. This cement (100 μm to 1 mm) is nonluminescent and presents lighter stable isotope values ($\delta^{18}O$ = -0.6 ‰, $\delta^{13}C$ = -2.4 ‰ PDB) suggesting a phreatic meteoric origin.

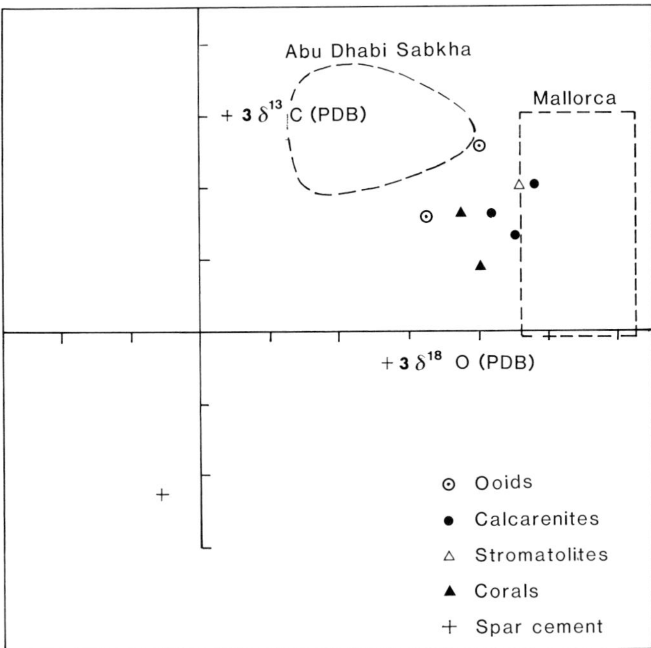


FIG. 13.—Cross plot of oxygen versus carbon stable isotopic data from different facies of the Terminal Complex Unit in the Santa Pola area. Dashed areas represent isotopic data from Mallorca after Oswald et al. (1990) and from Abu Dhabi after McKenzie (1981).

## CONCLUSIONS

1. The Late Miocene deposits in the Alicante-Elche basin consist of the following units from base to top: the Tabarca Unit, Torremendo Marls Unit, Reef Complex Unit, Terminal Complex Unit and the Gypsum and Marls Units. The age of these deposits varies from Upper Tortonian to Messinian.

2. The Late Miocene deposits in the studied area are divided into two depositional sequences here termed depositional sequence 1 (= Reef Complex Unit) and depositional sequence 2 (= Terminal Complex Unit).

3. The Reef Complex Unit comprises two reef morphologies: the fringing-belt morphology along the mainland in the Alicante-Elche area and the atoll-like platform morphology in offshore areas (Santa Pola area). The fringing reefs show coral diversity and morphological variations. The atoll-like reef consists of laminar and stick morphologies of *Porites*. The fringing reefs are older (Upper Tortonian?-Messinian) than the atoll-like reef of Santa Pola (Messinian).

4. The boundary between the Reef Complex Unit and the Terminal Complex Unit is an erosional surface modified by subaerial exposure which shows several platforms located at different levels (marine terraces), connected by steps (paleocliffs) in the Santa Pola hill.

5. The Terminal Complex Unit of the Santa Pola hill consists of a basal calcarenitic sub-unit (transgressive systems tract) and a cyclic stromatolitic sub-unit (highstand systems tract). The cyclic stromatolitic sub-unit in the Santa Pola platform shows restricted deposits (mainly massive sands) in the interior and high-to-moderate energy deposits (oolite shoals, massive sands with traction-current structures, and patch reefs) along the margins albeit in different palaeogeographic settings. The cyclic stromatolitic sub-unit consists of four main meter-scale shallowing-upward cycles bounded by erosional surfaces. The stromatolites occur at the base of each cycle and are interpreted as subtidal deposits in the intervals 0, 1 and 2 and as subtidal-intertidal deposits in interval 3. They grade upwards into different facies (oolites, patch reefs and sandy deposits).

6. The most conspicuous diagenetic feature of the Reef Complex Unit and the Terminal Complex Unit at Santa Pola is the intensive and pervasive but nondestructive dolomitization. The heavy stable isotope values ($\delta^{18}O$ = +3.3 to +4.7 ‰; $\delta^{13}C$ = +0.9 to +2.5 ‰) of the Santa Pola dolomites suggest a hypersaline influx, which is related to the Messinian evaporite event.

7. The geometric relationship between the carbonate platform deposits and the basinal marly-evaporitic deposits shows a wedge-on-wedge disposition. Similar relationships have been reported in different basins: the Silurian Michigan Basin, the Permian Zechstein Basin and the Messinian Western Mediterranean Basin.

## ACKNOWLEDGMENTS

The authors are deeply indebted to Anna Trave for the isotopic analysis, to Craig Docherty and George von Knorring for revision of the English text, and to Josep Agulló for drafting the figures. We are grateful to Mateo Esteban for numerous discussions in the field and for comments on the manuscript. The manuscript was considerably improved by the reviews of E. K. Franseen, R. Riding and D. S. Ulmer-Scholle. This work was supported by DGICYT Grants 91-0097/C01 to I.Z. and 91-0801 to F.C.

## REFERENCES

BELPEIRO, A. P., GOSTIN, V. A., CANN, J. H., AND MURRAY-WALLACE, C. V., 1988, Sediment-organism zonation and the evolution of Holocene tidal sequences in southern Australia, *in* de Boer, P. L., Van Gelder, A. and Nio, S.D. eds., Tide-Influenced Sedimentary Environments and Facies: Dordrecht, D. Reidel Publishing Company, p. 475-497.

BERNET-ROLLANDE, M. C., MAURIN, A. F., AND MONTY, CL., 1980, *Porites* reef vs. stromatolites at Santa-Pola, Spain: a Miocene sedimentological puzzle: Paris, 26e Congrès Géologique International, Résumés, v. 2, p. 435.

BIZON, G., BIZON, J.-J., AND MONTENAT, C., 1972, Le Miocène terminal dans le Levant espagnol: Revue de l' Institut Français du Pétrole, v. 27, p. 831-857.

CALVET, F., ZAMARREÑO, I., AND TRAVE, A., 1991, Los sistemas arrecifales del Mioceno superior en la cuenca de Alacant-Elx: Vic, I Congreso del Grupo Español del Terciario, Comunicaciones, p. 52-54.

CALVET, F., ZAMARREÑO, I., AND TRAVE, A., 1994, The Upper Miocene reef system of the Alicante-Elche Basin (Southeast, Spain). Miocene reefs ans carbonate platforms of the Mediterranean: Marseille, Interim Colloquium Regional Committee on Mediterranean Neogene Stratigraphy, p. 11.

CONIGLIO, M., JAMES, N. P., AND AISSAOUI, D. M., 1988, Dolomitization of Miocene carbonates, Gulf of Suez, Egypt: Journal of Sedimentary Petrology, v. 58, p. 100-119.

DABRIO, C., ESTEBAN, M., AND MARTIN, J. M., 1981, The coral reef of Níjar, Messinian (Uppermost Miocene), Almería Province, S.E.

Spain: Journal of Sedimentary Petrology, v. 51, p. 521-539.

Dill, F., Shinn, E. A., Jones, A. T., Kelly, K., and Steinen, R. P., 1986, Giant subtidal stromatolites forming in normal salinity waters: Nature, v. 324, p. 55-58.

Esteban, M., 1977, El arrecife de Santa Pola, *in* Salas, R., ed.,1er Seminario práctico de Asociaciones arrecifales y evaporíticas: Barcelona, Universidad de Barcelona, p. 4.11-4.51.

Esteban, M., 1979, Significance of the Upper Miocene coral reefs of the Western Mediterranean: Palaeogeography, Palaeoclimatology, Palaeoecology, v. 29, p. 169-188.

Esteban, M. and Giner, J., 1977, Field-guide to Santa Pola reef, UNESCO-IUGS IGCP Project nº96 Messinian Correlation: Málaga, Messinian Seminar nº 3, Field Trip 5, p. 23-30.

Esteban, M. and Giner, J., 1980, Messinian coral reefs and erosion surfaces in Cabo de Gata (Almería, SE Spain): Acta Geológica Hispánica, v. 15, p. 97-104.

Esteban, M. and Pray, L. C., 1981, A guidebook to the Tertiary reef carbonates and associated facies. Southeastern Spain and the Balearic platform: London, Internal Report, Erico, Company Ltd, 74 p.

Feldmann, M. and McKenzie, J. A., 1993, Neogene Stromatolites from Santa Pola, SE-Spain: Strasburg, EUG VII, Abstract supplement nº 1 Terra Nova, v. 5, p. 278.

Franseen, E. K. and Mankiewicz, C., 1991, Depositional sequences and correlation of middle (?) to late Miocene carbonate complexes, Las Negras and Níjar areas, southeastern Spain: Sedimentology, v. 38, p. 871-898.

Garcin, M., 1989, Le bassin de San Miguel de Salinas (Alicante, Espagne). Relations entre contexte structuro-sédimentaire et dépôts évaporitiques et carbonates au Messinien: Unpublished Ph. D. Dissertation, University of Paris-Sud, Orsay, 279 p.

Goreau, T. F. and Goreau, N. I., 1973, The ecology of Jamaican reefs. II. Geomorphology, zonation and sedimentary phases: Bulletin of Marine Sciences, v. 23, p. 399-464.

Gray, D. I., 1981, Lower Carboniferous Shelf Carbonate Palaeoenvironments in North Wales: Unpublished Ph.D. Dissertation, University of Newcastle upon Tyne, Newcastle upon Tynes, 321 p.

Hine, A. C., 1977, Lily Bank, Bahamas: History of an active oolite sand shoal: Journal of Sedimentary Petrology, v. 47, p. 1554-1581.

Kampschuur, W. and Simon, O. J ., 1969, Sur la géologie de l'île de Tabarca (Prov. d'Alicante, Espagne) et sa position tectonique dans la zone bétique (Cordillères bétiques): Comptes Rendus Sommaires de la Société Géologique de France, v. 2, p. 37-38.

Land, L. S. and Moore, C. H. Jr., 1977, Deep forereef and Upper Island Slope, North Jamaica, *in* Frost, S. H., Weiss, M. P. and Saunders, J. B., eds., Reefs and Related Carbonates-Ecology and Sedimentology: Tulsa, American Association of Petroleum Geologists Studies in Geology 4, p. 53-65.

Martin, J. M., Braga, J. C., and Riding, R., 1993, Siliciclastic stromatolites and thrombolites, Late Miocene, S.E. Spain: Journal of Sedimentary Petrology, v. 63, p. 131-139.

Martinez Diaz, C. and Perconig, E., 1977, Observaciones micropaleontológicas y estratigráficas sobre el corte de San Miguel de Salinas cerca del Km. 16 de la carretera de Orihuela: Málaga, UNESCO-IUGS IGCP Project nº96 Messinian Correlation, Messinian Seminar 3, Field Trip 5, p. 31-38.

McKenzie, J. A., 1981, Holocene dolomitization of calcium carbonate sediments from the coastal sabkhas of Abu Dhabi, UAE: a stable isotope study: Journal of Geology, v. 89, p. 185-198.

Meyers, W. and Lu, F., 1994, Dolomitization of a Miocene Reef Complex, Nijar, Spain. Miocene reefs and carbonate platforms of the Mediterranean: Marseille, Interim Colloquium Regional Committee on Mediterranean Neogene Stratigraphy, p. 34.

Montenat, C., 1973, Les formations néogènes et quaternaires du Levant espagnol: Unpublished Ph.D. Dissertation, University of Paris-Sud, Orsay, 1170 p.

Montenat, C., 1977, Les bassins néogènes du Levant d'Alicante et de Murcia (Cordillères Bétiques orientales-Espagne). Stratigraphie, paléogéographie et évolution dynamique: Documents des Laboratoires de Géologie de la Faculté des Sciences de Lyon, v. 69, p. 1-345.

Montenat, C., Ott d'Estevou, P., Larouziere, P. D., and Bedu, P., 1990, Originalité géodynamique des bassins Néogènes du Domaine Bétique Oriental (Espagne), *in* Berthon, J. L., Burollet, P. F., and Legrand, Ph., eds., Gènese et évolution des bassins sédimentaires: Notes et Mémoires, nº 21, p. 11-49.

Monty, Cl., 1981, Le Miocène supérieur de la région de Benejuzar (Province d'Alicante), et stromatolites associés: Annales de la Societé Géologique de Belgique, v. 104, p. 109-114.

Monty, Cl., Maurin, A. F., and Bernet-Rollande, M. C., 1980, Miocene stromatolitic reefs from SE Spain: Paris, 26e Congrès Géologique International, Résumés, v. 2, p. 521.

Orti, F. and Shearman, D. J., 1977, Estructuras y fabricas deposicionales en las evaporitas del Mioceno superior (Messiniense) de San Miguel de Salinas (Alicante, España): Publicaciones del Instituto de Investigaciones Geológicas de la Diputación de Barcelona, v. 32, p. 5-54.

Oswald, E. J., Meyers, W. J., and Pomar, L., 1990, Dolomitization of an Upper Miocene Reef Complex, Mallorca, Spain: Evidence for a Messinian dolomitizing Mediterranean Sea (abs.): American Association of Petroleum Geologists Bulletin, v. 74, p. 735.

Pomar, L., 1991, Reef geometries, erosion surfaces and high-frequency sea-level changes, Upper Miocene Reef Complex, Mallorca, Spain: Sedimentology, v. 38, p. 243-269.

Pomar, L., Fornos, J. J., and Rodriguez-Perea, A., 1985, Reef and shallow carbonate facies of the Upper Miocene of Mallorca, *in* Milà, M. D. and Rosell, J., eds., 6th European Regional Meeting Excursion Guidebook: Lleida, International Association of Sedimentologists and Universitat Autònoma de Barcelona, p. 495-518.

Pomar, L., Rodriguez-Perea, A., Sabat, F., and Fornos, J., 1990, Neogene stratigraphy of Mallorca Island, *in* Agustí, J., Domènech, R, Julià, R. and Martinell, J., eds. Iberian Neogene Basins: Sabadell, Paleontologia i Evolució, Memòria Especial 2, p. 269-320.

Riding, R., Awramik, S. M., Winsborough, B. M., Griffin, K. M., and Dill, R., 1991c, Bahamian giant stromatolites: microbial composition and surface mats: Geological Magazine, v. 128, p. 227-234.

Riding, R., Braga, J. C., and Martin, J. M., 1991a, Oolite stromatolites and thrombolites, Miocene, Spain: analogues of Recent giant Bahamian examples: Sedimentary Geology, v. 71, p. 121-127.

Riding, R., Martin, J. M., and Braga, J. C., 1991b, Coral-stromatolite framework, Upper Miocene, Almería, Spain: Sedimentology, v. 38, p. 799-818.

Rouchy, J. M., Saint-Martin, J. P., Maurin, A., and Bernet-Rollande, M. C., 1986, Evolution et antagonisme des communautés bioconstructrices animales et végétales à la fin du Miocène en Méditerranée occidentale: biologie et sédimentologie: Bulletin du Centre de Recherches, Exploration-Production Elf-Aquitaine, nº 10, p. 333-348.

Saint-Martin, J. P., 1990, Les formations récifales coralliennes du Miocène supérieur d'Algérie et du Maroc: Paris, Mémoires du Muséum National d'Histoire Naturelle, Sciences de la Terre, v. 56, 351 p.

Saint-Martin, J. P. and Rouchy, J. M., 1990, Les plates-formes carbonatées messiniennes en Méditerranée occidentale: leur importance pour la reconstitution des variations du niveau marin au Miocène terminal: Bulletin de la Société Géologique de France, t. 6, p. 83-94.

Santisteban, C. and Taberner, C., 1983, Shallow marine and continental conglomerates derived from coral reef complexes after dessication of a deep marine basin: the Tortonian-Messinian deposits of the Fortuna Basin, S.E. Spain: Journal of the Geological Society, London, v. 140, p. 401-411.

Shearman, D. J. and Orti, F., 1976, Upper Miocene gypsum: San Miguel de Salinas, S.E. Spain: Memoria della Societa Geologica Italiana, v. 16, p. 327-339.

Shinn, E. A., Hudson, J. H., Robbin, D. M., and Lidz, B., 1981, Spurs and grooves revised: construction versus erosion Looe Key Reef, Florida: Manila, Proceedings Fourth International Coral Reef Symposium, v. 1, p. 475-483.

Smona, R. A., 1984, Sedimentary facies and the early diagenesis of an oolite complex from the Silurian of West Virginia,*in* Harris, P. M., ed., Carbonate Sands - A Core Workshop: Tulsa, Society of Economic Paeontologists and Mineralogists Core Workshop 5, p. 20-57.

Valles, D., 1985, Sedimentologia dels materials carbonatats del Miocè Superior de l'area de Sta Pola: Unpublished M.S. Thesis, University of Barcelona, Barcelona, 121 p.

Valles, D., 1986, Carbonate facies and depositional cycles in the Upper Miocene of Santa Pola (Alicante, SE Spain): Revista d'Investigacions Geològiques, v. 42/43, p. 45-66.

# UPPER MIOCENE REEF COMPLEX OF THE LLUCMAJOR AREA, MALLORCA, SPAIN

LUIS POMAR,
*Departament de Ciencies de la Terra, Universitat de les Illes Balears, 07071 Palma de Mallorca, Spain*
WILLIAM C. WARD, AND DARRYL G. GREEN
*Department of Geology and Geophysics, University of New Orleans, LA 70148, USA*

Abstract: On all the Balearic islands (Mallorca, Menorca, Ibiza, and Formentera), the Upper Tortonian-Lower Messinian part of the post-orogenic sedimentary section is the Reef Complex. This unit is composed of carbonate rocks deposited as progradational reef-rimmed platforms with off-reef open-shelf, forereef-slope, reef-core and back-reef lagoon facies associations. The most extensive progradation of a Late Miocene carbonate platform in the western Mediterranean occurred on the Llucmajor Platform in the area of present-day southwestern Mallorca. Here as much as 20 km of basinward progradation took place during Late Tortonian and Early Messinian time. Core-hole data from the Llucmajor area show that the Reef Complex limestone and dolomite are up to 100 m thick and are spread over a platform of about 15 km by 20 km. The southern and western portions of the Llucmajor Platform coral-reef complex are superbly exposed in high vertical sea cliffs.

The lithofacies units of the Reef Complex are defined on the basis of their lithology, constituents, stratification and geometric relationships. There are two main types of open-shelf lithofacies: (1) dolomitized grainstone-packstone with abundant red algae and (2) packstone-wackestone with planktonic foraminifers. Interfingering landward with the finer-grained open-shelf lithofacies are dolomitized skeletal grainstone, packstone, and wackestone of the reef-slope deposits. These slope deposits are characterized by basinward-dipping clinoforms of variable thickness and lateral extent, depending on configuration of the forereef platform. The slope rocks interfinger landward with massive coral-reef limestone and dolostone. The reef rocks interfinger landward with flat-lying lagoon lithofacies composed of partly dolomitized packstone, wackestone, and grainstone.

The reef framework on the Llucmajor Platform is constructed mainly of only one to two genera of corals, *Porites* or *Porites* and *Tarbellastraea*. Along the southern part of the platform, depositional strike of the reef tracts was N35°W to N60°W, and progradation was toward the southwest. From the perspective of sea-cliff outcrops in this area, there are three main types of reef tracts: (1) discontinuous, mound-like *Porites* and *Tarbellastraea* reefs cropping out in the Vallgornera area, (2) more continuous *Porites* and *Tarbellastraea* reefs cropping out from Cala Beltrán to Els Bancals and (3) *Porites* reef tracts cropping out from Els Bancals to Cap Blanc. The youngest reefs of the Llucmajor Platform, which crop out along the western coast, are constructed of *Porites* and *Tarbellastraea,* and core data show that most reefs throughout the platform contain both these genera. The most complete sequence of the *Porites*-framework reef crops out on the high Cap Blanc sea cliff, where the reef is part of an aggradational sequence. Here there are three zones of coral morphology: (1) a lower zone of "dish coral", (2) a middle zone of "branching coral" and (3) an upper zone of "massive coral".

The Reef Complex has a complicated stratigraphy of accretional units, reflecting several orders (probably 4th through 7th) of high-frequency oscillations in relative sea level. These fluctuations in sea level produced the most characteristic facies relationship within the Reef Complex: progradation with vertical shifts (upward and downward) of the reef-core and associated lithofacies. The complex architecture of the Llucmajor carbonate complex can only be adequately defined from the reef-core facies stacking patterns in the dip direction. Changes in stacking patterns allow definition of four systems tracts: "low stillstand," "aggradational," "high stillstand" and "offlapping." These systems tracts are identified in the accretional units of every scale, except the basic accretional unit, the "sigmoid." The fore-reef slope and off-reef open-shelf facies are mainly built up by aggradational systems tracts separated by condensed intervals of fine-grained distal slope and open-platform carbonates. These distally condensed intervals correlate landward with the high-stillstand, offlapping, and low-stillstand systems tracts. Progradation of the reef systems on the southern part of the platform was more extensive during sea-level falls on a gentle depositional profile. The subsequent sea-level rises created wide lagoons, which apparently enhanced carbonate production and downslope shedding of sediment. On steeper topographic gradients, relatively minor reef progradation took place during sea-level falls, and only small back-reef lagoons were created during the subsequent sea-level rises. Barrier reefs with extensive lagoons and patch reefs formed during relative sea-level rises of different orders of magnitude; fringing reefs developed during sea-level falls.

The most conspicuous diagenetic features of this reef complex are: (1) extensive secondary porosity produced by dissolution of originally aragonitic constituents and (2) pervasive dolomite in much of the complex.

## INTRODUCTION

The most extensive progradation of a Late Miocene carbonate platform in the western Mediterranean occurred on the Llucmajor Platform in the area of present-day southwestern Mallorca (Fig. 1). Core-hole data from the Llucmajor area show a belt of reef and shelf limestone and dolomite about 20 km wide and up to 100 m thick. The southern portion of the Llucmajor Platform coral-reef complex is superbly exposed in the high vertical sea cliffs along the southwestern coast of Mallorca (Fig. 1A). These clean, three-dimensional exposures reveal in uncommon detail the complexity of facies geometries in a generally progradational reef complex (Pomar, 1991). Furthermore, this reef complex is noteworthy because its mid-sea location, away from continental-margin tectonic influences and terrigenous influx, may reflect paleo-oceanographic conditions more purely than do most other Upper Miocene reefs of the western Mediterranean. In this paper, we describe the facies associations within the aggradational and progradational stages of the different high-order depositional sequences cropping out along the southwestern coast of Mallorca.

Although Pomar (1991, 1993) and Pomar and Ward (1994,

Models for Carbonate Stratigraphy from Miocene Reef Complexes of Mediterranean Regions,
SEPM Concepts in Sedimentology and Paleontology #5, Copyright © 1996,
SEPM (Society for Sedimentary Geology), ISBN 1-56576-033-6, p. 191-225.

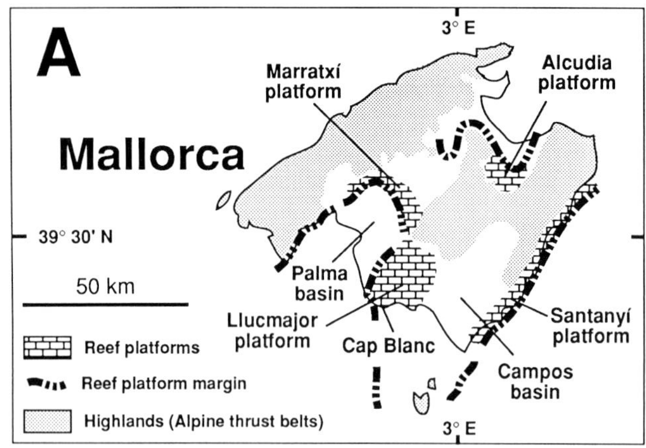


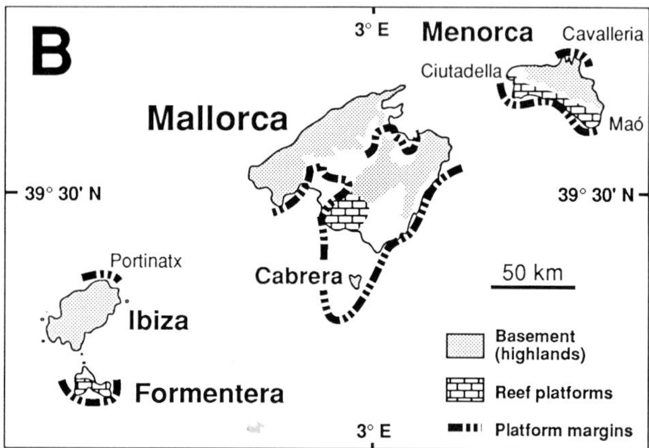


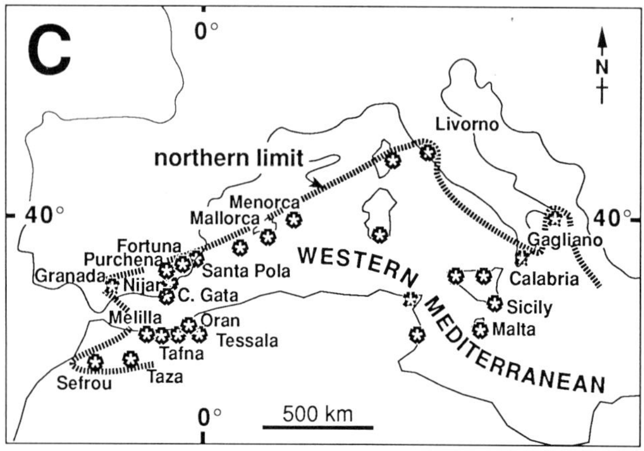


FIG. 1.—Late Miocene carbonate platforms in Mallorca (A), all the Balearic Islands (B), and the western Mediterranean Sea (C). In C, stars show location of Upper Miocene reefs, and "northern limit" marks the approximate northernmost position of Late Miocene reef growth (modified from Esteban, this volume).

1995) already discussed the controls on the characteristic stratigraphic architecture of the Llumajor Platform, this paper presents a more comprehensive, and partly new, interpretation of the sea-cliff exposures. We emphasize lithofacies components, lithofacies distribution and stratal geometries as they relate to the main factors which produced the stratigraphic architecture of this Upper Miocene carbonate buildup. In addition, this paper includes new descriptions of the stratigraphy of back-reef-lagoon lithofacies and of the major diagenetic features of the rocks exposed in the sea cliffs.

## REGIONAL SETTING

Upper Miocene reefs of the Balearic archipelago (Mallorca, Menorca, Ibiza, Formentera and smaller islands; Fig. 1B) developed near the northern limit of western Mediterranean reef growth (Fig.1C). During the Late Miocene, reef complexes grew in shallow submerged areas on all the Balearic platforms, but the most extensive carbonate platforms developed on the southern margins. On Menorca, the largest Upper Miocene carbonate platform crops out in the Cintadella-Maó area on the south, whereas smaller areas of Miocene limestone crop out near Cavalleria on the northern shore (Fig. 1B). Similarly, on Ibiza-Formentera, an Upper Miocene carbonate buildup lies along the southern part (most of Formentera Island), but there are only small areas of equivalent limestones on the northern side. On Mallorca, the Llucmajor Platform, on the southwestern side, is the most extensive area of Miocene carbonate rocks; smaller carbonate buildups make up the Marratxí Platform on the northeastern flank of the Palma Basin, the Alcudia Platform on the northern side of the island, and the Santanyí Platform on the eastern side (Fig. 1A).

The Balearic archipelago is the emergent part of the "Balearic Promontory," the northeastward extension of the Betic Range External Zone. The Balearic Islands are characterized by ranges of folded and thrusted Mesozoic, Paleogene, and Middle Miocene rocks that are flanked by downdropped areas covered with only slightly deformed Upper Miocene to Pleistocene sedimentary rock (Fallot, 1922; Rangheard, 1972; Bourrouilh, 1973; Colom, 1975; Sàbat, 1986; Sàbat et al., 1988; Fontboté et al., 1983; Rodriguez-Perea et al., 1990). Considered as a whole, the Balearic Promontory experienced extension and thinning during Mesozoic time; however, it underwent compression and Alpine-type thrusting during Cenozoic time. The major compressional events occurred during Mid Miocene time, the same as in southeastern Spain (Betic Range) and northern Africa (Maghrebides Ranges).

Miocene rocks comprise two broad groups of units relative to Middle Miocene tectonics (Pomar et al., 1983b). The lower group (Lower and Middle Miocene) is pre- and synorogenic. It onlaps the Paleogene and Mesozoic rocks and is folded and thrusted (except on Menorca Island). The Upper Miocene rocks are post-orogenic and overlie the lower group and the deformed Mesozoic and Paleogene rocks on all of the Balearic islands. The Upper Miocene rocks are fairly flat-lying, having undergone only slight tilting and flexure associated with normal and strike-slip faulting during the late Neogene to middle Pleistocene time.

## UPPER MIOCENE STRATIGRAPHY OF THE BALEARIC ISLANDS

Post-orogenic Upper Miocene rocks are extensively exposed on the Balearic Islands. They were referred to in the earlier literature as "Vindobonian molasse" (Colom, 1975; Bourrouilh, 1973). The Upper Miocene section can be divided into three major sedimentary units, which are tentatively considered (Pomar and Ward, 1994) as third-order depositional sequences (Haq et al., 1988) based on stratigraphic relationships and probable ages of these post-tectonic lithostratigraphic units. The lower sequence crops out on Mallorca (Unidad Calcisiltitas con *Heterostegina* or *Heterostegina* Calcisiltites Unit; Pomar, 1979; Pomar et al., 1983b; Barón and Pomar, 1985; Fig. 2A) and Menorca (Unidad Inferior de Barras or Lower Bar Unit; Obrador et al., 1983b; Fig. 2B). It includes extensive rhodalgal biostromes without coral reefs, and it is attributed to Early Tortonian time (N16). The middle sequence contains well-developed progradational coral reefs on all the islands (Complejo Arrecifal or Reef Complex Unit), and it is attributed to Late Tortonian time (N17) (Bizon et al., 1973; Alvaro et al., 1984) or Late Tortonian-Early Messinian time (Pomar et al., 1983b). The upper sequence, generally assigned to the Messinian, consists of a variety of lithologies, including oolites and stromatolites (Formación Calizas de Santanyí or Santanyí Limestone, equivalent to the Terminal Complex of Esteban, 1979), evaporite, dolomite and marls (Gypsum and Grey Marls Unit), and fan-delta conglomerates and marls (Unidad Margas de la Bonanova or Bonanova Marls Unit);(Fig. 2A).

### *Lower Tortonian*

*Mallorca.—*

The *Heterostegina* Calcisiltites Unit (Fig. 2A) is a 200-m-thick unit that was defined principally from borehole data in southwestern Mallorca (Garcia Yagüe and Muntaner, 1968; Fuster, 1973; Pascual and Barón, 1973; Barón, 1977). The name of this unit comes from the most abundant rock type in the subsurface, but at the outcrop the unit includes coarser-grained carbonate and siliciclastic rocks, particularly in the upper part. The boundaries of the *Heterostegina* Calcisiltite on Mallorca are poorly established. The unit unconformably overlies the folded basement rock of Jurassic and Cretaceous limestones, Oligocene marls and conglomerates, and middle Miocene marls. In the limited exposures on the southwestern side of Palma Bay, the upper boundary is an erosional surface (angular unconformity in some places) overlain by the Upper Miocene Reef Complex (Pomar et al., 1983a), but toward the center of the paleobasin the upper boundary apparently is conformable and is marked by lag deposits of *Heterostegina* (Barón, 1977).

On outcrop, this unit is predominantly soft fine-grained packstone and grainstone containing interbeds of terrigenous sandstone and conglomerate in the upper part and toward the paleoshoreline. Most of the carbonate section is moderately to thickly bedded, but definition of beds is obscured by extensive networks of ophiomorpha, thalassinoides and other burrows. Trough cross bedding, normal-graded bedding and channel-scour surfaces are well preserved in the sandstone and conglomerate (fan delta deposits) of the upper part. Common megafossils are irregular echinoids, scaphopods and thin-shelled bivalves with abundant rhodoliths and large oysters in the upper part of the unit. The terrigenous layers are composed of fragments of Mesozoic, Oligocene, and lower Neogene limestone, dolomite, sandstone, and conglomerate. Matrix material in the carbonate rocks is fine detritus of benthic foraminifers, red algae, and echinoids with planktonic foraminifers, ostracodes, molds of bivalves and terrigenous silt and very fine sand composed of quartz and carbonate-rock fragments. Hard irregular nodules occur in zones of patchy preferential cementation and dolomitization. Many of these indurated areas are related to burrow zones, and some are along fractures. In the characteristic weakly indurated ("marly") beds, the small moldic and intergranular pores are lined with irregular crusts of finely crystalline calcite cement. Typical porosity in these rocks is estimated to be 20-30%, with moldic and microvuggy porosity as high as 40% in patches of finely crystalline dolomite. Depositional setting for this unit on Mallorca is interpreted to be an open carbonate platform, along the landward margin of which were fan deltas deposited by streams draining from highlands of deformed Mesozoic and Paleogene rocks. The predominantly fine size of the carbonate sediments and the common assemblage of planktonic foraminifers, scaphopods, rhodoliths, echinoids and large oysters suggest that much of this carbonate platform was several tens of meters deep. No coral reefs are known in this unit.

*Menorca.—*

The Lower Bar Unit (Obrador and Pomar, 1983; Obrador et al., 1983a, b; Jurado, 1985; Fornós, 1987; Obrador et al., 1992) is a well-exposed unit of progradational rhodalgal ramp deposits (Figs. 2B, 3), up to 500 m thick in the subsurface. The lower boundary is a major erosion surface over the older Neogene units and the Mesozoic rocks. The upper boundary also is an erosion surface, locally with paleokarst and phosphatic crusts.

Obrador et al. (1992) described the following major lithofacies of the Lower Bar Unit (numbers below refer to numbers in Fig. 2B):

1. Up to 3 m of nearshore large-scale cross-bedded pebbly sandstones that onlapped onto an erosion surface over older Neogene and Mesozoic rocks.

2. From 30-150 m of highly burrowed fine-grained peloid-foraminifer packstone with plane-parallel to slightly undulating stratification and abundant ophiomorpha burrows grading upward to cross-bedded medium-grained red algae-bryozoan-mollusk packstone-grainstone, which is capped by cross-bedded coarse-grained rhodolith-bryozoan packstone to grainstone. These are interpreted as shoaling-upward subtidal deposits which accumulated on a shallow ramp.

3. As much as 150 m of large-scale clinoforms of coarse-grained rhodolithic grainstone (Fig. 3) interlayered with coarse- to medium-grained packstone rich in encrusting and branching red algae, echinoids, bryozoans, mollusks, and foraminifers.

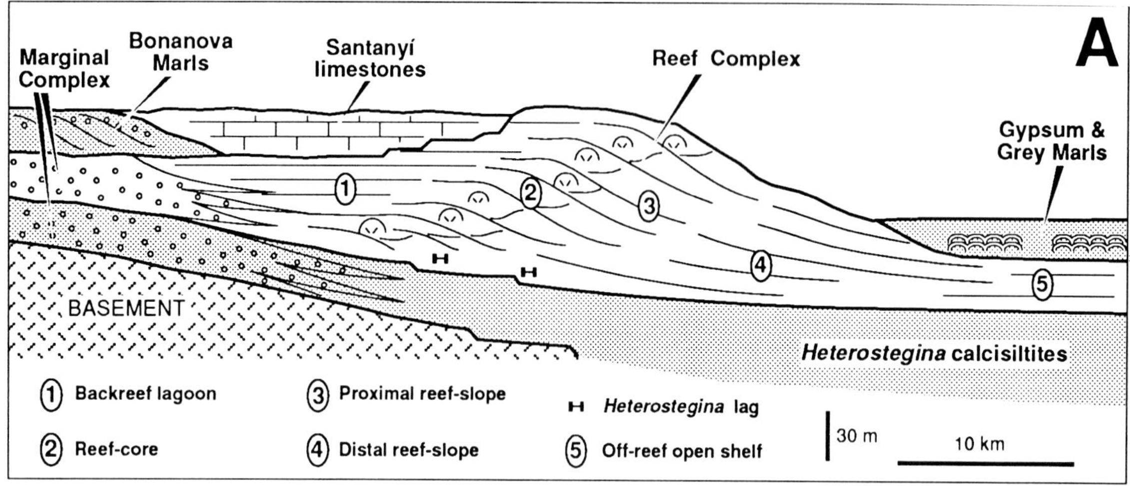


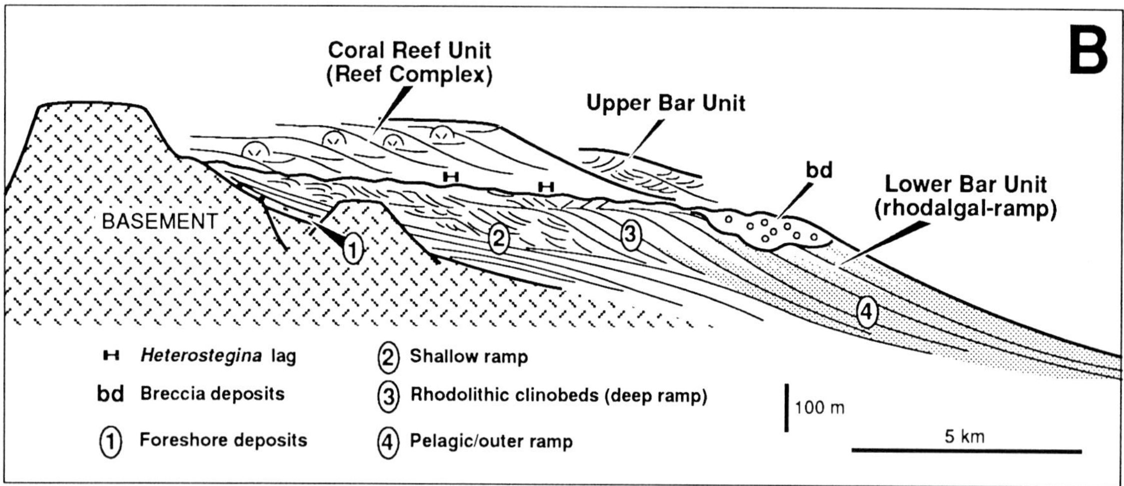


**STRATIGRAPHIC UNITS — C**

| SERIES | STAGES | CYCLES (3rd ORDER) | PLANKTONIC FORAMS BIOCHRONOZONES | MALLORCA | MENORCA | IBIZA and FORMENTERA |
|---|---|---|---|---|---|---|
| HOLOCENE | | | | | | |
| PLEISTOCENE | | 3.10 | N23 | Palma Silts & eolianites | eolianites | eolianites |
| | | 3.9 | N22 | | | |
| PLIOCENE | | 3.8 | N21 | ? — — — — — ? | | |
| | | 3.7 | | | | |
| | 3.5 my | 3.6 | N20 | Sant Jordi Calcarenites | | |
| | | 3.5 | N19 | Son Mir Calcisiltites | ? — — — — — ? | |
| | 5.2 | 3.4 | N18 | Santanyí Lm ? Gypsum & | Upper Bar Unit | |
| UPPER MIOCENE | MESSINIAN 6.3 | 3.3 | N17 | Bonanova Marls ? Grey Marls | ? — — — — — ? | Terminal Complex |
| | TORTONIAN | 3.2 | | Reef Complex Unit | Reefal Unit | Reef Complex |
| | | | N16 | — — — — — | | |
| | | 3.1 | N15 | Heterostegina Calcisiltites Unit | Lower Bar Unit | |

FIG. 2.—(A) Stratigraphic framework of major Upper Miocene lithologic units of Mallorca. (B) Stratigraphic framework of major Upper Miocene lithologic units of Menorca. (C) Stratigraphic column of Upper Miocene to Pleistocene lithologic units of the Balearic Islands. "Palma Silts" includes the Quaternary eolianites which commonly overlie the Reef Complex on the Llucmajor Platform. Third-order cycles according to Haq et al. (1988).

These clinoforms show sigmoidal and oblique beds, ranging from 100 to 200 m in length and some 5 to 8 m in thickness, and containing channels (3 m deep and 15 m wide) infilled with skeletal grainstone and rhodolith-rudstone. Sets of clinoforms are up to 40 m thick and 2 km long in some outcrops. These are interpreted to be deeper-ramp deposits.

4. The clinoforms interfinger basinward with thinly bedded fine-grained dolopackstone containing abundant foraminifers

FIG. 3.—(A) Clinoform beds of rhodolithic grainstone-packstone of the Lower Bar Unit (Fig. 2B) cropping out in Cala Biniparraix, 500 m west of Cala Binidalí on the southern coast of Menorca. (B) Lower Bar Unit exposed in Cala Binidalí, Menorca. Strata are steeply inclined toward the south (lower left). (C) Detail of rhodolithic grainstone-packstone shown in B. 14-cm pencil in center of photo.

and whole-shell pectinids. These deeper-ramp beds are gently undulating with low-angle cross-stratification, some 100 m in length and 1-2 m in maximum thickness.

### *Upper Tortonian-Lower Messinian*

On all the islands, the Upper Tortonian-Lower Messinian consists of the Coral Reef Unit or Reef Complex (Fig. 2C), deposited as progradational reef-rimmed platforms with similar facies associations: off-reef open-shelf, forereef-slope, reef-core and back-reef lagoon. These well-developed and extensively progradational reef complexes contrast with the underlying Lower Tortonian units, which, in the Balearic Islands, contain no coral reefs.

The Reef Complex is well known from the exceptional sea-cliff outcrops and from numerous boreholes on Mallorca (Fuster, 1973; Garcia Yagüe and Muntaner, 1968; Pascual and Barón, 1973; Barón, 1977; Simó and Ramón, 1986; Esteban et al., 1977, 1978; Esteban, 1979; Pomar et al., 1983a, b, 1985; Pomar, 1991, 1993; Pomar and Ward, 1994, 1995) and from good exposures on Menorca (Obrador and Pomar, 1983; Obrador et al., 1983a, b; Jurado, 1985) and Ibiza and Formentera (Simó, 1982; Simó and Giner, 1983). The first references to Upper Miocene corals in the Balearic Islands appeared in Rangheard (1972), who mentioned *Siderastraea crenulata* and *Porites cf. collegniana* on the northern coast of Ibiza and *Tarbellastraea cf. reussiana* on Formentera.

The lower boundary of the Reef Complex crops out on Menorca and locally on Mallorca. In landward settings, the reefs overlie an erosion surface on the Lower Tortonian and, locally, the Mesozoic basement. In more basinward settings this erosion surface shows phosphatic crusts and is overlain by the basinal facies of the Reef Complex unit (as in Menorca). In basinal settings, the lower boundary shows no erosive truncation and is overlain by a bioturbated *Heterostegina*-rich layer with large irregular echinoids; this boundary has been recognized in outcrop on Menorca and in boreholes on Mallorca. The upper boundary also is an erosion surface where it crops out on Formentera, Ibiza, and Mallorca. Paleocaves at the top of the Reef Complex on Mallorca are covered and filled by nearshore deposits of the Santanyí Limestones unit (later Messinian Terminal Complex). These major boundaries are considered to be the sequence boundaries which separate the three third-order depositional sequences of the Upper Miocene of the Balearic Islands.

### *Messinian*

Three major groups of lithofacies with different paleogeographic distribution and poorly known lateral geometry are attributed to the Messinian units on Mallorca. These units are the Bonanova Marls, Santanyí Limestones and Gypsum and Grey Marls (Figs. 2A, C). On Menorca, the Upper Bar Unit (Obrador et al., 1992) overlying the Reef Complex is of uncertain age, but it may be Messinian age because of its stratigraphic position (Figs. 2B, C). Simó (1982) and Simó and Giner (1983) attribute a stromatolitic unit on Ibiza and Formentera to the Terminal Complex of Esteban (1979), thus implying a Messinian age (Fig. 2C).

*Bonanova Marls.—*

This unit is composed of 35 m of fan-delta marls and conglomerates along the northern margins of the Palma Basin. It was first described by Garcia-Yagüe and Muntaner (1968) as "Margas Ocres" from a borehole and later by Barón and Pomar (1985), Fornós and Pomar (1983), and Pomar et al. (1983b) as "Margas de la Bonanova." This unit unconformably overlies the Reef Complex and is composed of marls with scallops, oysters, gastropods and small corals (shallow shelf) in the lower part, grading upward to conglomerates and red clays (alluvial fan) at the top. The upper boundary is not exposed, but this unit is overlain by the Santanyí Limestones at Palma (Pomar et al., 1983b; Fornós and Pomar, 1983; Barón and Pomar, 1985).

*Santanyí Limestones.—*

This limestone unit crops out near Palma and, more extensively, along the southeastern coast of Mallorca. It was first described by Garcia-Yagüe and Muntaner (1968) as "Calizas Pont d'Inca" from a borehole and later called the Terminal Carbonate Complex by Esteban (1979). A detailed study from the sea-cliff outcrops of southeastern Mallorca was carried out by Fornós (1983) and Fornós and Pomar (1983). Later, Fornós and Pomar (1984) re-defined the unit as "Calizas de Santanyí Formation" (Santanyí Limestones). The lower boundary is an erosion surface over the Reef Complex. The upper boundary is not exposed. The Santanyí Limestone rarely exceeds 30 m thickness and is composed of four major lithofacies, from base to top: (1) miliolid packstones and grainstones with vertical root traces (mangrove swamps), (2) stromatolitic boundstones and mudstones (intertidal-subtidal), (3) cross-bedded oolitic grainstones with giant thrombolites and stromatolites (oolitic shoals) and (4) skeletal grainstones and stromatolites (restricted subtidal). Near Palma, giant cyanobacteria domes (thrombolitic and stromatolitic) several meters in diameter are capped by one-meter cyanobacteria domes and worm "reefs," which, in turn, are overlain by cross-bedded oolitic grainstone.

*Gypsum and Grey Marls.—*

This unit occurs only within the Palma Basin and was first described by Fornós and Pomar (1983, 1984) as a lithofacies within the Santanyí Limestones. The Gypsum and Gray Marls Unit overlies the open-platform facies of the Reef Complex and is overlain by Pliocene Son Mir Calcisiltites and Sant Jordi Calcarenites (Fig. 2C), which partially filled the Palma Basin. The Gypsum and Grey Marls Unit is about 10 m thick in outcrop and is composed of dolomite, gray marls with stromatolites, marine mollusks and fish debris. These rocks may be correlated with massive gypsum cored in the center of the Palma Basin (Pomar et al., 1983b). Alvaro et al. (1984) and Simó and Ramón (1986) mentioned the presence of laminated marls with diatomites, fishes, fresh- to brackish-water ostracodes and chara in this unit.

*Correlation Problems.—*

The stratigraphic relationship of the Gypsum and Grey Marls of the Palma Basin with the Santanyí Limestones and Bonanova Marls (Fig. 2A) is uncertain because of the lack of chronostratigraphic data and the uncertain relationship of the Bonanova Marls and Santanyí Limestones of the western Palma Basin to the Santanyí Limestones of the southeastern coast of Mallorca. The Gypsum and Grey Marls Unit represents restricted shallow-marine to freshwater(?) deposits which conformably overlie the deeper-water deposits of the Reef Complex and, thus, records a lower relative sea level after deposition of the Reef Complex (presumably this fall in sea level is related to a major drawdown of sea level during the Messinian Salinity Crisis). The Gypsum and Grey Marls Unit, then, might be time equivalent to the alluvial upper part of the regressive Bonanova Marls, but it is unknown whether the Gypsum and Grey Marls Unit was deposited during this major lowering of sea level or during the subsequent rise.

The erosional surface on top of the Reef Complex and underlying the Santanyí Limestones may represent erosion during the relative fall of sea level when the Gypsum and Grey Marls Unit and, possibly, the upper Bonanova Marls Unit were deposited. If this unconformity is time equivalent to a major lowstand of the Messinian sea level, then the shallow-platform carbonate rocks of the Santanyí Limestones above this surface probably were deposited after the time of extensive evaporite deposition in the western Mediterranean. However, deeper-platform and basinal equivalents of the Santanyí Limestones are unrecognized, making any correlation difficult to substantiate.

## LLUCMAJOR REEF COMPLEX

### *Previous Work*

The extensive Upper Miocene coral-reef complexes of Mallorca remained unknown until the mid 1970's. The joint work of M. Esteban, L. Pomar, A. Barón and F. Calvet started in 1974, although only brief accounts of this original work were published (Esteban et al., 1977, 1978; Esteban, 1979). A more detailed facies analysis of the Upper Miocene reef complex of Mallorca was published in Pomar et al. (1983a, 1985). The work of Alvaro et al. (1984) and Simó and Ramón (1986) corroborated most of the results of Pomar et al. (1983a, 1983b) and added some valuable stratigraphic precision but also presented significant differences in the interpretation of facies geometries and stratigraphy. Recently, Pomar (1988, 1991, 1993) and Pomar and Ward (1994, 1995) described the high-frequency depositional sequences of the entire complex, whereas Green (1993) concentrated on depositional cyclicity of lagoon units. Pomar et al. (1990), Jenkins (1992), Lancaster (1993) and Bosence et al. (1994) used computer modelling to determine controls on the stratigraphic architecture of the Cap Blanc complex. Controls on carbonate accumulation and stratal geometry over the whole Llucmajor Platform are discussed by Pomar and Ward (1995). Perry (1996) studied the distribution and abundance of macroborers. Oswald (1992) made the first comprehensive study of dolomitization of this reef complex.

### *Age*

Regional considerations allow attributing the Cap Blanc reef to the late Tortonian-early Messinian (Pomar et al., 1983b) and more specifically (Pomar, 1991) to the late Tortonian global cycle TB3.2 of Haq et al. (1988). This is consistent with Bizon et al. (1973) and Alvaro et al. (1984), who determined that samples taken in the Cap Blanc area are from the N17 biochronozone (late Tortonian). Core data from water wells show that the Cap Blanc reefal unit progrades over the lower Tortonian (N16 in Alvaro et al., 1984) *Heterostegina* unit, and basinward it is partly onlapped by the Messinian gypsum unit and Lower Pliocene limestone. Age estimations from Sr isotopes (Oswald, 1992) give a late Tortonian age for the reef complex. K-Ar dates on well-preserved sanidine and biotite phenocrysts from a bentonite layer recently found in the back-reef lagoon deposits near Cap Blanc are 7.0±0.2 million years for the biotite (7.4% K) and 6.0±0.2 million years for the sanidine (9.75% K). The final stages of progradation, therefore, most likely took place during early Messinian time.

### *Database*

Extraordinarily well-exposed outcrops on the high sea cliffs of southwestern Mallorca (Fig. 1A) provide the main database for this study. Detailed stratigraphic relationships are obtained by numerous measured sections and by mapping lithologic units on photomosaics of the sea cliffs. Changes in trends of the sea cliffs allow study of three-dimensional relationships of the reef-complex stratigraphy. In addition to outcrop data, there is subsurface control from water-well cores and cuttings on the distribution of the Upper Miocene reef complexes. Field descriptions are supplemented and enhanced by petrographic study of over 500 thin sections.

### *Lithofacies*

This carbonate complex comprises several lithofacies that represent the various depositional environments associated with reef growth: off-reef open shelf, forereef slope, reef and back-reef lagoon (Fig. 4). Lithofacies are defined on their lithology, constituents, stratification, and geometric relationships. Overall the reef complex is about 100 m thick, but thicknesses of individual units of the various lithofacies are highly variable.

*Off-reef open-shelf lithofacies.—*

There are two main types of open-shelf lithofacies: (1) grainstone-packstone with abundant red algae and (2) packstone-wackestone with planktonic foraminifers. The first type crops out along the western margin of the Llucmajor Platform and is found in cores from the central, southern and western parts of the platform (Pomar and Ward, 1995). Whether this lithofacies is a basinward facies of the Reef Complex or whether it is underlies the Reef Complex as a separate lowstand unit can not be demonstrated at present. The second type of open-shelf lithofacies crops out along the southern coast and is in several cores from across the central platform. This lithofacies commonly overlies the red-algae-rich open-shelf deposits in cores and in outcrops along the western coast of the Llucmajor Platform, and it can be traced in those outcrops into distal-slope strata of the reef complex.

*Red algal lithofacies.*—The coarse-grained red algae-rich lithofacies is a poorly sorted dolograinstone-dolopackstone composed mostly of 1- to 15-cm rhodoliths and sand-size fragments of red algae (Fig. 5A). Clumps and layers of branching red algae are common in some intervals. Other skeletal constituents include large oysters, pectinids, the foraminifer *Heterostegina* and the coral *Tarbellastraea*. High-frequency depositional cyclicity in this lithofacies is recorded in generally coarsening-upward sequences a few meters thick. In some places, thin layers of *Tarbellastraea* mark the top of the cycles. Probably these are shoaling-upward cycles which record high-frequency sea-level fluctuations. Total thickness of this lithofacies is unknown, but it is about 30 m thick at the outcrop and at least 40 m thick in a well on the northwestern flank of the Llucmajor Platform. Based on the abundance of red algal and the scarcity of corals and other shallow-water biota, it is presumed that deposition of this lithofacies took place in water depths from a few tens to several tens of meters. Coralline-algal buildups are forming in the modern Mediterranean at depths of 20 to 160 m, with the deeper accumulations in clear water (Bosence, 1985; Ros et al., 1985).

*Planktonic-foraminifer lithofacies.*—The fine-grained open-platform lithofacies is characterized by flat-lying beds of packstone and wackestone several decimeters thick, with bed boundaries and internal stratification obscured by extensive burrowing. Along the southern coast of the Llucmajor Platform this lithofacies is mostly dolomitized, but the most distal beds on the western coast are mostly calcitic. The predominant components in finer-grained portions of this lithofacies are planktonic foraminifers, ostracodes and silt- to very-fine-sand-size detritus of oysters, other bivalves, echinoids and red algae (Fig. 5B). Characteristic megafossils are abundant oysters (*Neopycnodonte navicularis*), large irregular echinoids, rhodoliths and branching red algae, pectinid bivalves, small scaphopods and scarce sclerosponges. Depositional "cycles" on the scale of several decimeters in the packstone-wackestone units are defined by the repetitious alternation of mega-fossil-rich layers and mega-fossil-poor layers or by burrow-rich and burrow-poor layers. Thickness of this lithofacies is poorly constrained, but in the sea-

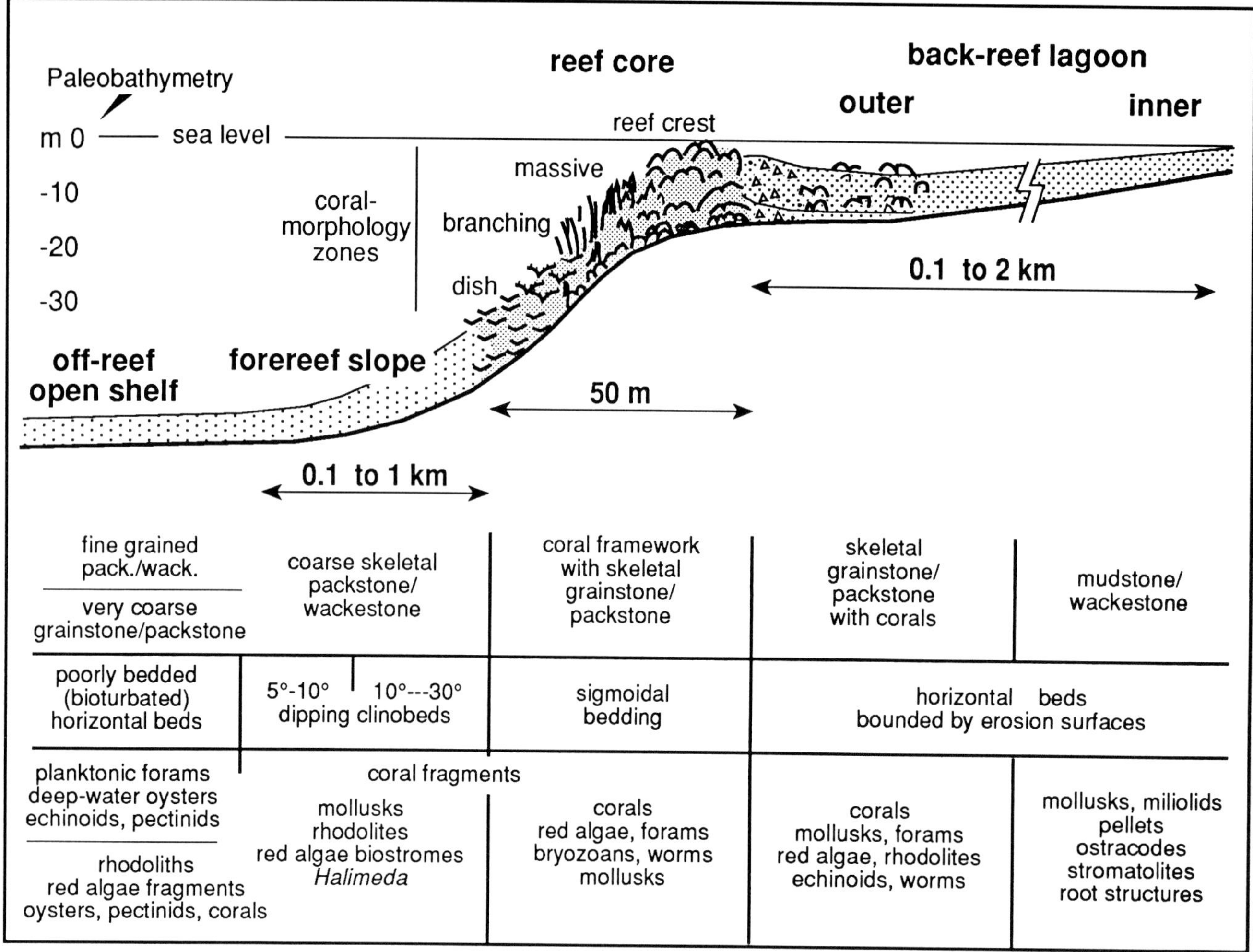


FIG. 4.— Depositional model and main facies characteristics of the Upper Miocene Reef Complex of the Llucmajor Platform.

cliff exposures it is at least 25 m thick. Judging from the depositional topography exposed in outcrop, this lithofacies was deposited in 50-70 m of water.

*Forereef-slope lithofacies.—*

Overlying and interfingering with the finer-grained open-shelf lithofacies are dolomitized skeletal grainstone, packstone and wackestone characterized by basinward-dipping clinobeds. Thickness and lateral extent of the slope deposits are highly variable, depending on configuration of the forereef platform and rate of sediment production. Maximum thickness of slope beds measured on outcrop and in cores is about 70 m.

*Distal-slope lithofacies.*—The more basinward reef-slope deposits are gently inclined (less than 10°) layers of poorly stratified, extensively burrowed fine-grained red algae-mollusk dolopackstone and dolograinstone. Gently dipping distal-slope beds equivalent to aggradational reefs can be traced at least 1.5 km along the southern sea cliffs. Characteristic megafossils in distal-slope beds are sand- to pebble-sized rhodoliths, whole-shell bivalves and large oysters, and in some places there are bioherms of branching and encrusting red algae up to several decimeters thick and several decimeters in diameter. In some places, the distal-reef-slope and open-shelf units are cut by 1- to 8-cm-wide near-vertical fractures ("neptunian dikes") filled with laminated dolomudstone and dolowackestone.

Much of this lithofacies lacks well-defined stratification because of extensive burrowing, but the small-scale depositional units can be defined on textural changes. These units are about 3 m thick, and they, in turn, are composed of 0.5-3-m sequences bounded by less-distinct discontinuities. Typically, these sequences coarsen upward, and the tops of many of the depositional "cycles" are marked by concentrations of large rhodoliths. Discontinuity surfaces separating both the larger-scale and smaller-scale units lack discernible borings and encrustations.

*Proximal-slope lithofacies.*—Proximal-reef-slope deposits on this platform are steeply dipping (10°-30°) layers of dolomitized skeletal and intraclastic grainstone, packstone, rudstone, and floatstone that interfinger landward with coral reefs. Lenticular layers of coral rubble and skeletal debris-flow deposits are common. Blocks of reef rock up to 80 cm in diameter occur in the upper layers. Fossils of the proximal reef slope are red algae fragments and rhodoliths, coral fragments, bivalves (some whole),

FIG. 5.—(A) Densely packed rhodoliths in the red-algal packstone-grainstone open-shelf lithofacies cropping out at Na Segura on the western coast of the Llucmajor Platform. (B) Photomicrograph of open-shelf skeletal wackestone cropping out at Ses Olles on the northwestern margin of the Llucmajor Platform. Fauna includes large echinoids (E), abundant planktonic foraminifers (PF), and worm tubes (WT). Bar scale represents 1 mm. Plane-polarized light. (C) Steep clinoforms in slope beds cropping out at Pas des Verro on the western margin of the Llucmajor Platform. Sea cliff is 90 m high. (D) Gently dipping clinoforms in slope beds cropping out at Cala Carril on the southern coast of the Llucmajor Platform. Sea cliff is 60 m high.

gastropods (including vermetids), echinoids, bryozoans and the green alga *Halimeda*. Although *Halimeda* fragments occur throughout the forereef and perireef deposits, this alga is a principal component only in the forereef slopes of the youngest reef tracts.

Proximal-slope clinobeds may be high-amplitude or low-amplitude, reflecting the depth of the forereef platform adjacent to the reef. Long, steep slope beds with 25-50 m of depositional relief and dipping up to 30° (Fig. 5C) crop out on the western and northwestern margins of the Llucmajor Platform, where the reef platform apparently was adjacent to an open shelf many tens of meters deep. In exposures on the southern sea cliffs, however, proximal slope beds show only several meters of depositional relief (Fig. 5D). Here sequences of reef and slope deposits apparently built up on a flat-lying open platform no more than a few tens of meters deep.

*Reef lithofacies.—*

Massive coral-reef limestone and dolostone overlie and interfinger basinward with the forereef-slope lithofacies (Fig. 4). The reef framework is constructed mainly of only one to two genera of corals, *Porites* or *Porites* and *Tarbellastraea*. Secondary framework components are encrustations of red algae, foraminifers, bryozoans, worm tubes and vermetid gastropods as well as scattered small colonies of the coral *Siderastraea* and microcrystalline rinds and crusts (cyanobacteria; Riding et al., 1991). The Upper Miocene coral reefs of Mallorca are predominantly framework reefs. Only locally is the reef-core lithofacies composed predominantly of rudstone and floatstone.

The coral colonies are vertically zoned in the reef core according to depth-controlled growth morphology (Fig. 6). In general, deeper-water corals show platy forms to platy with finger-like projections, intermediate-depth corals are branching, and shallower corals are hemispheroidal to columnar. Reef-crest growth

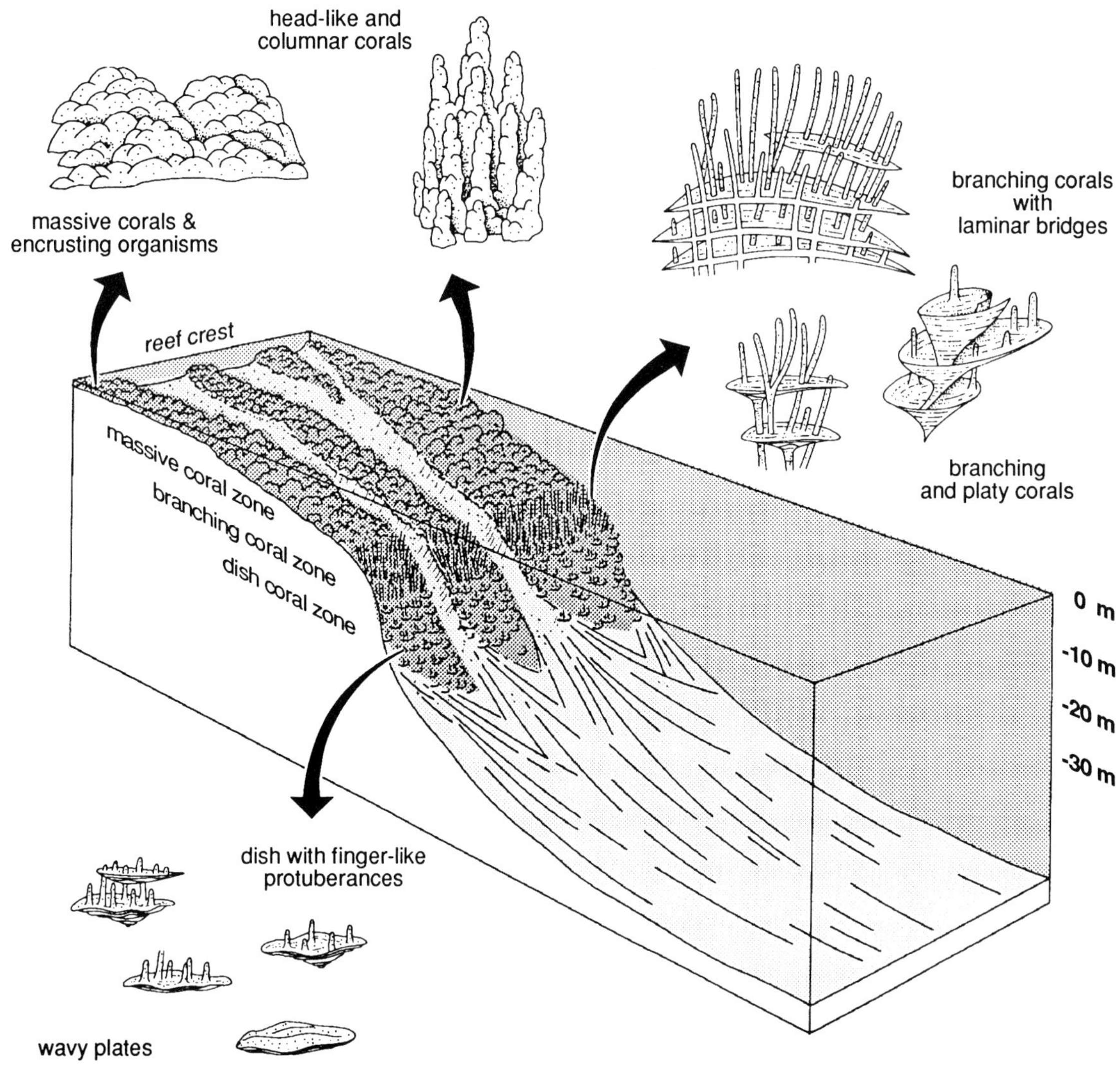


FIG. 6.—Reef-core depositional model showing zonation of coral-colony morphologies with respect to paleo-depth. This model is inferred from outcrops on the southern coast of the Llumajor Platform from Cap Blanc to Punta Negra (Fig. 7C).

forms are domal and massive laminar. Sponge and pholad borings as well as encrustations of red algae and cyanobacteria are more abundant in the shallow-zone reefs. Intercoral spaces commonly are filled with coarse skeletal grainstone-packstone and/or wackestone, but locally primary intercoral voids remain empty of sediment.

Along the southern part of the Llucmajor Platform, depositional strike of the reef tracts was N35°W to N60°W, and progradation was toward the southwest (Figs. 7A-C). From the perspective of sea-cliff outcrops, there are three main types of reef tracts: (1) discontinuous, mound-like *Porites* and *Tarbellastraea* reefs cropping out in the Vallgornera area, (2) more continuous *Porites* and *Tarbellastraea* reefs cropping out from Cala Beltrán to the middle of the Els Bancals coast and (3)

FIG. 7.—A. Major Upper Miocene reef tracts of Mallorca. (B) Main trends of reef tracts cropping out in the sea cliffs of the Llucmajor Platform. (C) Trends and types of reef tracts exposed from Vallgornera to Cap Blanc. Letters (e.g., CB) designate location of measured sections used to construct cross section in D. Lines of projection, aligned parallel to the direction of reef-tract progradation, show segments used to construct composite cross section in D. (D) Cross section (drawn along lines of projection shown in C) showing distribution of major lithofacies and types of reef tracts cropping out in sea cliffs of southwestern Mallorca. Letters (e.g., CB) correspond to measured sections located on map in C.

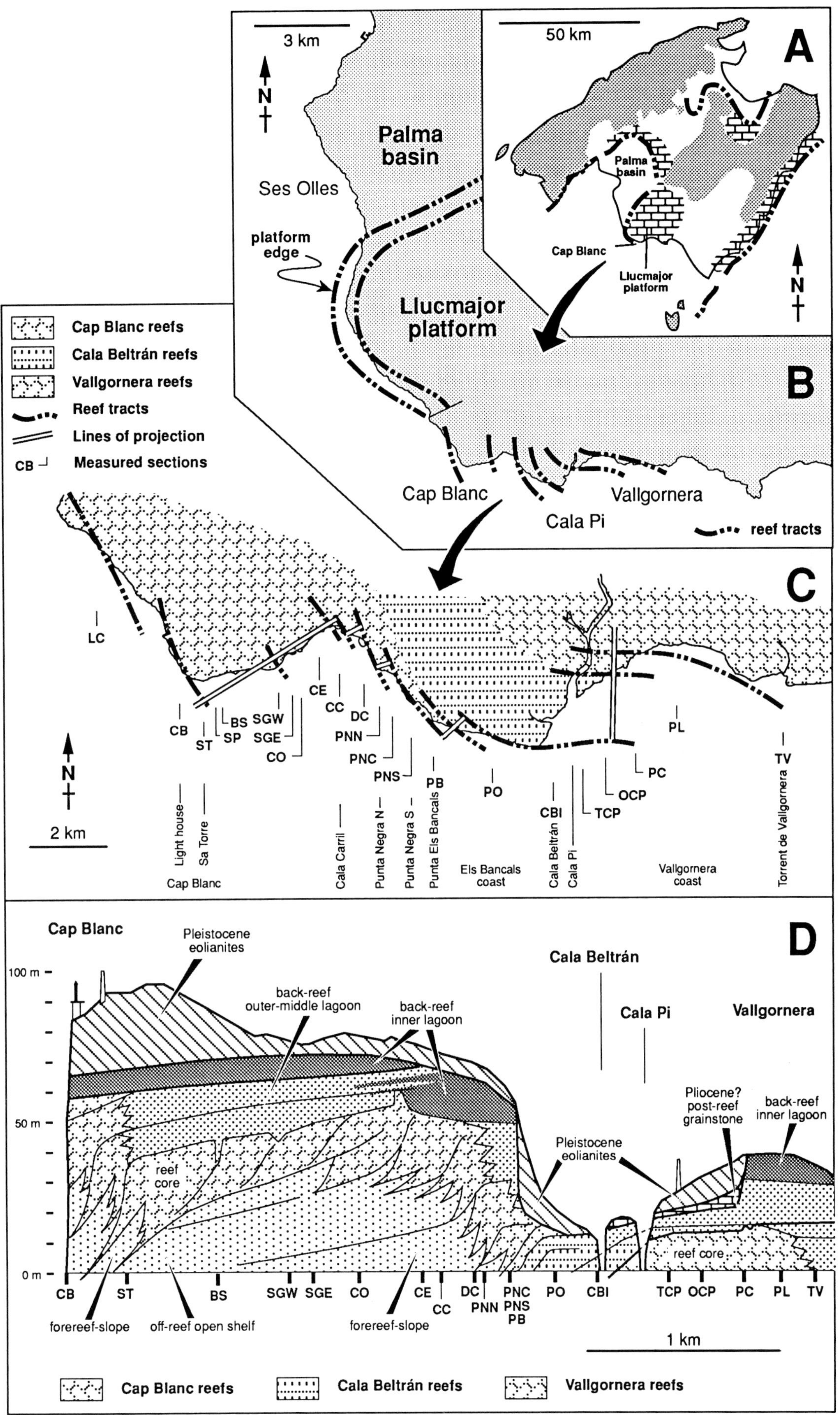

A
50 km
Palma basin
Cap Blanc
Llucmajor platform
N
B
3 km
N
Palma basin
Ses Olles
platform edge
Llucmajor platform
Cap Blanc
Cala Pi
Vallgornera
reef tracts
Cap Blanc reefs
Cala Beltrán reefs
Vallgornera reefs
Reef tracts
Lines of projection
CB Measured sections
C
LC
CB
ST
BS
SP
SGW
SGE
CO
CE
CC
DC
PNN
PNC
PNS
PB
PO
CBI
TCP
OCP
PC
PL
TV
N
2 km
Light house
Sa Torre
Cap Blanc
Cala Carril
Punta Negra N
Punta Negra S
Punta Els Bancals
Els Bancals coast
Cala Beltrán
Cala Pi
Vallgornera coast
Torrent de Vallgornera
D
Cap Blanc
Pleistocene eolianites
back-reef outer-middle lagoon
back-reef inner lagoon
Cala Beltrán
Cala Pi
Vallgornera
100 m
50 m
0 m
reef core
Pliocene? post-reef grainstone
back-reef inner lagoon
Pleistocene eolianites
reef core
CB
ST
BS
SGW
SGE
CO
CE
CC
DC
PNN
PNC
PNS
PB
PO
CBI
TCP
OCP
PC
PL
TV
forereef-slope
off-reef open shelf
forereef-slope
1 km
Cap Blanc reefs
Cala Beltrán reefs
Vallgornera reefs

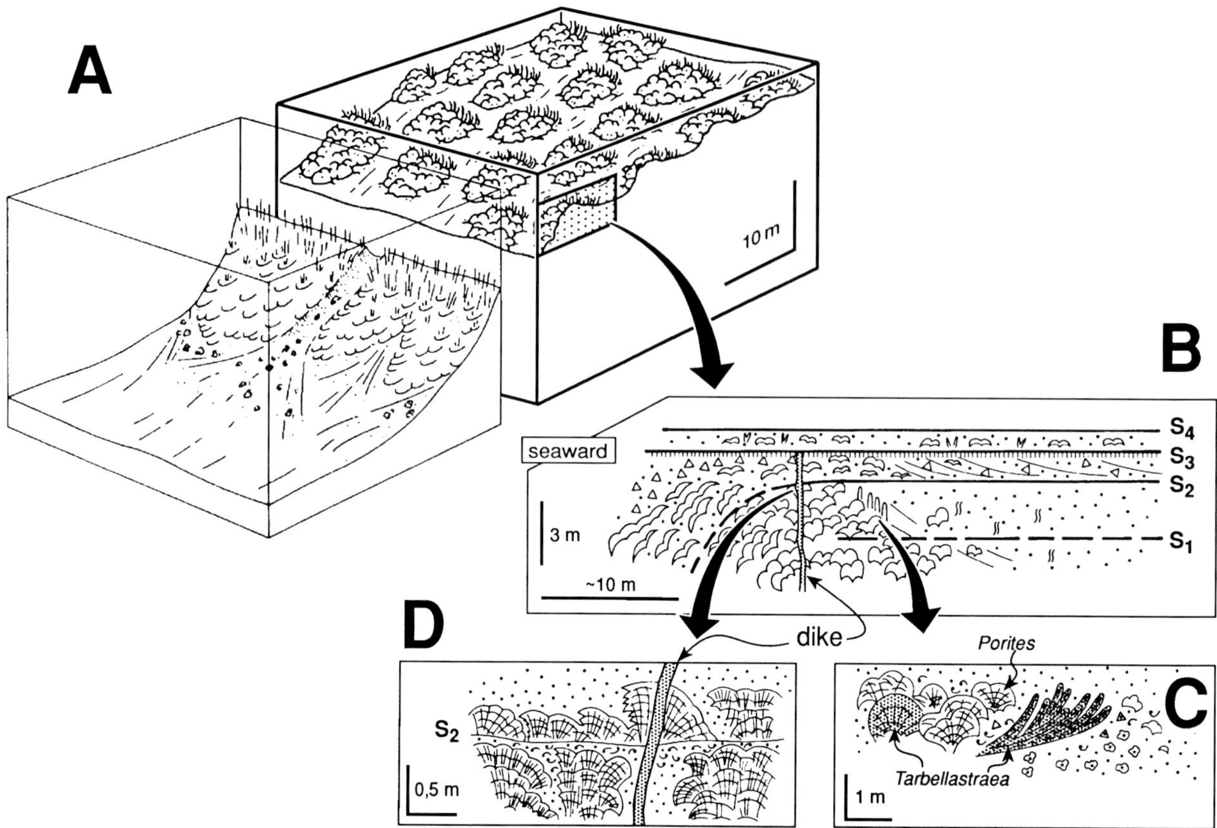


FIG. 8.—Vallgornera reefs. (A) Block diagram of mound-reef tract. Right part of the block (thick lines) is based on outcrop data, and left part of the block (thin lines) shows speculated seaward projection of tract. (B) Field sketch of one mound. Above discontinuity $S_1$ (which is traced into several mounds) there is reduction in lateral extent of the mounds. Surface $S_2$ (which represents a non-erosional hardground) is overlain by massive coral colonies in the mound and coral and mollusk rudstone landward of the reef mound. Surface $S_3$ is an erosion surface with worm borings (Fig. 9F). This surface truncates corals and lithified sediment as well as sedimentary dikes in the reef mounds and can be traced along the sea cliffs for at least one km. $S_3$ is overlain by an outer-lagoon unit, in turn, overlain by an erosional discontinuity ($S_4$). (C) and (D) Details of B.

*Porites* reef tracts cropping out from Els Bancals to Cap Blanc (Fig. 7C). The youngest reefs of the Llucmajor Platform, which crop out along the western coast south of Ses Olles (Fig. 7B), are constructed of *Porites* and *Tarbellastraea,* and core data show that most reefs throughout the platform contain both these genera. This indicates that the *Porites*-framework reefs are only a small portion of the Llucmajor Platform.

*Vallgornera reefs.*—"Mound" reefs characterize the oldest outcropping reef complex, exposed between Torrent de Vallgornera and Cala Pi (Figs. 7C, 7D, 8A). The base of these mound-shaped buildups is below present sea level; consequently it is impossible to determine the total thickness or assess the vertical zonation of coral morphology. The visible part of the buildups are up to 8 m thick and 10-12 m across, and individual mounds are spaced several meters apart. They are constructed of massive and columnar colonies of *Tarbellastraea* and *Porites* with minor heads of *Siderastraea*. The largest single *Tarbellastraea* and *Porites* colonies are up to 3 m in diameter and 1.5 m high. The cores of the mounds (Figs. 8B-D) are made up of closely spaced massive corals with minor amounts of detrital material. *Porites* is the most abundant lower in the reef core, with *Tarbellastraea* equally abundant in upper portions. The seaward side of the mounds are massive coral encrustations dipping at angles up to 45°, presumably in the paleo-seaward direction (S15°W-S25°W). Some of the lateral flanks low in the reef are near-vertical walls of encrusting *Porites*. On the landward side, columnar coral colonies predominate (Fig. 8D), and most of these corals, especially the *Tarbellastraea*, are highly bored by bivalves, worms and a variety of sponges (Figs. 9A, B). In some "mounds", the upper part of the landward flank is encrusted with laminar and domal masses of red algae up to about 0.7 m thick. In addition, micritic (cyanobacteria?) encrustations are thicker at the top of the reef. Coarse grainstone between the coral colonies contains whole-shell mollusks (pectinid bivalves and *Haliotis* gastropods are characteristic), red algae fragments and crabs.

The inter-mound grainstone is rich in mollusks, red algae fragments, small rhodoliths (1-cm diameter), coral debris and

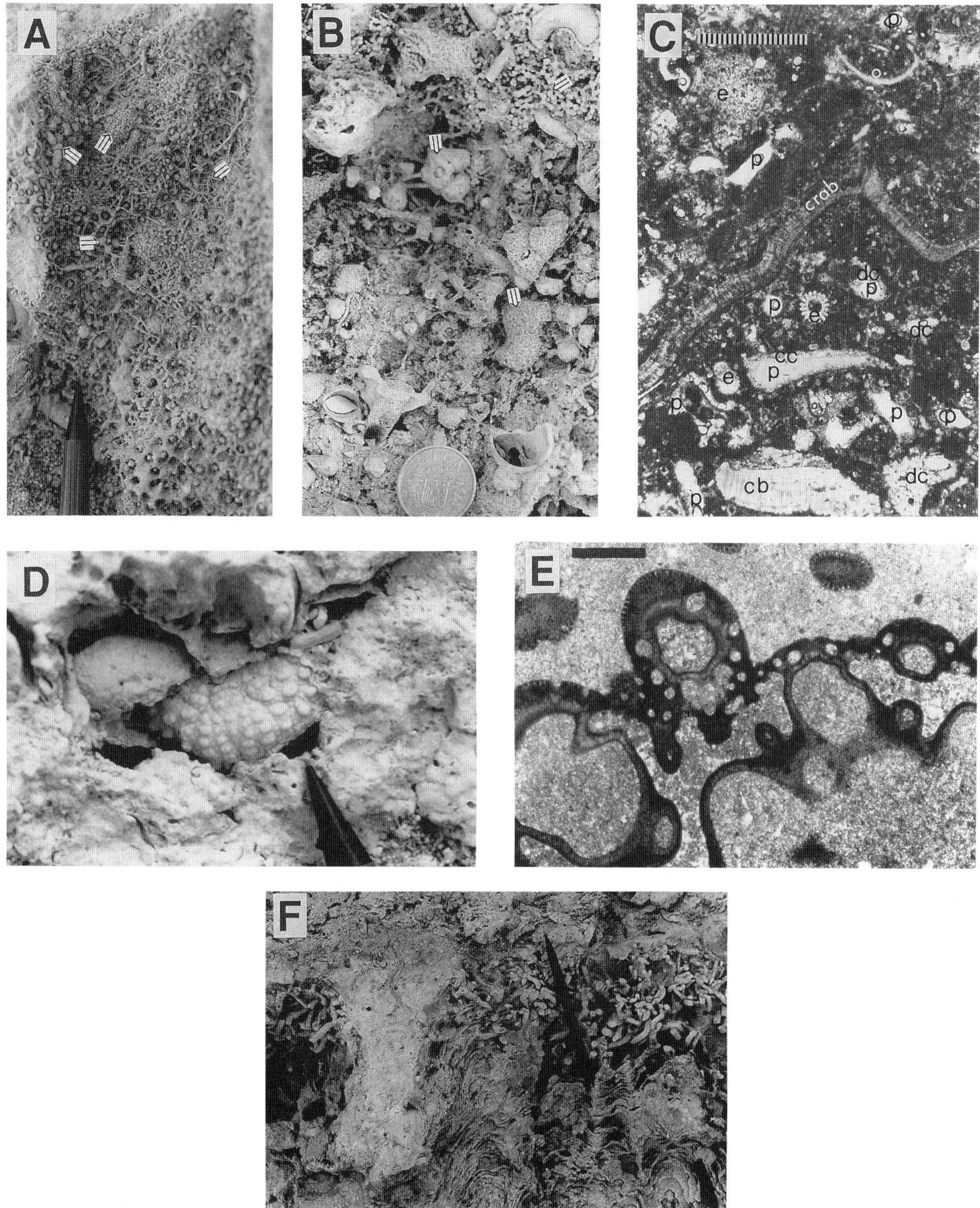


FIG. 9.—Vallgornera reefs. (A) Casts of different types of sponge borings (arrows) in mold of *Tarbellastraea*. As many as seven different types of sponge borings have been found in this reef. Black part of pencil is 18 mm long. (B) Casts of borings of different types of sponges (arrows) in mold of *Porites*. Coin is 2.5 cm in diameter. (C) Photomicrograph of inter-reef skeletal packstone, Vallgornera coast. Numerous dark grains are red algae. Other constituents are echinoids (e), calcitic bivalves (cb), and ostracodes (o). Common pore type (p) are molds of aragonitic mollusks and corals. Some moldic pores contain dolomite cement (dc), and others contain calcite cement (cc). Crab and some red algal fragments are dolomitized. Bar scale represents 1 mm. Plane-polarized light. (D) Crab *Daira speciosa* (center of photo) within intermound grainstone-packstone. Black part of pencil is 18 mm long. (E) Photomicrograph of crab-shell microstructure. Bar scale represents 1 mm. Plane-polarized light. (F) Pencil points to extensively bored (polychaete worms?) erosion surface ($S_3$, Fig. 8B). Borings penetrate coral, encrusting red algae, and lithified sediment. Boring casts stand in relief within mold of truncated *Porites* heads. Pencil is 14 cm long.

the crab *Daira speciosa* (Figs. 9C-E). Seaward flanks of the reef are overlain by coral rudstone, while lateral margins pass abruptly into red algal grainstone with abundant ophiomorpha burrows. Channels cutting perpendicular to the reef tract are filled with red algal grainstone containing abundant rhodoliths as well as crab shells, small coral heads, pectinid bivalves and *Haliotis*

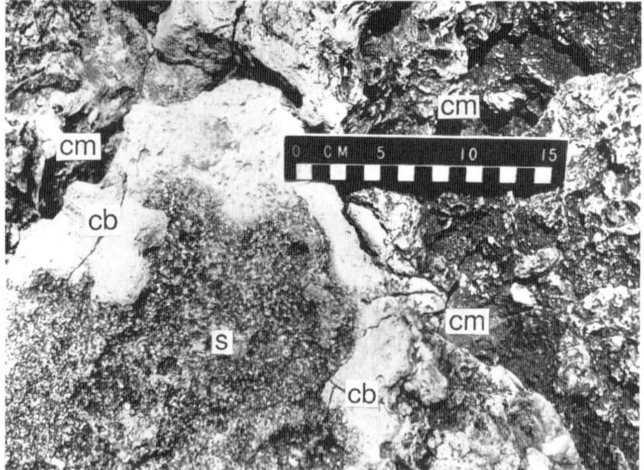


FIG. 10.—Looking down on microcrystalline cyanobacteria coatings (cb) on corals (now molds) (cm) at Punta Els Bancals (Fig. 7C). Sand- and gravel-sized skeletal debris (s) fills inter-coral voids.

gastropods. The crests and upper flanks of the mounds are covered with coarse grainstone containing abundant whole-shell bivalves, large gastropods (*Strombus*), coral fragments, heads of *Siderastraea* and red algae. Erosional discontinuities at the crest of the reefs are encrusted with corals, red algae, and micritic rinds (cyanobacteria?); some surfaces are highly bored by worms (Fig. 9F) and bivalves. Near-vertical neptunian dikes, which cut some mound reefs (Fig. 8D), are filled with fine sediment, and some apparently are lined with microcrystalline crusts (cyanobacteria?). Most of the reef rock in this area is calcitic, with only small dolomitized patches.

*Cala Beltrán reefs*.— The reef tracts from Cala Beltrán to the middle of the Els Bancals coast (Fig. 7C) also are composed of both *Porites* and *Tarbellastraea*, but they lack the mound morphology of the older Vallgornera reefs and all the reef rock is pervasively dolomitized. In this area, too, the base of the reefs is not exposed; however, as much as 4 m of reef crop out along the Els Bancals sea cliffs. Presumably these reefs are elongated perpendicular to the direction of progradation.

At Cala Beltrán, the reef is mainly composed of thick near-vertical columnar colonies of *Porites* and *Tarbellastraea* 0.5-2 m long and a few decimeters wide. In some places, massive and platy *Porites* occur among the long columns. Toward Els Bancals, successive reefs are constructed of massive and columnar *Porites* and *Tarbellastraea*, with *Tarbellastraea* apparently increasing in abundance upward. Some individual colonies reach 1 m wide and 2 m high, but the largest colonies decrease in size to about 1 m high toward the top of the reef. Corals are encrusted with red algae and cyanobacteria? rinds, especially in the upper part of the reef (Fig. 10). Several kinds of sponge borings as well as mollusk and fungal borings riddle the periphery of upper-reef corals. Some original cavities between the coral columns still remain only partly filled by internal sediment. Intra-reef rock is red algal-mollusk dolograinstone with pectinid bivalves, *Haliotis*, echinoids and crab shells. Accretion surfaces within the reef core are steeply inclined (up to 25°) in the paleo-seaward direction (S15°W), but they become horizontal landward, bounding the reef-crest subfacies at the top of the reef core. These flat-lying erosional discontinuities are bored by pholad bivalves and/or encrusted with red algae, micritic cyanobacteria crust and corals.

The reef crest is well preserved on Els Bancals coast. It is composed of large colonies of *Porites* and *Tarbellastrea*, some as much as 1 m high and 2 m across (Figs. 11A, B). Small domal heads of *Siderastraea* also are typical of this zone. Inter-coral rocks are mostly dolomitized packstone, grainstone and wackestone rich in red algae fragments and rhodoliths and also containing coral debris, mollusks, benthic foraminifers, and echinoids. In some units, the reef-crest corals occur as buildups 5-10 meters wide separated by "channels" oriented perpendicular to the reef tract. These channels are 2-3 m wide and are filled with coarse dolograinstone with rhodoliths and mollusks. At the margins of these channels, coral breccia covers the flanks of the coral buildups. Near the middle of the Els Bancals coast is a 1-m-thick sheet of densely stacked large rhodoliths (Fig. 11C) deposited on a back-reef flat or in a broad reef-crest channel. These are mostly 10-12 cm in diameter, with some as large as 20 cm across. Typically the centers of these large rhodoliths are concentric laminae wrapped around fragments of reef rock, and outer layers are densely branched (Fig. 11D). Also in a back-reef position there is a 1.5-m-thick layer of organ-pipe-shaped *Tarbellastraea* and *Porites*. Individual coral colonies are predominantly near-vertical columns 0.5-3 m long and several centimeters wide.

*Cap Blanc reefs*.— From Els Bancals to Cap Blanc (Fig. 7C), reefal buildups are constructed mainly of *Porites* with minor amounts of *Tarbellastraea* and *Siderastraea*. This change in the coral-framework reef-builders is sharp and occurs across a minor stratigraphic boundary without change in coral morphologies. Most of the reef rock from Els Bancals to Cap Blanc is pervasively dolomitized, except in the vicinity of Punta Negra N (Figs. 7C, D), where dolomite is patchy. The extensive exposures along this coast reveal a well-developed vertical zonation of the growth forms of *Porites* (Esteban, 1979; Pomar et al., 1983a, 1985; Pomar, 1991). The most complete vertical sequence of this reef type crops out on the high Cap Blanc sea cliff, where the reef is part of an aggradational sequence. Here there are three zones of coral morphology: (1) a lower zone of "dish coral," (2) a middle zone of "branching coral" and (3) an upper zone of "massive coral" (Figs. 6, 12).

*Dish-coral zone*.—The dish-coral zone is commonly about 10 m thick (ranging from 5-12 m). In some localities, as at Cap Blanc, it rests on a sharp surface above the fore-reef slope. At other localities, this lower zone conformably and transitionally overlies reef-slope beds of coral rubble and skeletal dolograinstone-packstone. The sharp boundaries are below progradational-downstepping reefs, and the transitional boundaries are at the base of aggradational-progradational reefs. At Cap Blanc, there are two parts of the dish-coral zone. A lower part is about 6 m thick and is composed of wedges of reef detritus and dish corals

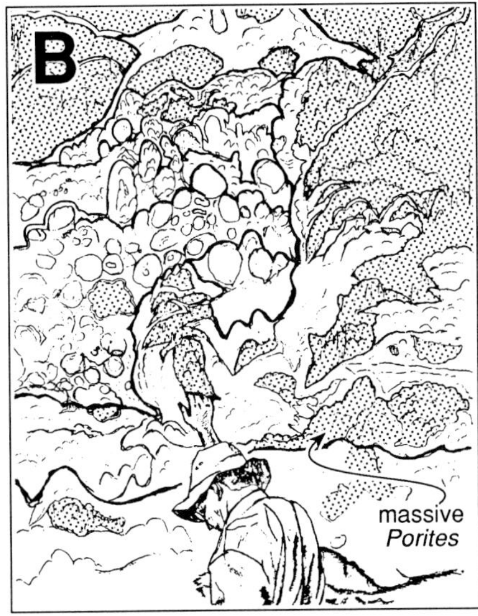

FIG. 11.— Reef-crest facies, middle of Els Bancals coast (Fig. 7C). (A) Sketch of massive coral colonies encrusted with thick red-algae and micritic cyanobacteria rinds, with pockets of rhodoliths. Shaded areas are coral molds; light areas are encrusting red algae, micrite rinds and intercoral sediment. (B) Detail of A (area of box) showing intergrowth of massive corals and red-algae rinds (bottom and right of picture), a pocket with rhodoliths and red-algal coated coral rubble (central area), overlain by massive corals (upper part). Note vertical elongation in many rhodoliths as well as red-algae pavements linking rhodoliths and corals. (C) Layer of large spherical rhodoliths 1 m thick adjacent to outcrop sketched in A, interpreted to have been deposited in a back-reef flat or in a broad reef-crest channel. (D) Detail of rhodolith from layer shown in C. Coin is 3 cm in diameter.

in living position, with thin layers of medium to coarse skeletal dolograinstone separating these wedges. The upper part is about 8-10 m thick and is composed of coral buildups which pass laterally into coral breccia.

Characteristic coral morphology in the dish-coral zone is wavy plates up to 30 cm in diameter. Platy forms with finger-like vertical protuberances become more abundant upward, and finger corals up to 0.5 m long are the dominant form in the upper part of this zone. Borings in the dish-coral zone are rare. This is in contrast to the highly bored coral fragments in the underlying reef-slope deposits, which were derived from shallow-water corals. Intra-coral cavities are filled with dolowackestone and dolopackstone containing fine detritus of echinoids, red algae, foraminifers, mollusks, and ostracodes (Fig. 13A). The amount

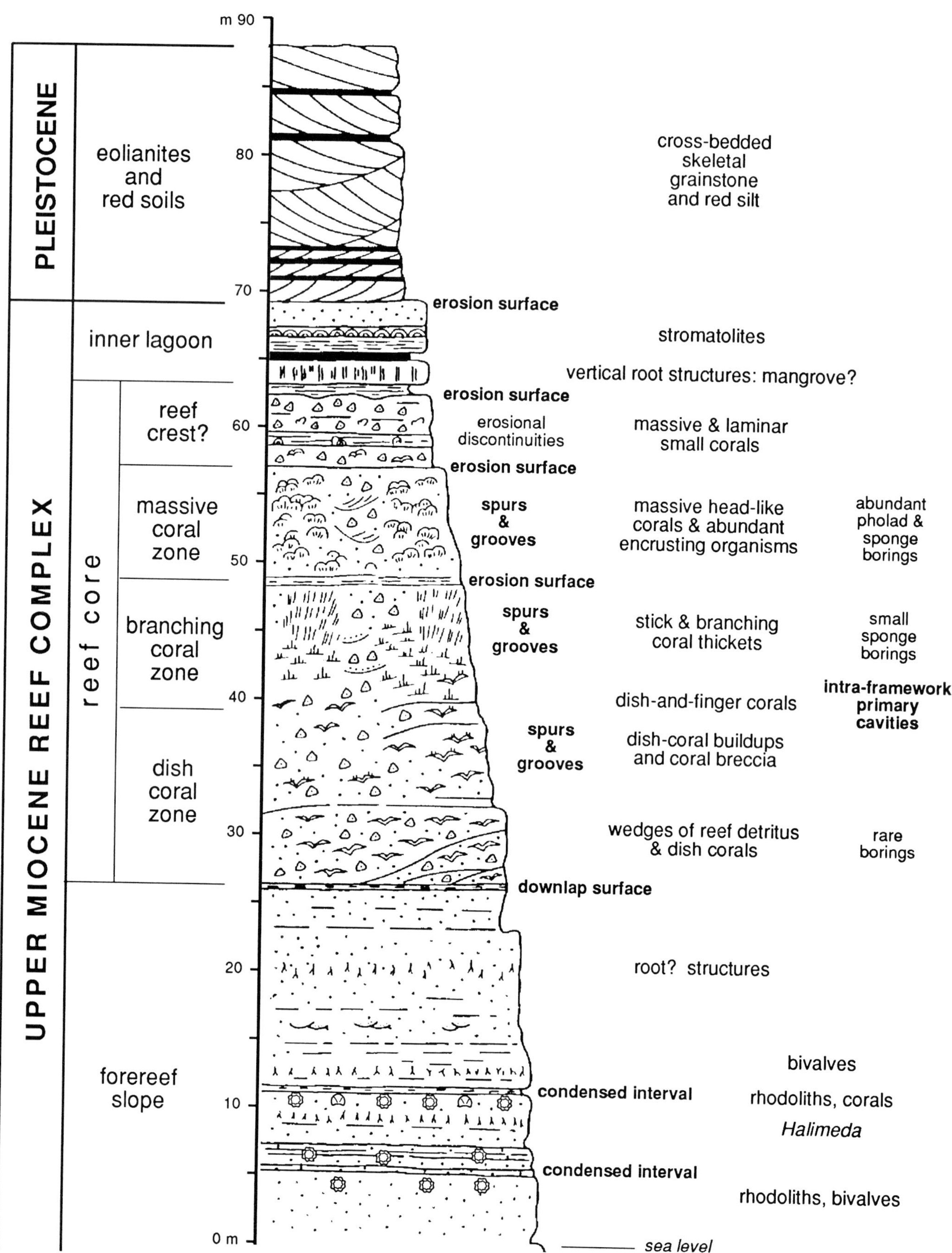

PLEISTOCENE
eolianites and red soils
UPPER MIOCENE REEF COMPLEX
inner lagoon
reef core
reef crest?
massive coral zone
branching coral zone
dish coral zone
forereef slope
m 90
80
70
60
50
40
30
20
10
0 m
cross-bedded skeletal grainstone and red silt
erosion surface
stromatolites
vertical root structures: mangrove?
erosion surface
erosional discontinuities
massive & laminar small corals
erosion surface
spurs & grooves
massive head-like corals & abundant encrusting organisms
abundant pholad & sponge borings
erosion surface
spurs & grooves
stick & branching coral thickets
small sponge borings
dish-and-finger corals
intra-framework primary cavities
spurs & grooves
dish-coral buildups and coral breccia
wedges of reef detritus & dish corals
rare borings
downlap surface
root? structures
bivalves
condensed interval
rhodoliths, corals
Halimeda
condensed interval
rhodoliths, bivalves
sea level

of allochemical material decreases up section, so that partly empty cavities in the coral breccia and coral framework are more numerous in the upper part of this zone (Fig. 13B). Commonly these primary framework cavities are 2-25 cm across and several centimeters high.

*Branching-coral zone.*—Transitionally overlying the dish and finger *Porites* is a zone characterized by coral colonies with stick and branch shapes (Fig. 12). The thickness of this zone may vary from 2 to 7 m, but preservation is irregular. Large-scale depositional dips in the paleosea direction are up to 35°.

At Cap Blanc, the largest branched *Porites* form thickets 2-3 m across and up to 4 m high (Fig. 13C). Between some of the branching-coral buildups are channel deposits of coral dolorudstone and coarse skeletal dolograinstone with intergrown laminar corals. Large pieces of branching coral colonies lie at the margins of some of the channels. These constructional reefs and associated channels are comparable to spurs and grooves characteristic of many modern reef fronts. Within the branching framework, layers of internal sediment alternate with laminar coral bridges (Fig. 13D). Typically, the intra-coral deposits coarsen upward, with dolowackestone and dolopackstone overlain by dolograinstone. Unfilled original intra-framework pores may be numerous and large in this zone. Most of these cavities are several decimeters in diameter, but several larger cavities are 2 m wide and 1.3 m high (Fig. 13E), and one exceptionally large primary cavity is 6 m across and 2.5 m high (Fig. 14). These are partly filled with multiple cones and fans of coarse skeletal dolograinstone filtered in from above. Allochemical components are red algae, corals, worm tubes, mollusks, and echinoids. Encrusting organisms within the coral framework in this zone are red algae, bryozoans, worms, foraminifers and cyanobacteria. The only borings in the stick and branching corals are small sponge borings.

Commonly, the branching-coral zone is truncated, and it may be overlain by deeper-water dish corals or by shallower-water massive corals, depending on high-frequency changes in sea level during buildup of successive reefs.

*Massive-coral zone.*—The upper morphology zone (Fig. 12) is characterized by massive head-like *Porites*, with *Tarbellastraea* and *Siderastraea* as minor constituents. The massive coral zone is up to about 8-12 m thick, but it is preserved only in aggradational reefs. Preservation of this shallower part of the reef core is unlikely during periods of falling sea level and progradational offlapping, because of removal by erosion.

FIG. 12.—Stratigraphic section below Cap Blanc lighthouse, showing vertical zonation of the Cap Blanc-type reefs. Distal to proximal reef-slope facies contains two condensed intervals mainly composed of red algae and rhodolith packstone with small coral colonies. The reef is made up of two accretional units: (1) the dish- and branching-coral zones, belong to a lower accretional unit that is truncated at the top by an erosional surface and (2) the massive-coral zone, which is part of a younger accretional unit. Coral breccia at top of the section, previously interpreted as reef-crest deposits, may be breccia related to an erosion surface (sequence boundary). Back-reef inner-lagoon deposits compose the uppermost part of the Miocene section.

Commonly, the massive coral colonies are several decimeters in diameter; however, large heads may be up to 1.5 m in diameter and columnar forms up to 2 m high. In the upper part of the reef, massive *Porites* colonies are thickly encrusted by red algae and riddled with pholad and sponge borings (Fig. 13F). Other encrusting organisms in this zone are foraminifers, cyanobacteria(?), worms, bryozoans and vermetid gastropods. Common allochemical components are mollusks, red algae and echinoids. Channels filled with dolograinstone and dolorudstone trend perpendicular to the reef tract, suggesting that the spur-and-groove system extended into the upper morphology zone. In some localities (e.g., Punta Negra), large pieces of reef rock, up to 4-m diameter, are overturned and lying at the bottom of some of these channels.

At the lighthouse section of Cap Blanc, this coral zone is about 8 m thick and overlies an erosional surface truncating the branching coral zone (Fig. 12). The upper boundary is also an erosional surface, which is overlain by 6 m of dark-gray and dense dolomite characterized by small massive and laminar *Porites*, plus a few spherical heads of *Siderastraea crenulata* (P. Busquets and G. Alvarez, pers. commun.). Several erosional discontinuities within this dark-gray unit truncate both coral colonies and penecontemporaneously lithified sediment, which have been interpreted as reef-crest deposits (Pomar 1991; Pomar et al., 1983b, 1985).

*Paleobathymetry.*—The Cap Blanc section and other reef exposures have been used to infer the paleobathymetry of the different coral-morphology zones (Figs. 6, 12) by measuring the vertical distance of each zone below the crest of these high-framework reefs (Pomar, 1991). This does not take into account possible variation in "drowning depth" (Hallock and Schlager, 1986) and assumes that the top of the massive-coral zone built up approximately to sea level. Although the Cap Blanc section (Fig. 12) does not represent a single and complete vertical sequence of the reef core, it allows estimation of the approximate water depth at which each zone developed. If these estimates are valid, the dish-coral zone was developed at depths between 20 and 30 m, the branching-coral zone, at depths between 10 and 20 m and the massive-coral zone, at depths shallower than 10 m. Even if the estimated depths are imprecise, this relative range of paleobathymetric distribution of the different coral-morphology zones is an important tool for determining the relationship between the facies architecture and relative sea-level fluctuations (Pomar, 1991) and for analyzing the sequence-stratigraphic framework (Pomar and Ward, 1994, 1995).

*Lagoon lithofacies.—*

The reef rocks are overlain by and interfinger landward with flat-lying lagoon lithofacies (Fig. 4). Thickness and extent of lagoon units are unequal in different parts of the reef complex, depending on the relative sea-level changes that controlled each reef-building episode. Lagoons were a few kilometers wide during sea-level rises and aggradation of the reef but were narrow or nonexistent during falling sea level and offlapping of the reef. Maximum thickness of lagoonal lithofacies in the sea-

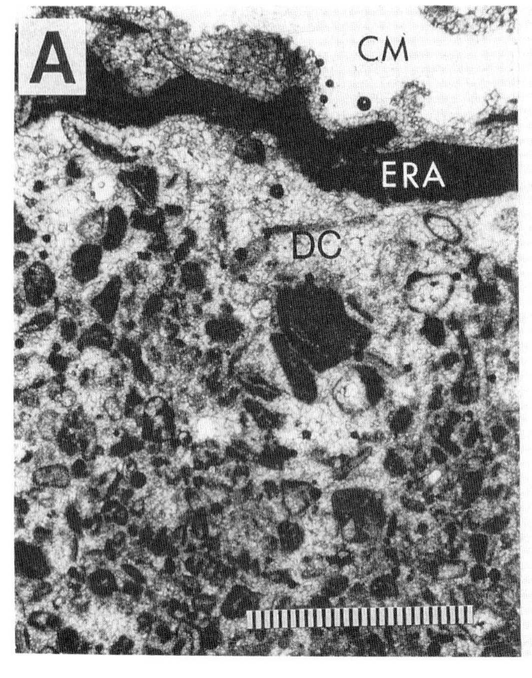
A
CM
ERA
DC

B

C

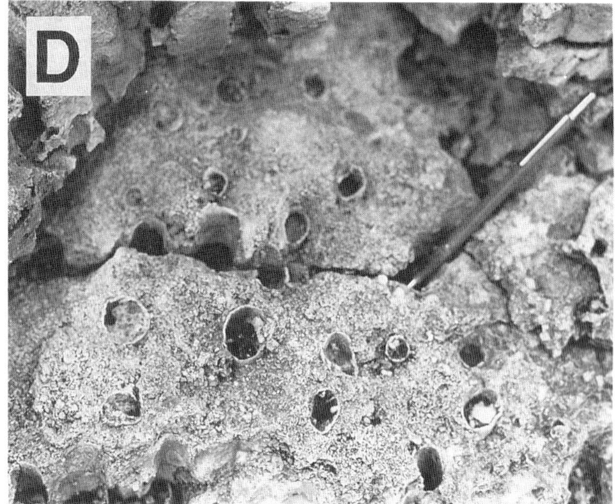
D

E

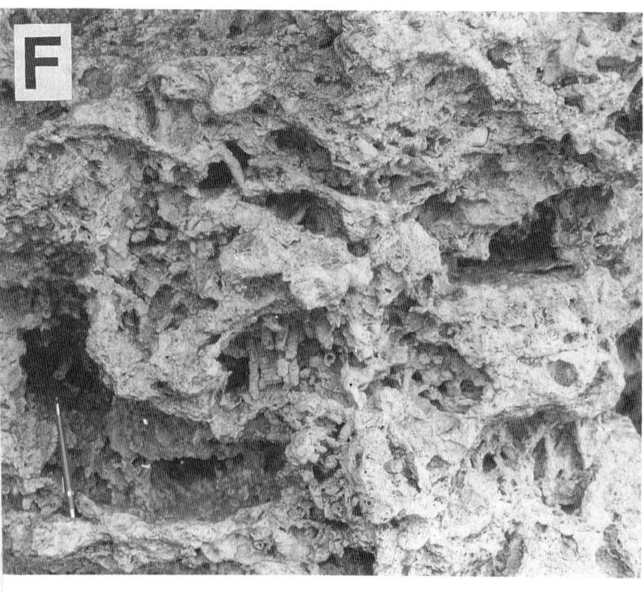
F

cliff exposures is about 30 m. These lagoonal rocks can be differentiated into "outer-lagoon," "middle-lagoon" and "inner-lagoon" lithofacies. For this reef complex, the outer-lagoon is the portion of the lagoon adjacent to the reef tract (up to 0.5-1 km landward from the reef). The middle lagoon included the area about 0.5-1.5 km landward of the reef, and the inner lagoon is the most distal part of the lagoon, in places extending at least 2-3 km behind the reef tract. Outer-lagoon rocks contain normal-marine fossils, while inner-lagoon rocks have mostly "restricted-marine" fossils (Fig. 15). Middle-lagoon units contain a mixture of restricted and normal-salinity components. The lagoon units are composed of centimeter- to meter-scale depositional sequences, mostly shallowing-upward cycles.

*Outer-lagoon lithofacies.*—Outer-lagoon rocks behind the reef crest are characterized by horizontal layers of skeletal grainstone and packstone with coral-patch reefs and lenses of coral breccia, rhodoliths and mollusks. In some places these rocks are partly to completely dolomitized. Typically there is a lack of internal stratification because of extensive burrowing. Most of the beds are bounded by erosional discontinuities and hardgrounds, and some of these surfaces are perforated by mollusk and worm(?) borings.

Coral patch reefs are common in the outer-lagoon units. Larger patch reefs are 5-10 m in diameter (Fig. 16A) with inter-reef areas about the same width. Massive domal and columnar colonies, some as large as 2 m in diameter and 0.5-1.5 m high, characterize the outermost-lagoon lithofacies, whereas smaller hemispheroidal heads are typical farther inland. Most corals are commonly riddled with borings of pholads, sponges and worms (Perry, 1996). In the Cala Pi area, patch reefs are composed of *Porites* and *Tarbellastraea*, but in the Cala Carril-Cap Blanc area, they are mostly *Porites*.

Outer-lagoon grainstone and packstone characteristically contain abundant red-algae (crustose and branching), echinoids, mollusks, benthic foraminifers and coral fragments (Figs. 15, 16B). Other constituents include serpulid worm tubes, *Halimeda*, planktonic foraminifers, bryozoans, peloids and intraclasts.

*Middle-lagoon lithofacies.*— Middle-lagoon units are generally poorly stratified packstone and grainstones with scattered small coral patches. Common components are red algae, benthic foraminifers including common miliolids, mollusks and echinoids (Fig. 15). Other grains are pellets and peloids, *Halimeda*, bryozoan, intraclasts and cerithid gastropods. Some middle-lagoon units include few-meter-thick sequences of crossbedded skeletal grainstone. The sedimentary structures and textures in these grainstones represent an upward progression from lower shoreface through upper-shoreface to foreshore environments. Constituent grains of these "beach-ridge" deposits are mostly mollusks, pellets, peloids and intraclasts with lesser amounts of benthic foraminifers (especially miliolids), echinoids and red algae (Fig. 15). These grainstones indicate that shoals and barrier bars accumulated in inner parts of some middle lagoons. Near Cap Blanc most of the middle lagoon units are pervasively dolomitized, but near Cala Carril these beds are only partly dolomitic, and near Vallgornera they are calcitic.

*Inner-lagoon lithofacies.*—Inner-lagoon lithofacies are thin- to medium-bedded grainstones, packstones and mudstones containing mainly miliolids, bivalves, pellets, peloids and cerithid gastropods (Fig. 15). Some zones also contain abundant alveolinids, soritids and trochamminids (Fig. 16C). Ostracodes, dasyclad algae, echinoids, *Favreina* pellets, calcispheres and red algae are less common constituents. Ooids are a major constituent in only a few grainstone layers.

In general, the inner-lagoon lithofacies is made up of 1- to 3-m-thick sequences of shallow-subtidal limestones passing upward into intertidal limestones. Many of these shallowing-upward units contain abundant burrows and root traces. Some of these high-frequency depositional cycles are capped by root zones, which are overlain by discontinuous wavy-laminated caliche crusts a few centimeters thick. A meter-thick root zone (Fig. 15) within the upper inner-lagoon unit from the Cap Blanc area to near Cala Carril (Figs. 7C, D) was attributed to mangrove swamp deposits (Pomar, 1991). Some depositional cycles are capped by stromatolites. A meter-thick stromatolite blanket (Fig. 15), overlying the mangrove(?)-root layer in the Cap Blanc-Cala Carril area, is composed of flat-lying wavy laminae and domal structures up to several centimeters high. Some subaerial-exposure surfaces are microkarsts, with dissolution and brecciation zones several centimeters thick. Typically just overlying the subaerial surfaces are transgressive deposits 5- to 30-cm-thick. These layers of packstone-grainstone with abundant cerithid gastropods, whole-shell bivalves, miliolids and other benthic foraminifers and pellets (Fig. 15). These layers also commonly contain small blackened lime-mudstone

FIG. 13.—Parts of the Cap Blanc-type reef, Cap Blanc lighthouse section. All the rocks in these photos are dolomite. (A) Photomicrograph of internal sediment within the dish-coral zone. Dark layer (ERA) is red algae which encrusted *Porites* (now moldic pore, CM). Most dark grains are red-algae fragments. Mollusk and coral fragments are represented by micrite envelopes around cement-filled molds. All cement is dolomite (DC). Bar scale represents 1 mm. Plane-polarized light. (B) Coral rubble within upper part of the dish-coral zone showing primary inter-clast porosity as well as coral-moldic porosity. Tape is marked in 10-cm divisions. (C) Sketch of thickets of inclined *Porites* sticks with laminar morphologies bridging base of the thickets (drafted from a slide by M. Esteban). This pattern of in-place thickets (4-6 m wide and 4 m high) with inter-thicket rubble zones (4-8 m wide, 4 m high) occurs repeatedly along an exposure face parallel to the reef tract, and it is interpreted to be part of a spur-and-groove system. (D) Looking down on internal sediment caught on laminar *Porites* bridges within stick-coral thickets shown in C. Holes are stick-coral moldic pores, which are surrounded by crusts (thin cylinders standing in relief) composed of dolomitized intraskeletal microcrystalline carbonate (marine cement?) and extraskeletal finely crystalline dolomite cement in addition to some encrusting red algae and worm tubes. Pencil is 14 cm long. (E) Megaporosity within the lower part of the branching-coral zone. The large pores are both original framework pores and coral-moldic pores. Just below the E on the upper left is an original pore floored with fine-gravel-size internal sediment that flowed into the cavity. Pencil is 14 cm long. (F) Massive-coral zone. *Porites* heads (now moldic pores) are riddled with pholad-bivalve borings (the casts of which project into the moldic pores) and encrusted by thin layers of red algae (laminar layers standing in relief). Pencil is 14 cm long.

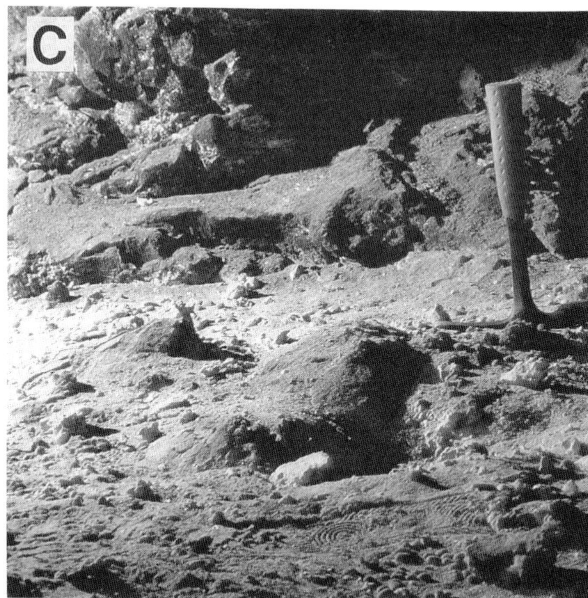


FIG. 14.—"Ballroom" porosity within lower branching-coral zone about 0.5 km northwest of Cap Blanc section (Fig. 12). All rocks in these photos are dolomite. (A) Large original pore within coral-reef framework. Walls of cavity are constructed of dish-and-finger and branching *Porites* (now moldic pores). Hammer rests on fan of coarse internal sediment that partly fills cavity. (B) Hammer lies on steep depositional surface of one of several fans of coarse internal sediment that was deposited along the margins of this original submarine cavity. Fan sediment was extensively cemented with submarine cement. (C) Conical mound ("hour-glass deposit" in foreground) on floor of the "ballroom" pore is one of several piles of internal sediment that filtered in from the roof of the submarine cavity.

intraclasts. Locally, scalloped dissolution surfaces are encrusted by large oysters.

In the Cap Blanc area, most of these beds are dolomitized, although in the Vallgornera area the inner lagoon contains only minor patches of dolomite associated with subaerial-exposure surfaces.

| COSTITUENTS | OUTER LAGOON | MIDDLE LAGOON | INNER LAGOON | ROOT ZONE | STROMATOLITE | TRANSGRESSIVE GRAINSTONE | BEACH RIDGE |
|---|---|---|---|---|---|---|---|
| worm tubes | r | | | | | | |
| corals | r | r | | | | | |
| *Halimeda* | r | r | | | | | |
| bryozoans | r | r | | | | | |
| red algae | XXXXX | CCC | r | | | r | CCC |
| echinoids | XXXXX | CCC | r | r | | r | r |
| benthic forams | CCC | CCC | r | | | r | r |
| mollusks | CCC | CCC | XXXXX | | r | XXXXX | XXXXX |
| miliolids | r | CCC | XXXXX | | r | CCC | CCC |
| peloids | r | r | CCC | XXXXX | CCC | XXXXX | XXXXX |
| pellets | | r | XXXXX | XXXXX | | XXXXX | XXXXX |
| cerithid gast. | | r | CCC | r | | CCC | |
| ooids | | | r | | | | r |
| ostracodes | | | r | | | | |
| calcispheres | | | r | | | | |

XXXXX: > 25% CCC: 25% - 5% r: < 5%

FIG. 15.—Major constituents of lagoon lithofacies. Modified from Green, 1993. Relative abundances of grains (X - abundant, C - common, and r - rare) are based on estimated volume percentages in thin section.

*Stratigraphic Architecture*

The areal distribution of depositional facies of the Upper Miocene carbonate rocks on Mallorca (Fig. 1) is relatively well known both from sea-cliff exposures and from borehole data (Pomar and Ward, 1995). The common vertical succession from open-shelf to reef-slope to reef-core to lagoonal lithofacies is typical of progradational reef platforms. Within this simple vertical succession of lithofacies, however, the Reef Complex has a complicated stratigraphy of high-order accretional units, reflecting high-frequency oscillations in relative sea level (Pomar 1988, 1991; Pomar and Ward, 1994, 1995). The oscillations in sea level produced the most characteristic facies relationship within the Reef Complex: progradation with vertical shifts (upward and downward) of the reef-core and associated lithofacies (Fig. 17). A hierarchy of four different magnitudes of accretional units is recognized among these vertical shifts in reefal lithofacies, which allow for estimation of the amplitudes of Mediterranean sea-level oscillations during the late Miocene (Fig. 17). Sea-level cycles with estimated amplitudes of less than less than 15 m, 20-30 m, 60-70 m, and about 100 m have been interpreted to control the architecture of the Reef Complex in the Llucmajor platform (Pomar, 1991). If the Reef Complex of the Llucmajor Platform is considered to represent part of a 3rd-order depositional cycle, then the depositional cycles within the Reef Complex could represent from 4th- to 7th-order sea-level fluctuations.

The basic building block of this carbonate complex is the "sigmoid" (Fig. 17), which is interpreted to be the result of accretion during short-term rises in relative sea level and erosion during short-term falls in sea level (Figs. 17, 18). This apparently 7th-order basic accretional unit fractally stacks in different magnitudes of larger-scale accretional units: "sets," "cosets" and "megasets" of sigmoids (Figs. 17, 19), which reflect lower frequencies of sea-level fluctuation (probably 6th, 5th, and 4th, respectively). All these units have the same characteristics in terms of bedding geometries, bounding surfaces, facies distribution and stratal architecture in relation to sea-level fluctuations. All of them are: (1) sigmoidal in shape; (2) bounded landward by erosional surfaces which pass basinward to correlative conformities; and (3) composed of an inner belt of horizontal lagoonal beds, a middle belt of sigmoid-shaped reef-core lithofacies and an outer belt of reef-slope and open-shelf lithofacies. Thus, they can be considered to be depositional sequences.

Changes in stacking patterns allow definition of four systems tracts in all hierarchical units above the sigmoids: "low stillstand," "aggradational," "high stillstand" and "offlapping" (Figs. 20, 21). Low-stillstand systems tracts (LST), which formed during generally low sea levels as the sea rose after maximum fall, are composed of relatively thin prograding reef-core lithofacies and

A

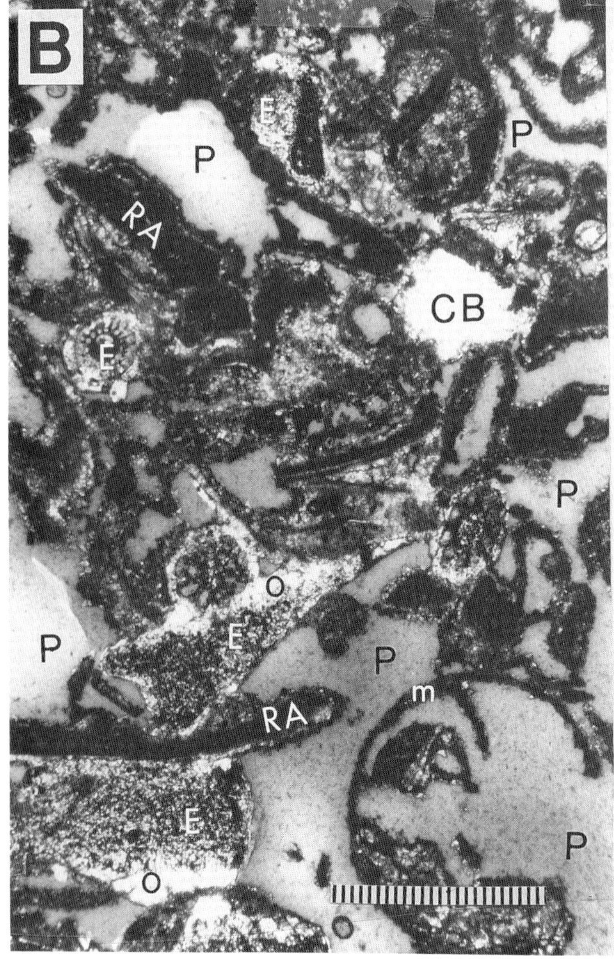
B
E
P
P
RA
CB
E
P
o
E
P
P
m
RA
E
P
o

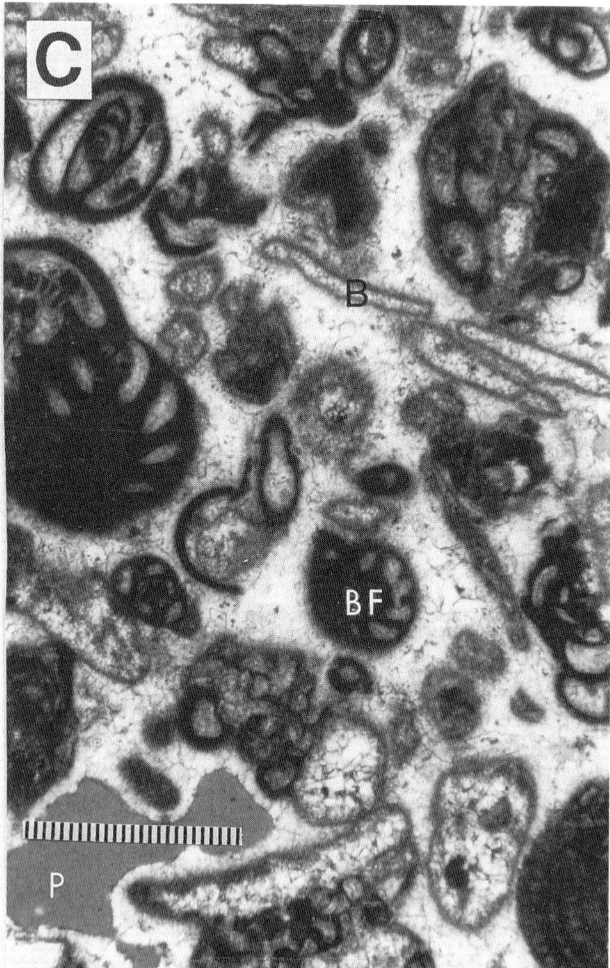
C
B
BF
P

poorly developed or absent lagoonal beds. The forereef-slope and open-shelf lithofacies thin basinward. Aggradational systems tracts (AST) are characterized by aggradation in all depositional systems from the lagoon to the open shelf. These were formed during the rise of sea level and are the most volumetrically important systems tracts in the whole system. Highstillstand systems tracts (HST), which formed during the high part of a sea-level cycle, are composed of thin progradational reef-core facies, forereef-slope facies wedging out basinward and volumetrically condensed open-shelf lithofacies. Lagoon beds are relatively thin, and erosion during sea-level falls has completely removed some of them. Offlapping systems tracts (OST) formed during falling sea level. These are composed of thin prograding and downstepping reef-core lithofacies, which downlaps on to the open-shelf facies of the previous systems tract, without significant forereef-slope facies. There is no lagoon lithofacies, and the open-shelf lithofacies are volumetrically condensed. The complex architecture of the Llucmajor carbonate complex can only be adequately defined from the reef-core lithofacies stacking patterns in the dip direction. At the outcrop scale, there are no lithologic criteria for defining the boundaries of the systems tracts nor the level of hierarchy of the discontinuities and erosion surfaces.

The fore-reef slope and off-reef open-shelf lithofacies are mainly built up by aggrading systems tracts separated by condensed intervals. These condensed intervals correlate landward with the high-stillstand, offlapping, and low-stillstand systems tracts (Fig. 21). Distal-slope and open-shelf deposits show different expressions of this high-frequency sea-level cyclicity. In the red-algae-rich lithofacies it is recorded in the few-meters-thick coarsening-upward sequences. The thin layers with dish-shaped and flat colonies of corals at the top of these cycles may be correlated with the condensed intervals. In the fine-grained open-platform lithofacies, rich in planktonic foraminifers, only changes in bioturbation density and the repetitious alternation of mega-fossil-rich and mega-fossil-poor layers may be tentatively attributed to this high-frequency cyclicity. In distal- to proximal-slope settings, in addition to the coarsening-up general character of the vertical sequence, the condensed intervals have a different composition. These zones show concentrations of

FIG. 16.—Back-reef-lagoon lithofacies. (A) Cliffs on the Vallgornera coast (Fig. 7C) showing outer-lagoon beds. Coral patch reefs (now mostly voids such as in middle zone on left of photo) with inter-reef grainstones (behind man). (B) Photomicrograph of outer-lagoon red algae and echinoid grainstone from the Punta Negra area (Fig. 7C). Most dark grains are red algae (RA); gray grains are echinoids (E), commonly with overgrowths (o). Mollusk molds are outlined by micrite envelopes and microcrystalline cement (m). CB designates a fragment of calcitic bivalve. Intergranular and moldic porosity (P) is abundant. Bar scale represents 1 mm. Plane-polarized light. (C) Photomicrograph of inner-lagoon benthonic foraminifer and mollusk grainstone from the Vallgornera coast (Fig. 7C). Foraminifers (BF) are well-preserved, but bivalves (B) are represented by micrite envelopes surrounding cement-filled molds. First-stage fibrous calcite cement is overlain by blocky to columnar sparry calcite cement, which occludes much of the intergranular (P) and moldic porosity. Bar scale represents 1 mm. Plane-polarized light.

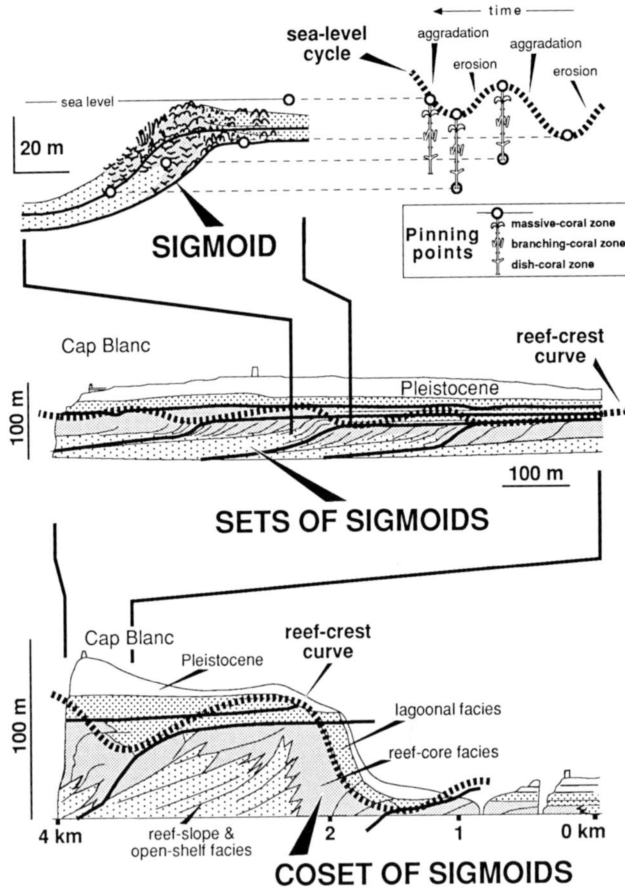


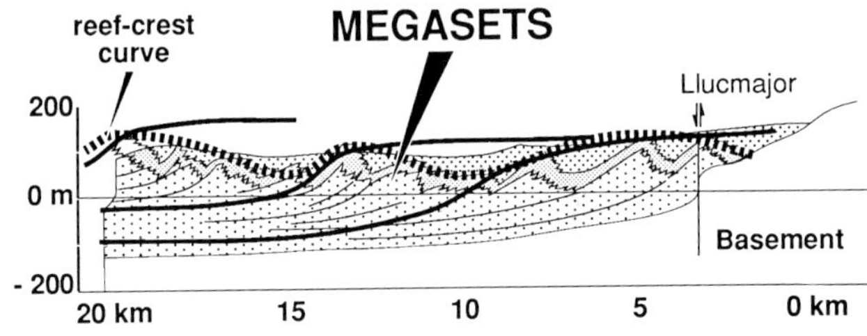


FIG. 17.—Stacking of high-frequency depositional sequences of the Llucmajor Platform as seen in sea-cliff exposures along the southern Llucmajor Platform. Basic sequence is sigmoid-shaped unit that formed during seventh-order sea-level cycle. Reef-crest curve (hachured line) is defined by the successive positions of the reef crest, measured or inferred from the coral-morphology zonation. The basic accretional units (sigmoids) are stacked in different magnitudes of increasingly larger-scale accretional units: sets, cosets, and megasets. Reef-crest curve (Pomar, 1991) reflects the amount of relative-sea-level fluctuation during progradation. Modified from Pomar and Ward (1995). Utilization of pinning points to track sea level uses method described by Franseen et al. (1993) and Goldstein and Franseen (1995).

whole-shell bivalves, rhodoliths and/or burrows. In proximal-slope settings, coral fragments and small coral heads also may in this condensed interval.

In reef core facies, the thickness of the vertical sequence and the preservation of the coral-morphology zones are distinctive of the different systems tracts (Fig. 21). Low stillstands systems tracts and aggradational systems tracts generally preserve the three coral-morphology zones, but aggradational systems tracts

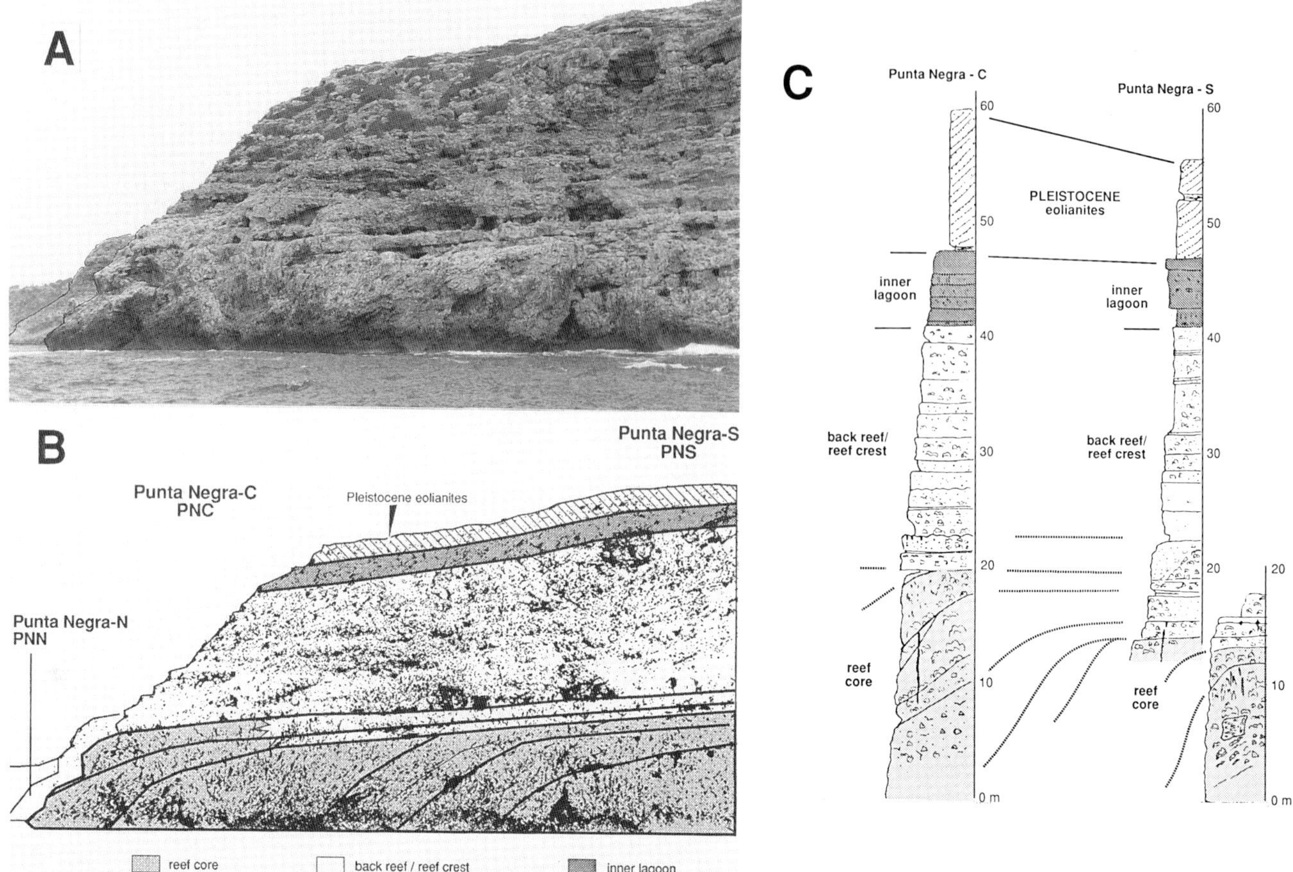


FIG. 18.—(A) Sea-cliff exposures in Punta Negra area (Fig. 7C) showing progradational sigmoids (seventh-order depositional sequences) composed of reef and back-reef lithofacies (at base of outcrop) overlain by gently inclined reef-crest and outer-lagoon units, in turn, overlain by flat-lying inner-lagoon beds. Paleoseaward direction was toward the left. (B) Tracing from photo in A, showing lithofacies and bedding geometries. (C) Stratigraphic sections at PNC and PNS (Fig. 7C, D) showing correlative erosional surfaces (hachured lines). Reef lithofacies on left passes landward (right) into outer-lagoon lithofacies. Truncated sigmoids are wedged-shaped.

have a thicker sequence of shallow-water corals (massive coral zone), presumably because of greater accommodation. High stillstand and offlapping systems tracts commonly have a thin reef-core sequence because of erosive truncation during fall and lowstand of sea level. They usually only have the dish- and part of the branching-coral zones.

The back-reef-lagoon facies belt is mainly aggrading systems tracts, bounded by erosion surfaces that are coeval with the offlapping and low-stand systems tracts (Figs. 21, 22). The lagoonal beds also hierarchically stack in different orders of units which correlate with the different accretional units in the reef core. In some units, basal layers of outer-lagoon lithofacies are overlain by coral patches. This records the flooding and subsequent submergence of a previous erosion surface during a rise of sea level. Some outer-lagoon cycles show an upward decrease in coral colonies and sediment size, indicating shallowing upward. Erosional truncations on top of such beds are related to the subsequent sea-level falls. These beds stack in sets of beds, with similar characteristics: the flooding over the subaerial surfaces is commonly recorded by a thin transgressive layer at the base of the cycle with stromatolites or with cerithid gastropod-rich packstone-grainstone. The subsequent submergence of the platform top is recorded by the predominantly subtidal character of the overlying lagoonal beds. Most lagoon units shallow upward but lack extensive intertidal caps. Intertidal and supratidal deposits, normally deposited at the top of shallowing-upward units during late transgression and highstand, may have been removed by erosion during subsequent falls in sea level. Such erosion events are recorded by the caliche crusts or microkarst surfaces commonly associated with the bed-set boundaries. Similar relationships can be seen at the level of the cosets of beds. The flooding of the platform top is recorded by sets of beds with restricted facies (stromatolites, root zones and paleosols), and the subsequent submergence, by the overlying open-marine grainstone-packstone.

FIG. 19.—Stacking patterns of lithofacies and boundaries which define the sets of sigmoids (A and B) and cosets of sigmoids (C and D) as traced on photomosaics of the sea cliffs from the Punta Negra area to Cap Blanc (Fig. 7C).

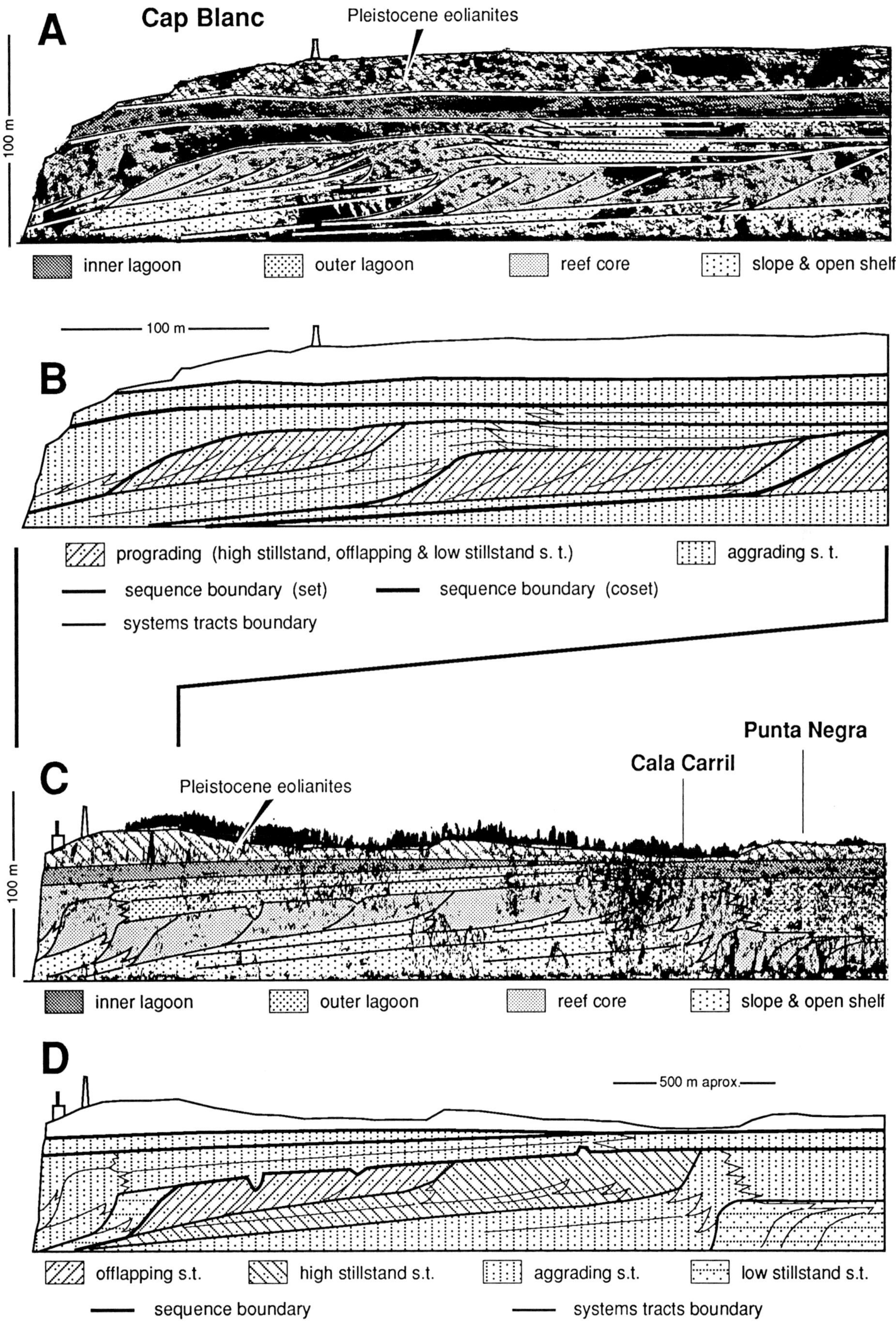
A
Cap Blanc
Pleistocene eolianites
100 m
inner lagoon
outer lagoon
reef core
slope & open shelf
100 m
B
prograding (high stillstand, offlapping & low stillstand s. t.)
aggrading s. t.
sequence boundary (set)
sequence boundary (coset)
systems tracts boundary
Punta Negra
Cala Carril
C
Pleistocene eolianites
100 m
inner lagoon
outer lagoon
reef core
slope & open shelf
D
500 m aprox.
offlapping s.t.
high stillstand s.t.
aggrading s.t.
low stillstand s.t.
sequence boundary
systems tracts boundary

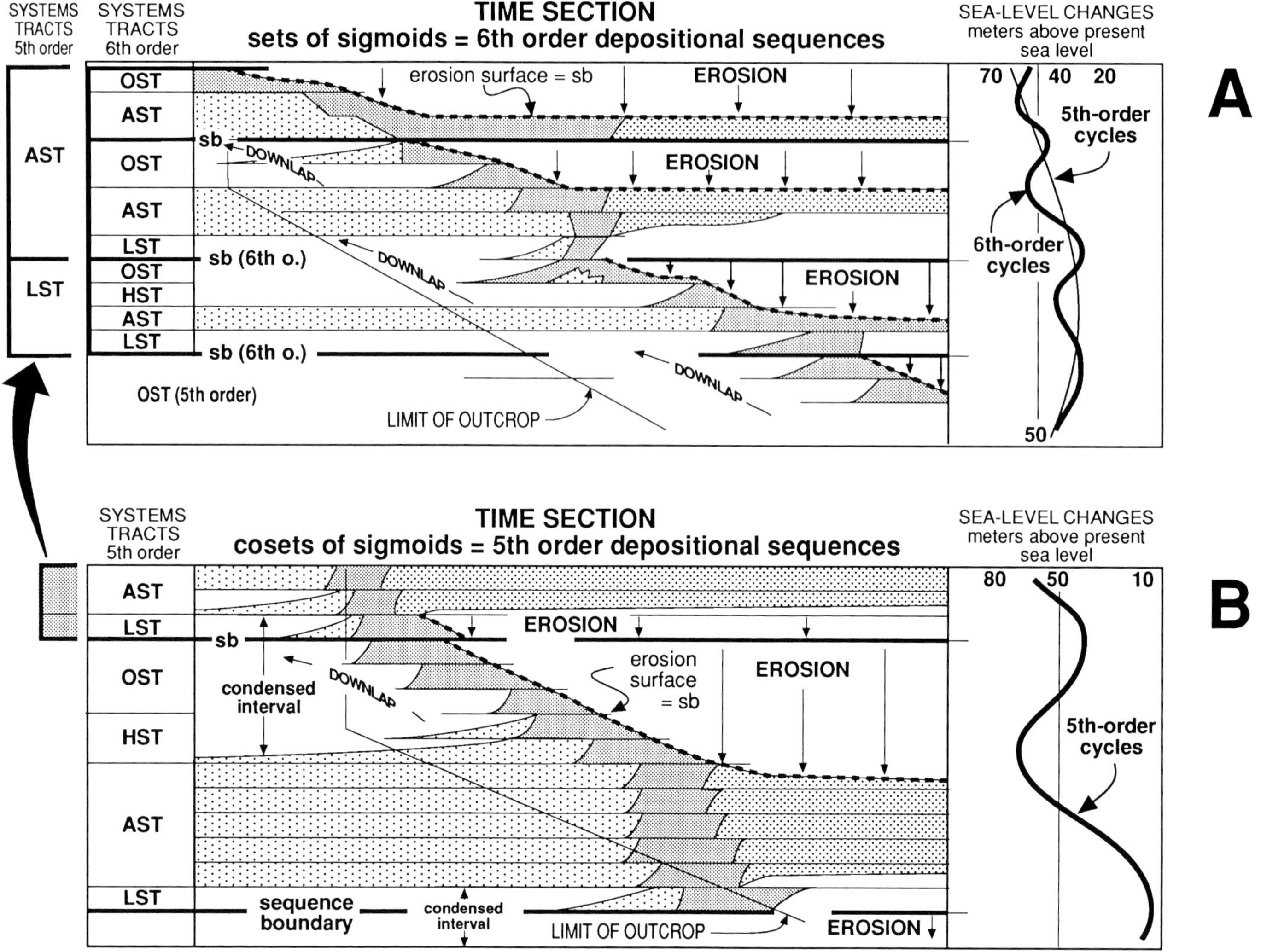


FIG. 20.—(A) Time sections showing lithofacies, condensed sections, erosion surfaces, downlap surfaces, sequence boundaries, and systems tracts of sets of sigmoids (sixth-order depositional sequences) in the same as area as shown in Figures 19A, 19B. Dark pattern shows reef lithofacies; coarse-dot pattern shows lagoon lithofacies; fine-dot pattern shows slope and platform lithofacies. AST - aggrading systems tract; LST - lowstand systems tract; OST - offlapping systems tract; HST - highstand systems tract. (B) Time sections of a coset (fifth-order sequence) in same area as shown in Figures 19C, 19D. Upper two fifth-order systems tracts (indicated by arrow) are the same as those of A. Systems tracts of a coset of sigmoids may be composed of one or several sets of sigmoids, within which the stacking of sigmoids defines the different systems tracts. Modified from Pomar and Ward (1994).

The progradation of the Reef Complex across as much as 20 km of the Llucmajor Platform reflects a post-tectonic depositional setting characterized by a broad shallow carbonate platform, low terrigenous influx and low rates of subsidence (Pomar, 1991). Although the progradation was extensive toward the southwest, where the carbonate platform was shallow and broad, progradation was less than 2 km along the margin of the Palma Basin, where the platform was relatively deep and steeper. Progradation of the reefal systems was more extensive during sea-level falls and downstepping across a gentle depositional profile, but relatively minor progradation took place during sea-level falls across a steep profile (Pomar and Ward, 1995). On the broad shelf, subsequent sea-level rises created wide back-reef lagoons, which probably enhanced carbonate production and downslope shedding of sediment. Direct petrographic evidence for this, however, is lacking, and there is a possibility that greater build up of slope and platform beds may be related to enhanced carbonate production throughout the entire system during sea-level rises. On steeper topographic gradients, only small back-reef lagoons were created during the subsequent sea-level rises. Barrier reefs with extensive lagoons and patch reefs occurred during relative sea-level rises of different orders of magnitude; fringing reefs were developed at times of sea-level falls. Preservation of the lagoon and the upper reef-core facies depends on the amount of erosional truncation, which was controlled by the magnitude of the sea-level falls.

## DIAGENETIC OVERPRINT

The most outstanding diagenetic features of this reef complex are: (1) extensive secondary porosity produced by dissolution of originally aragonitic constituents and (2) pervasive dolomite in

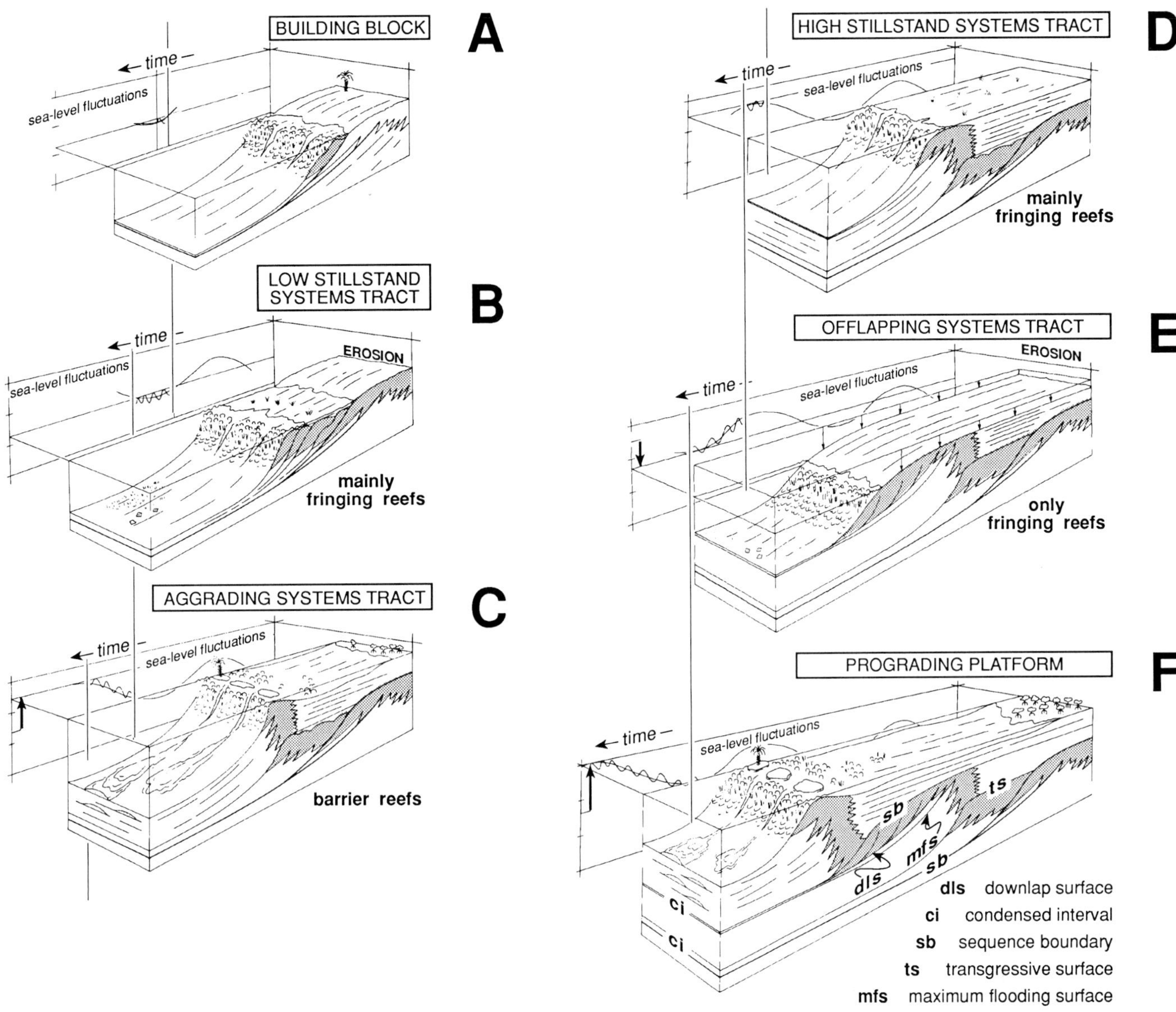


FIG. 21.—Block diagrams of four systems tracts recognized from stacking patterns of the accretionary units within sets and cosets. Dark shading shows location of reef lithofacies. Modified from Pomar and Ward (1994, 1995).

much of the complex. Other diagenetic components are syndepositional marine cements in some reef and lagoon units, various stages of phreatic calcite cement unevenly distributed through the complex and thin subaerial crusts and microkarst in some lagoon units.

### *Dissolution Features*

Virtually all originally aragonitic constituents were removed by dissolution. Most corals, mollusks, and green algae, therefore, are identified by reference to external and internal molds (e.g., Figs. 9A, 13E, 13F). Impressions of coral structure generally are best preserved on the periphery of colonies where lithification of infiltered sediment or microcrystalline submarine cement provided internal molds. Commonly the cores of coral colonies are completely removed, leaving large coral-shaped molds. Skeletons and cements that were originally Mg calcite or calcite are better preserved. Red algae, echinoids, ostracodes, benthic foraminifers, oysters and other calcitic bivalves generally are well preserved in both limestones and dolomites; although in some dolomites these calcitic constituents are partly to totally leached out. Some crusts of isopachous cement were dissolved, leaving cavities subsequently filled with dolomite cement (Figs. 23A, B).

Channels, caves and fractures related to Tertiary and Quaternary karsting also are common in the Upper Miocene reef complex. Most of these holes are partly filled with travertine and red sediment. Prominent features of large-scale dissolution in the Cap Blanc area are the paleocliffs of Upper Miocene limestone and cave-roof collapses (V-structures) affecting beds of the shallow-platform beds and upper parts of the reef core. Paleocliffs in the Els Bancals-Vallgornera area (Figs. 7C, D)

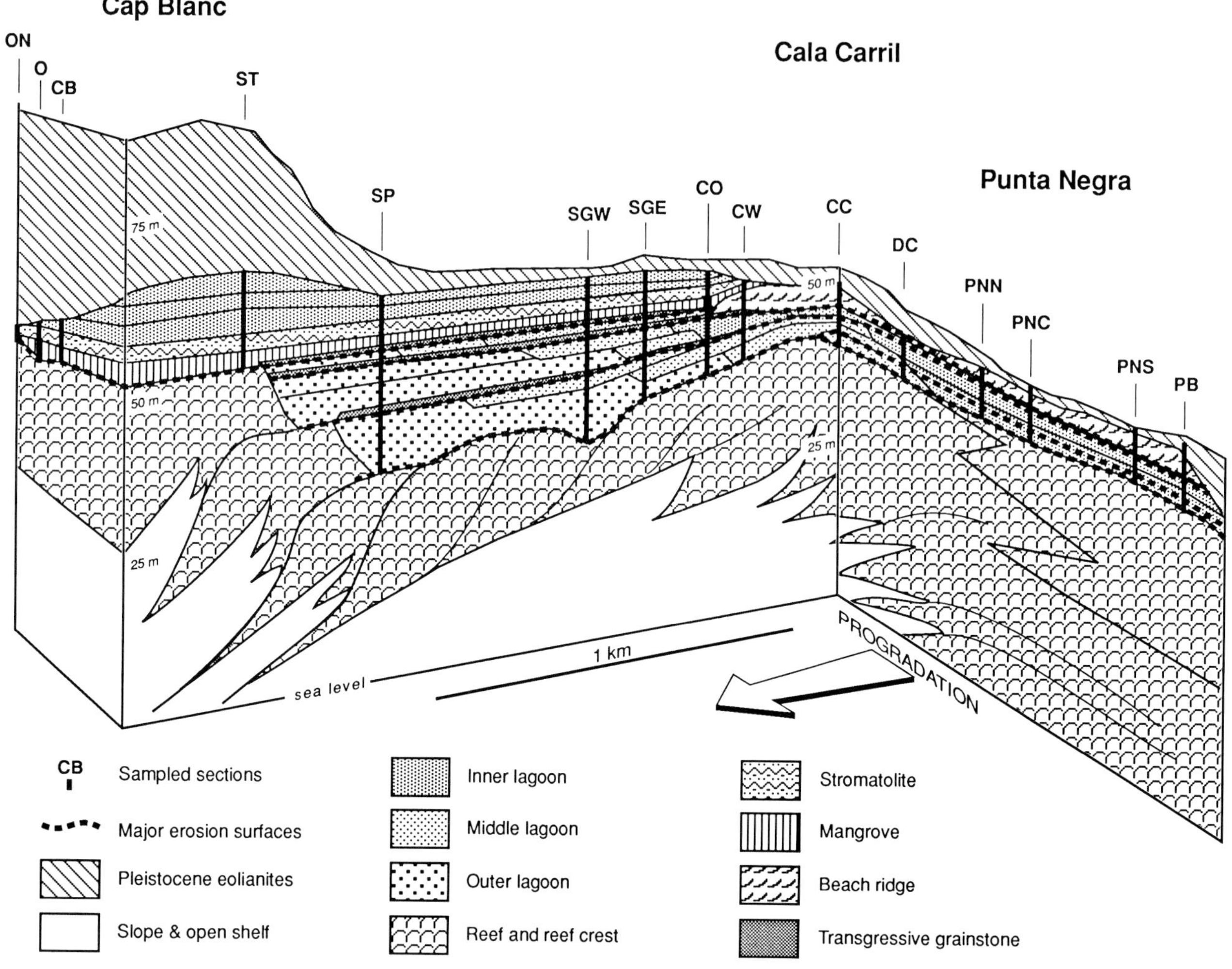


FIG. 22.—Lithofacies of back-reef lagoons from Punta Els Bancals to the Cap Blanc area (modified from Green, 1993). Letters denoting sampled sections are same as those in Figures 7C, 7D. Center part of cross section is perpendicular to depositional strike, and left and right parts are parallel to depositional strike.

were formed during subaerial exposure of the reef complex prior to deposition of the Pliocene limestones. The collapse structures, which occur extensively along the eastern coast of Mallorca (Esteban and Klappa, 1983; Fornós and Pomar, 1983; Pomar et al., 1985; Simó and Ramón, 1986; Fornós et al., 1988), clearly were produced during the sedimentation of the Messinian Santanyí Limestone.

## *Dolomite*

About two-thirds of the reef complex from the Vallgornera coast to Cap Blanc is extensively dolomitized (Fig. 24). Both replacement dolomite and dolomite cement are abundant (Figs. 23, 25). Carbonate mud and originally calcitic constituents are replaced by dolomite, but aragonitic constituents were leached out during and prior to dolomitization (Ward et al., 1993). Most replacement dolomite is very finely crystalline and, therefore, preserves much of the original skeletal structures and sediment

FIG. 23.—Submarine cements of the Llucmajor reef complex. (A and B) Plane-light (A) and cathodoluminescence (B) photomicrographs of red-algal grainstone with isopachous cement (IC; now dolomite), from reef-crest grainstone cropping out at Punta Els Bancals (Fig. 7C). All grains are dolomitized red algae (DRA). Cathodoluminescence shows that dolomite cement (mpdc) filled mold of inner layer of isopachous cement. Much of this mold-filling cement grew inward from the microcrystalline out layer of cement. Some dolomite cement (ipdc) also grew in the intergranular pores (P). Bar scale represents 0.5 mm. (C) Fibrous calcite cement in intergranular and intraskeletal pores (P) of grainstone from within the Punta Negra N (PNN of Figs. 7C, D) reef. Fibrous calcite cement is overlain by coarser sclenohedral calcite cement. Dark grain is red algae (RA). Mollusk molds are outlined by micrite envelopes. Intramoldic cement is dolomite (D). Bar scale represents 0.5 mm. Plane-polarized light. (D) Cathodoluminescence photomicrograph showing dolomite-replaced fibrous cement in dolograinstone from large cavity in Cap Blanc *Porites* reef (Figs. 7C, 13E). Skeletal grains such as dolomitized red algae (DRA) and mollusks (mold in lower left) are thickly coated with dolomitized isopachous fibrous cement (DFC). Luminescent-banded dolomite cement (DC) lines intergranular pores (P), and similar dolomite cement (DCm) fills moldic pores. Bar scale represents 0.2 mm. (E) SEM micrograph of dolomite-replaced fibrous cement (DFC) and pore-lining dolomite cement (DC) in same sample shown in D.

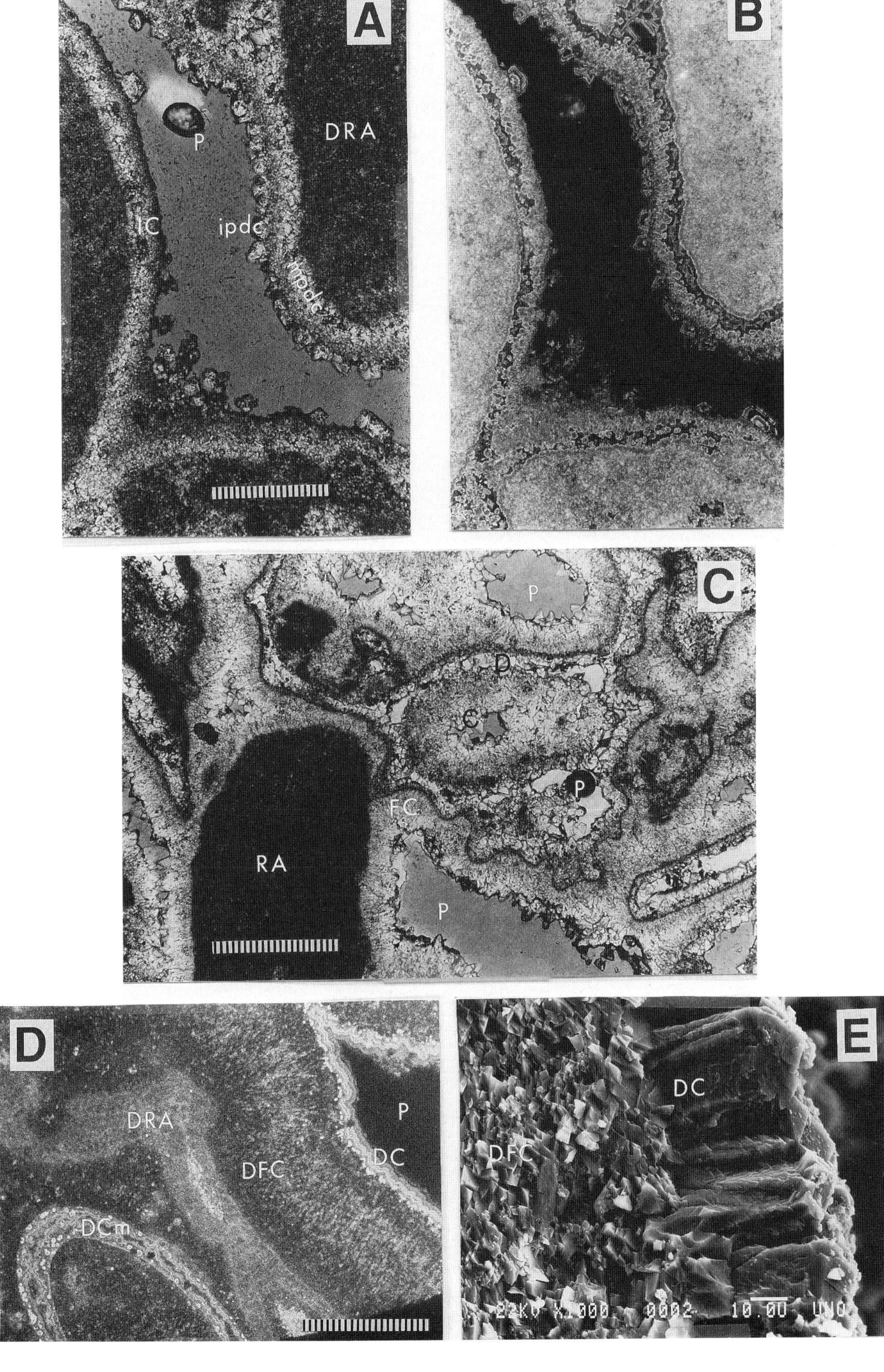
A
P
DRA
IC
ipdc
B
C
P
D
P
FC
RA
P
D
DRA
P
DFC
DC
DCm
E
DC
DFC
22KV X1000 0002 10.0U UNO

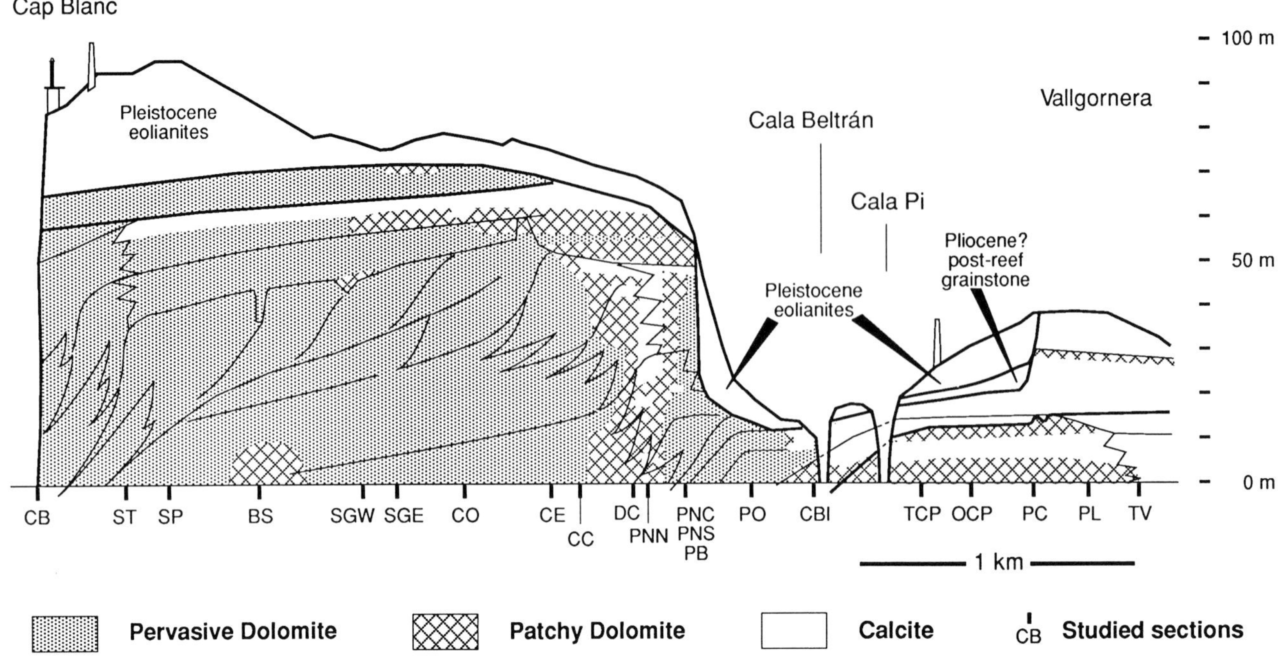


FIG. 24.—Distribution of dolomite in the sea-cliff exposures from Torrente Vallgornera to Cap Blanc (Figs. 7C, 7D). Letters designating sampled sections are same as those in Figure 7D.

fabric. Dolomitization was one of the earliest diagenetic events, and in pervasively dolomitized parts of the reef complex, both dolomitization and aragonite dissolution apparently preceded precipitation of non-marine calcite cements. Stratigraphic relationships show that much of the dolomitization took place during a highstand of Messinian sea level following a major fall in sea level. The evidence that earliest dolomite cement crusts precipitated directly on the coral skeletons (Fig. 25), therefore, shows that there was extensive preservation of metastable aragonite in the reef complex during the sea-level fall (Ward et al., 1993). This suggests that the southwestern portion of the Llucmajor Platform was removed from a source of freshwater during the Late Miocene and that the lowstand climate was arid (Sun and Esteban, 1994).

Dolomite cement is common in moldic pores left by dissolution of aragonite skeletons (Figs. 23C, 25A). Generally this cement occurs as one-crystal-wide fringes of clear euhedral-subhedral crystals 30 to 120 microns in diameter. Where dolomite cement completely occludes pores, it is a mosaic of anhedral to subhedral crystals. The dolomite cement exhibits complicated cathodoluminescent zonation (Fig. 25), which indicates that dolomite replacement and cementation began before aragonitic components were completely dissolved (Ward et al., 1993). Intracoral pores were lined with dolomite cement before wholesale dissolution of the coral (Fig. 25C). Following coral dissolution, precipitation of dolomite cements continued in the intraskeletal pores and in moldic pores (Fig. 25C).

Whole-rock analyses by Oswald et al. (1990), Oswald (1992) and Green (1994) show the Mallorca dolomites have $\delta^{18}O$ values of +4.5 to +6.3‰ and $\delta^{13}C$ values of -O.1 to +3.14‰ PDB. From 39 samples of Mallorca dolomite, Staudt et al. (1993) found Na ranged from 489-1,995 ppm; Cl, from <90-856; and $SO_4$, from 887-5,968. The "heavy" oxygen isotopic ratios and trace-element data indicate that the dolomitizing fluids were brines derived from evaporation of sea water (Oswald et al., 1990; Oswald, 1992; Sun, 1992; Staudt et al., 1993).

The distribution of dolomite changes from patchy or absent in the Vallgornera outcrops to pervasive in the Cap Blanc exposures (Fig. 24). The reef mounds on the Vallgornera coast (Fig. 8) contain small patches of dolomite associated with coral molds. The overlying outer-lagoonal beds in this area are mostly calcitic. The *Porites-Tarbellastraea* reefs and overlying lagoonal beds along the Els Bancals coast (Fig. 11) are pervasively dolomitized as is most of the *Porites* barrier tract from the western end of Els Bancals to Cap Blanc (Fig. 24). There are, however, areas of partial dolomitization. For example, at Punta Negra N (PNN, Fig. 24) the *Porites* reef and overlying lagoonal beds are only partly dolomitic. Much of the dolomite in the Punta Negra reef is cement lining aragonite-fossil molds (Fig. 23C) and coating calcitic grains, with only small areas of replacement dolomite. The lateral distribution of dolomite in the Punta Negra area suggests that the flow of dolomitizing fluid was partly controlled by vertical fractures. Inner-lagoon units throughout much of the reef complex are calcitic (Fig. 24), except for patches of dolomite in thin partly calichified zones at major stratigraphic breaks. This relationship suggests some dolomitization accompanied early stages of inner-platform flooding after subaerial exposure.

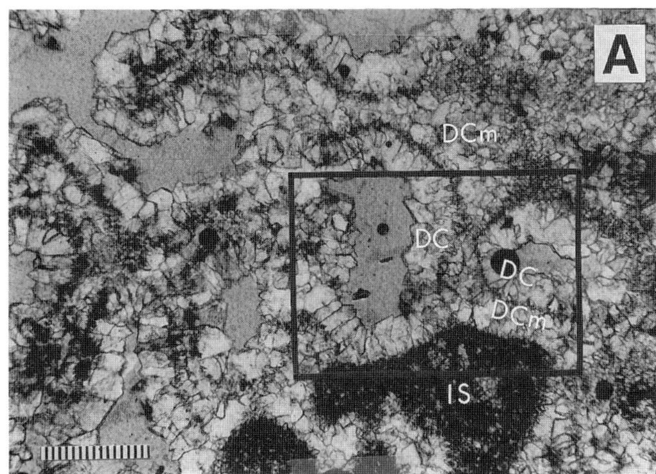


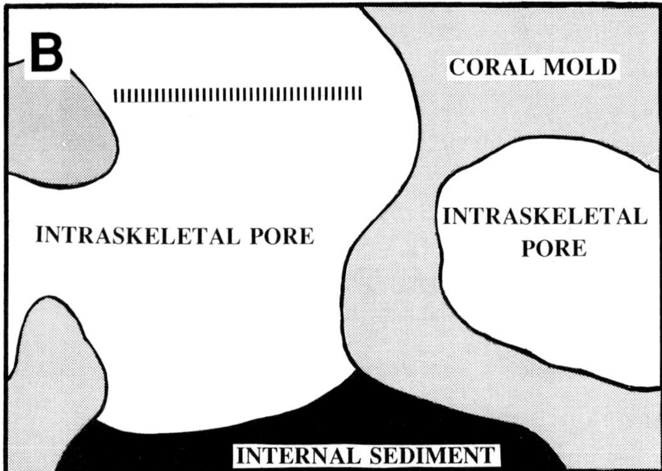


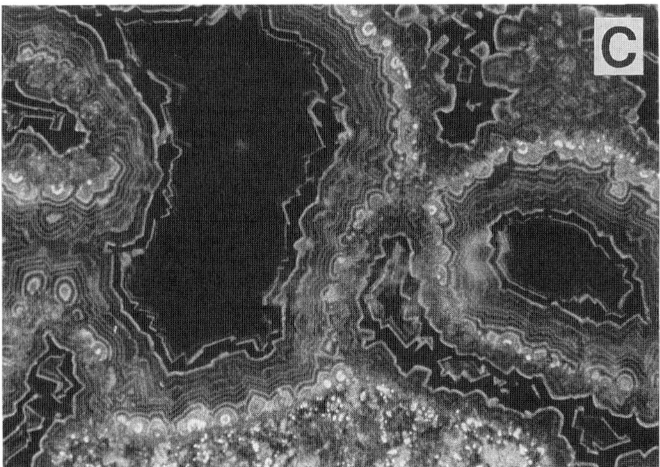


The general pattern of dolomite apparently becoming less abundant toward older parts of the reef complex suggests that the distribution of dolomite is partly related to proximity to the final platform margin (Oswald et al., 1990; Oswald, 1992; Sun, 1992). From the stratigraphic relationships and the oxygen- and carbon-isotope composition of the dolomites, Oswald et al. (1990) and Oswald (1992) concluded that dolomitization occurred during the Messinian when Mediterranean seawater with high Mg/Ca ratios invaded the reef complex following the Salinity Crisis drawdown.

### *Subaerial Crusts and Karst*

Caliche horizons and microkarst preserved in the lagoonal lithofacies are products of intermittent subaerial exposure of the shallow platform during deposition of the Upper Miocene reef complex. Superimposed over the Miocene carbonate rocks are calichified zones and larger-scale karst features (collapse structures and fracture-controlled paleo-seacliffs) that reflect subaerial exposure during late Tertiary and Quaternary time.

### *Calcite, Aragonite and Gypsum Cements*

In older parts of the reef complex, a few coral skeletons, especially *Tarbellastraea*, are replaced by medium-crystalline calcite. In these corals, intraskeletal cement (presumably originally aragonitic syndepositional cement) is preserved as fibrous-shaped inclusions within blocky calcite spar. Textures of originally Mg-calcitic marine cements also are retained in some reef and lagoon rocks. For example, intraskeletal pores in dolomitized and non-dolomitized reef rock commonly contain microcrystalline "pellets" like the peloidal Mg-calcite precipitates of modern reefs (e.g., Macintyre, 1985; Chafetz, 1986). In addition, intergranular pores in some grainstones and dolograinstones are lined with isopachous layers of fibrous crystals (Fig. 23) resembling Mg-calcite submarine cement (e.g., James and Ginsburg, 1979). Abundant microdolomite inclusions in calcitic layers of this fibrous cement are further indication of its origin as marine Mg-calcite cement (Lohmann and Meyers, 1977). Dolomite cement commonly is overlain by equant to fibrous calcite cement, representing various stages of post-dolomitization fresh-phreatic and vadose cementation.

Late-stage aragonite cement was precipitated locally along

FIG. 25.—Dolomite cement in Cap Blanc *Porites* reef. (A) Photomicrograph of clear subhedral dolomite cement (DC) lining intraskeletal pores and subhedral-anhedral dolomite cement (DCm) filling coral-moldic pores. Dark area is dolomite-replaced internal sediment (IS). Bar scale represents 0.25 mm. Plane-polarized light. Rectangle outlines area shown in B and C. (B) Diagram showing outlines of intraskeletal and moldic pores into which dolomite cement grew. Bar scale represents 0.25 mm. (C) Cathodoluminescence photomicrograph of area shown in B. Early-stage brightly luminescent cement occurs only in intraskeletal pores. Later-stage dark-and-light-banded crusts of dolomite cement were precipitated in both intraskeletal and moldic pores. This suggests that complete dissolution of the coral occurred after initial stages of dolomite cementation and after the final stages of dolomite cementation.

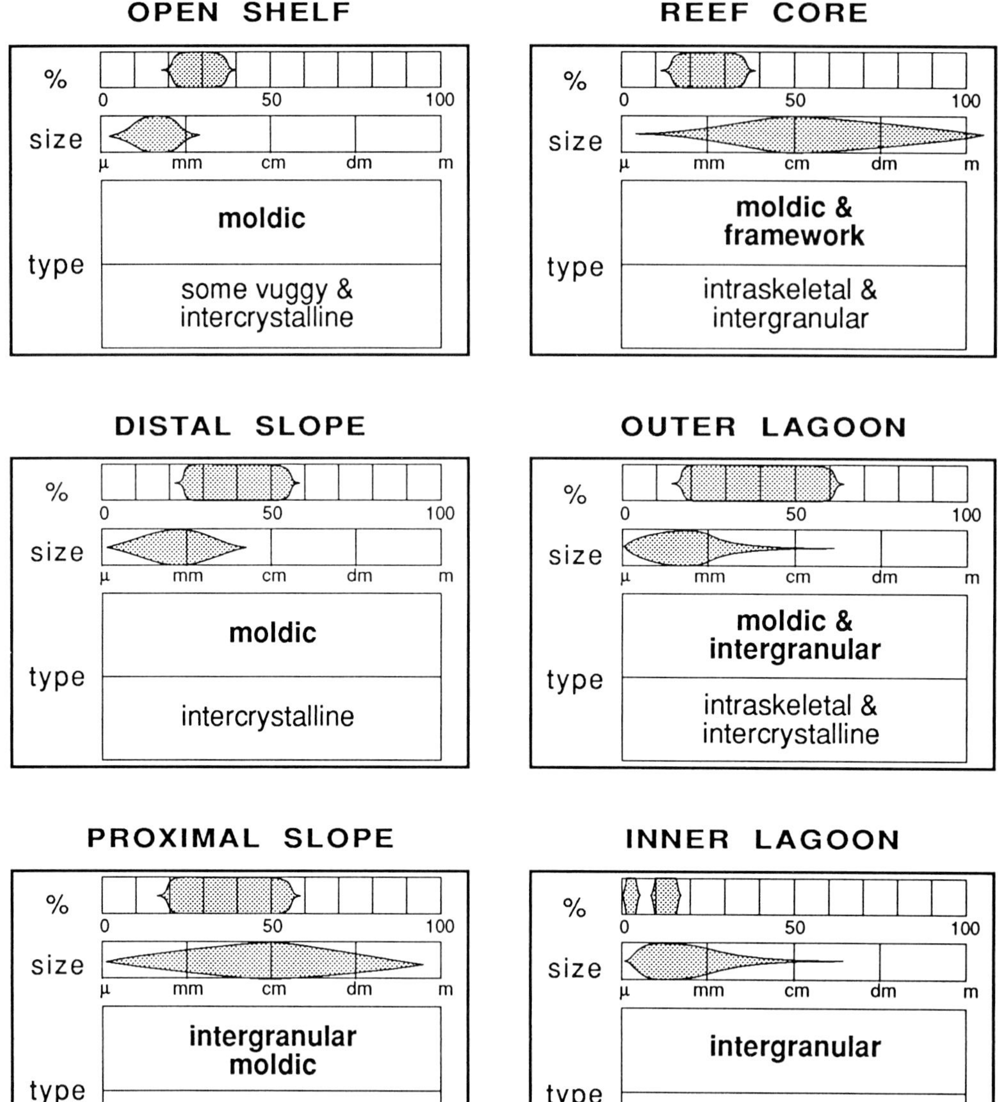


FIG. 26.—Estimated percentage and types of porosity of major lithofacies of the Reef Complex as well as estimated range of pore sizes. Under "type": bold lettering designates predominant porosity types for each lithofacies.

the Vallgornera and Els Bancals coast. Aragonite botryoids a few centimeters in diameter are found in open fractures about 3 m above present sea level, and irregular fringes of aragonite needles occur in some pores within the reef rock. These aragonite cements, which overlie dolomite and post-dolomite calcite cements, probably were precipitated within the coastal-karst system during a higher stand of the late-Neogene sea. Thin reddish-brown micrite crusts and Quaternary sediment overlie the large aragonite botryoids in the fractures. In many samples taken along the sea cliffs, the youngest diagenetic components are gypsum crystals. Their distribution is independent of stratigraphic level, and presumably they are precipitates from sea spray.

### *Porosity*

Most of the reef complex is highly porous (Fig. 26). Besides the abundant cavities left by dissolution of corals, many voids within the original reef framework were only partly filled with internal sediment. Moldic and framework pores within the reef tracts, therefore, range from a fraction of a centimeter to several meter in diameter. Aragonite-fossil molds as well as intergranular pores in many grainstones of the slope, reef, and lagoon lithofacies remain partly open (Figs. 9A, 9C, 9F; 11B; 13B, 13E; 14; 16B, 16C). The amount of porosity in wackestones and packstones is greatest in those rocks that contained abundant aragonite skeletons. Porosity is enhanced in some dolomitized rocks where calcitic constituents also are leached out. Neither syndepositional marine cement nor post-depositional calcite and dolomite cement is abundant enough to completely occlude porosity in most parts of the reef complex.

## CONCLUSIONS

The Llucmajor Platform was the site of the most extensive progradation (up to 20 km) of a Late Miocene coral-reef complex in the western Mediterranean. Excellent exposures along high sea cliffs reveal the depositional history during about 4 km of southwestward progradation on southern and western parts of the Llucmajor Platform. The Upper Tortonian-Lower Messinian Reef Complex overlies Lower Tortonian carbonate rocks (shallow-platform *Heterostegina* Unit of Mallorca or rhodalgal-ramp Lower Bar Unit of Menorca). Around the flanks of the Palma Basin, the slope and open-platform beds of the Reef Complex are overlain by shallow-marine marl (Bonanova Marls and Gypsum and Grey Marls), but the reef-platform beds are unconformably overlain by intertidal to shallow-subtidal limestones (Santanyí Limestones).

Four main lithofacies belts of the Upper Miocene Reef Complex are off-reef open-shelf planktonic foraminifer wackestone-packstone (rhodalgal packstone-grainstone also may be an off-reef open-shelf lithofacies); slope wackestone, packstone and grainstone; reef framestone, floatstone and rudstone; and back-reef lagoon grainstone, packstone, and wackestone. Much of the reef complex cropping out in this area is pervasively dolomitized.

Between Vallgornera and Cap Blanc on the southern part of the Llucmajor Platform, three types of reef rock crop out. Along the Vallgornera coast are mound reefs composed of *Porites* and *Tarbellastraea*. From Cala Pi to Els Bancals are well-developed reef tracts also constructed of those two corals, but from Els Bancals to Cap Blanc the reef frameworks are made up predominantly of *Porites*. The youngest reef tracts of the Reef Complex, which crop out along the western coast of the platform, are composed of both *Tarbellastraea* and *Porites*, and reefs penetrated in core holes throughout the platform contain both corals. This suggests that the *Porites*-framework reefs represent only a small segment of the Reef Complex.

In terms of scale of progradation and quality and size of the exposures, the reef complex cropping out in sea cliffs of the Llucmajor Platform appears as unique in the Upper Miocene of the Mediterranean. The spectacular progradation geometries of the Reef Complex reflect a post-tectonic depositional setting of (1) a broad shallow shelf, (2) low terrigenous influx and (3) low rates of subsidence (Pomar, 1991). Although the progradation was extensive toward the southwest, where the carbonate platform was shallow and broad, progradation was much less (less than 2 km) along the margin of the Palma Basin, where the platform was relatively deep with a steeper profile. Progradation of the reefal systems was more extensive during sea-level falls on a gentle depositional profile. The subsequent sea-level rises created wide lagoons, which apparently enhanced carbonate production and downslope shedding of sediment. On steeper topographic gradients, relatively minor reef progradation took place during sea-level falls, and only small back-reef lagoons were created during the subsequent sea-level rises. Barrier reefs with extensive lagoons and patch-reefs occurred during relative sea-level rises of different orders of magnitude; fringing reefs were developed at times of sea-level falls. Preservation of the lagoon and the upper reef-core facies depends on the amount of erosional truncation, which was controlled by the magnitude of the sea-level falls.

The basic element of the stratigraphic architecture is the sigmoid, with reef core, slope, open-shelf and back-reef lagoon facies. The pattern of stacking of sigmoids, sets of sigmoids, and cosets of sigmoids is delineated by facies shifts and erosional truncations, which reflect four orders of magnitude of sea-level oscillations of a frequency higher than the 3rd-order cycles of Haq et al. (1988). Geometries and facies distribution in the reef complex result from the pattern of stacking of these different accretional units, controlled by the corresponding high-frequency sea-level oscillations.

The Upper Miocene reef complex is highly porous because of the dissolution of aragonitic components, the generally low volume of pore-occluding cement, and the lack of compaction. About two-thirds of the reef complex from Vallgornera to Cap Blanc is extensively dolomitized. Both replacement dolomite and dolomite cement are abundant. Stratigraphic relationships indicate that dolomitization took place during the late Messinian. In most of these rocks, dolomitization was one of the earliest diagenetic events, preceding any sort of fresh-water diagenesis. This suggests that this part of the Reef Complex was removed from a source of fresh groundwater during the Late Miocene and probably reflects an arid climate. Much of the dolomitic rock retains high porosity because it remains free of extensive post-dolomite calcite cements.

## ACKNOWLEDGMENTS

This research was supported in part by grant RA-19-A from the Louisiana Education Quality Support Fund to Ward. We also are grateful to the Ministerio de Educación y Ciencia de España (Programa de Sabáticos) for providing living and travel expenses to Ward. Pomar gratefully acknowledges the support of Dirección General de Investigación Cientifica y Tecnica project PB87-0812. We are grateful for the travel funds provided by NATO Collaborative Research Grant 931472. Mateu Esteban provided much helpful discussion and advice during our study. E. Franseen, R. Goldstein and R. Handford made excellent suggestions to improve the manuscript.

REFERENCES

Alvaro, M., Barnolas, A., del Olmo, P., Ramirez del Pozo, J., and Simó, A., 1984, El Neógeno de Mallorca: Caracterización sedimentológica y bioestratigráfica: Boletín Geológico y Minero, v. 95, p. 3-25.

Barón, A., 1977, Estudio estratigráfico y sedimentólogico del Mioceno medio y superior postorogénico de la Isla de Mallorca: Palma, Premio Ciudad de Palma, unpublished, 180 p.

Barón, A. and Pomar, L., 1985, Stratigraphic correlation tables: area 2c Balearic Depression, *in* Steininger, F. F., Senes, J., Kleemann, K., and Rog, F., eds., Neogene of the Mediterranean, Tethys and Paratethys: Vienna, Institute of Paleontology, University of Vienna, v. 1, p. 17 and v. 2, p. 17.

Barón, A., Bayo, A., and Fayas, J. A., 1984, Valor acuifero del modelo sedimentario de plataforma carbonatada del Miocene de la isla de Menorca, *in* Obrador, A., ed., Libro Homenaje a L. Sanchez de la Torre: Barcelona, Cuadernos de Geologia, Universidad Autónoma Barcelona, v. 20, p. 189-207.

Bizon, G., Bizon, J. J., Bourrouilh, R., and Massa, D., 1973, Présence aux îlles Baléares (Méditerranée Occidentale) de sédiments "messiniens" dépossés dans une mer ouverte á salinité normale: Paris, Comptes Rendus des seánces de l'Academie des Sciences, v. 277, p. 985-988.

Bosence, D. W. J., 1985, The "coralligène" of the Mediterranean - a Recent analog for Tertiary coralline algal limestones, *in* Toomey, D. F., ed., Paleoalgology: Contemporary research and applications: Berlin, Springer-Verlag, p. 216-225.

Bosence, D. W. J., Pomar, L., Waltham, D. A., and Lankester, H. G., 1994, Computer modeling a Miocene carbonate platform, Mallorca, Spain: American Association of Petroleum Geologists Bulletin, v. 78, p. 247-266.

Bourrouih R., 1973, Stratigraphie, sédimentologie et tectonique de l'île de Minorque et du NE de Majorque (Baléares): Unpublished Ph.D. Thesis, University of Paris VI, Paris, 822 p.

Chafetz, H. S., 1986, Marine peloids: a product of bacterially induced precipitation of calcite: Journal Sedimentary Petrology, v. 56, p. 812-817.

Colom, G., 1975, Geologia de Mallorca: Palma, Gráficas Mallorca, 522 p.

Esteban, M., 1979, Significance of the Upper Miocene reefs of the Western Mediterranean: Paleogeography, Paleoclimatology, Paleoecology, v. 29, p. 169-188.

Esteban, M., Calvet, F., Dabrio, C., Barón, A., Giner, J., Pomar, L., and Salas, R., 1977, Aberrant features of the Messinian coral reefs, Spain (abs): International Geological Correlation Program, UNESCO, Messinian Seminar 3, Málaga, p. 41-45.

Esteban, M., Calvet, F., Dabrio, C., Barón, A., Giner, J., Pomar, L., Salas, R., and Permanyer, A., 1978, Aberrant features of the Messinian coral reefs, Spain: Acta Geológica Hispánica, v. 13, p. 20-22.

Esteban, M. and Klappa, C. F., 1983, Subaerial exposure environment, *in* Scholle, P. A., Bebout, D. G., and Moore, C. H., eds., Carbonate Depositional Environment:Tulsa, American Association Petroleum Geologists Memoir 33, p. 1-54.

Fallot, P., 1922, Etude géologique de la Sierra de Majorque: Unpublished Ph.D. Thesis, Polytechnique ch. Béranger, Paris, 480 p.

Fontboté, J. M., Obrador, A., and Pomar, L., 1983, Islas Baleares, *in* Geologia de España, Toma II (Libro Jubilar J. M. Rios): Madrid, Comisión Nacional de Geología e Instituto Geológico y Minero de España, p. 343-391.

Fornós, J. J., 1983, Estudi sedimentològic del Miocé terminal de I'illa de Mallorca: Unpublished M.S. Thesis, Universidad de Barcelona, Barcelona, 228 p.

Fornós, J. J., 1987, Les plataformes carbonatades de les Balears: Unpublished Ph.D. Thesis, Universidad de Barcelona, Barcelona, 954 p.

Fornós, J. J., Ginés, A., Ginés, J. and Pomar, L., 1988, Paleokarst collapse breccias in the uppermost Miocene of Mallorca Island, Spain (abs): Leuven, International Association Sedimentologists 9th Regional Meeting on Sedimentology, p. 76-77.

Fornós, J. J. and Pomar, L., 1983, Mioceno superior de Mallorca: Unidad Calizas de Santanyí ("Complejo Terminal"), *in* Pomar, L., Obrador, A., Fornós, J., and Rodriguez-Perea, A., eds., El Terciario de las Baleares (Mallorca-Menorca), Guía de las Excursiones del X Congreso Nacional de Sedimentología: Palma, Institut d'Estudis Baleárics and Universidad de Palma de Mallorca, p. 177-206.

Fornós, J. J. and Pomar, L., 1984, Facies, ambientes y secuencias de plataforma carbonatada somera (Formación Calizas de Santanyí) en el Mioceno Terminal de Mallorca (Islas Baleares), *in* Obrador, A., ed., Sánchez de la Torre: Barcelona, Cuadernos de Geologia, Universidad Autónoma de Barcelona, v. 20, p. 319-338.

Franseen, E. K., Goldstein, R. H., and Whitesell, T. E., 1993, Sequence stratigraphy of Miocene carbonate complexes, Las Negras area, southeastern Spain: Implications for quantification of changes in relative sea level, *in* Loucks, R. G. and Sarg, J. F., eds., Carbonate Sequence Stratigraphy: Recent Developments and Applications: Tulsa, American Association of Petroleum Geologists Memoir 57, p. 409-434.

Fuster, J., 1973, Estudio de los recursos hidráulicos totales de las Baleares: Informe de síntesis general: Ministerio de Obras Públicas, Ministerio de Industria y Ministerio de Agricultura, Comisión de Coordinación, v. 2, unpublished, 73 p.

Garcia-Yagüe, A., and Muntaner, A., 1968, Estudio hidrogeológio del Llano de Plama: Palma de Mallorca, Unpublished report, Ministerio de Obras Públicas, Dirección General Hidráulicas, Servicio Geológico de Obras Públicas, 79 p.

Goldstein, R. H. and Franseen, E. K., 1995, Pinning points: A method that provides quantitative constraints on relative sea-level history: Sedimentary Geology, v. 95, p. 1-10.

Green, D. G., 1993, High-frequency cyclicity recorded in back-reef lagoon units, Upper Miocene Reef Complex, Mallorca, Spain: Unpublished M.S. Thesis, University of New Orleans, New Orleans, 140 p.

Hallock, P. and Schlager, W., 1986, Nutrient excess and the demise of coral reefs and carbonate platforms: Palaios, v. 1, p. 389-398.

Haq, B. U., Hardenbol, J., and Vail, P. V., 1988, Mesozoic and Cenozoic chronostratigraphy and cycles of sea-level change, *in* Wilgus, C. K., Hastings, B. S., Kendall, C. G. St. C, Posamentier, H. W., Ross, C. A., and Van Wagoner, J. C., eds., Sea-level Changes: An Integrated Approach: Tulsa, Society of Economic Paleontologists and Mineralogists Special Publication 42, p. 71-108.

James, N. P. and Ginsburg, R. N., 1979, The seaward margin of Belize barrier and atoll reefs: International Association Sedimentologists Special Paper 3, 191 p.

Jenkins, K. F., 1992, The response of coral reefs to glacio-eustatic sea level cyclicity: An application of computer modelling: Unpublished Ph.D. Thesis, University of Manchester, Manchester, 296 p.

Jurado, M. J., 1985, Estudi sedimentològic del Neogen de l'àrea de Ciutadella: Consell Insular de Menorca, 144 p.

Lankaster, T. H. B., 1993, Computer modelling of carbonate platform stratigraphies: Unpublished Ph.D. Thesis, Royal Holloway, University of London, London, 199 p.

Lohmann, K. C. and Meyers, W. J., 1977, Microdolomite inclusions in cloudy prismatic calcites a proposed criterion for former high magnesium calcites: Journal Sedimentary Petrology, v. 47, p. 1078-1088.

Macintyre, I. G., 1985, Submarine cements - the peloidal question, *in* Schneidermann, N. and Harris, P. M., eds., Carbonate Cements: Tulsa, Society Economic Paleontologists and Mineralogists Special Publication 36, p. 109-116.

Obrador, A. and Pomar, L., 1983, El Neógeno del sector de Maó, *in* Pomar, L., Obrador, A., Fornós, J., and Rodriguez-Perea, A., eds., El Terciario de las Baleares (Mallorca-Menorca), Guía de las Excursiones del X Congreso Nacional Sedimentología: Palma,Institut d'Estudis Baleàrics and Universidad de Palma de Mallorca, p. 207-232.

Obrador, A., Pomar, L., Jurado, M. J., RodriguezPerea, A., and Fornós, J. J., 1983a, El Neogeno del sector de Ciutadella, *in* Pomar, L., Obrador, A., Fornós, J., and Rodriguez-Perea, A., eds., El Terciario de las Baleares (Mallorca-Menorca), Guía de las Excursiones del X Congreso Nacional Sedimentología: Palma, Institut d'Estudis Baleàrics and Universidad de Palma de Mallorca, p. 233-255.

Obrador, A., Pomar, L., Rodriguez-Perea, A., and Jurado, M. J., 1983b, Unidades deposicionales del Neógeno Menorquín: Acta Geológica Hispánica, v. 18, p. 87-97.

Obrador, A., Pomar, L., and Taberner, C., 1992, Late Miocene megabreccia of Menorca (Balearic Islands): A basis for the interpretation of a megabreccia ramp deposit, *in* Pedley, H. M., ed., Carbonate Ramps: Processes and Diagenesis: Sedimentary Geology, v. 79, p. 203-223.

Oswald, E. J., 1992, Dolomitization of a Miocene reef complex, Mallorca, Spain: Unpublished Ph.D. Thesis, State University of New York at Stony Brook, Stony Brook, 424 p.

Oswald, E. J., Meyers, W. J., and Pomar, L., 1990, Dolomitization of an Upper Miocene reef complex, Mallorca, Spain: evidence for a Messinian dolomitizing Mediterranean Sea (abs.): Bulletin American Association Petroleum Geologists, v. 74, p. 735.

Pascual, D. and Barón, A., 1973, Estudio hidrogeológico de la Unidad de Llucmajor-Campos: Unpublished Informe interno, Servicio Hidráulico de Baleares, Ministerio de Obras Públicas, 72 p.

Perry, C. T., 1996, Distribution and abundance of macroborers in an Upper Miocene reef system, Mallorca, Spain: Implications for reef development and framework destruction: Palaios, v. 11, in press.

Pomar, L., 1979, La evolución tectonosedimentaria de las Baleares: análisis crítico: Acta Geológica Hispánica (Libro Homenaje a Luis Solé Sabarís), v. 14, p. 293-310.

Pomar, L., 1988, Reef architecture and high-frequency relative sea-level oscillations, Upper Miocene, Mallorca, Spain (abs.): Leuven, Abstracts, International Association Sedimentologists 9th Regional Meeting on Sedimentology, p. 174-175.

Pomar, L., 1991, Reef geometries, erosion surfaces and high-frequency sea-level changes,Upper Miocene Reef Complex, Mallorca, Spain: Sedimentology, v. 38, p. 243-269.

Pomar, L., 1993, High-resolution sequence stratigraphy in prograding carbonates: Application to seismic interpretation, *in* Loucks, R. and Sarg, R., eds., Recent Advances and Applications of Carbonate Sequence Stratigraphy: Tulsa, American Association of Petroleum Geologists Memoir 57, p. 389-407.

Pomar, L., Bosence, D. W. J., and Waltham, D. A., 1990, Computer modelling of a Miocene carbonate platform, Mallorca, Spain (abs.): Nottingham, Abstracts, 13th International Association of Sedimentologists Congress, p. 431.

Pomar, L., Esteban, M., Calvet, F. and Barón, A., 1983a, La unidad arrecifal del Mioceno superior de Mallorca, *in* Pomar, L., Obrador, A., Fornós, J., and Rodriguez-Perea, A., eds., El Terciario de las Baleares (Mallorca-Menorca), Guía de las Excursiones del X Congreso Nacional Sedimentología: Palma, Institut d'Estudis Balearics and Universidad de Palma de Mallorca, p. 139-175.

Pomar, L., Marzo, M., and Barón, A., 1983b, El Terciario de Mallorca, *in* Pomar, L., Obrador, A., Fornós, J., and Rodriguez-Perea, A., eds., El Terciario de las Baleares (Mallorca-Menorca), Guía de las Excursiones del X Congreso Nacional Sedimentología: Palma, Institut d'Estudis Balearics and Universidad de Palma de Mallorca, p. 21-44.

Pomar, L., Fornós J. J., and Rodriguez Perea, A., 1985, Reef and shallow carbonate facies of the Upper Miocene of Mallorca, *in* Mila, M. D., and Rosell, J., eds., 6th European Regional Meeting Excursion Guidebook: Barcelona, International Association Sedimentologists and Universitat Autònoma Barcelona, p. 495-518.

Pomar, L. and Ward, W. C., 1994, Response of a late Miocene Mediterranean reef platform to high-frequency eustasy: Geology, v. 22, p. 131-134.

Pomar, L. and Ward, W. C., 1995, Sea-level changes, carbonate production and platform architecture: The Llucmajor Platform, Mallorca, Spain, *in* Haq, B. U., ed., Sequence Stratigraphy and Depositional Response to Eustatic, Tectonic and Climatic Forcing: Kluwer Academic Publishers, p. 87-112.

Rangheard, Y., 1972, Étude géologique des îles d'Ibiza et de Formentera (Baléares): Memoria del Instituto Geológico y Minero de España, Toma 82, 340 p.

Riding, R., Martín, J. M., and Braga, J. C., 1991, Coral-stromatolite reef framework, Upper Miocene, Almeria, S. E. Spain: Sedimentology, v. 38, p. 799-818.

Rodriguez-Perea, A., Gelabert, B., and Sàbat, F., 1990, An outline of the Serra de Tramuntana of Mallorca, Balearic Islands. The Valencia Trough: Geology and Geophysics: Terra Abstracts, v. 2, p. 9-10.

Ros, J. D., Romero, J., Ballesteros, E., and Gili, J. M., 1985, Diving in blue water: The benthos, *in* Margalef, R., ed., Key Environments - Western Mediterranean: Pergamon Press, p. 233-295.

Sàbat, F., 1986, Estructura Geologica de les Serres de Llevant de Mallorca (Balears): Unpublished Ph.D. Thesis, Universidad de Barcelona, Barcelona, 120 p.

Sàbat, F., Munoz, J., and Santanach, P., 1988, Transversal and oblique structures at the Serres de Llevant thrust belt (Mallorca Island): Geologische Rundschau, v. 77, p. 529-538.

Simó, A., 1982, El Mioceno Terminal de Ibiza y Formentera: Unpublished M.S. Thesis, Universidad de Barcelona, Barcelona, 165 p.

Simó, A. and Giner, J., 1983, El Neógeno de Ibiza y Formentera (Islas Baleares): Revista de Investigaciones Geológicas, v. 36, p. 67-81.

Simó, A. and Ramón, X., 1986, Análisis sedimentológico y descripción de las secuencias deposicionales del Neógeno postorogénico de Mallorca: Boletín Geológico y Minero, v. 97, p. 445-472.

Staudt, W., Oswald, E., and Schoonen, M., 1993, Determination of sodium, chloride and sulfate in dolomites: a new technique to constrain the composition of dolomitizing fluids: Chemical Geology, v. 107, p. 97-109.

Sun, S. Q., 1992, Skeletal aragonite dissolution from hypersaline seawater: a hypothesis: Sedimentary Geology, v. 77, p. 249-257.

Sun, S. Q. and Esteban, M., 1994, Paleoclimatic controls on sedimentation, diagenesis, and reservoir quality: Lessons from Miocene carbonates: American Association of Petroleum Geologists Bulletin, v. 78, p. 519-543.

Ward, W. C., Oswald, E. J., and Pomar, L., 1993, Dolomite cementation before and after aragonite dissolution, Upper Miocene coral reefs of Mallorca, Spain (abs.): American Association of Petroleum Geologists 1993 Annual Convention Program, p. 198.

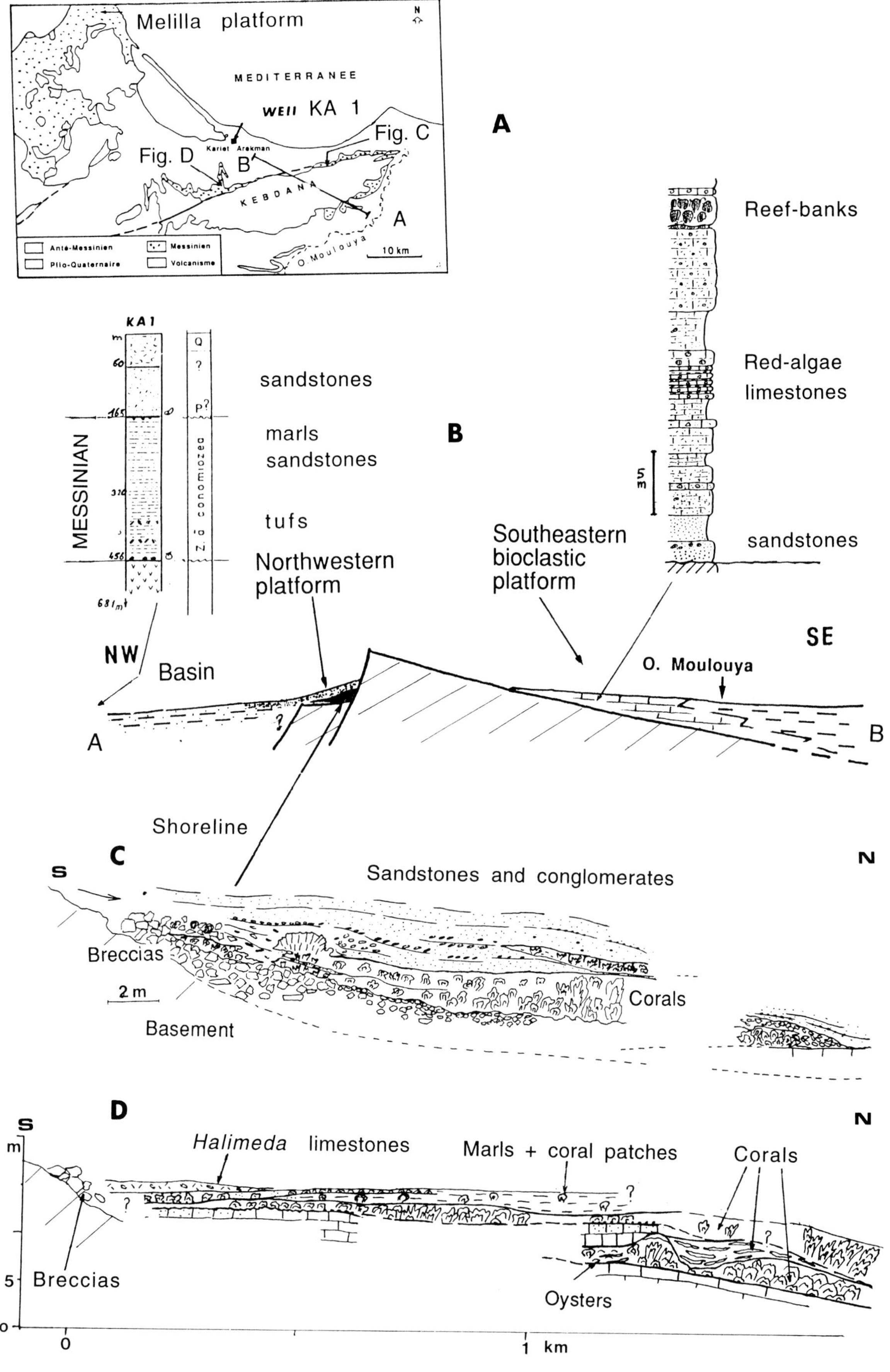

Melilla platform
MEDITERRANEE
well KA 1
Fig. C
Fig. D
B
Kariet Arekman
KEBDANA
A
O. Moulouya
10 km
Anté-Messinien
Plio-Quaternaire
Messinien
Volcanisme
A
KA1
MESSINIAN
sandstones
marls
sandstones
tufs
B
Reef-banks
Red-algae
limestones
5 m
sandstones
Southeastern
bioclastic
platform
Northwestern
platform
NW
Basin
SE
O. Moulouya
A
B
Shoreline
C
S
N
Sandstones and conglomerates
Breccias
2 m
Corals
Basement
D
S
N
m
Halimeda limestones
Marls + coral patches
Corals
Breccias
Oysters
5
0
0
1 km

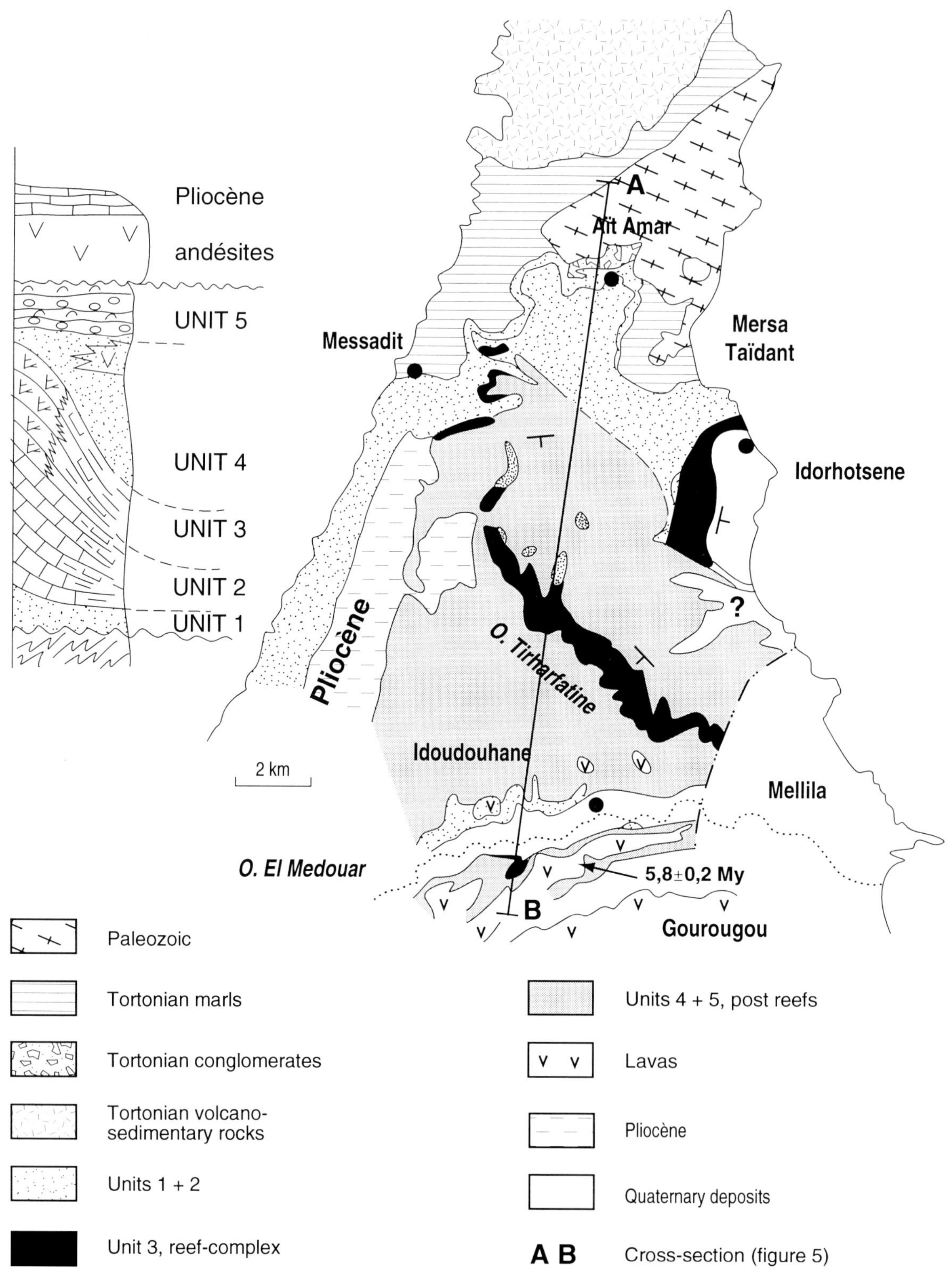


FIG. 3.—Generalized regional geologic map of the Trois Fourches Peninsula.

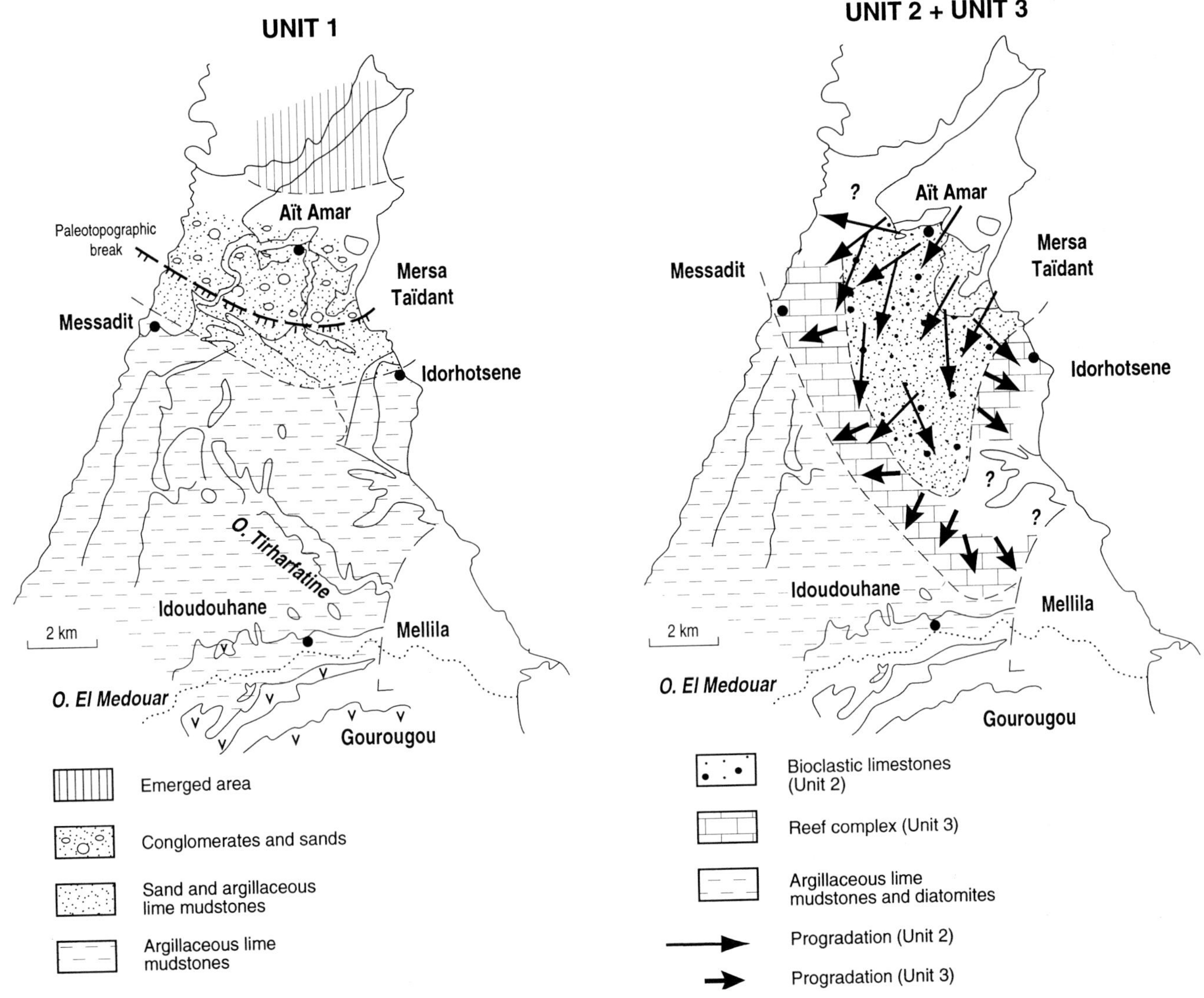


FIG. 4.—Synthetic Messinian lithofacies map of the Trois Fourches Peninsula.

rections of progradation diverge in a distal position (Fig. 4), and sediments become more fine-grained. More precisely, André et al. (1993) showed that Unit 3 was composed of three subunits separated by probably submarine erosional surfaces, deposited during three different periods of constant accomodation. To the south, the bioclastic deposits progressively change into argillaceous marls and diatomites (Fig. 5).

*Unit 3* consists of a reef complex mainly composed of *Porites* boundstones and *Halimeda* or red algae mixed skeletal packstones/ grainstones. The reef complex was established on the crest of a lobe-shaped prograding skeletal sand body (Unit 2; Fig. 3). The development of the reef complex mimicked the preexisting paleotopography of Unit 2 (Fig. 4), as shown by more divergent sedimentary slopes, from southwest to southeast (Saint Martin et al., 1991). Numerous cross sections throughout the platform show a complex geometrical pattern (Fig. 5). It is possible to separate, throughout the whole platform, first prograding reefs, then aggrading/retrograding reefs.

*Unit 4* begins with giant thrombolites-stromatolites which developed on the edge on the youngest agradational reef, above a limited amplitude erosional unconformity. Basinward facies change into thin oolitic limestones then into lime mudstones. Above these carbonate rocks, fine-grained yellow siliciclastic sandstones infill paleotopographic depressions between the previous buildups of the outermost part of reef complex. At the top of the reef buildups, we observed only local erosion defined by a limited truncation of some of the coral colonies. More detailed studies are required for an accurate determination of this unconformity, but it is probably not as important as the siliclastics conformably overlying the reef and coeval thrombolites-stromatolites distal deposits. The sandstones of Unit 4 locally show synsedimentary deformations which were related to seismic activity (Machhour et al., 1993).

Uni*t 5* unconformably rests upon Units 3 and 4 (Fig. 5). It is

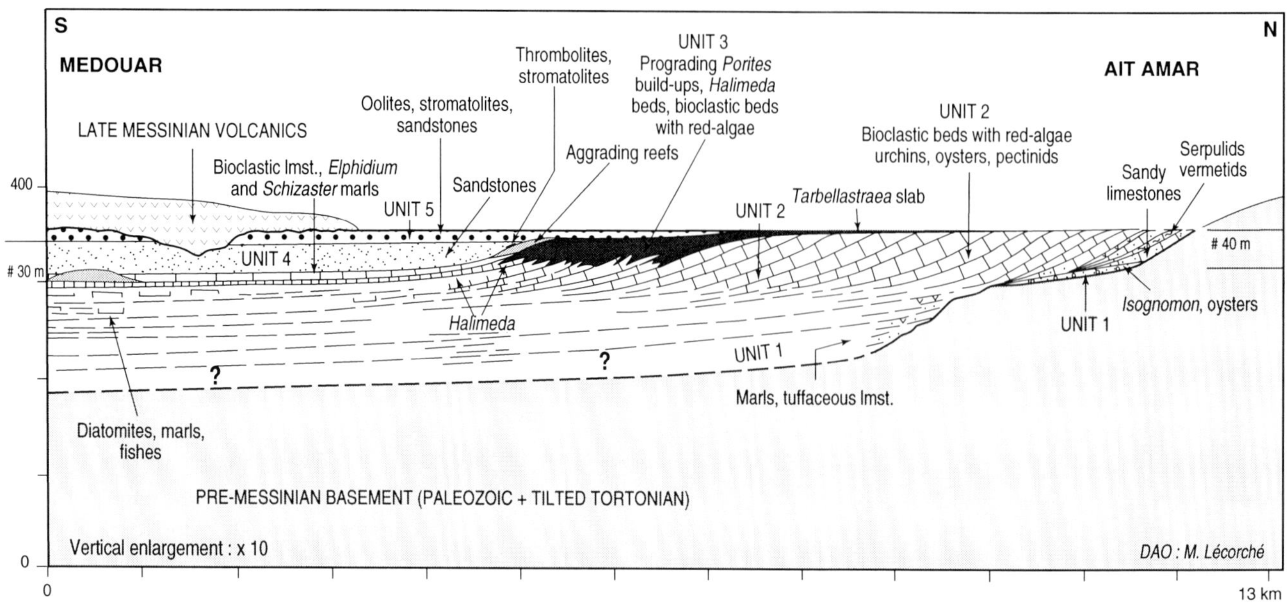


FIG. 5.—Structurally corrected cross-section through the Melilla platform (see Figure 3 for location) without post-Messinian deformations. Compaction was not removed.

composed of oolitic packstone/grainstone and/or stromatolitic boundstones deposited in a tidal environment. Upward, facies change into sandstones with interbedded conglomeratic micritic limestones, fluviatile conglomeratic channels and red paleosols. Southward, the whole sequence changes into volcanoclastic sediments near the Gourougou volcanic complex.

South, on the northern flank of the Gourougou volcanic complex, Unit 5 is overlain by a Late Messinian andesitic subaerial lava-flow (Hernandez and Bellon, 1985) which infills paleovalleys. Such paleovalleys indicate an important subaerial exposure at the end of the Messinian sedimentation in this region.

## THE CORAL REEF COMPLEX

### *Prograding Reefs*

*General features.—*

The main reefal outcrops are organized in a progradational pattern. In a general way, reefs are oligospecific *Porites* lenses of coral boundstones. Cunningham (1992) recorded seven prograding coral reefs in the Tirharfatine valley. Basinward, the oldest coral lenses laterally pass into red algae grainstones; the youngest coral lenses pass into *Halimeda* mixed-skeletal floatstone. Distally, these last *Halimeda* beds can be identified to 4 km from the reef complex (Fig. 5). Coral lenses and associated sediments show a general southward progradation which can be divergent in the detail.

*Reef settlement.—*

Pre-reef rocks of Unit 2 (André et al., 1993) range from skeletal grainstones to packstones to wackestones. To the north, Unit 2 consists of coarse molluscan packstones/grainstones/rudstones, rich in ostreids, pectinids, echinids and brachiopods. These prograding sedimentary bodies include red algal rudstone beds. The oldest *Porites* reefs rest directly upon the clinoforms at the top of the underlying bioclastic lobe (Unit 2), upon serpulid-rich mixed-skeletal wackestones with red algal and molluscan remains or upon mixed skeletal packstones with red algae, molluscs, and benthonic foraminifera on the slope of the lobe. Above, the reef displays mainly *Porites* corals with minor amounts of *Tarbellastraea* in its inner part. The reef is 1- to 2-m-thick bank resting above the foresets of the underlying bioclastic limestones on the flat top of the prograding lobe (Fig. 6A). At the margin of the underlying bioclastic lobe, the reef thickens and shows a progradational geometry reaching a thickness of tens of meters.

*Reef core facies.—*

In general, the prograding reefs are lense shaped, and the youngest ones are the thickest in the outer part of the complex (Fig. 6A). They rest upon slopes of bioclastics beds (red algae packstones/grainstones or *Halimeda* mixed skeletal floatstones) dipping 1° to 25° (Fig. 6B). Cunningham (1992) defined the complicated relationships between reefs and bioclastic slopes, with possible onlaps of bioclastic beds on some of the reefs. Reef cores are locally several hundred m in width and average 10 m thick. Smaller buildups occur as patch reefs on the *Halimeda* slopes in the outermost part of the prograding complex. The coral lenses are associated first with red algae-rich packstones/grainstones slopes, then with well-bedded *Halimeda* mixed-skeletal floatstones slopes.

The prograding reef framework (framestone), is composed of

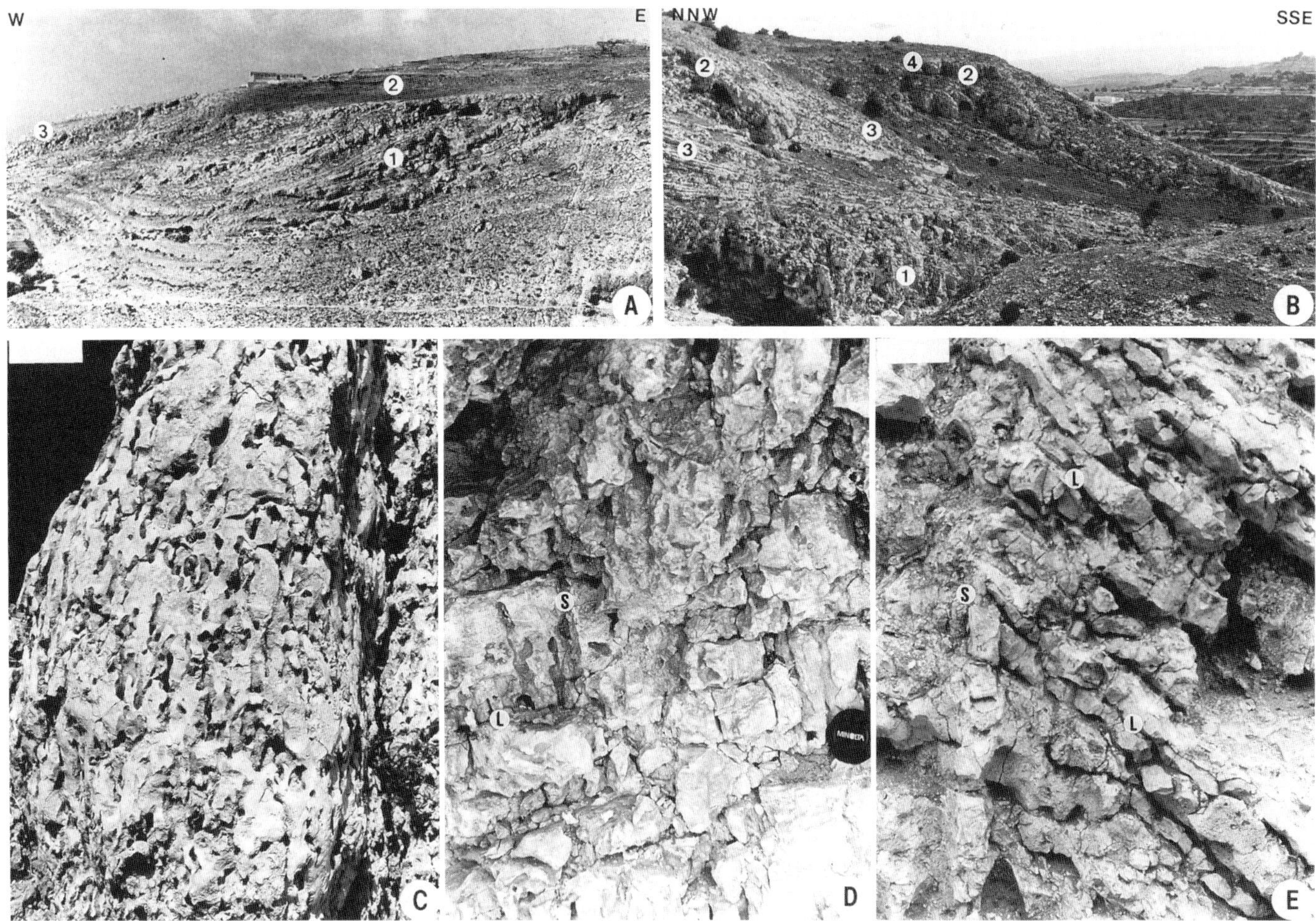


FIG. 6.—(A) Settlement of the reef complex: (1) bioclastic limestones of the prograding Unit 2; reef-bank part (2) and prograding lens-shaped reef part (3) of the first reef buildup. Irhzazene area. (B) View of the Tirharfatine valley: (1) prograding red-algae bioclastic limestones; (2) prograding *Porites* reefs; (3) prograding bioclastic slopes including red-algae and *Halimeda* plates; (4) aggrading reef-bank. (C) Prograding reef core framework made of *Porites* stick-like colonies associated with abundant white microbialites. Scale bar: 10 cm. (D) Combination of laminar-platy colonies (L) and stick- like colonies (S) of *Porites*. (E) *Idem* photo D. Scale bar: 5 cm.

a dense combination of branching and laminar growth-forms of *Porites* (Fig. 6C, D, E, 7B). The finger-type colonies may constitute masses that range 2 or 3 m (8-10 ft) high, forming very dense buildups of branching-coral boundstone (Fig. 7A) described by Saint Martin (1990). Finger-type buildups generally grow orthogonally from the base of the reef which consists of slopes of red algae bioclastic limestones or *Halimeda* mixed-skeletal floatstones.

The *Porites* colonies are rimmed by micritic and laminated coatings (Fig. 6C, 7C) that may be as thick as 10 cm (3 in). Micritic coatings are present throughout the whole reef complex, since the oldest prograding reef to the youngest aggrading one. This rim is mainly made of several types of pelloidal and laminated micrites. They are probably microbial in origin (Riding et al., 1991; Saint Martin et al., 1991; Saint Martin et al., 1993) having developed around dead and previously bored parts of corals. This micrite may represent 20 to 50% of the buildups facies. Encrusting red algae and encrusting foraminifera are also reported between branching corals and micritic coatings.

Various types of infilling sediments occur within the reef. Between finger coral colonies, the infilling sediment is a mixed skeletal packstone/grainstone with peloids, algae, echinoids and molluscs or a *Halimeda* floatstone/rudstone. In places, thin reef breccias occur around coral lenses. Sediments that separate reefs or that form flanking beds of reefs are composed of mixed skeletal grainstones/packstones with coralline algae and *Porites* constituents, including *Halimeda* plates which are abundant during the youngest prograding reefs building.

*Outer reef facies.—*

Geometric reconstructions and field observations show that the prograding reef complex is coeval in the outer platform with laminated diatomites, laminated argillaceous lime mudstones which only contain fossil fishes from shallow water environments (Gaudant et al., 1994) and rare marls with some planktonic foraminifera or rare carbonate platform-derived thin bioclastic limestones. The transitional zone from the reef complex to the outer platform is narrow (some of meters), and

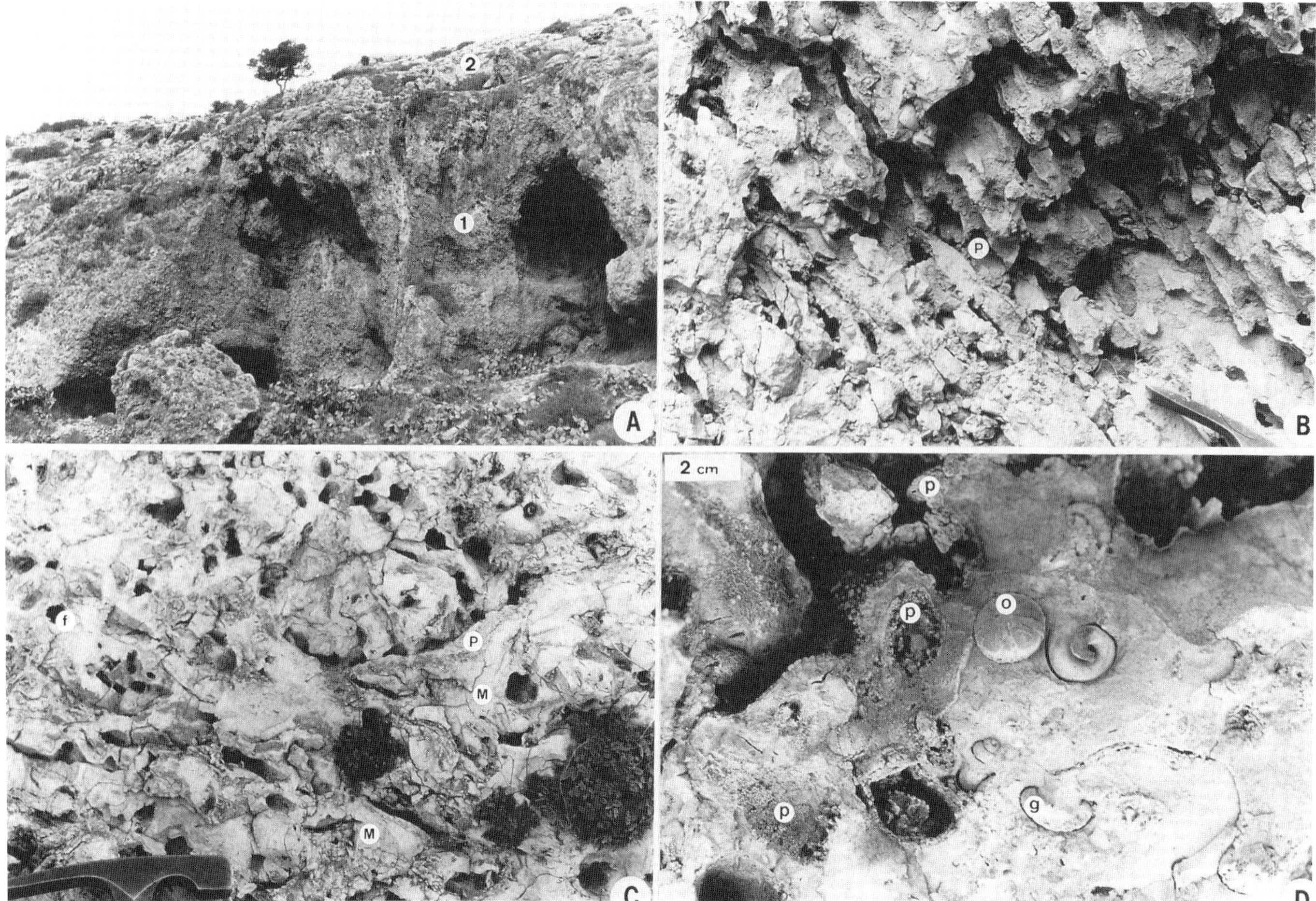

FIG. 7.—(A) Coral buildups in the outer part of the reef complex (height: 10 m): (1) branching colonies in the core of a prograding reef; (2) aggrading reef bank. Tirharfatine valley. (B) Finger-type *Porites* colonies. Corals are dissolved (p). (C) Combination of platy-like (p) and finger-like (f) colonies of Porit*es* with white microbial micritic coatings (m). (D) Example of reef-dwelling fauna between the *Porites* colonies (p), with the *Psammechinus* urchins (o) and gastropods (g).

reworked detritus from the platform on the slope are poorly developed.

### *Aggrading reefs*

*General features.—*

The last *Porites* reefs show aggradational/retrogradational pattern. The two oldest aggrading reefs are 2 to 3 m thick. These reef banks overlie a flat subaerial(?)/submarine unconformity (Figs. 6B, 7A) with substratum-derived pebbles, and they rest upon the top of the underlying prograding complex. The youngest reef is composed of small sized coral patches embedded in prograding oolitic beds in its innermost part, and of buildups reaching 10 m thick in its outer part.

*Reef core facies.—*

Aggrading/retrograding reefs at the end of the reef complex are still composed of *Porites* boundstones associated with oolitic limestones which mainly have gastropods fauna and may include local *Siderastraea* in marly coeval deposits. Aggrading reefs banks are made of platy or massive (cauliflower-like) *Porites* head coral boundstones, associated with red algae bindstones. Finger-like colonies are developed in the youngest reef. Locally, in small channels between large laminar coral colonies, oolitic peloid packstones were deposited.

Detailed mapping of the outer platform shows that aggrading/retrograding reefs developed also basinward as isolated platforms in the Medouar (Fig. 5) and Taourirt Medil areas far to the southwest, above the distal *Halimeda* beds of the prograding complex. These isolated reefal platforms rapidly change into lime mudstones and marls, and the transition zone is characterized by marls with isolated *Siderastaea* and *Porites* colonies or by bioclastic limestones with molluscs, red algae and serpulids.

As in the prograding reefs, microbialites develop around coral colonies.

*Outer reef facies.—*

Reefs basinward change into lime mudstones and marls which contain benthic foraminifera (as *Elphidium*) and numerous burrowing urchins (*Brissopsis*). The lime mudstones locally

TABLE 1.—MELILLA REEF-DWELLING FAUNA.

| SCLERACTINIA | GASTROPODA |
|---|---|
| *Porites lobatosepta* | *Diodora* sp. |
| *Porites calabricae* | *Haliotis tuberculata* |
| *Tarbellastraea* sp. | *Clanculus* sp. |
| | *Alvania* sp. |
| **BIVALVIA** | *Thericium vulgatum* |
| | Triphoridae |
| *Arca noae* | *Cerithiella* cf. *exbicarinata* |
| *Barbatia barbata* | *Cypraea* sp. |
| *Anadara turonica* | *Cymbium* cf. *pepo* |
| *Glycymeris bimaculata* | *Bulla* sp. |
| *Modiola adriatica* | |
| *Lithophaga lithophaga* | **ECHINOZOA, ECHINOIDEA** |
| *Chlamys pusio* | |
| *Aequipecten seniensis* | *Psammechinus* sp. |
| *Spondylus* sp. | |
| *Ostrea* sp. | |
| *Gastrana fragilis* | |
| *Diplodonta rotundata* | |
| *Codakia decussata* | |
| *Tellina planata* | |
| *Dosinia exoleta* | |
| *Venus multilamella* | |

suffered in situ brecciation ("Megabreccia" of Cunningham et al., 1994). Around the reef complex the carbonate platform-outer platform transition is sudden, without reworked material. That suggests low-energy environment and a very low-angle dipping sea bottom.

### *Associated Reef Fauna*

It is difficult to give a complete inventory of the entire reef-dwelling fauna because of its poor preservation. The faunal list of Table 1 (from Saint Martin, 1987, 1990), however, gives an idea of the different biotas. The reef fauna (Fig. 7D) are composed of numerous epibionts (encrusting bryozoans, serpulid worms, cerithid gastropods, and pelecypods of the families Pectinidae, Spondylidae, and Chamidae) and nestling organisms, such as echinoids and bivalvia (Arcidae). In sediment between coral colonies, one finds small burrowing bivalves, such as *Gastrana fragilis* and *Diplodonta rotundata*. Reef flank beds mostly contain burrowing bivalvia (*Glycymeris*, *Codakia*, *Tellina*, *Dosinia*, *Venus*)

### *Diagenetic Overprint*

The diagenetic evolution of the Melilla platform is presently under study. Only some preliminary data from thin sections and X-ray analysis will be discussed.

Aragonitic skeletons (especially corals) were dissolved probably early in the diagenetic history of the reef as their molds were coated by a thin lining of micrite (fine-grained internal sediment) along the walls of the dissolved skeleton with the central part of the colony usually completely missing. For example, *Porites* finger coral colonies end up looking like empty cylinders or "organ-pipes". Next, drusy calcite filled these cavities, accompanied by recrystallization of micrite through a process of microsparitization. This diagenetic history reflects the progressive emergence of the reef from the marine environment to a vadose environment of dissolution, then to a phreatic environment where drusy calcite was emplaced. Primary intragranular porosity is very important in the reef-complex (e.g. in coral skeletons and Ha*limeda* plates) and was enhanced by a Late Messinian to Recent karstification cutting through the whole Messinian carbonate platform.

However, the dissolution phase may be missing in certain reef lenses. In such cases, the original aragonitic framework of corals is there replaced by a mosaic of recrystallized calcite.

A late dolomitization was evidenced throughout the most elevated parts of Units 2, 3 and 5. In the present-day knowledge, dolomitization timing and distribution need more accurate studies.

### *Discussion*

It is commonly accepted that reefs are organized in a general southward downstepping pattern (Rouchy, 1982; Saint Martin and Rouchy, 1990; Saint Martin et al., 1991; Cunningham et al., 1994). Nevertheless, doubts about this geometry were recently discussed in Cornée et al. (1992), André et al. (1993), Benmoussa et al. (1994), Saint Martin et al. (1994) and Cornée et al. (1994b). From our point of view, downstepping geometry is only apparent. Our structural analysis clearly indicates, using index beds which were deposited on subhorizontal surfaces as oolites, stromatolitic beds, reef banks, diatomite beds and Quaternary marine terraces, that the Melilla platform suffered different late tectonic movements, especially differential tilting (1° to 3° southward tilting), late faulting, and compaction (the outer part of the carbonate platfom is much more argillaceous than the northern part). Consequently, a structurally corrected cross section (Fig. 5) shows that the reefs are not organized in a regular downstepping progradational pattern, but just in a progradational then aggradational pattern. Such a result was deduced in the Murdjadjo platform in Algeria (Cornée et al., 1994a). However, there are probably high-frequency relative sea-level changes during the reef complex development, as discussed by Cunningham (1992; not more than several meters amplitude from our observations).

## CONCLUSION

The Melilla *Porites* reef complex, in its main northern part, is horseshoe-shaped, because of the pre-existing prograding lobes, and extends laterally tens of kilometers (about 20 mi). It is first made of lens-shaped southward prograding fringing reefs then of aggrading reef banks and patches. The coral framework is largely built by laminar and finger-like *Porites* colonies, associated with microbialites. In the reef banks, platy and massive *Porites* colonies are dominant.

In general, coeval deposits of the prograding reefs consist of laminated marls, diatomites and platform derived bioclastic limestones. Coeval deposits of the aggrading reefs consist of brecciated shallow water lime mudstones. The transition zone from the reef buildups and the outer platform is always very narrow.

The reef complex can be included in the general relative sea-level history of the whole carbonate platform. Unit 1 indicates a major transgression (the "Upper Tortonian-Messinian transgression" of the Mediterranean). Unit 2 developed during a global stillstand, but there are high-frequency relative sea-level rises and drops. The reef complex (Unit 3) first developed during a stillstand. After limited relative sea-level drop with submarine planation, the youngest reefs were emplaced during several relative sea-level rises. The sandstones of Unit 4 then infilled the preexisting reefal paleotopography. The oolitic and stromatolitic sediments of Unit 5 were deposited during a new major transgression. Finally, the whole platform emerged and was eroded prior to the Pliocene transgression.

## ACKNOWLEDGMENTS

This work was accomplished with the help and the advice of Dr. J. Muller (mapping), A. Benmoussa, and Dr. J. P. André (sedimentological and diagenetic data). M. Lecorché and B. Guichané are thanked for technical support. The authors thank the anonymous reviewers for improvement of the manuscript.

## REFERENCES

André, J. P., Benmoussa, A., Cornée, J. J., Muller, J., Saint Martin, J. P., and Bessedik, M., 1993, Les corps bioclastiques progradants du Messinien d'Afrique du Nord (abs.): Paris, Réunion Spécialisée de la Société géologique de France, "Carbonates Intertropicaux", p. 8.

Benmoussa A., Demarcq, G., and Lauriat-Rage, A., 1987, Pectinidés messiniens du bassin de Melilla (N.E. Maroc), comparaisons inter-régionales et intérêts paléobiologiques: Revue de Paléobiologie, v. 6, p. 11-129.

Benmoussa, A., El Hajjaji, K., Cornée, J. J., Saint Martin, J. P., and Muller, J., 1994, Les unités biosédimentaires de la plate-forme carbonatée messinienne de Melilla (NE Maroc) (Abs.): Marseille, Interim-Colloquium of the Regional Committee on Mediterranean Neogene Stratigraphy, p. 5.

Benmoussa, A., El Hajjaji, K., Pouyet, S., and Demarcq, G., 1989, Les mégafaunes marines du Messinien de Melilla (Nord-Est Maroc): Palaeogeography, Palaeoclimatology, Palaeoecology, v. 73, p. 51-60.

Charlot, R., Choubert, G., Faure-Muret, A., Hottinger, L., Marçais, J., and Tisserant, D., 1967, Note au sujet de l'âge isotopique de la limite Miocène-Pliocène au Maroc: Comptes-Rendus de l'Académie des Sciences de Paris, série D, v. 264, p. 222-224.

Choubert, G., Charlot R., Faure-Muret, A., Hottinger, L., Tisserant, D., and Vidal, P., 1968, Note préliminaire sur le volcanisme messinien-"pontien" au Maroc: Comptes-Rendus de l'Académie des Sciences de Paris, série D, v. 266, p. 197-199.

Choubert, G., Faure-Muret, A., Hottinger, L., and Lecointre, G., 1966, Le Néogène du bassin de Melilla (Maroc septentrional) et sa signification pour définir la limite mio-pliocène: Berne, Processes of the International Congress of the Mediterranean Neogene, 3rd Sess., p. 238-249.

Cornée, J. J., Guieu, G., Muller, J., and Saint Martin, J. P., 1994b, Mediterranean Messinian carbonate platforms: some controlling factors (abs.): Géologie Méditerranénne, v. 21, no. 3-4, p. 45-48.

Cornée, J. J., Muller, J., and Saint Martin, J. P., 1992, Organization and dynamics of Messinian pericontinental carbonate platforms. Some examples from the North-African Mediterranean margins (abs.): Chichiliane, Platform Margins International Symposium, p. 33.

Cornée, J. J., Saint Martin, J. P., Conesa G., and Muller, J., 1994a, Geometry, paleoenvironments and relative sea-level (accomodation) changes in the Messinian Murdjadjo carbonate platform (Oran, western Algeria). Consequences: Sedimentary Geology, 89, p. 143-158.

Cunningham, K., 1992, Depositional sequence hierarchy in an upper Miocene carbonate complex, Cap des Trois Fourches, Morocco (abs.): Dijon, "Sequence Stratigraphy of European Basins", p. 308-309.

Cunningham, K., Franseen, E., Enos, P., Farr, M. R., and Rakic-El-Bied, K, 1994, High-resolution sequence stratigraphy and magnetostratigraphy for the Melilla Basin, northeastern Morocco: implications for regional and global controls on sequence development (abs.): Marseille, Interim-Colloquium of the Regional Committee on Mediterranean Neogene Stratigraphy, p. 19.

El Hajjaji, K., 1988, Les Bryozoaires messiniens du bassin de Melilla (N.E. Maroc) et leur signification biostratigraphique, paléobiogéographique et paléoécologique: Géologie Méditerranéenne, v. 15, p. 105-121.

Esteban, M., 1979, Significance of the Upper Miocene coral reefs of the western Mediterranean: Palaeogeography, Palaeoclimatology,

Palaeoecology, v. 29, p. 169-188.

Gaudant, J., Benmoussa, A., Cornée, J. J., El Hajjaji, K., Muller, J., and Saint Martin, J. P., 1994, L'ichthyofaune messinienne du bassin de Melilla-Nador (Nord-Est du Maroc) (abs.): Géologie Méditerranénne, v. 21, no. 1-2, p. 25-35.

Guillemin, M., 1976, Les formations néogènes et quaternaires des régions de Melilla-Nador et leurs déformations (Maroc nord-oriental): Unpublished Thesis, University of Orléans, Orléans, 220 p.

Guillemin, M. and Houzay, J. P., 1982, Le Néogène post-nappe et le Quaternaire du Rif nord-oriental (Maroc), Stratigraphie et tectonique des bassins de Melilla, du Kert, de Boudinar et du piedmont des Kebdana: Notes et Mémoires du Service Géologique du Maroc, n. 314, p. 7-238.

Hernandez, J. and Bellon, H., 1985, Chronologie K-Ar du volcanisme miocène du Rif oriental (Maroc): implications tectoniques et magmatologiques: Revue de Géologie dynamique et de Géographie physique, v. 26, p. 85-94.

Machhour, L., Cornée, J. J., Saint Martin, J. P., Lehman, P., and Muller, J., 1993, Enregistrement de l'activité séismique dans les sédiments: Exemple des plates-formes carbonatées messiniennes d'Afrique du Nord: Eclogae Geologicae Helvetiae, v. 86, p. 265-281.

Rampnoux, J. P., Angelier, J., Coletta, B., Fudral, S., Guillemin, M., and Pierre, G., 1979, Sur l'évolution néotectonique du Maroc septentrional: Géologie Méditerranéenne, t. vi, p. 439-464.

Riding, R., Martin, J. M., and Braga, J. C., 1991, Coral-stromatolite reef framework, Upper Miocene, Almeria, Spain: Sedimentology, v. 38, p. 799-818.

Rouchy, J. M., 1982, La genèse des évaporites messiniennes de Méditerranée: Mémoires du Muséum national d'Histoire naturelle de Paris (C), t. 50, 267 p.

Rouchy, J. M., Saint Martin, J. P., Maurin A., and Bernet-Rollande, M. C., 1986, Evolution et antagonisme des communautés bioconstructrices animales et végétales à la fin du Miocène en Méditerranée, biologie et sédimentologie: Bulletin des Centres de Recherches Exploration-Production Elf Aquitaine, v. 10, p. 333-348.

Saint Martin, J. P., 1987, Les formations récifales du Miocène supérieur du Maroc et d'Algérie, Aspects paléoécologiques et paléogéographiques: Unpublished Doctorat d'Etat Thesis, University Aix-Marseille 1, Marseille, 499 p.

Saint Martin, J. P., 1990, Les formations récifales du Miocène supérieur du Maroc et d'Algérie: Mémoires du Muséum national d'Histoire naturelle de Paris (C), t. 56, 351 p.

Saint Martin, J. P., André, J. P., Conesa, G., Cornée, J. J., Le Campion, T., Muller, J., and Riding, R., 1993, Les micrites d'origine microbienne dans les récifs messiniens de Méditerranée (abs.): Lyon, First European Palaeontological Congress, p. 115.

Saint Martin, J. P., Cornée J. J., Conesa G., Muller, J., and André, J.P., 1994, Les plates-formes carbonatées messiniennes de Méditerranée occidentale: une revue (abs.): Marseille, Interim-Colloquium of the Regional Committee on Mediterranean Neogene Stratigraphy, p. 55.

Saint Martin, J. P., Cornée, J. J., Muller, J., Camoin, G., André, J. P., Rouchy, J. M., and Benmoussa, A., 1991, Contrôles globaux et locaux dans l'édification d'une plate-forme carbonatée messinienne (bassin de melilla, Maroc): apport de la stratigraphie séquentielle et de l'analyse tectonique: Comptes-Rendus de l'Académie des Sciences de Paris, série II, t. 312, p. 1573-1579.

Saint Martin, J. P. and Rouchy, J. M., 1986, Intérêt du complexe récifal du Cap des Trois Fourches (Bassin de Nador, Maroc septentrional) pour l'interprétation paléogéographique des événements messiniens en Méditerranée occidentale: Comptes-Rendus de l'Académie des Sciences de Paris, série II, t. 302, p. 957-962.

Saint Martin, J. P. and Rouchy, J. M., 1990, Les plates-formes carbonatées messiniennes en Méditerranée occidentale: leur importance pour la reconstitution des variations du niveau marin au Miocène terminal: Bulletin de la Socété géologique de France, v. VI, p. 83-94.

Wernli, R., 1988, Micropaléontologie du Néogène post-nappes du Maroc septentrional et description systématique des Foraminifères planctoniques: Rabat, Notes et Mémoires du Service Géologique du Maroc, v. 331, 270 p.

# MESSINIAN CORAL REEFS OF WESTERN ORANIA, ALGERIA

JEAN-PAUL SAINT MARTIN
*CNRS URA 1208, Centre de Sédimentologie/Paléontologie, Université de Provence, 13331 Marseille Cedex 03, France*

ABSTRACT: Messinian reefs from western Orania occur along a narrow outcrop belt which is several hundred meters wide and tens of kilometers long. The reefs show generally a sequential organization of buildups in a backstepping pattern. Each step of reef-building is represented by a tabular fringing reef with poor internal zonation. Cyclic input of argillaceous sediment produced a rythmic response in coral growth. These well-bedded reefs developed during a general lower Messinian transgressive episode.

## INTRODUCTION

### *Previous Works*

Messinian coral reefs are widespread in northwestern Algeria where they crop out in the Tessala Mountains and at the margin of the Traras Massif to the west (Fig. 1). Miocene reef limestones are well exposed and form nearly continuous outcrop belts along paleogeographic limits of a Neogene basin in western Orania called the Tafna Basin.

Miocene coral reefs of Algeria were initially described in the late 1800's (Pouyanne, 1877; Pomel and Pouyanne, 1882), the stratigraphy having been established at the beginning of the century (Gentil, 1903, 1917). Sedimentary cycles of Miocene deposits were named with regional terminology: Cartennian (Early Miocene), Vindobonian with Helvetian and Tortonian (Middle and Late Miocene), and Sahelian (latest Miocene), with reefs belonging to the Sahelian cycle. More recent syntheses, including micropaleontological, sedimentologic, and structural data, have led to a general paleogeographic reconstruction of Algeria with special emphasis on relationships with Alpine thrusting and nappe development. According to Perrodon (1957) limestone units of the Tafna Basin belong to the upper part of the second post-nappe Miocene sedimentary cycle (latest Miocene or Sahelian).

Several papers have been devoted to the description of coral buildups of western Orania (Saint Martin and Chaix, 1981; Saint Martin, 1984; Saint Martin, 1990) and their associated faunas (Fréneix et al., 1987a, b, 1988; Guernet et al., 1984; Saint Martin et al., 1985; Moissette, 1985, 1988; Saint Martin, 1990).

### *Location*

Miocene outcrops of western Orania occur as a strip between the Mediterranean and the Sebaa Chioukh Hills to the south. They are limited southward by the Traras Inlier, and eastward they are exposed up to the vicinity of the Oran Sebkha (the "Great Sebkha" in Fig. 1). The main reef outcrops described in this paper are those of the Sebaa Chioukh Hills and those of the eastern limit of the Traras Inlier.

Reefs of the Sebaa Chioukh Hills correspond to the western extension of the Tessala Mountains, trending WSW-ENE and occur at an elevation of 500-600 m (1640-1970 ft). Miocene coral reefs there appear as limestone blocks, several kilometers (about 2 mi) in length and hundreds of meters (about 1000 ft) wide, easily recognizable on hilltops (Fig. 2).

Hills trending north-south, 300 to 400 m high, flank the Traras Massif and limit the Miocene basin to the west. The crest of these hills along a distance of about 10 km (6 mi) are composed of narrow and distinct limestone reef masses (Fig. 3).

### *Stratigraphic Sequence*

In western Orania, post-nappe sedimentation is divided into two depositional cycles (Guardia, 1975). The first one is characterized by continental detrital beds and was deposited over nappes or in grabens; these sediments were deformed before the second cycle began. The second cycle began with continental "red beds" of the Tafna Basin which are overlain by argillaceous lime mudstones of brackish to marine origin and contain vertebrate fossils (including rodents) of Turolian age. Above these beds are marine argillaceous lime mudstones, rich in planktonic microfauna which indicate a Tortonian age for the lower part and a Messinian age for the upper part of the second cycle. This second cycle ends with the deposition of limestones containing red algal beds (packstones, grainstones, rudstones and boundstones) and coral reefs. Thus, the whole second cycle seems to indicate an overall transgressive sequence. This reef formation completes the Messinian cycle in the western Orania region. The deposits of the second cycle thin toward the flanks of certain paleotopographic highs, with some reefs resting upon pre-Tortonian substrates (i.e., nappes or Mesozoic rocks).

### *Geologic Age*

Micropaleontological data infer a Messinian age for coral reefs of the western Orania region. Underlying marls contain an assemblage of planktonic foraminifera of the *Globorotalia mediterranea* zone (Guardia et al., 1974; Guardia, 1975; Saint Martin et al., 1983; Saint Martin, 1990).

### *Geologic Framework*

The trend of reef outcrops is closely superimposed with major structural trends of the Tessala and the Traras Mountains, suggesting that previous topographic expression has in fact

Models for Carbonate Stratigraphy from Miocene Reef Complexes of Mediterranean Regions,
SEPM Concepts in Sedimentology and Paleontology #5, Copyright © 1996,
SEPM (Society for Sedimentary Geology), ISBN 1-56576-033-6, p. 239-246.

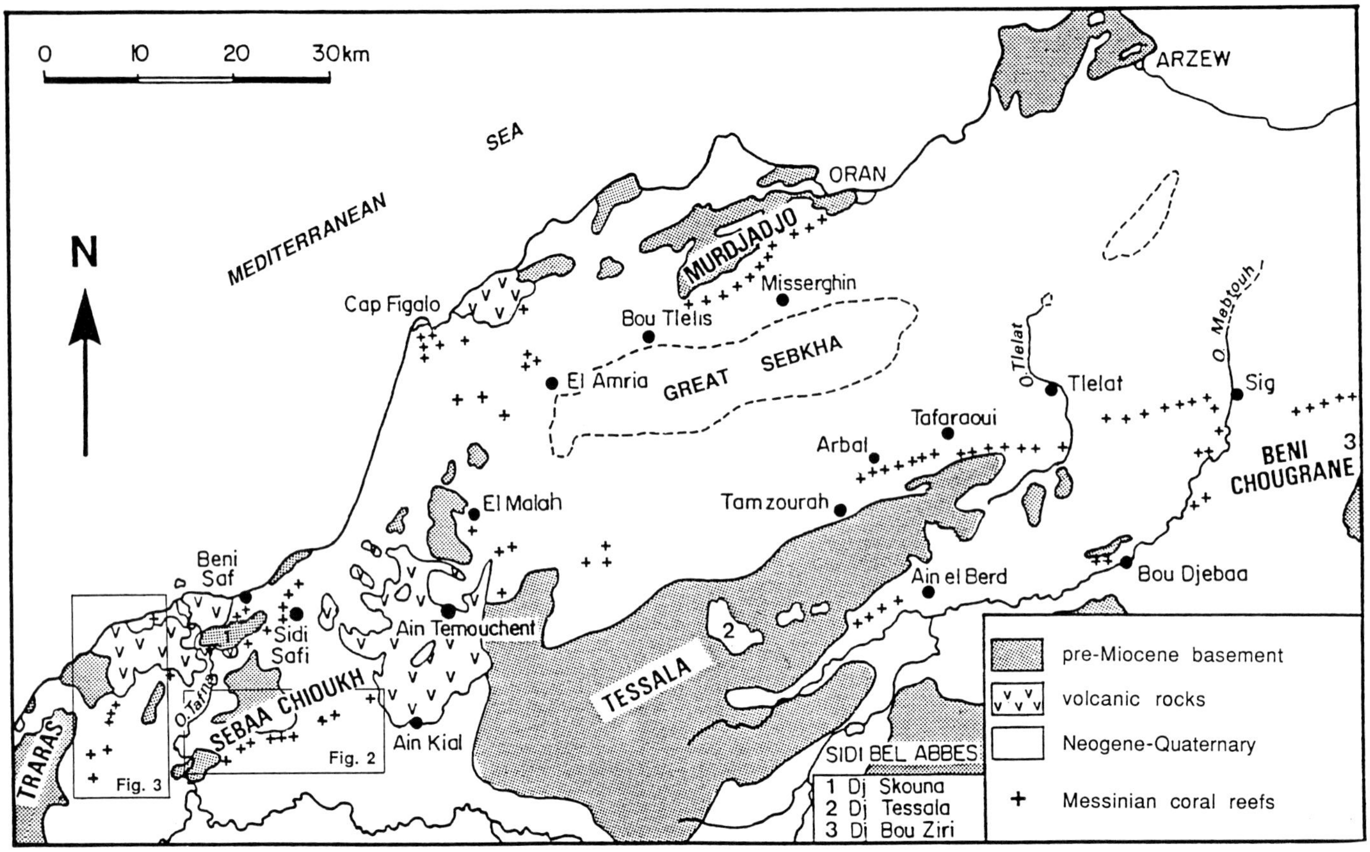

FIG. 1.—Location map of Messinian reefs in northwestern Algeria.

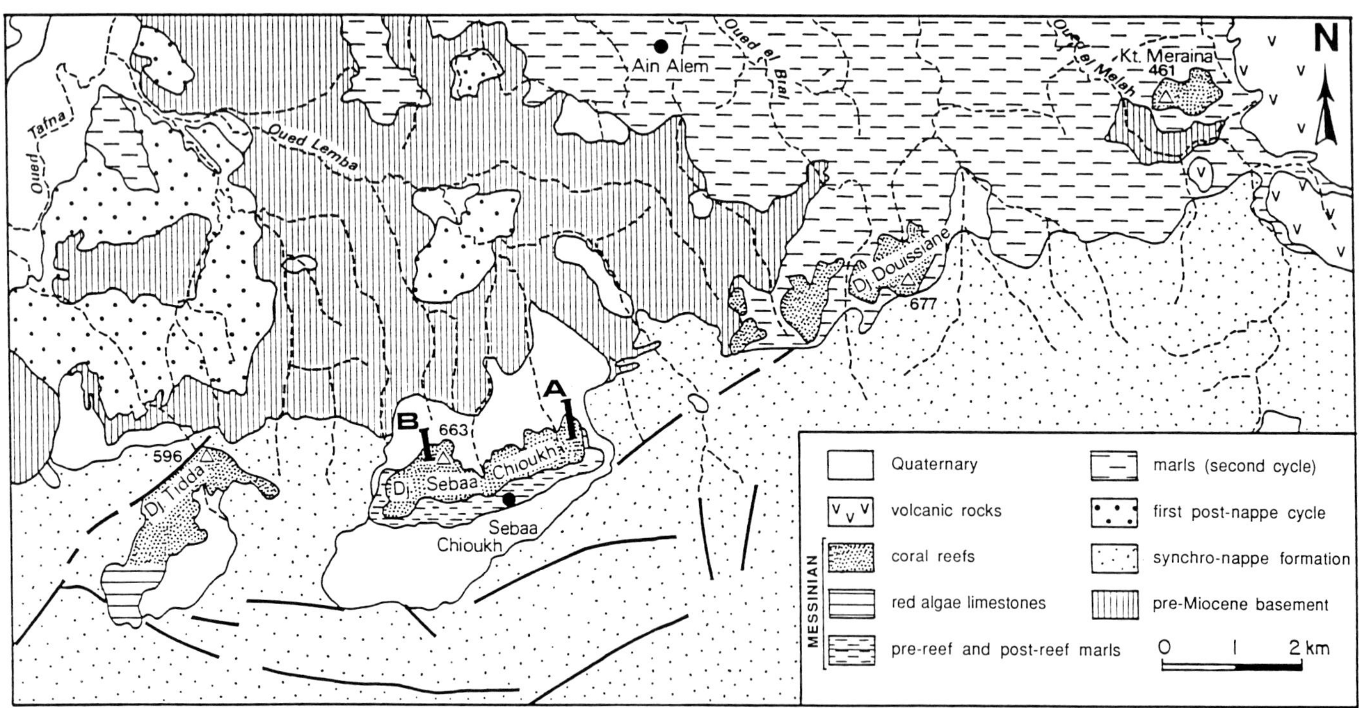


FIG. 2.—Regional geologic map of the Sebaa Chioukh Hills area. A and B indicate location of measured sections.

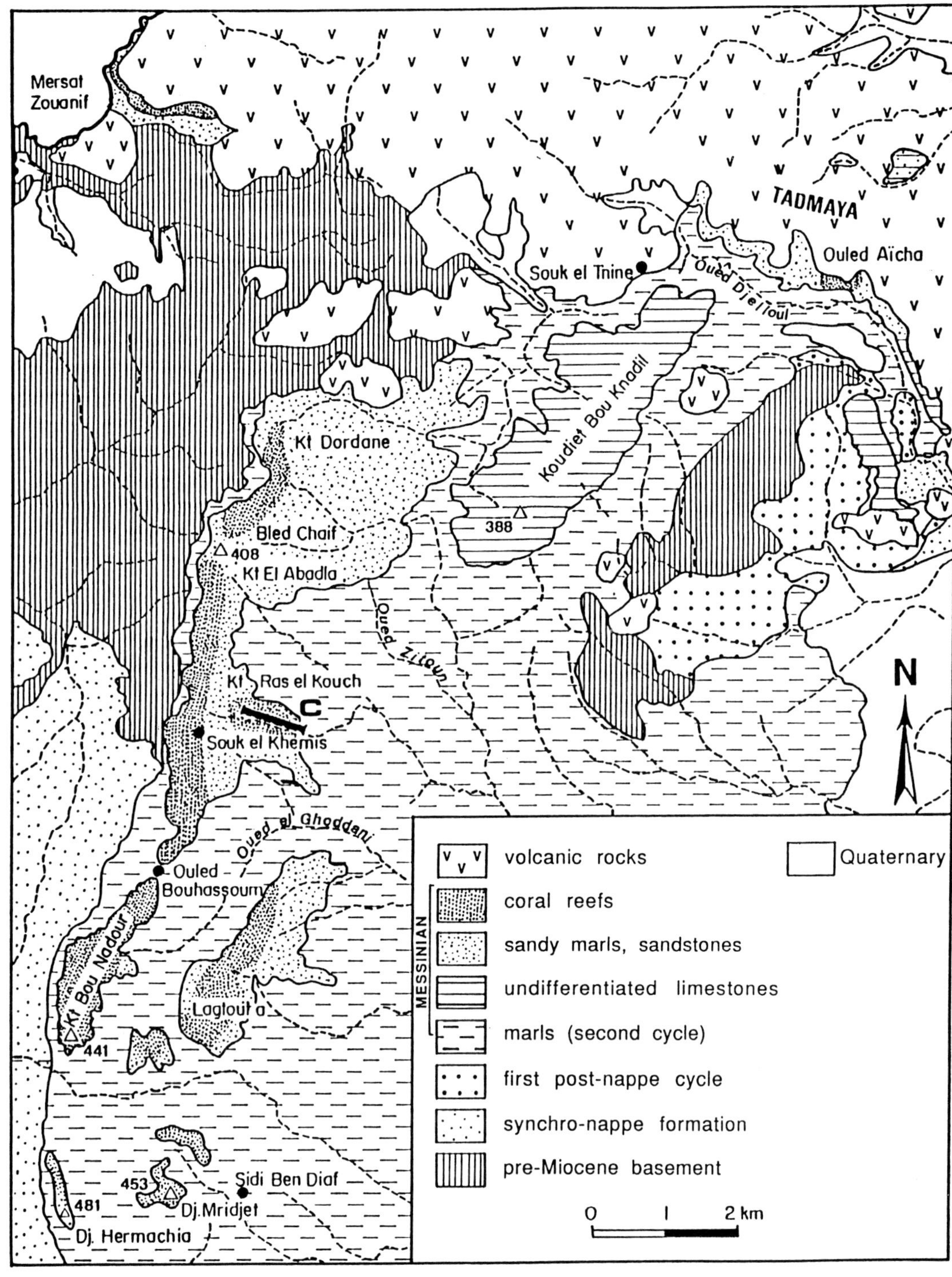


FIG. 3.—Regional geologic map of the Traras margin. C is the location of a measured section.

strongly influenced reef and shoal development, these reefs being founded upon previously formed folds of Upper Tortonian age (Thomas, 1985). On the other hand, one should notice that these reefs were uplifted 600 m (180 ft) above sea level in the Sebaa Chioukh Hills and 400 m (120 ft) in the Traras Mountains. Guardia (1975) demonstrated that paleotopographic relationships were inherited from nappe-emplacement during Middle Miocene times. The distribution of sedimentary facies is a good example of such a pattern (Figs. 2, 3), with argillaceous lime sediments being deposited in the depocenter of the basin, whereas red algae and coral rich limestones were deposited on shallow-water shoals.

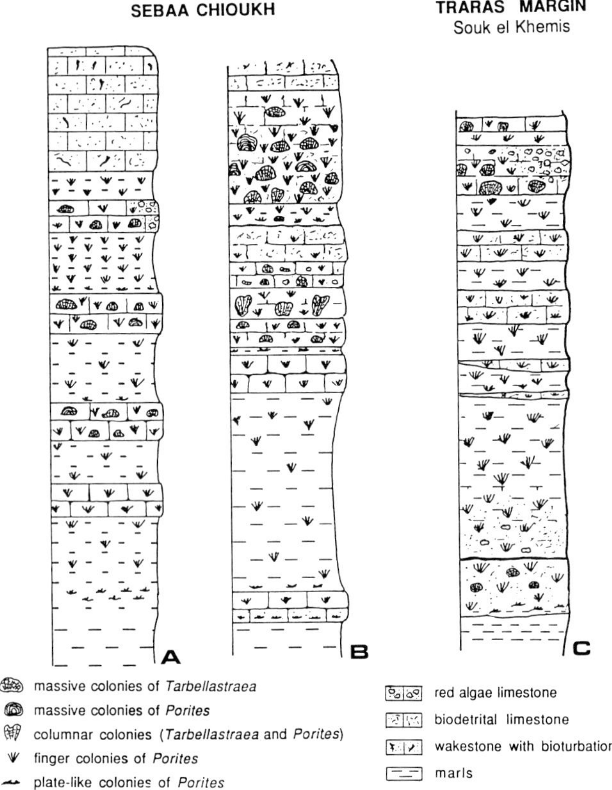


FIG. 4.—Reef-building sequences (location of three measured sections is shown in Figs. 2, 3).

## LITHOFACIES ASSOCIATED WITH THE REEF

### *Data Base*

Cross sections in the Sebaa Chioukh Hills and in the Traras Massif areas show a relatively uniform trend: a sequential organization of buildups in a stairstep pattern (Fig. 4). Unconformities separate reefal subunits and show changes in reef framework within these subunits. At the bottom of such an unconformity-bounded sequence, finger-type colonies of *Porites* (together with laminar colonies) initiated reef growth on a muddy sea floor. Generally, an argillaceous lime mudstone matrix is associated with a dominance of sheet-like finger colonies. When the suspended load of fine-grained terrigenous material became relatively low, reef fabrics were more dense, more massive, with interbedded lenses of reef-derived skeletal packstone. Corals here exhibit hemispheric, subspheric, and pillar-like growth-forms of *Porites, Tarbellastraea, Siderastraea,* and *Acanthastraea*.

In the Sebaa Chioukh Hills (Fig. 5), no horizontal zonation is reported in Miocene reefs, and little lateral change is observed as well. To the north, they are 10 m (33 ft) thick and rest upon pre-Messinian substrates; to the south, they reach a maximum thickness of 50 m (164 ft) and include numerous lenses of skeletal packstone/grainstone, foraminiferal packstones/ grainstones rich in *Heterostegina*, and, at the top, small reefs of coral boundstone. Similar observations, made near the Traras margin where reef sections are poorly exposed, indicate that Miocene reefs developed on a shoal parallel to the emergent Traras Massif with a narrow trough of lime mudstone deposits between the hills and the reefs. The back side of the reef faced the Traras Massif, whereas its front faced eastward (Fig. 6).

### *Depositional Facies*

*Pre-reef facies.—*
The reefs rest upon argillaceous mudstones or sandy (siliciclastic) argillaceous mudstones. Pioneer colonies of *Porites* developed in muddy environments and were scattered in the argillaceous lime mudstone matrix. Some buildups (e.g., those in the Sebaa Chioukh Hills) developed on pre-Messinian formations.

*Reef core facies.—*
Major framework builders of the Miocene reefs of western Orania include *Porites lobatosepta, Porites calabricae,* and *Tarbellastraea* cf. *reussiana.* Accessory corals include *Siderastraea crenulata, Acanthastraea* sp. and *Paleoplesiastraea* sp.

The abundance and growth-form of reef frame-builders changed as a function of the amount of lime mud matrix present. Build-ups with abundant framework elements display a complex mixture of branching, platy, and head-like colonies in a coral encrusting-red-algal framestone texture. In comparison, isolated finger colonies of *Porites* in mud-rich beds form a branching-coral bafflestone texture.

Sediment infilling of reef-core facies is composed of argillaceous lime mudstones, lime mudstones and mixed-skeletal wackestones with fragments of foraminifera, molluscs, red algae, and corals.

*Interreef facies.—*
Mixed-skeletal wackestones, packstones, grainstones, with foraminifera and molluscs, and locally rudstones (red-algae and coral debris) were deposited in interreef environments and formed lens-shaped interbeds. On the external margin of the Sebaa Chioukh Hills, abundant *Heterostegina* lime wackestones are recorded.

*Off-reef facies.—*
Reefs are laterally replaced basinward by argillaceous skeletal wackestones/mudstones rich in planktonic biota (Saint Martin, 1990). Slope deposits are not clearly identified.

*Post-reef facies.—*
Coral reefs represent the last Messinian sedimentation in western Orania. However, argillaceous lime mudstones, with oysters and a sparse microfauna, locally overlie the reefs.

### *Depositional Facies Distribution*

Facies distribution is more uniform laterally than vertically. Vertically, facies are characterized by several backstepping

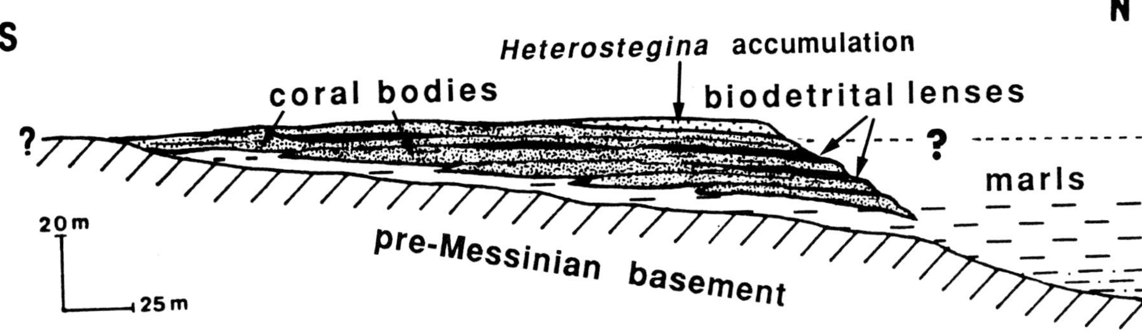


FIG. 5.—Diagram of reef structure in the Sebaa Chioukh Hills.

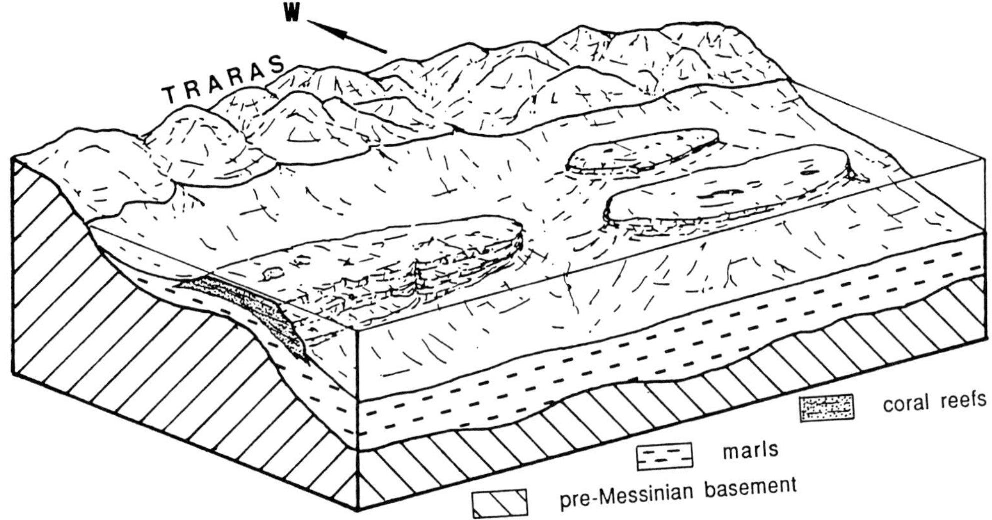


FIG. 6.—Reconstruction of the Traras margin reefs.

geometry sequences. Sequences generally show marls, then coral colonies embedded in marls, and finally carbonate coral reef banks. Each sequence, more or less complete, is interpreted as a shallowing upward sequence. In conclusion, the backstepping geometry and the repeated sequences indicate transgressive deposits. Poorly developed zonation is identified only at the top of the reef-complex with coral reef banks laterally passing into a narrow algal crest, then a bioclastic slope. All the features lead to interprete the reef complex as composed of a succession of reef-banks which developed in embayements, contemporaneous with fine-grained terrigeneous sedimentation.

### *Reef Associated Fauna*

The associated reef-dwelling fauna is rich and was the subject of a detailed study by Saint Martin (1990). More than 150 species from different groups (Bryozoa, Mollusca, Echinoidea, Crustacea and Brachiopoda) were distinguished (Table 1). The global fauna composition clearly indicates the diversity of reef biotopes. Various types of organisms associated with hard substrates are abundant and occur as cemented (molluscs and barnacles), encrusting (bryozoans), byssally-attached (bivalves), foot-attached (gastropods), and boring (bivalves). Nestling types of organisms (bivalves and echinoids) are associated with related sand-rich environments. Soft, muddy bottoms were intensely burrowed by infaunal molluscs.

## DIAGENETIC OVERPRINT

A comprehensive study concerning the diagenesis of the reefs of western Orania has not been conducted. However, thin-section examination of a limited number of samples reveals the following preliminary observations:

1. micritic infilling of almost all intraparticle pores;
2. recrystallization of skeletal aragonite into sparry calcite;
3. replacement of micritic matrix by calcitic microsparite;
4. dissolution of aragonitic skeletal elements, producing moldic porosity;
5. precipitation of mosaic calcite in moldic pores; and
6. late-stage dissolution of the previous sparry and mosaic calcite, producing vuggy porosity.

Apparently, drowning of these reefs by argillaceous lime mudstones favoured preservation of coral structures. Moreover, in some coral colonies, it is possible to observe dolomitized micrite rinds, convoluted micritic laminations and bituminous micro-rhombohedrons (Saint Martin, 1987). These phenomena

TABLE 1.—Reef-dwelling fauna, Sebaa Chioukh Hills.

| SCLERACTINIA | GASTROPODA |
|---|---|
| *Tarbellastraea* cf *reussiana* | *Haliotis tuberculata* |
| *Paleoplesiastraea* sp. | *Clanculus* cf. *cruciatus* |
| *Siderastraea crenulata* | *Astraea* sp. |
| *Acanthastraea* sp. | *Nerita emiliana* |
| *Porites lobatosepta* | *Rissoa* sp. |
| *Porites calabricae* | *Alvania* sp. |
| | *Architectonica* sp. |
| **BRYOZOA** | *Turitella unicarinata* |
| | *Turitella subarchimedis* |
| *Tubulipora* sp. | *Turitella* cf. *terebralis* |
| *Diaperoecia major* | *Vermetus* sp. |
| *Plagioecia* cf. *sarniensis* | *Thericium vulgatum* |
| *Lichenopora* cf. *grignonensis* | *Cerithiella* cf. *exbicarinata* |
| *Aplousina bobiesi* | Triphoridae |
| *Antrapora* sp. | *Epitonium* sp. |
| *Calpensia nobilis* | *Strombus coronatus* |
| *Cribrilaria radiata* | *Trivia* sp. |
| *Figularia figularis* | *Cypraea* sp. |
| *Unbonula monoceros* | *Cymatium affine* |
| *Schizoporella longirostris* | *Columbella* cf. *fallax* |
| *Escharina vulgaris* | *Oliva* sp. |
| *Schismoporella aculifera* | *Conus* cf. *mercati* |
| *Microporella ciliata* | *Conus vindobonensis* |
| *Schizotheca fissa* | *Terebra* sp. |
| *Celleporina costazi* | *Cylichnina* sp. |
| *Myriapora truncata* | |
| | **ECHINOZOA, ECHINOIDEA** |
| **BIVALVIA** | |
| | *Arbacina* sp. |
| *Barbatia barbata* | *Schizechinus duciei* |
| *Barbatia subhelbingi* | *Psammechinus* sp. |
| *Striarca lactea* | *Clypeaster* sp. |
| *Botula fusca* | |
| *Lithophaga lithophaga* | **CRUSTACEA, CIRRIPEDIA** |
| *Chlamys pusio* | |
| *Chlamys linguafelis* | *Creusia oraniensis* |
| *Aequipecten seniensis* | *Balanus concavus* |
| *Pecten aduncus* | *Balanus perforatus* |
| *Spondylus gaederopus* | *Balanus amphitrite* |
| *Lima lima* | |
| *Limaria tuberculata* | **CRUSTACEA, DECAPODA** |
| *Anomia ephippium* | |
| *Ostrea lamellosa* | *Dromia neogenica* |
| *Ostrea lamellosa* | |
| *Ostreola stentina* | **BRACHIOPODA** |
| *Hyotissa hyotis* | |
| *Diplodonta rotundata* | *Argyrotheca cf. neapolitana* |
| *Chama gryphoides* | |
| *Plagiocardium papillosum* | |
| *Gastrana fragilis* | |
| *Venus multilamella* | |
| *Pelecyora islandicoides* | |
| *Dosinia exoleta* | |
| *Lajonkairia rupestris* | |
| *Hiatella arctica* | |
| *Jouannetia tournoueri* | |

may have been related to the activity of filamentous organisms leaving pseudo-rhombohedrons of possible bacterial origin (Monty, 1986).

## OTHER MESSINIAN REEFS OF WESTERN ALGERIA

Throughout western Algeria, Messinian reefs are common in three main regions (Fig. 1): (1) western Orania and along the Mediterranean coast; (2) central Orania, extending along the coast from Cap Figalo to Oran; and (3) the lower Chelif Basin, 35 km (22 mi) southeast of Oran, along the Tessala and Beni Chougrane Mountains. The distribution of these reefs follows rather precisely the distribution of the present-day hills of Traras, Djebel Skouna, Djebel Murdjadjo, Tessala, and Beni Chougrane.

On a regional scale, Messinian coral reefs of Algeria vary in type, depending on their position in the Messinian cycle (Saint Martin, 1990; Saint Martin and Rouchy, 1990; Saint Martin et al., 1992; and Cornée et al., 1994). In the lowermost part of the Messinian times, reefs generally developed in a transgressive pattern and were commonly bedded. First, it was thought that a second Messinian reefal development occured after the maximum transgression; these reefs have been considered to be organized in general basinward downstepping pattern, because of a progressive sea-level drop (Rouchy, 1982; Rouchy et al., 1986; Saint Martin, 1990; Rouchy and Saint Martin, 1992). Nevertheless, recent studies indicate that these reef complexes developed through a progradational-aggradational process only (Saint Martin et al., 1992, 1995; Cornée et al., 1992, 1994).

Generally, reefs growing up to sea level ("shoal reefs") changed according to the shape and paleogeographic location of pre-existing Miocene ridges. Small isolated reefs or long reef belts occur along linear topographic highs (i.e., previous major structural trends) in the Traras, Tessala, and Beni Chougrane Mountains.

Platform reefs were developed over large flat, shallow-water areas (i.e., Beni Saf, Sidi Safi, and the southern margin of the Tessala Mountains). Near and especially at the top of these reefs, scattered, flat lense-shaped encrusting red-algal boundstones occured (i.e., larger than 10 $m^2$).

Fringing reefs are present on the flanks of partly emerged paleostructures (e.g. in the Skouna area). The reef complex of Murdjadjo is a single fringing-reef because of its large extension and its internal organization. This reef rests upon a pre-Neogene substrate that probably formed a narrow Miocene archipelago. In other places, the reef overlies a thick algal-limestone formation. The Murdjadjo reef is quite similar to the Melilla reef (Saint Martin and Cornée, this volume) in that it is composed of prograding lenses of *Porites* boundstone/ wackestone with interbedded mixed-skeletal *Halimeda* packstones/ grainstones (Rouchy, 1982; Rouchy et al., 1982, 1986; Saint Martin, 1990; Saint Martin and Rouchy, 1990; Rouchy and Saint Martin, 1992; Cornée et al., 1994).

It is also possible to distinguish reef types (Saint Martin, 1990) by their coral associations. The most diversified association, as observed in western Orania, central Orania, and along the southern margin of the Tessala Mountains, contains the genera *Porites*, *Tarbellastraea*, *Siderastraea*, *Acanthastraea*, and *Paleoplesiastrea*. A second, less diversified association includes *Porites* with local occurrences of *Siderastraea*.

## CONCLUSIONS

Messinian reefs from western Orania occur along a narrow outcrop belt, several hundred meters (approximately 1000 ft) wide by a few tens of kilometers (about 40 mi) long. Two reef trends are recognized: WSW-ENE in the Sebaa Chioukh and N-S in the Traras margin.

The linear reef that follows the crest of the Sebaa Chioukh Hills is in fact composed of several elliptical reef bodies, each approximately a kilometer in length and encased in argillaceous lime mudstones. The bedded internal structure and the stacking of coral-growth sequences indicate that reef growth was related to a transgression with backstepping geometry. Each step of reef building is represented by a tabular fringing reef with poor, but some, internal zonation. The whole sequence is about 50 m (164 ft) thick.

Reefs of the Traras margin developed on a narrow ridge which was a nearshore island in the Miocene sea. They display the same reef building sequences as observed in reefs of the Sebaa Chioukh Hills, except for basal conglomerates.

Repeated input of fine-grained terrigenous sediment produced a rythmic response in coral growth. When terrigenous input was high, *Porites* dominated reef faunas with both finger and platy growth-forms. When terrigenous input was low, the reef framework contains interbeds and lenses of mixed-skeletal packstone/ grainstone and shows higher diverity, including other coral genera. Reef diagenesis included an early stage of dissolution of aragonitic skeletons and a late stage of calcitic recrystallization. The coral reefs of western Orania were the last marine deposits in the region and were not associated with Messinian evaporites.

## ACKNOWLEDGMENTS

Dr. J. J. Cornée is thanked for help in the preparation of the manuscript and Dr. L. Montaggioni for advice about reef sedimentology and anonymous reviewers for improving the manuscript.

## REFERENCES

Cornée, J. J., Muller, J., and Saint Martin, J. P., 1992, Organization and dynamics of Messinian pericontinental carbonate platforms. Some examples from the North-African Mediterranean margins (abs.): Chichiliane, Platform Margins International Symposium, Abstracts, p. 33.

Cornée, J. J., Saint Martin, J. P., Conesa G., and Muller, J., 1994, Geometry, paleoenvironments and relative sea-level (accomodation) changes in the Messinien Murdjadjo carbonate platform (Oran, western Algeria). Consequences: Sedimentary Geology, v. 89, p. 143-158.

Fréneix, S., Moissette, P., and Saint Martin, J. P., 1987a, Bivalves Hétérodontes du Messinien d'Oranie (Algérie occidentale): Bulletin du Muséum national d'Histoire naturelle de Paris, série 4, C, v. 9, p. 425-453.

Fréneix, S., Moissette, P., and Saint Martin, J. P., 1987b, Bivalves Ptériomorphes du Messinien d'Oranie (Algérie occidentale): Bulletin du Muséum national d'Histoire naturelle de Paris, série 4, C, v. 9, p. 3-61.

Fréneix, S., Moissette, P., and Saint Martin, J. P., 1988, Huîtres du Messinien d'Oranie (Algérie occidentale) et paléobiologie de l'ensemble de la faune de Bivalves: Bulletin du Muséum national d'Histoire naturelle de Paris, série 4, C, v.1 0, p. 1-21.

Gentil, L., 1903, Etude géologique du bassin de la Tafna: Bulletin du Service de la Carte géologique de l'Algérie, v. 2, 425 p.

Gentil, L., 1917, Sur le Néogène de l'Algérie occidentale: Compte-Rendu Sommaire des séances de la Société géologique de France, v. 12, p. 168-169.

Guardia, P., 1975, Géodynamique de la marge alpine du continent africain d'après l'étude de l'Oranie occidentale: Unpublished Thesis, University of Nice, Nice, 289 p.

Guardia, P., Magne, J., and Moyes, J. 1974, Aperçu sur le Néogène autochtone de l'ouest oranais (Algérie occidentale): Mémoires du B.R.G.M., France, v. 78, p. 691-703.

Guernet, C., Poignant, A., and Saint Martin, J. P., 1984, Contribution à l'étude de la microfaune des récifs messiniens d'Oranie occidentale (Algérie): Géobios Lyon, v. 17, p. 155-161.

Moissette, P., 1985, Encrusting Bryozoans from two Messinian coral reefs of western Algeria, *in* Nielsen, C. and Larwood, G. P., eds., Bryozoa: Ordovician to Recent: Fredensborg, Olsen and Olsen, p. 205-212.

Moissette, P., 1988, Faunes de Bryozoaires du Messinien d'Algérie occidentale: Documents du Laboratoire de Géologie de Lyon, v. 102, 351 p.

Monty, C., 1986, Microbial dolomites (abs.): Canberra, 12th International Sedimentary Congress Australia, Session C4, p. 215-216.

Perrodon, A., 1957, Etude géologique des bassins néogènes sublittoraux de l'Algérie occidentale: Bulletin du Service de la Carte géologique de l'Algérie, v. 12, 323 p.

Pomel, A. and Pouyanne, J., 1882, Texte explicatif de la Carte géologique provisoire des départements d'Alger et d'Oran, Alger, Unpublished, p. 31-45.

Pouyanne, J., 1877, Notice explicative sur la subdivision de Tlemcen: Annales des Mines d'Algérie, v. 12, p. 81-155.

Rouchy, J. M., 1982, La genèse des évaporites messiniennes de Méditerranée: Mémoires du Muséum national d'Histoire naturelle de Paris (C), t. 50, 267 p.

Rouchy, J. M., Chaix, C., and Saint Martin, J. P., 1982, Importance et implications de l'existence d'un récif corallien sur le flanc sud du Djebel Murdjadjo (Oranie, Algérie): Comptes-Rendus de l'Académie des Sciences de Paris, série II, v. 294, p. 813-816.

Rouchy, J. M. and Saint Martin, J. P., 1992, Late Miocene events in the Mediterranean as recorded by carbonate-evaporite relations: Geology, v. 20, p. 629-632.

Rouchy, J. M., Saint Martin, J. P., Maurin A., and Bernet-Rollande, M. C., 1986, Evolution et antagonisme des communautés bioconstructrices animales et végétales à la fin du Miocène en Méditerranée, biologie et sédimentologie: Bulletin du Centre de Recherches Exploration-Production Elf Aquitaine, v. 10, p. 333-348.

Saint Martin, J. P., 1984, Le phénomène récifal messinien en Oranie (Algérie): Géobios Lyon, Mémoire spécial 8, p. 159-166.

Saint Martin, J. P., 1987, Les formations récifales du Miocène supérieur du Maroc et d'Algérie, Aspects paléoécologiques et paléogéographiques: Unpublished Doctorat d'Etat Thesis, University of Aix-Marseille 1, Marseille, 499 p.

Saint Martin, J. P., 1990, Les formations récifales du Miocène supérieur du Maroc et d'Algérie: Mémoires du Muséum national d'Histoire naturelle de Paris (C), t. 56, 351 p.

Saint Martin, J. P. and Chaix, C, 1981, Sur la paléoécologie des formations récifales du Miocène supérieur d'Oranie occidentale: Comptes-Rendus de l'Académie des Sciences de Paris, série II, v. 292, p. 1341-1343.

Saint Martin, J. P., Chaix, C., and Moissette, P., 1983, Le Messinien récifal d'Oranie, une mise au point: Comptes-Rendus de l'Académie des Sciences de Paris, série II, v. 297, p. 545-547.

Saint Martin, J. P., Cornée J. J., and Muller, J., 1995, Nouvelles données sur le système de plate-forme carbonatée du Messinien des environs d'Oran (Algérie). Conséquences: Comptes-Rendus de l'Académie des Sciences de Paris, série II, v. 320, p. 837-843.

Saint Martin, J. P., Cornée J. J., Conesa, G., Bessedik, M., Belkebir, L., Mansour, B., Moissette, P., and Anglada, R., 1992, Un dispositif particulier de plate-forme carbonatée messinienne: la bordure méridionale du bassin du Bas-Chelif (Algérie): Comptes-Rendus de l'Académie des Sciences de Paris, série II, v. 315, p. 1365-1372.

Saint Martin, J. P., Moissette, P., and Fréneix, S., 1985, Paléoécologie des assemblages de Bivalves dans les récifs messiniens d'Oranie occidentale: Bulletin de la Société géologique de France, v. I, p. 280-283.

Saint Martin, J. P. and Rouchy, J. M., 1990, Les plates-formes carbonatées messiniennes en Méditerranée occidentale, leur importance pour la reconstitution des variations du niveau marin au Miocène terminal: Bulletin de la Société géologique de France, v. VI, p. 83-94.

Thomas, G., 1985, Géodynamique d'un bassin intramontagneux, Le bassin du Bas-Chélif occidental durant le mio-plio-quaternaire: Unpublished Doctorat d'Etat Thesis, University of Pau, Pau, 594 p.

# MIOCENE REEF FACIES OF THE PELAGIAN REGION (CENTRAL MEDITERRANEAN)

MARTYN PEDLEY
*Department of Geology, Leicester University and School of Geography and Earth Resources, University of Hull, HU6 7RX, United Kingdom*

ABSTRACT: There are three depositional scenarios for Miocene reef development within the Pelagian Region as follows: Carbonate *ramps* provide an important setting for coral patch reefs, with the additional important development of coralline algal (often rhodolitic) biostromes immediately basinward of the lime-mud dominated patch reefs. *Delta tops and channels*, in total contrast to the clear water carbonates of gently sloping ramps, are the tectonically active zones associated with areas immediately to the south of the Sicilian Alpine fold belt. Here, in association with delta top channels and small offshore culminations above thrust-generated sea floor ridges, localised reefs developed whenever turbidity and siliciclastic sedimentation was reduced. Generally this was coincidental with marine onlap events of relatively short duration. Finally, *carbonate shelf slope breaks,* there are the monogeneric coral curtain reefs which developed under clear water conditions at basin-facing breaks in slope such as the submarine fault scarps bordering on to the central Mediterranean graben systems.

From a palaeoecological viewpoint, the limited range of coral genera present may most conveniently be catalogued on the basis of their respective growth form strategy. These range from slender rods in excess of 1 m long in the coral curtains, to vermiform and ramose bushy growths in the areas of highest siliciclastic sedimentation. Additionally, dome headed corals dominate upper fore-reef and reef crest locations; whereas, undulose growth forms typify lagoonal and ramp patch reefs.

Each peculiar reef type is the product of a variable range of parameters including substrate stability, eustatic fluctuations, tectonism and sedimentation rates. These are all considered before proposing several sedimentological models common to the central Mediterranean region and possibly of value elsewhere within Tertiary "Tethyan realm."

## INTRODUCTION

A resurgence of interest in Miocene reefs of the Central Mediterranean is reflected in the work of Sergre (1960) and Grasso and Pedley (1985, 1989) in Lampedusa; Chevalier (1961), Catalano and Esteban (1978), Grasso and Pedley (1988), Fois (1990) and Pedley et al. (1994) in western, central and northern Sicily; Pedley (1979, 1983, 1987b) in the Maltese Islands; Grasso et al. (1982) and Pedley (1983) in south-eastern Sicily; and Pedley and Grasso (1994) in Calabria. Reefs are widely distributed around the margins of the Pelagian basin (Fig. 1), a broad dish-shaped structure lying on the divide between the eastern and western Mediterranean basins.

These central Mediterranean Miocene reefs are generally small structures. Frameworks are often sparse and coral diversity is low (less than five species, sometimes only one). Nevertheless other biota may be diverse and clearly defined facies associations are observable.

Past workers have been hampered by inadequate knowledge of field associations and absence of coherent lithostratigraphy. Each of the accounts presented here is backed by extensive regional facies mapping, by biostratigraphic correlation and coherent lithostratigraphic correlation.

Other than local biostromes in the Aquitanian (Early Miocene) of south-east Sicily, the Central Mediterranean reefs are of Tortonian and Early Messinian age (N16 and 17 zones of Blow, 1969). It is their mode of occurrence and structure, rather than their size, which makes these bioherms significant. They record the subtle physical changes influencing the shallow threshold which divides the eastern and western Palaeomediterranean basins, and in particular they monitor water flow direction, eustasy and tectonic movements more precisely than does any other sediment association in the region (Fig. 1).

## DEPOSITIONAL SCENARIOS OF PELAGIAN REEFS

Reefs are developed in three depositional scenarios: (i) ramps, (ii) delta tops and channels, and (iii) shelf breaks in slope.

The principal Pelagian reef developments, indicated by black dots on Figure 1, generally are small constructions usually less than 100 m long and less than 10 m thick. Generally they face eastward-driven surface water currents upwelling from the deeper western Palaeomediterranean basin. This circulation pattern became set up by Burdigalian times (Pedley, 1987a), at which time the Indian Ocean connection was lost (Adams et al., 1983), and the eastern Palaeomediterranean Basin immediately became a site of net water loss due to evaporation. The most extensive reef developments were on sites exposed to a constant through flow (e.g., Malta and SE Sicily). Nevertheless, brief Late Miocene reef episodes became established under transgressive eustatic conditions which caused flooding of siliciclastic deltas associated with the Terravecchia Formation of north central Sicily. Here, the encouragement to growth was the temporary establishment of clear-water highstand conditions with better flushing within the embayments. During Late Tortonian-Early Messinian deposition abrupt breaks of slope (e.g., submarine fault scarps) also encouraged reef development.

## REEFS ON RAMPS

Two ramp carbonate successions occur in the Tortonian units of the eastern Pelagian Block (northern segment of the African Plate). That of south-east Sicily is of classic form (*sensu* Ahr, 1973). The carbonates were deposited on a gentle basin-facing paleoslope (see model in Buxton and Pedley, 1989). These developed under conditions of eastward water flow and infrequent tectonic activity. The Sicilian example has a history extending back into Early Paleogene times. The Maltese ramp

Models for Carbonate Stratigraphy from Miocene Reef Complexes of Mediterranean Regions,
SEPM Concepts in Sedimentology and Paleontology #5, 
SEPM (Society for Sedimentary Geology), ISBN 1-56576-033-6, p. 247-259.

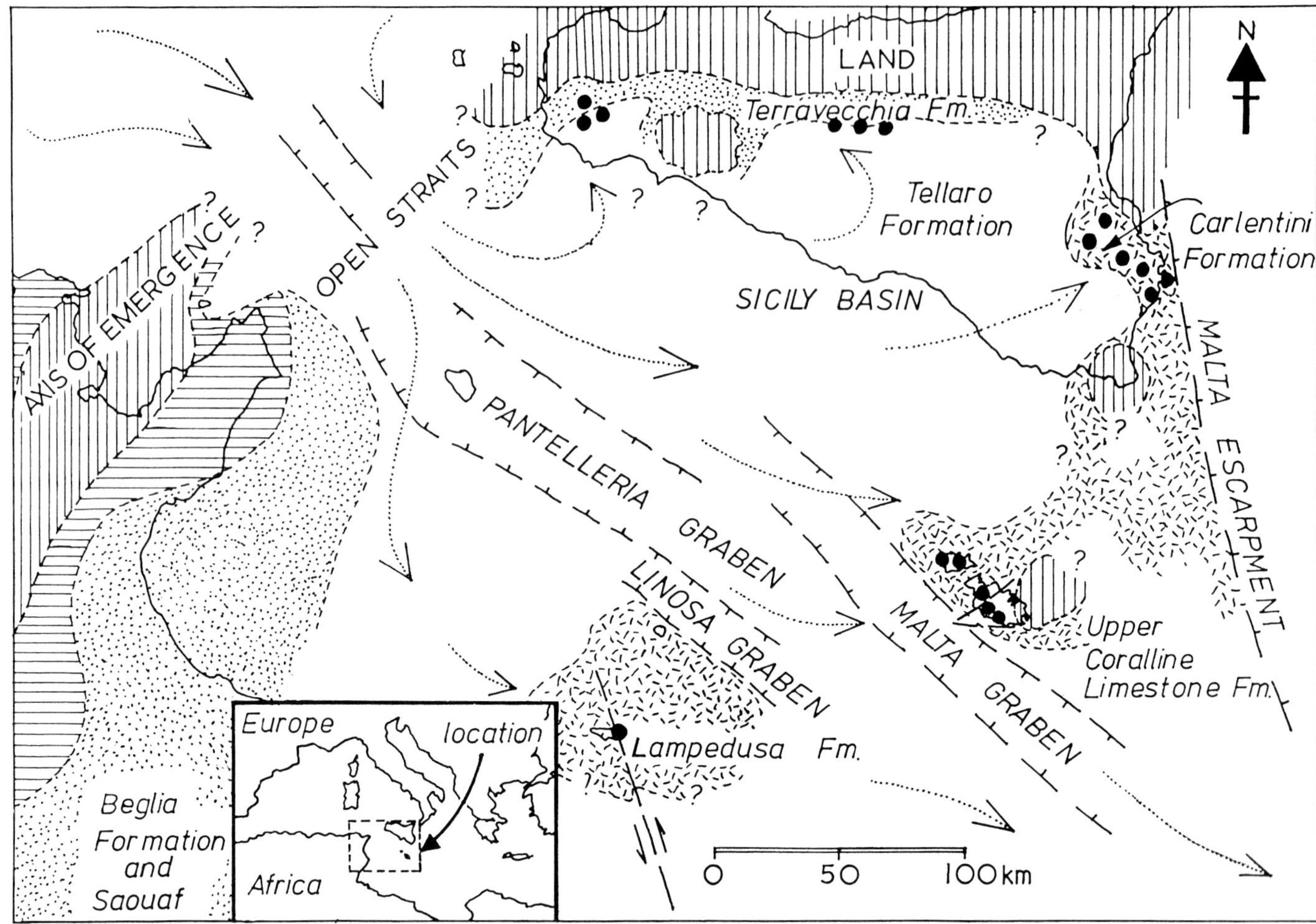


FIG. 1.—General geological map of the Pelagian Block during Late Miocene times. Vertical and horizontal bars = subaerially emergent and erosional areas; coarse stipple = fluvial, marine and deltaic siliciclastics; random strikes = marine bioclastic (reefal carbonates marked by solid black dots); blank areas = pelagic lime mudstones, marls and clays and arrows with dotted shafts = inferred Paleomediterranean marine circulation patterns.

is a good example of a leeward ramp. Incipient movement on the eastern margins of the Malta Graben (see Fig. 1) created a northeasterly regional paleoslope upon which the ramp developed. Leeward building was encouraged by the vigorous east flowing bottom current (Pedley, 1987a).

### *South-east Sicily: A Windward Ramp Case History*

This structure (area D in Fig. 2) faces the "Caltanissetta Basin," an area of deep water, grey pelagic marl sedimentation during Tortonian time. These marls (Tellaro Formation of Rigo and Barbieri, 1959) pass upramp into nodular to thinly bedded, bioclastic lime mudstones and wackestones (Palazzolo Formation of Rigo and Barbieri, 1959) containing a sparse echinoid macrofauna and extensive thalassinoidean bioturbation (C in Fig. 3). In inferred water depths of about 40 m, the wackestone facies gives way to a *Heterostegina*-rich coarse bioclastic packstone which is transitional with rhodolitic coralline algal biostromal beds (A in Fig. 3) of the Siracusa Limestone Member (Monti Climiti Formation of Pedley, 1981). Small patch reefs are developed in the shallowest eastern zone. They are up to 4 m high and 50 m wide. Their pale-grey micritic cores with scattered *Porites* and *Tarbellastraea* colonies and rare *Palaeoplesiastraea* and *Siderastraea,* crustose corallines, articulated burrowing bivalves and early lithification features contrast strongly with the coarser grained off-reef packstones and grainstones which are often *Halimeda*-rich. Evidence from the overlying Carlentini and Monte Carrubba Formations demonstrate that shoreward of the patch reefs lay an outer shoal zone and inner restricted lagoon (Pedley, 1983).

The ramp profile is about 30 km long and suggests a maximum depositional paleoslope of about 3°, however, the relatively external position of the *Heterostegina* belt indicates a convex cambered profile with a shallow, and only gently inclined, internal and steeper external zone (distally steepened ramp model of Read, 1982). The ramp extended at least 50 km along depositional strike and apparently maintained itself from (Eocene?) early Oligocene times as a result of continuous sediment supply from the area of attenuated sedimentation to the east of the Melilli-Siracusa Line (Pedley, 1981; Pedley et al.,

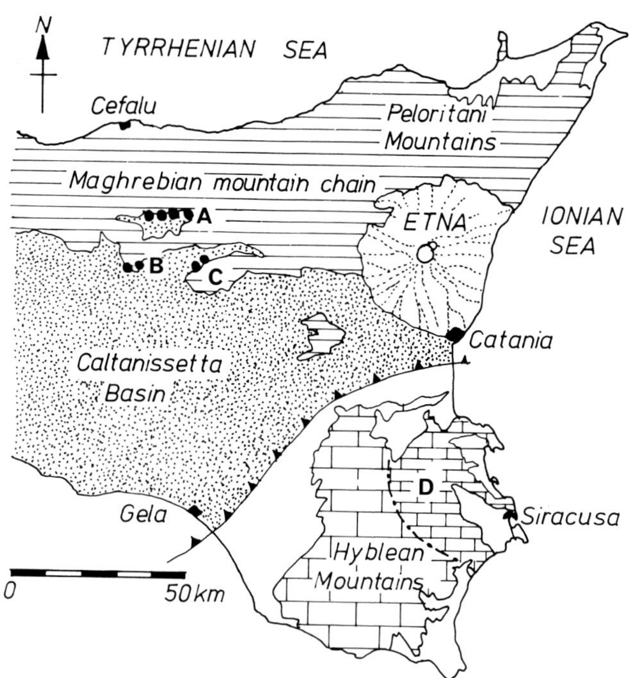


FIG. 2.—General geology of central and eastern Sicily during the Late Miocene. Horizontal bars = subaerial mountain belts; coarse stipple = marls and clays of the "Caltanissetta Basin;" coarse brick pattern = Tertiary outer ramp carbonates of the stable African Foreland; fine brick pattern = inner ramp carbonates; areas marked "A" = carbonates in proximal channels (Monte Corvo); area "B" = Landro group of channel-mouth reefs; area "C" = Villadoro reefs; area "D" = patch reefs of the upper ramp slope in S. E. Sicily. Unshaded areas in south east Sicily are Quaternary sediments. The toothed line extending from Gela to Catania represents the Gela Nappe and divides African autochthon (platform carbonates) from allochthonous European Plate components to the north.

1992). The early Miocene (Aquitanian) history of the ramp includes only localized patch reefs and associated coralline algae in the extreme east. The majority of the ramp carbonates from the shallowest areas consist of coarse bioclastic packstones and rudstones (Melilli Member of Monti Climiti Formation, B in Fig. 3). Rhodolitic coralgal patch reefs and lagoonal associations (Grasso et al., 1982; Pedley, 1983) continued developing into Early Messinian times (see Fig. 1).

*Malta: A Leeward Ramp Case History*

Although not so extensively exposed as the Sicilian ramp, the Upper Coralline Limestone Formation of the Maltese Islands illustrates well the morphology of a leeward ramp. Figures 4 and 5 show the facies distributions which in part are based on Pedley (1976, 1978). The deeper water area (D in Fig. 4) was the site of hemipelagic lime mudstones and bioclastic wackestones. This passes up-ramp into a coralline algal biostrome (Bosence and Pedley, 1979, 1982), area B in Figure 4. Several algal facies including algal marls, rhodolith and crustose pavements are recognized. Upramp, *Heterostegina* lime mudstones and wackestones replace the biostrome, and in the shallowest zone in the extreme west small *Porites* thickets associated with large pectinids and *Clypeaster* echinoids occur (E in Fig. 4).

Progradation of the ramp is illustrated by facies C in Figure 4. These later sediments consist of coarse grained bioclastic (especially rhodolitic) and peloidal packstones (Tal Pitkal Member, Pedley, 1979, 1993). The topset beds of these show colonization by scattered rhodoliths and the development of lenticular patch reefs identical in most respects to those described previously in south-east Sicily. *Porites* and *Tarbellastraea* are commonest but are associated with occasional *Palaeoplesiastraea* corals. In addition, sheet-like biostromes are developed in some beds. The reefs show early lithification phenomena (Pedley, 1979, 1983) and a rich association of encrusting, nestling and burrowing invertebrates. Marine planation terminated patch reef development, and the resulting surface was then buried in a thin (Messinian) ooid grainstone.

The Maltese and Sicilian ramp associations contain the most diverse faunas and floras including corals. These are documented in Pedley (1976, 1983) and Buxton and Pedley (1991). The faunal composition of each of the lithofacies (Figs. 4, 5) are all quite distinct and lateral facies boundaries are sharp.

In the deepest water areas of the ramp, hemipelagic wackestone facies, low in macrofauna but dominated by planktonic foraminifera, are developed. Other than a few irregular echinoids and considerable bioturbation (Pedley, 1992a), the dominant benthonic fauna is foraminiferal. A coralline algal biostrome (Bosence and Pedley, 1982) is developed in water depths of about 60 m (Pedley, 1976) on the Maltese ramp in a narrow gently subsiding corridor (Fig. 4, area B). The constructors here are both crustose and rhodolitic coralline algae. Bosence and Pedley (1982) believe that the environment was similar to the present day "biocoenose du detrique cotier" and "Biocoenose du coralligene de plateau" of Peres and Picard (1964). There are also middle Miocene analogs (e.g., southern Italy, Carannante et al., 1981).

The "Large Foraminifera facies" (Buxton and Pedley, 1989) is developed immediately up-ramp of the outer ramp association (probably in depths of 40-50 m, see Hottinger, 1977). Locally it is a monospecific *Heterostegina* packstone, especially in the Maltese Islands. In south-east Sicily, however, the foraminifers are usually disseminated through the first meter or two of the succeeding rhodolith packstone facies (between C and A in Fig. 4). The Large Foraminifer biofacies finds a close parallel in the Recent Aqaba ramp (Riess and Hottinger, 1984).

Upramp from the Large Foraminifera facies the macro-boring and encrusting community continues to be poorly represented due to the general absence of lithification. The rhodolitic algal facies (A in Fig. 4) is dominated by cheilostome bryozoans (mainly arborescent and articulated forms in the sheltered areas but with encrusters dominant in the shallower parts). In Malta, infaunal filter feeding molluscs, brachiopods, irregular echinoids and benthic foraminifera comprise the rest. It is within the rhodolitic facies that patch reef development occurs.

The patch reefs of SE Sicily and Malta (solid black lenses in Figs. 3, 4) are well defined bioherms occupying the shallower water areas of their respective ramps and usually associated with

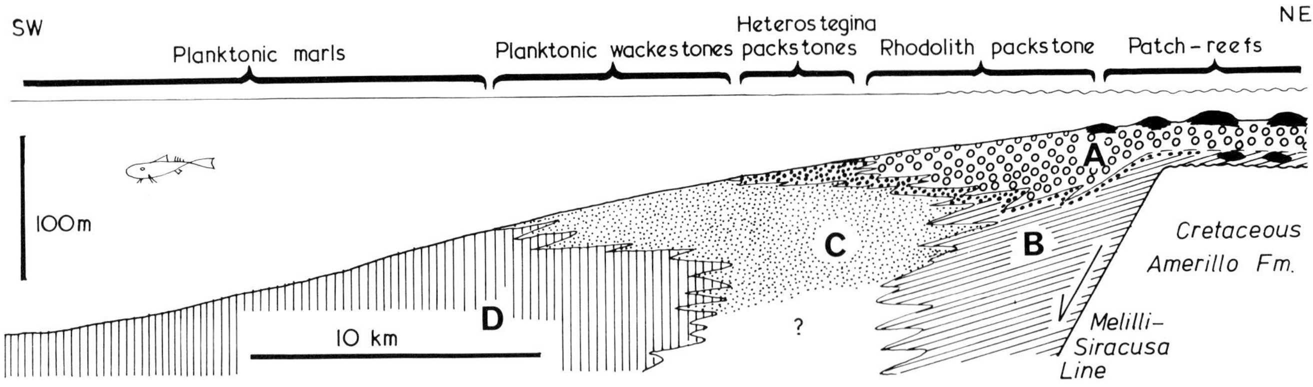


FIG. 3.—Profile down the Tortonian ramp of south east Sicily. The Melilli-Syracusa Line is defined here as a boundary fault dividing stable shelf with a Cretaceous rudistid reef development from subsiding basin underlaid by deepwater Cretaceous "Scaglia Facies". N. B. in this and all succeeding figures the fish faces the dominant current. Symbols: A = rhodalgal packstones with coral patch reefs in black; B = bioclastic packstones; C = hemipelagic wackestones of the Palazzolo Formation; D = Hemipelagic marls.

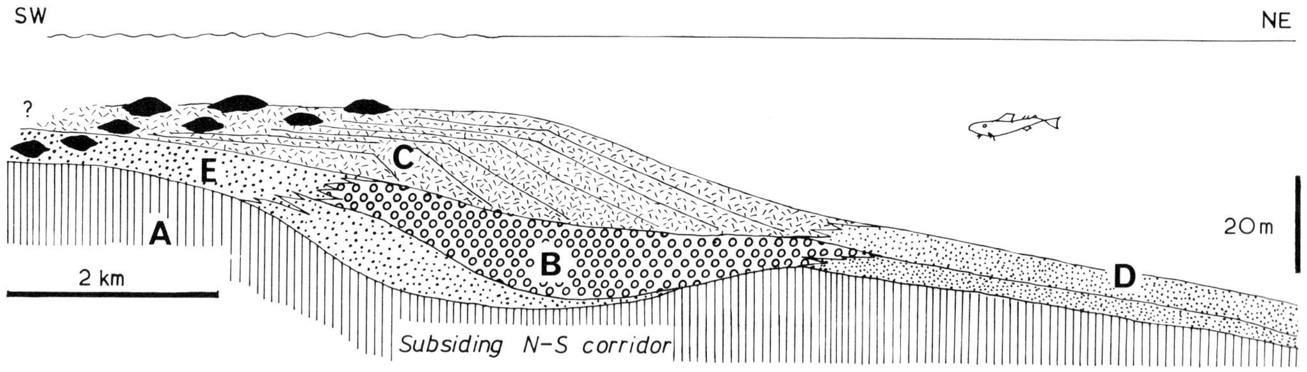


FIG. 4.—Profile of the Maltese Tortonian ramp: A = Tortonian Marls; B = Coralline Algal Biostrome (stippled area between A and B is *Heterostegina* packstone facies); C = prograding coralline algal-rich bioclastic intra-shelf ramp of the Tal Pitkal Member filling in the subsiding N-S corridor (see Fig. 3 for location); D = planktonic outer slope wackestones; E = bioclastic wackestones with rhodoliths, and local coral thickets and small patch reefs further west.

rhodalgal facies. The dominance here of the alveolinid foraminifer *Borelis* is taken to indicate water depths of about 25 m (Hottinger, 1977). Patch reefs are the principal bioconstructions found on ramps. However, they are identical to patch reefs from other Pelagian settings, consequently, they will be collectively considered. The majority have a lenticular form. Sheet-like biostromal developments are also common in Malta. Table 1 illustrates the diversity of the commoner marine macro-biota associated with the patch reefs of the Maltese Islands south-east Sicily and the coral curtains Lampedusa.

Typically, the coral growth forms are either sheets, low domes or stocky bushes (Fig. 6, see later). In all cases, colony diameter is typically about 20 cm. These primary frameworks are strengthened by an interlinking crustose coralline algal framework and by invertebrates such as homotrematid foraminifers, serpulid worms, vermetid gastropods and encrusting bryozoan colonies (Pedley, 1979). Colonization by both encrusters and nestlers alternates with interstitial sedimentation of micrite and bioclasts. Early submarine isopachous cement fringe development (Pedley, 1979, 1983) further links allochems and frameworks, encircling them with alternating micrite and spar fringes. These processes appear to have been operating concurrently to produce turbulence and bioerosion-resistant structures.

The abundance of bivalve borings in these reefs is considered to indicate rapid lithification of the sediments. Boring bivalve crypts can often be followed in from reef margins and clearly cross cut both coral and inter-framework sediment infills.

Volumetrically, the micritic areas dominate the patch reefs are commonly peloidal. The peloids are usually less than 70 μm long but may be aggregated into multinucleate 'clotted' masses. Generally, they have a clear spar fringe around their outer margins (cf. Macintyre, 1977; Lighty, 1985). Their abundance and uniformity of size argues for a single common origin and against them being produced as fecal pellets by grazers. The peloids might be inorganically precipitated (see review of peloid formation in Macintyre, 1985). However, there is a remarkable similarity between the Miocene marine peloids and freshwater biofilm associated peloids in tufas (Chafetz, 1986; Pedley,

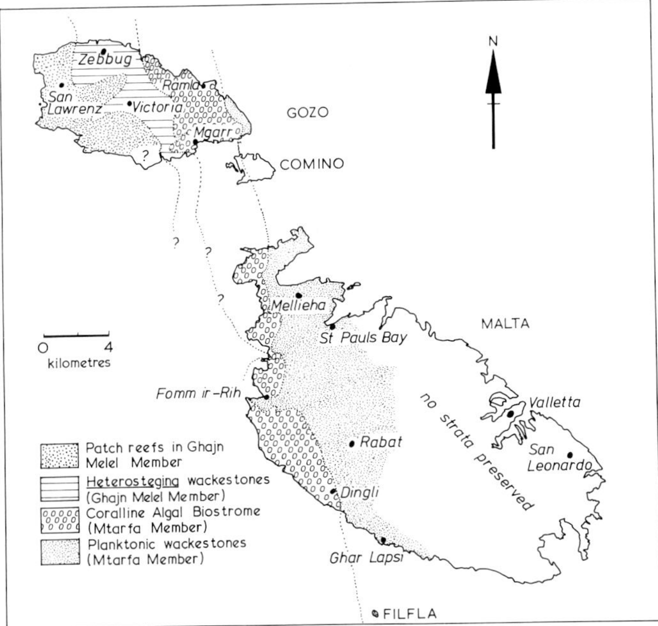


FIG. 5.—Facies distribution across the Tortonian ramp of the Maltese Islands. Note the marked N-S strike to the facies. This probably is caused by local synsedimentary subsidence over subparallel basement fractures.

1992b). Additionally, there is considerable encrustation of corals and coralline algae by stromatolite coatings ("compact crust" of Pedley, 1979). Collectively, peloids and micritic coatings in the ramp associated patch reefs are currently thought to be bacterial clump precipitates (see review in Pedley, 1992b). Tsien (1985) provides further ideas on algal-bacterial origins for other forms of micrite. Nevertheless, a ready supply of micrite, sourced by a plethora of borers and mechanical breakdown, must also have been available and must have remained unlithified for a sufficient time period to permit colonization by burrowers and semi-infaunal bivalves.

The diversity of filter feeders (see Table 1) in Miocene patch reefs is a reflection of the widespread availability of soft lime mud sediment within the growing Pelagian bioherms (cf. Bermuda cup-reefs of Ginsburg and Schroeder, 1973). *Panopea*, for example, up to 160 mm in length and commonly in life position, occurs in some Sicilian patch reefs. Undamaged bivalve shells (rarely articulated) are the most common bioclasts in many others (Pedley, 1979).

The absence of *Halimeda* from such sediment pockets but dominance in off-reef sediments (particularly in association with coral curtains to be described later) is puzzling. It suggests that these codiacean algae were not patch reef dwellers in these Miocene reef areas but belonged to inter-reef site communities.

TABLE 1.—COMMON INVERTEBRATE AND FLORAL ASSOCIATIONS OF THE CORAL REEFS

| | *PATCH REEFS* | *CORAL CURTAINS* |
|---|---|---|
| **Borers** | Lithophaga (M, L)<br>Gastrochaena (M, S)<br>Clionid Sponges (M, S)<br>Bryozoa (M)<br>Fungae (M) | Clionid Sponges (CS, L, M)- rare<br>Lithophaga (L)- rare |
| **Encrusters** | Homotrematid Foraminifera (M, S)<br>Bryozoa (M, S)<br>Vermetus (M, S, L)<br>Serpulid Worms (M, S, L)<br>Spondylus (M, S, L)<br>Ostraea (M, S) | Bryozoa (CS, L)<br>Serpulid Worms (CS)- rare<br>Vermetid Gastropods (CS)- rare |
| **Burrowers** | Glycymeris (M)<br>Cardium (M, L)<br>Cardita (M, L)<br>Chama (M)<br>Lucina (M, S, L)<br>Indet. Heterodonts (M, S, L)<br>Ostraea (S)<br>Astarte (M, S)<br>Panopea (S) | none |
| **Nestlers** | Arca (M, S)<br>Terebratula (M, S)<br>Megathyrid Brachiopods (M, S) | none |
| **Grazers** | Cidarid Echinoids (M, S, L)<br>Haliotis (M)<br>Echinid Echinoids (M)<br>Strombus (M, S)- off-reef<br>Patella? (M)<br>Clypeaster (M, S)- off-reef | Cidarid Echinoids (CS)- fore-reef, rare |
| **Predators** | Conus (M, S)- off-reef<br>Turritella (M, S)- off-reef<br>Natica (M, S)<br>Crustaceans (M, S, L) | none |
| **Nektonic** | Macrochlamis (M, S)- off-reef<br>Chlamys (M, S, L)- off-reef | Chlamys (CS)- fore-reef |
| **Vegetation** | Crusose Coralline Algae (M, S)<br>Rhodolitic Coralline Algae (M, S, L)- off-reef<br>Halimeda (M, S, L)- mainly off-reef | Halimeda (CS, L, M)- off-reef |

CS = central Sicily channel reefs
L = Lampedusa
M = Maltese Islands
S = S. E. Sicily

REEFS IN DELTA TOPS AND CHANNELS

### *Central Sicily Case History*

Thin bioclastic carbonate units occur within the argillaceous to sandy Terravecchia Formation (Schmidt di Freiberg, 1962; Ruggieri and Torre, 1982, 1984) around the northern margins of the "Caltanissetta Basin" in north-central Sicily. They are also known from more westerly areas (Catalano et al., 1976; Aruta and Buccheri, 1971; Catalano, 1979; Esteban et al., 1982, Fig. 2). Microfaunal analysis of the succession indicates that it spans a time range of Late Tortonian to Early Messinian (Grasso and Pedley, 1988). The carbonate beds vary from lenticular sheets of wackestone and packstone, rich in bryozoans, alveolinid and miliolinid foraminifers and echinoid debris, to thick but local

| | N. Central Sicily Channel-mouth reefs: talus | wall | N. Central Sicily Intra-channel fill: mud | sand | N. Central Sicily Intra-basin high patches: mud | sand | S.E. Sicily, Malta and Lampedusa Reef-curtains at slope break | S.E. Sicily Malta and Lampedusa Platform-top patch-reefs | Maltese Islands Coralline algal biostrome |
|---|---|---|---|---|---|---|---|---|---|
| A (1, 2) | | $P_1$ | | $P_2$ | | | $P_1$ | | |
| B (1, 2) | | $Pa_1, C_1$ | | $P_2, T_2$ | | $P_1, C_1$ | | $T_2$ | |
| C | | | P, T | | P | | | | |
| D (1, 2) | | $P_1, T_1$ | | | | $T_2$ | | | |
| E | | | | | | | | P | P |
| F | P, T | | | P | | | | T | T |
| G | | | | F | | | | | |

FIG. 6.—Distribution of coral growth forms within the various reef types of the Pelagian Region: A = slender rod; B = bushy (ramose); C = bone-like and vermiform; D = dome; E = undulose; F = sheet; G = cushion. Coral genera: P = *Porites*; T = *Tarbellastraea;* C = *Coeloria;* F = *Favites;* Pa = *Palaeoplaesiastraea.* Numbers to the right of the coral genera: 1. = colony height greater than 0.5 meter; 2 = colony height less than 0.5 meter.

reefal accumulations with abundant *in situ* coral growths (see Grasso and Pedley, 1985, 1989 for sedimentological and structural analysis). The dominant coral genera are *Porites*, *Tarbellastraea* and *Palaeoplesiastraea* but with subordinate *Siderastraea*, *Coeloria* and *Heliastraea* and rare *Meandrina.*

Much of the carbonate is the product of reworking and subsequent deposition of material within low sinuosity channels that were cut into the delta top shelf during preceding marine lowstands. The earliest deposits in the channels are cobbles, pebbles, and quartz sands. These prograde south and southwestwards down the steep sided proximal channels and are poorly fossiliferous (e.g., Manca di Corvo channel, A in Fig. 2). Thinner bedded packstones conformably succeed these marginal marine and fluvial siliciclastics. The packstones represent a transgressive marine event which flooded the delta top but are now only preserved within the preexisting incised channels. These carbonates contain abundant benthonic foraminifers, echinoid debris, Cheilostome bryozoan and bivalve material, (e.g., Petralia Channel).

Close to the shelf edge break of slope, broad channels with gently sloping lateral margins open into the deeper basin (e.g., Balza di Rocca Limata Channel, B in Figs. 2, 7A). The mouths of these channels are choked by thick reef breccia deposits. Areas immediately up-channel are filled by clays and sands. Well established reefs with prominent fore-reef talus slopes and diverse reef crest coral associations face the basin.

Pelagic clays and marls dominate the basin fill; however, these deposits are attenuated over submarine hills of older strata thrust into south-verging imbricate slices by synsedimentary tectonism. Here, small coralgal patch reefs are locally developed (e.g., Villadoro area, C in Fig. 2).

The Landro group of outcrops centered on Balza di Rocca Limata (Fig. 7A) well illustrates the characteristics of these channel-mouth reefs. Here, the channel mouth is about 2 km wide. A thickness of about 45 m of reef deposit is preserved in the axial region of the channel, thinning to a feather edge both southwestwards and to the northeastward. The reef deposit rests above pre-reefal clays and quartz sand channel fill which in turn lie directly on eroded, older Terravecchia Formation (E in Fig. 7A).

The fore-reef facies (A in Fig. 7A) consist of cobble to sand sized reef-derived clasts forming an ill stratified breccia, which passed directly into the reef core. Faunas are sparse but encrusting coral growth forms are present in the lower forereef slopes together with pectinid bivalves. No clearly discernible reef crest occurs but *in situ* coral stands are extensively developed on the upper forereef slope.

Further deposits of massive bedded breccia composed of lithified reef material occur up-channel from the forereef front (C in Fig. 7A). They are underlaid by clay and quartz sand deposits and further silty clay beds occur intercalated with the breccias higher in the sequence (D in Fig. 7A).

Erosion at outcrop has removed proximal areas of this channel deposit and any reef developments formerly existing beyond the channel margins. Data from other channels indicates that the reefoidal breccias give way to finer sand and clay fills up-channel from the mouth area. The paleoenvironment is summarized in Figure 7B.

Collectively, these deposits represent reef developments contemporaneous with patch reefs from south-east Sicily, Malta, and coral curtains to be described later from Lampedusa. They developed on the southern side of a rising Late Miocene landmass (the Maghrebian Mountain Chain, part of the Alpine fold belt system) created by Pelagian Block-European Plate collision.

In profile, these reefs appear to have the morphology of a fringing reef; indeed in the extreme west of Sicily, Catalano (1979) and Catalano and Esteban (1978) record true fringing reefs within the Baucinna Formation of the same age. In the north-central Sicilian examples, the lenticular developments have no lateral continuity. In situ *Porites* coral stands are, nevertheless, extensively preserved on the upper fore-reef slope towards the center of the channel fill. These *Porites* and *Tarbellastraea* rods appear to have formed an *in situ* build-up facing the basin (B in Fig. 7A) and are closely associated immediately up channel by a rich coral assemblage of up to five common genera of dome-headed and stocky branching coral growth forms, all well described by Chevalier (1961; see Fig. 6 for further details). This reef area contains a moderately diverse (but low actual numbers) invertebrate association of boring and encrusting bivalves, echinoids, bryozoans and crustose coralline algae. There is an abundance of benthonic foraminifera in the micritic matrices. Furthermore, there is considerable evidence here of penecontemporaneous reef framework destruction and recolonization.

Up-channel reef developments are dominated by siliciclastic silts and clays (Fig. 7A) and contain an abundance of bone-like and vermiform *Porites* growth forms. This growth form has also been commonly recorded from coral knolls associated with muddy sediments in intra-basin locations (Grasso and Pedley,

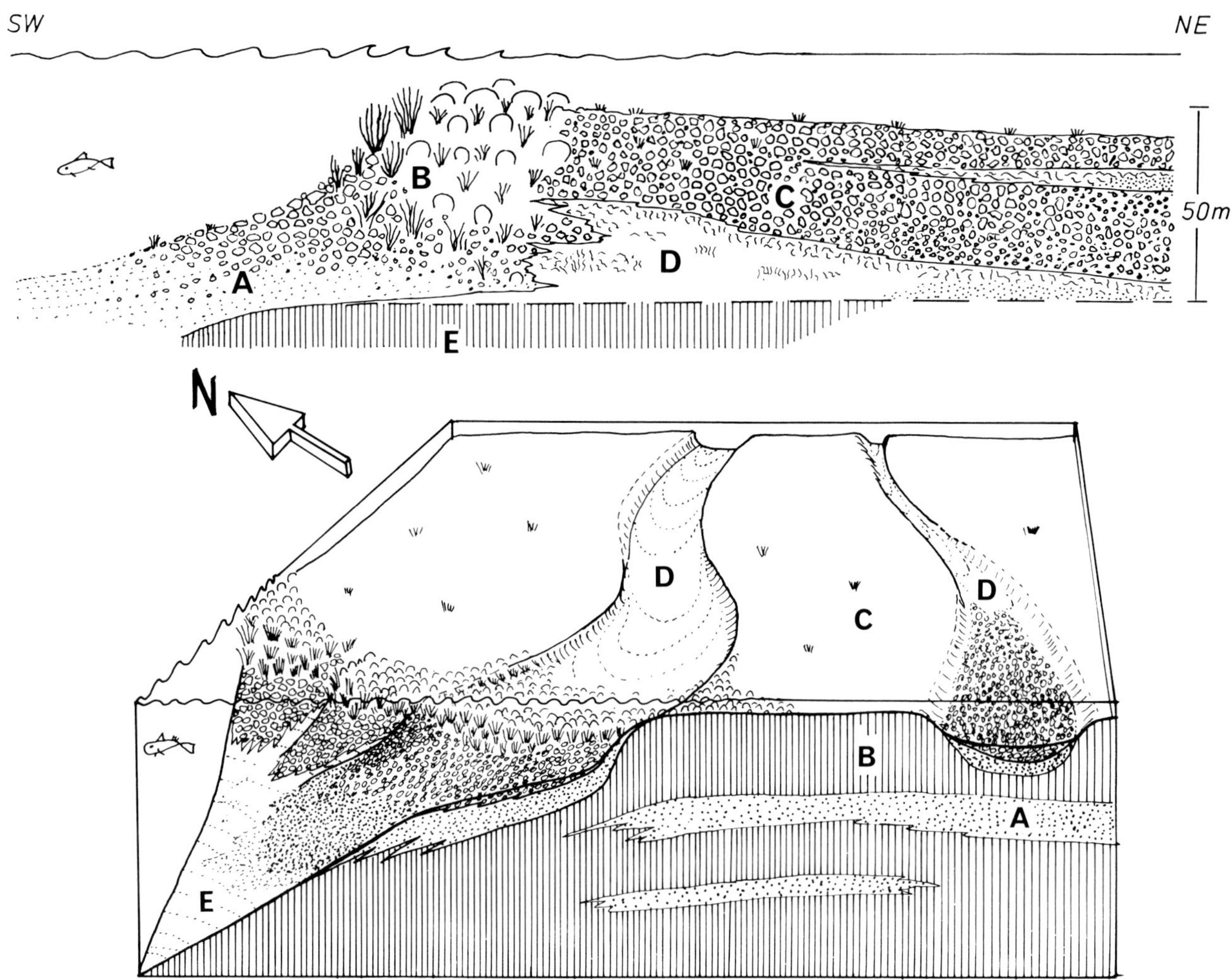


FIG. 7.—(A) Profile of the Late Tortonian-Early Messinian Landro channel-mouth reef, north Central Sicily: A = fore-reef talus of lithified coral wackestones and packstones; B = upper forereef with *Porites* colonies and in place *Tarbellastraea* and *Coeloria* colonies domination areas immediately up-channel; C = lithified reef cobble channel fill with siliciclastic silt and clay intercalations containing vermiform *Porites* growth forms; D = Siliciclastic basal channel silts and clays containing monospecific *Porites* (vermiform growth form); E = eroded Terravecchi Formation forming the channel base. (B) Block model of the channel-reef paleoenvironment: deltaic sheet sands (A) and clays (B) of the Terravecchi Formation; C = delta top with scattered coral patches and marginal reef; D = low sinuosity delta top channels occupied by channel-mouth reefs and more proximal reef rubble and siliciclastic deposits and E = delta front slope.

1988).

## REEFS IN SHELF BREAKS

An unusual bioherm type found in a variety of fully marine situations of Late Tortonian to Early Messinian age occurs in this setting and consists of monogeneric scleractinian coral build-ups (coral curtains). The slender rod (organ-pipe) growth form is dominant. Invariably these are *Porites* rods which are typically 2 or 3 cm in diameter and 1 to 3 m in length. The corals (now mainly *in situ* external moulds) are preserved in a lime mudstone or wackestone matrix. Other macrofaunas are scarce.

### *Malta Case History*

The Ghar Lapsi reef complex in western Malta (Pedley, 1987a) provides a good example of a shelf break reef. The *Porites* curtain (A in Fig. 8A) is orientated along the crest of a synsedimentary fault scarp (a splay fault of the Maghlaq Fault which is part of the offshore Malta Graben, Fig. 1). It is about 50 m wide but extends over 800 m along fault-strike before passing offshore. Immediately adjacent to the south occur crudely-bedded, cobble intraclast, reef-debris breccias (E in Figure 8A). Well-sorted *Halimeda*-rich wackestones and packstones lie close behind the reef curtain (B in Fig. 8A) and are locally cross stratified. These backreef deposits lack planktonic elements.

A similar situation occurs in Lampedusa Island lying 160 km west of Malta (Fig. 8B). Here the monogeneric *Porites* curtain (A) along the eastern side of the harbour separates a southwestern sequence of bioclastic rudstones (F) and a rhodolith pavement (E) from *Halimeda* packstones (d) and associated patch reefs (C) to the northeast.

The coral curtains (Figs. 8A, B) are in stark contrast to the ramp associated patch reefs (see Table 1). Slender rod ('organ-

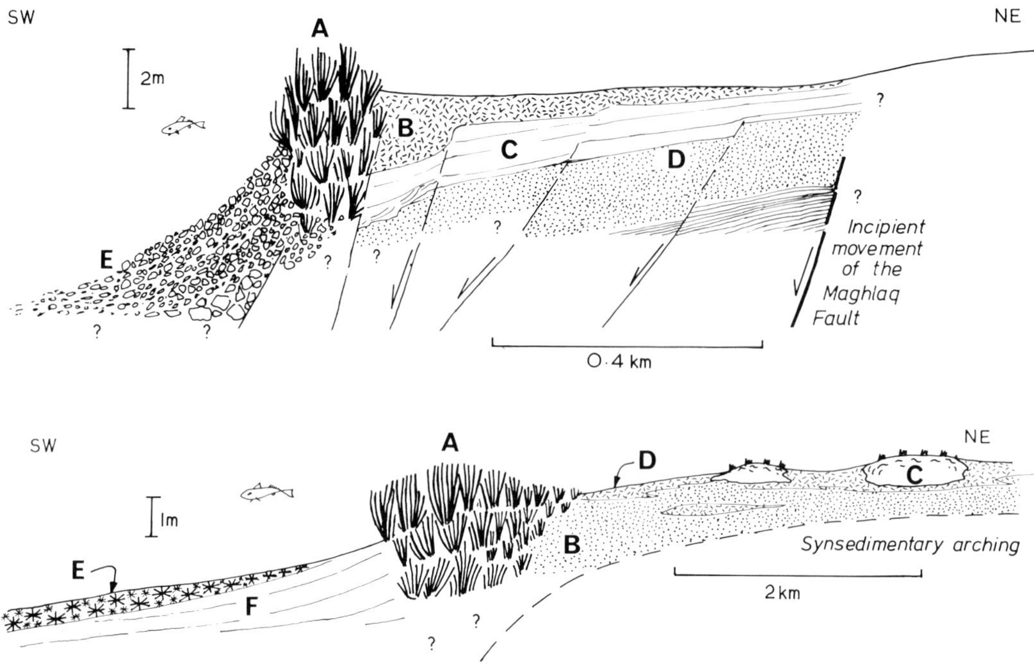


FIG. 8.—(A) Profile of the Late Tortonian-Early Messinian Ghar Lapsi reef complex of Malta: A = monospecific *Porites* stand (slender rod growth form); B = *Halimeda* - rich packstones of the back reef; C = cross stratified to planar shelf resediments involved in synsedimentary deformation; D = Basal Upper Coralline Limestone Formation and E = fore-reef talus composed of lithified cobbles of packstone, partly derived from the coral curtain. (B) Profile of the Late Tortonian-Early Messinian Lampedusa coral curtain: A = monospecific *Porites* coral curtain; B = back-reef wackestones; C = early cemented coralgal patch reefs; D = *Halimeda* -rich packstones; E = rhodolith packstone pavement and F = packstones and grainstones.

pipe') *Porites* coral growth forms frequently are the only forms present. The corals (now mainly *in situ* external moulds) are typically preserved in a lime mudstone or wackestone matrix. Other macrofaunas are rare.

The monogeneric coral fabric consist of unlinked frameworks without submarine isopachous fringe cements. Such interstitial micrite present is devoid of *in situ* fauna other than that which could be brought in postmortally. This is especially true of the *Halimeda* plates which are concentrated within the back reef areas and less so in the inter-coral micrites. All other macrofauna, including encrusters, are rare. There is also a virtual absence of borers, including clionid sponges, even in the coral heads. Some areas, however, have variable thicknesses of stromatolitic coatings around the coral rods

## CORAL GROWTH FORMS

For the purpose of this account only those build ups in which corals play a key role will be considered. Biostromes, and in particular the "Coralline Algal Biostrome" of the Maltese ramp, are considered elsewhere (this volume; Pedley, 1976; Bosence and Pedley, 1979, 1982).

Coral growth form study offers a fruitful source of data of value in the interpretation of environmental conditions. Many invertebrate genera typically are characterized by constancy in growth form, (e.g., bryozoans and brachiopods, see Tortonian Maltese ramp study in Pedley, 1976). The Mediterranean Miocene scleractinian corals, like their modern counterparts, however, (see Brackel, 1977; Geister, 1980; Frost, 1981), appear to have been capable of considerable growth form variation according to prevailing local conditions, (cf. Maltese ramp coralline algae in Bosence and Pedley, 1982.)

The distribution of growth forms associated with the commoner coral genera is presented in Figure 6. Seven principal growth forms are recognized, the majority of which occur within the central north Sicilian reef structures:

*A. Slender rods*- Those in excess of 1 m in length (10-30 mm diameter) occur as monogeneric *Porites* stands associated with the reefs at Ghar Lapsi, Lampedusa, and rarely in the Sicilian intra-channel reef curtains. Identical, but shorter rods are characteristic of the upper fore-reef talus slope in Sicilian channel-mouth reefs. They are absent elsewhere.

*B. Bushes*- These are a common multi-generic growth form both in patch reefs and channel reefs. Generally, these ramose colonies are under 0.2 m high and may be heavily encrusted with coralline algae. They are absent from the reef curtains.

*C. Bone-like or vermiform*- Only in sites associated with siliciclastic sedimentation do these *Porites* growth forms occur and always to the exclusion of other growth forms. Individuals tend to be up to 40-mm diameter and less than 0.5 m long. They are often enveloped in muddy sediment in which case the vermiform habit appears to be a growth response to toppling and recovery, or in the case of the bone-like growths is the result of inter-colonial competition for space. Favored sites appear to have been either deeper water locations or channel floors behind reef core areas. The growth form is absent from patch reefs and curtains.

*D. Dome*- This multi-generic growth form only appears in the largest reefs where about five genera dominate the shallowest water core zones (colony height about 1 m, diameter up to 1 m). Smaller domes occur in association with intra-basin coral patches. Domes are absent from typical shelf patch reefs and curtains.

*E. Undulose*- This is the characteristic *Porites* growth form in shelf patch reefs of Sicily, Malta and Lampedusa. Colonies may extend laterally for 0.15 m be 0.1 m high and have characteristically undulose tops. This growth form corresponds to the "Cavoli" head of Italian authors. They are developed as isolated colonies which are invariably bivalve drilled and rhodalgal encrusted.

*F. Sheets*- Thin (1 to 20 cm thick) sheets of *Porites* and *Tarbellastraea*, a few cm to several meters laterally) are widespread though uncommon in all bioherms except the reef-curtains, and are occasionally present in the "Coralline Algal Biostrome" of Malta. They are usually associated with local seafloor highs, talus slopes and patch reef crests.

*G. Small cushions*- Small (less than 10 cm) *Favites* cushions are only seen encrusting cobbles and pebbles in the central north Sicilian delta top channel deposits. These and the associated thin, crustose coralline algal laminae invariably provide the earliest evidence of marine incursion into these areas.

## ENVIRONMENTAL CONTROLS

### *Foreland Stability*

The development of both south east Sicilian and Maltese Tortonian ramps was characterized by environmental stability with only infrequent small scale tectonism (e.g., movement of the Melilli-Siracusa Line of Pedley, 1981 in Sicily and subsiding north-south corridor of Pedley, 1987a in Malta).

Salinities remained normal, deduced from the high faunal diversity, as did oceanic circulation (see water flow vectors in Fig. 1). Closure of the Indian Ocean connection by Burdigalian (Early Miocene) times (Adams et al., 1983) prevented the previously prevailing westwards equatorial Tethyan flow. This effectively removed tides from the late Miocene Paleomediterranean.

### *Eustasy*

All reef areas were subjected to small scale eustatic events but generally were encouraged to develop during highstands (see Pedley, this volume).

These events were the inevitable consequence of several restricted water flow events across the Betic and Rif straits (Atlantic connection) during Late Tortonian and Early Messinian time prior to the main Mediterranean desiccation event (Ruggieri and Sprovieri, 1976; Pedley, 1983). In particular, further water losses by evaporation in the eastern basin were now compensated by a steady eastward flow from the Moroccan and Spanish connections with the Atlantic (see Fig. 1 for Pelagian flow directions deduced from reef and bed facing directions in Pedley, 1976, 1979, 1981, 1983, 1987a, Grasso and Pedley, 1985, 1989).

Further plate closure and continuing desiccation, possibly tied in with minor eustatic falls in world sea level (Vail et al., 1977) interrupted inflow from the Atlantic and was responsible for the initiation of drawdown events which culminated in the Messinian salinity crisis.

The Channel mouth reefs of north central Sicily, suffered greatly from being perched precariously on steep delta front locations in tectonically active situations. Here, they were easily terminated by minimal offlap events. In contrast, relatively small sea-level rises effected rapid establishment of clear water conditions over wide areas previously choked by siliciclastic delta muds and silts of the Terravecchia Formation. Although short term in duration they permitted quasi-climax reef community development in drowned delta front locations (e.g., Landro) and pioneer communities in more proximal channels (e.g., Manca di Corvo channel, A in Fig. 2).

The most diverse biotal associations occur in the channel-mouth reef locations and the previously described patch reefs (shallow ramp setting). Coral growth form diversity is greater in the former except in up-channel locations where high siliciclastic input rates operated periodically. Here, vermiform *(Porites)* growth forms alone thrived.

### *Tectonism and Sedimentation*

A total contrast exists between the biotal diversity of the previously described reefs and those occurring at basin facing breaks of slope (i.e., coral curtains). This is illustrated in Table 1. Particularly striking is the almost total absence of faunas other than *Porites* in the coral curtains. Those elements that do occur are either nektonic (e.g., *Chlamys*) or epifaunal encrusters or scavengers. Significantly, there is a total absence of microborers and burrowers in these coral curtain biocoenoses.

The faunally impoverished Pelagian examples demonstrably are developed at the same time as the high diversity ramp associated patch reefs of Malta and southeastern Sicily and with the patch reefs of eastern Lampedusa (see Fig. 8B). Clearly, the coral curtains developed under normal salinity marine conditions where additional factors were in operation.

The differences cannot simply be explained as reflecting salinity variations concomitant with Late Miocene Mediterranean desiccation as postulated for similar low diversity reefs in Spain (Esteban, 1980; Dabrio et al., 1981).

In the case of the Ghar Lapsi and Lampedusa curtains, local syndepositional tectonism has played an important role. Both synsedimentary arching (Lampedusa) and small scale block faulting (Ghar Lapsi) were responsible for the creation and maintenance of a marked step to the shelf margin. Certainly in the Maltese case large allochthonous blocks of coral curtain several meters long lying within the forereef talus suggest contemporaneous earthquake development. Excessively high carbonate sedimentation rates appear to have prevailed as a result of soft sediment stripping from the backreef zones. This detritus was effectively held at the shelf edge by the coral curtains (thus the backreef *Halimeda* sands of the Ghar Lapsi curtain are never recorded in the forereef deposits, Pedley, 1987b).

Clearly there is some interrelationship between reef morphology, faunal diversity and tectonically active regions. Additionally, Frost (1977, 1981) and Reinhold (1995) suggests that high sedimentation rates may be a further factor as they appear to be responsible for low diversity *Porites* dominated pioneering communities in Recent, Oligocene and Miocene sequences elsewhere. The interrelationships are complex but are considered further in Pedley (this volume).

## DISCUSSION

### *The Paleomediterranean Late-Miocene Coral Niche*

From the foregoing account, it may be seen that Late Miocene coral diversity of the Central Mediterranean was much lower than in the Paleomediterranean Oligocene (cf. Geister and Ungaro, 1977; Frost, 1977). The genus *Porites* was the only coral sufficiently adaptable to cope with all scenarios. Today, it is also a frequent element in low diversity pioneer reef communities (Frost, 1977, 1981) and in deeper substrates not subject to wave surges. During Miocene deposition monogeneric *Porites* thrived under conditions of excessively high fine-siliciclastic sedimentation (vermiform and bone-like growth forms). Under conditions of high carbonate resedimentation the slender rod growth form developed. Furthermore, *Porites* also appears equally frequently in the quasi-climax communities of the channel-mouth reefs (low branching or bushy growth form) and on the ramp patch reefs (undulose and sheet growth forms).

Clearly, *Porites* was able to colonize and stabilize shifting substrates. This was to a large extent brought about by the perfection of a mechanism for shedding settling sediment (tentacular manipulation of Hubbard and Pocock, 1972) which allows living examples of the genus to inhabit inhospitable locations.

The genus *Tarbellastraea* is the most accommodating of the Faviid corals which otherwise are restricted to the more diverse quasi-climax communities of the channel-mouth reefs. It is the principal associate of *Porites* in the ramp patch reefs of Malta and Sicily. Today, the Faviids maintain their position under severe competition by digestive dominance interactions (Frost and Langenheim, 1974). This type of aggression displaces the pioneer community in modern reefs and must also have been equally as successful in the case of the Miocene examples.

*Favites* occurs as a pioneer encruster on cobbles in the basal deposits of the north central Sicily channels and in the main channel-mouth reefs.

### *Late Miocene Scenario*

The rich coral associations of Tethyan Paleogene strata are, by Miocene times, represented only by the most successful pioneer and quasi-climax genera. Although probably not able to cope with elevated salinities they were the product of selection for ability to colonize rapidly. The ramose forms clearly could withstand the highest sedimentation rates though this resulted in lower diversity assemblages (cf. Roy and Smith, 1971). The disappearance of tidal influence during Miocene sedimentation removed the need for turbulence resistant framework architecture. Genera such as *Porites* were freed to experiment with fragile growth forms which could then selectively evolve to cope with such aspects as high sedimentation rates and competition for space from algae and stromatolitic biofilms. Coral frameworks became more open generally and internal mud-grade sediments began to dominate patch reef structures in the more stable areas.

These conclusions, while drawn from the Pelagian region only could have basinwide applicability within the Mediterranean Miocene realm.

## CONCLUSIONS

Three recurring Miocene palaeoenvironmental scenarios are proposed (see Fig. 9) in order to encompass the biotal variability observed in central Mediterranean reef associated successions. These models may have wider applications within the Tertiary strata of the Tethyan belt:

### *The Ramp Model*

The facies sequence in both the south east Sicilian and Maltese ramps is very similar. The Large Foraminifera zone finds a precise analogue in older Pelagian ramp settings, it is *Lepidocyclina* in Chattian (Late Oligocene) sediments of Malta (Buxton and Pedley, 1989) and *Nummulites* in the Eocene strata of Tunisia and Sicily.

Clearly, it is possible to recognize distinct ramp facies associations in Palaeomediterranean Cenozoic ramps (basin to shore) as follows: planktonic wackestones zone -> large Foraminifera zone -> rhodolitic algal packstones zone -> coralgal patch reef zone. Furthermore, it is suggested that the Miocene Pelagian ramps are typical of Tethyan Cenozoic ramps, though few have been described.

In current-facing (windward) ramps, these zones are followed more interiorly by a barrier zone (often oolitic) and a partly restricted lagoonal zone of thinly bedded lime mudstones passing laterally into peritidal carbonates. In leeward ramps, these internal zones are starved of water circulation, thus biostrome/bioherm development, sediment transport and ramp build up is hindered and shoalwater deposits will be thin or absent.

The associated coral bioherms always appear to be coralgal patch reefs and invariably are characterized by high biotal diversity. These are to be located between a large foraminifer zone (basinward) and a partially restricted lagoon (with or without sand shoals) in more interior locations. Where strong bottom currents prevail a further coralline algal biostrome facies can develop down ramp of the large foraminifer belt (see Figs. 4, 5) in water depths of 50-70 m.

## CENTRAL MEDITERRANEAN REEF ENVIRONMENTS

*Stable*
RAMP

| | *PATCH-REEF ZONE* → | *RHODALGAL SLOPE* → | *LARGE FORAM. ZONE* → | *C.ALGAL BIOSTROME* |
|---|---|---|---|---|
| Framework | Coralgal | | | Algal |
| Corals | 2 Genera | | | —— |
| Growth Forms | Low Domes, Sheets | | | —— |
| Other Biota | Diverse | | | Diverse |
| Bioerosion | High | | | High |

*Tectonism & Nutrients*
SHELF MARGIN

*Eustasy & Sedimentation*
DROWNED DELTA TOP AND BASIN

| | *CORAL CURTAIN* | *INTRA-CHANNEL CURTAIN* → | *CHANNEL MOUTH REEF* → | *PINNACLE REEF* |
|---|---|---|---|---|
| Substrate | Lime Sand | Quartz Sand | Quartz Sand, Mud | Quartz Sand, Mud |
| Framework | Coral | Coral | Coral (algal) | Coral (algal) |
| Corals | *Porites* only | *Porites* only | 3-4 genera | 3-4 genera |
| Growth Forms | Rods | Rods | Mixed | Mixed |
| Other Biota | Low diversity | Low diversity | Low diversity | Low diversity |
| Bioerosion | Low | Low | Low | Low |

FIG. 9.—Summary of Late Miocene coral reef environments.

The southeastern Sicilian ramp in particular is proposed as a current facing (windward) ramp model to illustrate the most typical aspects of Paleomediterranean ramps. This ramp model is considered to have widespread application in Cenozoic carbonates throughout the entire Tethyan carbonate belt (Buxton and Pedley, 1989).

### *The Delta Top /Channel-Mouth Model*

Delta-channel reefs occur in other Mediterranean Miocene areas from Spain to Turkey but few accounts are published. These reefs can be expected to develop during eustatic rises in sea level when partly emergent deltas or paralic belts become inundated and brief periods of clear water occur. Subsequent sea-level fall may be responsible for their demise, but parts will be preserved in deeper channel sites. Basin cut-offs prone to desiccation are obvious candidates for the development of this reef association although glacio-eustatic events on a global scale could conceivably create the necessary rapid sea-level fluctuations.

### *Shelf Break (Coral Curtain) Model*

Coral curtains are examples of reef growth under tideless conditions which are, nevertheless, highly stressed on account of static, shallow water conditions with a potential for nutrient and toxin imbalance. Under potentially catastrophic conditions, most organisms are excluded. The exceptions are pioneer communities of opportunistic genera with high rates of standing crop turnover (e.g. *Halimeda*) or aberrant, rapidly growing, tall morphotypes (e.g., *Porites*).

To date, the monogeneric *Porites* rod coral curtain reef type is only recorded from the Mediterranean Miocene. Perhaps it is the product of a unique combination of circumstances involving substrate instability, random, rapid variations in sediment supply, and possibly also nutrient imbalances (also see Pedley, this volume). Only within the Alpine foldbelts do all these variables operate. Consequently, it is not surprising that monogeneric stands of coral rods find their greatest development here.

ACKNOWLEDGMENTS

Many of the ideas expressed in this article have benefited from lengthy discussions with Italian workers. In particular I am indebted to Professors Mario Grasso and Fabio Lentini. Mateu Esteban has also provided stimulating discussions on many occasions for which I am also most grateful. Assistance with fieldwork expenses has mainly been from The Royal Society of London and NERC grant GR3/8610.

REFERENCES

Adams, A. E., Gentry, A. W., and Wybrow, P. J., 1983, Dating the terminal Tethyan event, *in* Meulenkamp, J. E., ed., Reconstruction of marine palaeoenvironments by means of carbon isotope analysis: Utrecht Micropaleontological Bulletins, v. 30, p. 273-297.

Ahr, W. M., 1973, The carbonate ramp: an alternative to the shelf model: Transactions of the Gulf Coast Association Geological Society, v. 23, p. 221-225.

Aruta, L. and Buccheri, G., 1971, Il Miocene preevaporitico in facies carbonatico detritica dei dintorni di Baucina. Venmtimiglia di Sicilia, Calatafimi (Sicilia): Rivista Mineraria Siciliana, v. 188, p.130-132.

Blow, W. H., 1969, Late Middle Eocene to Recent planktonic foraminiferal biostratigraphy: Geneva, Proceedings of the First International Conference on Planktonic Microfossils,v. 1, p. 199-422.

Bosence, D. W. J. and Pedley, H. M., 1979, Palaeoecology of a Miocene coralline algal bioherm: Bulletin Recherches Exploratie Elf Aquitaine, v. 3, p. 463-470.

Bosence, D. W. J. and Pedley, H. M., 1982, Sedimentology and palaeoecology of a Miocene coralline algal biostrome from the Maltese Islands: Paleogeography, Palaeoclimatology, Palaeoecology, v. 38, p. 8-43.

Brackel, W. M., 1977, Corallite variation in *Porites* and the species problem in corals: Miami, Proceedings of the 3rd International Coral Reef Symposium, v. 1, p. 457-462.

Buxton, M. and Pedley, H. M., 1989, A standardized model for Tethyan Tertiary carbonate ramps: Journal of the Geological Society of London, v. 46, p. 746-748.

Catalano, R., 1979, Scogliere ed evaporiti Messiniane in Sicilia. Modelli genetici ed implicazioni strutturali: Palermo, Lavori dell 'Istituto geologica dell' Universita, v. 18, p. 3.-21.

Catalano, R. and Esteban, M., 1978, Messinian reefs of western and central Sicily: Rome, Abstracts, Messinian Seminar, v. 4.

Catalano, R., Renda, P., and Slaczka, A., 1976, Redeposited gypsum in the evaporitic sequence of the Ciminna Basin (Sicily), *in* Catalano, R., Ruggieri, G., and Sprovieri, R., eds., Messinian Evaporites in the Mediterranean: Palermo, Memorie Societa Geologica Italiana, v. 16, p. 83-93.

Carannante, G., Simone, L., and Barbera, C., 1981, "Calcari briozoi e litotamni" of southern Apennines. Miocene analogs of Recent Mediterranean rhodolitic sediments: Bologna, International Association of Sedimentologists, 2nd European Meeting, p. 17-20.

Chafetz, H. S., 1986, Marine peloids: a product of bacterially induced precipitation of calcite: Journal of Sedimentary Petrology, v. 56, p. 812-817.

Chevalier, J. P., 1961, Recherches sur les *Madreporaires* et les formations recifals miocenes de la Mediterranee occidentale: Memoires Societe Geologique France, v. 93, 562 p.

Dabrio, C. J., Esteban, M., and Martin, J. M., 1981, The coral reef of Nijar, Messinian (uppermost Miocene), Almera province, SE Spain: Journal of Sedimentary Petrology, v. 51, p. 521-539.

Di Stefano, E. and Catalano, R., 1976, Biostratigraphy, palaeoecology and tectonosedmentary evolution of the preevaporitic deposits of the Ciminna Basin (Sicily): Memorie della Societa Geologica Italia., v. 16, p.95-110.

Esteban, M., 1979/1980, Significance of the Upper Miocene coral reefs of the western Mediterranean: Palaeogeography, Palaeoclimatology, Palaeoecology, v. 29, p. 169-188.

Esteban, M., Catalano, R., and di Stefano, E., 1982,Scogliere Messiniane a *Porites* nella Sicilia sud-orientale: Rendiconti della Societa Geologica Italiana, v. 5, p. 61-64.

Fois, E., 1990, Stratigraphy and palaeogeography of the Capo Milazzo area (NE Sicily, Italy): clues to the evolution of the southern margin of the Tyrrhenian Basin during the Neogene: Palaeogeography, Palaeoclimatology, Palaeoecology, v. 78, p. 87-108.

Frost, S. H., 1977, Ecologic controls of Caribbean andMediterranean Oligocene reef coral communities: Miami, Proceedings of the 3rd International Coral Reef Symposium, Rosentiel School of Marine and Atmospheric Science, University of Florida, v. 2, p. 367-373.

Frost, S. H., 1981, Oligocene reef coral biofacies of the Vicentin, northeast Italy, *in* Toomey, D. F., ed., European Fossil Reef Models: Tulsa, Society of Economic Paleontologists and Mineralogists Special Publication 30, p. 483-539.

Frost, S. H. and Langenheim, R. L., 1974, Cenozoic Reef Biofacies: Tertiary Larger Foraminifera and Scleractinian corals from Chiapas, Mexico: DeKalb, Northern IllinoisUniversity Press, 338 p.

Geister, J., 1980, Morphologie et distribution des coraux dans les recifs actuels de la Mer des Caraibes: Annali dell' Universita di Ferrara, n. s. sez IX Sc., Geologiche e Paleontologiche, v. 6, p. 15-28.

Geister, J. and Ungaro, S., 1977, The Oligocene coral formations of the Colli Berici (Vicenza, Northern Italy): Ecologae Geologicae Helveticae, v. 70, p. 811-823.

Ginsburg, R. L. and Schroeder, J. H., 1973, Growth and submarine fossilization of algal cup reefs, Bermuda: Sedimentology, v. 20, p. 573-614.

Grasso, M., Lentini, F., and Pedley, H. M., 1982, Late Tortonia-Lower Messinian (Miocene) palaeogeography of S.E. Sicily: information from two new formations of the Sortino Group: Sedimentary Geology, v. 32, p. 279-300.

Grasso, M. and Pedley, H. M., 1985, The Pelagian Islands: A new geological interpretation from sedimentological and tectonic studies and its bearing on the evolution of the Central Mediterranean Sea (Pelagain Block): Geologica Romana, v. 24, p. 13-34.

Grasso, M. and Pedley, H. M., 1988, The sedimentology and development of the Terravecchia Formation carbonates (Upper Miocene) in North Central Sicily: possible eustatic influences on facies development: Sedimentary Geology, v. 57, p. 131-149.

Grasso, M. and Pedley, H. M., 1989, Palaeoenvironment of the Upper Miocene coral build-ups along the northern margins of the "Caltanissetta Basin" (Central Sicily), *in* Di Geronemo, I., ed., Proceedings of the 3rd Symposium on Ecology and Palaeontology of Benthonic Communities: Societa Paleontologica Italiana, p. 373-389.

Hottinger, L., 1977, Distribution of the larger Peneroplidae, Borelis and Nummulitidae in the Gulf of Elat, Red Sea, *in* Reiss, Z., Leutenegger, S., Hottinger, L., Feront, W. J. J., Meulenkamp, J. E., Thomas, E, Hansen, H. J., Buchardt, B., Larsen, A. R., and Drooger, C. W. , eds., Depth Relations of Recent Foraminifera in the Gulf of Aqaba-Elat: Utrecht, International Geological Correlation Program Project 1, Utrecht Micropaleontological Bulletins 15, 244 p.

Hubbard, J. A. E. B. and Pocock, Y. P., 1972, Sediment rejection by recent scleractinian corals: a key to palaeoenvironmental reconstruction: Geologische Rundschau, v. 161, p. 598-626.

Lighty, R. G., 1985, Preservation of internal reef porosity and diagenetic sealing of submerged early Holocene barrier reef, southeast Florida shelf, *in* Schniedermann, N. and Harris, P. M., eds., Carbonate Cements: Tulsa, Society of Economic Paleontologists and Mineralogists Special Publication 36, p. 23-149.

MacIntyre, I. G., 1977, Distribution and submarine cements in a modern Caribbean fringing reef, Galeta Point, Panama: Journal Sedimentary Petrology, v. 47. p. 503-516.

MacIntyre, I. G., 1985, Submarine cements-the peloid question, *in* Schniedermann, N. and Harris, P. M., eds., Carbonate Cements: Tulsa, Society of Economic Paleontologists and Mineralogists Special Publication 36, p. 109-116.

Pedley, H. M., 1976, A palaeoecological study of the Terebratula-Aphelesia Bed based on bryozoan growth-form studies and brachopod distributions: Palaeoclimatology, Palaeogeography, Palaeoecology, v. 20, p. 209-234.

Pedley, H. M., 1978, A new lithostratigraphical and palaeoenvironmental interpretation for the coralline limestone formations of the Maltese

Islands: Institute of Geological Sciences, Overseas Geology and Mineral Resources, v. 54, p. 273-291.

Pedley, H. M., 1979, Miocene bioherms and associated structures in the Upper Coralline Limestone of the Maltese Islands: their lithification and palaeoenvironment: Sedimentology, v. 26, p. 577-591.

Pedley, H. M., 1981, Sedimentology and palaeoenvironment of the southeast Sicilian Tertiary platform carbonates: Sedimentary Geology, v. 28, p. 273-291.

Pedley, H. M., 1983, The petrology and palaeoenvironment of the Sortino Group (Miocene) of S.E. Sicily: evidence for periodic emergence: Journal of the Geological Society London, v. 140, p. 335-351.

Pedley, H. M., 1987a, Controls on Cenozoic sedimentation in the Maltese Islands: review and reinterpretation, *in* Lentini, F., ed., Proceedings Societa Geologica Italia-II Systema Avampiese-Avanfossa lungo la catena Appeninico Maghrebide: Rome, Memorie della Societa Geologica Italiana, v. 38, p. 81-94.

Pedley, H. M., 1987b, The Ghar Lapsi limestones: sedimentology of a Miocene intra-shelf graben: Centro, v.1, p. 1-14.

Pedley, H. M., 1992a, "Bioretexturing": early fabric modifications in outer ramp settings- a case study from the Oligo-Miocene of the Central Mediterranean, *in* Pedley, H. M., Harwood, G., and Sellwood, B., eds., Ramps and Reefs: Sedimentary Geology, v. 79, p. 173-188.

Pedley, H. M., 1992b, Freshwater (Phytoherm) reefs: the role of biofilms and their bearing on marin esedimentation, *in* Pedley, H. M., Harwood, G., and Sellwood, B., eds., Ramps and Reefs: Sedimentary Geology, v. 79, p. 255-274.

Pedley, H. M., 1993, Oil Exploration Directorate, Office of the Prime Minister, Valletta, Malta, Geological map of the Maltese Islands (sheet 1, Malta; sheet 2 Gozo and Comino): Keyworth, British Geological Survey.

Pedley, H. M., Cugno, G., and Grasso, M., 1992, Gravity slide and resedimentation processes in a Miocene carbonate ramp, Hyblean Plateau, southeastern Sicily, *in* Pedley, H. M., Harwood, G., and Sellwood, B., eds., Ramps and Reefs: Sedimentary Geology, v. 79, p. 189-202.

Pedley, H. M. and Grasso, M., 1994, Upper Miocene peri-Tyrrhenian reefs in the Calabrian Arc: sedimentological, tectonic and palaeogeographic implications. Geologie Mediterraneenne, v. 21, p. 123-136.

Pedley, H. M., La Manna, F., and Grasso, M., 1994, A new record of Upper Miocne reef carbonates from Santo Stefano di Camastra-Caronia area (Northern Sicily) and its regional significance: Bollitino Societa Geologia Italiana, v. 113, p. 435-444.

Peres, J. M. and Picard, J., 1964, Nouveau manuel de bionomie benthique de la Mer Mediterranee: Recueil Travaux Station Marine d'Endoume-Marseille Bulletin, v. 31, 137 p.

Read, J. F., 1982, Carbonate platforms of passive (extensional) continental margins: types, characteristics and evolution: Tectonophysics, v. 81, p. 195-212.

Reinhold, C., 1995, Guild structure and aggredation pattern of Messinian Porites patch reefs: ecological succession and external environmental control (San Miguelde Salinas Basin, SE Spain): Sedimentary Geology, v. 97, p. 157-175.

Riess, Z. and Hottinger, L., 1984, The Gulf of Agaba: ecological micropalaeontology: Ecological Studies, v. 50, 354 p.

Rigo, M. and Barbieri, R., 1959, Stratigrafia pratica applicata in Sicilia: Bollettino Servizio Geologico Italia, v. 88, p. 1-98.

Roy, K. J. and Smith, S. V., 1971, Sedimentation and coral reef development in turbid water: Fanning Lagoon: PacificScience, v. 25, p. 234-248.

Ruggieri, G. and Sprovieri, R., 1976, The "dessication Theory" and its evidence in Italy and Sicily, *in* Catalano, R., Ruggieri, G., and Sprovieri, R., eds., Messinian Evaporites in the Mediterranean: Palermo, Memorie della Societa Geologica Italia 16, p. 165-169.

Ruggieri, G. and Torre, G., 1982, Il ciclo saheliano dintorni di Castellana Sicula (Palermo): Atti della Societa Italiana di Scienze Naturali e del Museo Civile di Storia Naturale di Milano, v. 123, p.425-440.

Ruggieri, G. and Torre G., 1984, Il Miocene superiore di Cozzo Terravecchia (Sicillia Centrale): Giornale Geologia, v. 46, p. 33-43.

Schmidt di Frieberg, P., 1962, Introduction a la geologie petroliere de la Sicile: Revista Institut Francais du Petrole, v. 17, p. 635-668.

Segre, A. G., 1960, Geologia, *in* Zavatiari, E , ed., Biogeografia delle Isole Palagie: Rendiconti Accademia Nazionale de XL, serie 4, v. 11, p. 115-162

Tsien, H. H., 1985, Algal-bacterial origin of micrite inmud mounds, *in* Toomey, D. F. and Nitecki, M. H., eds., Paleoalgology Contemporary Research and Applications: NewYork, Springer-Verlag, p. 290-296.

Vail, P. R., Mitchum, R. M. Jr., and Thompson, S., Iii., 1977, Seismic stratigraphy and global changes of sea level, part 3: Relative changes of sea level from coastal onlap, in Payton, C. E., ed., Seismic Stratigraphy — Applications to Hydrocarbon Exploration: Tulsa, American Association of Petroleum Geologists Memoir 26, p. 83-97.

# RHODOLITH FACIES IN THE CENTRAL-SOUTHERN APENNINES MOUNTAINS, ITALY

GABRIELE CARANNANTE
*Dipartimento di Scienze della Terra, University of Napoli, Largo San Marcellino 10, 80138 Napoli, Italy*
AND
LUCIA SIMONE
*Dipartimento di Scienze della Terra, University of Cagliari, Via Trentino 51, 09127 Cagliari, Italy*

Abstract: Miocene sequences composed of bioclastic limestones rich in red algal concretions (rhodoliths) and bryozoans (Bryozoan and Lithothamnium Limestones- BLL) crop out in the central-southern Apennines (Italy). In general, these limestones document the evolution of an open carbonate shelf, the upper portion of which shows evidence of a drowning event. The facies consist of rhodalgal-type grain associations which typically are indicative of carbonate platforms developing in temperate seas or in subtropical to tropical areas that are characterized by anomalous (e.g., cooler, eutrophic, upwelling) water conditions. Most of the facies constituents were derived from organisms adapted to low-light intensities (sciaphile assemblages) in the cooler deep euphotic zone, whereas components derived from organisms living in the photic zone (photophile assemblages) and warmer waters are subordinate. These facies appear to be analogous to Modern bioclastic sediments that cover large sectors of the middle to outer shelf in the Mediterranean Sea, described by Pérès and Picard (1964) as *Détritique Côtier* biocenosis sediments. Within the *Détritique Côtier* deposits, the *Faciès à Pralines* (cf. Pérès and Picard, 1964) which is characterized by red algae concretions, match well with the rhodolith-rich Apennine Miocene facies. BLL skeletal-rich deposits were locally (and partially) stabilized to form complex agglomerates. These are analogous to present-day concretionary hard grounds dominated by encrusting red algae (*Coralligène* bottoms in Pérès and Picard, 1964).

The facies identified in this study were likely formed as a result of the upwelling of cold, nutrient-rich water on the outermost-shelf sectors which produced anomalous water conditions on the Apennine Miocene subtropical shelves. As a consequence, rhodalgal-type associations were favored and a "temperate-type" platform developed, where growth potential was relatively low. A significant Burdigalian transgressive event resulted in increased water depths and exposure of the uppermost bioclastic level of the BLL to the sediment-water interface thereby creating a sheet of relict sediment. A further increase in water depth and enlargement or lateral shifting of the anoxic levels connected with an upwelling maximum resulted in exposure of the shelves to nutrient-rich and oxygen-poor waters and phosphatization of the sediments. The drowning event culminated in deposition of planktonic-rich sediments that were mixed with the relict neritic bioclasts resulting in a complex basal depositional interval (palimpsest interval *sensu* Swift et al., 1971) that pass upwards into hemipelagic globigerinid-rich wackestones).

## INTRODUCTION

### *Location and Sratigraphic Sequence*

Lower Miocene carbonate rocks, characteristically rich in algal concretions (rhodoliths), crop out in the central and southern Apennines (Italy, Fig. 1) with discontinuous but sometimes extensive exposures with thicknesses that generally do not exceed 60 m. These are known in Italian literature as *Calcari a Briozoi e Litotamni* (Bryozoan and Lithothamnium Limestones-BLL). They cover a Mesozoic-Paleogene substrate which passes upward into hemipelagic and pelagic deposits rich in planktonic foraminifera. The transitional interval is commonly marked by abundant phosphatic and glauconitic grains. A clear drowned sequence has been recognized (Carannante, 1982b; Simone and Carannante, 1985, 1988). The BLL are rhodolith rudstones-floatstones in a matrix of bioclastic grainstones and have been reported as organogenic reefs (Cocco, 1971). Sporadically, algal bindstones are present with limited areal extent (up to several square meters wide) and thickness (a few decimeters thick).

The BLL are essentially composed of encrusting coralline algae, bryozoans, mollusks and benthic and encrusting foraminifera. They almost completely lack non-skeletal grains (ooids, lumps), micrite, green algae and hermatypic corals. Due to the dominance of encrusting coralline algae (rhodophyta), they may be defined as rhodalgal deposits (*sensu* Carannante et al., 1988a).

In some areas (M. Soprano, Fig. 1), Lower Miocene limestones crop out with characteristics (on the basis of grain association and evolutionary trend) that may be referred to as BLL despite scarce abundances of rhodoliths and coralline algae. Because of the dominance of mollusks, echinoids and benthic forams, these bioclastic limestones may be defined as molechfor deposits (*sensu* Carannante et al., 1988a) that rapidly pass upward into resedimented Upper Miocene mixed arenites.

### *Geologic Age*

Dating of the BLL is not easy. When miogypsinids are present, they allow good stratigraphic dating (Barbera et al., 1980; Schiavinotto, 1985). If miogypsinids are absent, the mollusks help in dating these limestones (Barbera, 1979; Barbera et al., 1978). The basal sediments of the BLL are Aquitanian age in the more southern areas (e.g., M. Soprano) where *Miogypsina socini* and *Miogypsina globulina* characterize the basal ostreid banks (Carannante et al., 1988b). They become younger to the north (e.g., early Burdigalian at M. Camposauro) where *Miogypsina intermedia* are found in pelecypod banks with *Flabellipecten burdigaliensis* Lk (Barbera et al., 1980; Schiavinotto, 1985). In the Matese Group, the *Miogypsina* are only locally present; however, the presence of *Pecten pseudobeudanti* Dep. Rom. may support a Burdigalian age for the basal interval. The age of the base of the overlying

Models for Carbonate Stratigraphy from Miocene Reef Complexes of Mediterranean Regions,
SEPM Concepts in Sedimentology and Paleontology #5, Copyright © 1996,
SEPM (Society for Sedimentary Geology), ISBN 1-56576-033-6, p. 261-275.

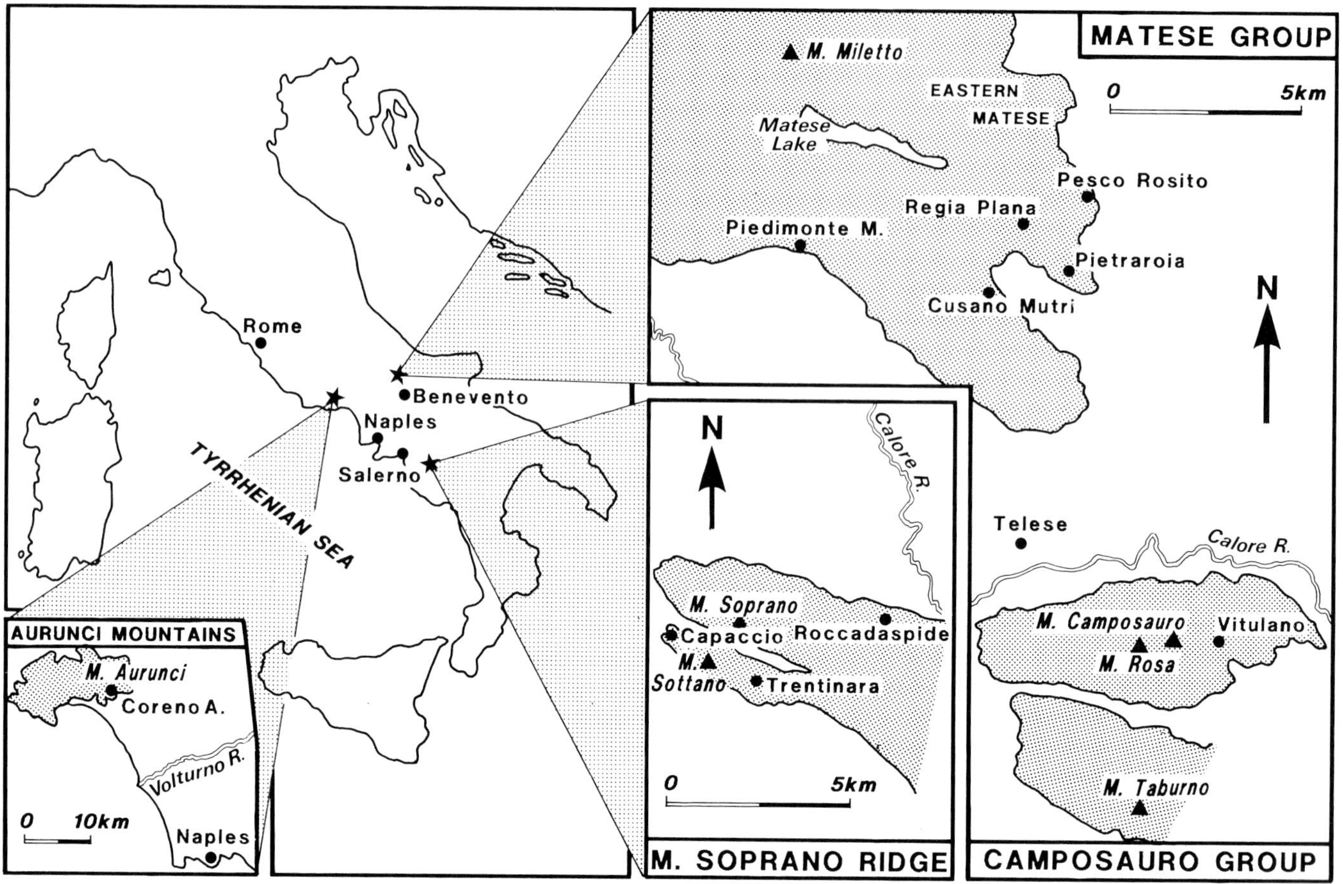


FIG. 1.—Location map of the studied outcrops mentioned in the text.

hemipelagic/pelagic deposits also varies from Burdigalian age in the southern area to Messinian age in the northern area. Diachronism of the Miocene sequences suggests a progressive drowning of the Cenozoic carbonate platform towards the north-northeast.

### *Geological Framework*

In the literature, the central-southern Apennines are considered to be part of a thrusted range, the complex internal geometry of which has been progressively modified since the Late Tertiary. The studied outcrops are part of the Structural-Stratigraphic Units (cf. D'Argenio and Scandone, 1969) resulting from the deformation of a paleogeographic domain characterized by carbonate sedimentation from Triassic to lower Miocene age. The sedimentary model commonly accepted for Mesozoic limestones is that of a carbonate platform with scarps of variable inclination and height. Following an important stratigraphic break (unconformity), the lower Miocene limestones were deposited on the Mesozoic-Paleogene substrate and then subsequently deformed during late Miocene tectogenetic events. The BLL sediments may be considered as the last neritic lithofacies deposited before the drowning of increasingly larger sectors of the Miocene foreland which is related to migration of the foredeep at the front of the Apennine nappes (Carannante et al., 1987). The lower Miocene limestones were subsequently dissected and overthrusted and are now part of the Apennine chain.

## LITHOFACIES ASSOCIATED WITH RHODOLITH DEPOSITS

### *Measured Sections*

We selected detailed sections (Figs. 1, 2A-G) that show different characteristics and/or evolution of these organogenic limestones: Coreno Ausonio (Aurunci Mountains), Pietraroia, Cusano Mutri and Pesco Rosito (Matese Group), M. Rosa (Camposauro Group) and Capaccio and Roccadaspide (M. Soprano Ridge).

*Aurunci Mountains.—*

In the Aurunci Mountains (Coreno Ausonio, Fig. 1), the BLL crop out with a maximum thickness of 60 m above a Paleocene calcareous substrate (Fig. 2A). BLL consist of rhodolith floatstones and rudstones in bioclastic grainstones and/or packstones (Fig. 3A). Locally, silty mud filled the inter- and intragranular cavities of the bioclastic sediment in which the rhodoliths are characterized by a predominance of bryozoans (*Holoporella sp.*). Algal crusts, decimeters thick and up to

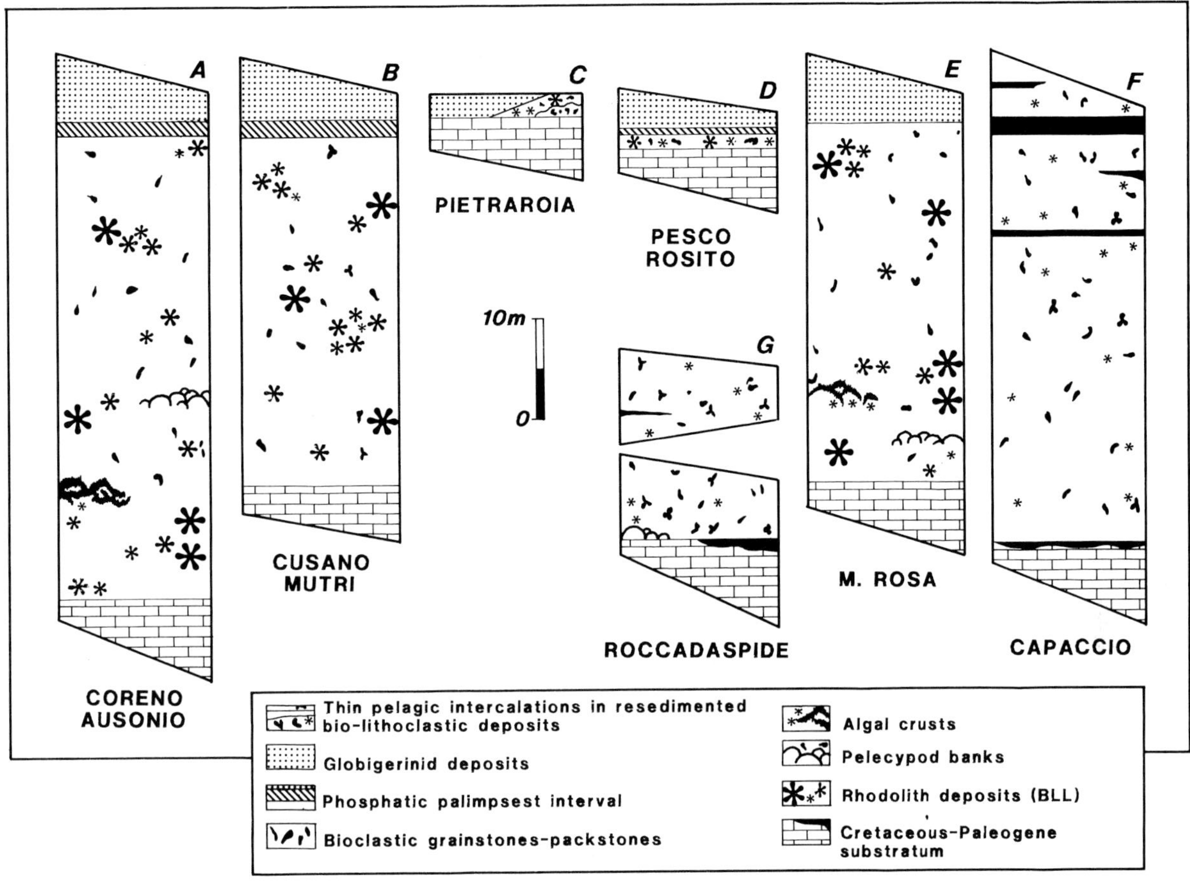


FIG. 2.—Stratigraphic sections of the drowned Miocene sequences in the study area.

several square meters wide, are present. The binding and encrusting by coralline algae formed a hard bottom stabilizing a previously loose sediment. Thick coralline algae-dominated crusts cover rhodoliths, pelecypods and other skeletal fragments (Fig. 3B).

The sharp transition with the overlying hemipelagic sediments is characterized by a complex depositional interval (1 m thick). Here, planktonic-rich matrix contains deeply bioeroded and reddened benthonic bioclasts, abundant phosphatic grains and rare glauconitic grains. This complex interval was derived from the mixing of neritic bioclasts that remained exposed for long periods to the water/sediment interface (relict sediment), with the subsequently deposited globigerinid ooze (palimpsest interval *sensu* Swift et al., 1971).

*Matese Group.—*

In the Matese area (Cusano Mutri, Pesco Rosito, Pietraroia, Fig. 1), the BLL crop out with thicknesses ranging from a few cm to 40 m (Figs. 2B, C, D, 4A, B). Locally, they may be absent in some channeled areas (e.g., Pietraroia) where hemipelagic deposits directly overly the Cretaceous substrate covered by ferriferous crusts. In those areas, the BLL only infill some pelecypod borings in Cretaceous units (Fig. 4A). The BLL are essentially rhodolith floatstones and rudstones in a matrix of bioclastic grainstones. The BLL grades upwards through a transitional palimpsest interval, very rich in phosphatic grains (Figs. 5, 6) and up to 150 cm thick (e.g., Pesco Rosito) into hemipelagic sediments.

*Camposauro Group.—*

In the Camposauro Group (Monte Rosa, Fig. 1), the BLL crop out with a thickness of 40 m covering a rocky lower Cretaceous substrate (Fig. 2E). The BLL are rhodolith rudstones and floatstones in a bioclastic grainstone matrix. Well-developed pelecypod banks (Fig. 4C), several decimeters thick and several meters wide, characterize the basal interval. Small algal crusts, up to 15 cm thick, are also present. The BLL passes upwards through a palimpsest interval, consisting of scattered rhodoliths and/or benthonic grains in a matrix of planktonic packstones, into pelagic sediments.

FIG. 3.—Lower Miocene rhodolith deposits of the "Bryozoan and Lithothamnium Limestones", central-southern Apennines, Italy: (A) Rhodolith rudstones in a matrix of bioclastic grainstones from Coreno Ausonio (Aurunci Mountains); (B) Coralline algae bindstones associated with bioclastic packstones from Coreno Ausonio (Aurunci Mountains).

*M. Soprano Ridge.—*

The M. Soprano area (Capaccio and Roccadaspide, Fig. 1) contains exposures of lower Miocene glauconitic limestones that range in thickness from a few m to 50 m ("Formazione di Roccadaspide," Selli, 1957). These glauconitic limestones are biolithoclastic grainstones with compositional constituents similar to that of previously described BLL. The limestones of the "Formazione di Roccadaspide" lie on a Paleogene substrate (Figs. 2F, G, 7), locally intensely bioeroded, and on discontinuous lenses of red residual clay (up to 10 m thick) associated with lenses of dark gray clay rich in *Cerithium* and freshwater ostracods. Rhizocretions and *Microcodium* are common on the top of the Paleogene substrate associated with the continental deposit. On this substrate, well-developed ostreid banks reach up to 50 cm in thickness (Fig. 4D). The beds of the "Roccadaspide Formation" limestones are massive and several decimeters thick, becoming thinner upwards into the sequence that grades basinward into beds characterized by numerous intercalations of mixed turbidites. The "Roccadaspide Formation" limestones are devoid of glauconite grains in the lowermost intervals. The glauconite grains become abundant in the upper beds where reworked *Miogypsina globulina-intermedia* occur concomitant with the increase of coralline algae and bryozoa fragments.

## *Depositional Facies*

*Substratum facies.—*

The BLL sediments overlie the Mesozoic-Paleogene substrate with a sharp contact commonly marked by stylolitic surfaces. In many locations, lithophagous organisms have intensely bored the substrate (Figs. 4A, D, 8A). The substrate surface shows, in addition to well preserved pelecypod bores (*Parapholas* and *Pholadidea*, Galdieri, 1913), small borings and traces of polycheta and sponges. The dimensions and morphology of the preserved bore-holes suggest they are part of previously larger perforations. Fragments of the substrate, produced by intensive bioerosion, are common in the basal part of the overlying BLL (Fig. 4B). Thus, the substrate limestones acted as a rocky bottom repeatedly colonized by boring and sessile organisms before the deposition of the bioclastic grainstones.

*Rhodolith facies.—*

BLL are massive and crop out with thicknesses ranging from a few cm to 60 m. They show a rather constant biogenic composition and may be considered as a lithofacies association illustrated by Figures 8 (A, C), 9 (A-D) and 10 (A-D). The lithology commonly shows sutured grain to grain contacts. Micrite is not present. Only locally (Aurunci Mountains), a significant silty mud component may be observed.

Normally, coralline algae (*Sporolithon*, *Lithothamnium*, *Lithophyllum*, *Mesophyllum*), associated with encrusting foraminifera and membraniporiform bryozoa, form rhodoliths ranging from a few cm to 20 cm in diameter, with an average size of 8 cm (Figs. 3A, 9A, B, 10C, D). The shape of the rhodoliths is essentially subspherical but ellipsoidal forms are locally abundant. The rhodoliths are massive and compact, with low primary porosity. The internal structure is laminar although columnar forms are observed rarely; the nuclei are composed of coralline algae or bryozoan fragments and less frequently barnacles and pelecypods. In addition in the rhodoliths, some epibionts have been recognized, such as barnacles and serpulids. Rhodoliths typically show one or more stages of bioerosion by sponges and pelecypods; these bores may show internal fillings with discordant geopetal orientations which provide evidence of early diagenesis during growth stages (Fig. 9C).

Only locally (Camposauro Group, Aurunci Mountains) do the

FIG. 4.—Lower Miocene rhodolith deposits of the "Bryozoan and Lithothamnium Limestones", central-southern Apennines, Italy: (A) Complex network of bioeroded cavities (white areas) filled by bioclastic sediment rich in bryozoa and coralline algae (see Fig. 8A) from Pietraroia (Matese Group); (B) Basal interval of the rhodolith deposits (M in photo) above the Cretaceous substrate (C in photo) from Pietraroia (Matese Group). The arrow indicates a large fragment of deeply bioeroded substrate. (C) Pelecypod (pectinid) rudstones that form banks in the lower part of the rhodolith limestones from the M. Rosa (Camposauro Group). (D) Pelecypod (ostreid) rudstones that form banks directly upon the deeply bioeroded Paleogene substrate (arrows) from Roccadaspide (M. Soprano Ridge).

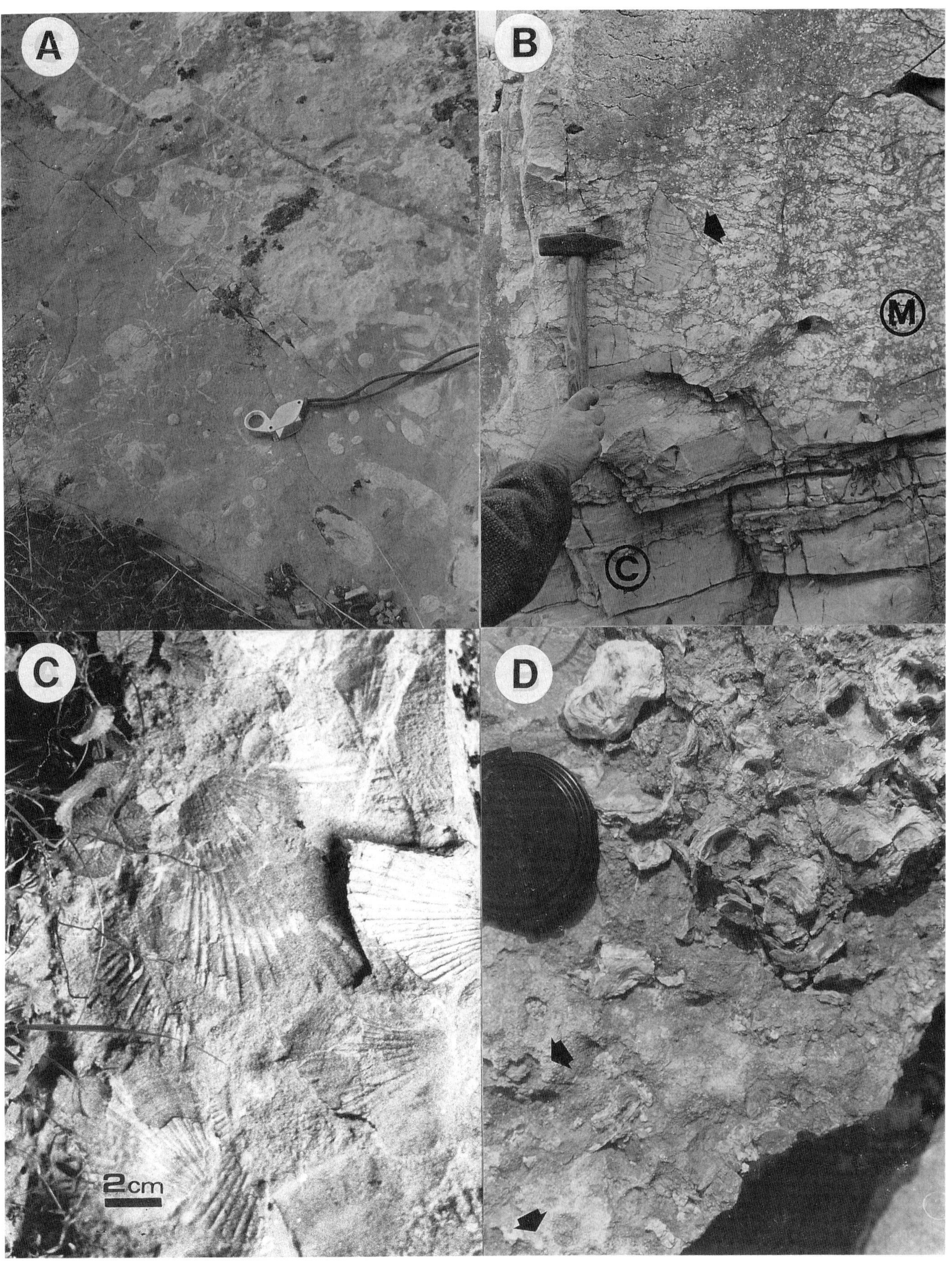
A
B
M
C
C
D
2cm

FIG. 5.—"Bryozoan and Lithothamnium Limestones" (BLL) from Pesco Rosito (Matese Group, central-southern Apennines, Italy). The Lower Miocene rhodolith deposits of the BLL rest unconformably upon the Mesozoic substrate with a sharp contact (arrows) and pass upward through a phosphatic palimpsest interval (P) into thin bedded hemipelagic deposits (see Fig. 6).

FIG. 7.—Lower Miocene limestones of the "Roccadaspide Formation" from Roccadaspide (M. Soprano Ridge, Southern Apennines, Italy). The resedimented bioclastic deposits of the Roccadaspide Formation (M) rest upon a bioeroded, locally pedogenized Paleogene substratum (P) that is overlain by a lense of dark gray claystone (V) rich in *Cerithium* and freshwater ostracods.

FIG. 6.—Phosphatic palimpsest interval at the upward transition from the "Bryozoan and Lithothamnium Limestones" neritic deposits to the hemipelagic, globigerinid-rich deposits from Pesco Rosito (Matese Group, central-southern Apennines, Italy).

coralline algae, associated with bryozoans and sessile foraminifera, form large crusts a few decimeters thick and a few square meters wide (Fig. 3B) with local intensive bioerosion. More frequently, the rhodoliths are interconnected by several millimeters-thick discontinuous algal crusts.

FIG. 8.—Lower Miocene carbonate deposits, central-southern Apennines, Italy: (A) Bryozoa and coralline algae grainstones (Lithofacies 4) of the BLL filling a pholad bore-hole in the Cretaceous substrate from Pietraroia (Matese Group); (B) Bryozoa and coralline algae grainstones (Lithofacies 4) of the Roccadaspide Formation (M. Soprano Ridge), showing sessile forams, pelecypod and echinoid fragments, and micritized coralline algae; (C) Bryozoa and coralline algae packstones with a muddy matrix (Lithofacies 5 of the BLL) from Coreno Ausonio (Aurunci Mountains ) showing large bryozoa and barnacles fragments; (D) Bryozoa and coralline algae floatstones rich in small fluorapatite grains in a matrix of globigerinid packstones at the transition between the BLL and overlying hemipelagic deposits from Cusano Mutri (Matese Group); (E) Bioclastic packstones rich in planktonic foraminifera (Globigerinids) from Cusano Mutri (Matese Group); (F) Phosphate-rich bioclastic grainstones from Pesco Rosito (Matese Group) characterizing the palimpsest interval at the transition with the overlying hemipelagic deposits. Note the phosphatized fillings (p) in sporangic cavities.

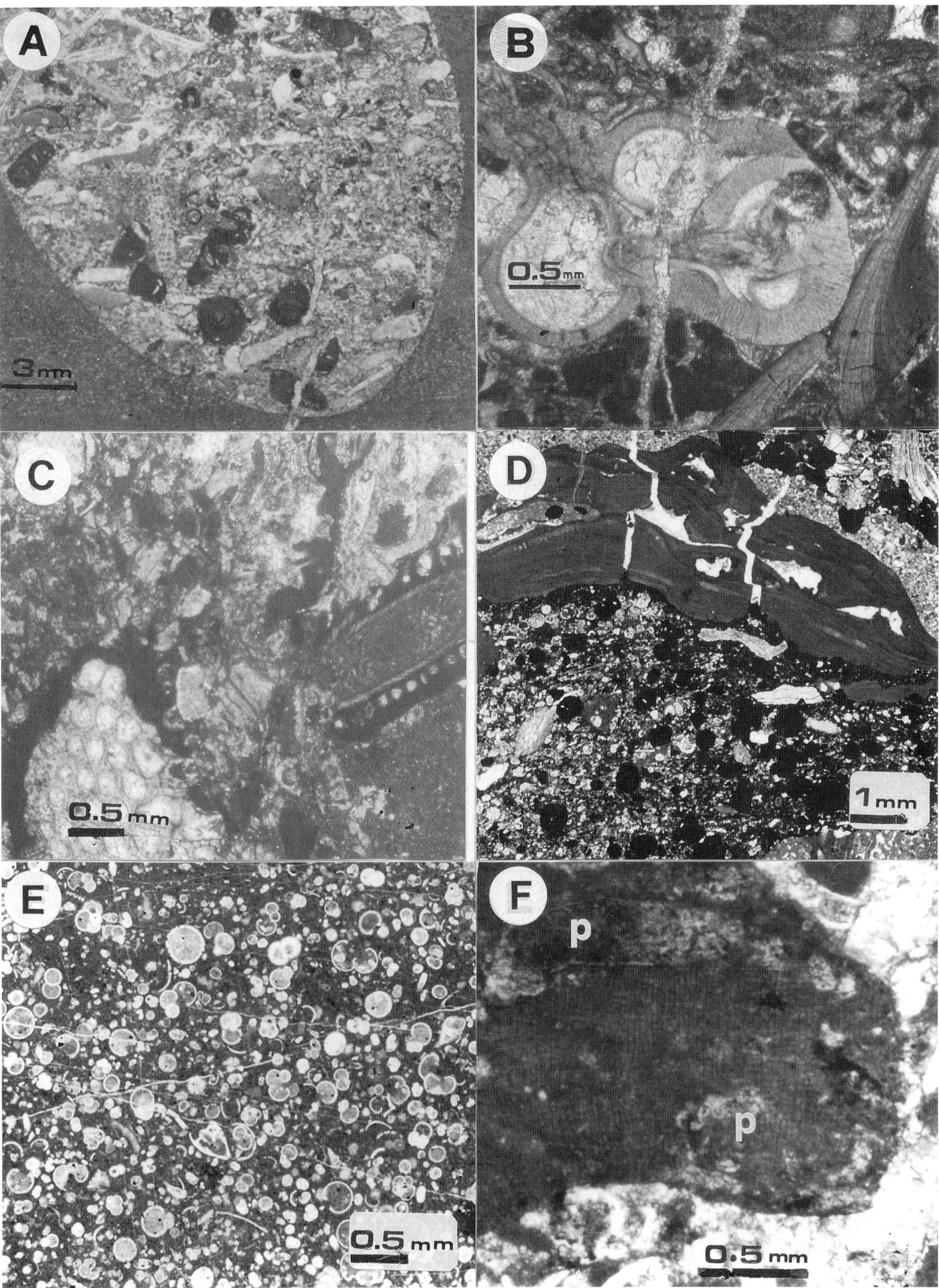
A
3mm
B
0.5mm
C
0.5mm
D
1mm
E
0.5mm
F
p
p
0.5mm

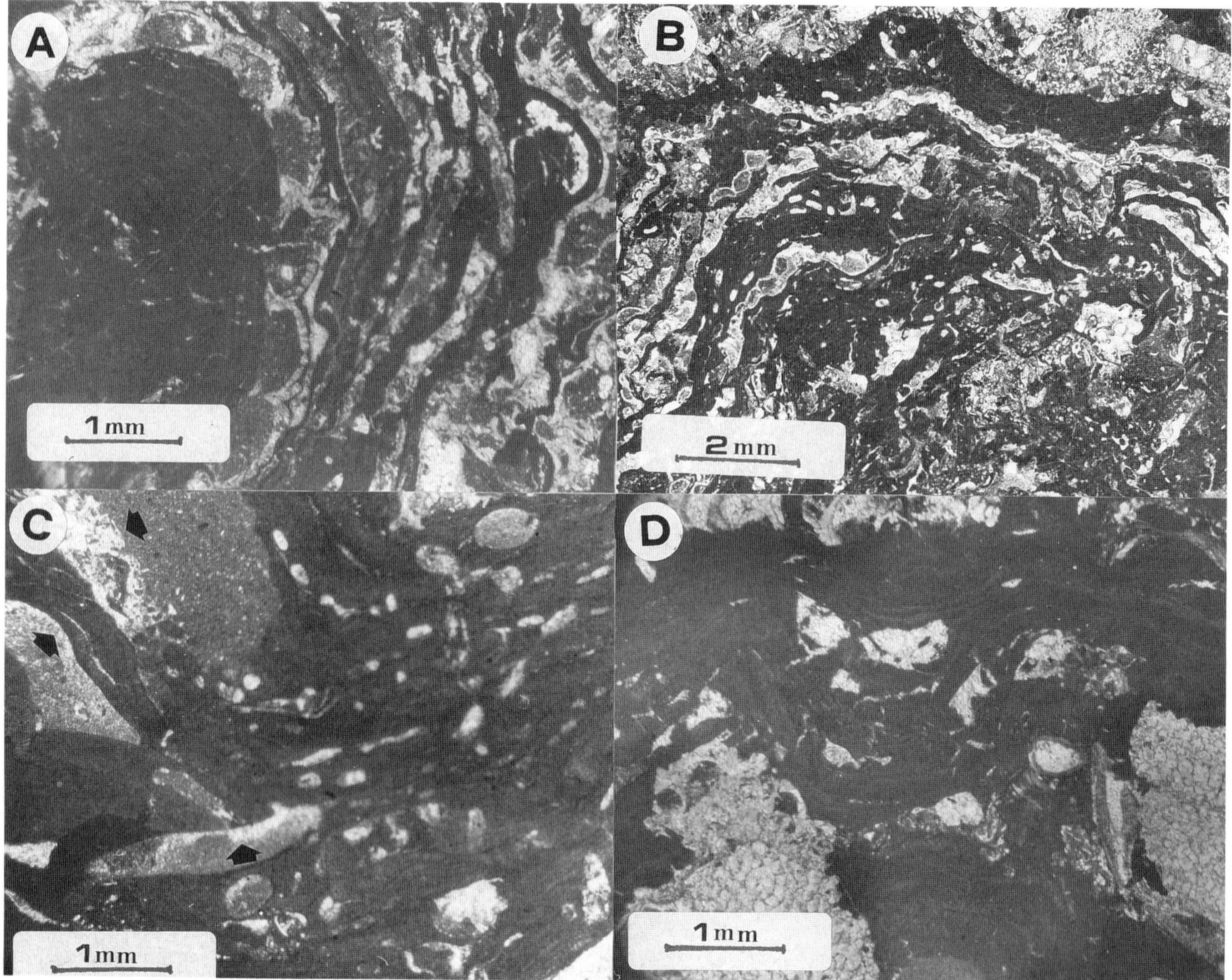


FIG. 9.—Lower Miocene rhodolith deposits of the Bryozoan and Lithothamnium Limestones (BLL), central-southern Apennines, Italy: (A) and (B) Competitive overgrowth of encrusting coralline algae, bryozoans and forams in rhodolith envelops from (A) M. Rosa (Camposauro Group ) and (B) Cusano Mutri (Matese Group ); (C) Encrusting coralline algae in rhodolith envelops from Coreno Ausonio (Aurunci Mountains ). Note sporangic cavities and bioerosional bore-holes with variously oriented geopetal fillings (arrows); (D) Coralline algae bindstones that trapped large benthic forams (*Miogypsina*) from M. Rosa (Camposauro Group).

Benthic forams are very abundant; among them Barbera (1979) recognizes: *Textularia sp.*, *Bigenerina* sp., Miliolids, *Cyclamina* sp., *Elphidium* sp., Lagenids, Buliminids, Rotalids, *Operculina* sp., *Eterostegina* sp., *Amphistegina* sp., *Gypsina* sp., *Miogypsina* sp., *Miniacina* sp., and *Homotrema* sp. Some of them occur only locally, such as Miogypsinids, forming distinct macroforaminifera levels in some localities.

Planktonic forams occur only sporadically in BLL. Globigerinids increase in number and diversity in the uppermost levels of the sequence at the sharp transition to hemipelagic-pelagic sediments.

Because of the strong lithification of the sediments, bryozoa colonies could not be isolated. Based on the zoarial form, adeoniform and vinculariform bryozoa have been identified in thin section (Barbera et al., 1978, 1980).

The mollusks occur as inarticulated valves, bored by clyonids (*Enthobia* bores), or they are articulated and in life position forming banks (Fig. 4C, D). Barbera (1979) recognizes: *Pecten pseudobeudanti* Dep. Rom., *P. hornensis* Dep. Rom., *Flabellipecten besseri* (An.), *F. burdigaliensis* (Lk.), *F. benedictus* (Duj), *F. fraterculus* (Sow), *Chlamys holgeri* (Gein.), *C. latissima* (Brocchi), *C. scabrella* (Mich.), *C. opercularis* (L.), *Ostrea lamellosa* (Brocchi) and *Neopycnodonta* sp.

Fish teeth (Teleostea, Selacia) are common at the top of the BLL. Among them, Franco (1960) recognized: *Carcharodon megalodon* Agassiz, *Odontaspis cuspidata* Agassiz, *O. contortidens* Agassiz, *Oxyrhina desori* Agassiz, *O. hastalis* Agassiz, *Miliobatus crassus* Gervais, *Hemipristis serra* Agassiz and *Chrysophrys cincta* Agassiz.

On the whole, the BLL and the "Roccadaspide Formation"

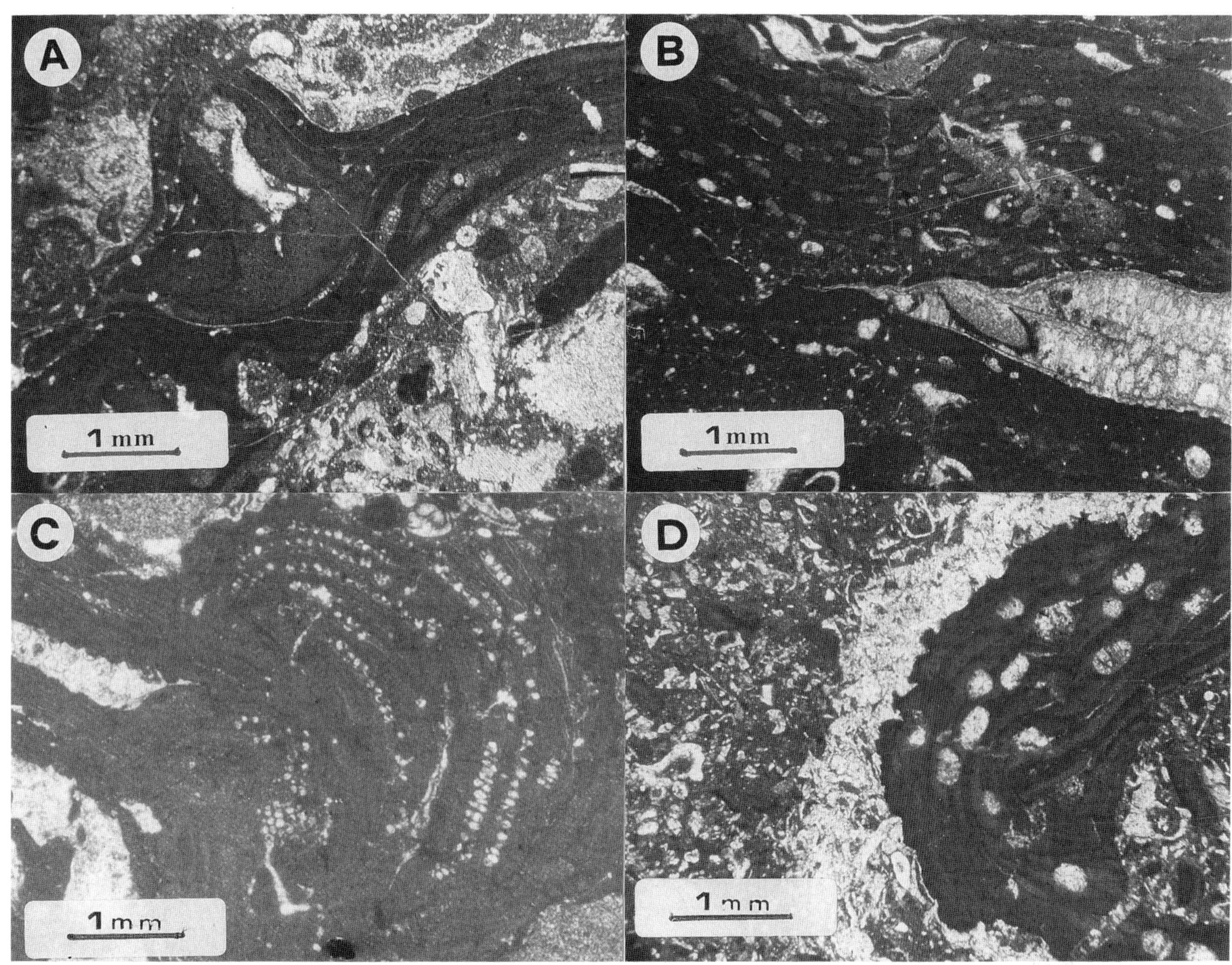


FIG. 10.—Lower Miocene rhodolith deposits of the Bryozoan and Lithothamnium Limestones (BLL), central-southern Apennines, Italy: (A) Thin coralline algae layers stabilizing bioclastic packstones from Coreno Ausonio (Aurunci Mountains.); (B) Coralline algae bindstones that trapped a large mollusk fragment from Coreno Ausonio (Aurunci Mountains); (C) and (D) Close-up of algal envelops in rhodoliths from (C) Cusano Mutri (Matese Group) and (D) Coreno Ausonio (Aurunci Mountains).

limestones (at least in the portions rich in skeletal grains) can be subdivided into six lithofacies:

Lithofacies 1. Rhodolith and/or pelecypod clasts in a matrix of bryozoa and coralline algae grainstones generally similar to that of lithofacies 4 and locally lithofacies 5.

Lithofacies 2. Rhodolith rudstones (Fig. 3A).

Lithofacies 3. Pelecypod rudstones (Fig. 4C, D).

Lithofacies 4. Bryozoa and coralline algae grainstones. Fragmented bryozoans and coralline algae are the predominant skeletal constituents, and benthonic forams and bioeroded pelecypod fragments also are abundant (Figs. 8A, B).

Lithofacies 5. Bryozoa and coralline algae packstones. Silty mud fills the inter- and intragranular cavities of the bioclastic sediment (Fig. 8C).

Lithofacies 6. Coralline algae bindstones (algal crusts). The sediment is made up of encrusting coralline algae associated with sessile and encrusting foraminifera and bryozoa (Figs. 3B, 9D, 10B). The intrabiolithitic cavities are filled with bioclastic sediments with a silty mud similar to that in lithofacies 5 (Fig. 10A).

Relationships among these lithofacies suggest a fundamental lithology of skeletal (algae- and bryozoa-rich) grainstones (Lithofacies 4), more or less enriched in algal concretions (rhodoliths) and pelecypod shells (Lithofacies 1). Where rhodoliths or pelecypods become more abundant, the texture gradually changes to a rhodolith rudstone (Lithofacies 2; sometimes stabilized by encrusting coralline algae) or a pelecypod rudstone (Lithofacies 3) forming banks made up of ostreids and pectinids.

The ostreid banks seem to characterize the basal interval directly covering the Mesozoic-Paleogene substrate since these bivalves needed a hard bottom (Fig. 4D). In the sequence, they are limited to patches of stabilized substrate. Banks of pectinids (Fig. 4C) are repeatedly developed in the lower part of the sequence, reaching 20-30 cm in thickness and extending for about one hundred square meters.

Rhodolith rudstones may occur as rhodolith pavements but, more commonly, the larger rhodoliths are packed and form lenticular bodies a few meters in diameter that are surrounded by coarse bioclastic sediment. In some localities (e.g., Pietraroia), the rhodolith accumulation forms a mounded morphology, several meters in diameter and up to 2 m thick, with surfaces encrusted by ostreids and serpulids (Carannante, 1982a). Due to the dominance of rhodophyta among the constituents, these deposits are classified as rhodalgal sediments (Carannante et al., 1988a).

*Transitional facies to the hemipelagic facies.—*
BLL generally changes gradually upward through an up to 150-cm-thick phosphatic interval (Fig. 5) to thin bedded hemipelagic and pelagic limestones. This interval is characterized by abundant fluorapatite grains, subspherical to ellipsoidal in shape, having an average diameter of 2 mm, and scattered smaller glauconite grains (Fig. 6). Both these grain types occur with bioclastic fragments very similar to those of the underlying organogenic limestones to form rudstones or floatstones with a matrix of globigerinid packstones (Fig. 8D).

The bioclastic grains typically show reddish brown surfaces and sharp fracture margins that offset internal sediments. Complex grains, formed by several skeletal fragments bound together by encrusting organisms, are very common; these complex intraclasts are cemented by micrite. The muddy matrix is scarce in the lower part of this facies and increases upward to form a wackestone with only a few scattered skeletal calcareous fragments and abundant phosphatic grains. Glauconite grains occur as internal sediment within some planktonic foraminifera. These grains became increasingly abundant upward in the hemipelagic (locally marly) and pelagic limestones (Fig. 8E). Phosphatized sediments fill macroboring and/or internal cavities of planktonic as well as benthonic organisms. These phosphatic grains seem to be a result of several different processes. Whereas most are large elliptical grains interpreted as fecal pellets (Zalaffi, 1963), other phosphatic grains may have been formed (Carannante, 1982b) by substitution of muddy calcareous sediments that occluded coralline algae sporangia (Fig. 8F) and/or by a gradual alteration of benthonic foraminifera as suggested by Manheim et al. (1975) for the marine phosphorite formation off Peru.

Where the phosphatic interval is missing, reddish brown ferriferous crusts separate BLL from the overlying hemipelagic deposits (Bergomi and Damiani, 1976; Barbera et al., 1978). In some localities (e.g., Pietraroia), these crusts, 2-3 cm thick, directly coat the Cretaceous substrate, fossilizing the lower Miocene bioclastic sediments that fill macroboring cavities.

*Resedimented bioclastic facies.—*
In the M. Soprano Ridge, the biolithoclastic, partially glauconitic grainstones (Fig. 8B) of the "Roccadaspide Formation" are dominated by large benthic forams, mollusk and echinoid fragments (molechfor association, *sensu* Carannante et al., 1988a). Planktonic forams, which are scarce at the base of the section, increase upwards in the sequence to form thin millimeter laminations composed of globigerinid grainstones intercalated with the biolithoclastic wackestones. In addition, the globigerinids may be the major components of the overlying hemipelagic deposits which preceded the mixed arenite turbidite deposits. Although they appear macroscopically homogeneous, microscopically the "Roccadaspide Formation" limestones display repeated textural, compositional and biotic characteristics (Carannante et al., 1988b). In particular, older miogypsinid species alternate with younger species suggesting repeated reworking of the sediments. The glauconite limestones of the "Roccadaspide Formation" were previously described by Selli (1957) as shallow homogeneous neritic deposits. More recently, these limestones have been interpreted by Carannante et al. (1988b) as sediments accumulated on the upper slope or on distal portions of the ramp. The glauconitic limestone formation involved complex processes of transport and deposition ranging from simple storm-controlled grain resuspension and deposition to grain falls and/or cohesionless gravity flow deposition. The nature and characteristics of these remobilized deposits suggest that the source of the bioclastic grains was an open carbonate shelf comprising rhodalgal and/or molechfor assemblages (foramol *sensu lato* platform).

## DIAGENETIC OVERPRINT

In the BLL, neither originally aragonite skeletal debris, nor evidence of moldic solution has been found, and calcite is the primary mineral of most of the organic components (Barbera et al., 1978). Coralline algae, barnacles, echinoids, serpulids, as well as most bryozoa and encrusting foraminifera tests, are characterized by high-Mg calcite. The ostreid and pectinid shells and the tests of most of the planktonic forams are low-Mg calcite and aragonite is present only in the thin prismatic layers of some pectinids.

### *Complex Grains*

In the BLL, some complex grains are recognized that were formed by several skeletal fragments bound together by the competitive overgrowth of encrusting organisms. These complex grains underwent an early submarine diagenesis documented in the topmost interval of the neritic sediments by the presence of limiting surfaces that offset both the rigid organic structures and the silty intragranular filling. Similar organogenic intraclasts become dominant in the phosphatic interval. In addition, bore holes in the rhodoliths show geopetal fillings with a variable orientation (Fig. 9C). Therefore, we deduce that the internal sediment was, at least partially, cemented on the sea floor. The cementation became more significant at the top of the bioclastic deposits and occurred during a non-depositional stage which preceded deposition of the hemipelagic sediments. According

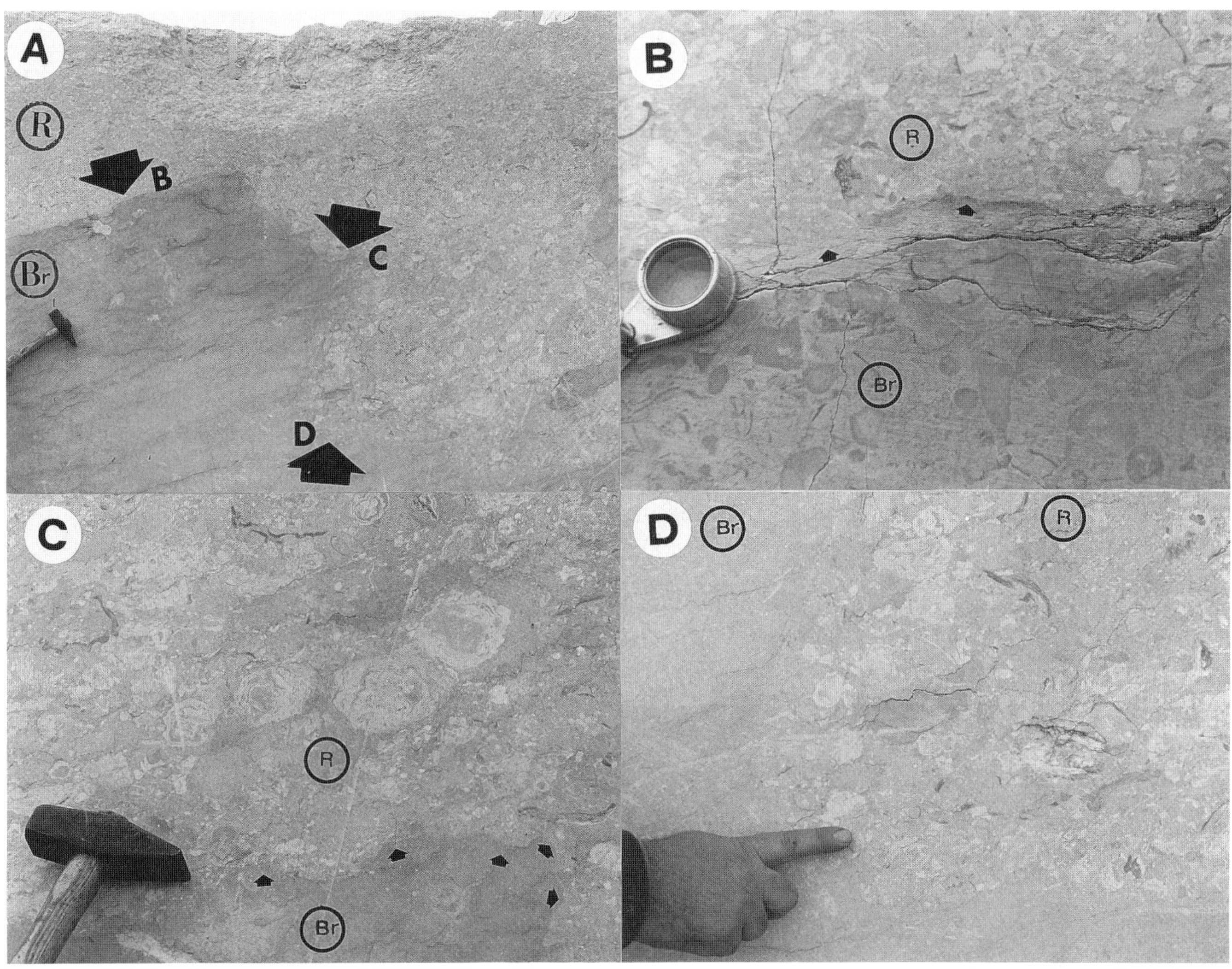


FIG. 11.—Lower Miocene Bryozoan and Lithothamnium Limestones (BLL), Pietraroia (Matese Group, central-southern Apennines, Italy): (A) Erosional surface (arrows) within the BLL between bryozoa-rich sediment (Br) and rhodolith-rich sediment (R). On the left side of the figure the top of the (Br) shows a hardened level partially preserved (arrows B and C in the photo). This level grades downward to poorly cemented deposits, and was the product of early submarine diagenesis during the deposition of the BLL (submarine hard ground); (B) and (C) Close-up of (A) showing the sharp contact between the well cemented bryozoa-rich grainstones (Br) and the overlying rhodolith floatstones (R); (D) Close-up of (A) in the eroded area (arrow D in the figure A) where the poorly cemented bryozoa-rich deposits (Br) mixed with the following rhodolith-rich sediment (R), producing a complex transition boundary.

to Barbera et al. (1978), the deposits of BLL were progressively stabilized during their accumulation, and an incipient cementation occurred in the stabilized sediments immediately below the loose and mobile sediments of the more superficial layers. This submarine cementation was formed by sea waters introduced into the pore system through porous sediments. Well-cemented surfaces characterize the BLL in some limited areas (e.g., Pietraroia). In those areas, eroded submarine hardgrounds (Figs. 11A-D) are present within the bioclastic sediment of the BLL. They appear as decimeters thick hardened levels grading downward into poorly cemented sediment.

Environmental and hydrological conditions changed with increasing depth within the sediments. The topmost level, left unburied," was exposed for a long time to the water/sediment interface. It became the surface for phosphate enrichment processes and subsequently a place for the accumulation of hemipelagic sediments, thus generating a palimpsest sediment (*sensu* Swift et al., 1971).

### *Phosphate Grains*

In the transitional facies to the overlying hemipelagic deposits, the selective phosphatization of some grains indicates anoxic interstitial pore conditions with fluids richer in phosphates than the overlying water (Carannante, 1982b). Under these conditions, the phosphorous was concentrated in particularly favorable microenvironments (e.g., foraminifera tests, fecal pellets or micritic filling of intragranular cavities such as coralline algae sporangic cavities)

### *Glauconite Grains*

Further evidence of low sedimentation rates, at the transition between the bioclastic calcarenites and the overlying hemipelagic-pelagic deposits, is evidenced by scattered green glauconite grains (Carannante et al., 1984). They are comparable to the marine green grains of the glaucony facies *sensu* Odin and Matter (1981). As a consequence, conditions of semiconfinement may be hypothesized for the microenvironment where the glauconitization processes took place within a relict sediment sheet exposed for long periods to the sediment-water interface.

### *Reddened Grains*

Bioclasts with reddish surfaces are quite common at the BLL transitional interval in our study area. Reddish goethite grains, resulting from the partial oxidation of glauconite and phosphatic grains, are reported from the Peru-Chile and eastern Australian continental shelves (Burnett, 1980; Marshall and Cook, 1980). The reddening is attributed to the change from poorly oxygenated waters to waters progressively enriched in dissolved oxygen. According to Burnett (1980), this change might be due to a lateral shifting and/or a contraction of the oxygen minimum layer, caused by a shift of the maximum upwelling locus.

### *Reddish-brown Crusts*

A distinct diagenetic phase seems to have led to the reddish-brown ferriferous crusts which locally separate the hemipelagic-pelagic sediments from the BLL. Formation of the ferriferous crusts required low sedimentation rates associated with a strong current regime and availability of hard substrata. The hard substrates were represented by the exposed Cretaceous rocky substrate which was deeply bored by lithophagous organisms (e.g., Pietraroia) and by the stabilized Miocene bioclastic sediments that formed hardgrounds.

## DISCUSSION OF THE DEPOSITIONAL AND DIAGENETIC ASPECTS

### *BLL Characteristics*

Barbera et al. (1978), Carannante et al. (1981) and Simone and Carannante (1985) pointed out the inadequacy of the tropical Bahamian rimmed shelf model to explain carbonates with BLL characteristics and instead found them to be characterized by: (a) the predominance of constituents derived from organisms adapted to low-light intensities (sciaphile assemblages) in the deep-euphotic zone where circalittoral bottoms are deeper than 50 m and (b) the presence and local abundance of constituents derived from organisms living in well-lit areas (photophile assemblages) of infralittoral bottoms and warmer waters.

The composition of the BLL was first attributed to a foramol-type association (see Lees, 1975) by Barbera et al. (1978, 1980) and Carannante et al. (1981, 1984). Later, it was redefined as a rhodalgal association by Carannante et al. (1988a) on the basis of the dominance of encrusting coralline algae, bryozoans and variables amounts of large benthic forams and barnacles. According to these authors, rhodalgal sediments are particularly important on temperate shelves, at the transition between chlorozoan-chloralgal lithofacies (dominated by zoantharia and chlorophyta) and the colder and deeper molechfor lithofacies. The general trend in the zonation of sediment types, primarily controlled by latitude and water depth, may show variations due to local environmental factors (e.g., upwelling, freshwater input, cold current regime).

According to Carannante (1982b) and Simone and Carannante (1985), upwelling waters seem to have played a major role in favoring the development of these Burdigalian rhodalgal assemblages and, as a consequence, in producing a temperate-type carbonate platform. In the shallower areas characterized by more turbulent waters, which are only laterally affected by upwelling processes, the development of some organisms may be favored by induced eutrophic conditions. Red algae and bryozoa are particularly phosphorous-rich (Milliman, 1974). Examples of rhodalgal facies and extensive algal crusts are known from areas with nutrient-rich and cold upwelling waters (Summerhayes et al., 1976). The presence of active upwelling may also be inferred by the subsequent phosphatization of the uppermost level of the BLL. Phosphates are now widely forming in areas of intensive upwelling (Summerhayes et al., 1976; Baturin, 1982), where very low values of dissolved oxygen are often found in association with the cold and nutrient-rich upwelled waters. When the BLL depositional area became drowned, the bioclastic deposits underwent conditions favorable for phosphatization, resting at the periphery of nutrient-rich and low oxygen water bodies directly connected with upwelled waters. In the BLL-dominated platform true reefs are not present. Instead, the association of the BLL lithofacies indicate an open carbonate ramp that graded laterally into deeper environments, where resedimentation processes were particularly active.

The Cretaceous-Paleogene rocky substrate locally acted as a distally steepened ramp on the bottom of the Miocene sea. In these areas, sea-level oscillations created distinctive features which are characteristic of the response of the temperate-type carbonate platforms to sea-level changes (Carannante and Simone, 1988). During rapid transgressive phases, these carbonate platforms easily drown because of their relatively low rate of growth and inability to keep pace with rising sea level (Simone and Carannante, 1985, 1988; Carannante et al., 1986); whereas, during lowstands, neritic sediments may be more easily eroded and redeposited by gravitational processes. According to the age of the reworked *Miogypsina* sp. in the deposits of the "Roccadaspide Formation" (Carannante et al., 1988b), a major resedimentation episode seems to have occurred during the upper Aquitanian sea-level drop which corresponds to the end of the TB1 supercycle of Haq et al. (1987), and additional resedimentation episodes occurred during the sea-level drop that corresponds to the TB2 supercycle of Haq et al. (1987). These latter resedimentation episodes included sediments previously glauconitized during the rapid Burdigalian transgressive phase.

### *Modern Analogues*

Modern analogues of the Miocene rhodolith limestones are well known. Present-day Mediterranean sediments provide some of the best analogues, as pointed out by many authors (Walther, 1885; Walther and Schirlitz, 1886; Studencki, 1979; Carannante et al., 1981; Bosence, 1985; Simone and Carannante, 1985). The classic zonation of Mediterranean benthic assemblages of Pérès and Picard (1964) allows a better understanding of the Miocene lithofacies of this study and a more complete paleoenvironmental interpretation of the rhodolith limestones.

In the Mediterranean Sea, large areas of the deeper infralittoral-circalittoral zones contain accumulations of bioclastic sediments (*Détritique Côtier* in Pérès and Picard, 1964). These deposits, characterized by low sedimentation rates, display a textural composition similar to Lithofacies 4 of the Miocene limestones of this study. The bioclastic constituents of the *Détritique Côtier* are derived mostly from bioerosion and fragmentation of: (a) sciaphile assemblages living on localized hard bottoms (biocenosis of the *Coralligène de Plateau* in Pérès and Picard, 1964) as well as on the local mobile sedimentary sheet (biocenosis of the *Détritique Côtier* ) and (b) photophile assemblages from infralittoral areas less than 50 m deep, such as benthic forams living within the grass meadows. The organisms of this second type of assemblage live mainly as epiphyta on phanerogam leaves and may be carried in as isolated grains and/or fixed on plant fragments.

In the BLL, lithofacies displaying large rhodoliths (Lithofacies 1 and 2) coincide with the sediments of present-day rhodolith-rich *Faciès à Pralines* of the above biocenosis of the *Détritique Côtier* (Figs. 12A, B). This facies is known in deep infralittoral-circalittoral (30-80 m of depth) open shelf areas.

Modern rhodolith deposits, that frequently characterize areas with an elevated topography on which an active current regime impinges, cannot be considered as true buildups. They are passive accumulations even if they may be more or less stabilized and may show a positive relief (Fig. 12C). In the rhodolith accumulations, the rhodoliths are subject to periodic movements related to water dynamics and to the activity of grazing organisms (Pérès and Picard, 1964; Prager, 1987; Prager and Ginsburg, 1989). During phases of stasis, the algal concretions may be stabilized by a covering of non-calcified filamentous red algae (*Précoralligène* bottoms in Pérès and Picard, 1964) and/or by the presence of large brown algae, the holdfasts of which are frequently fixed on rhodoliths.

Additionally, the loose sediment may be stabilized by the trapping and binding action of some encrusting organisms of the *Détritique Côtier* biocenosis (e.g., coralline red algae, membraniporiform bryozoa, encrusting foraminifera). As a consequence, hard substrata develop in accordance with the evolution of the *coralligène* bottoms, the climax of which is represented by the biocenosis of the *Coralligène* itself. These processes of aggregation lead to the formation of massive and cavernous agglomerates (Fig. 12D). They may undergo bioerosion and fragmentation by boring and rasping organisms which produce complex skeletal debris.

The biocenosis of the *Coralligène de Plateau*, as defined by Pérès and Picard (1964), produces only small buildups where intensive bioerosion releases a large amount of skeletal debris. These buildups may be considered as sediment producers with a low preservation potential, except for some structures in which coralline algae are dominant (Lithofacies 6, Figs. 9D, 10B). This coralline algal facies which are more compact and resistant with respect to those dominated by invertebrates, may form buildups a few meters high on the sea floor, with a major preservation potential.

## CONCLUSIONS

1. The BLL accumulated at rates as low as 10 m/Ma (Barbera et al., 1980) over circalittoral bottoms in open shelf areas that correspond to the middle-outer sectors of a distally steepened ramp. This ramp developed in the still undeformed central-southern Apennine Miocene foreland. These deposits are dominated by rhodalgal associations. This facies, typical of transitional areas from tropical to temperate regions, may also form in subtropical-tropical regions where various factors preclude the development of chlorozoan associations such as: salinity changes, cooler waters, lack of suitable substrata, and excessive nutrient content (Simone and Carannante, 1985; Carannante et al., 1986; Hallock and Schlager; 1986).

2. During the deposition of the BLL rhodolith sediments, the outermost sectors of the shelf were subjected to upwelling of cold and nutrient-rich waters which favored rhodalgal-type associations and, consequently, the development of a "temperate-type" platform (Carannante, 1982b). At increased depth, a sheet of relict sediments was formed. Due to a further deepening or to a widening (and/or lateral shifting) of the anoxic layers connected with upwelling maximum, the relict cover remained exposed for long periods to nutrient-rich and poorly oxygenated waters and subsequently underwent phosphatization. Planktonic-rich deposition followed and these deposits were mixed with the relict grains, thus forming a palimpsest sediment (*sensu* Swift et al., 1971).

3. The skeletal-rich deposits of the BLL were partially (and locally) stabilized to form complex cavernous agglomerates analogous to the present day *Coralligène* bottoms. They formed massive organogenic accumulations that graded laterally to hemipelagic deposits on distal portions of the ramp.

4. On the whole, the BLL accumulated during a significant Burdigalian transgressive event that appears to correspond to the transgressive system tract of the TB2.1 cycle of Haq et al. (1987) following the global sea-level drop at the Aquitanian-Burdigalian boundary. During this time interval, the Mediterranean Sea was likely to have had an estuarine-type water circulation and the climate was subtropical with wet and relatively cool conditions (Esteban, this volume). Deep Atlantic waters entered the Mediterranean to form deep Mediterranean waters which were raised locally along the shelf as active upwelling waters, whereas the

FIG. 12.—Recent Mediterranean coralline algae sediments: (A) Bioclastic sediments composed of rhodoliths, mollusks and bryozoan fragments dredged at a depth of 60 m from the Gulf of Taranto (Ionian Sea); (B) Rhodoliths dredged at a depth of 50 m on the Cajola Bank from the Bay of Naples (Tyrrhenian Sea); (C) Rhodolith mound partially stabilized by thin coralline algae laminae at 35 m of depth from the Bay of Naples (Tyrrhenian Sea); (D) Coralligène bottom at 40 m of depth from the Bay of Naples (Tyrrhenian Sea).These organogenic frameworks, dominated by the growth of encrusting coralline algae, were formed from repeated aggregation processes and contain a network of cavities, that led to the formation of complex cavernous agglomerates (photo F. Toscano).

warmer and less saline surface Mediterranean waters were discharged into the Atlantic Ocean. Large diffusion of Burdigalian temperate-type open shelves, inhabited by rhodalgal/molechfor (foramol *sensu lato*) assemblages in the circum-Mediterranean region, fits well in the scenario in which upwelling processes were common.

Resedimentation of foramol bioclastic sediments occurred during the sea-level drop at the end of the TB1 supercycle, and additional main resedimentation episodes were active during the TB2 supercycle when tectonic controls interacted with sea-level oscillations (Carannante et al., 1988b).

The palimpsest interval that marks the upwards transition from the BLL neritic deposits to the overlying hemipelagic deposits represents a drowning unconformity (*sensu* Schlager, 1989) due to the low growth rate of the temperate-type carbonate platforms. These latter types of platforms are unable to continue to develop during rapid and significant relative sea-level rises (Simone and Carannante, 1988). The resulting drowned successions may be related to destructive sea-level rises (*sensu* Follmi et al., 1994).

5. Later, tectonic subsidence took away increasingly larger sectors of the central-southern Apennine Miocene foreland from the neritic domain, and the area became the location of turbiditic deposition related to the migration of the foredeep at the front of the Apennine nappes (Carannante et al., 1987).

ACKNOWLEDGMENTS

We thank D. Bosence, M. Esteban and J.-M. Rouchy who critically read the manuscript. Financial support for this research was provided by the Centro Nazionale della Ricerche of Italy and by the Ministero della Publica Istruzione of Italy.

REFERENCES

BARBERA, C., 1979, Lamellibranchi Miocenici della "Formazione di Cusano" provenienti da Cusano Mutri (Matese Orientale, Benevento): Bollettino della Società dei Naturalisti in Napoli, v. 86, p. 193-211.

BARBERA, C., CARANNANTE, G., D'ARGENIO, B., AND SIMONE, L., 198O, Il Miocene calcareo dell'Appennino meridionale: contributo della paleoecologia alla costruzione di un modello ambientale: Annali Università di Ferrara, v. 6, p. 281-299.

BARBERA, C., SIMONE, L., AND CARANNANTE, G., 1978, Depositi circalittorali di piattaforma aperta nel Miocene Campano. Analisi sedimentologica e paleoecologica: Bollettino della Società Geologica Italiana, v. 97, p. 821-834

BATURIN, G. N., 1982, Phosphorites on the sea floor. Origin, composition and distribution: Elsevier, Amsterdam, Developments in Sedimentology, v. 33, 343 p.

BERGOMI, C. AND DAMIANI, A. V., 1976, Diagenesi precoce nei depositi Serravalliano-Tortoniani del Lazio e considerazioni sulla evoluzione strutturale del Bacino di sedimentazione miocenico: Bollettino del Servizio Geologico d'Italia, v. 97 (1978), p. 35-66.

BOSENCE, D. W. J., 1985. The "Coralligène" of the Mediterranean - a Recent analog for the Tertiary coralline algal limestones, *in* Toomey, D.F. and Nitecki, M.H., eds, Paleoalgology: Contemporary Research and Applications: Berlin, Springer-Verlag, p. 216-225.

BURNETT, W. C., 1980, Apatite-glauconite associations off Perù-Chile: paleo-oceanographic implications: Journal Geological Society London, v. 137, p. 757-764.

CARANNANTE, G., 1982a, La Valle del Canale (Civita di Pietraroia, Matese). Una incisione miocenica riesumata sul margine della Piattaforma carbonatica Abruzzese-Campana: Geologica Romana, v. 21, p. 511-521.

CARANNANTE, G., 1982b, Modello deposizionale e diagenetico di un livello fosfatico nel Miocene carbonatico dell'Appennino Campano: Rendiconti della Società Geologica Italiana, v. 5, p. 15-20.

CARANNANTE, G., DI GERONIMO, I., AND SIMONE, L., 1984, "Relict sediments" dans les séquences carbonatées miocènes de l'Apennin méridional et leur analogues récents dans la Mer Méditerranée (abs.): Marseille, 5th International Association of Sedimentologists Regional Meeting, p. 90-91.

CARANNANTE, G., ESTEBAN, M, MILLIMAN, J., AND SIMONE, L., 1988a, Carbonate facies as paleolatitude indicators: problems and limitations: Sedimentary Geology, v. 60, p. 333-346.

CARANNANTE, G., MATARAZZO, R., PAPPONE, G., SEVERI, C., AND SIMONE, L., 1988b, Le calcareniti mioceniche della Formazione di Roccadaspide (Appennino Campano-Lucano): Memorie della Società Geologica Italiana, v. 41, p. 775-789.

CARANNANTE, G., PESCATORE, T., SENATORE, M. R. AND SIMONE, L., 1987, L'annegamento delle aree di Avampaese miocenico e la migrazione delle facies a briozoi e litotamni (foramol) nell'Appennino meridionale (abs.): Naxos, Convegno della Società Geologica Italiana, p. 22-23.

CARANNANTE, G. AND SIMONE, L., 1988, Foramol carbonate shelves as depositional site and source area: recent and ancient examples from the Mediterranean Region. (abs.): American Association of Petroleum Geologists Bulletin, v. 72, p. 993-994.

CARANNANTE, G., SIMONE, L. AND BARBERA, C., 1981, "Calcari a Briozoi e Litotamni" of Southern Apennines. Miocenic analogous of recent Mediterranean rhodolitic sediments (abs.): Bologna, 2nd International Association of Sedimentologists Regional Meeting, p. 17-20.

CARANNANTE, G., SIMONE, L. AND NEUMANN, C., 1986, Miocene drowning of a temperate (foramol) carbonate platform: upper Miami Terrace (abs.): American Association of Petroleum Geologists Bulletin, v. 70, p. 571.

COCCO, E., 1971, Note illustrative della carta geologica d'Italia, Foglio 161, Isernia: Roma, Servizio Geologico d'Italia, 38 p.

D'ARGENIO, B. AND SCANDONE, P., 1969, Jurassic facies pattern in the Southern (Campania-Lucania) Apennines: Annales Hungarian Geological Institut, v. 54, p. 383-396.

FOLLMI, K. B., WEISSERT, H., BISPING, M. AND FUNK, H., 1994, Phosphogenesis, carbon-isotope stratigraphy, and carbonate-platform evolution along the Lower Cretaceous northern Tethyan margin: Geological Society of America Bulletin, v. 106, p. 729-746.

FRANCO, D., 196O, Su alcuni ittioodontoliti rinvenuti nei calcari Terziari a pettinidi di Pietraroia (Benevento): Bollettino della Società dei Naturalisti in Napoli, v. 67, p. 187-200.

GALDIERI, A., 1913, Osservazioni sui calcari di Pietraroia in provincia di Benevento: Napoli, Rendiconti Real Accademia Scienze Fisiche e Matematiche, Fasc. 6° a 10°, 10 p.

HALLOCK, P. AND SCHLAGER, W., 1986, Nutrient excess and the demise of coral reefs and carbonate platforms: Palaios, v.1, p. 389-398.

HAQ, B.U., HARDENBOL, J. AND VAIL, P. R., 1987, Chronology of fluctuating sea level since the Triassic: Science, v. 235, p. 1156-1167.

LEES, A., 1975, Possible influence of salinity and temperature on modern shelf carbonate sedimentation: Marine Geology, v. 19, p. 159-198.

MANHEIM, F., ROWE, G.T., AND JIPA, D., 1975, Marine phosphorite formation off Perù: Journal of Sedimentary Petrology, v. 45, p. 243-251.

MARSHALL, J. F. AND COOK, P. J., 1980, Petrology of iron and phosphorous rich nodules from the E. Astralian continental shelf: Journal Geological Society London, v. 137, p. 765-771.

MILLIMAN, J. D., 1974, Marine Carbonates: Berlin, Springer-Verlag, 375 p.

ODIN, G. S. AND MATTER, A., 1981, De Glauconiarum origine: Sedimentology, v. 28, p. 611-641.

PÉRÈS, J. M. AND PICARD, J., 1964, Nouveau manuel de bionomie benthique de la Mer Méditerranée: Marseille, Travaux Station Marine d'Endoume, v. 31, Fasc. 47, 137 p..

PRAGER, E. J., 1987, The growth and structure of calcareous nodules (for-algaliths) on Florida's outer shelf: Unpublished Master Thesis, University of Miami, Miami, 64 p.

PRAGER, E. J. AND GINSBURG, R. N., 1989, Carbonate nodule growth on Florida's outer shelf and its implications for fossil interpretations: Palaios, v. 4: p.310-317.

SCHIAVINOTTO, F., 1985, Le Miogypsinidae alla base della trasgressione miocenica del Monte Camposauro (Appennino Meridionale): Bollettino della Società Geologica Italiana, v. 104, p. 53-63.

SCHLAGER, W., 1989, Drowning unconformities on carbonate platforms, *in* Crevello, P. D., Wilson, J. L., Sarg, J. F. and Read, J. F., eds, Controls on Carbonate Platforms and Basin Development: Tulsa, Society of Economic Paleontologists and Mineralogists Special Publication. 44, p. 15-25.

SELLI, R., 1957, Sulla trasgressione del Miocene nell'Italia meridionale: Giornale Geologia di Bologna, Serie 2, v. 26, p. 1-54.

SIMONE, L. AND CARANNANTE, G., 1985, Evolution of a carbonate open shelf up to its drowning: the case of the Southern Apennines: Napoli, Rendiconti Accademia Scienze Fisiche e Matematiche, Serie IV, v. 53, p. 1-43.

SIMONE, L. AND CARANNANTE, G., 1988, The fate of foramol ("temperate-type") carbonate platforms: Sedimentary Geology, v. 60, p. 347-354.

STUDENCKI, W., 1979, Sedimentation of algal limestones from Busko-Spa Environs (Middle Miocene, Central Poland): Palaeogeography, Palaeoclimatology, Palaeoecology, v. 27, p. 155-165.

SUMMERHAYES, C. P., DE MELO, U. AND BARRETO, H. T., 1976, The influence of upwelling on suspended matter and shelf sediments off Southeastern Brazil: Journal of Sedimentary Petrology, v. 46, p. 819-828.

SWIFT, D. J. P., STANLEY, D. J. AND CURRAY, J. R., 1971, Relict sediments, a reconsideration: Journal Geology, v. 79, p. 322-346.

WALTHER, J., 1885, Die gesteins bildenden Kalkalgen des Golfs von Neaplel und die Intstehung structurloser kalke: Zeitschift d. Deutschen Geologischen Gesellschaft, p. 1- 28.

WALTHER, J. AND SCHIRLITZ, P., 1886, Sdudien zur Geologie des Golfes von Neapel.:Zeitschrift d. Deutschen Geologischen Gesellschaft, p. 295-341.

ZALAFFI, M, 1963, Segnalazione di un livello con piccole coproliti fosfatiche e glauconitiche nel Miocene del Lazio meridionale: Geologica Romana, v. 2, p. 331-334.

# ROSIGNANO REEF COMPLEX (MESSINIAN), LIVORNESI MOUNTAINS, TUSCANY, CENTRAL ITALY

ALESSANDRO BOSSIO
*Dipartimento di Scienze della Terra, Università di Pisa, via S. Maria 53, 55126 Pisa, Italy*
MATEU ESTEBAN
*Carbonates International Ltd., Vilanova 70, Esporles, Mallorca, Spain*
RENZO MAZZANTI
*C.N.R., Dipartimento di Scienze della Terra, Università di Pisa, via S. Maria 53, 55126 Pisa, Italy*
ROBERTO MAZZEI, AND GIANFRANCO SALVATORINI
*Dipartimento di Scienze della Terra, Università di Siena, via delle Cerchia 3, 53100 Sienna, Italy*

ABSTRACT: The Rosignano Reef Complex ("Calcare di Rosignano") consists of two fringing coral reef units with small lagoons, the Acquabona (lower) and the Castelnuovo (upper), separated by restricted marine carbonates and conglomerates and associated unconformities. The Acquabona reef is exclusively made-up of *Porites*; whereas, the overlying Castelnuovo reef shows a higher diversity (3-5 coral species). Large stromatolitic domes associated with marine conglomerates also are present in Castellina. Red algal mounds are common in association with coral reefs but never reach significant geometries to be described separately. Basinal sections consist of basal lacustrine deposits overlain by restricted marine marls with three evaporitic units evolving into another unit of marine to brackish-freshwater facies post-dating the Castelnuovo reef. Litho- and biostratigraphic correlation provides a complete framework for the northernmost coral-reef complex of the Upper Miocene of the Mediterranean.

## INTRODUCTION

The term "Calcare di Rosignano" (Rosignano Limestone) has been used since the last century (Capellini, 1874, 1878; De Bosniaski, 1878) to refer to the Upper Miocene skeletal-rich limestones in the area of the Livornesi Mountains (Livorno, Fig. 1) in Tuscany (northern Italy). De Bosniaski (1879) was the first to mention the presence of a variety of reef-building corals in the Rosignano Limestone. Coral reefs occur in the Livornesi Mountains with other scattered, small outcrops in Casaglia and Volterra (Fig. 2; Bossio et al., 1978).

While the reefs in Tuscany are not the most spectacular outcrops in the Mediterranean, they are of great importance because of their stratigraphic complexity and precise correlation with several units of Messinian evaporites. Furthermore, these reefs are paleogeographically significant as the northernmost outcrops of the Upper Miocene coral reef belt in the Mediterranean (see final note).

Although this paper was prepared in its present form in 1991, continuing research work by the authors in the region (Bossio et al., 1991, 1992, 1993, 1994) fully corroborates the interpretations presented here.

### *Previous Work*

Chevalier (1961) summarized the extensive classical paleontological work in the region and corroborated the findings of eight coral species of De Bosniaski (1878), with *Porites lobatosepta* as dominant builder and local presence of *P. calabricae, Siderastraea crenulata, Tarbellastraea, Heliastraea*, and several varieties of *Cladocora* cf. *michelotti*. Chevalier (1961) also stressed the exclusivity of *Porites* in many of the outcrops, interpreted as an expression of difficult environmental conditions. He considered the Rosignano Limestone as a barrier reef complex trending northwest-southeast, with three backstepping ridges, across the Livornesi Mountains. Giannini (1962) presented a detailed map of this area recognizing a variety of lithologies (oyster, bivalve beds, corals, algal limestones, conglomerates, sands and clays) in the unit mapped as Rosignano Limestones. Esteban et al. (1978) and Bossio et al. (1978) presented new sedimentological and biostratigraphic data, correlating these outcrops with other Messinian coral reefs in the western Mediterranean (Esteban, 1979). Bossio et al. (1981, 1986) and Bartoletti et al. (1986) refined the stratigraphic and cartographic data in the area of the town of Rosignano. This paper summarizes our present understanding of the Rosignano Reef Complex and points out a correlation hypothesis with other Upper Miocene units.

### *Geologic Setting*

The Livornesi Mountains are part of the Ligurian nappe overthrusting the Toscana unit, one of the stacked structural units that form the northern Appennines Alpine foldbelt (Fig. 1). The Ligurian nappe contains intensively deformed Jurassic radiolarites, serpentinites, basalts, gabbros, deep-water Cretaceous sediments and Paleogene flysch. These pre-Upper Miocene rocks are collectively termed "basement" in the sketches of Figures 2 and 3. The paroxysmal orogenetic phase occurred during Serravallian-Tortonian time and was followed by block-faulting and rifting during Late Miocene and Pliocene-Pleistocene times, with predominant north-northwest/south-southeast trends (Trevisan, 1952). Semi-allochtonous Langhian, Serravallian and Lower Tortonian sandstones occur to the east (Mazzanti et al., 1981), but in the Livornesi Mountains the first autochtonous Neogene marine sediments are dated as Messinian age (Ruggieri, 1956; Giannini, 1962). However, extensive lacustrine deposits (500-900 m thick in adjacent basins to the

Models for Carbonate Stratigraphy from Miocene Reef Complexes of Mediterranean Regions,
SEPM Concepts in Sedimentology and Paleontology #5, Copyright © 1996,
SEPM (Society for Sedimentary Geology), ISBN 1-56576-033-6, p. 277-294.

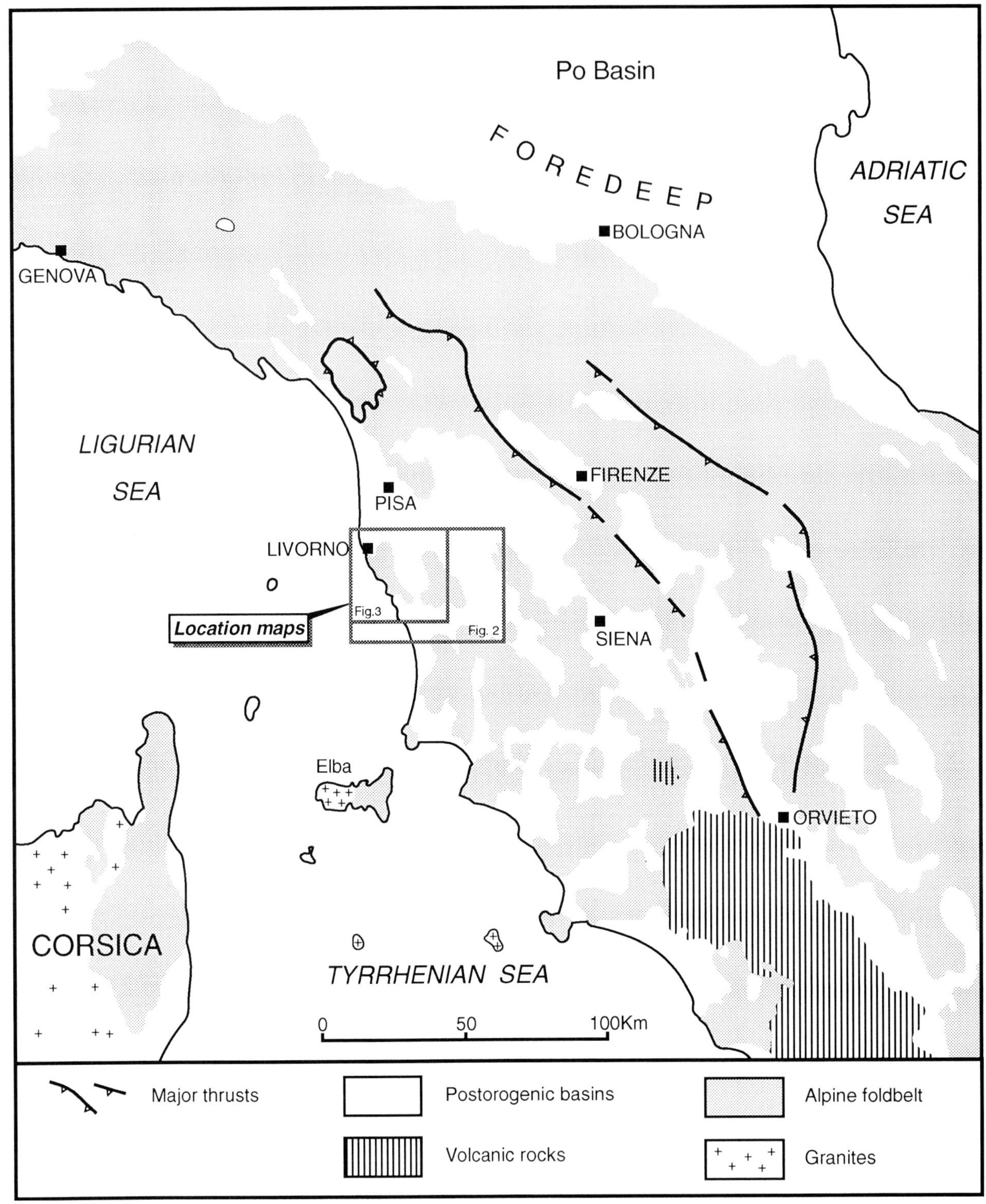


Fig. 1.—Location of the Rosignano Reef Complex. Neogene and Pleistocene post-paroxysmal basins of the northern Appennines (modified from Pieri, 1975).

southeast) below the marine Messinian section (Bossio et al,. 1978, 1981) are dated as "Turolian" age (Lazzarotto and Sandrelli, 1979); according to Giannelli et al. (1981), and Mazzanti et al., (1981), these lacustrine deposits belong to a Turolian interval equivalent to the Late Tortonian time.

The block-faulting that started in Late Miocene time divided the region into multiple basins (Fine-Tora, Cécina, Volterra and others) limited by emerging topographic highs that approximately correspond to present-day "monti" (Fig. 2). Some outcrops of Calcare di Rosignano have suffered evident tectonic

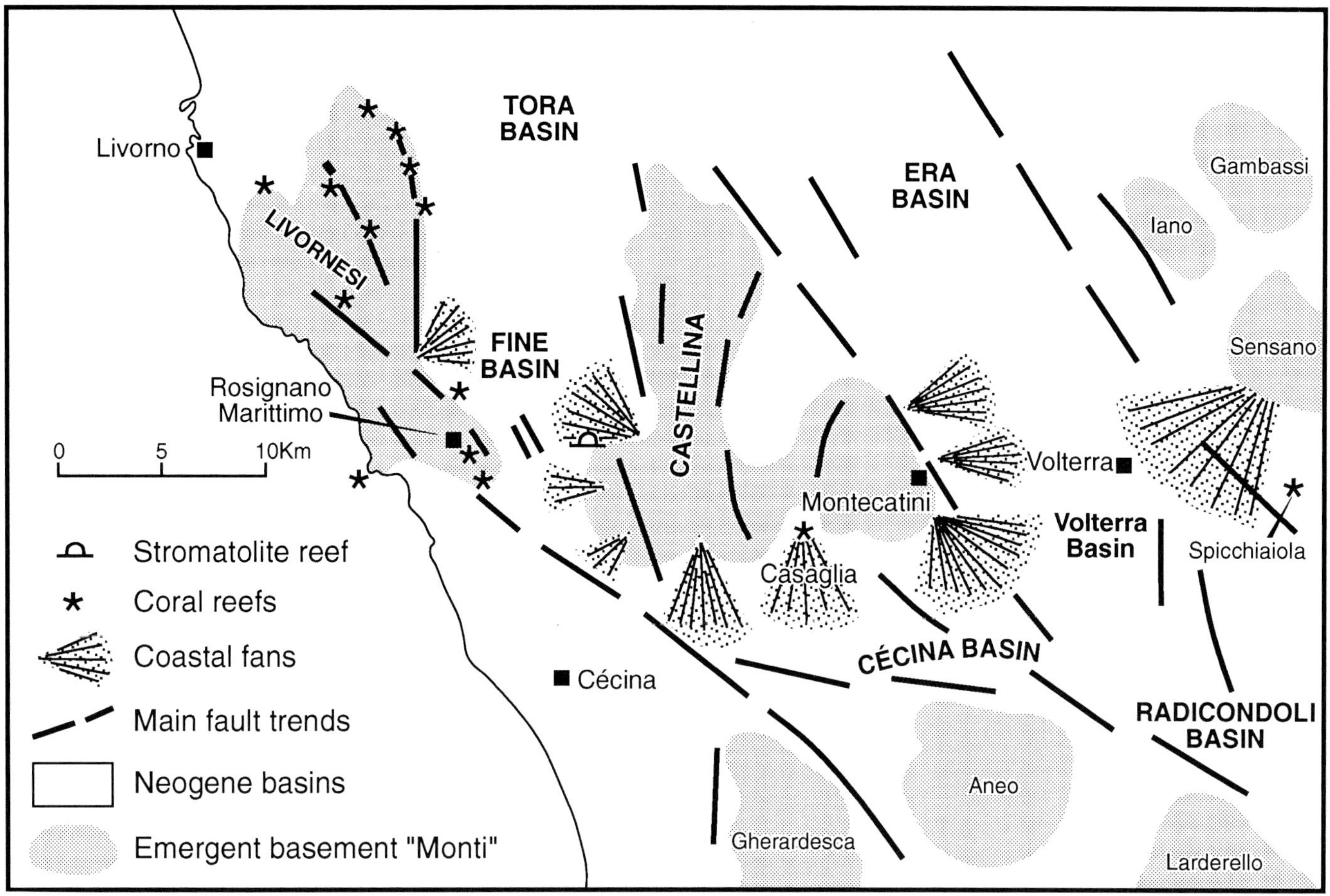


Fig. 2.—Schematic paleogeography of the Neogene and Pleistocene basins between Livorno and Volterra, showing locations of the reef outcrops, coastal fans and emergent areas ("monti"). Modified from Bossio et al. (1978).

tilting, but, in general, most of the depositional dips and slopes are essentially preserved with only minor modifications. The total thickness of Neogene sediments can be up to more than 1,000 m, but the distribution is very irregular. The basal lacustrine sediments ("Serie Lignitifera") are followed by a thick (up to more than 800 m) marl sequence with intercalations of evaporitic units (including halite in the Volterra Basin), sandstones and conglomerates representing a wide range of fresh-water, brackish, restricted marine and hypersaline environments. This marly sequence is much thinner towards the margins of the basins and locally grades into alluvial fans, delta fans and shallow-water carbonates, commonly less than 100 m thick.

### *Stratigraphy of the Rosignano Area*

The Rosignano Limestones occur directly on the basement in most outcrops around the Livornesi Mountains, but in adjacent basins to the east (Cecina, Volterra), it also overlies the basal lacustrine unit. The best outcrops occur along the western margins of the Fine Basin, near the town of Rosignano Marittimo (Figs. 2, 3), where the reference section (Figs. 4, 5) is here established (modified from Bossio et al., 1978, 1981; Bartoletti et al., 1986) with the following units:

1. Jurassic basalts (doleritic, ophitic) and serpentinites;
2. Rosignano Limestones, consisting of three informal units (Acquabona Dolomite, Lagoonal Carbonate Interval and Castelnuovo Limestone) separated by subaerial-exposure surfaces and shallow coastal-marine conglomeratic wedges ("Cantine Conglomerates" and "Villa Mirabella Conglomerates"). These conglomerates (with minor carbonates) are extensively developed on the eastern margin of the basin ("Sant'al Poggio Conglomerates"). Although part of the Rosignano unit is dolomitized, the term "Rosignano Limestones" is here maintained for consistency with the literature;
3. Marine marls, commonly laminated and capped by marine diatomitic marls ("tripoli of Paltratico");
4. Lower gypsum unit, with marine marls and local wedges of sandstones and conglomerates ("Rio Sanguigna Formation" of Bartoletti et al., 1986);
5. Lacustrine marls and gypsarenites ("upper gypsum unit"), with a mappable conglomeratic intercalation ("Sabbie e conglomerati della Villa di Poggio Piano"); this unit has the characteristic "Lac Mer" facies; and
6. Shales, silts and sandy shales ("Argille Azzurre") with a well-dated basal Pliocene section (*Sphaeroidinellopsis seminulina* s.l. zone, *Discoaster variabilis* s.l. zone).

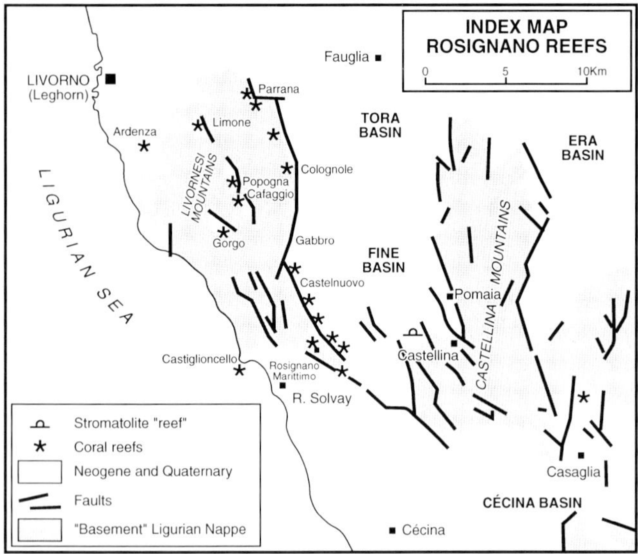


Fig. 3.—Index map of the main reef localities in the Livornesi Mountains. Structural sketch simplified from Lazzarotto and Mazzanti (1978).

Along the western margin of the Fine Basin, the section is predominantly marly and reduced to 30-75 m in thickness. In the middle of the Fine Basin, the marly section can be up to 500 m thick, capped with a few hundred meters of Pliocene shales (Bossio et al., 1981). The micropaleontological analysis (Bossio et al., 1986) of the marls reveals a complex paleoecological succession and lateral variations.

The Rosignano Limestone, cropping out in the Livornesi Mountains, is capped by conglomerates, recrystallized breccias, evaporitic dolomites and stromatolites, with only few occurrences of gypsum wedges. This implies a major pinchout of the thick basinal section of marls and gypsum.

### *Age of the Rosignano Reef Complex*

Ruggieri (1956), re-examining previous paleontological data, attributed a Messinian age to the Rosignano Limestones. Bossio et al. (1986) confirmed the occurrence of Early Messinian microfauna in the Castelnuovo Limestone, with the presence of dextral *Globorotalia acostaensis* and *Bulimina echinata*. The correlation with the type sections in Sicily and elsewhere in Italy indicates a late Early Messinian age. As stressed in Mazzanti et al. (1981), the "tripoli" in Tuscany is correlated (by planktonic foraminifers and calcareous nannofossils) with the upper part of the pre-evaporite "tripoli" of the Sicilian type section, and, as a consequence, it is likely that all the Rosignano reef complex is Early Messinian age. The ostracods in the slopes of the Acquabona Dolomite also are consistent with, and supports, a Messinian age (*Pokornyella italica, Capsacythere sicula, Cletocythereis minor, Loxoconcha punctatella*). Hypothetical correlation with the 4th-order sequence stratigraphy in the western Mediterranean (Esteban, this volume) could suggest the possibility of a fini-Tortonian age for the Acquabona Dolomite.

## REEF FACIES ANALYSIS

In the Rosignano Limestones of the Livornesi Mountains, there are two distinctive types of coral reefs in terms of lithofacies, composition and geometries: the Acquabona and the Castelnuovo. This was already noted in Bossio et al. (1978), who considered the Castelnuovo type as a series of patch reefs (locally fringing) and the Acquabona as a barrier reef. However, Bossio et al. (1978, 1981), Esteban et al. (1978) and Bartoletti et al. (1986) insisted that these two types were separated by a major erosional surface and that there was no evidence of a frontal barrier reef during Castelnuovo time and little evidence of a lagoon during Acquabona time. Figure 5 shows the inferred relationship between the Acquabona and Castelnuovo reefs, with the Lagoonal Carbonate Interval and the overlying marls.

### *Acquabona Reef*

The best exposures of the Rosignano Limestones, offlaping basaltic and serpentinitic rocks, are in the quarry of Acquabona (Rosignano Marittimo). Patches, less than 10 m thick, of flat-lying reefal limestone, attributed to the Acquabona unit (Bartoletti et al., 1986), occur in the flat-top basement hills around the quarry. They are associated with a basal conglomerate, up to 2 m thick ("Cantine Conglomerates"), of basement pebbles grading up into a Miocene coral breccia, skeletal packstones and typical in-place *Porites* stick framework. These conglomerates occur as pockets or channels in a scoured erosion surface.

*Reef framework facies.—*
The Acquabona reef framework (Fig. 6A) is exclusively constructed by *Porites lobatosepta* (Chevalier, 1961), with variable amounts of encrusting red algae, bryozoans, serpulids (Fig. 6G), encrusting forams and extensive micritic crusts 2-8 cm thick (Figs. 6E, F). The colonial morphology is dominated by 1- to 3-m-long, 2- to 3-cm-thick sticks, with only a few patches of horizontal laminar colonies in the lower parts or at the base of the vertical bushes of sticks. Most of the coral skeletons have been intensively leached. Organic boring is not intense. Sediments are wackestones and packstones (Figs. 6B, D), with abundant fragments of *Halimeda*, echinoids, *Pecten* and other bivalves, few rare gastropods, red algae, peloids, miliolids, *Textularia, Amphistegina,* and *Elphidium*. The coral framework forms massive wedges (up to 8 m thick) in the upper part of the quarry, thinning down and intercalated into talus-slope sediments. The coral sticks are not vertical but inclined towards the depositional slope. It is not possible to ascertain the amount of truncation of the reef framework under the subaerial exposure surface. The flat-lying patches of reef framework around the quarry may represent shallower reef-flat areas, suggested by the presence of *Porites* fragments up to 10 cm in diameter, few vertebrate bones, intense sponge boring, and abundant serpulids, bryozoans and gastropods.

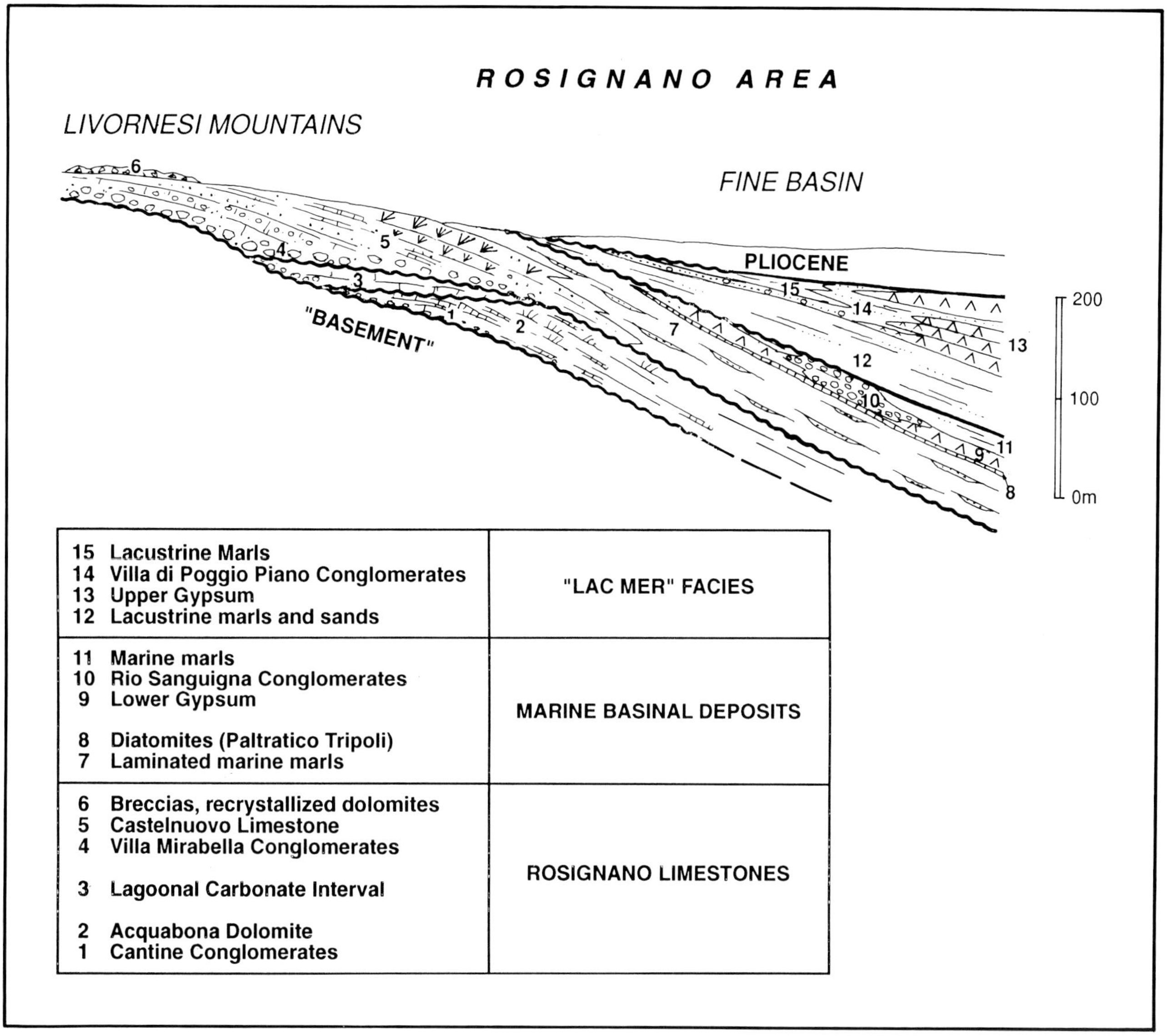

ROSIGNANO AREA
LIVORNESI MOUNTAINS
FINE BASIN
PLIOCENE
"BASEMENT"
1
2
3
4
5
6
7
8
9
10
11
12
13
14
15
200
100
0m
15 Lacustrine Marls
14 Villa di Poggio Piano Conglomerates
13 Upper Gypsum
12 Lacustrine marls and sands
"LAC MER" FACIES
11 Marine marls
10 Rio Sanguigna Conglomerates
9 Lower Gypsum
8 Diatomites (Paltratico Tripoli)
7 Laminated marine marls
MARINE BASINAL DEPOSITS
6 Breccias, recrystallized dolomites
5 Castelnuovo Limestone
4 Villa Mirabella Conglomerates
3 Lagoonal Carbonate Interval
2 Acquabona Dolomite
1 Cantine Conglomerates
ROSIGNANO LIMESTONES

Fig. 4.—Regional Upper Miocene stratigraphic units and correlation of the Rosignano reef complex.

Maestá
Acquabona Quarry
W - E
Castelnuovo Limestone
Lagoonal Interval
Acquabona dolomite
Le Melette
Section Figure 10
BASEMENT
QUATERNARY
30m
approx.
200m
PLIOCENE
Gypsum
Tripoli
laminated marine marls
STRATIGRAPHIC RELATIONSHIPS OF THE ROSIGNANO LIMESTONE IN ROSIGNANO MARITTIMO

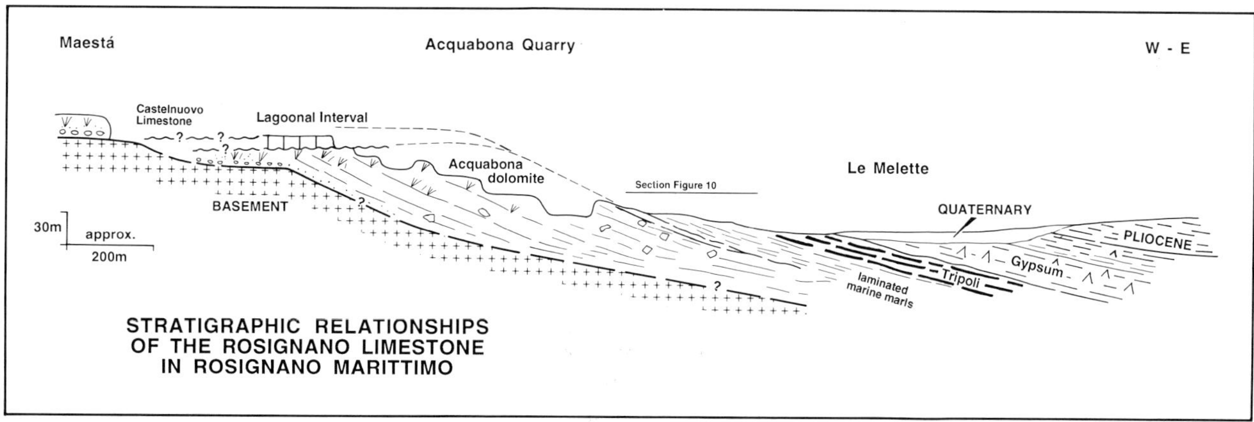

Fig. 5.—Stratigraphic relationships of the Acquabona Dolomite, Lagoonal Carbonate Interval and Castelnuovo Sandy Carbonate units of the Rosignano Limestone in the type locality of Rosignano Marittimo.

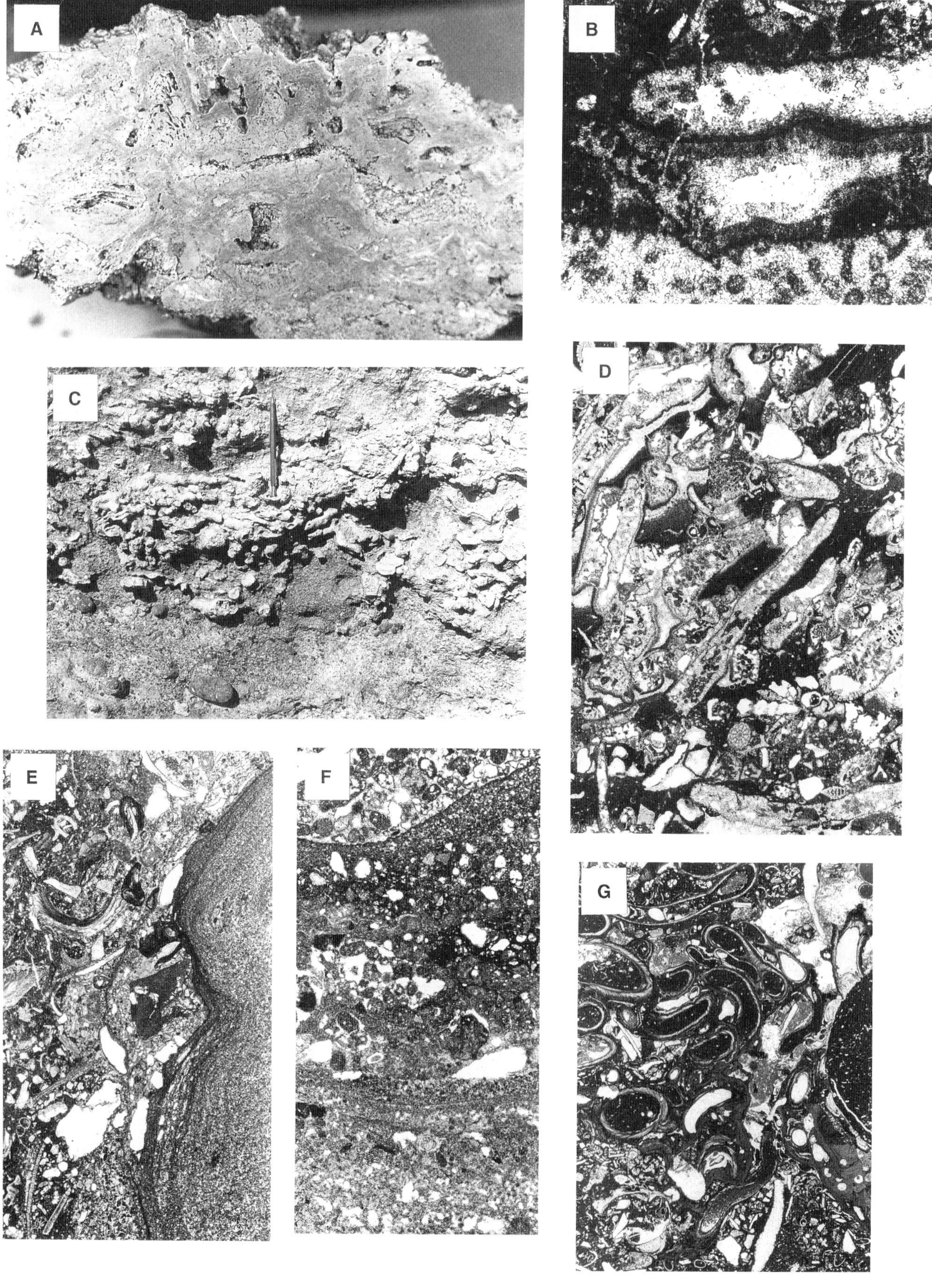
A
B
C
D
E
F
G

Fig. 7.—Partial view of the quarry at Acquabona, with reef slope sediments (Acquabona and Castelnuovo) and part of the Fine Basin and its Pliocene cover in the background. Massive beds (left) are *Porites* breccias and *Halimeda* packstones overlain by bedded skeletal wackestones and packstones (central and right) with a general fan-array disposition to the right with red algae, molluscs and blocks of *Porites* framework. These dips are considered to be close to the depositional angle of slope, with only minor modifications by post-reef fractures.

*Off-reef talus slope facies.*—

Most of the quarry corresponds to a proximal reef-talus slope prograding to the east (varying from east-northeast to east-southeast) with dips up to 25° and maximum vertical thickness of about 60 m (Fig. 7). The quarry is affected by numerous macro-microfractures that do not seem to have disturbed significantly these essentially depositional clinoforms. The upper part of this clinoform is truncated by a flat-lying, undulating erosion surface overlain by horizontal Messinian carbonates (Lagoonal Carbonate Interval). The upper thick beds thin downward and interfinger downslope with poorly lithified, fine-grained carbonate sediments; locally 3- to 10-cm-thick lamination is the most conspicuous structure. Scattered, round pebbles of the basement rocks are present in the lower part of the clinoform. The general trend of the reef framework and slope facies indicates downstepping progradation, with each successive coral framework colonizing the previous upper slope.

The reef slopes show three main types of lithologies:

A. *Halimeda* packstones, in white thick strata, with abundant *Halimeda* plates showing preferred grain orientation parallel to bedding, and vague cm-scale lamination. Matrix is peloidal, with fine-sand size fragments of echinoids, molluscs, red algae, bryozoans, rare planktonic forams and few pebbles of *Porites* fragments. Floored interstices in intergranular pore spaces are common, moldic *Halimeda* porosity is pervasive (Figs. 6B, D). This lithology is well developed in the upper part of the quarry and wedges downslope.

B. Laminated crusts and skeletal grainstones, forming layers 3-30 cm thick, are intercalated in the *Halimeda* packstones and in the coral framework. The laminated crusts contain layers of peloidal micrite (Figs. 6E, F) and layers of encrusting forams, serpulids and red algae. These peloidal micrite crusts are very common in the Messinian reefs of the western and central Mediterranean and are interpreted as cyanobacterial, stromatolitic (Pedley, 1973; Riding et al., 1991). The skeletal grainstones are medium-sorted and dominated by fragments of molluscs, echinoids and red algae, with scattered serpulids, bryozoans and miliolids.

C. Poorly lithified, fine-grained, mud-supported carbonate sediments, locally marls, with thick layers of coarse-grained red algae wackestones and packstones. Sporadically, there are concentrations of fragments of *Porites* branches (2 cm) and blocks of *Porites* framework (up to 3 m in diameter). Intense burrowing has destroyed most of the lamination. Layers of bivalves (*Pecten, Spondylus*, oysters, and others) and rhodoliths, up to 30 cm thick, are common. Other fossils include a few entire echinoids and branching bryozoans. Chevalier (1961) mentioned *Pecten dunkeri, P. stazzanensis*, and *Flabellipecten besseri.* However, most skeletal grains are reduced to fine detritus mixed with low amounts of quartz silt, and poorly preserved microfauna (benthic forams, few planktonic forams and ostracodes). Bossio et al. (1978) listed *Neoconorbina terquemi, Cibicides lobatulus, Florilus boueanus, Elphidium sp., Quinqueloculina sp., Cythereidea acuminata, Callistocythere pallida, Ruggieria tetraptera, Aurila interpretis, A. philippi, A. convexa, Cletocythereis minor, Heliocythere magnei, Loxoconcha punctatella, L. variesculpta, L. cristatissima, Xestoleberis reymenti.*

Distal slope-basin transition facies and basinal equivalents of the Acquabona reef are not exposed in the Fine Basin. Outcrops of basement rocks in the basin are covered by evaporitic and

Fig. 6.—Some aspects of the Rosignano coral reefs. (A): Lower part of coral framework in Acquabona; laminar *Porites* with incipient vertical coral sticks and some broken coral fragments. Coral framework is intensively coated by laminated-peloidal micritic crusts. (B): *Halimeda* packstones in the Acquabona reef slopes. Some *Halimeda* plates are partly leached; note fibrous intergranular cement (dolomitized) followed by a more calcitic fringe. (C): Basal contact of the Castelnuovo reef framework with the basal Mirabella conglomerate, Rosignano Marittimo. (D): Partly leached *Halimeda* packstone from Acquabona reef slopes with micrite flooring interstices in intergranular porosity. (E, F): Laminar micrite-peloidal coatings on corals and sediments of the Acquabona reef framework. Some encrusting forams are trapped in E; in F the micritic coatings incorporated abundant skeletal and peloidal grains. (G): Serpulid-rich knobs. Serpulids are common in the Rosignano Limestone, particularly abundant in the Lagoonal Interval.

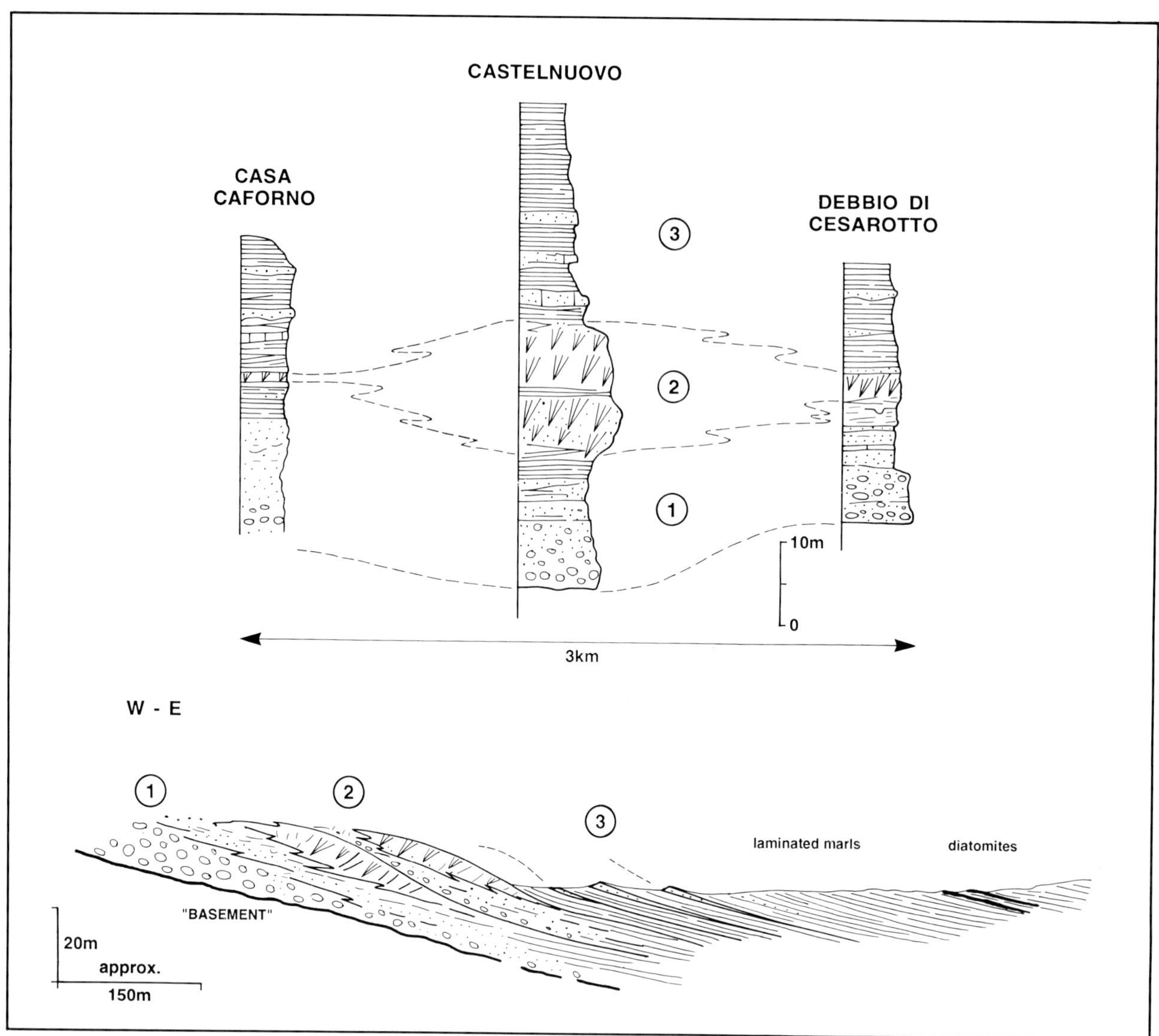


Fig. 8.—The Castelnuovo reef in the type locality of Castelnuovo della Misericordia with the transition into the basinal marls and diatomites. 1: conglomerates, sands, marls, 2: sandy and marly carbonates with coral reef framework, 3: marls and thin sandy carbonates.

stromatolitic carbonates and conglomerates, suggesting important erosional processes or non-deposition.

*Post-reef facies.—*

The Acquabona reef clinoforms are truncated and overlain by the horizontally bedded Lagoonal Carbonate Interval in the upper part of the Acquabona quarry. The erosional surface has an undulose morphology, with vertical relief up to 3 m and crests separated by 6-15 m. This could represent a karstic topography, possibly inheriting a depositional buttress morphology in the eroded reef. The Lagoonal Carbonate Interval consists of horizontal, well-bedded, well-lithified shallow-water limestones, less than 30 m thick. The basal 20-40 cm contain very abundant serpulids directly encrusting the surface, followed by 3 m of well-bedded wackestones and mudstones with fragments of bryozoans, serpulids, echinoids, red algae (including rhodoliths up to 4 cm in diameter), gastropods, oysters, *Pecten* and few small *Porites* fragments. This is capped by one layer of hemispheroidal stromatolites 1 m in diameter, partly truncated by a minor erosion surface. The rest of the section consists of skeletal packstones and wackestones, with serpulids, *Halimeda*, red algae, bryozoans, *Pecten*, echinoids, oysters, miliolids, peneroplids and few small *Porites* fragments. This sequence is repeatedly (2-3 times) interbedded with 20-to 60-cm-thick layers of abundant serpulids, oolitic-peloidal grainstones and small oncoliths. The upper 5 m of the Lagoonal Carbonate Interval are intensively recrystallized and capped by a thin caliche crust with abundant rhizocretions. This crust is overlain by 5 m of poorly

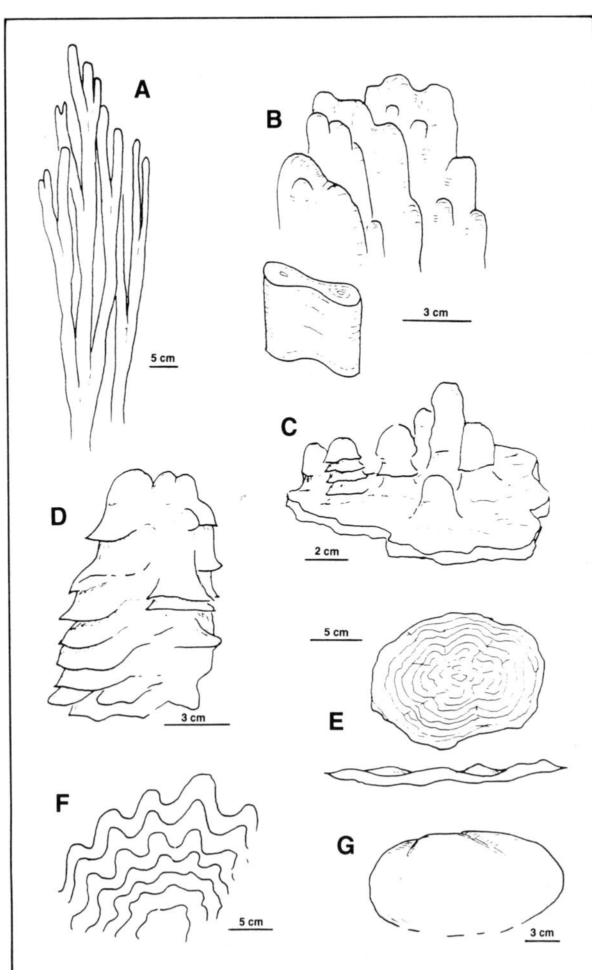


Fig. 9.—Main variations of coral colonial morphology in the predominant reef builders (*Porites* and some *Tarbellastraea*) of the Castelnuovo reef. (A): thick branching sticks, (B): flat branches or blades, (C): laminar colonies with incipient sticks, some with minor aprons (tendency to laminar growth), (D): knobby, columnar colony with aprons, (E): laminar, dish-like colony; view of the underside with concentric growth lines and lateral view, (F): knobby, cauliflower or botryoidal, and (G): subspherical to elliptical head, also common in other corals (*Siderastraea* and others). From Bossio et al. (1978).

exposed whitish marls and fine-grained skeletal packstones with rare planktonic forams. The Lagoonal Carbonate Interval is interpreted as a shallow-marine lagoon, with periodic recurrences of more restricted (brackish? hypersaline?) marginal facies; this unit has many similarities with the facies of the Terminal Complex (Esteban, 1979) of the Messinian reefs in the western Mediterranean. The nature of the barrier defining the lagoon of the Lagoonal Carbonate Interval cannot be demonstrated (shoals?, reefs?). Detailed mapping (Bartoletti et al., 1986) proves that the Castelnuovo reef overlies the Lagoonal Carbonate Interval and the Acquabona reefs.

*Acquabona reef distribution.—*

There are no other well documented Acquabona reef outcrops in the Livornesi Mountains and the rest of the studied area. The small outcrop of Monte Tignoso (Ardenza, North of the Livornesi Mountains, Fig. 3), including large massive colonies of *Porites*, could be tentatively considered as part of an Acquabona coral talus breccia. This breccia is capped by a karstic erosion surface and oolitic-serpuliditic carbonates similar to the Lagoonal Interval. Some of the outcrops and quarries in the Casa Limone-Limoncino (Fig. 3) also expose possible Acquabona reef talus, with massive *Porites* (at least 1m in diameter) and capped by oolitic-oncolitic carbonates and intense microsparitization. These limited and problematic outcrops are insufficient to define the paleogeography of the Acquabona reef.

## *Castelnuovo Reef*

The Castelnuovo reef is characterized by abundant terrigenous-siliciclastic mixing (Fig. 6C) and wider variety of corals and colonial morphology than in the Acquabona reef. The very long coral sticks, thin micritic crusts and *Halimeda* packstones, typical of the Acquabona reef, are absent or uncommon. Most corals are well preserved, without leaching, and the reef rock is poorly cemented. The Castelnuovo reefs are much smaller (less than 15 m thick) and do not show well-developed, prograding clinoforms. The Castelnuovo reef type is extensively distributed as laterally discontinuous mounds in the Livornesi Mountains (Fig. 3), occupying an apparently higher stratigraphic position than the Acquabona reefs in most outcrops.

*Pre-reef facies.—*

A sandy conglomerate ("Villa Mirabella Conglomerates" of Bartoletti et al., 1986), 2-100 m thick, occurs at the base of the Castelnuovo reefs, with very gradual vertical and lateral transitions between the two lithologies. Basement pebbles are up 40 cm in diameter. Plant debris (lignite), ferruginous coatings on the pebbles, and coarse fibrous and prismatic calcite cements are common in some intervals, and, towards the upper part, organic boring and red algae coatings become more common. The sandy conglomerates grade upwards into sandy marls and, locally, into coral reefs. In some localities, these conglomerates contain scattered skeletal grains without traces of coral reefs and form most of the "Rosignano" mapping unit below the basinal marls. These conglomerates are interpreted as delta fans in coastal areas, locally supporting mounds and patches of coral reefs.

In some localities around Rosignano Marittimo, the Castelnuovo reefs seem to occur above the white marls of the upper part of the Lagoonal Carbonate Interval. However, observations are limited by the quality of the exposures in cultivated fields and complex fracturing; therefore, a completely clear relationship cannot be established.

*Reef framework facies.—*

The outcrops around Castelnuovo della Misericordia and Castelvecchio provide the type locality (Fig. 8). The reef

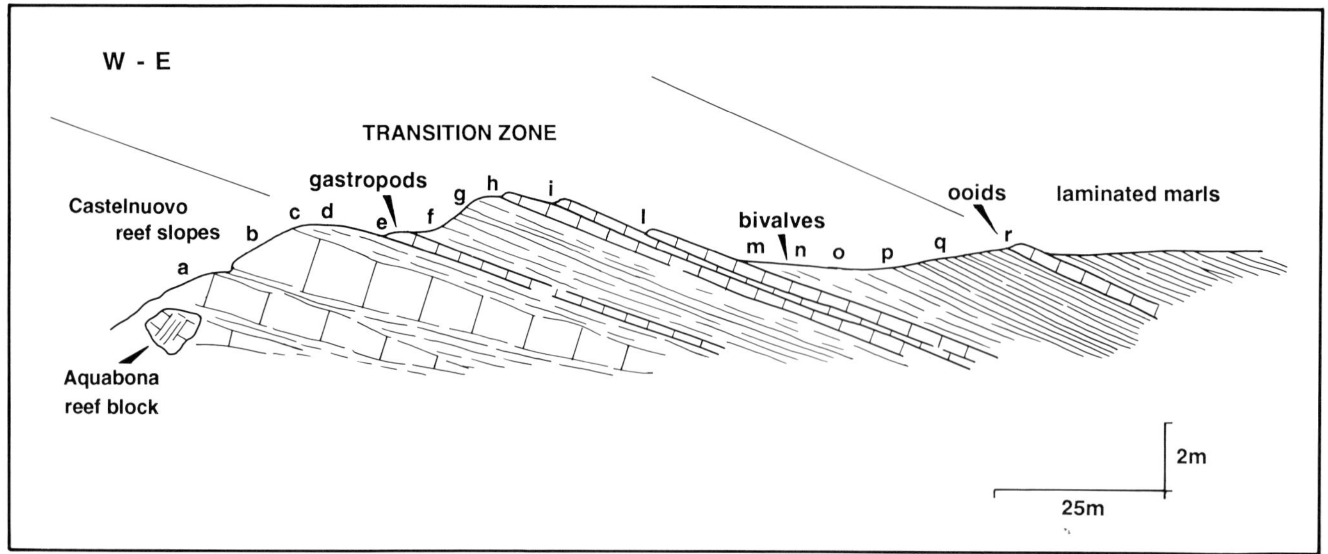


Fig. 10.—The transition zone between the Castelnuovo reef slopes and the overlying laminated marls (modified from Bossio et al., 1981). Letters (a-d) indicate samples described in the text.

framework is up to 15 m thick, with a predominance of *Porites lobatosepta* in a sandy marly matrix. Chevalier (1961) also mentions *Heliastraea, Siderastraea crenulata, Porites* cf. *calabricae, Porites* sp and abundant red algae, associated to *Pecten dunkeri, P. stazzanensis, P. marvilensis, Chlamys multistriata*, crustaceans and bryozoans. Some horizons contain abundant rhodoliths, *Arca*, oysters and serpulids. The reef framework contains a wide variety of colonial morphologies in *Porites* and *Tarbellastraea* (Fig. 9), indicating different responses to local environmental conditions. The sediment contains up to 40% quartz sand and silt. The extensive micropaleontological analysis of Bossio et al. (1986) reveals abundant and varied ostracofauna (*Pokornyella, Capsacythere, Tenedocythere, Loxoconcha, Aurila, Xestoleberis, Cytheridea, Callistocythere, Cletocythereis, Loxocorniculum, Neonesidea,* and others) and less diversified benthic foraminifera association (miliolids, *Ammonia, Rosalina, Glabratella, Neoconorbina, Elphidium,* and others). The Castelnuovo reefs are interpreted as laterally discrete, elongated mounds, less than 15 m high and a few hundred meters wide; due to poor exposure, their lateral extent is difficult to evaluate (0.5-2 km?). These coral reef mounds colonized and interacted with siliciclastic deposits interpreted as episodically active, abandoned channels and bars in small delta fans.

*Off-reef facies.—*
The Castelnuovo reef is flanked by marls and sandy marly limestones that could be considered as a poorly developed slope in shallow waters (lobes, abandoned bars in a delta-front setting?), with *Lucina, Turritella, Conus* (Figs. 10A-D), rapidly grading upwards into the laminated marls and diatomites. In many localities, the flanks of the Castelnuovo reef are bounded by fractures and tectonically tilted down to the marly section. This could give the impression of steep reef slopes and makes it difficult to evaluate the paleodepths around the Castelnuovo reefs. The benthic forams are more diversified than in the reef framework with abundant buliminids, bolivinitids, cassidulinids, *Textularia, Buccella, Calcarina, Planorbulina, Cibicides*, etc. The abundance of *Bulimina echinata, Bolivina dilatata* and *B. detellata* is suggestive of conditions of poor bottom oxygenation and tendency to hypersalinity (Bossio et al., 1986). The ostracofauna is enriched with abundant *Hiltermannicythere, Leptocythere* and *Ruggieria*. Planktonic forams (*Orbulina suturalis, O. bilobata, Globorotalia acostaensis, Globigerina bulloides, G. nepenthes, G. quinqueloba, Globigerinoides obliquus extremus, G.* gr. *quadrilobatus, Globorotalia pseudoobesa,* and *Globigerinita bradyi*) and calcareous nannoplankton (*Calcidiscus leptoporous, C. macintyrei, Discoaster* cf *asymmetricus, D. brouweri, D. intercalaris, D. mendomobensis, Helicosphaera carteri, Pyrocyclus* spp., *Sphenolithus abies, Syracosphaera* sp. and others) are present but not abundant (see Bossio et al., 1986, for extended description).

*Post-reef facies.—*
The flanks of the Castelnuovo reef are overlain by, and partly grade into, the laminated marls. Laminae are 1-3 mm thick with alternating light-gray calcareous laminae and thicker dark-gray shaly laminae. Bossio et al. (1986) mentioned the presence of scattered radiolarians, sponge spicules, bryozoans, echinoid debris, molluscs, microgastropods, crustaceans, fish and plants (lignite). The micropaleontological analysis reveals the marked shallow, oligotypic character for most of the succession below the diatomitic marls and the similarity with the flanks of the Castelnuovo reefs. The predominance of *Bulimina* and *Bolivina*, immature, impoverished ostracofauna and the rare and dwarfed planktonic forams and nannoplankton are considered indicators

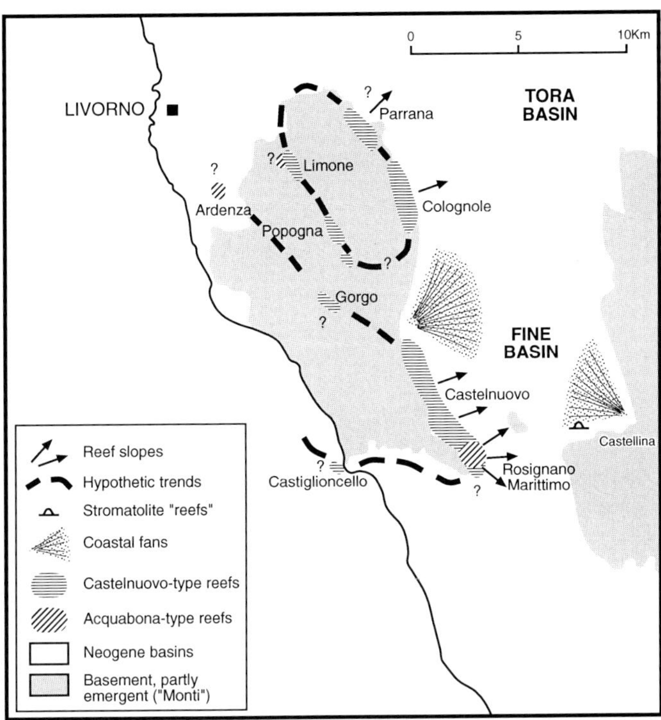


Fig. 11.—Distribution of Acquabona and Castelnuovo reefs in the Livornesi Mountains, with hypothetic grouping of two paleogeographic trends.

of higher-salinity, restricted marine conditions (Bossio et al., 1986).

A distinct post-reef facies occurs on the distal slope of the Castelnuovo reef (Fig. 10). A 6-m-thick transitional zone (samples e-r, Fig. 10) of gray silty marls with few intercalations of 20-to 40-cm-thick limestones occurs between the typical Castelnuovo reef-slope facies and the basinal laminated marls and diatomites. The basal bed (e) of this transitional zone shows abundant, small brownish gastropods, partly phosphatized. The marly sections have abundant, blackened fragments of oysters and *Chlamys*. The carbonates in the middle part of the section (i) are poorly-sorted skeletal grainstones (bivalves, gastropods, *Halimeda*, echinoids, peloids, benthic and rare planktonic forams). The primary and secondary moldic porosity is pervasively cemented by large crystals of poikilotopic gypsum. The marly carbonate beds of the upper part (r) contain packstones of micritized ooids, with abundant debris of molluscs, ostracods, serpulids, miliolids and other benthic forams, rare planktonic forams, and less developed late gypsum cements. These lithologies are clearly absent in the Acquabona and Castelnuovo reef-slope facies and indicate an important change in environmental conditions. The presence of oolitic grains and abundant gastropods could suggest the possible occurrence on top of the Castelnuovo reef (now eroded away) of a carbonate unit similar to the Lagoonal Carbonate Interval (Fig. 4). The micropaleontological analysis (Bossio et al., 1978) shows similar associations to the marly intercalations in the reef-slope facies, except for a wider variety of benthic forams (including *Bulimina, Rosalina, Cibicides, Elphidium, Calcarina, Fursenkoina, Spirillina, Bolivina, Patellina* and *Astegerinata*) and rare planktonic forams (*Orbulina* spp., *Globorotalia acostaensis* and *Globigerina bulloides*). The upper part grades into the typically impoverished laminated marine marls (*Bolivina dilatata, Bulimina echinata, Protoelphidium granosum, Criboelphidium decipiens* and *Cythereidea acuminata*) and partly covered diatomitic marls.

In a landward position, the post-reef facies is poorly exposed. There are thin patches of a variety of lithologies (recrystallized stromatolitic and oolitic carbonates, mudstones with evaporite molds, fresh-water gastropods, conglomerates, coarsely crystalline dolomite, breccias, silicified carbonates, etc.), without a clear relationship with the underlying Castelnuovo Limestone. These lithologies are common in the Terminal Complex of the Messinian reefs in the western Mediterranean (Esteban, 1979; this volume).

*Regional distribution.*—

The northeastern flank of the Livornesi Mountains, between Parrana and Colognole (Fig. 3) contains the northernmost Upper Miocene coral reef in the entire Mediterranean (see final note). Facies, size and geometries are very comparable to the Castelnuovo reef in the type locality, but the reef framework is practically monospeciphic (*Porites lobatosepta*). In the area of Popogna, the Castelnuovo reef is tilted 40°, the framework is less than 5 m and contains abundant *Siderastraea* and *Porites* and few *Heliastraea*. Locally, the matrix around the framework contains up to 70% quartz sand and silt. The upper part of the reef in Popogna is a sandy marl, with abundant small branching ahermatypic corals (*Cladocora* cf. *michelotti, C. michelotti* var. *popognae, C. liburnensis*; Chevalier, 1961), followed by skeletal sandy layers with gastropods and stromatolites. The benthic forams are characterized by the abundance of miliolids, *Elphidium, Cibicides, Ammonia, Nonion, Amphistegina*, peneroplids and textularids; these are more conspicuous than in the Castelnuovo-Parrana area. The basal conglomerates contain *Pecten stazzanensis, Arca diluvii, Ostraea lamellosa, Vulgocerithium costatum* var. *longoutriculata, Venus plicata, Aloidis gibba, Turritella turris, Conus pirula*, etc.(Chevalier, 1961). Overall, the Popogna outcrops indicate a more marginal-marine, restricted and physically stressed environment than the Castelnuovo-Parrana trend, as noted by Chevalier (1961).

Other isolated outcrops are more difficult to attribute to the Castelnuovo reef trend. In the Gorgo-Cafaggio area, reefs and conglomerates are clearly interstratified, locally with ferruginous hard-ground surfaces. *Porites* and a few *Siderastraea* are the only corals encountered with concentrations of *Halimeda*, serpulid and peneroplid layers and some large (1m) hemispheroidal heads of *Porites*. These lithologies are somewhat similar to the Lagoonal Carbonate Interval, but the regional context favors a Castelnuovo trend. The outcrops of the Limone area contain lithologies similar to the Lagoonal Interval and Castelnuovo reef, with sandy *Porites* framework dominated by *Porites*. In the area of Castiglioncello, there are abundant small

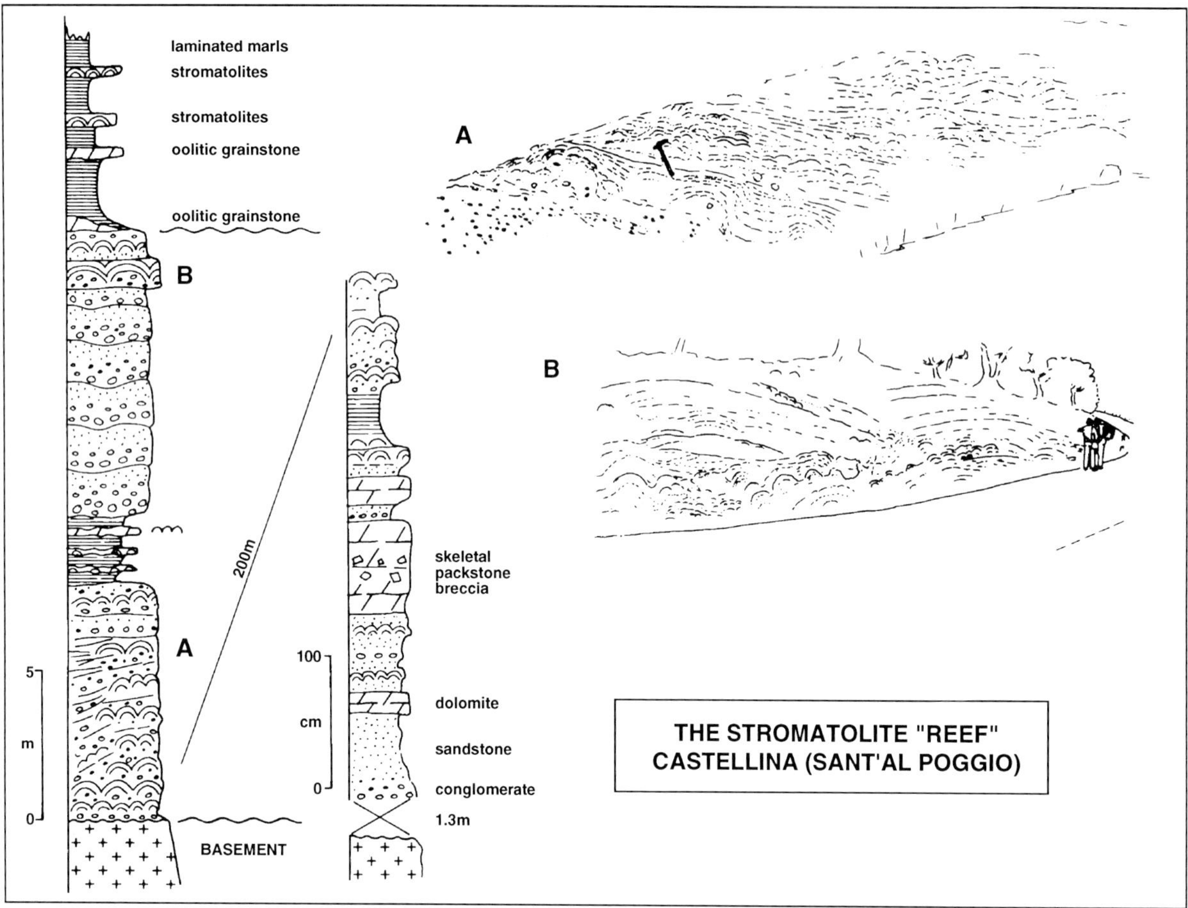


Fig. 12.—The stromatolite reef of Castellina (Sant'al Poggio), intergrowing with conglomeratic sand deposition. (A, B): Details of the complex stromatolitic dome or mound structure. Hammer in A is 50 cm long.

branching *Porites* and red algae in a siliciclastic-rich matrix, also tentatively attributed to the Castelnuovo trend. In summary, the outcrops attributed to the Castelnuovo reef trend in the Livornesi Mountains are tentatively grouped into two areas (Fig. 11): the northern (Parrana) and the southern (Castelnuovo-Rosignano), separated by a central area (Gorgo-Cafaggio) with slightly more restricted (and deeper?) environments.

Outside the Livornesi Mountains, Castelnuovo-type reefs are located in Casaglia (Cecina Basin) and Spicchiaiola (Volterra Basin, Figs. 2, 3). These reefs occur on marls, sandstones and conglomerates of brackish-lagoonal environments overlying the basal lacustrine unit, and are capped by marine marls and gypsum. These observations and the general pattern in the Livornesi Mountains suggest a peculiar feature of the Castelnuovo reef trend. Reef development occurred in shallow, restricted and stressed marine conditions prior to the deposition of the diatomites and gypsum.

## STROMATOLITE "REEFS"

Coral reefs are undetected in the eastern part of the Fine Basin. The Rosignano Limestone in the area of Castellina contains abundant stromatolites (Fig. 12) associated with a predominantly sandy-conglomeratic sequence ("Sant'al Poggio Conglomerates"). This sequence occurs directly on the basement and grades upwards into laminated marls and gypsum beds. Most stromatolites are 10-30 cm in diameter, but some horizons show large domes (1 m in diameter), locally integrated into large mounds up to 5 m wide (Figs. 12A, B) informally termed "reefs." Basement pebbles are common in the lower part of the section, but carbonate pebbles (1-8 cm in diameter) are dominant. The quartz sand matrix shows fibrous calcitic cements in some horizons. Two major conglomeratic units (8-10 m thick each) are well exposed at the base of the section, separated by marly intercalations and thin stromatolitic carbonate layers (Figs.

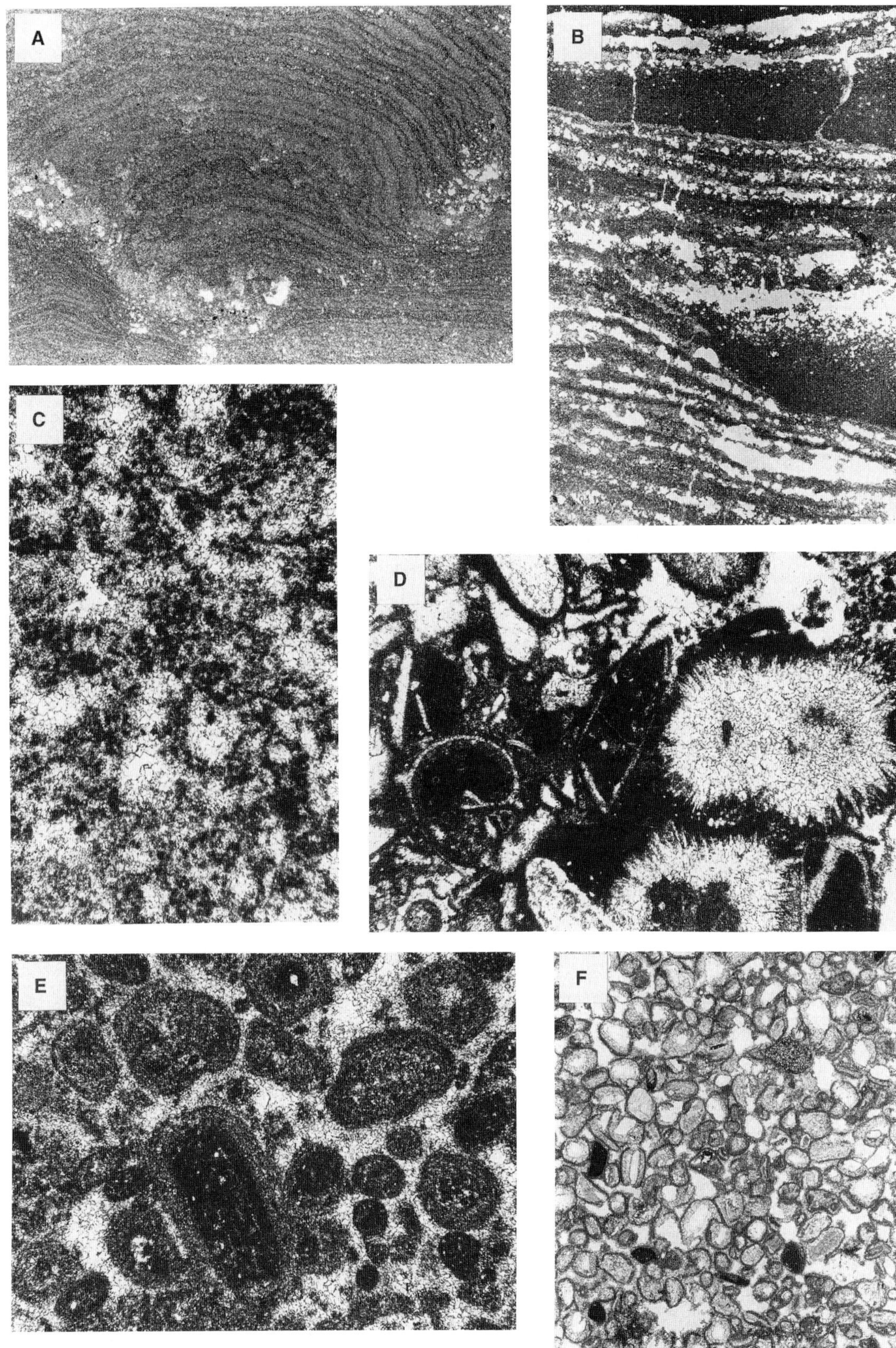


Fig. 13.—Stromatolitic "reef" of Castellina (Sant'al Poggio) and associated facies. (A): Densely laminated, micritic stromatolite. (B): Vuggy-laminar porosity in stromatolite. Some vugs could be enlarged evaporite-moldic porosity. (C): Partly cemented spongy structure in stromatolite. (D): Spherulitic and micritic coatings in skeletal packstone. (E): Oolitic grainstone, some peloids resemble crustaceans fecal pellets. (F): Intensively leached oolithic grainstone, some nuclei are quartz sand grains.

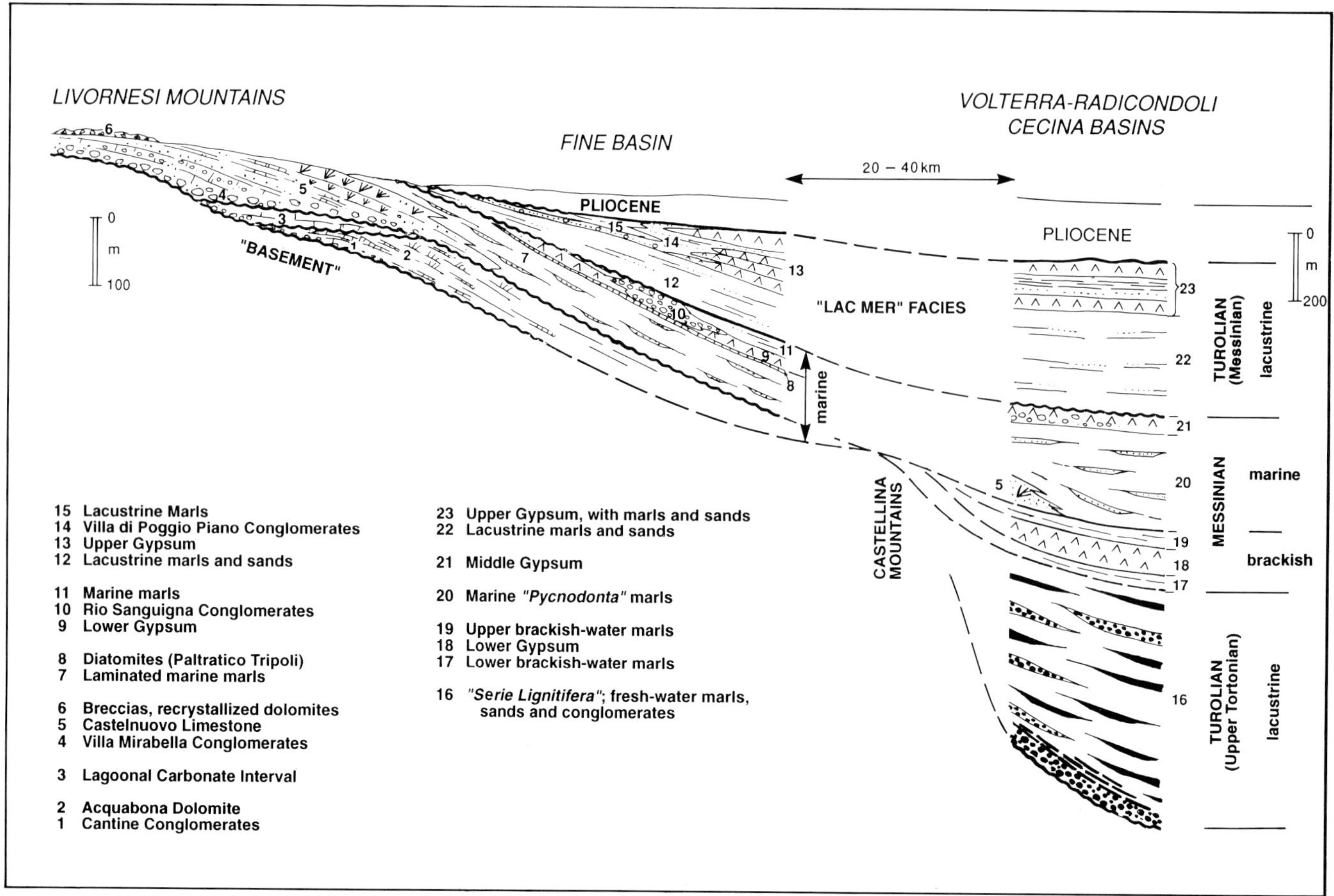


Fig. 14.—Schematic time-correlation of the Upper Miocene stratigraphic units in Tuscany.

13A-C). A 2-m-thick sequence of sandy conglomerates with small stromatolites intercalated with strata of skeletal limestone (Fig. 12) is locally preserved at the base of the lower conglomeratic unit. These intercalated skeletal carbonates are poorly sorted packstones and grainstones with fragments of red algae, echinoderms, serpulids, *Pecten, Pycnodonta*, bryozoans, textularids, miliolids, microgastropods and fragments of stromatolite (Fig. 13D). Some skeletal grains are coated by micritic cements and fibrous calcite (Fig. 13D). The top of the second conglomeratic unit contains large stromatolitic domes capped by thin well sorted oolitic grainstones (Figs. 13E, F) and gray laminated marls with thin layers of oolitic limestones and stromatolites. All these lithologies are common in the Terminal Complex defined in the western Mediterranean (Esteban, 1979, this volume).

The Castellina stromatolitic "reef" is interpreted as having been deposited on a small delta fan, probably in part reworked as a beach, in the margins of a restricted shallow basin evolving from marginal marine to stressed, hyperhaline conditions. The stratigraphic position and lithofacies associated with the Sant'al Poggio conglomerates leads us to tentatively consider them as a lateral equivalent of the Castelnuovo reefs and the basal Villa Mirabella Conglomerates in the western margin of the Fine Basin (only 5 km to the west). It is quite remarkable that sizes, geometries and style of association with conglomerates are very similar for the Castelnuovo coral reefs and the Castellina stromatolite reefs; only the organic composition is different. This further illustrates the complexity of the changing water chemistries in the Fine Basin.

## DIAGENETIC PATTERNS

The two types of coral reefs in the Rosignano Limestones show different diagenetic patterns. The Acquabona reef is well lithified with intensively leached corals and low matrix porosity in the reef framework lithofacies. The Castelnuovo reefs show, in general, excellent preservation of the corals, poor lithification and high matrix porosities in areas with low clay content. Fibrous cements, interpreted as marine, have only been found in some of the Acquabona slope facies. The abundant micritic (cyanobacterial?) crusts of the Acquabona framework and slopes are absent in the Castelnuovo reefs. The extensive skeletal-moldic porosity of the Acquabona reef probably occurred before the deposition of the Castelnuovo reef (and the Lagoonal Interval?), and could be related to the important erosion surface (karst?) truncating the reef or to the basal Castelnuovo unconformity. The Castellina stromatolite reef shows extensive matrix and vuggy porosity (Figs. 13B, C), not occluded by locally important calcitic cements.

The majority of the Acquabona reef is intensively dolomitized, although most of the original textures are preserved (Fig. 6). This dolomitzation event appears diagenetically early, after or during a phase of early leaching of originally aragonitic constituents, and it is followed by pervasive calcite cementation. These fabrics are very similar to other Messinian dolomitized reefs in the western Mediterranean (Oswald et al., 1991). Oswald (1992) developed a model explaining this dolomitization in relation to basinal brines flushing the metastable reef during the transgressive events. On the other hand, the Castelnuovo reefs and Lagoonal Carbonate Interval appear to be mostly calcitic (without systematic sampling). The abundance of fresh-water influx, diluting brine concentration, could explain the absence of dolomitization in the upper units of the Rosignano Limestones. Finally, dolomitization also occurs in the poorly exposed carbonate breccia pockets on the landward top of the Castelnuovo reefs. These breccias are associated with coarsely recrystallized and silicified fabrics and are interpreted as diagenetic evidence of an eroded, hypothetic evaporite cover.

## RELATIONSHIPS OF ROSIGNANO LIMESTONE AND EVAPORITES

The Rosignano Limestones provide important clues to the interpretation and correlation of coral reefs and evaporites in the entire Mediterranean. In the Fine Basin, there are two distinctive coral reef units and two evaporite units; in the southeastern basins (Volterra, Radicondoli and Cecina Basins) there are three evaporite units and only one reef unit. Their relationship (Fig. 14) is supported by detailed mapping and precise, regionally consistent biostratigraphy based on associations of planktonic foraminifers, calcareous nannofossils and ostracods. A summary of the phases of sedimentary evolution follows:

1. Deposition of lacustrine fresh-water deposits of the "Serie Lignitifera" (unit 16 in Fig. 14) in the southeastern Basins during the Late Tortonian? (Turolian) times. This event is poorly recorded in the Fine Basin (not represented in Fig. 14).

2. Earliest Messinian marine transgression, with development of the oligotypic Acquabona coral reef (unit 2 in Fig. 14) on the margins of the Fine Basin. This transgression included at the base minor patches of coastal-marine sandstones and conglomerates (Cantine Conglomerates, unit 1 in Fig. 14). Only brackish-water marls (unit 17 in Fig. 14) were deposited on top of the lacustrine deposits in the southeastern basins without major sedimentary discontinuity.

3. Subaerial exposure and partial erosional truncation of the Acquabona reef in the Fine Basin. This phase of evolution is correlated with the deposition of the Lower Gypsum unit in the southeastern basins (unit 18). Most of this gypsum unit is clastic (gypsarenites) and resedimented in brackish-water conditions.

4. Deposition of the Lagoonal Carbonate Interval (unit 3) in the Fine Basin, with shallow-marine to restricted hyperhaline facies. This is correlated with the continued deposition of brackish-water marls (unit 19) in the southeastern basins.

5. Major subaerial exposure and erosional truncation in the Fine Basin. This event eroded the record of a hypothetic barrier (reefs? shoals?) corresponding to the Lagoonal Carbonate Interval and it is correlated to a paraconformity at the top of unit 19 in the southeastern basins.

6. Major Early Messinian marine transgression in the Fine Basin and southeastern basins. This transgression is recorded by the laminated marls (unit 7) and "*Pycnodonta* "marls (unit 20) with generalized open-marine conditions and indications of stressed and restricted environments. The margins of all the basins are extensively onlapped with abundant coarse terrigenous deposits (Villa Mirabella Conglomerate, unit 4) and the mixed carbonate-siliciclastic Castelnuovo coral reefs (unit 5). The eastern margin of the Fine Basin shows a predominance of restricted marine conglomerates and stromatolite reefs (Sant'al Poggio Conglomerates). The basinal sections are capped by diatomites (Paltratico Tripoli, unit 8), clearly correlated with the upper part of the pre-evaporitic tripoli in Sicily (Bossio et al., 1986).

7. Deposition of marine evaporites as the Lower Gypsum in the Fine Basin (unit 9) and the Middle Gypsum in the southeastern basins (unit 21). Commonly these gypsum units show conglomeratic and turbiditic textures. In the Fine Basin, there are local clastic-marine wedges (Rio Sanguigna Conglomerates; unit 10) and local deposition of marine marls (unit 11). During the deposition of the marine gypsum in the basins, most of the paleogeographic highs (such as Monti Livornesi) were subaerially exposed. However, localized patches of recrystallized breccias of restricted marine carbonates and evaporitic dolomites (unit 6) on top of the Castelnuvo reef and the substrate suggest the possibility of synchronous carbonate deposition during this major evaporitic event.

8. Deposition of brackish to fresh-water marls ("Lac Mer") in the Fine Basin (unit 12) and southeastern basins (unit 22). There is evidence of erosional truncation of underlying units along the basin margins suggesting that most of the paleogeographic highs ("monti") were subaerially exposed.

9. Deposition of lacustrine gypsum and gypsarenites as the Upper Gypsum in the Fine Basin (unit 13) and southeastern basins (unit 23). Lacustrine conglomerates (unit 14) and marls (unit 15) were deposited on the margins. This phase of evolution continued with the fresh-water "Lac Mer" character in all the studied area.

10. Earliest Pliocene marine transgression.

The stratigraphic framework in the Rosignano area reflects the complexity of the sedimentary evolution in Messinian time in the Mediterranean, particularly in partly restricted, marginal basins.

## CONCLUDING REMARKS

As stressed by Esteban et al. (1978) and Bossio et al. (1978, 1981), the Rosignano Reef Complex shows close similarities with other reefs of the same age interval in the western Mediter-

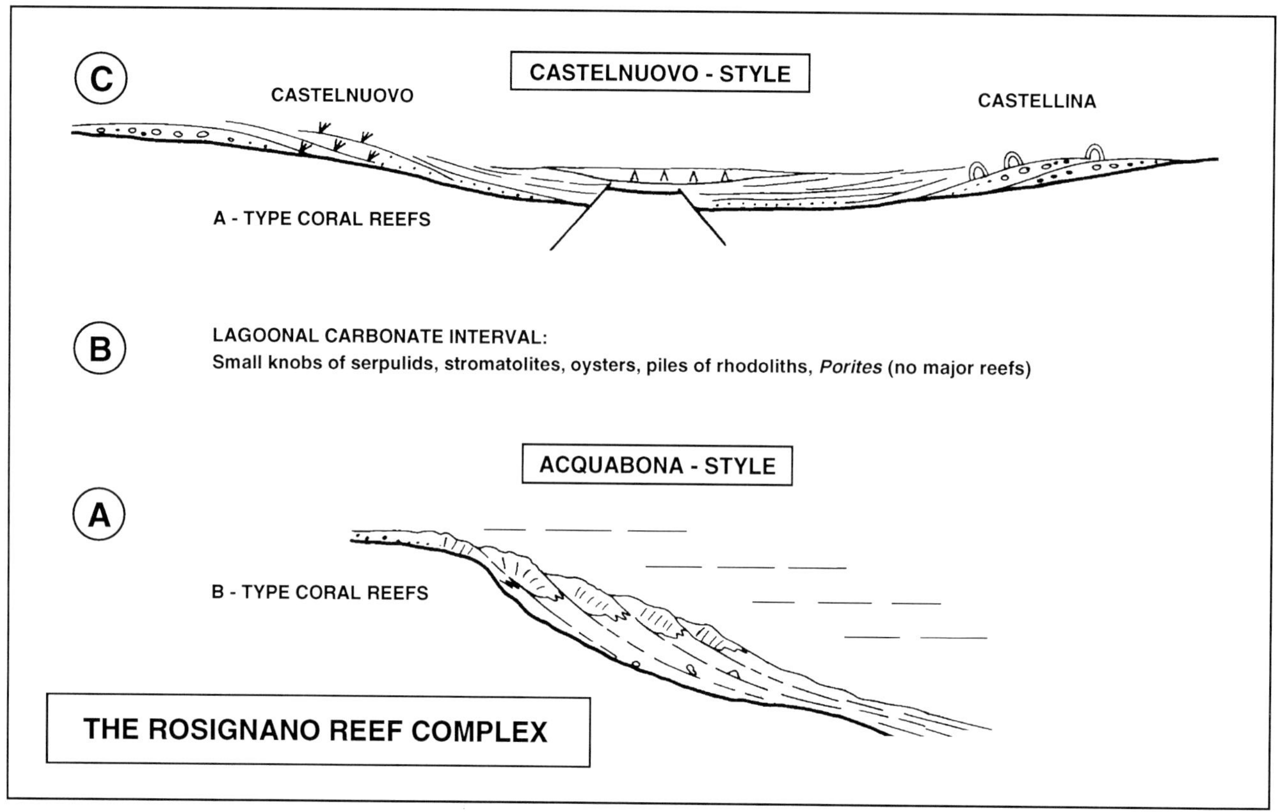


Fig. 15.—Types of reef development in the Rosignano Reef Complex. Types A and B (Esteban, 1979) refer to coral diversity (A, high; B, low).

ranean. The Acquabona reef is a type B reef (monogeneric, in the sense of Esteban, 1979, this volume), whereas the Castelnuovo reef is a type A reef (3-5 coral genera). Figure 15A shows the basic aspects of the Acquabona reef sequence. In the early phase of progradation, water depths at the basin transition could have been on the order of 60-100 m, and progressively decreased by a relative sea-level fall. The truncation and karstification of the top of the Acquabona reef framework is related to this sea-level fall. These types of offlapping (downstepping) and monogeneric reef complexes are characteristic of the Lower Messinian reefs in the western Mediterranean (Esteban, this volume; also Saint-Martin and Rouchy, 1990), but there are also indications of their possible occurrence in the Uppermost Tortonian reefs just below the *G. mediterranea/G. conomiozea* datum (Santisteban, 1981). The Lagoonal Carbonate Interval shows no major reef development, only small stromatolites, accumulations of serpulids, piles of rhodoliths and occasional *Porites* colonies.

The Castelnuovo reef is a classic type A reef (Esteban, 1979, this volume) interbedded with coarse terrigenous deposits. It grew after a major erosional event (and block faulting?). The Castelnuovo reefs are clearly expansive on the basement, indicating a higher relative sea level than for the underlying Acquabona reef; this is interpreted as the result of the generalized subsidence in the area. The presence of relatively high-diversity type-A reefs could be explained by the influx of terrigenous sediments (Santisteban, 1981), which also could explain the cessation of reef growth in the early phases of progradation and the lack of well developed downstepping. In any case, the accommodation space and water depths for the Castelnuovo reefs were small, and the basin rapidly evaporated. Paleoecological interpretations of Bossio et al. (1986) indicate a shallow, stratified water body in the Fine Basin at the end of the Castelnuovo reef growth. In our correlation, the Castelnuovo coral reefs occurred at the same time as the stromatolite reefs (Castellina) on the eastern margin of the Fine Basin. A buried ridge in the middle of the Fine Basin (Val di Perga) could have contributed to isolating a more restricted eastern margin (Fig. 15B). This would be consistent with the above mentioned tectonic event at the base of the Castelnuovo sequence, which apparently changed the bathymetry and configuration of the Fine Basin. No more reef growth is recorded in the Fine Basin after the Castelnuovo reef event.

The Rosignano reef complex shows a clear reversal of the common trend of decreasing coral diversity in many Messinian reefs in the Mediterranean (Esteban, 1979, this volume). The older Acquabona unit is the typical oligotypic *Porites* reef, whereas the younger Castelnuovo unit shows higher diversity in corals and associated communities. This clearly indicates that the drop of coral diversity responds to local conditions and cannot be used as a correlation tool in the Messinian reefs.

Furthermore, each reef unit (Acquabona and Castelnuovo) preceded the development of evaporites in the basin and exhibit similar geometric relationships; but only the older Acquabona reef displays the peculiar features of oligotypia, colonial morphology and micritic crusts (Esteban, 1979, this volume), indicating the interaction with local environmental or tectonic conditions.

The Rosignano Limestone also illustrates the problems of correlation of lithologic units widely termed Terminal Carbonate Complex (Esteban, 1979, this volume). Lithologies and geometries characteristic of the Terminal Complex of many Messinian reefs of the Mediterranean regions are observed in: (i) the Lagoonal Carbonate Interval (unit 3 in Fig. 14) capping the Acquabona reef; (ii) Castellina stromatolite reef, considered a lateral equivalent of the Castelnuovo reef unit; and (iii) the recrystallized carbonate breccias (unit 6), capping the Castelnuovo reef, show lithologic features and geometries common in the Terminal Carbonate Complex. Consequently, there could be up to three distinctive Terminal Complexes in the Rosignano area, with different ages and stratigraphic relationships.

The Lower Gypsum unit of the southeastern basins (unit 18 in Fig. 14) is a local development in marginal, restricted basins and should not be correlated with the Mediterranean-scale Messinian evaporitic units represented by the other two evaporites (units 9-21 and 13-23 in Fig. 14). Local evaporite developments, although thick and extensive, reflect local restriction in marginal basins and cannot be used as markers for inter-basinal correlations of Messinian reef complexes.

In spite of the stratigraphic complexity, the example of the Rosignano Limestones demonstrates detailed biostratigraphic correlation with Mediterranean-scale events, an essential tool for the understanding of the Messinian carbonates and reefs. The Rosignano example provides a reference for refined interpretation of other areas with a more limited stratigraphic record (i.e., only one reef unit or one evaporite unit or one Terminal Complex).

Recently, Barrier et al. (1994) have mentioned the presence of blocks attributed to the Uppermost Tortonian-Lower Messinian coral reefs in Vigoleno (near Piacenza in the Po Basin) at about 200 km to the north-northeast of Rosignano. In that locality, sticks and laminar colonies of *Porites lobatosepta* and colonies of *Siderastraea crenulata* are associated with abundant red algae. Although the age and structural setting of these small outcrops require corroboration, it could demonstrate reefal development further north than previously considered. However, the Rosignano reefs studied here continue to be the best developed outcrops preserved in the northern zone of the Mediterranean Messinian coral reefs. Most of their features are closely comparable to other more meridional ones. The abundance of *Cladocora* mixed with hermatypic corals in the Popogna outcrops (Castelnuovo reef) could be the only indication of proximity to this northern limit.

ACKNOWLEDGMENTS

The authors thank the reviews and comments by L. Pomar, Q. Sun, J. Jiménez, E. G. Purdy, M. Pedly and D. S. Ulmer-Scholle.

REFERENCES

BARRIER, P., CAUQUIL, E., RAFFI, S., RUSSO, A., AND TRAN VAN HUU, M., 1994, Signification du plus septentrional des récifs messiniens à Algues et Porites connus en Méditerranée (Vigoleno, Piacenza, Italie) (abs.): Marseille, Interim Colloquium Regional Committee of Mediterranean Neogene Stratigraphy, p. 2-3.

BARTOLETTI, E., BOSSIO, A., ESTEBAN, M., MAZZANTI, R., MAZZEI, R., SALVATORINI, G., SANESI, G., AND SQUARCI, P., 1985, Studio geologico del territorio comunale di Rosignano Marittimo in relazione alla carta geologica alla scala 1:25.000: Livorno, Quaderno di Museo di Storia Naturale di Livorno, v. 6, p. 33-127.

BOSSIO, A., BRADLEY, F., ESTEBAN, M., GIANNELLE, L., LANDINI, W., MAZZANTI, R., MAZZENI, R., AND SALVATORINI, G., 1981, Alcuni aspetti del Miocene superiore del Bacino del Fine: Pisa, IX Convegno della Società Paleontogica Italiana: v. 3-8, p. 21-54.

BOSSIO, A., CERRI, R., COSTANTINI, A., GANDIN, A., LAZZAROTTO, A., MAGI, M., MAZZANTI, R., MAZZEI, R., SAGRI, M., SALVATORINI, G., AND SANDRELLI, F., 1992, I bacini distensivi Neogenici e Quaternari della Toscana, Field Trip Guidebook, Excursion B4, 76 Reunione Estiva Società Geologica Italiana: Firenze, Università di Firenze, p 199-277.

BOSSIO, A., COSTANTINI, A., FORESI, L. M., MAZZANTI, R., MAZZEI, R., MONTEFORTI, B., SALVATORINI, G., SANDRELLI, F., AND TESTA, G., 1994, Note preliminari sul Neoautoctono dell'area di Sassa (settore SW del Bacino di Volterra) Province di Pisa e Livorno: Studi Geologici Camerti, Volume Speciale, p. 33-43.

BOSSIO, A., CONSTANTINI, A., LAZZAROTTO, A., LIOTTA, D., MAZZANTI, R., MAZZEI, R., SALVATORINI, G., AND SANDRELLI, F., 1993, Rassegna delle conoscenze sulla stratigrafia del neoautoctono toscano: Memoria della Società Geologica Italiana, v. 49, p. 17-98.

BOSSIO, A., ESTEBAN, M., GIANELLI, L., LONGINELLI, A., MAZZANTI, R., MAZZEI, R., RICCI LUCCHI, F., AND SALVATORINI, G., 1978, Some aspects of the upper Miocene in Tuscany: Pisa, Messinian Seminar 4, Intenational Geological Correlation Project 96 (fasc. spec.), 88 p.

BOSSIO, A., ESTEBAN, M., MAZZANTI, R., MAZZEI, R., AND SALVATORINI, G., 1991, Ipotesi di correlazione tra facies sedimentarie del Miocene superiore dei bacini compresi tra il Valdarno inferiore e la val di Cécina, *in* Cioni, R. and Sbrana, A., eds., Evoluzione dei bacini Neogenici e loro rapporti con il magmatismo Plio-Quaternario nell'area Tosco-Laziale: Pisa, Università di Pisa, p. 70-72.

BOSSIO, A., MAZZANTI, R., MAZZEI, R., AND SALVATORINI, G., 1985, Analisi micropaleontologiche delle formazioni mioceniche, plioceniche e pleistoceniche dell'area del Comune di Rosignano: Livorno, Memorie degli Quaderni di Museo di Storia Naturale di Livorno, v. 6, p. 129-170.

CAPELLINI, G., 1874, La formazione gessosa di Castellina Marittima e i suoi fossili: Bologna, Memoria della Accademia di Scienci Naturale di Bologna, Serie 3, v. 4, p. 525-602.

CAPELLINI, G., 1878, Il calcare di Leitha, il Sarmatiano e gli strati a Congeria nei Monti di Livorno, di Castellina Marittima, di Miemo e di Monte Catini: Atti Rendicompti della Accademia Lincei, Serie 3, Memoria Scienza fisica matematica e naturale, v. 2, p. 1-19.

CHEVALIER, J. P., 1961, Recherches sur les Madreporaires et les formations récifales miocènes de la Méditerranée occidentale: Paris, Published Ph. D. Thesis, Mémoires de la Societé Géologique de France 93, 562 p.

DE BONIASKI, S., 1878, Sui fossili miocenici del Gabbro: Atti della Societa Toscana di Scieza Naturali: Atti Società Toscana Scienze Naturali, v. 1, p. LII-LIX.

ESTEBAN, M., 1978, Significance of the Upper Miocene reefs in the western Mediterranean (abs.): Rome, Messinian Seminar, v. 4.

ESTEBAN, M., 1979, Significance of the Upper Miocene coral reefs of the western Mediterranean: Palaeogeology, Palaeoclimatology, Palaeoecology, v. 29, p. 169-188.

ESTEBAN, M., BOSSIO, A., GIANNELLI, L., MAZZANTI, R., MAZZEI, R., AND SALVATORINI, G., 1978, The Messinian reef complex (Calcare di Rosignano) in Tuscany (abs.): Rome, Messinian Seminar, v. 4.

GIANNINI, E., 1962, Geologia del Bacino della Fine (provincie di Pisa e Livorno): Bolletino della Società Geologica Italiana, v. 81, p. 99-224.

LAZZAROTTO, A. AND MAZZANTI, R., 1978, Geologia dell'alta Val di Cècina: Bolletino della Società Geologica Italiana, v. 95, p. 1365-1487.

LAZZAROTTO, A. AND SANDRELLI, F., 1979, Stratigrafia e assetto tettonico delle formazioni neogeniche nel bacino del Casino (Siena): Bolletino della Società Geologica Italiana, v. 96, p. 747-761.

MAZZANTI, R., MAZZEI, R., MENESINI, E., AND SALVATORINI, G., 1981, L'Arenaria di Ponsano: nuove precisazioni sopra l'età: Pisa, IX Convegno della Società Paleontologica Italiana, p. 135-159.

OSWALD, E. J., 1992, Dolomitization of a Miocene reef complex, Mallorca, Spain: Unpublished Ph.D. Dissertation, State University of New York at Stony Brook, Stony Brook, 437 p.

OSWALD, E. J., FRANSEEN, E. K., AND MEYERS, W. J., 1991, Similarities in the dolomitization of upper Miocene reef complexes in Mallorca and the Las Negras area, Spain: Possible evidence for a Mediterranean dolomitizing event during the Messinian (abs.): American Association of Petroleum Geologists Bulletin, v. 74, p. 735.

PEDLEY, H. M., 1983, The petrology and palaeoenvironment of the Sortino Group (Miocene) of SE Sicily: evidence for periodic emergence: Journal of the Geological Society of London, v. 140, p. 335-350.

RIDING, R., MARTIN, J., AND BRAGA, J. C., 1991, Coral-stromatolite reef framework, Upper Miocene, Almería, Spain: Sedimentology, v. 38, p. 799-818.

RUGGIERI, G., 1956, I lembi miocenici del Livornese nel quadro della tettonica dell'Appennino: Atti Accademia di Scienza di Istituto di Bologna, Serie 11, v. 3, p. 1-12.

SAINT-MARTIN, J. P. AND ROUCHY, J. M., 1990, The Messinian carbonate platforms in the western Mediterranean: their importance for the reconstruction of the late Miocene sea level variations: Bulletin de Géologie de la Societé de France, v. 8, p. 83-94.

SANTISTEBAN, C, 1981, Petrología y sedimentológia de las materiales del Mioceno superior de la cuenca de Fortuna (Murcia) a la luz de la Teoría de la Crisis de Salinidad: Unpublished Ph.D. Thesis, Universitat de Barcelona, Barcelona, 725 p.

TREVISAN, L., 1952, Sul complesso sedimentario del Miocene superiore e del Pliocene della Val di Cécina e sui movimenti tettonici tardivi in rapporto ai giacimenti di lignite e di salgemma: Bolletino della Società Geologica Italiana, v. 70, p. 65-78.

# TECTONIC CONTROLS ON MIOCENE REEFS AND RELATED CARBONATE FACIES IN CYPRUS

EDWARD J. FOLLOWS
*UEDN/711, Shell UK Expro, 1 Alten's Farm Road, Aberdeen, AB9 2HY, United Kingdom*

ALASTAIR H. F. ROBERTSON AND TERENCE P. SCOFFIN
*Department of Geology and Geophysics, Grant Institute, West Mains Road, Edinburgh EH9 3JW, United Kingdom*

Abstract: Reefs of Early Miocene (Aquitanian/Burdigalian) and Late Miocene (Tortonian) age are exposed in Cyprus, mainly around the periphery of the Troodos ophiolitic massif. Reef growth followed early Tertiary deep-water sedimentation and localized tectonic uplift. The Aquitanian/ Burdigalian Terra Member formed on isolated, stable basement highs in southeast and west Cyprus. Up to 500 m- by 80 m-sized, faunally diverse patch reefs grew in relatively shallow, calm seas. Fore-reef and basinal facies are exposed in western Cyprus. Reefs of the Tortonian Koronia Member reflect local tectonic settings. Those in southern Cyprus developed on linear highs, related to earlier crustal compression. These reefs are preserved almost entirely as channelised talus in adjacent basins. Around the western and, particularly the northern margins of the Troodos ophiolite, reef growth was influenced by active crustal extension. Patch reefs in the north formed on uplifting fault blocks and large volumes of talus were shed down steep slopes into basins to the north. In west Cyprus, the reefs grew on both flanks of a gradually subsiding graben, with local preservation of back-reef facies. The Late Miocene reefs were dominated by Porites, first largely domal, then mainly as sheet-like encrustations. Growth of both the Early and Late Miocene reefs was preceded by erosion and was then followed by transgression, associated with a relative sea-level rise. Reef growth was finally brought to an end by Messinian desiccation of the Mediterranean. The reefs were cemented by Mg - calcite and botryoidal aragonite during and shortly after growth. Subsequently, during exposure to meteoric and brackish waters, the fabric of the reef was modified by dissolution, neomorphism, calcite spar and locally gypsum cementation. Uplift and extensional faulting caused fracturing of the, by then, brittle reefs (Terra Member) and fissures were locally enlarged by karst-forming solutions. A marine transgression in the ensuing Early Pliocene time then filled the fissures with fine-grained carbonate sediments, which were prone to mixing-zone dolomitization.

## INTRODUCTION

The objective of this paper is to summarize the environments of deposition, tectonic setting, field relations and diagenesis of Miocene reefs and related carbonate sediments of Cyprus (Fig. l). The limestones overlie a basement of Late Cretaceous ophiolitic rocks and deep-sea sediments of Late Cretaceous-Early Tertiary age (Robertson and Hudson, 1974; Fig. 2). The overall tectonic setting involved initial genesis of oceanic crust and mantle in the Tethys ocean, followed by localized tectonic uplift by Early Miocene time, allowing reef growth in some areas (Robertson, 1977a). The later stages of this uplift are believed to relate to northward subduction of Tethyan oceanic crust beneath Cyprus during the Neogene (Kempler and Ben-Avraham, 1987; Robertson, 1990). Different areas of the island underwent compression, extension, or were relatively stable during the Miocene. There is, thus an opportunity to document the influences of local tectonics on reef growth, in relation to other factors, including relative sea-level change, climate and sedimentation. The Cyprus Miocene reefs (Fig. 2) establish links between the Central and Eastern Mediterranean reefs through comparison with Pliocene reefs in Israel (Buchbinder, 1979), Turkey (Hayward, 1982; Hayward et al. this volume) and the Red Sea (James et al., 1988; Burchette, 1988). These limestones also shed light on regional paleo-oceanographic changes during closure of the Tethys ocean.

## EARLIER WORK

Early contributions to the study of the Miocene units of Cyprus were mainly biostratigraphical (e.g., Bellamy and Jukes-Browne, 1905). Cowper-Reed (1929, 1930, 1932, 1933, 1935, 1939) established comprehensive faunal lists of Helvetian-Tortonian age and drew comparisons with Malta, Cilicia and north Syria.

A comprehensive stratigraphy of the Cyprus sediments first appeared in 1949 (Henson et al., 1949). After mapping in the 1960s, a reef unit was recognized along the northern margin of the Troodos ophiolite, comprising a massive core and bedded flanks, overlying older sediments and lavas (Carr and Bear, 1960; Bear, 1960). Bagnall (1960) and Pantazis (1967) also gave early petrographic descriptions, while Bear (1960) briefly discussed diagenesis. Turner (1968) and Kluyver (1969) also mentioned Miocene limestones in western Cyprus, including those on the Akamas Peninsula.

In the early 1970s, attempts were being made to integrate the Miocene limestones into tectonic interpretations of the Eastern Mediterranean area (Cleintaur et al., 1977; Robertson, 1977b). Research was further stimulated by drilling of DSDP sites 375 and 376 (Leg 42A) off southwestern Cyprus (Hsü et al., 1978). In 1979, the Geological Survey Department published a map of Cyprus (1;250,000) with all the main outcrops of Miocene limestones. New work on the Miocene carbonates in southern Cyprus was documented by Eaton (1987) and Eaton and Robertson (1993). Recently, detailed accounts have been published on the local field relations and structure (Follows and Robertson, 1990), of the diagenesis (Follows, 1992) and the role of local tectonics versus eustatic sea-level change in the Neogene of Cyprus (Robertson et al., 1991). However, this is the first overall summary of the relatively small, but locally well exposed reefs and related facies in Cyprus. Little subsurface

Models for Carbonate Stratigraphy from Miocene Reef Complexes of Mediterranean Regions,
SEPM Concepts in Sedimentology and Paleontology #5, Copyright © 1996,
SEPM (Society for Sedimentary Geology), ISBN 1-56576-033-6, p. 295-315.

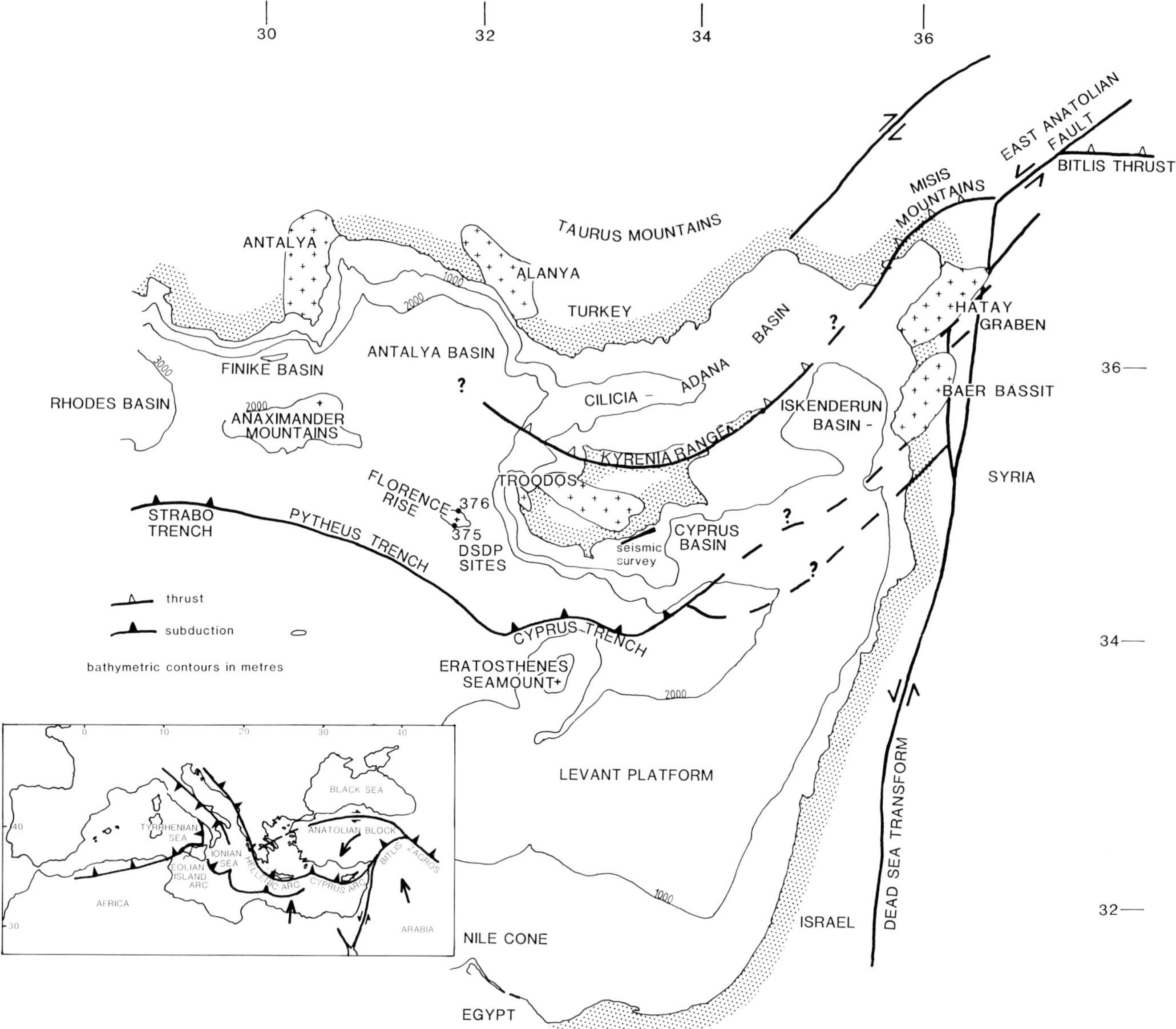


FIG. 1.—Outline tectonic map of the Eastern Mediterranean showing the present day setting of Cyprus above a northwards-dipping subduction zone. This subduction is believed to have been active since Early Miocene time and to have influenced Miocene reef development in Cyprus. Inset: plate tectonic framework of the Eastern Mediterranean region. From Robertson et al. (1991).

data exist, and the area is not believed to have any hydrocarbon potential.

## STRATIGRAPHY AND DISTRIBUTION OF REEF LIMESTONES

Reef limestones are found in four areas of Cyprus: first, along the north margin of the Troodos ophiolite; secondly, in southeast Cyprus; thirdly, in south Cyprus; and fourthly, in west Cyprus (Fig. 2). Minor occurrences of reef limestones in north Cyprus, along the southeast flank of the Kyrenia Range, are not considered here (Fig. 2). All of these occurrences are buildups exhibiting a distinctive framework mainly controlled by the growth of corals. The reef carbonates are now mainly limestone, in contrast to the off-reef facies that are more chalky.

The stratigraphy of the Tertiary sediments overlying the Troodos ophiolite is mapped out in Figure 2. Above the Troodos ophiolite, a thin (tens of meters-thick) sequence of Late Cretaceous metalliferous mudstones, radiolarian cherts and volcanogenic facies typically passes abruptly upwards into pelagic carbonates of Maastrichtian-Early Tertiary age (Lefkara Formation). By Early Miocene time, the basement was uplifted sufficiently to allow local reef growth (Terra Member). Later reef growth (Koronia Member) was mainly around the periphery of the ophiolitic basement, associated with shallow- to deeper-marine carbonate sedimentation (Pakhna Formation; Robertson, 1977b; Eaton, 1987; Eaton and Robertson, 1993). Overlying

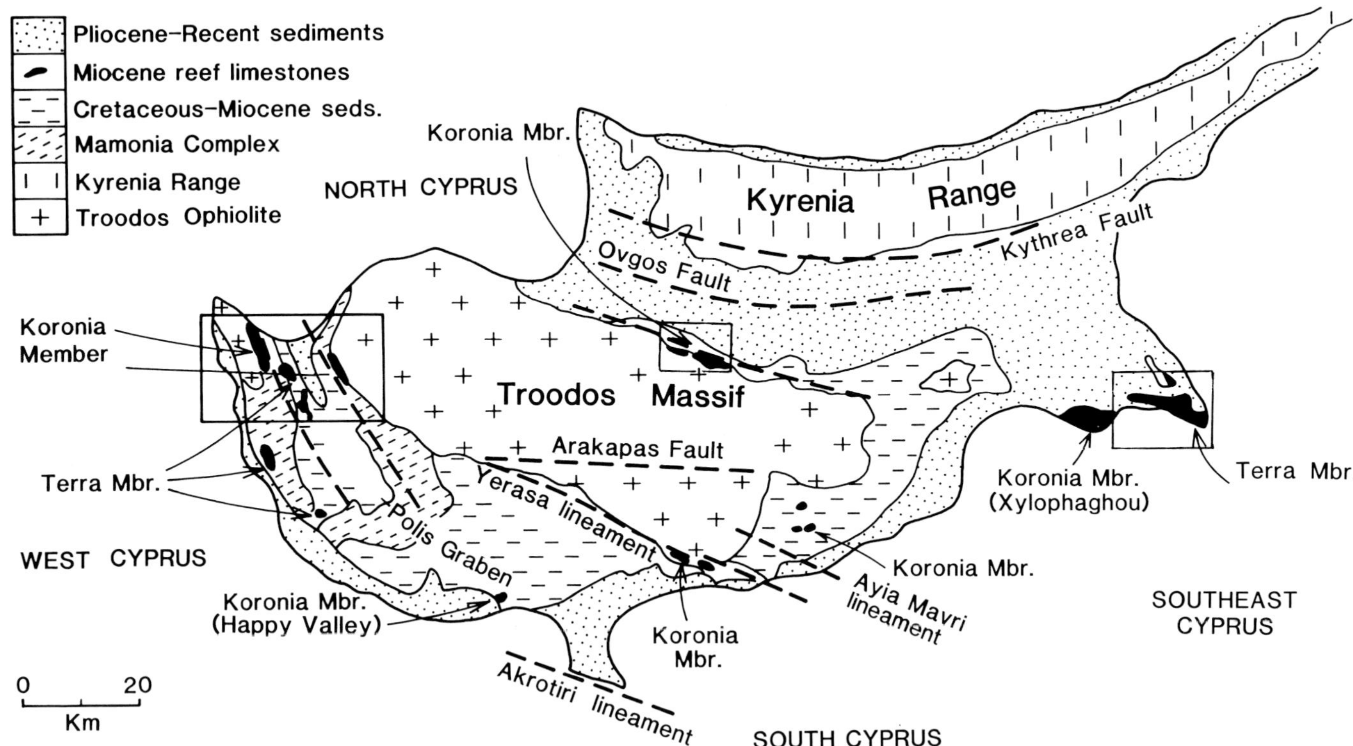


FIG. 2.—Location of the Miocene reefs in relation to the main structural lineaments active in Miocene time. Boxes mark areas shown in more detailed maps.

evaporites (Kalavasos Formation) precipitated during desiccation of the Mediterranean (Orszag-Sperber et al., 1989; Robertson et al., 1995). Transgressive marine silts, muds and minor conglomerates then accumulated in the Pliocene (Nicosia Formation; McCallum and Robertson, 1990, 1995), followed by accelerated uplift and deposition of Quaternary alluvial fans, fan deltas, coastal and subaerial deposits (Poole et al., 1990; Poole and Robertson, 1991). The dominant control of this drastic uplift was initial continental collision and underthrusting of Eurasia by Africa (Gass and Masson-Smith, 1963).

## BIOSTRATIGRAPHY

In southeastern Cyprus, nannofossil dates indicate the Terra Member reefs are Early Miocene age. The reefs rest on Lefkara Formation chalks (NP15) and are in turn directly overlain by Middle Miocene pelagic carbonates of an NN6 age (see Martini, 1971; O. Varol pers. commun., 1989; Fig. 3). Unconformably overlying chalks were dated as Late Miocene age (NN10), while pelagic carbonates still higher in the succession were dated as Early Pliocene. O. Varol (pers. commun., 1989) dated the Koronia Member reefs along the north Troodos margin. A Middle Miocene age (NN6) was assigned to off-reef facies (debris-flows) at the base of the sequence and higher up, marls of the Pakhna Formation, laterally equivalent to the reefs, are of Late Miocene age (NN10). Marls of the Nicosia Formation above this are Early Pliocene (NN15).

Lists of pelagic and benthonic foraminifera (Mantis, in Pantazis, 1967; Baroz and Bizon, 1977) present severe difficulties in dating the Miocene reefs: (i) reworking is common; (ii) benthonic foraminifera are localized and of low diversity in the stratigraphically higher reef facies; and (iii) published microfossil locations cannot be easily matched with the stratigraphy used in this study. However, the benthonic assemblages *Eulepidina*, *Spiroclypeus*, *Miogypsinoides*, *Lepidocyclina*, *Miogypsina*, *Nephrolepidina* and *Miolepidocyclina burdigalensis*, when taken together, confirm an Aquitanian to mid Burdigalian (Early Miocene) age for the Terra Member (Table 1, Fig. 3). The Koronia Member is dominated by the benthonic assemblage *Operculina-Heterostegina-Borelis*, suggestive of a Middle to Late Miocene age, as is the presence of the planktonic foraminiferan *Orbulina universa*, which extends down to and including N9 (Blow, 1969; M. Simmons and T. Wonders, pers. communs., 1990).

The coral taxa of the Terra Member are a subset of the Western Mediterranean fauna, described by Chevalier (1961) and are thus consistent with an Aquitanian-Burdigalian age. Also, the abundance of poritid corals in the Koronia Member is characteristic of Tortonian to Messinian strata in the Western Mediterranean (e.g., Esteban et al., 1977; Esteban, 1979; Rouchy et al., 1986). The echinoid *Clypeaster altus*, common within the Koronia Member, is widespread in other Mediterranean Tortonian sediments (Challis, 1980; Rose and Poddubiuk, 1987).

In summary, the limited biostratigraphic evidence (Table 1, Fig. 3) is taken to indicate that the reefs of the Terra Member are Aquitanian-mid Burdigalian (Early Miocene) age while the

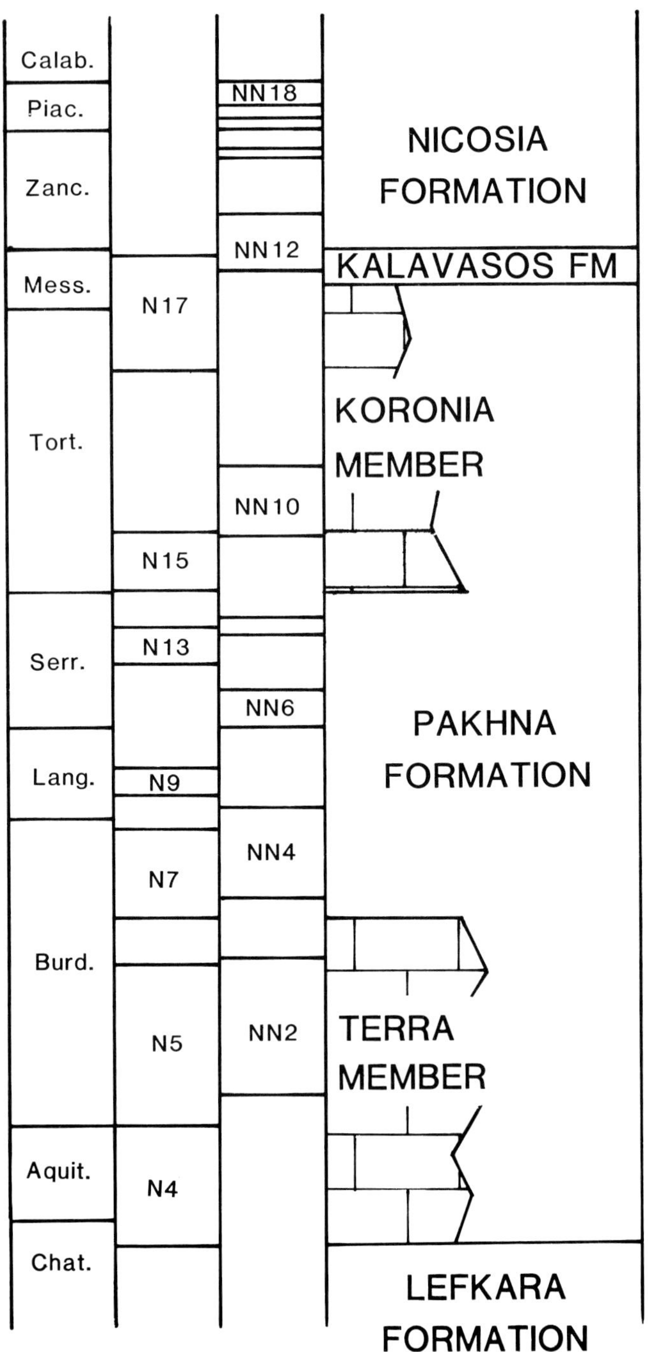


FIG. 3.—Summary of the inferred age of the reef units in Cyprus in relation to the foraminiferal and nannofossil zonations. The zonation is essentially that used by the Deep Sea Drilling Project in the Mediterranean by Hsü et al. (1978). See Figure 2 for the mapped distribution of the reef units.

Koronia Member is Tortonian (Late Miocene). Regional stratigraphical correlations are discussed later in the paper.

TABLE 1.—SUMMARY OF NEW FORAMINIFERAL BIOSTRATIGRAPHIC DATA FROM THE REEF-BEARING TERRA AND KORONIA MEMBERS, DETERMINED BY M. SIMMONS AND T. WONDERS. THESE CONFIRM AN EARLY MIOCENE (AQUITANIAN-BURDIGALIAN) AGE FOR THE TERRA MEMBER, AND A LATE MIOCENE (TORTONIAN) AGE FOR THE KORONIA MEMBER BOTH WITHIN THE PAKHNA FORMATION (FROM FOLLOWS, 1990).

| LOCATION | FORAMINIFERA IDENTIFIED |
|---|---|
| *Koronia Member* | |
| **N. of Mitsero** | *Operculina* |
| | *Orbulina universa* (p) |
| | *Globigerinoides* (p) |
| **N. of Mitsero** | *Operculina* |
| | *Amphestegina* |
| | *Schitzula* (p) |
| Kotaphi Hill | *Amphistegina* |
| W. of Kreatos | Miliolids |
| | Triloculinids |
| | Quinqueloculinids |
| | *Ophilium cristeum* |
| | *Borelis* |
| N. central area | Miliolids |
| Polykantho | *Borelis* |
| | Quinqueloculinids |
| **Southern and eastern Cyprus** | |
| Happy Valley | *Operculina* |
| | keeled planktics |
| Xylophagou | *Amphistegina* |
| | *Orbulina universa* |
| | *Spherogypsina* |
| *Terra Member* | |
| **Southeastern Cyprus** | |
| Ayia Napa | *Miolepidocyclina burdigalensis* |
| | *Amphistegina* |
| Ayios Antonios | *Eulepidina* |
| | *Miogypsinoides* |
| | *Spiroclypeus* |
| **Western Cyprus** | |
| Neokhorio | *Eulepidina* |
| | *Operculina* |
| Androlikou | *Miogypsina globulina* |
| Androlikou | *Lepidocyclina* |
| | *Miogypsina globulina* |
| | *Miolepidocyclina* |
| Inia | *Lepidocyclina* |
| | *Nephrolepidina* |
| | *Miogypsina* |

## STRUCTURAL SETTING

Early Miocene reefs (Terra Member) are present in southeast and west Cyprus, while the Late Miocene reefs (Koronia Member) developed in north, south, west and locally in southeast Cyprus (Fig. 2). The structural setting of Cyprus is critical to understanding the Miocene reefs and will now be summarized.

### *Southeast Cyprus*

Southeastern Cyprus is presently a flat area, marked by a number of low, limestone-capped hills (Figs. 4A, B). A locally exposed basement of Late Cretaceous melange is overlain by a thin succession of sub-horizontal pelagic and reef carbonates. The Middle and Late Miocene intervals are separated by a low-angle unconformity, thought to correspond with the fissuring of limestones and the development of chert dykes in the underlying pelagic carbonates, both features that reflect crustal extension (Follows and Robertson, 1990). Reef-bearing limestones in the Cape Greco area were later exposed in the Messinian, then partly eroded and terraced during Late Pliocene-Quaternary time (Poole and Robertson, 1991).

By contrast, 20 km further west, at Xylophaghou (Fig. 2), an isolated exposure of Koronia Member is composed of roughly horizontally-bedded bioclastic detritus. This limestone is cut by low-angle normal faults that downthrow to the west.

In summary, the reef-bearing limestones in southeast Cyprus developed on a relatively stable, flat-topped structural high, probably bounded by faults.

### *North Troodos Margin*

The northern margin of the Troodos ophiolite is characterized by a range of low hills dropping northwards onto the Mesaoria Plain. In this area, the location and lithofacies of the Miocene reefs of the Koronia Member were largely controlled by extensional faulting (Follows and Robertson, 1990). The faulting was active during Middle and Late Miocene and Early Pliocene times and dissected the northern area into two main blocks, bounded by NNE-SSW trending transverse faults (Figs. 5A, B). Igneous basement, cover sediments and the Miocene reefs are exposed in rotated fault blocks, typically 1 km across.

Several lines of evidence show that crustal extension was active during reef development: (i) wedge-shaped bodies of Miocene age thin northwards away from mapped faults; (ii) coarser-grained sediments (e.g., debris-flows) accumulated in more tectonically active areas; (iii) coeval successions are seen on either side of growth faults, whereas overlying facies (e.g. debris-flows) of the Pakhna Formation were not affected (Follows and Robertson, 1990); (iv) abundant debris-flows in the upper part of the Pakhna Formation relate to mass wasting from steep, active scarps, near the present edge of the ophiolite outcrop; and (v) dip patterns, within and between, individual fault blocks and small-scale structures (e.g., slumps, intraformational breccias) also indicate fault movement during sedimentation. The resulting basement uplift also had the effect of introducing clastic sediment and thus modifying the nature of reef growth.

Uplift and erosion removed much of the pre-Miocene sedimentary cover of the ophiolite prior to reef development. Growth-faulting was active by the time reef growth began. Faulting then intensified, as shown by the presence of large channelised and wedge-shaped debris-flow deposits that overstep earlier, smaller-scale fault blocks. Extension was again active during Messinian evaporite precipitation (Kalavasos Formation) as shown by the presence of gypsum debris-flows and tilted selenitic gypsum (Follows and Robertson, 1990; Robertson et al., 1995). In earliest Pliocene time, pelagic carbonate again accumulated in a relatively tectonically quiescent setting. Later, during Early Pliocene time, block faulting was active, with large-scale debris-flow deposition. Overlying Middle and Late Plio-Quaternary sediments are undeformed, and there is no evidence of later faulting (McCallum and Robertson, 1990; McCallum, 1990).

### *Southern Cyprus*

The development of Late Miocene reefs in southern Cyprus was preceded by uplift and crustal compression that created northwest-southeast-trending lineaments. Several such lineaments, Yerasa, Ayia Mavri and Akrotiri, were important in localizing reef growth (Fig. 6). Undeformed Miocene limestones on the Yerasa lineament unconformably overlie Late Cretaceous melange (Robertson, 1977b) or Troodos pillow lavas directly. Thus compression had ended prior to the onset of Koronia Member reef growth (Robertson et al., 1991).

Within the predominantly carbonate Miocene deposits in south Cyprus, very little in situ reef limestone is now preserved. Abundant reef-derived talus, along with ophiolite detritus, is exposed mainly in channels within the Pakhna Formation. From this evidence, it is clear that reefs were originally widely developed south of the Troodos Massif. Reef facies limestones are now preserved at only one locality, Happy Valley (Fig. 2). These reefs apparently developed on a local basement high within an adjacent, structurally controlled basin (Robertson et al., 1991). Adjacent basins, although mainly carbonate-dominated, also received clastic detrital sediment from the uplifted Troodos Massif (e.g., in the Maroni sub-basin).

### *West Cyprus*

The Terra Member in west Cyprus unconformably overlies pelagic carbonates of the Lefkara Formation, in places overstepping onto the basement of the Mesozoic Mamonia Complex. This area had already experienced considerable differential uplift and erosion, followed by relative tectonic quiescence during Early Miocene reef growth. An arcuate pattern of outcrops reflects the local paleo-topography of northwestern Cyprus. Reconstruction of the reef morphology is, however, complicated by later faulting and block-rotation during formation of the Polis graben (Payne and Robertson, 1995).

The topography of western Cyprus is dominated by the Polis-Paphos graben and the Akamas Peninsula, both elongate N-S trending structural features (Figs. 7A, B). The graben came into existence in the Late Miocene-Early Pliocene and has been episodically active until Recent time. In this area, the Early Miocene Terra Member predates the crustal extension, while the Koronia Member developed during active extension and graben formation.

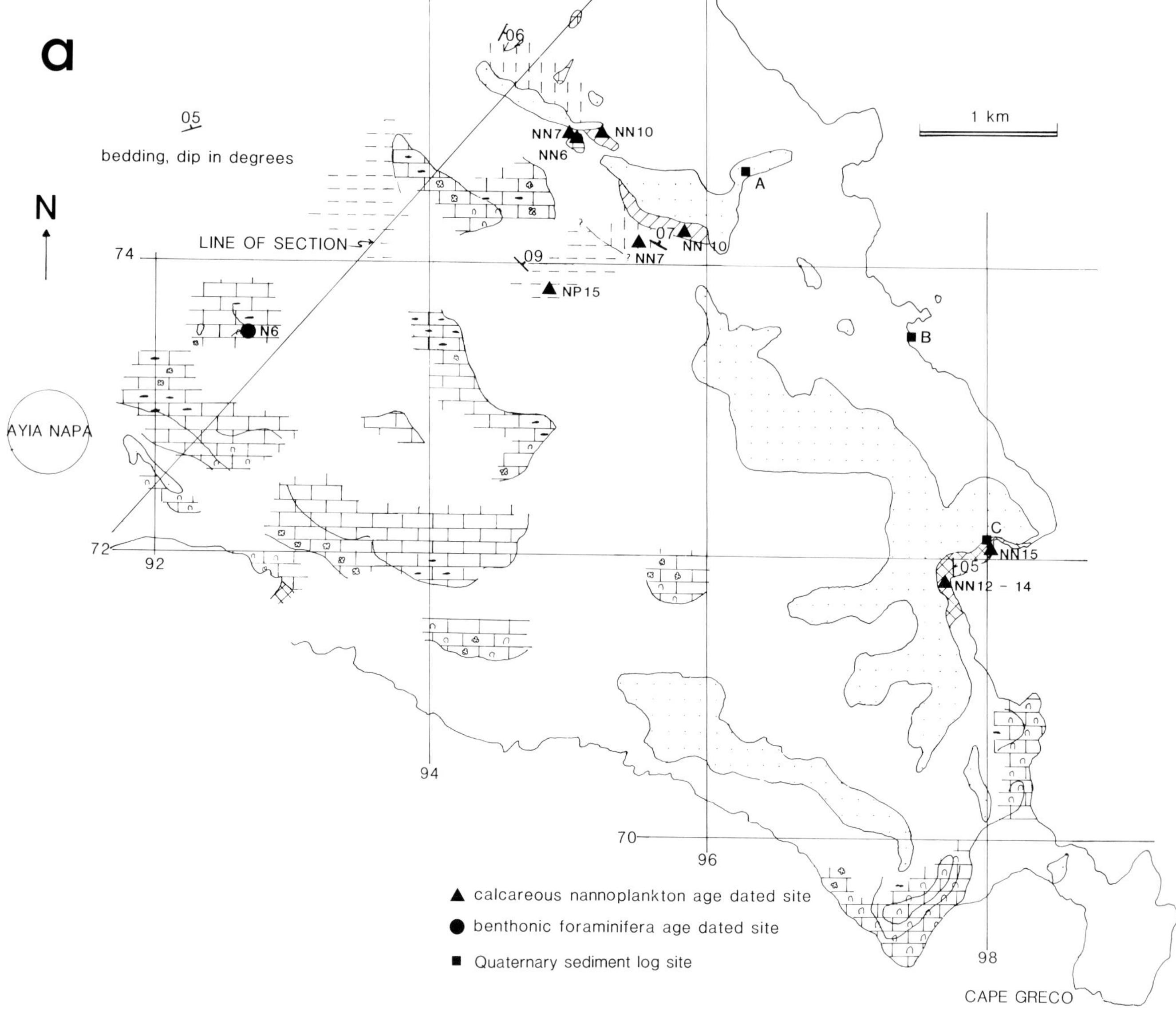


FIG. 4.—Geological setting of the Early Miocene Terra Member in southeast Cyprus. (A) Geological map. (B) Cross section. Note the locations of sites dated. Modified after Follows and Robertson (1990). See Figure 2 for location.

The reefs of the Late Miocene Koronia Member developed on both flanks of the northeast-southwest-trending Polis-Paphos graben. The resulting basin was bounded by uplifted ophiolitic basement of the Troodos Massif to the east and the Akamas Peninsula to the west. This basin, however, contains no clastic input with the exception of basal clasts on the Akamas Peninsula. The graben filled with basinal facies of the Pakhna Formation, Messinian evaporites (Weisgerber, 1978; Dupoux, 1983; Elion, 1983; Robertson et al., 1995) and Plio-Quaternary terrigenous sediments (Poole and Robertson, 1991; Payne and Robertson, 1995). Major faults bound the basin and these also helped to localize reef growth on the structural highs.

*Regional Summary*

In general, the Terra Member reefs developed with little tectonic influence. In contrast, the location and facies of the later Koronia Member reefs were influenced by local tectonics. The greatest effect was along the north Troodos margin, where uplift gave rise to slope facies and a significant clastic input, but reef growth in western Cyprus was also influenced by differential tectonic subsidence. In southern Cyprus, later uplift and erosion has preserved mainly slope and basinal reef-derived facies.

The regional tectonic setting is thought to have been an active continental margin (Kempler and Ben-Avraham, 1987), related to northward subduction of Tethyan oceanic crust beneath

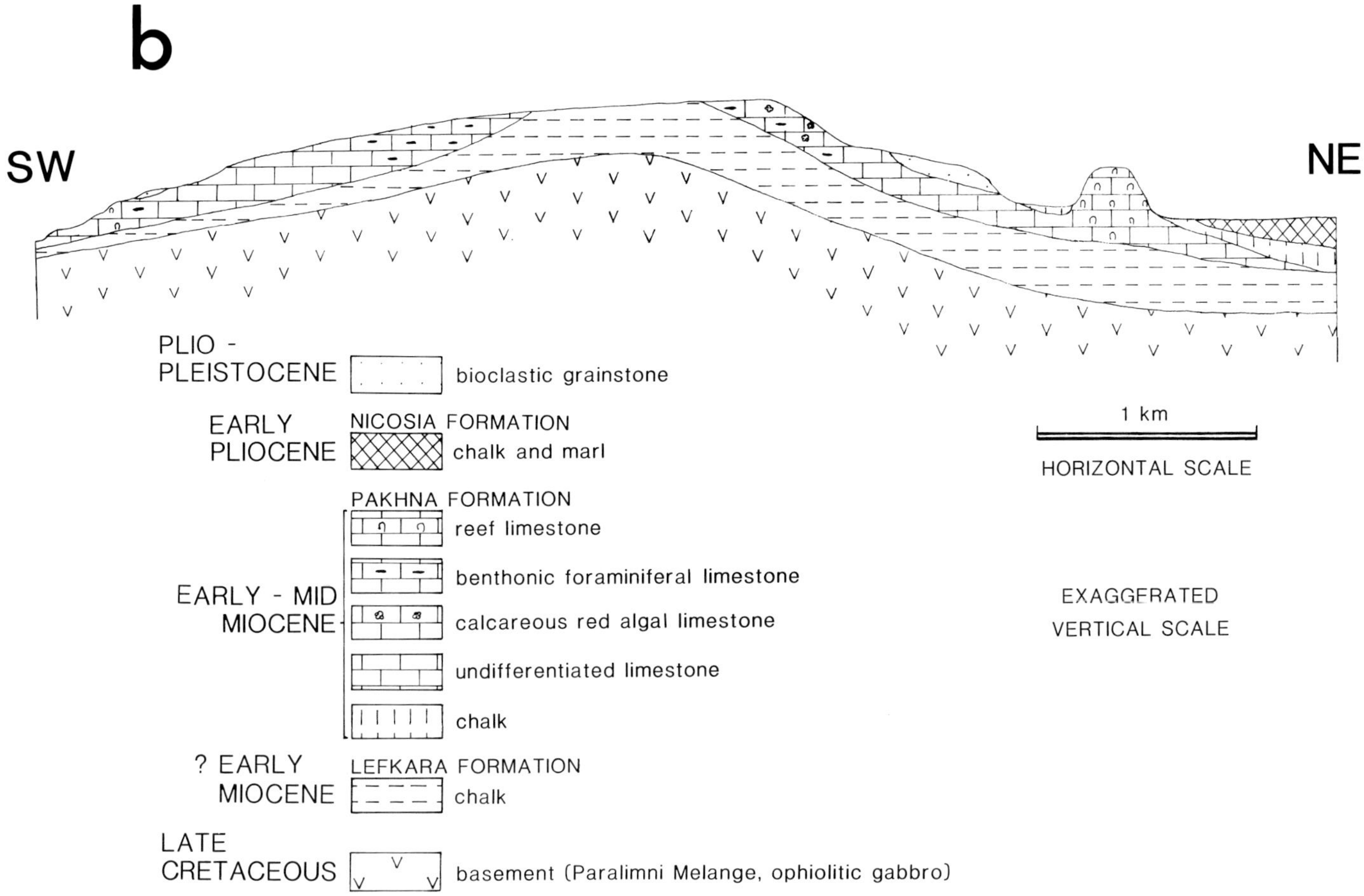

b

SW NE

PLIO - PLEISTOCENE — bioclastic grainstone

EARLY PLIOCENE — NICOSIA FORMATION — chalk and marl

PAKHNA FORMATION

EARLY - MID MIOCENE — reef limestone; benthonic foraminiferal limestone; calcareous red algal limestone; undifferentiated limestone; chalk

? EARLY MIOCENE — LEFKARA FORMATION — chalk

LATE CRETACEOUS — basement (Paralimni Melange, ophiolitic gabbro)

1 km

HORIZONTAL SCALE

EXAGGERATED VERTICAL SCALE

FIG. 4.—Continued.

Cyprus (Robertson, 1990; Payne and Robertson, 1995). In general, areas near the trench in the south underwent compression and uplift, while extension dominated further north and west. Tectonic movements exploited pre-existing lines of structural weakness, particularly in southwest Cyprus (Robertson et al., 1991).

## LIMESTONE LITHOFACIES

The nature of the exposures and the gradational facies relationships result in some problems when distinguishing between intra-reef and inter-reef particulate sediments. However, within the Miocene limestones it is possible to recognize: (1) in situ framestones, (2) particulate sediments intimately associated with framestones and (3) off-reef limestones.

### *Early Miocene Reefs (Terra Member)*

In situ framestones, together with the associated intra-reef sediment constitute the reef facies, which was found to be rather similar in both southeast and west Cyprus (Table 2).

*Reef Facies.—*

*Morphology and Distribution.*—Many reef outcrops approximate the shape and size of the original reef as post lithification erosion has selectively removed softer reef-margin facies, leaving the more lithified reef limestones intact (Fig. 8A). Typically, the reefs are circular to crescent-shaped in plan view, with a shallow, domed profile, often about 10 m in diameter and a few meters thick, but sometimes attaining a width of 500 m and a thickness of 80 m, particularly in the southeast. The individual structures are scattered about 1 km apart and resemble modern lagoonal patch reefs in size, shape and distribution (e.g., Scoffin, 1987).

*Framestone Components.*—The reef facies (Figs. 9, 10A, B) comprise small framestone structures scattered amongst lime mud intra-reef sediments, with a ratio of framestone to particulate sediments of about 1 to 10. The structure is made up of a primary framework of massive and branching corals (Figs. 8B, C, D), plus a secondary framework composed of encrusting organisms, including solitary corals, calcareous red algae, bryozoans, serpulids, foraminifera and molluscs. The main frame-building corals are *Porites* sp., *Favia aquitainensis* and *Favites*

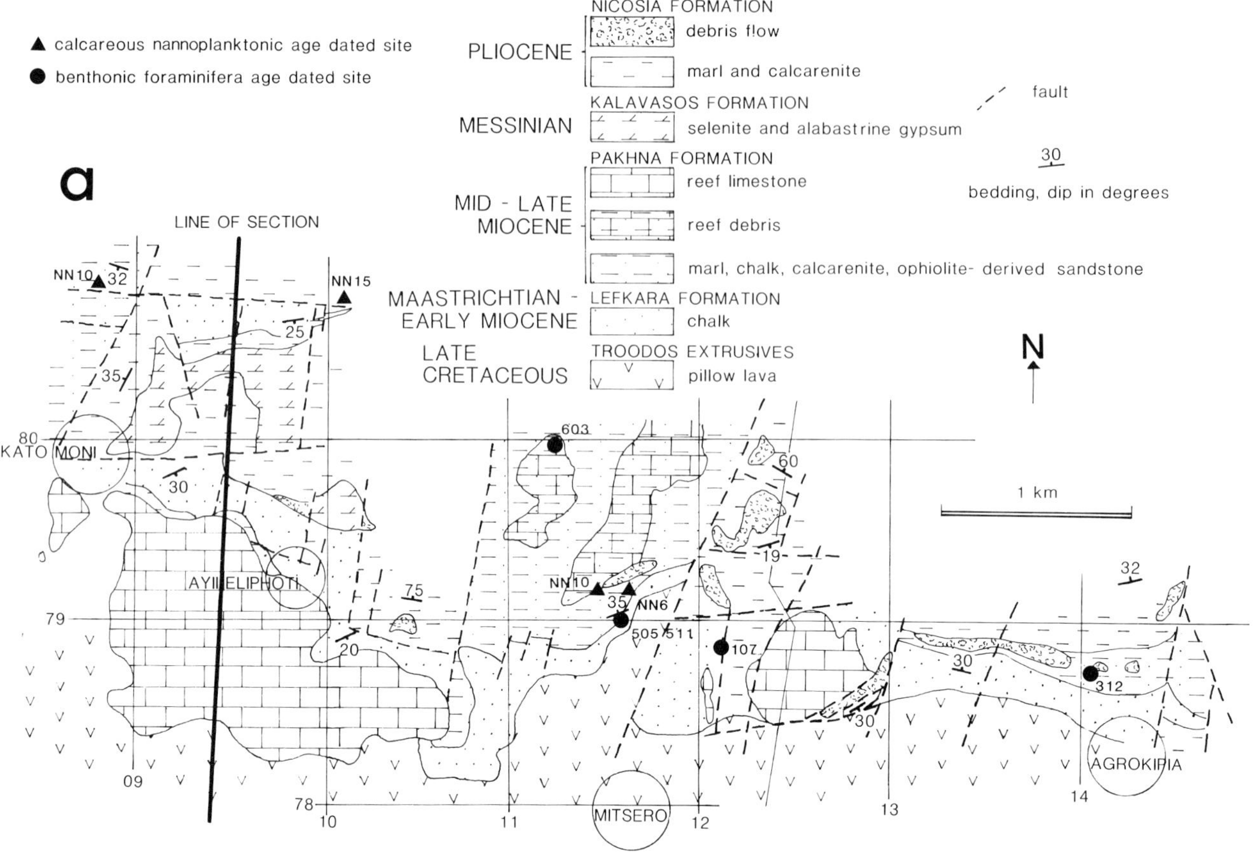


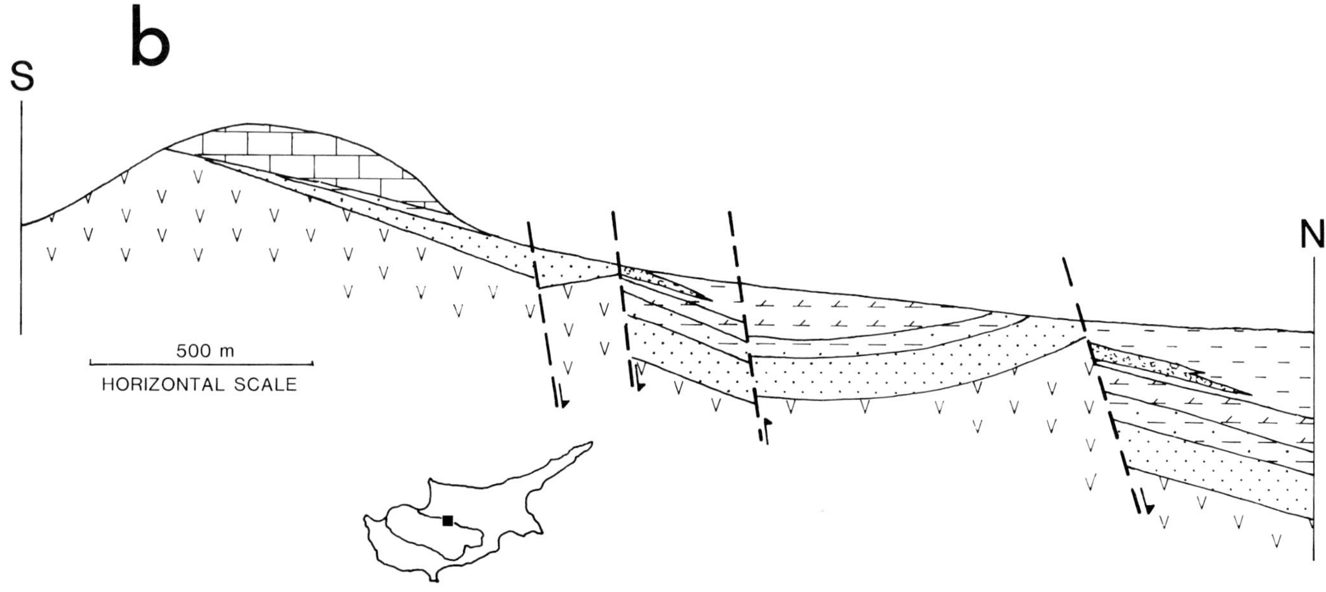


FIG. 5.—Geological setting of the Late Miocene Koronia Member along the northern margin of the Troodos ophiolite. (A) Geological map. Note the two major fault trends, the locations of dated samples and the line of section. (B) Cross section. Note the onlap of the reef limestones over sediments, locally directly onto the Troodos ophiolitic basement. The normal faults were active both during and after reef growth. Modified after Follows and Robertson (1990). See Figure 2 for location.

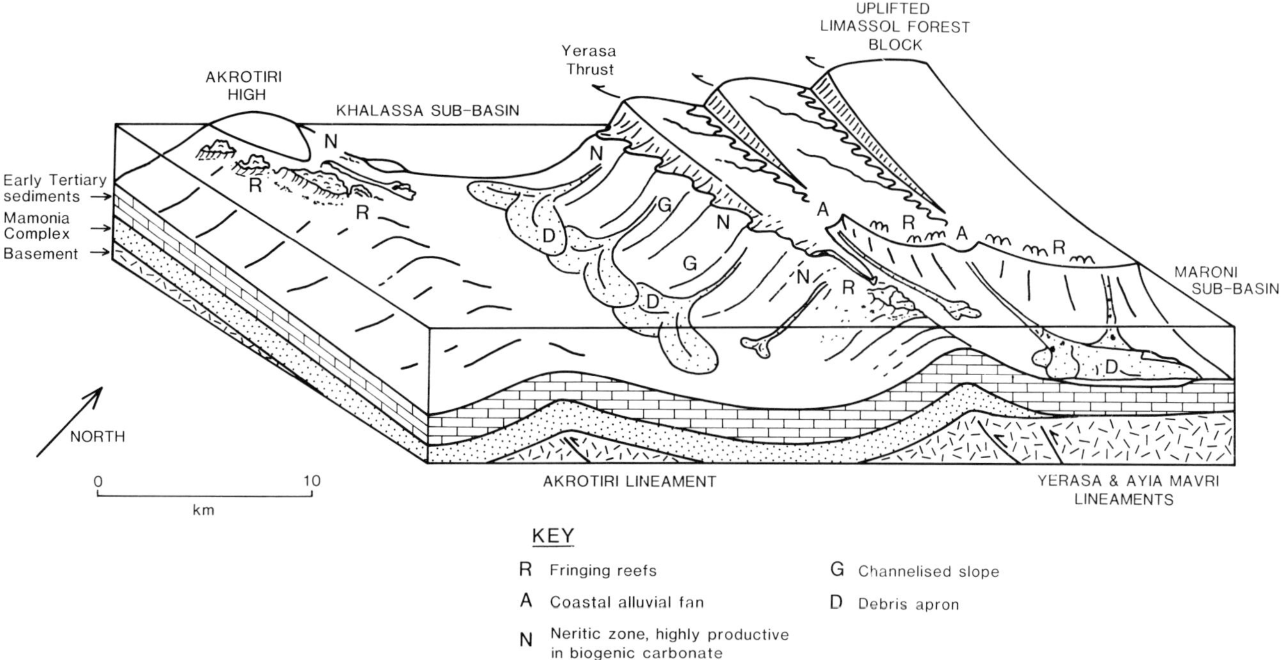


FIG. 6.—Reconstruction of the sedimentary environment and tectonic setting of Miocene limestones in southern Cyprus. The reefs developed on topographic lineaments, created by earlier crustal compression. Detritus accumulated in structurally formed, intervening basins. Modified after Eaton (1987).

TABLE 2.—CORALS IDENTIFIED IN THE EARLY MIOCENE TERRA MEMBER AND THE LATE MIOCENE KORONIA MEMBER. THE TERRA MEMBER IS FAUNALLY DIVERSE, WHILE THAT OF THE KORONIA MEMBER IS VERY RESTRICTED. NUMBERS: 1 RARE, TO 4, ABUNDANT. FROM FOLLOWS (1990).

| TERRA MEMBER CORALS | KORONIA MEMBER CORALS |
|---|---|
| **Frame builders** | |
| 4 *Porites* | 4 *Porites* |
| 4 *Favia aquitainensis* | 2 *Tarbellastraea reussiana* |
| 4 *Favites neugeboreni* | |
| **Reef dwellers** | |
| Faviids | |
| 2 *Favia melitae* | |
| 2 *Hydnophora (Monticulastraea) provincialis* | |
| 1 *Goniastraea* | |
| Heliastraeids | |
| 2 *Heliastraea grandis* | 1 *Montastraea* |
| 1 *Solenastraea tizeroutinensis* | 1 *Tarbellastraea carryensis* |
| 2 *Tarbellastraea carryensis* var *maj. nov var.* | |
| Mussids | |
| 2 *Lithophyllia michelotti* | |
| 1 *Leptomussa falloti* | |
| Archeocaeniids | |
| 3 *Actinastraea tarbellensis* | |
| 2 *Stylophorida reussiana* | |
| Acroporids | |
| 3 *Acropora* | |
| Poritids | |
| 2 *Goniopora* | |
| Agariciicids | |
| 1 *Siderasterea* | |

*neugeboreni*, which grew in massive domed and phaceloid forms, ranging from 10 cm to 1.5 m in diameter (Fig. 9). Other morphologies of frame-building corals include finger-like upright branches (*Porites* sp., *Acropora* sp.); dendritic branches (*Actinastrea* sp., Faviids) and sheet-like forms (*Porites* sp.). An interesting aspect of the Terra Member primary framework structure is the local occurrence of alternations of laminar sheet-like forms (normally *Porites*) and upright dendritic types (usually *Actinastrea* and/or *Acropora* colonies).

The secondary framework coats individual coral colony surfaces with encrustations up to 1 cm in thickness. These framework additions presumably strengthened the structure, but regular sequences of encrusters, which might signify changes in environment during growth, were not discerned.

Boring, mainly by Lithophagids (less than 1 cm in diameter) is common in poritid and faviid colonies. *Actinastrea* was particularly susceptible to microborers. These structures consist of fibres, less than 0.4 mm in diameter, connecting small 'swellings', which were revealed as a lacy network, formed by diagenetic solution. These morphological characteristics match the descriptions of *Entobia* (Bromley and d'Alessandro, 1984; Pleydell and Jones, 1988).

*Intra-Reef Facies.*—The sediment in the hollows between individual coral colonies is coarse bioclastic debris comprising skeletal fragments of reef-frame builders and reef dwellers. These include the following groups of organism: echinoderms (*Echinus sp.*), molluscs, gastropods (*Trochus sp.*, *Voluta. sp.*, *Conus sp.*), bivalves (of the venerid, carditid and glycimerid

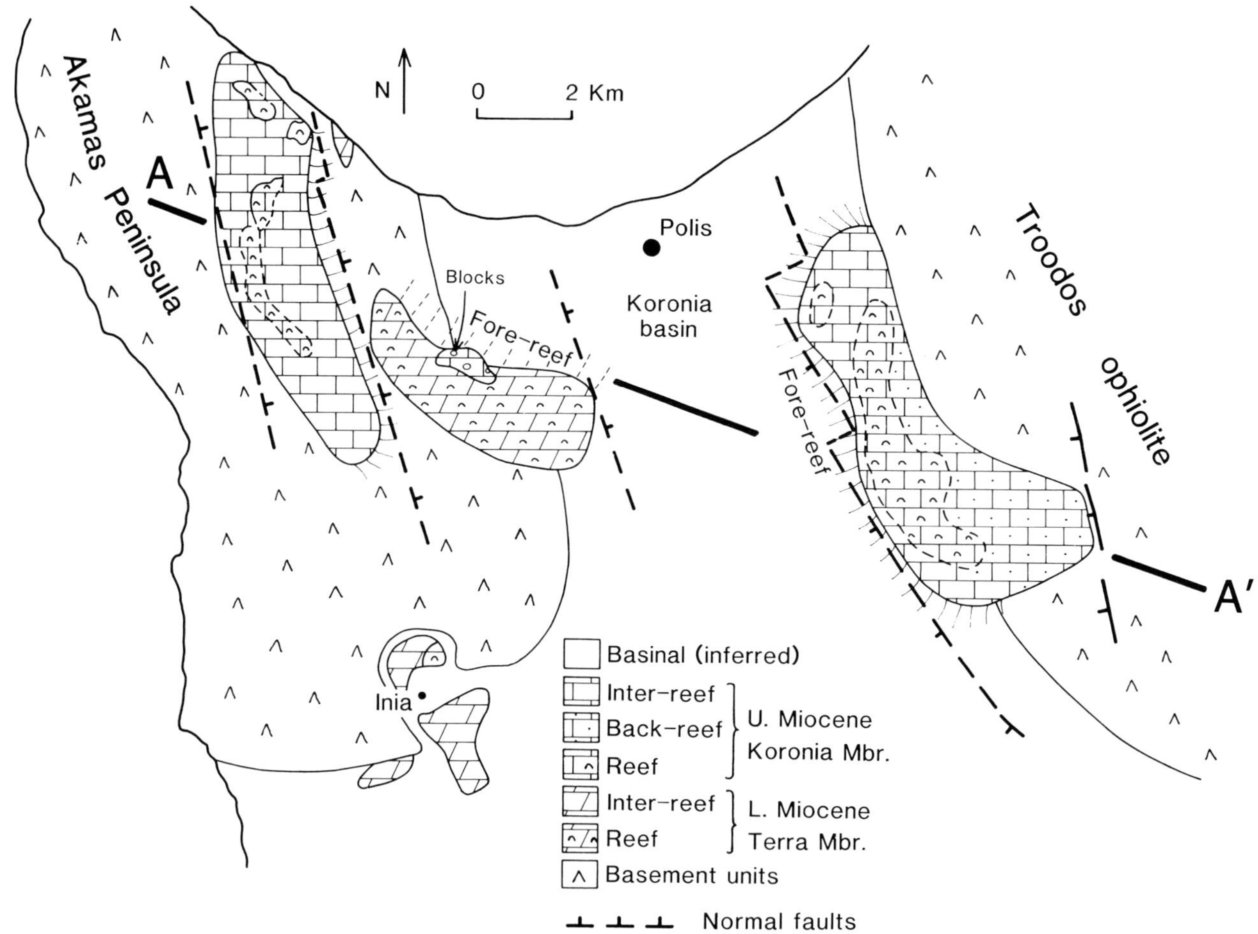


FIG. 7.—Geological setting of the Terra and Koronia Members in northwest Cyprus. Note the location of the reefs and their sediments in relation to the inferred trends of active faults. The Polis-Paphos graben was initiated during the Late Miocene and influenced the deposition of the Koronia Member. See Figure 2 for location.

families), bryozoans, benthonic foraminiferans and crustose coralline algae. Additional coarse components include aggregate grains and 1- to 2-cm-sized rhodoliths. On a smaller scale, in interstices between coral branches, line mud and benthonic foraminiferans are the main constituents. Within small geopetals, such as borings, several generations of void-filling micrite occur at various angles to the horizontal. The evidence from these cavities points to a history of early lithification of lime mud within the reef, and local tilting of derived reef blocks.

*Off-Reef Facies.*—In southeast Cyprus, off-reef facies are distributed in belts in which coral-dominated patch reefs passed outwards, first into algal-dominated deposits, then into benthonic foraminiferal packstones and grainstones and massive lime mud and then into pelagic carbonates. Individual beds of packstone-grainstone are locally up to 30 cm thick, cross-bedded and overlie basement clays. Benthonic foraminifera are abundant and include forms with inflated tests less than 1 cm in size. Fragmented coralline algal encrusters (*Lithophyllum*) and rhodoliths (*Mesophyllum* and *Lithophyllum*) are also common. These limestones represent inter-reef sediments. No examples of slope facies (for example, debris-flow deposits) are found, and it is clear that the sea floor topography around the reefs was subdued in this area of southeast Cyprus.

In west Cyprus, fore-reef facies are better developed and form a discrete, elongate east-west-trending zone, parallel to the trend of patch reefs, with a palaeoslope dipping northwards (Fig. 8E). The slope facies comprise abundant debris-flow deposits, made up of large benthonic foraminifera (Fig. 8F) and rhodoliths, which grade into pelagic carbonates (Fig. 10B). These sediments are similar to deep ramp facies reported in Miocene units of the Central Mediterranean (Buxton and Pedley, 1989). Talus deposits are also locally present between the reef and reef-slope but are restricted to a narrow zone 5 m wide. Blocks in this slope breccia are composed of coral and detrital grainstone.

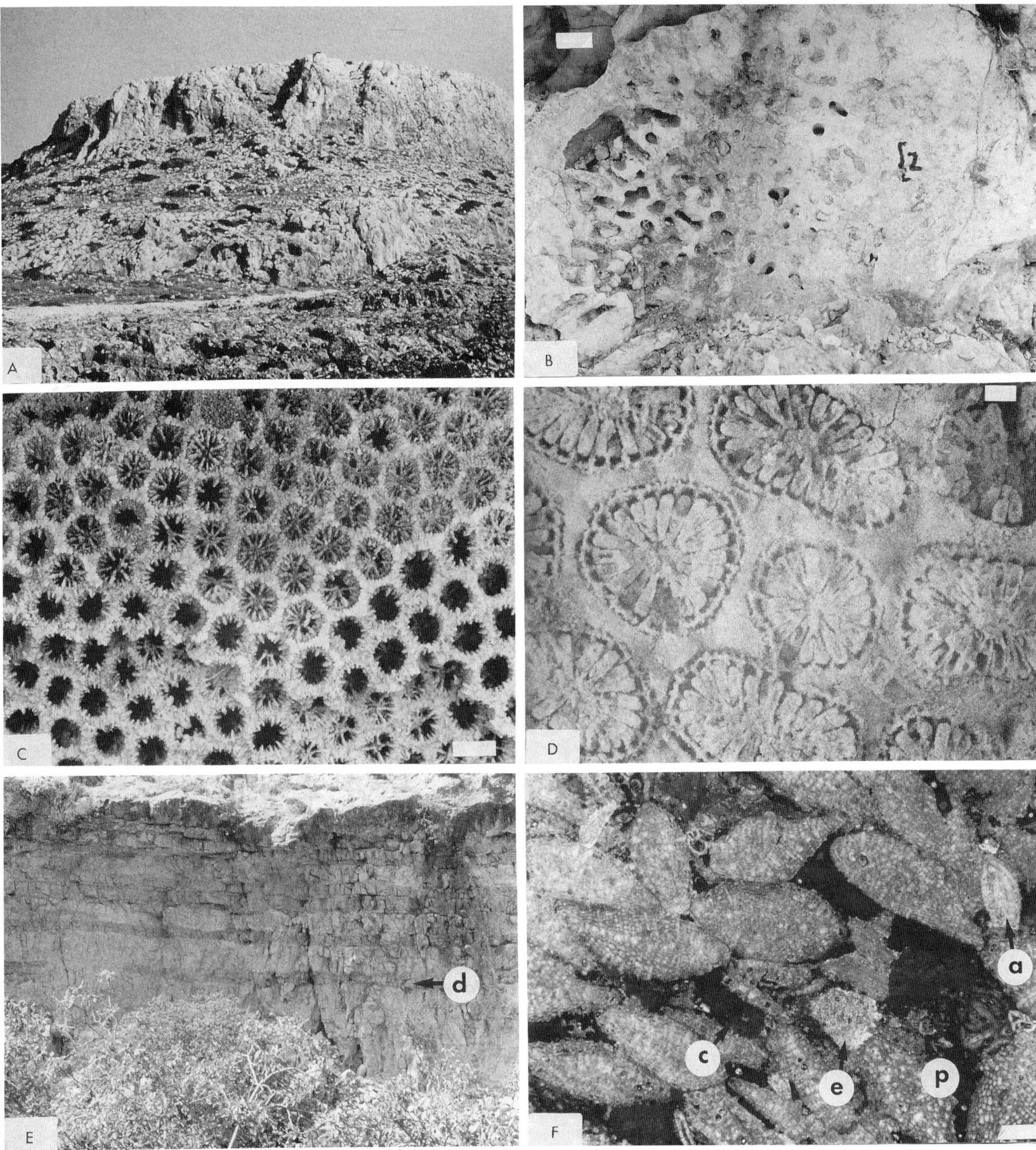


FIG. 8.—(A) 500-m-long patch reef exposed at Cape Greco, southeast Cyprus; the two escarpments total 80 m in height. (B) Effect of dissolution of the corallites of a ramose faviid leaving behind the surrounding lithified interstitial micritic sediments. Similar corals are occasionally observed within the larger upstanding patch reefs. Scale bar = 2 cm. (C) *Tarbellastraea carryensis* var. major nov. This coral has undergone dissolution and cementation, the septa were almost completely dissolved in the lower part of the plate. Scale bar = 0.5 mm. (D) Detail of *Favia aquitainensis*, illustrating the state of preservation of some of Terra Member colonial corals. Scale bar = 0.15 mm. (E) Plane bedded inter-reef facies; dolomitized horizons are darker (d). A typical exposure of inter- and fore-reef facies in southwest Cyprus. Height of section = 10 m (F) Benthonic foraminiferal grainstone dominated by *Miolepidocyclina burdigalensis*. Minor amounts of echinoid (e) and crustose coralline red algal (c) fragments are present, in addition to scattered *Amphistegina* (a) and planktonic foraminifera (p), Ayia Napa quarry. Scale bar = 0.5 mm.

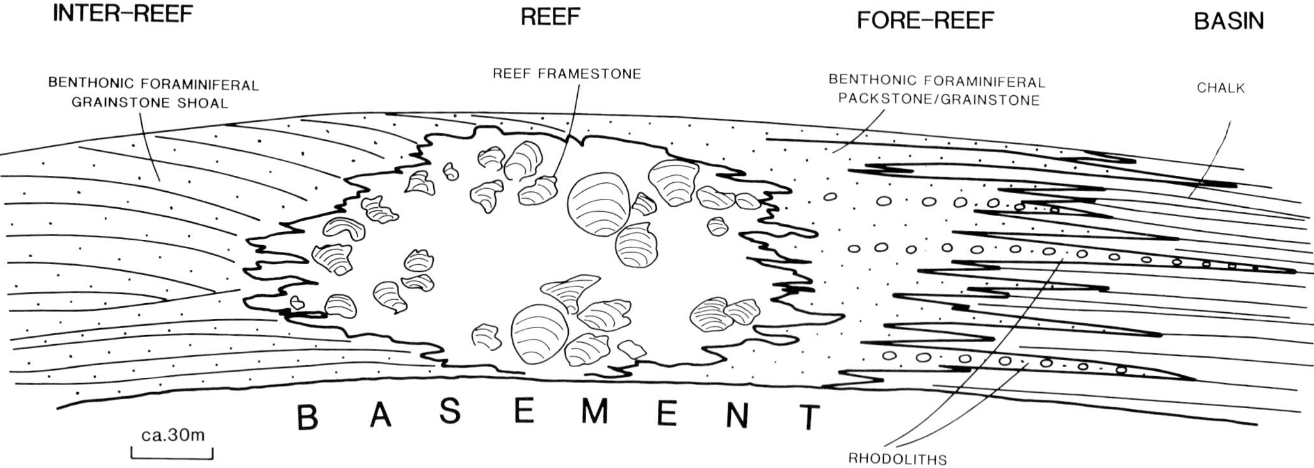


FIG. 9.—Schematic reconstruction of the Early Miocene Terra Member. See text for explanation. Individual coral types are not differentiated (see Fig. 8).

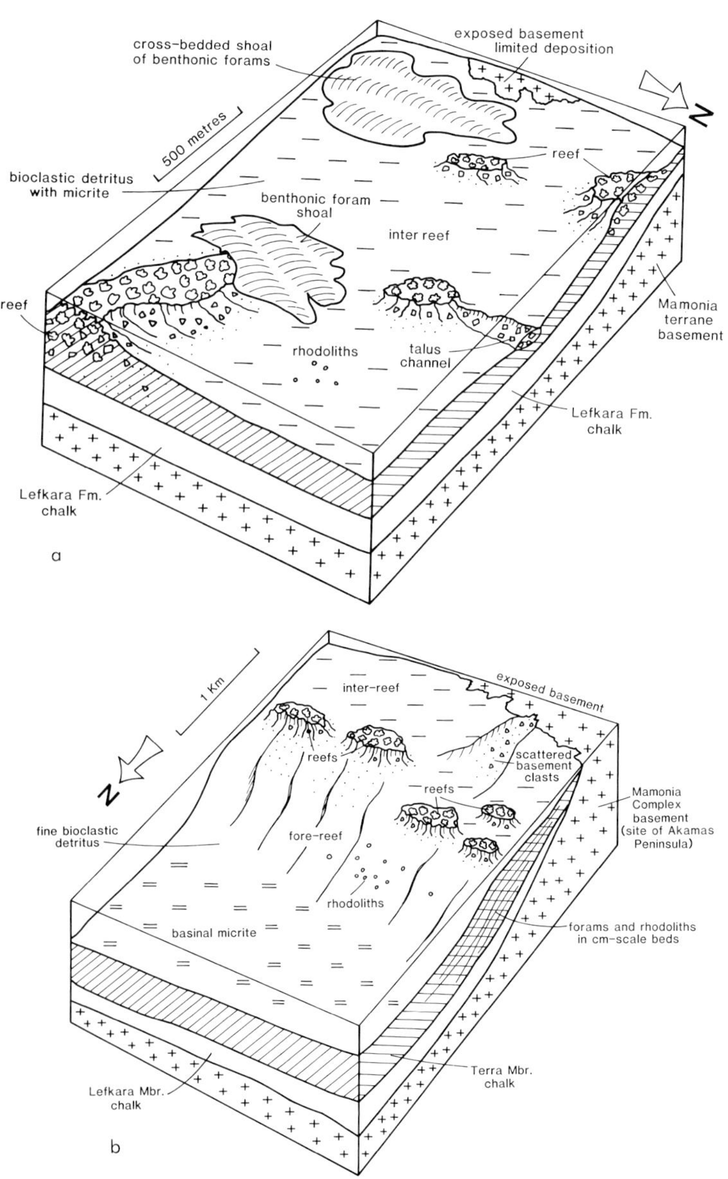


FIG. 10.—Reconstruction of the Early Miocene Terra Member. (A) In southeast Cyprus, there is little evidence of active tectonics during growth of these reefs. See Figure 4 for map and cross section of this area. (B) In west Cyprus, reef growth was influenced by gentle subsidence, which predated rifting of the Polis graben. See Figure 7 for map of this area.

### *Late Miocene Reefs (Koronia Member)*

The Koronia Member represents the second phase of reef growth on Cyprus, occurring in the upper part of the Pakhna Formation (Fig. 3). This later phase of reef growth is comparable with similar fringing reefs of monospecific, poritid coral fringing reefs, commonly developed around the Mediterranean basin in the Tortonian-early Messinian time (e.g., Esteban, 1979; Rouchy et al., 1986; Esteban et al., this volume; Pedley et al., this volume; Table 2). These Koronia Member reefs are marked by a greatly reduced diversity of coral taxa, relative to the Terra Member.

*Reef Facies.—*

In contrast to the Terra Member reefs, the poritid reefs generally did not form discrete patch reefs, but linear trends of low mounds, or sheet-like structures, intimately associated with particulate bedded sediments. Also, the Late Miocene reefs usually cap local topographic highs and considerable volumes of reef-derived debris were shed down fore-reef slopes into neighboring basins.

*Morphology and Distribution.*—The nature of the Koronia Member reef outcrops differs in each main area of exposure, largely controlled by local structural setting, as summarized earlier. In northern Cyprus, the reefs unconformably overstepped basement lava and chalks. Fringing reef tracts reached over 7 km long and 2 km wide, interrupted by transverse faults (Figs. 5A, B). Extensional faulting continued during reef growth and reefs were localized on the upstanding fault blocks.

In western Cyprus, north-south trending, roughly linear poritid

FIG. 11.—Illustrations of the Koronia Member. (A) Koronia Member reefs exposed at Happy Valley, south Cyprus (example arrowed). Note the block-like patch reefs weathered out from the vegetation covered hillside (Pakhna Fm.). (B) A stained (Alizarin red) cut sample of laminated poritid coral (p). The stain picks out additional cryptic laminae of algal coatings (a) of lithified micritic sediment between the coral layers. Note the porosity present in the thicker poritid layer (v). Scale bar = 0.5 mm. (C) Molds of glycimerids in packstone. This is a common in situ growth assemblage, Koronia Member. Scale bar = 1.5 cm. (D) Angular to sub-angular chalk clasts in a chalk matrix with some sandstone present. This facies is typical of the base of channels in the Koronia Member. Scale bar = 2.5 cm.

reefs developed on both the east (Troodos Massif) and west (Akamas Peninsula) flanks of the Polis-Paphos graben (Fig. 7). Exposures of the Koronia Member are aligned parallel to the graben-bounding faults, with fore-reef facies dipping towards the graben axis (Follows, 1990). Locally, along the western flank of the graben, lithified blocks of Koronia Member limestone were shed into pelagic carbonates, above limestones of the Terra Member, and this event may mark the start of extensional faulting in the area. The topographically lowest reefs crop out nearest the graben axis, whereas higher ones are located further away, on the rift flanks. These reefs are distributed *en echelon*; they were initially up to 8 m thick but decreased in size upwards and laterally, reaching decimeter scale, comparable to thicknesses of the off-reef beds. These unusual relationships are explained by syn-depositional subsidence of the graben axis, causing outwards migration of the reef facies with time. In contrast to northern Cyprus, however, there is a general absence of structurally controlled facies, such as debris-flows, and this suggests that the seafloor topography remained subdued during reef growth. The development of the present steep-sided graben, offset by transverse faults (e.g., west end of Peristerona valley) took place in Messinian to Early Pliocene time, only after reef growth ended.

In southern Cyprus, the reefs are mainly represented by talus, largely confined to channels. The composition and location of the debris suggests the reefs were dominated by poritids, that constituted linear reef tracts on the paleohighs formed earlier. At one location, Happy Valley, early Koronia Member reefs remain *in situ*, exposed on limestone bluffs as groups of corals surrounded by packstones, which pass laterally into channelised calciturbidites and debris-flows (Fig. 11A).

*Reef Framework.*—The earliest reefs do not show the development of a contiguous framework, instead discrete poritid, tarbellastraeid and montastreid coral colonies occur separately

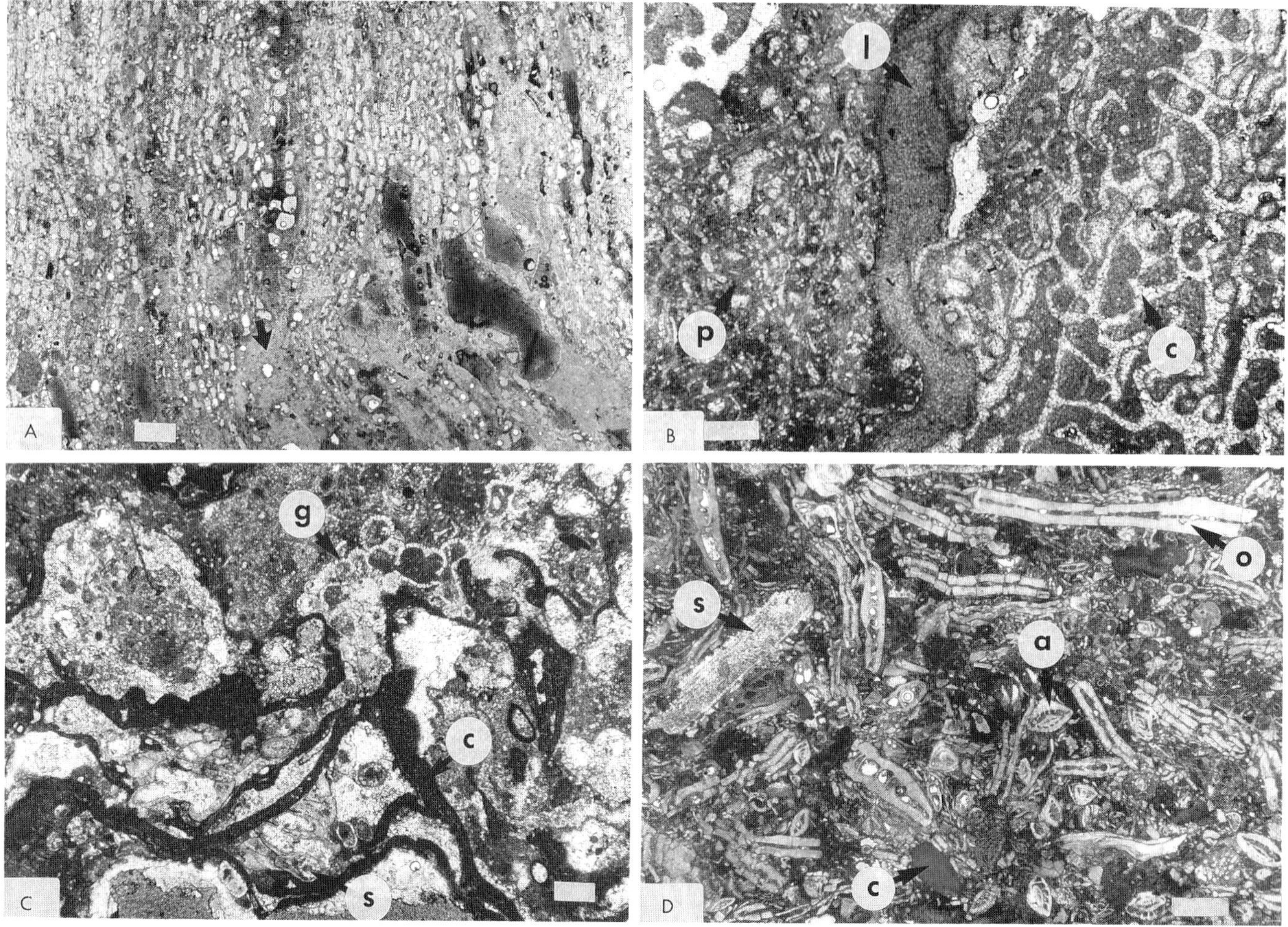


FIG. 12.—(A) The Anascan cheilostrome bryozoan, *Calpensia nobilis* (Michelin) with a multilamellar "celleporiform" colony (P. D. Taylor, pers. commun., 1987). The micritic patches between the structure (example arrowed) may represent replacements after soft-bodied fauna associated with the structure. Most of the cells within the sample are empty. Scale bar = 1 mm. (B) The crustose coralline alga (l), *Lithothamnium*, has colonized the surface of a poritid coral (c) before being overlain by fine-grained packstone (p). Note that internal sediment fills have created geopetals in the intra-coral void space, also the retention of void space between algal and coral surfaces. Scale bar = 0.5 mm. (C) Crustose coralline red algae (unidentified) (c) that plays a binding role within the Koronia Member reef. Encrusting foraminifera (g), *Gypsina* and serpulids (s) contribute to the structure. Scale bar = 0.5 mm. (D) Grainstone from the base of a talus channel on the north margin of the Troodos Massif, comprising *Operculina sp.* (o), *Amphistegina sp.* (a), crustose coralline red algal fragments (c) and echinoid spines (s) in addition to planktonic foraminifera, and other detritus. The bioclasts, particularly large foraminifera, illustrate compaction with limited cementation except for overgrowths on the echinoid fragments. Scale bar = 1 mm.

and upright within packstone sediments. These domal colonies decrease in abundance up sequence and are replaced by laminar poritids. The poritid layers (0.5 to 1 cm thick; Fig. 11B) are commonly parallel but also developed finger-and ridge-like projections up to 20 cm high. Projections were later roofed over to create "shelter cavities", which were later filled by internal sediment. The sheet-like growth forms of coral are emphasized by encrustations, calcareous red algae (*Lithothamnium*; Fig. 12A) bryozoans (*Calpensia nobilis*; Fig. 12B), encrusting foraminifera (e.g., *Gypsina*; Fig. 12C) and encrusting molluscs and serpulids. As in the Terra Member corals, *Lithophaga* and *Entobia* borings are relatively abundant. The typical reef morphology is reconstructed in Figure 13.

*Inter-Reef Facies.*—The sediment intimately associated with the reef framework occurs mainly as internal sediment within "shelter cavities", roughly 10 cm in size. These sediments vary from wackestones to grainstones and commonly show a coarsening-upward sequence. The grains consist of bioclastic fragments of reef framework and reef dwellers, including echinoderms (e.g., *Echinus*), molluscs (patellids, trochids, *Haliotis*, cerithids, oysters, glycimerids; Fig. 11C), bryozoans, benthonic foraminifera (*Amphistegina*, *Operculina*; Fig. 12D) and crustose algae. The dominant non-skeletal grains present are peloids.

*Off-Reef Facies.*—The Koronia Member reefs differ from those of the Terra Member in having distinctive fore-reef slope sediments associated with them. The fore-reef transition is best seen in the northern area, where syn-depositional faulting created a rugged seafloor. Proximal reef talus dominates, largely

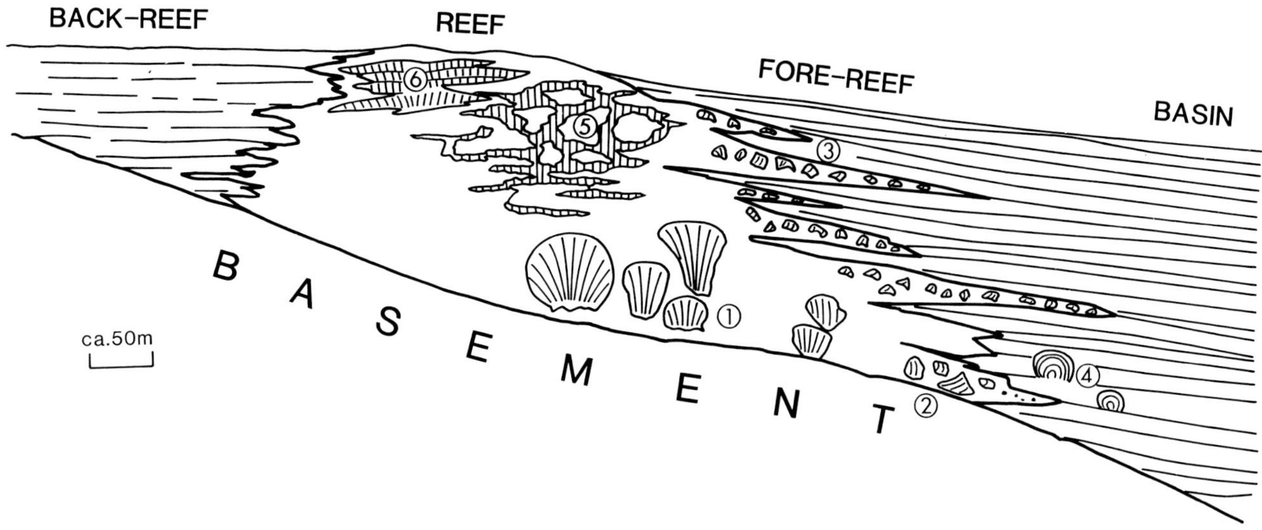


FIG. 13.—Schematic reconstruction of the Late Miocene Koronia Member. The reef framework is mainly composed as poritid corals. There is considerable variation in the scale and geometry of the poritid colonies, as discussed in the text. 1-domal colonial coral growth; 2-marginal domal coral fragments; 3-flanking reef talus; 4-isolated domal corals; 5- poritid growth forming cavities infilled with internal sediment; and 6-laminated poritid growths.

deposited by mass flow. Off-reef deposits lower in the succession are mainly channelised calciturbidites and debris-flows. These sediments pass upwards into lenses and sheets of fore-reef detritus (Fig. 11D), including both bioclastic debris and terrigenous, basalt-derived sand, intercalated with marl. Proximal fore-reef facies contain abundant whole bivalves (e.g., glycimerids), gastropods, echinoderms (*Schizaster*), benthonic foraminifera, rhodoliths and serpulids, whereas more distal facies mainly comprise graded beds of bioclastic sands. In this northern area, local back-reef exposures (Fig. 14A) include miliolid foraminifera. Also, the reefs are flanked by patchily distributed oolitic grainstones.

Off-reef facies are also widely distributed in southern Cyprus, where they accumulated in structurally controlled basins between the linear topographic highs capped by reefs (Eaton, 1987; Fig. 6). Such off-reef facies are particularly well developed in the Khalassa sub-basin (Fig. 6). Proximal basin slope facies exposed along the northern margin of this basin ("gullied slope-facies" of Eaton, 1987) are dominantly channelised calcarenites, rich in poritid coral fragments that become more distal southwards, passing through thin-bedded calciturbidites, into pelagic chalk. The patch reefs that were developed on isolated highs within this basin (Happy Valley, see above) disintegrated to form a variety of sediments, including interbedded calcarenites and chalks, overlain by debris-flows containing wackestone and packstone blocks. Hardgrounds developed beneath the calcirudites, as shown by the presence of well cemented limestone surfaces, with attached intact oysters and other bivalves, including *Spondylus*. The input of carbonate debris was episodic, possibly controlled by storms, local tectonics and/or relative sea-level changes.

In other areas, the reef to off-reef facies transition was less tectonically controlled, for example at Polykantho in the north (west of Fig. 5A), and in southeast Cyprus. In the southeast, the most reef-proximal sediments are mainly detrital, but also contain discrete colonies of *in situ* corals, up to 1.5 m across. Associated sediments include massive wackestone-packstones, in layers, 1-5 m thick. These are interstratified with thinner layers (20 cm thick) rich in reworked bivalves and corals (now moldic) including cylindrical, Tarbellastreids, up to 15 cm across. Rhodoliths form up to 40% of the rock. Packstones (up to 4 cm thick) contain burrowing echinoids (e.g., *Clypeaster altus*) and benthonic foraminifera. Individual sharp, basal units reflect sudden influxes of coarse-grained sediments, interpreted as storm deposits.

In west Cyprus, off-reef facies are extremely variable, largely controlled by local tectonics. On the east flank of the Polis-Paphos graben (Fig. 14B), for example, linear outcrops of reef facies are bordered by fore-reef facies in the west and by back-reef facies in the east. The fore-reef sediments dipping into the graben are thickly-bedded packstones, containing abundant rhodoliths and coarse fragments of crustose coralline algae, echinoderms and bivalves. The back-reef facies, at a higher structural level, are thin-bedded marls (containing scattered whole bivalves), chalks and packstone, with comminuted bivalves, crustose coralline algae and miliolids.

## DIAGENESIS

### *Terra Member*

The earliest cement coating the reef-building skeletons is a thin isopachous fringe of acicular crystals that is overgrown by

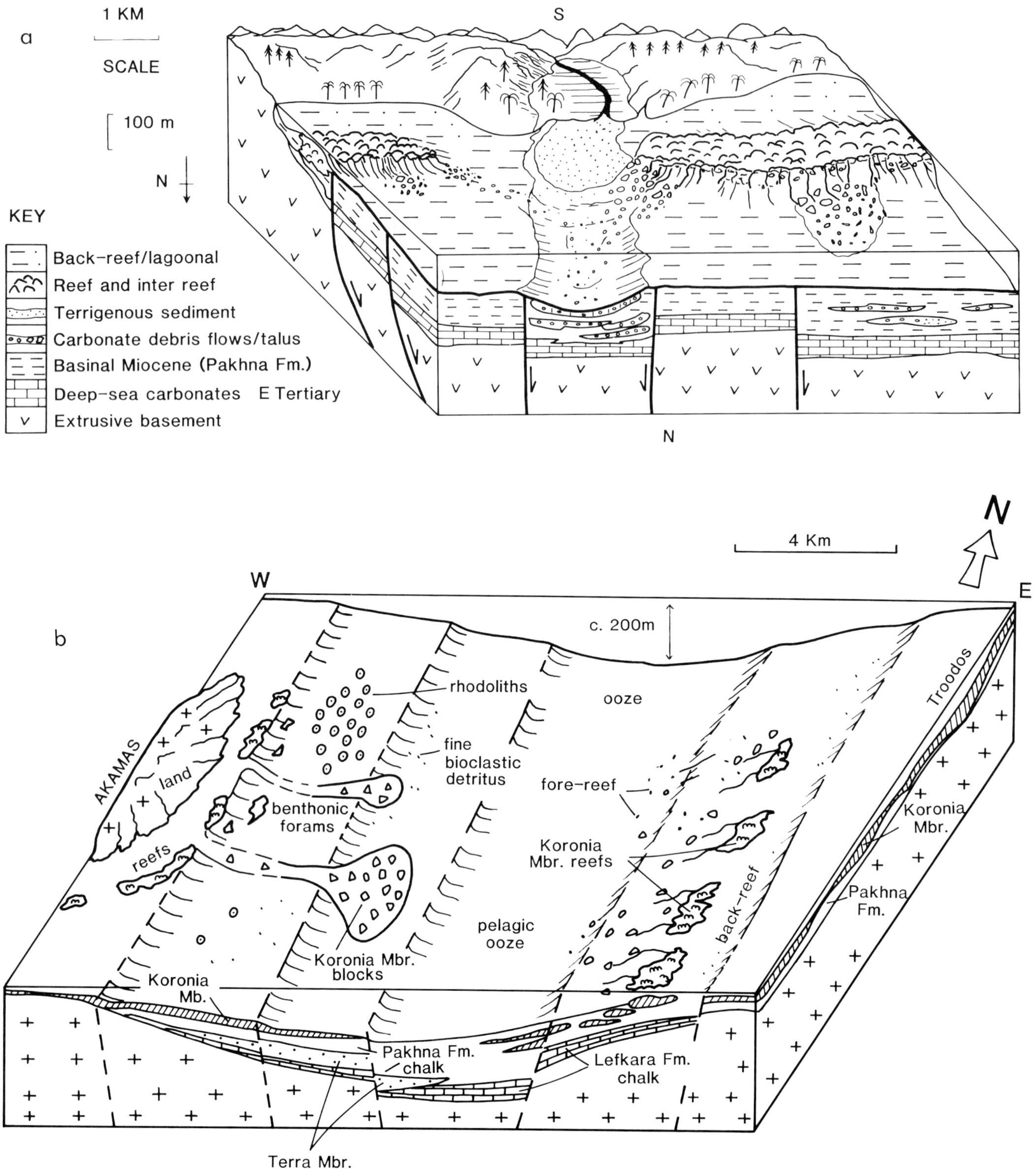

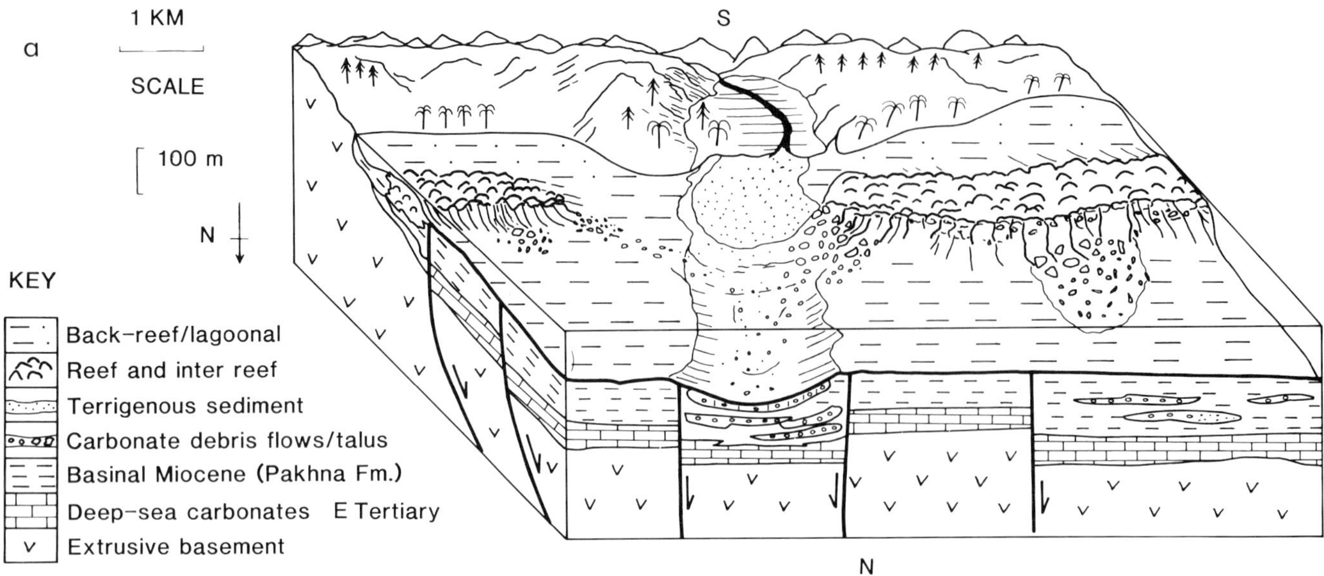




FIG. 14.—Reconstruction of the Late Miocene Koronia Member. (A) In northern Cyprus, note that reef development was mainly in the form of small poritid patches, developed in shallow, turbulent seas, with variable terrigenous input from the south. The reefs developed on fault blocks during active extensional tectonics. Reef talus, largely debris-flows accumulated in channels in the downfaulted basin further north. Debris was funneled along transverse faults. (B) In northwest Cyprus, the reef growth was controlled by active crustal extension and development of the Polis-Paphos graben. Sea-floor relief, however, was never steep in contrast to the southern and northern areas.

larger tabular cement crystals. Both are now calcite; however, these early cements are believed to have been marine in origin, as they commonly precede micritic internal sediment. Subsequent flushing by meteoric fluids led to the dissolution of aragonitic skeletons. Micrite envelopes of poritid coral walls support an acicular fringe cement on one side (primary pore) and a blocky calcite spar on the other (secondary pores) after dissolution of corals. The moldic voids were coated with tabular calcite and, in places, gypsum crystals, which also grew within the internal sediment micrite. The gypsum crystals were later pseudomorphed by calcite. The fact that the first cement is a fibrous calcite, rather than void-filling spar or pseudomorphed gypsum, suggests that its composition was unlike that of corals and was most probably originally Mg-calcite. The final cements were coarse calcite spar, which contain bands of abundant inclusions. The coarse, late cements have $\delta^{18}O$ and $\delta^{13}C$ isotopic compositions in the range for calcite precipitated from waters of meteoric origin ($\delta^{18}O$: -3 to -5‰; $\delta^{13}C$ -3 to -8 ‰; e.g., Allan and Matthews, 1982). The bands of late inclusions are composed of Mg-rich micrite and could represent a return to marine conditions at the time of the Pliocene transgression.

Numerous fissures cut the reef framework, typically one or two centimeters across and extending several meters downwards. These fissures are filled with micrite in two stages, first in the Messinian and then prior to marine transgression in Early Pliocene time. The micrite was commonly replaced by fine euhedral crystals of dolomite, but this dolomite has only rarely replaced the host reef facies limestones next to the fissures.

The off-reef limestones of the Terra Member are rich in micrite; consequently, calcite spar cements are not well developed. The two initial cement phases seen in the reef frameworks are also present in these calcarenites, especially within foraminiferal tests and as epitaxial overgrowths on echinoderm grains; there are also some local patches of late coarse spar. Gypsum is absent. The benthonic foraminifera-rich beds of the Terra Member off-reef facies seem to have been particularly susceptible to dolomitization and, in some low stratigraphical horizons, to silicification also. The dolomite crystals, in most cases, have replaced the micritic matrix and only partially attacked the foraminiferal grains. In some beds, however, foraminiferal grains have dissolved, leaving molds in a fine dolomitic matrix. The dolomitic crystals are euhedral rhombs approximately 60 mm in diameter with inclusion-rich cores and limpid rims, in which the Mg/Ca ratio increases from 0.54 to 0.60 from core to rim. The $^{87}Sr/^{86}Sr$ ratios of two samples of dolomitized Terra Member inter-reef facies, indicate that the dolomitizing fluids were less than 15.4 million years old (Follows, 1992).

### *Koronia Member*

As with the Terra Member, neomorphism and dolomitization have obscured the details of the depositional and early diagenetic fabrics. Within the reefs, two early cement phases, aragonite fringes and botryoids (with original mineralogy preserved), and acicular fringes of Mg-calcite, were synchronous with internal sedimentation of micrite. Some reef builders encrust this micrite indicating these early cements are marine in origin. The aragonite fringes are arranged as crusts coating grains whereas the botryoids form spherical splays up to 2 cm in diameter. The splays resemble marine botryoidal aragonite from the Belize reef front, as described by James and Ginsburg (1979). The aragonitic crystals have a high Sr content (0.85 to 1.2 wt% SrO and a typical marine isotopic signature ($\delta^{18}O$ of +1.7 to +3.0‰ and $\delta^{13}C$ of +1.9 to +3.4‰). The internal micrite sediment has a MgO wt% of 1.5 and only trace amounts of Sr, suggesting it was originally Mg-calcite. The Mg-calcite cements (3 wt% MgO), in places, line cavities within micrites and are themselves overlain by peloidal cements, which are thought to be of marine origin (Macintyre and Marshall, 1989).

The late-stage, coarsely crystalline calcite spar at the center of voids is similar in composition to calcite found in veins and is thought therefore, to be vein-fed in origin. As in the Terra Member, dolomitization is patchy, though especially common in those Koronia Member limestones close to the ophiolitic basement. Isotopic analysis of entirely dolomitized samples reveal values of $\delta^{18}O$ between +3.5 and +6.1 and $\delta^{13}C$ between +0.65 and +2.2. These values support a dolomitizing mechanism of mixed marine-meteoric water replacement, with only a minor component of freshwater. The presence of limpid (lower Ca) rims to euhedral dolomites suggests that the meteoric / salt water ratio decreased with time. Chemical analysis did not reveal any differences between the dolomites of the Terra, Koronia, or Pliocene fissure-fill limestones, and thus argue against the dolomites being a direct result of the salinity crisis in the Messinian.

### *Diagenetic Interpretation*

The burial history of the Terra and Koronia Member limestones is summarized in Figure 15. The limestones of both members, especially the reefs, were lithified by aragonite and Mg-calcite cements while in contact with sea water. A phase of uplift followed which resulted in regression and exposure of the reefs to fresh water. The meteoric fluids dissolved aragonite skeletons, particularly corals, although some aragonite cements that were enveloped in micrite escaped this dissolution. The molds were filled by sparry calcite cements with freshwater chemical signatures. Locally in southeast Cyprus, gypsum was precipitated within reef voids and internal sediments, possibly during drawdown of the sea in Messinian time. During crustal extension that ensued, rigid reefs fractured on a yielding bentonitic substrate. These enlarged solution voids (up to 1 m) were later finally filled by pelagic grains during the Early Pliocene marine transgression. It must have been this marine phase that permitted the formation of micritized sparry cements in the Terra Member, if the fine Mg-calcite crystals owe their origins to algal boring and subsequent infill, as suggested by Bathurst (1966) and Kobluk and Risk (1977). The mixing of fresh and marine waters during this time was responsible for patchy dolomitization, which partly obliterated earlier textures. Plio-

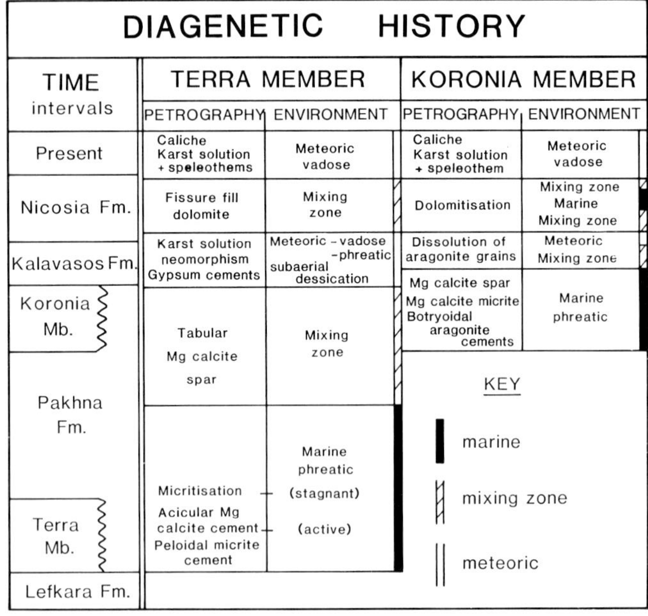


FIG. 15.—Summary of the diagenetic environments of the Early Miocene Terra Limestones and the Late Miocene Koronia Limestones. Modified from Follows (1990). See text for further explanation.

Quaternary uplift has again exposed these limestones to a freshwater regime in their present outcrop.

### *Porosity Development*

Some primary porosity is preserved within both limestone members, for example between layers of calcareous red algal skeletal growth and within incompletely cemented intergranular pores. Dissolution of aragonite skeletons and the incomplete filling of these voids has generated a high moldic porosity, especially in reef samples. The fissuring led to the development of an irregular vuggy porosity which cross-cuts earlier fabrics. Late dolomitization has created a moderate intercrystalline porosity between the euhedral rhombs. Also, preferential dolomitization of micrite matrix in wackestones was followed by local dissolution of relic calcitic foraminifera skeletons to create a late-stage secondary moldic porosity.

## REGIONAL COMPARISONS

Cyprus is strategically situated to facilitate comparisons with reefs in the central and western Mediterranean and Middle Eastern areas. In Cyprus, the Early Miocene corals are markedly lower in diversity than those of the Oligocene reefs in the western Mediterranean (Chevalier, 1961, 1977). The Early Miocene Terra Member appears to have survived longer (to mid Burdigalian) in Cyprus relative to counterparts in the western and central Mediterranean (e.g., Pedley et al., this volume). Reefs failed to develop throughout the Mediterranean as a whole between the Early and Late Miocene growth phases.

The Koronia Member reefs are somewhat similar to the Tortonian reefs of the Ziglaq Formation, fringing the coastal plain of Israel (Buchbinder, 1975, 1979, this volume). These reefs in Israel have recently been reassigned to a Langhian rather than Tortonian age (M. Esteban, pers. commun., 1993).

Terrigenous input played little role in Cyprus, on the shallow Malta shelf (Pedley, 1987), or in Israel, in contrast to southwest Turkey (Hayward, 1982; Hayward et al., this volume; Flecker et al., 1995), north central Sicily (Grasso and Pedley, 1988) and southeast Spain (Martín et al., 1989). The Gulf of Suez has a similar absence of terrigenous sediment, but there the reef front was probably much steeper, favoring extensive sediment bypassing (James et al., 1988). However, *Stylophora*, reported by James et al. (1988) and Buchbinder (1979), columnar poritids (in the Red Sea) and common Tabellastraeids (in southwest Turkey; Hayward, 1982) are all absent, consistent with a younger, Tortonian age. Esteban (1979) suggested that any Messinian reefs in the eastern Mediterranean should have a diverse fauna, but this is not seen on Cyprus. The Tortonian Koronia Member is instead most similar to the Miocene reefs in the Pelagian Islands, central Mediterranean (Grasso et al., 1985) and the Gulf of Suez (James et al., 1988; Burchette, 1988), two other areas where local tectonics greatly influenced reef development.

Several controls of reef growth inferred for other Mediterranean areas do not appear to be applicable in Cyprus. These include, first, the concept of multiple erosion surfaces as an indication of sea-level oscillation (e.g., in north central Sicily; Grasso and Pedley, 1988). In Cyprus, erosion surfaces are only visible at the base of the Terra and Koronia Members. Secondly, the notion that the reefs became topographically lower due to Messinian sea-level fall (e.g., southeast Spain; Esteban, 1979; Santisteban and Taberner, 1983) does not apply this area. The locations of the Koronia Member reefs were controlled by local faulting, rather than by progressive sea-level fall. In contrast, no such fault influence is reported in southeast Spain (Dabrio et al., 1981) or in Israel (Buchbinder, 1979). The small Miocene reefs in southwest Turkey were developed in association with alluvial fans and also without any obvious local syndepositional fault control.

## CONTROLS OF REEF GROWTH IN CYPRUS

In this final section, we briefly discuss the controls of reef development in Cyprus, in the light of regional comparisons.

### *Local Tectonics*

Tectonic effects were clearly most important in controlling Miocene reef development. First, regional tectonics resulted in sufficient uplift of the basement for local reef development. The reefs of the Early Miocene Terra Member grew on relatively stable, uplifted areas in west and southeast Cyprus. These reefs were floored by the highly deformed Mamonia Complex and related units, while the more structurally intact Troodos ophiolite remained submerged. By contrast the Late Miocene Koronia Member reefs were localized by active tectonic movements, during and/or immediately prior to reef growth. Some of these reefs developed on uplifted lineaments related to Early Miocene compression in south Cyprus, and debris accumulated in adjacent structurally controlled basins. The slope morphology was somewhat similar to the distally steepened ramp model of Read (1982). Elsewhere, reefs grew during active extensional faulting (i.e., growth faulting) in north and west Cyprus. This extension was most active in the north, where the reefs were bordered by submarine scarps, as in the gullied slope by-pass model of Read (1982). In the west, by contrast, sedimentation largely kept pace with subsidence and the rift floor topography remained subdued. The setting was thus somewhat akin to the rimmed shelf accretionary model of Read (1982). In this area, the poritid reefs were displaced from the depocentre to the rift flanks as subsidence continued. In general, most of the reef-capped fault lineaments were inherited from earlier lines of crustal weakness. On a more regional scale, the overall tectonic setting of reef development is inferred to have been the upper plate of an active, northward-dipping subduction zone (Robertson et al., 1991).

### *Sea-level Change*

Both the Terra and Koronia Members appear to have been preceded by erosion, caused by relative sea-level fall. The reefs developed on smooth, planed surfaces following transgression (i.e., relative sea-level rise). A possible explanation of the absence of mid Miocene (Langhian) reefs in Cyprus is that reef growth was terminated by a relative sea-level fall, following an earlier rise, during which the Terra Member was transgressed by pelagic carbonates. Certainly, the Late Miocene reef growth was again preceded by erosion, locally cutting down to basement. With further transgression, poritid reefs grew: first, as upstanding patch reefs dominated by domal morphologies; then, as relative sea-level fell, by smaller patches with mainly encrusting poritids. Reef growth was finally terminated by desiccation, with unconformably overlying evaporite deposition, followed by exposure and widespread development of caliche and local karst, prior to Pliocene transgression.

### *Regional Plate Tectonics*

By Early Miocene time, collision of Africa and Eurasia east of Cyprus had virtually eliminated the Tethys ocean (e.g., Dercourt et al., 1986), leaving only shallower seaways connecting the Mediterranean Sea and the Indian Ocean (McCall et al., 1989). In consequence, climate, currents and nutrient availability are all likely to have changed; the combined effect of these environmental changes being to reduce the diversity of corals, particularly framework reef builders, from Late Oligocene time onwards (Chevalier, 1977; Frost, 1977).

### *Sedimentation*

During Early Miocene time, reefs began to develop on pre-existing gentle highs made up of Mesozoic rocks of the Mamonia Complex in west Cyprus. Discrete patch reefs, up to tens of meters high, grew in calm, relatively shallow seas, surrounded by benthonic foraminiferal sands. Fore-reef sediments are exposed in west Cyprus. By contrast, in response to active tectonics, a new, more accentuated relief developed in the Late Miocene time. The ideal sites of poritid reef growth were fringing basement highs. The poritid coral shape changed through time from mainly domal, to more sheet-like encrustations. In the most active extensional areas, in northern Cyprus, the reefs were bounded by fault scarps and large volumes of coarse debris were shed down steep slopes into adjacent basins. Terrigenous clastic input was also significant in Late Miocene time of the northern and western areas. In west Cyprus, discrete reef tracts grew on the uplifted rift flanks and debris was shed into intervening rift basins, while back-reef facies locally developed between the rift flanks and the Troodos ophiolite basement to the east. In southern Cyprus the reefs developed on a recently stabilized lineament and shed large volumes of talus into basinal areas to the south.

## ACKNOWLEDGMENTS

This work was carried out under the tenure of a Natural Environmental Research Council Studentship held at Edinburgh University by E. J. Follows. We thank the Cyprus Geological Survey Department, particularly Dr. C. Xenophontos for advice and encouragement. Of the many who assisted with this project, we particularly thank J. McCallum, A. Poole and S. Eaton of the Grant Institute, and T. Fallick and G. Rogers of the Scottish Universities Research and Reactor Centre at East Kilbride. The manuscript benefited from comments by M. Esteban, R. Loucks, W. C. Ward and D. S. Ulmer-Scholle.

## REFERENCES

ALLAN, J. R. AND MATTHEWS, R. K., 1982, Isotope signatures associated with early meteoric diagenesis: Sedimentology, v. 29, p. 797-817.

BAGNALL, P. S., 1960, The Geology and Mineral Resources of the Pano Lefkara-Larnaca area: Nicosia, Geological Survey Department, Cyprus, Memoir 5, 116 p.

BAROZ, F. AND BIZON, G., 1977, La couverture Tertiaire du flanc nord du massif du Troodos et de la partie meridionale de la Mésaoria: étude stratigraphique et micropaléontologique: Révue de l'Institute Français du Pétrole, v. 32, p. 719-759.

BATHURST, R. G. G., 1966, Boring algae, micritic envelopes and lithification of molluscan biosparites: Geological Journal, v. 5, p. 15-32.

BEAR, L. M., 1960, The Geology and Mineral Resources of the Akaki-Lythrodondha area: Nicosia, Geological Survey Department, Cyprus, Memoir 3, 122 p.

BELLAMY, C. V. AND JUKES-BROWNE, A. J., 1905, The Geology of Cyprus: Plymouth, W. Brendon and Son, 72 p.

BLOW, W. H., 1969, Late Middle Eocene to Recent planktonic foraminiferal biostratigraphy, *in* Proceedings of the First International Conference on Planktonic Microfossils: Geneva, Brill, Leiden, p. 199-421.

BROMLEY, R. G. AND D'ALESSANDRO, A., 1984, The ichnogenus Entobia from the Miocene, Pliocene and Pleistocene of southern Italy: Rivista Italiana di Paleontologia e Stratigrafia Milan, v. 90, p. 227-296.

BUCHBINDER, B., 1975, Lithogenesis of Miocene reef limestones in Israel with particular reference to the significance of red algae: Geological Survey of Israel, Report OD.3.75, 173 p.

BUCHBINDER, B., 1979, Facies and environments of Miocene reef limestones in Israel: Journal of Sedimentary Petrology, v. 49, p.

1323-1344.

Burchette, T. P., 1988, Tectonic control on carbonate platform facies distribution and sequence development: Miocene, Gulf of Suez: Sedimentary Geology, v. 59, p. 155-178.

Buxton, M. W. N. and Pedley, H. M., 1989, A standardised model for Tethyan Tertiary Carbonate ramps: Journal of the Geological Society of London, v. 146, p. 746-748.

Carr, J. M. and Bear, L. M., 1960, The Geology and Mineral Resources of the Peristerona-Lagoudhera area: Nicosia, Geological Survey Department, Cyprus, Memoir 2, 79 p.

Challis, G. R., 1980, Palaeoecology and taxonomy of mid-Tertiary Maltese echinoids: Unpublished Ph. D. Thesis, Bedford College, University of London, London, 401 p.

Chevalier, J. P., 1961, Récherches sur les Madréporaires et les Formations Récifales Miocènes de la Méditerranée Occidentale: Mémoirs de la Societé de Géologie de France, v. 40/93, 563 p.

Chevalier, J. P., 1977, Aperçu sur la faune coralliene récifale du Néogène: Mémoires du Bureau de Récherches Géologiques et Minières, v. 89, p. 359-366.

Cleintaur, M. R., Knox, G. J., and Ealey, P. J., 1977, The geology of Cyprus and its place in the east Mediterranean framework: Geologie en Mijnbouw, v. 56, p. 66-82.

Cowper-Reed, F. R., 1929, Contributions to the geology of Cyprus: Geological Magazine, v. 66, p. 435-447.

Cowper-Reed, F. R., 1930, Contributions to the geology of Cyprus: Geological Magazine, v. 67, p. 241-271.

Cowper-Reed, F. R., 1932, New Miocene faunas from Cyprus: Geological Magazine, v. 69, p. 511-517.

Cowper-Reed, F. R., 1933, Notes on the Neogene faunas of Cyprus, I: The Annals and Magazine of Natural History, Series 10, v. 12, p. 225-244.

Cowper-Reed, F. R., 1935, Notes on the Neogene faunas of Cyprus, II: The Annals and Magazine of Natural History, Series 10, v. 15, p. 1-37.

Cowper-Reed, F. R. 1939, A Miocene limestone from Cyprus: Geological Magazine, v. 76, p. 310-311.

Dabrio, C. J., Esteban, M., and Martìn, J. M., 1981, The coral reefs of the Nìjar, Messinian, Almeria Province, Southeast Spain: Journal of Sedimentary Petrology, v. 51, p. 521-539.

Dercourt, J., Zonenshain, L. P., Ricou, L.-E., Kazmin, V. G., Le Pichon, X., Knipper, A. L., Grandjacquet, C., Sbortshikov, I. M., Geyssant, J., Lepvrier, C., Pechersky, D. H., Boulin, J., Sibuet, J.-C., Savostin, L. A., Sorokhtin, O., Besthal, M., Bazhenov, M. L., Lauer, J. P., and Biju-Duval, B., 1986, Geological evolution of the Tethys Belt from the Atlantic to the Pamirs since the Lias: Tectonophysics, v. 123, p. 241-315.

Dupoux, B., 1983, Étude comparée de la tectonique Néogène des bassins du sud de Chypre et du bassin d'Antalya (Turquie): Unpublished Ph. D. Thesis, Université de Paris Sud, Centre d'Orsay, Paris, 116 p.

Eaton, S., 1987, The sedimentology of Mid-Late Miocene carbonates and evaporites in southern Cyprus: Unpublished Ph. D. Thesis, Edinburgh University, Edinburgh, 259 p.

Eaton, S. and Robertson, A. H. F., 1993, The Miocene Pakhna Formation, southern Cyprus and its relationship to the Neogene tectonic evolution of the Eastern Mediterranean: Sedimentary Geology, v. 86, p. 273-296.

Elion, P., 1983, Étude structurale et sédimentologique du bassin Néogène de Pissouri (Chypre): Unpublished Ph. D. Thesis, Université de Paris Sud, Centre d'Orsay, Paris, 235 p.

Esteban, M., Calvet, F., Dabrio, C. J., Barón, A., Giner, J., Polmar, L., and Salas, R., 1977, Messinian (Upper Miocene) reefs in Spain: morphology, composition and depositional environments: Miami, 3rd International Coral Reef Symposium Abstracts.

Esteban, M., 1979, Significance of the Upper Miocene coral reefs of the Western Mediterranean: Palaeogeography, Palaeoclimatology, Palaeoecology, v. 29, p. 169-188.

Flecker, R., Robertson, A. H. F., Poisson, A., and Muller, C., Facies and tectonic significance of two contrasting Miocene basins in south coastal Turkey: Terra Nova, v. 7, p. 221-232.

Follows, E. J., 1990, Sedimentology and tectonic setting of Miocene reefs and related sediments in Cyprus: Unpublished Ph. D. Thesis, University of Edinburgh, Edinburgh, 385 p.

Follows, E. J., 1992, Patterns of reef sedimentation and diagenesis in the Miocene of Cyprus, *in* Sellwood, R. W., ed., Ramps and Reefs: Sedimentary Geology, v. 79, p. 225-253.

Follows, E. J. and Robertson, A. H. F., 1990, Sedimentology and structural setting of Miocene reefal limestones in Cyprus, *in* Malpas, J., et al., eds., Ophiolites, Oceanic Crustal Analogues: Nicosia, Proceedings of the International Symposium, p. 207-216.

Frost, S. H., 1977, Cenozoic reef systems of the Caribbean-prospects for paleoecological synthesis, *in* Frost, S. H., Weiss, M. P., and Saunders, J. B., eds., Reefs and Related Carbonates-Ecology and Sedimentology: Tulsa, American Association of Petroleum Geology Studies in Geology 4, p. 93-110.

Gass, I. G. and Masson-Smith, D., 1963, The geology and gravity anomalies of the Troodos Massif, Cyprus: Philosophical Transactions of the Royal Society of London, v. A255, p. 417-467.

Grasso, M., Pedley, H. M., and Reuther, C.-D., 1985, The geology of the Pelagian Islands and their structural setting related to the Pantelleria rift (central Mediterranean Sea): Centro, University of Malta Press, v. 1, p. 1-19.

Grasso, M. and Pedley, H. M., 1988, The sedimentology and development of Terravecchia Formation carbonates (Upper Miocene) of North Central Sicily: Possible eustatic influence on facies development: Sedimentary Geology, v. 57, p. 131-149.

Hayward, A. B., 1982, Coral reefs in a clastic sedimentary environment: fossil (Miocene, S.W. Turkey) and Modern (Recent, Red Sea) analogues: Coral Reefs, v. 1, p. 109-114.

Henson, F. R. S., Browne, R. W., and McGinty, J., 1949, A synopsis of the stratigraphy and geological history of Cyprus: Quarterly Journal of the Geological Society of London, v. 105, p. 1-41.

Hsü, K. J., Mondadert, L., Bernoulli, D., Cita, M. B., Erickson, A., Garrison, R. E., Kidd, R. B., Mélières, F., Müller, C., and Wright, R., 1978, History of the Mediterranean salinity crisis: Washington, United States Government Printing Office, Initial Report of the Deep Sea Drilling Project, v. 42, p. 1053-1078.

James, N. P. and Ginsburg, R. N., 1979, The Seaward Margin of Belize Barrier and Atoll Reefs: Oxford, Blackwell Scientific Publications, International Association of Sedimentologists Special Publication 3, 191 p.

James, N. P., Coniglio, M., Aïssaoui, D. M., and Purser, B. H., 1988, Facies and geologic history of an exposed Miocene rift-margin carbonate platform: Gulf of Suez, Egypt: American Association of Petroleum Geology Bulletin, v. 72, p. 555-572.

Kempler, D. and Ben-Avraham, Z., 1987, The tectonic evolution of the Cyprean Arc: Annales Tectonicae, v. 1, p. 58-71.

Kluyver, H. M., 1969, Report on a regional geological mapping in Paphos District: Geological Survey Department Cyprus Bulletin, v. 4, p. 21-36.

Kobluk, D. R. and Risk, M. J., 1977, Micritisation and carbonate-grain binding by endolithic algae: American Association of Petroleum Geology Bulletin, v. 61, p. 1069-1082.

MacIntyre, I. G. and Marshall, J. F., 1989, Submarine lithification in coral reefs: some facts and misconceptions: Townsville, Proceedings of the 6th International Coral Reef Symposium, v. 1, p. 263-272.

Martìn, J. M., Braga, J. C., and Rivas, P., 1989, Coral successions in Upper Tortonian reefs in SE Spain: Lethaia, v. 22, p. 271-286.

Martini, E., 1971, Standard Tertiary and Quaternary calcareous nannoplankton zonation: Rome, Proceedings of the II Planktonic Conference, Edizioni Tecnoscienza, p. 739-785.

McCallum, J. E., 1990, Sedimentation and tectonics of the Plio-Pleistocene of Cyprus: Unpublished Ph. D. Thesis, Edinburgh University, Edinburgh, 263 p.

McCallum, J. E. and Robertson, A. H. F., 1990, Pulsed uplift of the Troodos Massif; evidence from the Plio-Pleistocene Mesaoria basin, *in* Malpas, J., eds., Ophiolites, Oceanic Crustal Analogues: Nicosia, Proceedings of the International Symposium, p. 217-230.

McCallum, J. E. and Robertson, A. H. F., 1995, Late Pliocene-early Pleistocene Athalassa Formation, north central Cyprus: carbonate sand bodies in a shallow seaway between two emerging landmasses: Terra Nova, v. 7, p. 265-277.

McCall, G. J. H., Rosen, B., and Darrell, J. G. 1989, Miocene reefs of Iran: Marseille, I.S.R.S. Meeting Abstracts, p. 103.

ORSZAG-SPERBER, F., ROUCHY, J.-M., AND ELION, P., 1989, The sedimentary expression of regional tectonic events during the Miocene-Pliocene transition in the southern Cyprus basin: Geological Magazine, v. 126, p. 291-299.

PANTAZIS, T. M., 1967, The Geology and Mineral Resources of the Pharmakas-Kalavasos area: Nicosia, Geological Survey Department Cyprus Memoir 8, 190 p.

PAYNE, A. S. AND ROBERTSON, A. H. F., 1995, Neogene supra-subduction zone extension in the Polis graben system, west Cyprus: Journal of the Geological Society of London, v. 152, p. 613-628.

PEDLEY, H. M., 1987, The Ghar Lapsi limestones: sedimentology of a Miocene intra-shelf graben: Centro, University of Malta Press, v. 2, No. 3, p. 1-14.

PLEYDELL, S. M. AND JONES, B., 1988, Boring of various faunal elements in the Oligocene-Miocene Bluff Formation of Grand Cayman, British West Indies: Journal of Paleontology, v. 62, p. 348-367.

POOLE, A. J., SHIMMIELD, G. B. AND ROBERTSON, A. H. F., 1990, Late Quaternary uplift of the Troodos ophiolite, Cyprus: Uranium series dating of Pleistocene coral: Geology, v. 18, p. 894-897.

POOLE, A. J. AND ROBERTSON, A. H. F. 1991, Uplift and sealevel change at an active plate boundary: evidence from the Quaternary of Cyprus: Journal of the Geological Society of London, v. 148, p. 909-921.

READ, J. F., 1982, Carbonate platforms of passive (extensional) continental margins: types, characteristics and evolution: Tectonophysics, v. 81, p. 195-212.

ROBERTSON, A. H. F., 1977a, Tertiary uplift history of the Troodos Massif, Cyprus: Bulletin of the Geological Society of America, v. 88, p. 1763-1772.

ROBERTSON, A. H. F., 1977b, The Moni Melange, Cyprus: an olistostrome formed at a destructive plate margin: Journal of the Geological Society of London, v. 133, p. 447-466.

ROBERTSON, A. H. F., 1990, Tectonic evolution of Cyprus, *in* Malpas, J., ed., Ophiolites, Oceanic Crustal Analogues: Nicosia, Proceedings of the International Symposium, p. 235-252.

ROBERTSON, A. H. F., EATON, S., FOLLOWS, E. J., AND MCCALLUM, J. E., 1991, Role of local tectonics versus global sealevel change in the Neogene evolution of Cyprus, *in* MacDonald, D. I. M., ed., Sealevel Changes at Active Plate Margins: Processes and Products: Oxford, Blackwell Scientific Publications, International Association of Sedimentologists Special Publication 12, p. 331-369.

ROBERTSON, A. H. F., EATON, S., FOLLOWS, E. J., AND PAYNE, A. S., 1995, Deposition processes and basin analysis of Messinian evaporites, Cyprus: Terra Nova, v. 7, p. 233-253.

ROBERTSON, A. H. F. AND HUDSON, J. D., 1974, Pelagic sediments in the Cretaceous and Tertiary history of Cyprus, *in* Hsü, K. J. and Jenkyns, H. C., eds., Pelagic Sediments; on Land and Under the Sea: Oxford, Blackwell Scientific Publications, International Association of Sedimentologists Special Publication 1, p. 403-436.

ROSE, E. P. F. AND PODDUBIUK, R. H., 1987, Morphological variation in the Cenozoic echinoid Clypeaster and its ecological and stratigraphical significance: Annals of the Institute of Geology Publications, Hungary, v. 70.

ROUCHY, J.-M., SAINT-MARTIN, J.-P., MAURIN, A., AND BERNET-ROLLANDE, M.-C., 1986, Évolution et antagonisme des communautés bioconstructrices animales et végétales à la fin du Miocène en Méditerranée occidentale: biologie et sédimentologie: Bulletin des Centres de Récherches, Exploration et Production, Elf-Aquitaine, v. 10, p. 333-348.

SANTISTEBAN, C. AND TABERNER, C., 1983, Shallow marine and continental conglomerates derived from coral reef complexes after dessication of a deep marine basin: the Tortonian-Messinian deposits of the Fortuna basin, south east Spain: Journal of the Geological Society of London, v. 140, p. 401-411.

SCOFFIN, T. P., 1987, An Introduction to Carbonate Sediments and Rocks: Glasgow, Blackie and Son Ltd., 274 p.

TURNER, W. M., 1968, A progress report on the geology of western Cyprus including the Akamas Peninsula: Albuquerque, Department of Geology, University of New Mexico, 42 p.

WEISGERBER, F., 1978, Stratigraphie et sédimentologie du Miocène terminal et du Pliocène inférieur au sud de Chypre: la bordure méridionale du massif du Troodos: Rapport Interne, Institute Français du Pétrole, Reference 26, 194 p.

# MIOCENE PATCH REEFS FROM A MEDITERRANEAN MARGINAL TERRIGENOUS SETTING IN SOUTHWEST TURKEY

ANTHONY B. HAYWARD,

*Executive Office, British Petroleum Exploration Co Ltd, D'Arcy House, 146 Queen Victoria Street, London, EC4 4BY, United Kingdom*

ALASTAIR H. F. ROBERTSON, AND TERENCE P. SCOFFIN

*Department of Geology and Geophysics, University of Edinburgh, West Mains Road, Edinburgh, EH9 3JW, United Kingdom*

ABSTRACT: Small patch reefs, up to 8 m high and 40-50 m across, occur locally within Early Miocene terrigenous clastic sediments in the Kasaba basin of Southwest Turkey. The patch reefs are located within a prograding fan delta-type succession, interpreted as part of a foreland basin, underlain by a collapsed Mesozoic carbonate platform. The Miocene foreland basin developed in response to southeasterly thrusting of a thick pile of allochthonous thrust sheets, the Lycian Nappes, representing Tethyan continental margin and oceanic units. Well preserved, undeformed patch reefs are present in the Upper Miocene Kasaba Formation; poorly preserved patch reefs also occur in the Lower Miocene Kemer Formation.

The patch reefs developed directly on gravel and coarse sand fans, without pioneer coral development. The primary framework builders (*Favites, Tarbellastraea, Montastraea, Porites*) progressively changed in morphology upwards from dish-shaped, tabular corals, to large branching colonies. The patch reefs are asymmetrical in plan view, with more extensively developed off-reef facies on the landward flanks, where reworked reef-derived talus interfingers with terrigenous sediment. The primary coral framework was encrusted by a secondary framework of coralline algae and encrusting foraminifera. The reefs were modified by a variety of boring and grazing organisms (e.g., bivalves, sponges, bryozoa), producing abundant sediment that accumulated in areas between individual coral colonies and as off-reef flanking facies.

The patch reefs are interpreted to have formed on the abandoned, submarine toes of coastal alluvial fans, following switching in sediment supply and/or change in relative sea level. In Southwest Turkey, wave and storm activity in the microtidal Mediterranean Sea was mainly onshore, reworking reef-derived material landwards. The patch reefs were later buried by alluvial fans, overthrust during final emplacement of the Lycian Nappes, then exhumed following uplift and erosion during Plio-Quaternary time.

## INTRODUCTION

The present objective is to describe and interpret occurrences of mainly Upper Miocene patch reefs in a marginal, terrigenous palaeoenvironment in Southwestern Turkey. Many Mediterranean Miocene reefs formed away from sites of significant terrigenous sediment input (e.g., Cyprus; Follows et al., this volume; Sicily and Pelagian islands; Pedley, this volume). Terrigenous-dominated settings were once thought to be unfavorable to reef growth. However, it has now become clear that this is not always the case, as the present example from Southwest Turkey will illustrate. The reefs in Southwest Turkey are located in a sequence of fan delta-type clastics that were shed from advancing thrust sheets, including ophiolites. The depositional site was a foreland basin that overlies a foundered Mesozoic carbonate platform (Fig. 1).

## EARLIER WORK

The stratigraphy of the area in Southwest Turkey with reefs was first described by Poisson (1977, 1981) and by Önalon (1980), followed by a detailed sedimentological study (Hayward and Robertson, 1981; Hayward, 1982a, 1983, 1984, 1985). Hayward (1982b) briefly described Upper Miocene patch reefs from this area and compared them with counterparts in the modern Red Sea. The geology and tectonics of the adjacent allochthonous thrust system, the Lycian Nappes, was investigated by Poisson (1977) and discussed, in its regional context, by Robertson and Woodcock (1980, 1984), amongst others. Similar Miocene reefs developed in fan delta clastics were recently described to the east of the study area in Turkey (Flecker et al., 1995)

### *Regional Geological Framework*

The Upper Miocene patch reefs are located within the upper part of a thick succession of dominantly ophiolite-derived clastic sediments, interpreted as a foreland basin (Hayward, 1982a), related to southeasterly thrusting of the Lycian nappes onto a Mesozoic carbonate platform and represented by the Susuz Dağ and Bey Dağları (Fig. 1). The Lycian nappes originated as a Mesozoic continental margin, deep-water basins and oceanic crust, sited northwest of a Bahama-type carbonate platform (Poisson, 1977). This platform was located along the northern Mesozoic margin of Gondwana (Robertson and Woodcock, 1984). The Lycian nappes were first deformed in the Upper Cretaceous time, but were not thrust to their present position until the Miocene time. During this final emplacement the foreland, represented by the Mesozoic carbonate platform, underwent flexural subsidence, initiating a foreland basin, which progressively filled with an overall shallowing-upward succession of, first submarine fan, then fan delta-type clastics. More northwesterly areas of the basin, including the Upper Miocene patch reefs, were overthrust by the Lycian nappes, while areas farther southeast remained ahead of the final thrust front. Despite the overthrusting, the successions, including the reefs, have remained intact and are relatively undeformed.

The regional situation is complicated by the presence, in the west, of another system of allochthonous units, the Antalya Complex or Antalya Nappes. These units are reconstructed as

Models for Carbonate Stratigraphy from Miocene Reef Complexes of Mediterranean Regions,
SEPM Concepts in Sedimentology and Paleontology #5, Copyright © 1996,
SEPM (Society for Sedimentary Geology), ISBN 1-56576-033-6, p. 317-332.

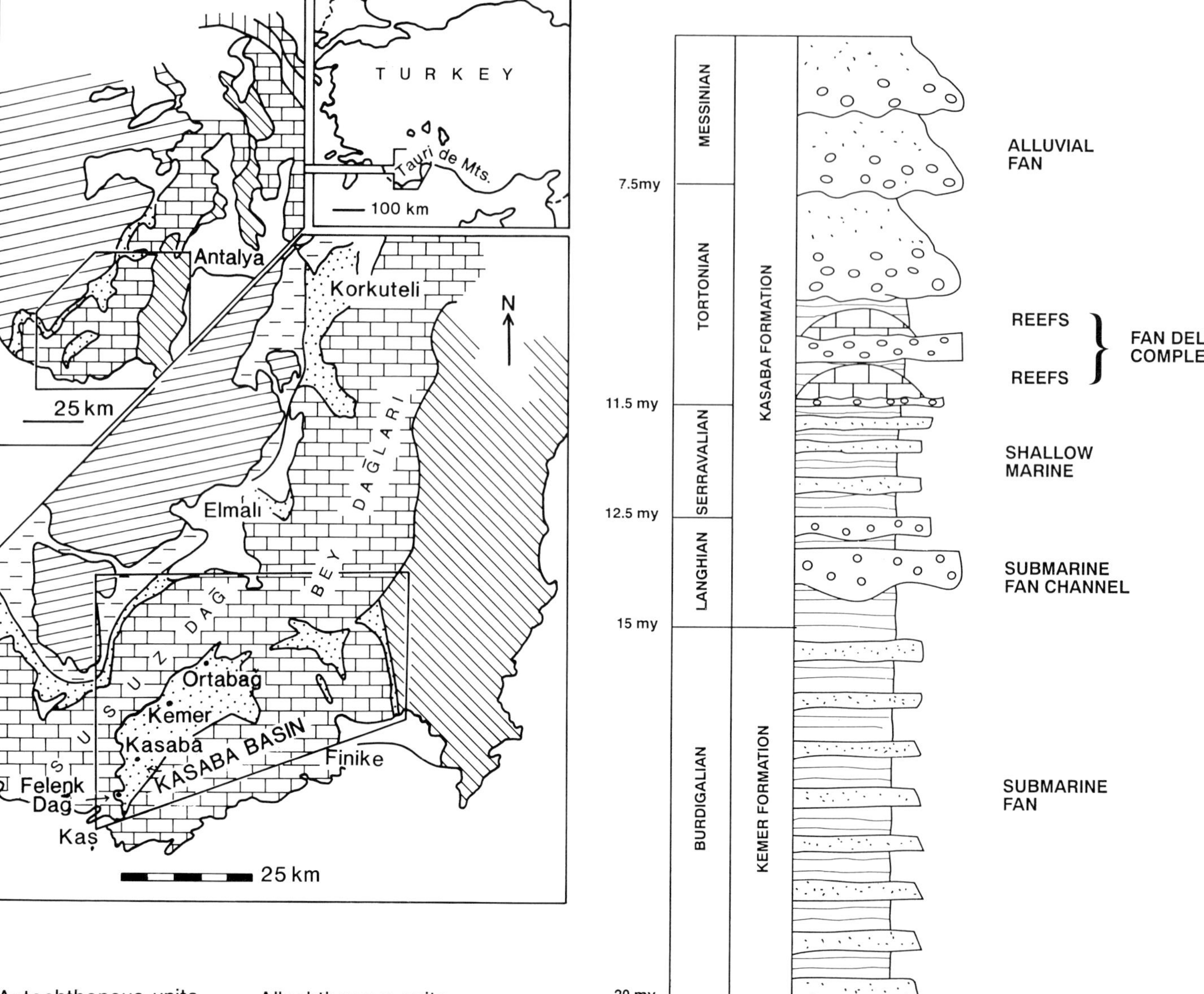


FIG. 1.—Location and geologic maps showing the distribution of Miocene clastic sediments associated with patch reefs in Southwest Turkey.

FIG. 2.—Stratigraphy of the Miocene clastic sediments in the Kasaba basin. The reefs are located within the Kasaba Formation; see Figure 1 for location.

a passive margin and ocean basin that lay to the east of the Bey Dağları carbonate platform during the Mesozoic. The Antalya Complex was finally emplaced, by dominantly strike-slip and transpressional processes, westwards over the easterly margin of the Bey Dağları carbonate platform during Miocene time (Robertson and Woodcock, 1984). However, the Upper Miocene reefs lay well to the west of areas of clastic input from this second allochthonous system, which will not be mentioned further here.

## LITHOFACIES ASSOCIATED WITH THE REEF

### *Reef Morphology*

The reefs are bilaterally symmetrical mounds, up to 8 m high and as much as 40-50 m across. They are generally oval or subspherical in plan morphology. The basal surfaces of the reefs are roughly horizontal, paralleling bedding in the underlying sediments. The exhumed tops are convex.

TABLE 1.—BIOSTRATIGRAPHY OF THE REEFS OF THE KASABA FORMATION. AGES QUOTED IN MY.

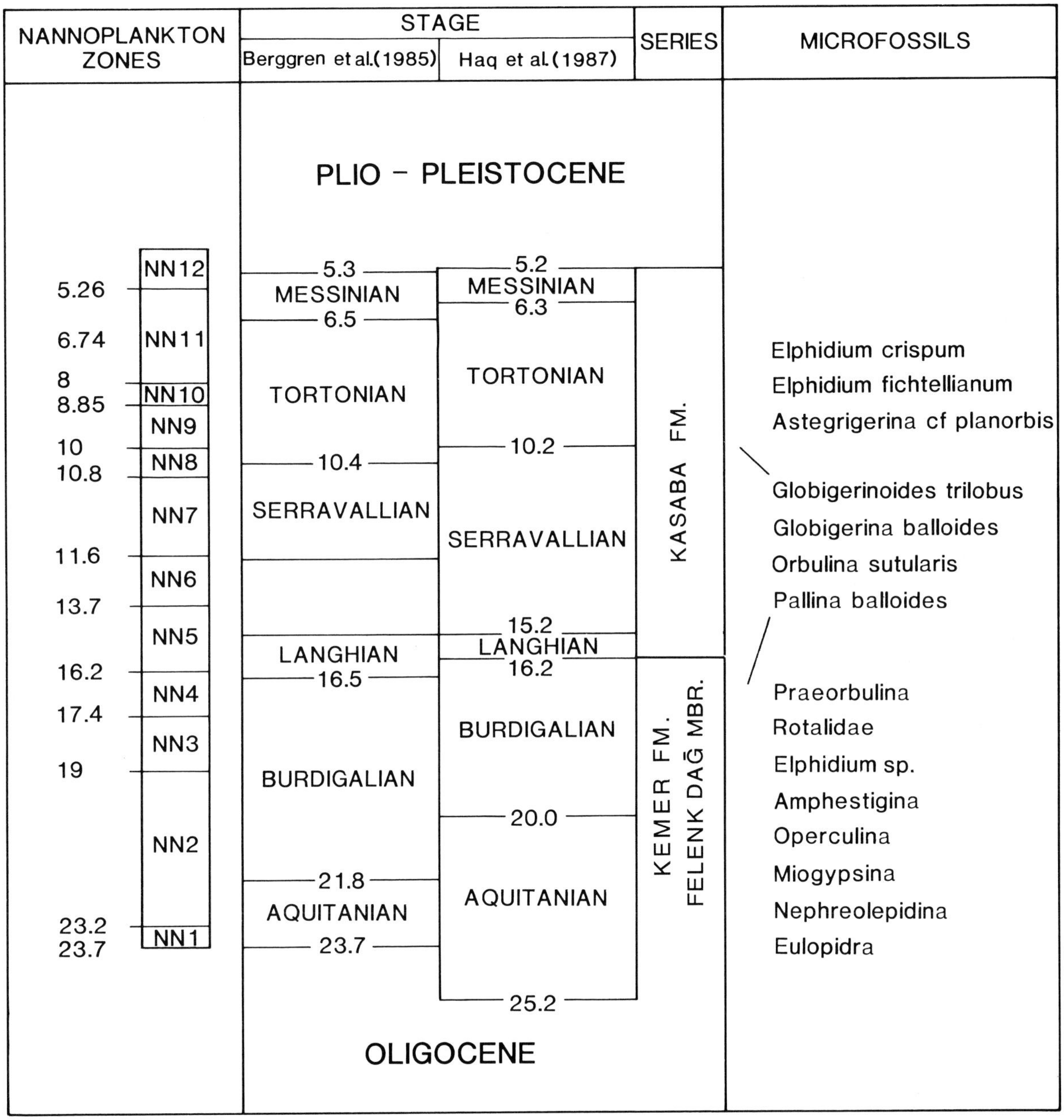

| NANNOPLANKTON ZONES | | STAGE | | SERIES | MICROFOSSILS |
|---|---|---|---|---|---|
| | | Berggren et al. (1985) | Haq et al (1987) | | |
| | | PLIO - PLEISTOCENE | | | |
| | NN12 | 5.3 | 5.2 | | |
| 5.26 | | MESSINIAN | MESSINIAN | KASABA FM. | |
| | | 6.5 | 6.3 | | |
| 6.74 | NN11 | | | | Elphidium crispum |
| 8 | | TORTONIAN | TORTONIAN | | Elphidium fichtellianum |
| 8.85 | NN10 | | | | Astegrigerina cf planorbis |
| | NN9 | | 10.2 | | |
| 10 | NN8 | 10.4 | | | Globigerinoides trilobus |
| 10.8 | NN7 | SERRAVALLIAN | | | Globigerina balloides |
| 11.6 | NN6 | | SERRAVALLIAN | | Orbulina sutularis |
| 13.7 | NN5 | | 15.2 | | Pallina balloides |
| | | LANGHIAN | LANGHIAN | | |
| 16.2 | NN4 | 16.5 | 16.2 | KEMER FM. FELENK DAĞ MBR. | Praeorbulina |
| 17.4 | NN3 | | BURDIGALIAN | | Rotalidae |
| 19 | | BURDIGALIAN | | | Elphidium sp. |
| | NN2 | | 20.0 | | Amphestigina |
| | | 21.8 | AQUITANIAN | | Operculina |
| | | AQUITANIAN | | | Miogypsina |
| 23.2 | NN1 | | | | Nephreolepidina |
| 23.7 | | 23.7 | | | Eulopidra |
| | | | 25.2 | | |
| | | OLIGOCENE | | | |

## *Stratigraphic Sequences*

The biostratigraphy of the reefs and associated sediments is summarised in Table 1. Reefs of two contrasting ages are present in the study area. The first comprises scattered disoriented blocks that will not be discussed in detail here. These reef blocks are known as the Felenk Dağ Member and form part of the Kemer Formation. The Felenk Dağ Member is dated as Early Miocene (Burdigalian) by the presence of *Rotalidae, Amphestigina sp., Miogypsina sp.* and *Nephreolepidina sp.* The occurrence of *Praeorbulina sp.* from just below the highest calcareous sandstone within the member confirms a Langhian age for the top of the Felenk Dağ Member.

The better developed reefs, discussed in this paper, are located at a number of locations with the Kasaba Formation (Fig. 2). At present, they can be dated only by association with these generally interbedded sediments. A Serravallian age for the base of the formation is suggested by the presence of planktonic foraminifera *Globorotalia meyeri, Globorotalia periphero rondo, Globigerina trilobus* and *Orbulina suturalis* (Poisson, 1977). *Globigerinoides trilobus* is taken as the biozonal marker of Upper Burdigalian-Lower Langhian boundary although occur-

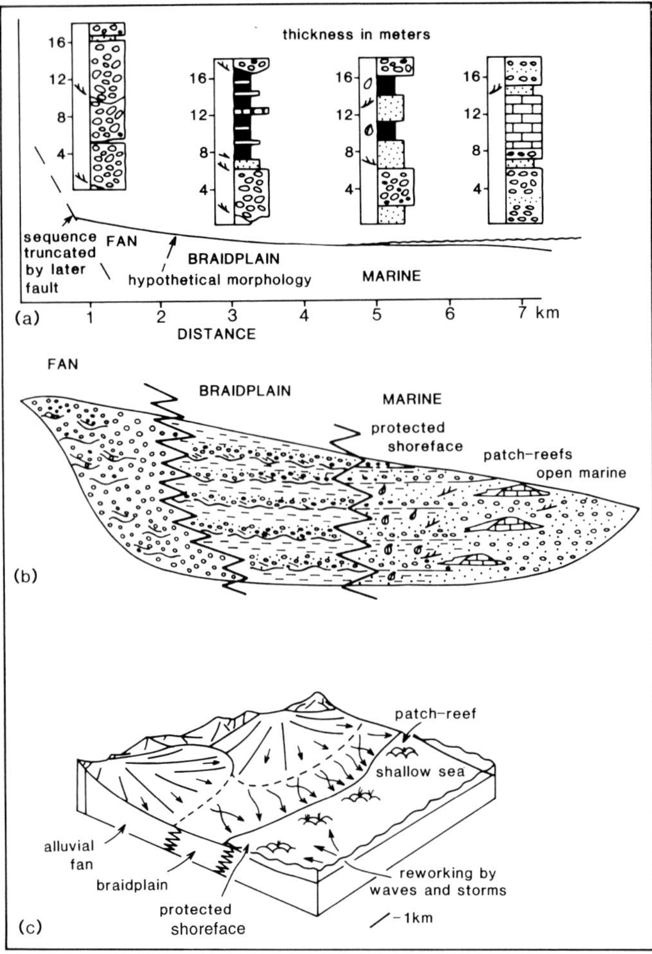


FIG. 3.—General sedimentary model for the Kasaba Formation (that contains the patch reefs) showing lateral and vertical sedimentary facies variations. (A) Down-fan change in vertical sequences - from alluvial fan to braidplain, to nearshore marine, to offshore marine; (B) Interdigitation of the three facies associations: alluvial fan passing downslope into a fluvial braidplain sequence and then into a shallow marine sequence; (C) Block diagram illustrating the deposition of the Kasaba Formation. Alluvial fans prograded over a fluvial braidplain into a shallow sea. During periods of low sediment supply, patch reefs developed on the submarine toes of the alluvial fan.

rences ranging from Lower Aquitanian-Serravallian? to Pliocene time are recorded in the literature (Bizon et al. 1974). *Globigerina balloides* ranges from the top of NN4 to NN5. Using the Berggren et al.'s (1985) time scale, the first occurrence of *Orbulina sp.* is dated at 15.2 Ma. Önalon (1980) recorded an abundant benthonic foraminiferal assemblage from the highest fossiliferous horizons within the Kasaba Formation, which include *Rotalia breccarii, Elphidium crispum, Elphidium fictellianum* and *Astegrigerina cf. planorbis*, but these do not define a specific stage.

The reefs of the Kasaba Formation are associated with abundant terrigenous clastics (Fig. 3). Relative sea-level change was a major factor influencing clastic sediment input and thus reef growth. A growing body of information (e.g., Grasso and Pedley 1988; Pedley, this volume) indicates that a major intra-Mediterranean onlap event (ca 3.2 highstand of Haq et al. (1987) caused

TABLE 2.—LIST OF CORAL SPECIES AND THEIR APPROXIMATE ABUNDANCES IN PATCH REEFS OF THE KASABA FORMATION.

| CORAL SPECIES | APPROXIMATE ABUNDANCE IN REEFS |
|---|---|
| *Montastrea* spp. | ~30% |
| *Tarbellastraea silciliae* } | ~40% |
| *Tarbellastraea* spp. } | |
| *Favites* spp. } | ~20% |
| *Favites neglecta* } | |
| *Favites neglecta* (Michelotti) } | |
| *Porites* spp. | ~10% |

two (or in places one) highstands of Late Tortonian-Early Messinian age, over much of the Mediterranean region; this would have inhibited clastic input but favored reef establishment.

In summary, it appears likely that, in the study area, as elsewhere in southern Turkey (Flecker et al., 1995) and in Cyprus (Follows et al., this volume), the reefs are mainly of Early Miocene (Burdigalian) and Late Miocene (Tortonian-Messinian?) age. However, only the Late Miocene reefs will now be discussed.

The Late Miocene reefs are not confined to one horizon but are distributed throughout the sequence (Figs. 4, 5). They are, however, restricted to the shallow near-shore zone, seaward of the shoreface. Along strike, the reefs did not form a continuous fringing or barrier reef, but rather occurred as isolated build ups paralleling the contemporary shoreline, resulting in an approximately east-west alignment. The reefs occur as isolated patches that are completely surrounded by coarse clastic terrigenous sediment.

### *Internal Structure and Facies Distribution*

Corals form the major frame-building organisms, comprising between 50% and 90% by volume of the reefs. Table 2 lists the dominant coral species, with an indication of their approximate abundance. Coralline algae are the other important frame-building organisms.

The reefs can be subdivided into three facies types (Fig. 6) which are discussed below.

*Pre-reef facies and basal unit.—*
All reefs overlie a cobble-pebble terrigenous conglomerate, which passes gradationally into a very coarse calcarenite (i.e., rudstone), over a thickness of approximately 1 m (Fig. 7A). This passes rapidly upwards into a bioclastic breccia, composed of coral fragments generally encrusted by algae, algal encrusted shell debris and foraminifera. The thickness of this bioclastic breccia varies between 0.5 and 1 m. The lowest unequivocally in situ corals are large (up to 0.7 m across) dish-shaped tabular

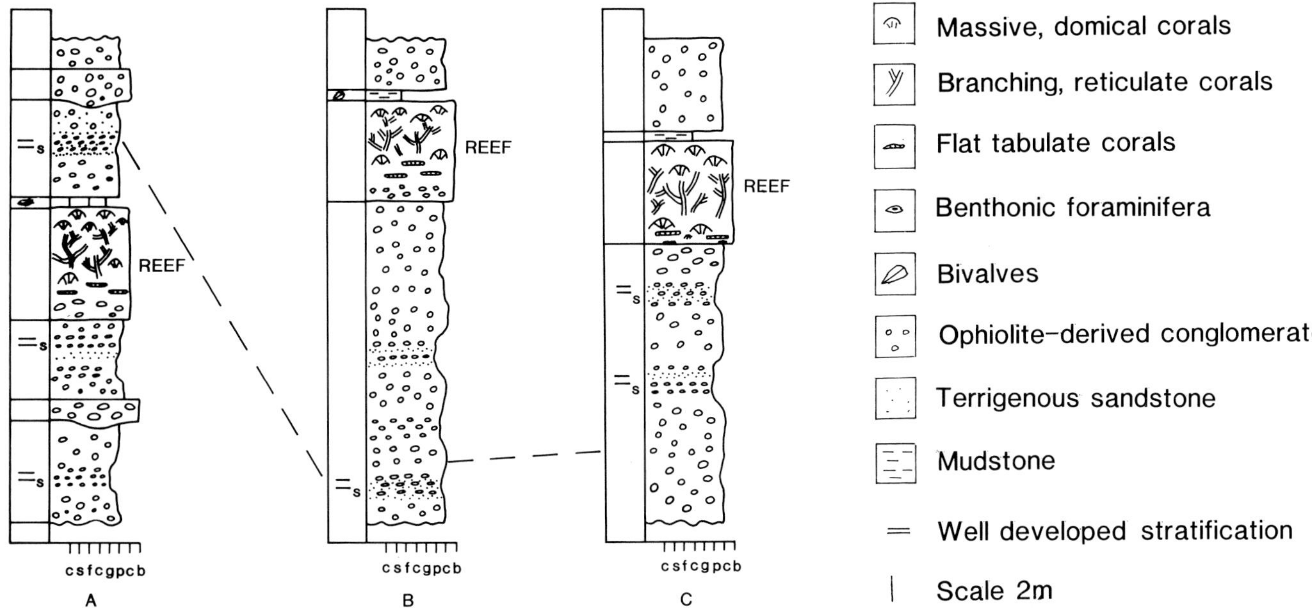


FIG. 4.—Sedimentary sections showing the distribution of reefs within coarse clastic terrigenous sediments. See Figure 5 for location. The correlation between the sections is tentative. Note the location of the reef limestone units depositionally overlying coarse clastic sediments.

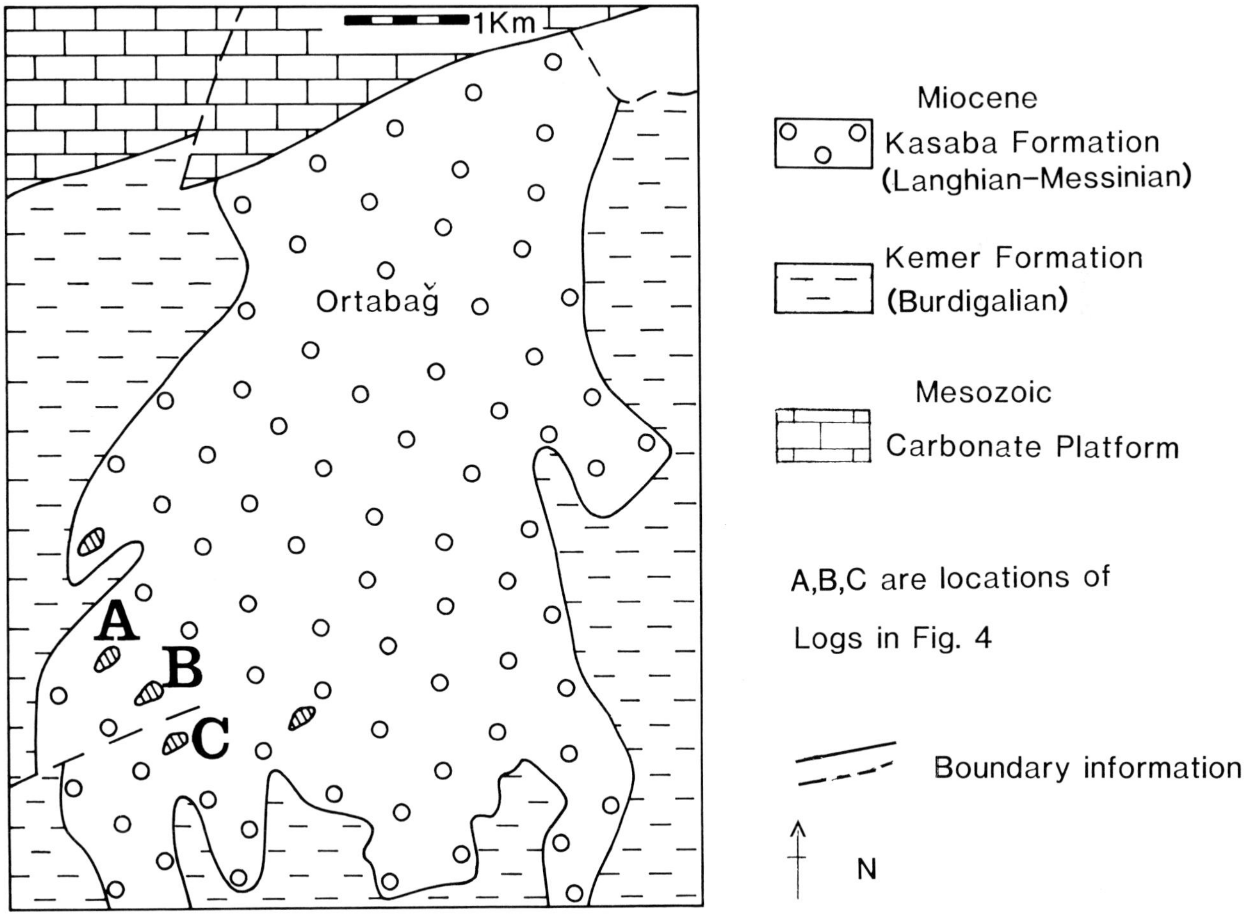

FIG. 5.—Simplified geological sketch map showing the distribution of the Kasaba Formation and the location of the patch reefs. See Figure 1 for regional setting.

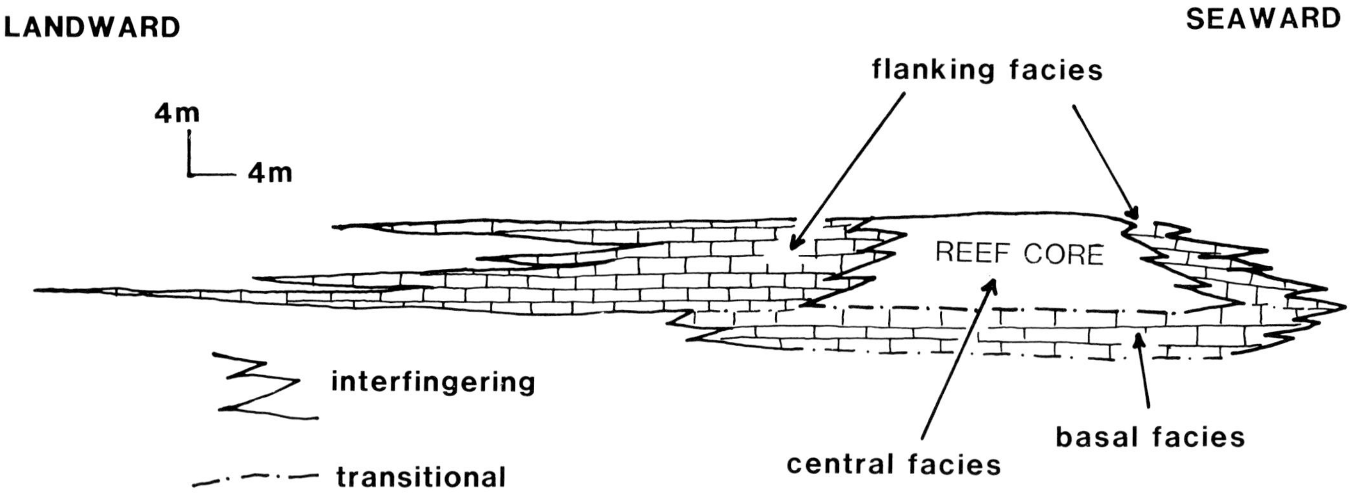

LANDWARD

SEAWARD

flanking facies

4m

4m

REEF CORE

interfingering

basal facies

central facies

transitional

FIG. 6.—Summary of the main facies types in the patch reefs of the Kasaba Formation.

FIG. 7.—Field photographs of the patch reefs. (A) Small patch reef (R) overlying terrigenous conglomerate (C); note that fore-reef breccia is mainly composed of disoriented coral heads (B); (B) Central framework of reef composed of a branching colony of *Tarbellastraea sp.* (T) overlain by more massive colonies of *Montastraea sp.* (M).

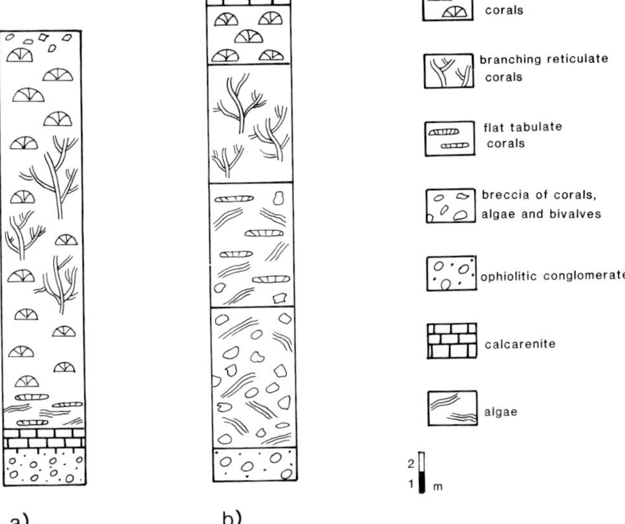


FIG. 8.—Progressive change in coral morphology upwards through the central reef framework (see text for details): (A) General overall zonation; (B) more complex zonation observed in some reefs.

forms, mainly *Favites sp.* and *Montastraea sp.* and a few scattered hemispherical forms. This zone, dominated by tabular compound forms, is rarely greater than 1 m thick and passes laterally into poorly sorted coral calcarenite breccia (mudstone-grainstone).

*Reef core facies.—*

The central framework of the reef is dominated by large branching colonies of *Tarbellastraea sp.* (Fig. 7B). The colonies have an average height of 1.5 m, maximum width of 3 m and commonly increase in width upwards (Fig. 8). Individual branches are up to 1 m tall, 0.1 m in diameter and variably encrusted by algae. Tight packing of the structure, in some areas, is demonstrated by immature small branches wedged between larger ones. Voids between branches are often filled with terrigenous mudstone. The other main *in situ* components of the central framework are large hemispherical coral colonies. The original dimensions of these amalgamated colonies are difficult to estimate as they now grade transitionally into zones of coral breccia. Zonation in coral type and form are described below. Inter-colony areas are composed of a mixed breccia of coral fragments and algal-bound reef debris. Broken and disoriented coral fragments range from large foundered blocks of *Tarbellastraea sp.* up to 2 m across, which have tumbled in locally, to small coral fragments 20-30 mm in size. Branches of *Tarbellastraea sp.* form up to 70% of the inter-colony breccia; the remainder is composed of broken and disoriented coral heads, bivalves and blocks of algal-bound reef material. The latter attest to breakage of colonies during growth and subsequent algal binding. The breccia is unstructured with large blocks being scattered randomly throughout.

*Off-reef flanking facies.—*

Laterally, the margins of the reefs grade into talus breccias, which themselves merge with surrounding conglomerate and sandstone. The core facies to talus breccia, transition typically occurs over a 4-5 m zone. Coral heads and colonies become progressively more disoriented away from the reef. The beds on the sides of the reef were apparently fairly steeply inclined at an angle of approximately 30-40°. Within the coarser, more proximal parts of the talus breccia, solitary corals and colonies are rarely found in growth positions. Away from the core, over a distance of about 50 m, the breccias become thinner and finer-grained, passing gradually into a very coarse calcarenite with weakly developed fining-upward, grading and parallel laminations. Farthest from the reef (up to 70 m from the reef core), these calcarenites become progressively thinner and interfinger with a mixed terrigenous calcareous sandstone. The percentage of coral debris within the flanking beds decreases away from the reef and is progressively replaced by more abundant algal and foraminiferal components. Analysis of geopetal sediment fills within oriented fragments suggest a depositional dip of between 10° and 15° for the flanking beds against the core areas. This decreases outwards over a distance of 10-20 m, to angles less than 4°.

The flanking facies are well developed only on the landward (back reef) margin of the reef, as determined from associated terrigenous clastic sedimentary facies and palaeocurrent analysis. On the seaward (forereef) margin, very coarse coral-dominated breccias pass abruptly into terrigenous clastic material, over distances of less than 15 m (Fig. 9). The unsorted, coarse-grained nature of the flanking beds and separation, in distal areas, into discrete depositional events, suggest rapid transportation and deposition, probably by periodic storms. The marked difference in flanking facies between seaward and leeward sides of the reef is probably the result of the prevailing environment. Wave and storm activity transported reef material into the lee of the reef, where it remained relatively undisturbed. Similar asymmetric development of reef-derived deposits has been described from the ancient by Lowenstam (1957) and, more recently, from studies of modern Australian reefs (Maiklen, 1970). Deposition of the coarse-grained debris on the seaward margin of the reef was probably by the steady accumulation of fallen blocks. The asymmetrical flank facies development of the Kasaba Formation reefs, therefore suggests a dominantly northerly (onshore) wind.

*Post-reef facies.—*

In some areas, reefs are overlain by a thin veneer of calcarenite composed of foraminiferal, algal and shell fragments and rare coral debris (Fig. 4). Elsewhere, this calcarenite is absent and the reefs are overlain by very coarse terrigenous pebbly sandstone. The basal 0.5 m of the sandstone contains abundant reef-derived material, suggesting some minor reworking of the reef top. Above this, the sandstone is devoid of reefal material.

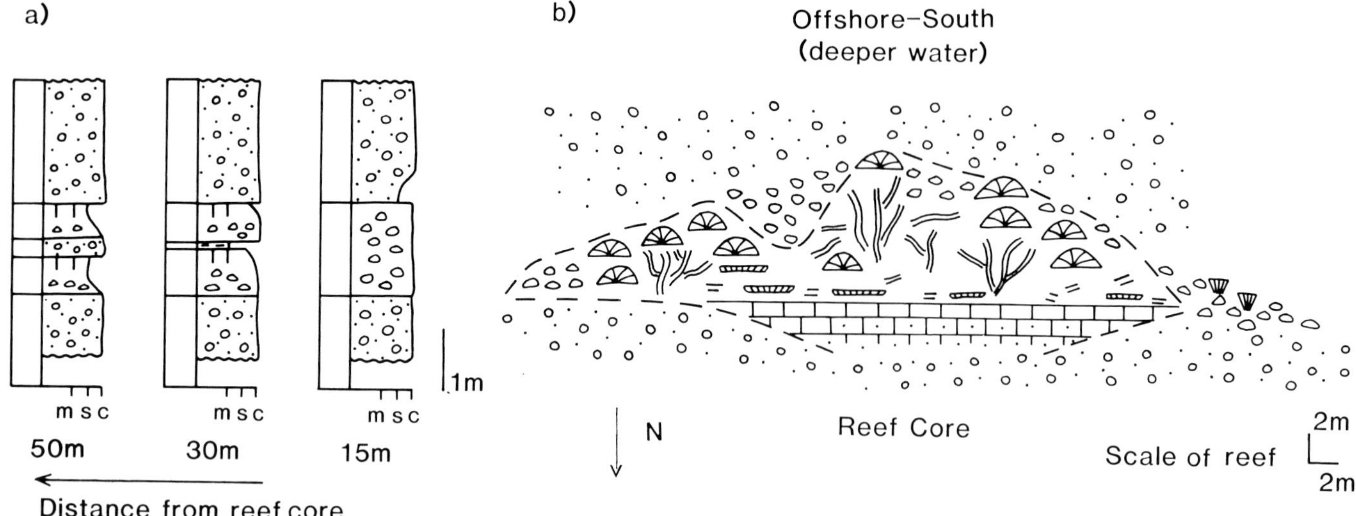


FIG. 9.—Reef flank facies: (A) sedimentary sections measured at increasing distances from the reef core; note that bioclastic breccias pass progressively into coarse calcarenites that are normally graded; (B) plan view of the reconstructed flank facies; note the lateral change in coral types and of associated facies. See Figure 8 for key to symbols.

*Depositional facies distribution.—*
The Kasaba Formation reefs formed on a firm, although unlithified, substrate (cobble and pebble gravel; Fig. 9). The reefs show no "pioneer community" that is significantly different from the overlying central framework of the reef. The only difference observed is an increase in the proportion of corals above the basal unit. The types and proportions of encrusting organisms and molluscs remain relatively constant throughout the reef core.

*Coral morphology.—*
The reefs exhibit a distinct change in coral morphology from the base upwards (Fig. 8). The general overall zonation is from flat, tabular dish forms (*Favites sp.*, some species of *Montastraea*), through branching reticulate forms (*Tarbellastraea sp.*), to massive domal forms (mainly *Montastraea*). In some reefs this zonation is particularly apparent. Elsewhere, it is not so well developed, and a complex relationship exists in the upper parts of the reef, with branching and domal forms being intimately intergrown.

Reefs developed in a similar environmental setting in southern Spain show a similar variation in coral morphology (Esteban, 1979; Santisteban and Taberner, 1980, 1983). In this instance, dish-shaped corals pass into globate forms and then into *Tarbellastraea* colonies in the upper part.

In summary, the vertical zonation in coral forms within the Kasaba Formation reefs was probably caused by changes due to increasing hydrodynamic stress. The absence of a pioneer community reflects development on an already firm substrate. Having become established, the gradual upward growth of the coral colonies took place in increasingly shallow-water, with a corresponding increase in wave energy (i.e., hydrodynamic stress).

## SECONDARY PROCESSES AND DIAGENETIC OVERPRINT

A variety of processes interact to form and preserve a reef (e.g., Schroeder and Zankl, 1974; Scoffin and Garrett, 1974). The initial stage of primary framework building by corals is subject to later processes that are both constructional: the addition of a secondary framework by encrusting organisms, sedimentation and cementation; and destructional: boring, rasping and grazing.

### *Encrusting Organisms Within The Kasaba Formation Reefs*

Within the Kasaba Formation reefs, encrusting organisms form an important element of the framework (Fig. 10). Prior to encrustation, many coral substrates show evidence of having been extensively bored.

FIG. 10.—Photomicrographs of thin sections (A, C and D) and acetate peel (B) of encrusting sequences in the Kasaba Formation reefs. (A) Crust of mixed composition: *Montastraea* (M) encrusted by a thin layer of *Lithophyllum* (L), subsequent crust consists of highly contorted, interlaminated *Mesophyllum* (Me) and an encrusting foraminifera *Planorbulina* (P). Crust growth has been terminated by reef sediment (biomicrite). Note extensive borings (b) in lower parts of crust. Scale bar = 2 mm. Plane polarized light. (B) Mixed crust of interlaminated *Mesophyllum* (Me), *Gypsina plana* (G) and *Planorbulina* (P): much of the lower parts of the crust has been destroyed by borers and subsequently infilled with fine reef sediment. Scale bar 2 mm. (C) Crust of mixed composition: coral substrate (c) is overlain by interlaminated *Lithophyllum* (L) and reef sediment (r). Successive crusts consist of interlaminated *Mesophyllum* (Me) and *Homotrema rubrum* (H). Crust growth was terminated by reef sedimentation. Scale bar = 2 mm. Plane polarized light. (D) Crust of *Mesophyllum* (Me) that grew downwards in the coral framework. This is overlain (i.e., upward) by a thin crust of fine bioclastic debris and peloids (b). P is *Porites*. In the remaining space, a late stage equant blocky sparite cement (S) has developed. Scale bar = 2 mm. Plane polarized light.

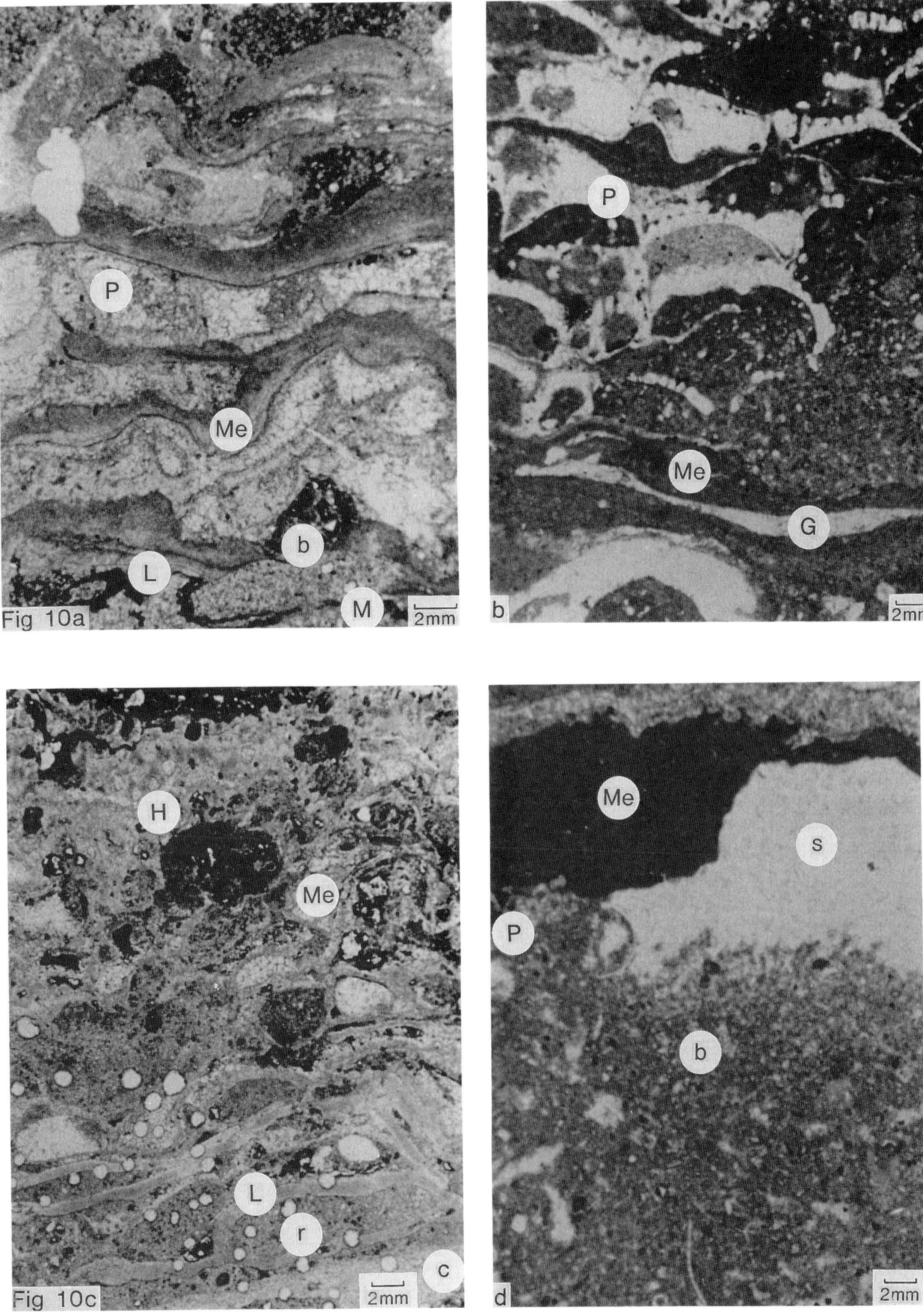
P
Me
b
L
M
2mm
Fig 10a
P
Me
G
b
2mm
H
Me
L
r
c
2mm
Fig 10c
Me
s
P
b
d
2mm

Several types of encrusting sequence are recognized and from comparison with studies of modern reefs, they can be directly related to their environment of formation.

*Mixed crusts.—*

Coral specimens taken from the base of colonies and inter-colony debris areas often show a thick (up to 8 mm), complex, crustal development.

An initial photophyllic crust of *Lithophyllum* is overgrown by successive encrustations of interlaminated *Mesophyllum* (and subordinate *Lithophyllum*) and encrusting (*Gypsina plana, Planorbulina, Homotrema rubrum*; Figs. 10A, B, C). Away from the coral substrate, *Mesophyllum* becomes progressively thinner and more contorted, supporting a number of encrusting foraminifera. The upper part of the encrusting sequence consists solely of foraminifera, which are overlain and interlaminated with reef sediment (i.e., biomicrite).

In Recent reef environments (Martindale, 1976, 1992; Scoffin, 1987) crusts of mixed composition record the progressive change from a photophyllic environment, through photophyllic/sciaphyllic, to a sciaphyllic shade-living environment as the structure builds and light is reduced.

Following initial encrustation by *Lithophyllum* in a photophyllic environment, upward and outward growth of the coral colony resulted in the lower (dead) areas of the colony becoming shaded. Cavities developed beneath the living reef surface by overgrowth, and within these cryptic environments, wave surge and currents prevented reef-derived sediment from being deposited. Such circulation aided the growth of intermediate crust types resulting in interlaminated *Mesophyllum* (and other photophyllic/sciaphyllic coralline alga) and foraminifera.

Continued upward growth of the colony, or an increase in reef debris overlying the cavity further decreased light intensity and the crusts became progressively more sciaphyllic. Ultimately there was insufficient light for the growth of crustose coralline algae and their position was taken by encrusting foraminifera, such as *Gypsina plana* and *Planorbulina*. The encrusting sequences were terminated by biomicrite reef sediment. A drop in the velocity of circulating water resulted in the cavities finally becoming choked with sediment with restricted growth. The similarity between sequences described from the Recent and those in the Kasaba Formation reefs suggests a very similar origin.

*Constant-composition crusts.—*

Corals taken from presumed life position, high within the reef core show a crust of variable thickness (0.40-2.5 mm) of the coralline algae *Lithophyllum sp.* Differential growth paths and rates of growth resulted in individual crusts frequently showing onlap and offlap relationships with adjacent crusts. Voids created by varying growth rates and borings were filled with reef sediment which locally forms discontinuous lenses, or by blocky microspar calcite cement. On the underside of coral colonies, crusts of *Lithophyllum* are absent or form discontinuous, very thin (40 µm) crusts. In some instances, large *in situ* coral colonies do not have algal crusts on their upper surfaces. By comparison with Recent reefs, the thick photophyllic crusts of *Lithophyllum* formed in a shallow well-lit environment that is consistent with their position at the top of the reef. The thickness of the crust is related to the time spent in that environment and to the degree of illumination and hydrodynamic exposure to which the coral was subjected (Martindale, 1976, 1992). Algal crust growth was halted by an influx of terrigenous sediment, terminating reef growth. Algal crusts are not present on some large corals, suggesting that they were buried by terrigenous sediment prior to the establishment of any encrusting organisms. The absence of photophyllic crusts on the underside of corals is consistent with their shaded position in the colony during growth.

Some in situ corals, toward the base of the reef, are encrusted by thick photophyllic crusts on their upper surface. The crusts, up to 6 mm thick, generally consist of one or two species of interlaminated coralline algae (e.g., *Mesophyllum, Lithophyllum*, Fig. 10D) and rarely encrusting foraminifera (*Gypsina sp., Planorbulina*). These crusts are always overlain by reef sediment.

Thick constant-composition algal crusts, from lower areas of the reef core, indicate that, upon death, the coral remained in essentially the same environment, and encrusting organisms were exposed to similar conditions through time. In present-day reefs, the destructive activity of boring organisms and rise in level of surrounding sediment results in burial for the coral (Martindale, 1976, 1992). A similar interpretation is favored for the Kasaba Formation reefs.

*Summary of encrusting sequences.—*

A summary of several types of encrusting sequences is shown in Figure 11. Although no regular zonation occurs, several associations are present:

1. Crusts of mixed composition occur dominantly in inter-colony debris areas.
2. *In situ* corals towards the base of the reef are frequently encrusted by thick photophyllic crusts on their upper surface.
3. *In situ* corals towards the top of the reef core are not encrusted, or only by a very thin photophyllic crust.

### *Bio-erosion*

Borings are found both in the primary (coral) framework and within the overlying encrusting sequences. Borings within coral skeletons consist dominantly of two types: (i) rounded to oval, smooth-walled cavities between 0.5 mm and 2.5 mm in diameter, which rarely occupy greater than 15% of the coral. The cavities are often lined with a microspar rim and invariably infilled with micritic reef sediment (Fig. 12A). In shape and size, they are similar to borings produced by bivalves and worms (Bromley, 1970). In several examples, the bivalve shells are preserved within borings; (ii) Slightly irregular walled, ramose, branching networks, restricted to the outer 5 mm of the coral skeleton. The borings are of variable length, up to 1 mm in

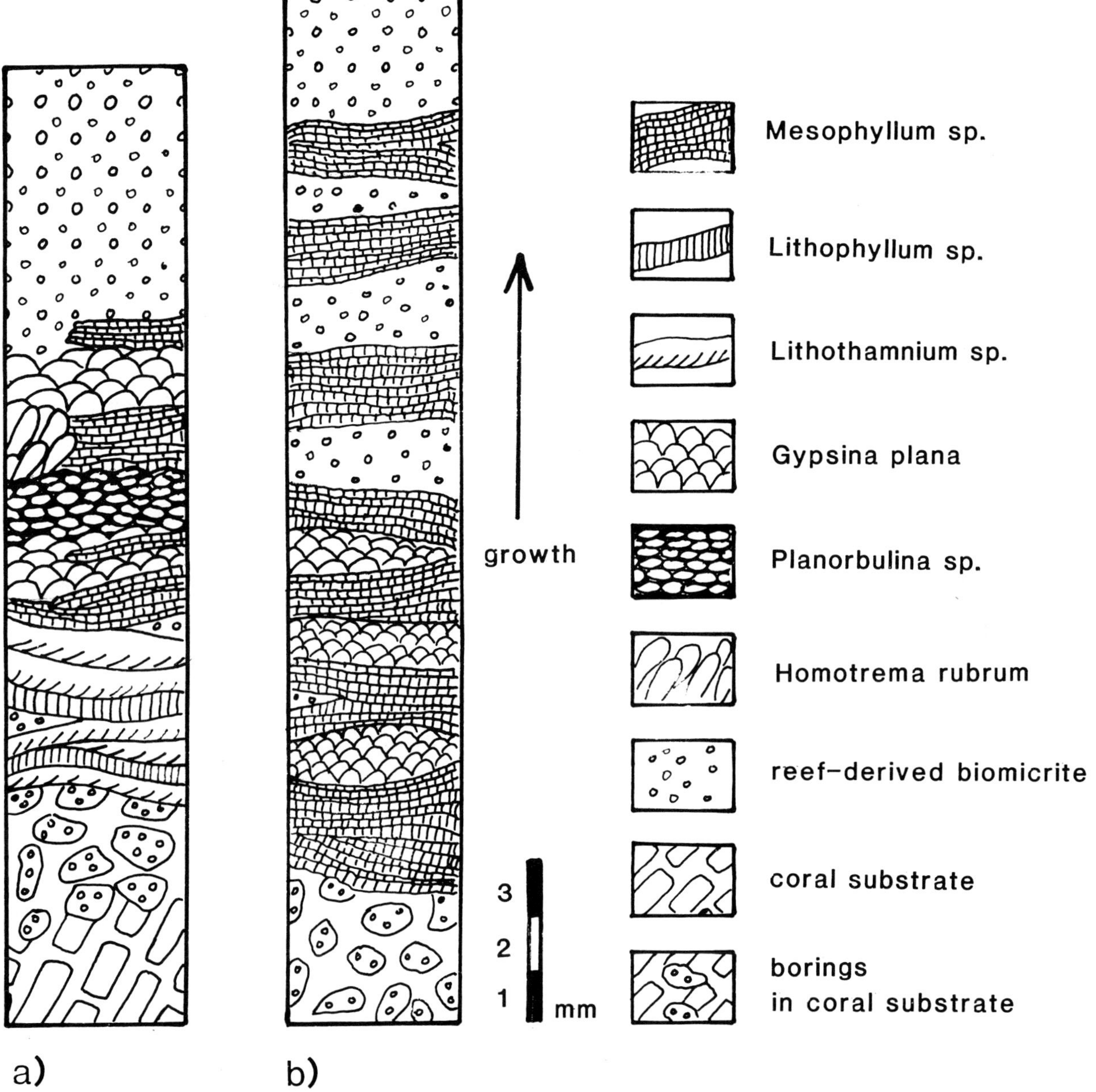


FIG. 11.—Diagrammatic cross sections of crusts: (A) Mixed-composition crust: A thick photophyllic crust of *Lithophyllum* and *Lithothamnium* is overlain by interlaminated *Mesophyllum*, and the encrusting foraminifera *Gypsina plana* and *Homotrema rubrum*. (B) Constant-composition crust: An initial thick crust of *Mesophyllum* is overlain by interlaminated *Mesophyllum* and reef sediment. Crust on *Favites neglecta* (Michelotti) from inter-colony debris area.

diameter. They often form a bored skin to the coral skeleton. They were probably produced by either sponges or bryozoans. Cavities are now infilled with fine-grained reef sediment.

The types of boring do not differ significantly in corals taken from life position and those taken from inter-colony debris areas. However, the density of borings increases markedly in corals and crusts from inter-colony shaded environments, where crust growth is slower and destructive processes are more dominant.

The boring activity of sponges, polychaetes, algae, bivalves and other organisms, outlined briefly above (see Bromley, 1970) within primary and secondary frameworks, resulted in formation of a variety of open and closed cavities. These, combined with non-bored skeletal and inter-skeletal voids, in both corals and encrusters, provided a suitable environment for the precipitation of carbonate cement and accumulation of reef-derived sediment.

### *Sedimentation*

Reef sediment within cavities consists of subangular, generally poorly sorted fragments of coral, coralline algae, benthonic foraminifera and shell fragments in a matrix of micrite (Fig. 12B). Grain size varies from 10 μm to 5 mm. Grading is rarely

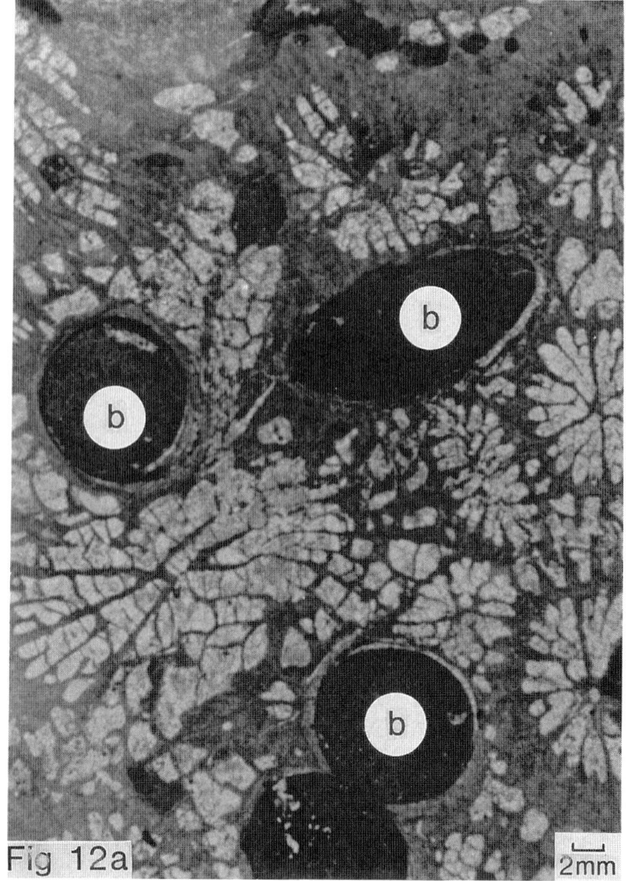


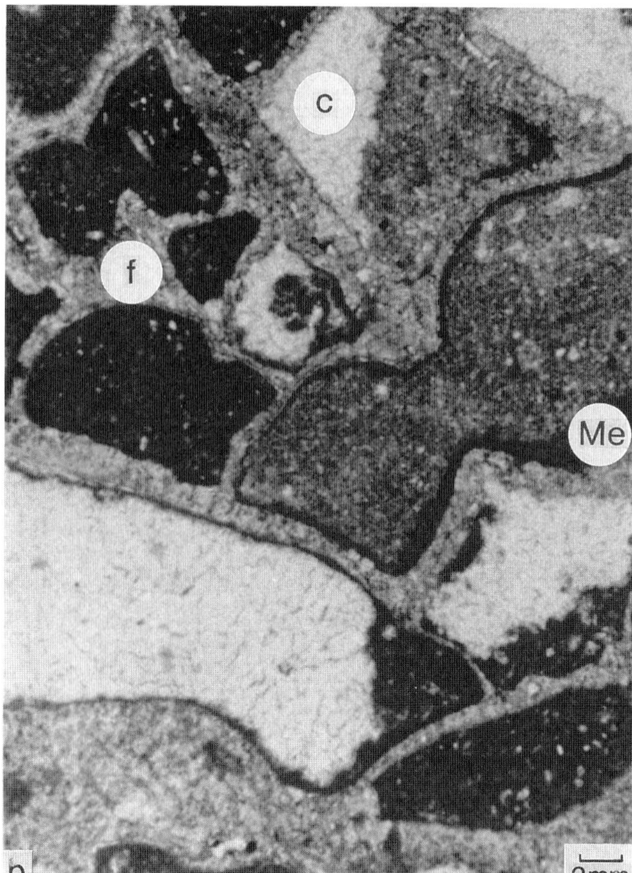


FIG. 12.—Borings and encrusting sequences in the Kasaba Formation reefs. Both scale bars = 2 mm; plane polarized light. (A) Bivalve borings (b) infilled with reef sediments. The remains of the shell are clearly visible: (B) isopachous cement fringes (f) are overlain by reef derived biomicrite sediment, some forming geopetals in intraskeletal voids. Voids are filled with late stage sparite cement (c). Note presence of thin crust of *Mesophyllum* (Me).

present. The sediment is essentially the same in cavities in both corals and encrusters. Despite all the terrigenous sediment in the area, such sediment is rare within the reef cavities. However, scattered terrigenous cobbles and pebbles are locally seen within the reef core.

The biomicrite sediment was probably produced by the organic breakdown of the frame. Wave action sucked water out of the cavities and voids, and the resultant turbulent inflow transported suspended sediment back into the frame. In areas where the initial sedimentary fill was subsequently bored by polychaetes or bivalves, early synsedimentary cementation of the sediment must be invoked. These second generation borings are also frequently infilled with reef sediment.

### *Submarine Cements*

Submarine cements occur as both peloidal cements and as isopachous fringes of calcite needles, up to 70 μm thick, around the margins of some intraskeletal cavities and voids within coral skeletons. In some cases the cavities subsequently have been infilled by micritic reef sediment.

The restriction of early cements to intraskeletal voids may indicate that interskeletal sedimentation was too high to allow the formation of aragonite, or high-Mg calcite cements, on surfaces exposed to external processes.

### *Interaction and Sequential Development*

Following establishment of the primary framework encruster growth, cementation and sedimentation were processes which continued the construction of the reef and reduced the porosity by the addition of a secondary framework. At the same time, boring organisms and mechanical breakdown by physical mechanisms (e.g., wave action) destroyed both primary and secondary frameworks, creating new surfaces on which the sequence of events may be repeated.

Studies of modern reefs (Garrett et al., 1971; Schroeder and Zankl, 1974; Scoffin and Garrett, 1974; Martindale, 1976) reveal that constructional processes dominate on the upper well-lit surface, whereas on the undersurface destructive processes are frequently more important. Similar relationships are observed in the Kasaba Formation reefs. The detailed examination of several specimens from the underside of large colonies and inter-colony debris areas reveals a complex interaction of constructive and destructive processes during the growth and partial, or complete, destruction of the primary or secondary framework. By comparison, corals taken from life position high in the reef framework, show only one generation of boring and subsequent photophyllic encrustation. The interaction of destructive and constructive processes continued until boring either completely destroyed the primary framework and the secondary frame, or until the frame was removed from the environment of growth (Martindale, 1976, 1992). In this case, growth was terminated by the influx of terrigenous clastic material.

### *Burial and Diagenesis*

Following burial, the reefs were subject to a variety of diagenetic effects.

*Primary framework.—*

Coral skeletons which comprise the primary framework consist of either dusty, equant, blocky calcite or fibrous calcite that mimics aragonite crystal forms, both formed by the alteration of aragonite to calcite shortly after burial (Bathhurst, 1971; James, 1974). The resultant structure is formed around a mold composed of voids and cavities that are filled with cemented micritic sediment. Where skeletal micro-architecture is partially preserved by fibrous calcite, a gradual neomorphism of aragonite to calcite is suggested without significant intermediary void stage. X-ray diffraction reveals that a large proportion of the coral skeletons have retained up to 50% of their original aragonite mineralogy. None of the rocks analysed contained dolomite.

*Secondary framework.—*

The encrusting sequences demonstrate well-preserved skeletal micro-architecture. This is because the skeletons of crustose coralline algae and foraminifera consist of high-Mg calcite which undergoes little or no structural alteration and is preserved as the stable polymorph low-Mg calcite.

*Cements.—*

Excluding the initial submarine cement, which is patchily developed (see above), two generations of cement are present. The first consists of patchy developed fringes of microspar approximately 40 μm thick which lines the rims of voids and cavities (Fig. 12B) and in some areas overlies reef sediment. The second generation of cement consists of equant blocky sparite that fills intra-skeletal voids and cavities in both primary and secondary frameworks and also intraskeletal voids.

## REEFS IN A COARSE CLASTIC SEDIMENTARY ENVIRONMENT: COMPARISON WITH RED SEA REEFS

Classically, reefs were not thought to be associated with high terrigenous clastic sedimentation input. However, studies of ancient and modern reefs and their environment of formation have shown that this is not necessarily the case. Present-day coral reefs are found in close association with terrigenous clastic sediments from a number of areas (e.g., Jamaica, Wescott and Ethridge, 1980; Red Sea, Gvirtzman and Buchbinder, 1978; Hayward, 1982b,1985; Sicily, Grasso and Pedley, 1988; Pedley and Grasso, 1991; Pedley, this volume, and elsewhere in Southern Turkey, Flecker et al., 1995). Those in Sicily are most comparable to the reefs discussed here as they are associated with a flexural foredeep tectonic setting.

Fringing coral reefs are extensively developed around the margins of coastal alluvial fans along the coast of the Red Sea and Gulfs of Elat and Suez. However, these areas differ in their overall tectonic-environmental setting from the Miocene sequence of Southwest Turkey, in that they are related to rifting and crustal extension, rather than compression and foreland basin development; however, these reefs do provide a useful modern analogue.

The coarse gravel sediments of coastal alluvial fans provide an ideal substrate for coral planulae to settle. In the Red Sea, small coral colonies grow directly on non-lithified terrigenous pebble clasts greater than 60 mm in diameter. Material finer than this is not normally colonized by corals. However, material as fine as granule gravel may be initially bound by coralline algae, upon which coral colonization can follow.

The grain size of a potential growth substrate, therefore, exerts a very strong control on reef location in the Red Sea. In the Kasaba Formation (Fig. 13), reefs are only found where claystone forms a minor part of the sequence, suggesting a similar control. The presence of claystone in the nearshore, back-reef sequences is probably the result of the protection of this area by the reef from wave action. Mud provides an unsuitable substrate for colonization by coral planulae (or coralline algae), thus landward progradation of the reef is prevented.

The Kasaba Formation was deposited just prior to the Messinian desiccation event, recorded throughout the Mediterranean area (Hsü et al., 1973), as a semi-arid environment characterized by low precipitation and low seasonal run-off. The critical factor in reef development was the localized and episodic sedimentation patterns typical of semi-arid alluvial environments. In recent examples of this climate, sedimentation is restricted to one or two flash-flood events every year (Gvirtzman and Buchbinder, 1978; Hayward, 1982b, 1985). In addition, these periodic sediment influxes are confined to the active portion of a fan and, depending on the prevailing marine currents, have little effect on the marine sediment adjacent to inactive areas of the fan. With low episodic sedimentation rates, a reef would have ample opportunity to re-establish itself, should it be swamped by an influx of terrigenous material. In addition, small-scale eustatic changes in sea level prior to and possibly during the Mediterranean desiccation event may well have influenced reef growth (e.g., Follows et al., this volume).

In the Red Sea, catastrophic flood events during the winter result in terrigenous material being carried onto the reef. Marine

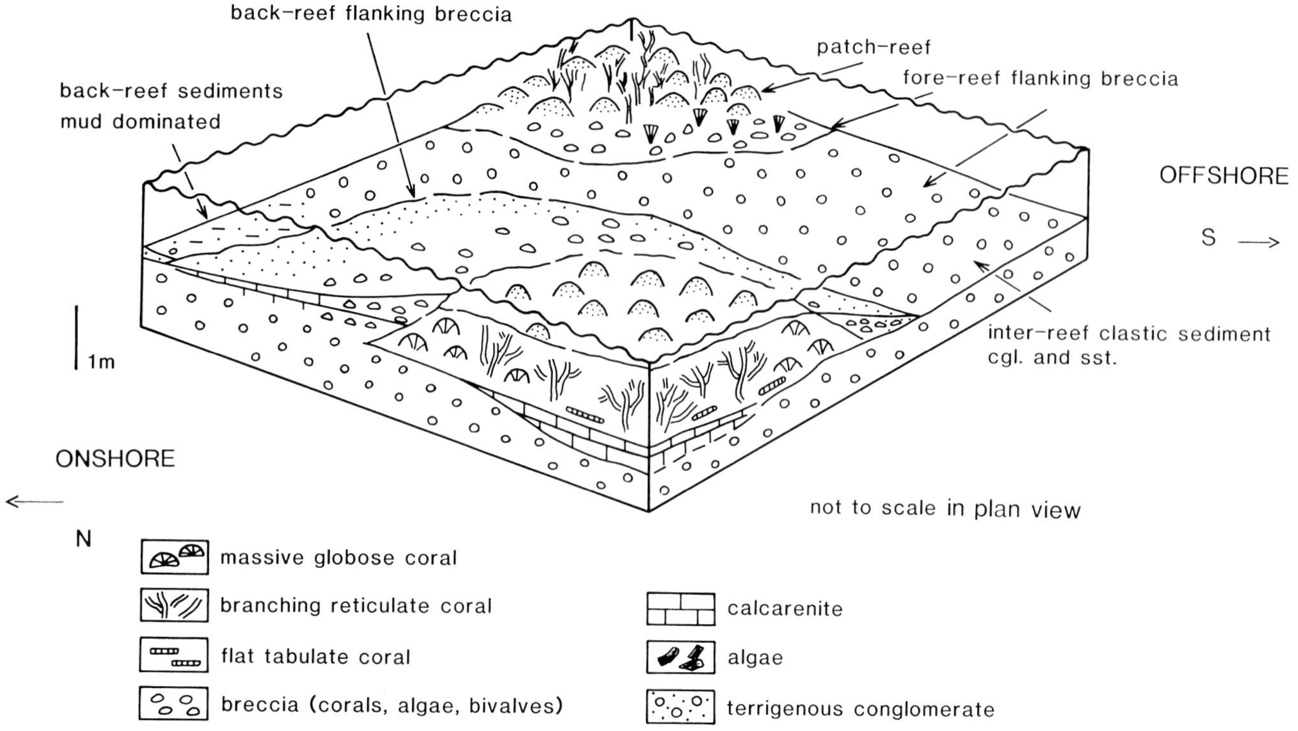


FIG. 13.—Generalized sedimentological model for the Kasaba Formation patch reefs showing greater development of flanking sediments on back-reef (landward margin). Inter-reef sediment is dominated by coarse terrigenous clastic sediment.

currents, mainly longshore drift and waves, take up to ten days to move the finer material away from the living reef area. The effect of such influxes on the reef, however, is minimal (B. Buchbinder, pers. commun., 1981). Much of the terrigenous material coarser than pebble gravel is retained in the reef with coralline algae, other encrusting organisms and corals binding the clasts into the reef. In some areas, terrigenous material bound in this way forms up to 50% of the reef framework (Hayward, 1982b, 1985).

In the Kasaba Formation, the restriction of terrigenous clastic sediment to rare terrigenous cobbles and pebbles within the reef core suggests that clastic influx rarely swamped reefs with enough debris to terminate growth. However, the scattered development of reefs within the clastic sequence indicates that conditions for growth were only satisfied sporadically, probably as the locus of terrigenous sedimentation switched. This type of control on reef location is also evident in the Red Sea where entrenched fluvial channels that formed during a Pleistocene sea-level lowstand, pass seawards into narrow incised canyons. Reefs are developed away from the active fluvial-marine channels, but are not growing in the channel areas as a result of the high sedimentation rates formed by lateral confinement of the fluvial system.

The Kasaba Formation sedimentary sequences, however, were developed in a very different tectonic regime than the Red Sea. Rapid subsidence associated with the regional emplacement of allochthonous units (Lycian Nappes) and extensive migration of the active sedimentation tract across the fan surface resulted in relatively high sedimentation rates over the entire alluvial fan surface. Although in the short term, sedimentation rates were probably similar to the Red Sea, subsidence and active progradation of the alluvial fan in the Kasaba Formation sequences was far more extensive than in the Red Sea examples, which are in a more or less equilibrium situation at the present day.

## CONCLUSIONS

Late Miocene reefs developed in patches parallel to the palaeoshoreline on the submarine toes of coastal alluvial fans, in a tectonically active foreland basin setting. Gravel and coarse sand of the fans provided an ideal substrate on which coral planulae could settle and grow. As a result no pioneer community is present. Primary framework builders consist dominantly of the corals *Favites sp., Tarbellastraea sp., Montastraea sp.* and subordinate *Porites sp.* Coral morphology changed progressively from tabular to branching as the reef grew upwards towards the surf zone. There is no evidence of surf effects and the tops of the reefs were probably several meters below the sea surface during life, consistent with growth in a micro-tidal sea.

The primary coral framework was encrusted by a secondary framework of coralline algae and encrusting foraminifera. The effects of boring and grazing by a number of organisms (bivalves, sponges, bryozoa) produced abundant debris which accumulated in inter-colony areas. Dominantly onshore wave and storm activity periodically redeposited some of this material landward into the lee of the reefs where it remained relatively undisturbed.

Flanking facies are only well-developed on the landward (back reef) margins of patch reefs where they comprise layered breccias passing outwards into bedded calcarenites. By contrast, on the seaward (forereef) margin very coarse breccias, rich in coral clasts, pass directly into terrigenous sediment. The patch reefs are generally overlain by pebbly terrigenous sandstones, or more locally by thin calcarenite.

The patch reefs developed as part of a prograding fan delta-type succession within a foreland basin setting related to the final stages of tectonic emplacement of a regional nappe complex.

## ACKNOWLEDGMENTS

This study is based part of a Ph. D. Dissertation by A. B. Hayward of a Miocene foreland basin succession in Southwest Turkey. The first author was funded by a Natural Environmental Research Council Studentship. The second author's fieldwork was supported by a N.E.R.C. research grant. Helpful reviews were provided by R. Flecker, M. Pedley and E. Franseen.

## REFERENCES

Bathhurst, R. B. C., 1971, Carbonate Sediments and Their Diagenesis: Amsterdam, Elsevier, 620 p.

Berggren, W. A., Kent, D. V., and Van Couvering, J. A., 1985, Neogene geochronology and chronostratigraphy, *in* Snelling, N. J., ed., The Chronology of the Geologic Record: London, Geological Society of London Memoir 10, p. 211-260.

Bizon, G., Biju-Duval, B., Letouzay, J., Minod, O., Poisson, A., Ozer, B., and Oztumer, E., 1974, Nouvelles précision stratigraphiques concernant les bassins tertiaires du sud de la turquie (Antalya, Mnt, Adana): Paris, Revue del'Institute Français du Petrole, v. 19, p. 305-320.

Bromley, R. G., 1970, Borings as trace fossils and Entobea cretocea Portlock as an example: Geological Journal, v. 3, p. 49-90.

Esteban, M., 1979, Significance of the Upper Miocene coral reefs of the Western Mediterranean: Palaeogeography, Palaeoclimatology, Palaeoecology, v. 29, p. 169-188.

Flecker, R., Robertson, A. H. F., Poisson, A., and Muller, C., 1995, Facies and tectonic significance of two contrasting Miocene basins in Southern Central Turkey: Terra Nova, v. 7, p. 221-232.

Garrett, P., Smith, D. L., Wilson, A. O., and Patriquin, D., 1971, Physiography ecology and sediments of two Bermuda patch reefs: Journal of Geology, v. 79, p. 647-668.

Grasso, M. and Pedley, H. M., 1988, The sedimentology and development of Terravecchia Formation carbonates (Upper Miocene) of North Central Sicily: Possible eustatic influence on facies development: Sedimentary Geology, v. 57, p. 131-149.

Gvirtzman, G. and Buchbinder, B., 1978, Recent and Pleistocene coral reefs and coastal sediments of the Gulf of Elat: Jerusalem, Post Congress Guidebook, Tenth International Congress of Sedimentology, p. 163-189.

Haq, B. U., Hardenbol, J., and Vail, P. R., 1987, Chronology of fluctuating sea levels since the Triassic: Science, v. 235, p. 1156-1167.

Hayward, A. B., 1982a, Tertiary ophiolite-related sedimentation in S.W. Turkey: Unpublished Ph.D. Dissertation, University of Edinburgh, Edinburgh, 420 p.

Hayward, A. B., 1982b, Coral reefs in a clastic sedimentary environment: Fossil (Miocene, Southwest Turkey) and Modern (Recent, Red Sea) analogues: Coral Reefs, v. 1, p. 109-114.

Hayward, A. B., 1983, Coastal alluvial fans and associated marine facies in the Miocene of Southwest Turkey, in Collinson, J. D. and Lewin, J., eds., Modern and Ancient Fluvial Systems: Oxford, Blackwell Scientific Publishers, Special Publication of the International Association of Sedimentologists 6, p. 323-336.

Hayward, A. B., 1984, Miocene clastic sedimentation related to the emplacement of the Lycian Nappes and the Antalya Complex, Southwest Turkey, *in* Dixon, J. E. and Robertson, A. H. F., eds., The Geological Evolution of the Eastern Mediterranean: London, Special Publication of the Geological Society of London 17, p. 237-300.

Hayward, A. B., 1985, Coastal alluvial fans (fan deltas) of the Gulf of Aqaba (Gulf of Eilat), Red Sea: Sedimentary Geology, v. 43, p. 241-260.

Hayward, A. B. and Robertson, A. H. F., 1981, Direction of ophiolite emplacement inferred from Cretaceous and Tertiary sediments of an adjacent authochthon, the Bey Dalarl, southwest Turkey: Geological Society of America Bulletin, v. 93, p. 68-75.

Hsü, K. J., Ryan, W. B. F., and Cita, M. B., 1973, Late Miocene desiccation of the Mediterranean: Nature, v. 242, p. 240-244.

James, N. P., 1974, Diagenesis of scleractinian corals in the subaerial vadose environment: Journal of Paleontology, v. 48, p. 785-799.

Lowenstam, H. A., 1957, Nigarian reefs in the Great Lakes area, *in* Hedgepeth, J. W. and Lad, H. W., eds., Treatise on Marine Ecology and Paleoecology 2: Boulder, Geological Society of America Memoir 67, p. 215-248.

Maiklen, W. R., 1970, Carbonate sediments in the Capricorn Reef Complex, Great Barrier Reef, Australia: Journal of Sedimentary Petrology, v. 40, p. 55-80.

Martindale, W., 1976, Calcareous encrusting organisms of the Recent and Pleistocene reefs of Barbados, West Indies: Unpublished Ph. D. Thesis, University of Edinburgh, Edinburgh, 141 p.

Martindale, W., 1992, Calcified epibionts as palaeoecological tools: examples from the Recent and Pleistocene reefs of Barbados: Coral Reefs, v. 11, p. 167-177.

Önalon, M., 1980, Elmali-Kas (Antalya) Arasirdake Bolgerin jeologisi: Istanbul Universitesi, Fen Fakulltesi Monografileri, Sayi 29, 140 p.

Pedley, M. and Grasso, M., 1991, Sea level change around the margins of the Catania-Gela Trough and Hyblean Plateau, southest Sicily (African-European plate convergence zone): a problem of Plio-Quaternary plate buoyancy, *in* MacDonald, D. I. M., ed., Sedimentation, Tectonics and Eustasy. Sea-level Changes at Active Margins: Oxford, Blackwell Scientific Publishers, Special Publication of the International Association of Sedimentologists 12, p. 451-464.

Poisson, A., 1977, Recherches géologiques dans les Taurides occidentales (Turquie). Lithothamnium pseudoramossissimum nouvelle espèce d'algue rouge de la formation de Karabayir: Bulletin of Mineralogical Research, Exploration Institute of Turkey, v. 82, p. 67-71.

Poisson, A., 1981, Miocene transgression forms in the Gökbuk-Catallan basin (Western Taurus range, Turkey): T.C. Petrol. Isleri Genel Müddurlüü Dairesi, v. 25, p. 217-221.

Robertson, A. H. F. and Woodcock, N. H., 1980, Strike-slip related sedimentation in the Antalya Complex, southwest Turkey, *in* Ballance, P. F. and Reading H. G., eds, Sedimentation in Oblique-Slip Mobile Zones: Oxford, Blackwell Scientific Publishers, Special Publication International Association Sedimentologists 4, p. 127-145.

Robertson, A. H. F. and Woodcock, N. H., 1984, The Southwest segment of the Antalya Complex as a Mesozoic-Tertiary Tethyan continental margin, *in* Dixon, J. E. and Robertson, A. H. F., eds., The Geological Evolution of the Eastern Mediterranean: London, Special Publication of the Geological Society of London 17, p. 251-272.

Santisteban, C. and Taberner, C., 1980, Reefs in the U. Miocene of S.E. Spain: Cambridge, Reefs Past and Present, Abstracts, p. ?.

Santisteban, C. and Tabener, C., 1983, Shallow marine and continental conglomerates derived from coral reef complexes after desiccation of a deep marine basin: the Tortonian-Messinian deposits of the Fortuna basin, south east Spain: Journal of the Geological Society of London, v. 140, p. 401-411

Schroeder, J. H. and Zankl, H., 1974, Dynamic reef formation: A sedimentological concept based on studies of Recent Bermuda and Bahama reefs: Brisbane, Proceedings of the 2nd International Coral Reef Symposium 11, Great Barrier Reef Committee, p. 413-428.

SCOFFIN, T. P., 1987, Introduction to Carbonate Sediments and Rocks: Glasgow, Blackie and Son, 174 p.

SCOFFIN, T. P. AND GARRETT, P., 1974, Processes in the formation and preservation of internal structure in Bermuda patch reef: Brisbane, Proceedings of the 2nd International Coral Reef Symposium 11, Great Barrier Reef Committee, p. 429-449.

WESCOTT, W. A. and ETHRIDGE, F. A., 1980, Fan-delta sedimentology and tectonic settings-Yallahs Fan-Delta: American Association of Petroleum Geology Bulletin, v. 64, p. 374-399.

# MIDDLE AND UPPER MIOCENE REEFS AND CARBONATE PLATFORMS IN ISRAEL

BINYAMIN BUCHBINDER
*Geological Survey of Israel, 30 Malkhe Yisrael Street, 95501 Jerusalem, Israel*

ABSTRACT: During most of the Miocene epoch, deposition in the Mediterranean coastal plain and offshore areas of Israel was characterized by hemipelagic sediments of fine-grained siliciclastics. These were deposited along the ancient continental slope as a result of terrigenous influx into the eastern Mediterranean from river systems preceding that of the present Nile River. Miocene reefs and carbonate platforms developed during two short periods when sea level rose and flooded the shelf, in early Middle Miocene (upper part of N8 to N9) and in Late Miocene times (N17). Sediments of the first event are known as the Ziqlag Formation and of the second, as the Pattish Formation. The early Middle Miocene Ziqlag phase is characterized by two carbonate platforms which occupy different topographic levels in the Shefela (foothills) area. They are a result of two successive transgressive events corresponding to third-order cycles 2.3 and 2.4 combined with a continued westward tilting of the shelf. During the younger 2.4 cycle, the distal part of the first platform was tectonically elevated and an abrasional sea-cliff was formed marking the boundary between the higher and lower platforms. The higher platform is represented by a rhodalgal-ramp lithofacies and constitutes small-scale cycles of shoreface storm deposits of grainstones/packstones with hummocky cross stratification and of graded shell-beds (floatstones), alternating with claystones of backshore lagoon environment with large oysters. The stacking pattern of the small-scale cycles exhibits progradational regressive characteristics. The lower platform is poorly exposed. The Ofaqim reef at its western edge shows a relatively low diversity coral assemblage (faviids, stylophorids and poritids). This together with the "temperate" rhodalgal lithofacies of the higher platform may either reflect the beginning of the gradual cooling towards the Upper Miocene or the lowering of the salinity of the Mediterranean surface layer in Middle Miocene times (Serravallian crisis). Late Miocene carbonates (Pattish Formation), corresponding to 3rd-order cycle 3.2, predate the Messinian desiccation. They did not penetrate far beyond the shelf edge area except along the Gaza Beer Sheva canyon. In Ofaqim, they truncate the seaward part of the Ziqlag reef, showing prograding clinoforms of alternating dolomitized rhodoliths and claystones, topped by a veneer of branching *Porites* colonies. Debris flow deposits of mixed reef-derived clasts and Cretaceous lithoclasts accumulated in the slope along the submarine canyons at that time. Messinian evaporites of deep marine origin overlie these sediments. Landwards, the evaporites onlap older formations, reflecting a sea-level rise during cycle 3.3.

## INTRODUCTION

Miocene reefal limestones in Israel form patchy outcrops in the foothills area and along the country's eastern coastal plain (Fig. 1) where they are also found in drillholes. These limestones have been referred to as the Ziqlag Formation (Reiss and Gvirtzman, 1966; Gvirtzman and Buchbinder, 1969; Gvirtzman, 1970; and Buchbinder, 1975 and 1979). Buchbinder et al. (1993) showed that these sediments represent two phases of reef and carbonate platform development: an early Middle Miocene phase (upper part of N8 to N9) designated as the Ziqlag Formation and a Late Miocene phase (upper part of N16 to N17), designated as the Pattish Formation. The coastal plain and shelf area of Israel are marked by the development of a post-Eocene wedge of fine clastic sediments of the Saqiye Group, which is up to 2000 m thick. These sediments accumulated along the continental slope of the northeastern margin of the Arabian platform as the result of a significant influx of terrigenous sediments into the eastern Mediterranean (Figs. 1, 2) from river systems preceding that of the present Nile. The eastward-flowing circum-Mediterranean current carried these sediments toward the Israeli coast where they accumulated on the slope in front of a well-developed shelf break. The process was interrupted by a few erosional events, resulting in river incision onshore, in the formation of a system of submarine canyons along the continental slope, and in the accumulation of evaporites during Messinian time. Erosional processes ceased after the Messinian time and thick Pliocene clays forming the Yafo Formation filled up pre-existing canyons and prograded westward to form the present continental shelf. The development of reefs and carbonate platforms within this predominantly clastic regime was possible only during short periods when the effect of clastic influx was minimized due to high sea levels.

## STRATIGRAPHY AND GEOLOGICAL FRAMEWORK

Figure 3 depicts the spatial and temporal distribution of Oligocene-Neogene sediments across the ancient slope and shelf. The scheme is superimposed on Haq et al.'s (1987, 1988) chronostratigraphy, sequence stratigraphy, and global sea-level curve. The stratigraphic scheme is based on drillhole data from the shelf edge and slope (coastal plain area) and on outcrop data from the shelf (Judean foothills area).

The Oligocene to Pleistocene section (Saqiye Group) was divided into the Bet Guvrin, Ziqim, Ziqlag, Mavqiim, and Yafo formations (Gvirtzman, 1970 and Gvirtzman and Buchbinder, 1978). Deposition was mostly confined to the slope and shelf edge areas. However, high sea levels during Early Oligocene and early Middle Miocene resulted in marine deposition on the shelf platform area. Deposition on the slope was occasionally interrupted by erosion phases resulting in the formation of submarine canyons, some of which extended onto the shelf area (Buchbinder et al., 1993). Debris flows, turbidite deposits, and deep-water sandstones are found in canyons in the slope area, in sediments dating from Oligocene to Late Miocene times. The most significant erosion phases occurred in the following times: (1) Middle Oligocene (P21); (2) Late Burdigalian (between zones N6 and N8); (3) Late Serravallian (N13 to lower part of N14); and (4) Late Messinian, after the deposition of Messinian evaporites. The third erosional event (in Late Serravallian time)

Models for Carbonate Stratigraphy from Miocene Reef Complexes of Mediterranean Regions,
SEPM Concepts in Sedimentology and Paleontology #5, Copyright © 1996,
SEPM (Society for Sedimentary Geology), ISBN 1-56576-033-6, p. 333-345.

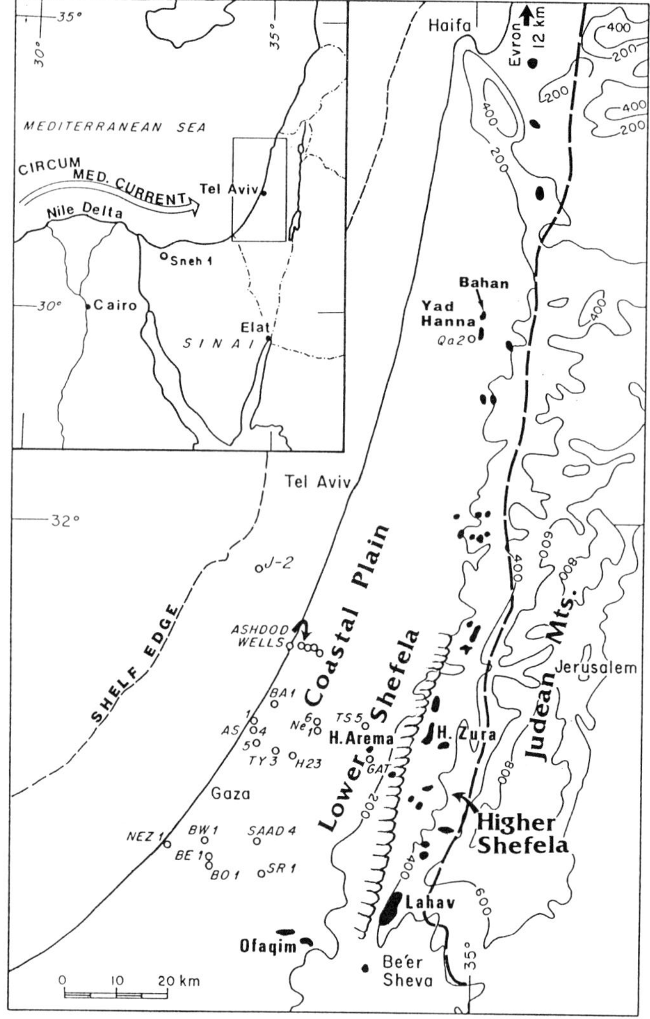


FIG. 1.—Location map of study area, showing physiographic units, topographic contours (in meters), outcrops of Miocene limestones (black patches) and drillholes (circles) noted in the text. The thick dashed line marks the easternmost extension of Miocene reef limestones. The jagged line marks the scarp separating the higher Shefela from the lower Shefela. Borehole abbreviations: AS- Ashqelon, BA- Barnea, BE- Beeri, BO-Beeri Old, BW- Beeri west 1, H- Helez, J- Joshua, Ne- Negba, NEZ- Nezarim, SR- Sharsheret, TS- Tel Safit, TY- Talme Yafe, Qa- Qaqun.

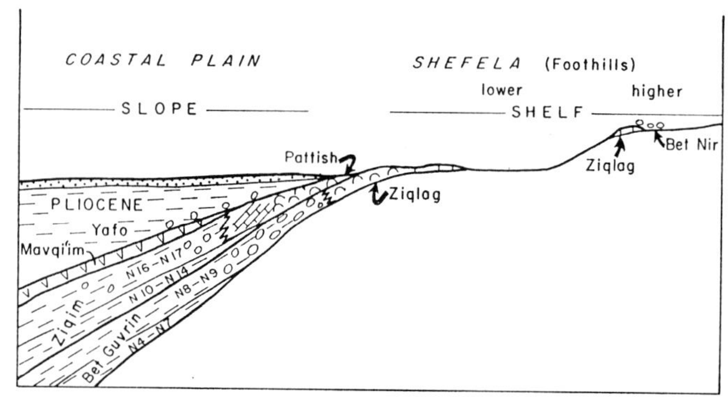


FIG 2.—A schematic E-W section across the Miocene shelf and slope in Israel showing the Neogene sedimentary wedge. Note the relationships between the Middle Miocene Ziqlag reef complex and its slope sediments and the Late Miocene Pattish reef complex and its slope sediments.

was the most dramatic because of its extensive seaward continuation, its deep incision, and its well developed drainage system (Gvirtzman, 1970, Gvirtzman and Buchbinder, 1978). This event corresponds to a major sea-level drop at the beginning of the TB3 super cycle. The fourth event is superimposed on the previous one and is highly pronounced on seismic sections because channel outlines are enhanced by the evaporite infilling (Fig. 4).

Miocene carbonate deposition occurred during highstands of sea level in early Middle Miocene (Ziqlag Formation) and Late Miocene times (Pattish Formation) when the shelf was fully or partially flooded. During periods when shelf areas were emergent, hemipelagic sedimentation took place mainly in slope settings. It is possible, however, that at intermittent sea-level stages shelf areas were covered with shallow marine or continental deposits which were later completely eroded.

Outcrops of the Middle Miocene Ziqlag Formation are found on Israel's eastern coastal plain and in the Shefela foothills (Fig. 1). The easternmost outcrops mark the ancient shoreline which existed during the peak of the transgression that deposited the Ziqlag sediments. The limestones unconformably overlie Upper Cretaceous to Oligocene formations and occupy the synclinorial area of the Shefela foothills adjacent to the Judean and Samarian anticlinorial backbone which separates the Mediterranean coastal plain to the west from the Jordan Rift Valley to the east. The unconformity at the base of the formation forms two distinct morphological units: (1) the eastern part of the coastal plain and the lower Shefela Plain, which rises gradually from an elevation of 100 m in the west to 350 m in the east; and (2) the higher Shefela plain, at elevations of about 300-480 m above sea level, forming a wide platform which abuts the Judean Mountains to the east. West of the Ziqlag outcrops, the Coastal Plain area constitutes the ancient continental slope.

The western-most outcrops of Horvat Arema and Evron (Figs. 1, 4), which are close to the slope, include planktonic foraminifera of units N8 and N9 (Buchbinder et al., 1993). Slope sediments of debris flows with reef-derived lithoclasts and pebbles of Cretaceous limestones and dolostones embedded in calcareous claystones and locally with contorted bedding are found in drillholes west of the outcrops (e.g., in Ashqelon 5 well at 1369 m depth).

The carbonates of the Late Miocene Pattish Formation formed mainly along the shelf edge in front of the Ziqlag reef. The only outcrops are found along the Gaza Beer Sheva canyon, in Ofaqim and in the town of Beer Sheva. These carbonates are also found in drillholes on the Gaza-Beer Sheva canyon's shoulders west of Ofaqim (Fig. 4). Conglomerates containing coral debris and pebbles of Cretaceous limestones and dolostones of the Judea Group embedded in calcareous claystones that include planktonic foraminifera of units N16-N17 (Tortonian to Messinian) are found in many deep drillholes in the slope area, especially along the courses of the ancient canyons of Ashdod, Ashqelon, and Barnea (Fig. 4). These conglomerates are considered as debris flows which were derived from Pattish reefs on the shelf edge and from submarine outcrops of the Judea limestone

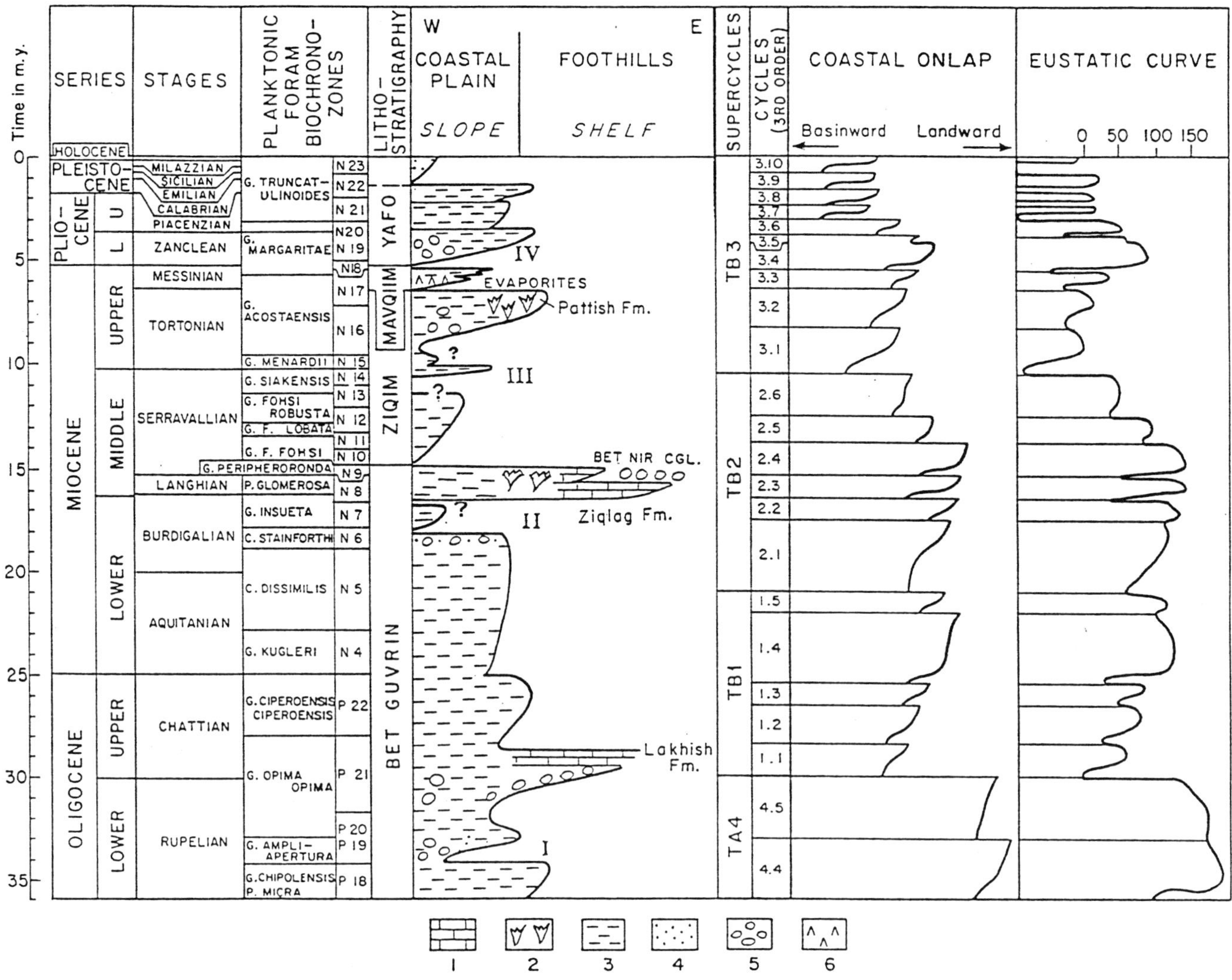


FIG. 3.—Stratigraphic scheme of the Oligocene - Neogene marine section in Israel correlated with the chronostratigraphy, sequence stratigraphy, and eustatic curve of Haq et al. (1987, 1988). Roman numerals I - IV denote major erosion events. Lithology: (1) platform limestones; (2) coral reefs; (3) calcareous claystones; (4) sandstones; (5) conglomerates; and (6) anhydrite or salt.

along the upper continental slope. The record provided by these deep-water deposits of reef debris is particularly important because most of the reefs at the shelf edge were completely eroded in the Messinian. These debris sheets deposited in slope environments were later overlain by Messinian evaporites of the Mavqiim Formation.

## FACIES DISTRIBUTION

### *Early Middle Miocene Ziqlag Phase*

Sediments of the Ziqlag phase are included within the sedimentary package of units N8 (upper part) and N9, corresponding to the uppermost part of the Bet Guvrin Formation (Fig. 3). They reflect a relatively short time span of about 1.5 my. (Fig. 3). In terms of the global sea-level curve and sequence stratigraphy of Haq et al., (1988), this time-span is characterized by sharp fluctuations in sea level of third order cycles 2.2 (upper part), 2.3 and 2.4 (lower part).

Sediments associated with the Ziqlag phase were deposited in distinct facies belts in shelf platform, shelf edge reef and slope settings. Most outcrops (Figs 1 and 4) are on the carbonate platform belt of the higher Shefela surface. Exceptions are the Ofaqim and Yad Hanna outcrops, which represent a shelf edge reef facies, and the Horvat Arema outcrop, which represent, an upper slope facies. There is no evidence for a significant backreef lagoon behind the reef facies. The sediments of the Lahav platform reflect a storm-effected open shelf setting.

*Carbonate platform facies.—*

The outcrops in the higher Shefela and the lower Shefela areas (Fig. 1) represent two carbonate platforms. The higher Shefela platform is separated from the lower Shefela platform by a 80-m-high morphological step which is probably of abrasional origin. Most outcrops are located in the upper Shefela platform; their thicknesses reach up to 50 m. Outcrops of the lower Shefela

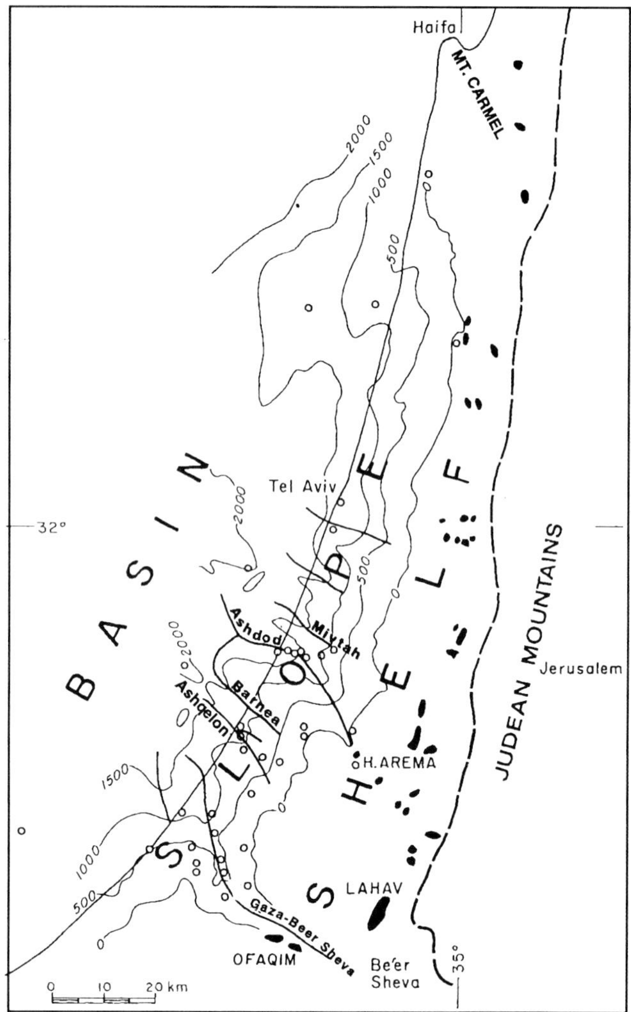


FIG. 4.—Map showing subsurface contours (in meters) of base Pliocene Yafo Formation (modified from Gelberman and Grossowicz, 1990). Major canyons along the continental slope are marked by a heavy line. Black patches denote outcrops of Miocene reef limestones. Small circles denote drill holes.

platform are very scarce. The small relicts left are too thin to allow a meaningful study of lithofacies and depositional environments. Generally, platform outcrops show bioclastic packstones and grainstones with occasional lithoclast pebbles and up to 10% quartz sand, especially towards the eastern (landward) side. Particle grain-sizes vary between medium to very coarse sand. The common coralline-algae grains and broken branches indicate that the algae lived on the loose sandy substrate as semi-free crusts and branches. Platform facies are also characterized by rich molluscan and echinoderm faunas, especially clypeasterid echinoderms and burrowing pelecypods such as cardiids, lucinids, and arcriids. Among the gastropods, *Strombus* is common, especially in areas with coral colonies. *Halimeda,* though not so common, is locally present in the Lahav outcrops. A few coral colonies of faviids and poritids were found, especially near the western rim of the higher Shefela platform in H. Zura (Fig. 1). Benthonic foraminifera are quite

FIG. 5.—The upper Shefela platform in the Lahav area. The Ziqlag Fomation (about 40 m thick) unconformably overlying Middle Eocene chalk (white patch below the cliff on the left hand side).

common and include miliolids, *Amphistegina, Archaias, Operculina, Heterostegina,* peneroplids, *Borelis melo,* and rarely *Borelis melo curdica.*

Outcrops in the Lahav area of the higher Shefela platform have the best exposures of carbonate platform facies (Figs. 1, 5, 6). The Ziqlag Formation is 50 m thick, unconformably overlies Middle Eocene chalks (Maresha Formation) and is composed of repeating cycles of carbonate units of skeletal packstones or grainstones with shell beds (molluscan rudstones) alternating with calcareous claystones with large oyster shells (Fig. 6). The lower 32 m of the section includes four such cycles, and the rest of the section is too weathered to allow clear cycle distinction. The lowermost cycle, which is also the thickest (15 m), is well preserved. The upper cycles are highly weathered.

The sharp flat contact of the Ziqlag platform with the underlying Eocene pelagic chalk is characterized by a dense network of *Thalassinoides* burrows and is overlain by 0.5-1.0 m of pelecypod packstone, with a few phosphate grains. The packstone locally exhibits faint wave ripples (Figs. 6, 7) and is overlain by 3-5 m of arenaceous skeletal packstones and grainstones showing parallel laminations, gently dipping trough cross stratification and hummocky and swaley cross stratification (Figs. 7 ,8). The hummocks and swales are amalgamated and are 1-3 m in diameter and about a 25 cm deep. Theses are overlain by up to 5-m ledge of pelecypod echinoid rudstone with most of the echinoids belonging to the genus *Clypeaster.* This lithofacies first was interpreted as a "mollusc bank" by Gvirtzman and Buchbinder (1969) and later as a beach deposit by Buchbinder (1975, 1979). Some fresh outcrops reveal a succession of upward-fining cycles, 20-30 cm (8-12 inches) thick, with shell accumulations at the bottom, gradually passing upward to finer skeletal packstones showing parallel lamination and locally hummocky cross stratification. The base of each cycle is sharp, and the graded shell layer, in places, contains pebble-size lithoclasts. The pelecypod rudstone ledge is overlain by 2.5 m of finer grainstone/packstone layer showing parallel lamina-

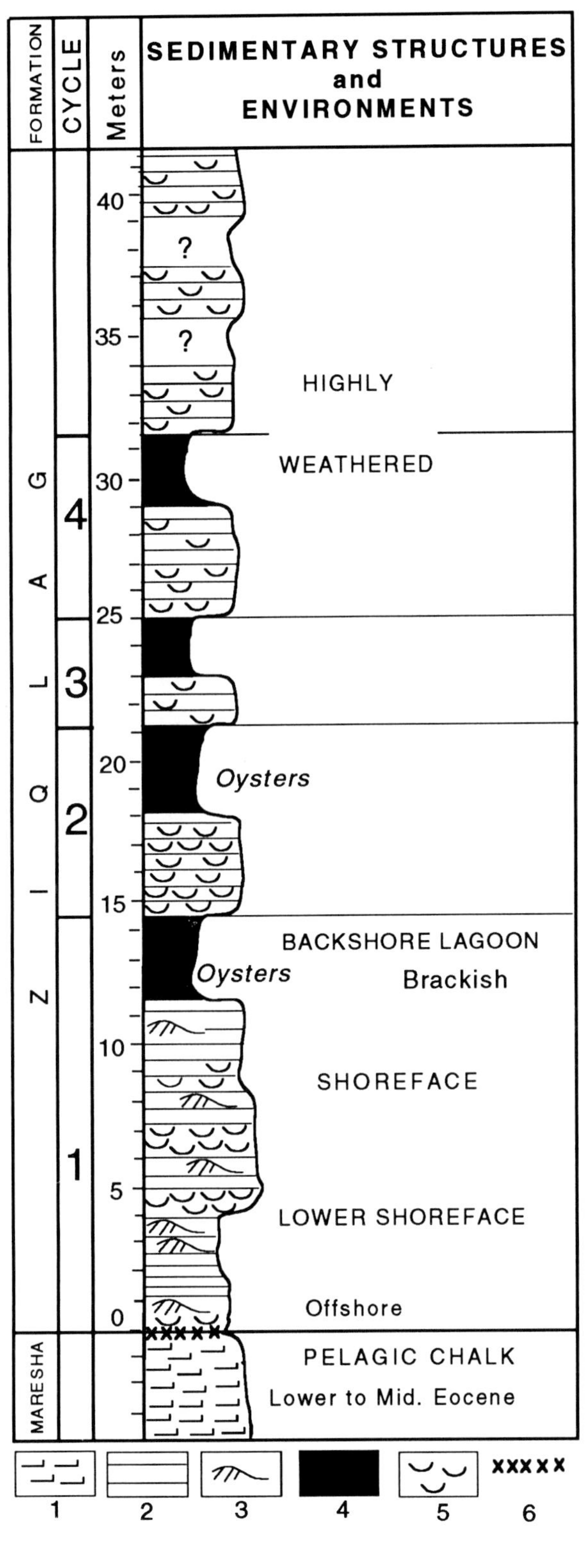


FIG. 6.—Columnar section of platform carbonates of the Ziqlag Formation in the Lahav area, showing small-scale cycles of shoreface/backshore deposits. (1) pelagic chalk; (2) parallel bedding in grainstones/packstones; (3) hummocky cross stratification; (4) calcareous claystone; (5) mulluscan and echinoderm shells; and (6) *Thalassinoides* burrows.

FIG. 7.—The lower part of the Ziqlag Formation in the Lahav area. Note the pelecypod packstone bed with faint ripple structures sharply overlying Middle Eocene chalks and overlain by arenaceous molluscan grainstones showing parallel lamination and swaley cross lamination.

tions and hummocky cross lamination.

The hummocky cross stratified arenaceous packstones are interpreted as storm deposits (Hamblin and Walker, 1979; Dott and Bourgeois, 1982; McCroy and Walker, 1986). Upward-fining cycles also characterize storm deposits; occasional storms produce intense bottom-shear conditions well below normal wave base, concentrating shells of living and dead organisms on the sea floor. Winnowing and suspension of the finer sediment by storm turbulence thus accounts for upward-fining cycles (Aigner, 1982; Kreisa, 1981; and Kumar and Sanders, 1976).

The sedimentary signature of storms, whether few or numerous, may erase most of the record of the fair-weather deposits (Kumar and Sanders, 1976). Such deposits in the area were probably composed of burrowed skeletal wackestones and packstones with a diverse fauna of soft-bottom burrowing pelecypods, grazing echinoderms, and detached to semi-attached coralline algae. The lowermost bed of the Lahav section may represent these deposits; it commonly shows faint wave-ripple laminations (Fig. 7). The storm deposits above the lowermost bed indicate a lower shoreface environment. The absence of

FIG. 8.—Hummocky cross stratification (storm deposit) in fine arenaceous molluscan grainstones of the Ziqlag Formation in the Lahav area.

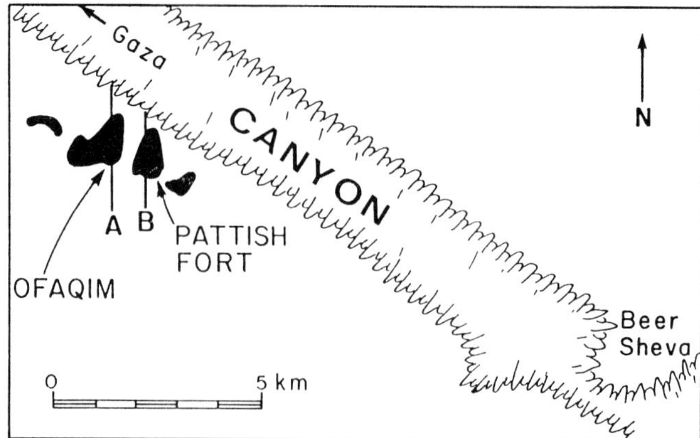


FIG. 9.—Sketch map showing Miocene reef outcrops of the Ziqlag and the Pattish Formations in the Ofaqim area on the southern bank of the Gaza-Beer Sheva canyon.

bioturbated wackestone to mudstone beds excludes the possibility of a much deeper environment of deposition (McCroy and Walker, 1986). The lower cycle is terminated by a recessive 3-m thick unit of terrigenous claystone. Except for large oyster shells, the unit is devoid of marine skeletal grains. It is interpreted to represent a backshore lagoon environment of brackish water.

The marine carbonate units in the overlying cycles are thinner and range between 1.5-3 m, thus indicating increasing regressive conditions.

*The shelf-edge reef facies.—*

The shelf edge is cut by erosional canyons that originated in the Oligocene and were rejuvenated in late Burdigalian times. Reef development along the shelf edge is patchy, and though most reefs were probably destroyed later by Late Miocene erosion, it is assumed that the reefs never formed a continuous belt.

The main outcrops of true coral reef are found in Ofaqim, on the southern bank of the Gaza-Beer Sheva canyon. A small relict outcrop is found in Yad Hanna. Other reefal sediments were found in a few boreholes along the shelf edge in Negba and Qaqun. However, because of poor micropaleontological control, it is impossible to distinguish between these reefs and the Late Miocene Pattish reef.

Thick talus aprons of reef debris and lithoclasts are found along the southeastern edge of the Ashdod canyon. The Hurvat Arema outcrop at the eastern end of the Ashdod canyon (Figs. 1, 4) shows boulders of faviid coral colonies, interbedded in *Globigerina* packstones. In the TS-5 well, a 313-m thick sequence of interbedded skeletal packstones and mudstones was penetrated. Shallow boreholes on the platform adjacent to the canyon (Gat A, B, and G) penetrated skeletal grainstones and packstones. Only a few coral specimens were found. Therefore, it is believed that the shelf edge in this area did not sustain a significant reef growth. However, the skeletal accumulation with occasional coral thickets was significant enough to provide ample material for a thick talus apron along the canyon slopes. These sediments were included in the Middle Miocene Ziqlag event based on the identification of N8-N9 planktonic foraminifera from the Horvat Arema outcrop, which is the farthest inland occurrence of talus apron sediments in the Ashdod Canyon.

The thick accumulation of corals and coarse clastics of the Upper Cretaceous Judea limestone in the Helez 23 (770 to 850-m depth) and Mivtah 1 (1017 to1167-m depth) wells (Figs. 1, 4) may also represent talus apron and not true reef buildup, as was previously thought (Buchbinder, 1975). Unfortunately, only cutting samples were available from these wells, and therefore it is difficult to ascertain the true nature of the section. The identification of *Borelis melo curdica* from the Helez 23 well (Reiss and Gvirtzman, 1966) indicates that the sediments were derived from the Middle Miocene Ziqlag limestones and not from the younger Late Miocene Pattish phase (Buchbinder et al., 1993). In the absence of clear biostratigraphic control, the age of the section in the Mivtah 1 well is uncertain. It may also belong to the Late Miocene Pattish reef.

*The Ofaqim Reef.*— In Ofaqim (Figs. 1, 9), the Middle and Late Miocene reef complexes are found in juxtaposition wherein the Late Miocene Pattish reef truncates the seaward part of the Ziqlag reef (Fig. 10). This interpretation is based on a sedimentological and micropaleontological study of outcrops and drillholes in the area (Buchbinder et al. 1993).

The Middle Miocene Ziqlag reef belt (Figs. 9, 10) in Ofaqim constitutes a series of outcrops on the southern bank of the Gaza-Beer Sheva canyon. The reef is breached by a series of channels running northward into the canyon. The exposed thickness is about 12 m, and the reef overlies Late Eocene to Oligocene marlstones. The width of the exposed reef varies between 100 and 150 m. The southern (landward) side of the reef is truncated by Pliocene to Pleistocene sandstones, and the northern (seaward) side is truncated by the Pattish reef system (Fig. 10), which removed all forereef deposits.

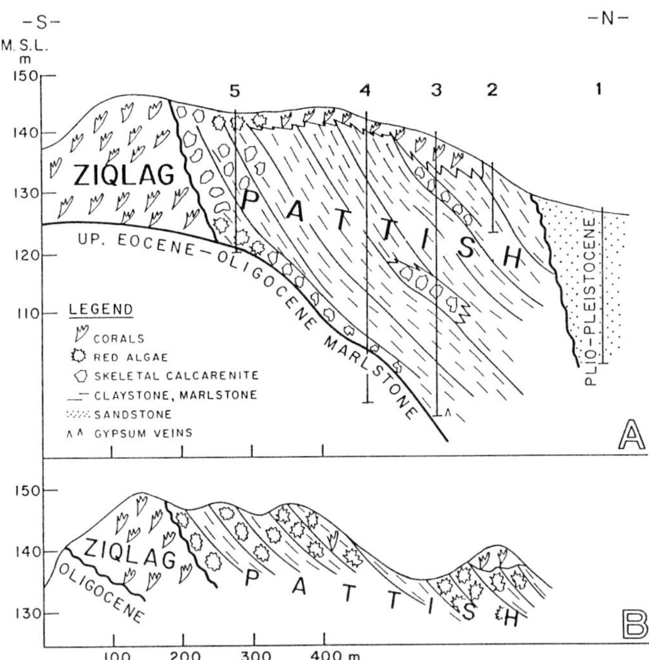


FIG. 10.—Sections A and B of Figure 9 across the Ziqlag and the Pattish reefs in the Ofaqim area. Coreholes drilled into the Pattish reef complex are marked 1 to 5. The Pattish reef truncates the forereef side of the Ziglag reef. Note progradational clinoforms of the Pattish cycle overlain by a thin veneer of poritid corals (in Section A), representing the highstand systems tract of the Pattish cycle.

Massive head-shaped faviids are the most common frame builders in the Ofaqim reef (Fig. 11A). Branching stylophorids (Fig. 11B) form local concentrations in certain areas. Poritids (Fig. 11 C) are relatively rare and occur as massive colonies. This indicates a relatively low coral diversity. In most cases, corals exhibit original growth position and are well preserved. They occupy 40-80% of the surface area of individual outcrops. Coralline algae form a significant part of the reef volume. They occur as crusts which bind and connect coral colonies and other skeletal constituents. Separate branches and rhodoliths are also found. *Lithophyllum* and *Lithoporella* are the most abundant algal genera in the reef. Encrusting foraminifera locally form alternating crusts with coralline algae; bryozoans are much less common. The space between frame builders is occupied by skeletal packstones with *Amphistegina, Operculina, Heterostegina, Borelis,* rare *Archaias* and *Elphidium.*

*Slope facies.—*

Slope sediments of the Ziqlag cycle generally consist of calcareous claystones, rich in planktonic foraminifera which are usually included in biostratigraphic units N8-N9. Slope sediments reach a maximum thickness of 240 m in the interfluve areas between the canyons (e.g., in the Nezarim 1 drillhole), but are usually much thinner, about 60-m. In the latter cases, the sediment may be very rich in planktonic foraminifera tests of unit N8, in places forming *Globigerina* ooze with glauconite grains (e.g., Beeri-west 1 well, Druckman et al., 1983). This sediment apparently represents a condensed section (Loutit et al., 1988), reflecting starved conditions during the peak of the Ziqlag cycle which probably correlates with cycle 2.3 of Haq et al. (1987, 1988).

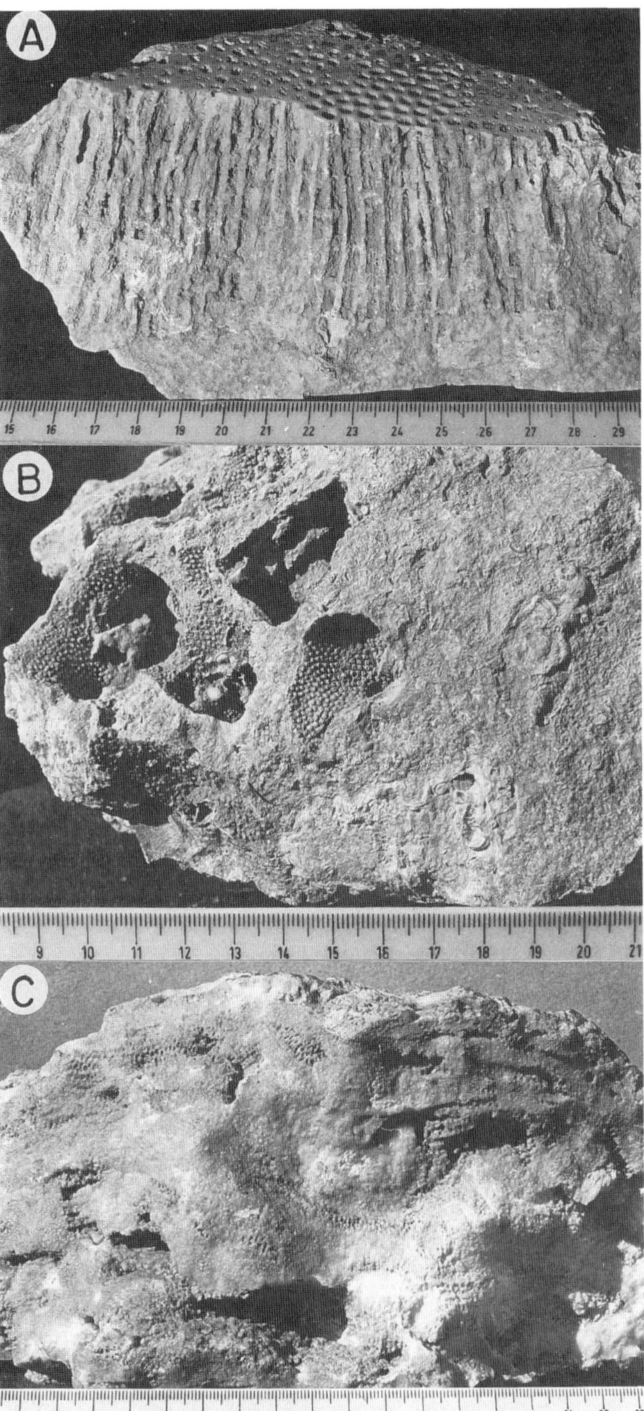


FIG. 11.—Corals from the Ziqlag Formation in Ofaqim: (A) faviid coral, (B) stylophorid coral and (C) poritid coral

A section with mass-transported sediments was found in cores taken from the Ashqelon 1 and Ashqelon 5 drillholes in the Ashqelon channel. It is composed of alternations of thinly bedded claystones and sandy claystones, highly contorted in

FIG. 12.—Clinoform beds of red algae debris and rhodoliths from the Pattish reef (Pattish Fort).

place. Boulders and pebbles of Cretaceous limestones, rhodoliths, and shallow-water carbonates are also present.

*Late Miocene Pattish Reef Phase*

The Pattish reef complex is part of the sedimentary package of units N16-N17 (the upper part of the Ziqim Formation). Sediments of this sedimentary cycle are confined to the shelf edge area and locally penetrate further inland through the erosional canyons that were rejuvenated prior to this cycle. Most of the outcrops attributed to the Pattish reef complex form either clinoforms at the upper shelf edge or thick debris flows at the slope. Only small-size coral reefs were found in outcrops. However, judging from the abundant corals in the debris sheets in the slope area, extensive coral buildups probably characterized the shelf edge and upper slope, but were stripped by later erosion.

*The shelf-edge belt.—*
The Pattish reef complex crops out in Ofaqim and in the Pattish Fort area (Fig. 9). The exposures are poor; however, a few shallow core holes were drilled in the Ofaqim outcrop (Figs. 10A), penetrating through the Oligocene substrate and providing valuable stratigraphic information. A residential complex was recently built on the outcrop; thus it is totally concealed today. On the surface of the outcrop, branching colonies of dolomitized *Porites* were found. Circular undulating outlines of individual branches (10-25 cm in diameter) could be detected on the rock surface. Coralline-algae crusts and skeletal grainstones were also found. The *Porites*-rich unit is up to 5 m thick. Below this, the drillholes penetrated a wedge of calcareous claystones (up to 42 m thick) with occasional beds of skeletal debris showing gentle dips (up to 10°) to the north. Two coreholes (4 and 5) penetrated a few meters of coralline-algae crusts and coral debris in the lower part of the sequence. Rare planktonic foraminifera from the claystone wedge in corehole 3 typify the

FIG. 13.—Calcareous claystone bed overlain by a bed of rhodolith rudstone from the Pattish reef clinoform-complex in the Pattish Fort.

N17 zone (upper Tortonian-Messinian). Notably, the 2-m-thick interval at the very bottom of core hole 3 is rich in gypsum veins and gypsum crystals.

The Pattish Fort exposure shows (Fig. 10B) steeply dipping (about 20°) clinoforms consisting of dolomitized rhodolith grainstones and packstones alternating with calcareous claystones (Figs. 12, 13). Rare poritid corals were found at the top of the inclined bed. The coralline algae in the clinoform beds consists of a single species *of Mesophyllum laffitei* (Buchbinder, 1977a).

Although the exposures are limited in lateral extent and vertical dimension, the clinoform prograding style seems to fit highstand systems tract of Esteban's et al. (this volume) model of Miocene reef sequences in the Mediterranean area. This sedimentary style is similar also to Messinian carbonate platforms in Spain and North Africa (Franseen and Mankiewicz, 1991, this volume; Saint Martin and Rouchy, 1990, Rouchy and Saint Martin, 1992). The two sections (Fig. 10) display the highstand systems tract of a 3rd- or 4th-order eustatic cycle, probably corresponding to 3rd-order cycle 3.2 of Haq et al. (1987, 1988). Clinoform progradation (Figs. 10, 12) toward the Gaza-Beer Sheva canyon is clearly evident. The thin veneer of the poritid corals on the top of the clinoforms (Fig. 10A) fits the coralgal reef phase of the highstand systems tract (Esteban, this volume) or the "middle coral reef unit" of Rouchy and Saint Martin (1992). The claystone beds in the progradational clinoform reflect some terrigenous contribution from the land in the east and possibly from a pre-Nile river system in the west.

Steeply inclined clinoforms (10°-15°) composed of sandy echinoderm grainstones were also found in an old quarry in the town of Beer Sheva. They prograde towards the center of the canyon and are overlain by a mollusk lithoclast floatstone bed with large boulders and with scouring grooves at its base. This bed is probably a debris-flow deposit. It was previously referred to as "Ziqlag Tongue" (Gvirtzman and Buchbinder, 1969, Martinotti et al., 1978). Based on biostratigraphic considerations, however, Buchbinder et al. (1993) attributed the section of the Beer Sheva old quarry to the Late Miocene Pattish cycle.

Other occurrences of the Pattish reef unit are known from drillholes on the banks of the Gaza-Beer Sheva canyons (Saad 4, Sharsheret 1, Beeri 1, and Beeri-old). Only cutting samples are available from these drillholes. Poritid corals, coralline algae, and skeletal packstones and grainstones were identified in thinsections. Most samples are highly dolomitized, and the thickness of the reef units varies from 10 m in Sharsheret 1 to 90 m in Saad 4.

The Pattish reef phase is also known from northern Sinai where it forms a continuous belt up to 200 m thick a few hundred meters below sea level (see also Jenkins, 1990). A core sample from the Sneh 1 drill hole (Fig. 1) shows rhodolith pavement facies with a few poritid corals.

*The slope belt.—*

The Pattish reef is included within the hemipelagic marlstone unit (upper part of the Ziqim Formation) spanning N16 to N17 in the slope area. Unit N15 is absent in most wells and unit N16 unconformably overlies sediments of unit N14. The thickness of the N16-N17 sediments ranges from up to 50 m in the erosional canyons to 210 m outside the canyons, pointing to submarine erosion processes in the canyons. Cutting samples from drillholes in the canyons generally yield an assortment of coral, algal and skeletal bioclasts together with lithoclasts of the Cretaceous Judea limestones; whereas, outside the canyons the unit consists of hemipelagic claystones only. The mixture of bioclast-lithoclast deposits and pelagic marlstone in the canyons are considered debris flows channeled by mass flows through the submarine canyons. They are found along the Ashqelon canyon system in the Ashqelon 4 and Talme Yafe 4 wells, in the Barnea canyon (Barnea 1 well), along the Ashdod canyon in the Ashdod 1, 2, 3 and 5 wells, in the Hof-Ashdod 1 well and the offshore Joshua 2 well. These deposits were previously regarded as part of the shallow marine reef facies of the Ziqlag Formation (Derin and Reiss, 1973, Buchbinder, 1975, 1979; Gvirtzman and Buchbinder, 1978). Considering the "Ziqlag" sediments on the slope as an organic *in-situ* reef implies dramatic up and down "yo-yo tectonics" of 2000-m amplitude within a short geologic time span and the formation of a continuous shallow marine reef sheet along 2000-m slope (Gvirtzman and Buchbinder, 1978). However, upon reconsideration, such a process is highly inconceivable. The present interpretation considers the slope area as a predominant feature in the Miocene times (see also Tibor et al. 1992) and the canyons cutting it are considered a result of submarine erosion processes followed by clastic deposition of mass flows.

## LATE MIOCENE EVAPORITES

Late Miocene evaporites of the Mavqiim Formation are mostly confined to the slope area (Fig. 2), where they either sharply overlie debris sheets of the Pattish cycle in the Ashdod, Ashqelon and Barnea canyons or hemipelagic marlstones of N16-N17 age. They, therefore, must have been deposited in a relatively deep-water basin (Buchbinder et al. 1993). Gypsum deposits are also found on the platform along the southern bank of the Gaza-Beer Sheva canyon in Beeri. Dolomitization of the Pattish reef on the shoulders of the Gaza-Beer Sheva canyon could be attributed to the evaporation phase (see also Oswald et al., 1990). However, reef debris along the canyons in the slope area, although directly overlain by anhydrites, are not dolomitized. The gypsum crystals and veins in the lowermost part of the Pattish reef complex in Ofaqim could be related to a marginal occurrence of the Messinian evaporites on the canyon's flanks. However, additional borehole information in the area is needed to verify this.

Cohen (1987, 1993) and Cohen and Parchamovsky (1986) indicated an onlap of the evaporites on to older formations. Although they related the onlapping relations to a westward tectonic tilt, sea-level rise during cycle 3.3 as postulated by Saint-Martin and Rouchy (1990), Rouchy and Saint Martin (1992) and by the 4th-order oscillations suggested by Esteban (this volume) could also explain the onlap features of the Late Miocene evaporites in the subsurface of the coastal plain of Israel.

## DIAGENETIC OVERPRINT AND POROSITY

### *Biogenetic Constituents*

*The early Middle Miocene Ziqlag carbonates.—*

The microfabric of both high magnesium calcite constituents (such as coralline algae, benthonic foraminifera, and echinoderms) and of calcite constituents (such as ostreids, pectinids, and bryozoans) is well preserved. Aragonite constituents such as pelecypods and gastropods are usually leached and their molds usually filled with blocky calcite. Aragonite shells of pelcypods and gastropods from the platform remained empty or partially filled, resulting in very high moldic porosity.

Corals reflect different diagenetic histories, following their taxonomic grouping (Buchbinder, 1977b, 1979). Colonies of branching stylophorids are completely leached and appear as empty molds of whole colonies (Fig. 11B). In the poritids, aragonite walls were completely leached and refilled with blocky calcite, and their detailed outlines are preserved by micritic envelopes. In other cases, (e.g., in coral debris of the Barnea 1 well), poritid coral walls exist as empty molds while the primary coralline voids are filled with micrite, exhibiting an extremely high porosity. Faviid-coral walls commonly show dissolution and reprecipitation on a fine scale, preserving some relict wall structures. In other occurrences, the coral wall is completely leached and is either partially or totally filled with blocky calcite. Calcite cement, filling primary voids of the faviids, commonly displays two generations: (1) calcitized needle cement lining the coral's primary cavities and (2) late blocky cement filling space left by the needle-cement, though locally leaving some porosity at the center of the cavities.

*The Late Miocene Pattish carbonates.—*
The Pattish reef is characterized by dolomitization which becomes more intensive toward the basinal side in the Ofaqim area. $\delta^{18}O$ values of the dolomites range between +2.2‰-+3.5‰, averaging +3‰ (Magaritz, 1973). $\delta^{13}O$ values range between 0‰- +3‰. The oxygen isotope values indicate dolomitization from evaporated sea water (Magaritz, 1973). Evidently, the higher (landward) part of the clinoform succession escaped dolomitization, presumably because of its exposure to a freshwater phreatic or vadose environment, whereas the lower part was subjected to hypersaline fluids of the post-reef evaporative phase, supporting the model of Oswald et al. (1990). Coralline algae and micritic envelopes are preferentially dolomitized. In some cases the outlines of the pore system of poritid corals is well preserved by dolomitized micritic envelopes, whereas primary cavities and leached wall cavities show late-stage filling by calcite. Dolomitization of coralline algae mimics the original fine skeletal architecture, which is better preserved than in calcified algae (Buchbinder, 1979). However, in advanced stages of dolomitization, original microstructures of biogenic constituents are almost totally obliterated and the rocks show fine to coarse mosaics of dolomite with intercrystalline porosity and moldic porosity after molluscan shells.

## DISCUSSION AND CONCLUSIONS

Miocene carbonate deposition on the eastern Mediterranean margin occurred during two short-term high sea-level events in the early Middle Miocene and Late Miocene times. Both carbonate phases were confined to the periphery of the basin, though the early Middle Miocene Ziqlag carbonates penetrated much further inland. The latter was due to notably high sea levels. High sea level in early Middle Miocene times is also implied by a condensed section of planktonic-foraminifera "siltstone" (of zone N8) with glauconite grains in the upper reaches of the slope (Beeri-west 1 well, Druckman et al., 1983).

Clastic sedimentation prevailing in the slope area during the deposition of the two reefal phases, originated from two sources: the Nile precursors, which provided mostly prodelta clays and silts; and a local continental drainage system, which provided coarser clastics. Sediments from the latter source moved mainly as gravity flows along the submerged canyons and were deposited in the distal parts of the canyons as debris sheets developing basinward into submarine fans. In this fashion clastic sediments bypassed the shelf area, thus facilitating the development of coral reefs.

The Ziqlag platform does not exhibit typical tropical chloralgal (or chlorozoan) characteristics. The common grains of coralline algae, pelecypods, echinoderms and benthic foraminifera fit rhodalgal to molechor lithofacies (Carannante et al., 1988). Although this facies co-existed with chloralgal facies in Ofaqim, coral diversity in the Ofaqim reef is relatively low and may already reflect the gradual cooling trend towards Late Miocene times (Chevalier, 1977). Other authors postulated both cooling (Bizon, 1981, Demarcq's, 1989 Serravalian crisis) and lowering of surface water salinity (Van der Zwaan and Gudjonsson, 1986) in the Mediterranean in Middle Miocene (Serravalian) times. This could be the reason for the predominace of "temperate" rhodalgal lithofacies on the platform. The stratigraphic unit overlying the Ziqlag unit in the slope area (the lower part of the Ziqim Formation) spans the N10-N13 zones (Figs. 2, 3). These zones correspond to the *G. foshi* lineage zone. Derin (in Horowitz and Derin, 1987, in Goldsmith et al., 1988) noted the absence of the N10-N13 zones in Israel. Martinotti (1981, 1990) claimed that although the *G. foshi foshi* lineage is absent in Israel, a parallel biostratigraphic unit, bounded at its base by the last occurrence of *G. foshi peripheroronda* and at its top by the last Middle Miocene occurrence of *G. ruber*, can be distinguished. This unit is characterized by reduced salinities, as evidenced by nannoplankton distribution (Moshkovitz and Ehrlich, 1982) and by the reduction in number and size of *Cassigerinella chipolensis* (Martinotti, 1989). In fact, a freshening episode of the Mediterranean surface layer in this period was proposed based on stable isotopes (van der Zwaan and Gudjonsson, 1986). Such an episode could also explain the interrupted evolution of the *G. foshi* lineage.

The Ziqlag platform is marked by abundant storm deposits which dominate the lower part of the section in Lahav. This indicates an open shelf setting, where reefs at the shelf edge, if they existed, did not obstruct storm impact. Thus, the higher Shefela may be regarded as a distally steepened ramp, similar to other rhodalgal ramps which are quite extensive in the Miocene of the Mediterranean (Pomar and Rodriguez-Perea, 1983; Carannante, this volume; Pedley, this volume).

### *Sequence Stratigraphy and Geodynamics*

Judging from the distribution of planktonic foraminifera in the transitional shelf-slope outcrops (Upper N8-N9, Buchbinder et al., 1993), the Ziqlag carbonates represent a relatively short time-span of 1-1.5 my. In terms of sequence stratigraphy this time span corresponds to Haq's *et al.* (1987, 1988) third order cycles 2.3 and the lower part of cycle 2.4 (Fig. 3). It is possible that the two platforms of the higher and lower Shefela represent these two third-order cycles, of which the higher Shefela platform reflects the older (2.3) cycle. The section of the higher Shefela platform in Lahav shows at least four smaller-scale cycles (Fig. 6). Similar cycles are considered as fundamental elements of carbonate platforms (Schwarzacher, 1987, Wanless, 1991). They typically have a duration of $10^4$-$10^5$ years (Goldhammer, 1990). The small-scale cycles in the Lahav section are therefore possibly of 4th or 5th order. The reduction in the thickness of the marine (shoreface) carbonate units in the upper cycles (Fig. 6) signifies increasing regressive conditions and progradational stacking pattern in the platform, thus representing the highstand systems tract of the third-order cycle. The section in the Lahav area begins with a dense network of *Thalassinoides* burrows. The superimposed nature of the burrows and their concentration at this stratigraphic level only, indicates conditions of sediment starvation at the base of the section. The overlying pelecypod packstone bed lacks sedimentary structures of traction transport. The overlying sediments show wave ripples (Fig. 7) which are followed by hummocky cross stratification, indicating shallowing to storm-wave base. These features indicate that the base of the Lahav section marks the maximum flooding surface of the third-order cycle and that the *Thalassinoides* network represents a condensed section. Sediments of the transgressive systems tract may have been distributed west of Lahav, but were probably abraded by the successive cycle.

The 80-m elevation difference between the higher and lower Shefela platforms is explained by the combination of (1) a continuous tectonic process of westward tilting and (2) cyclic eustatic processes of two transgressive events of different 3rd-order cycles. During the first cycle (2.3), a single platform (or distally steepend ramp) probably existed in the Shefela area. Sea-level drop at the end of the first cycle resulted in the exposure of the platform (or ramp), while its eastern distal part was further uplifted because of westward tectonic tilt. When sea level rose again at the beginning of cycle 2.4, the higher Shefela remained emerged and an abrasional sea cliff was formed marking the boundary between the lower and higher Shefela (see also Buchbinder et al., 1993) and eroding the proximal part of the higher Shefela ramp. Thus, although the storm-dominated Ziqlag outcrops imply a ramp setting, the sediments at the proximal end of this ramp were eroded and therefore, cannot, be studied to verify this assumption. The reef in Ofaqim, however, which is situated on the western (proximal) side of the lower Shefela platform, corresponds, probably, to the shelf edge of this younger platform.

The limited exposures of Late Miocene carbonates makes the reconstruction of this phase difficult. Based on the distribution of planktic foraminifera (Tortonian-Messinian, N16-N17), the Pattish sediments in Ofaqim and Beer Sheva areas corresponds to third-order cycle 3.2. This cycle predated the Messinian desiccation of the Mediterranean and is characterized by an increase in terrigenous supply, by a strong progradational deposition and by limited eastward distribution compared to the previous Ziqlag cycle. These characteristics may be explained by (1) the continuous uplift of the basin margins, resulting in increased clastic shedding into the basin and (2) by lower sea levels compared to the Middle Miocene cycles. It is not clear whether the sections in Ofaqim and the Pattish Fort (Fig. 10) represent a third-order cycle, or a higher-order cycle of a series of down-stepping reefs, as known from Late Miocene reefs in the Western Mediterranean (Pomar, 1991, Rouchy and Saint Martin, 1992). The section in Beer Sheva with the steep clinoforms and debris-flow floatstone deposits indicates the existence of a shallower water platform adjacent to the Gaza-Beer Sheva canyon, which provided the skeletal material to the prograding clinoforms and to the debris flows that gradually filled the canyon.

Comparison between the spatial distribution of Miocene sediments in the coastal plain and the foothills area (as depicted in Figure 3) and Haq et al.'s (1987, 1988) coastal onlap curve show many differences. The absence of marine deposits from the shelf (foothills) area in Lower Miocene epoch and in most of the Serravalian (Middle Miocene) age is striking in the light of Haq's et al. (1987, 1988) relatively high sea levels at these time intervals. On the other hand, global sea levels during cycle 3.2 which correspond to the deposition of the Pattish reef on the shelf are lower than those corresponding to the Early and Middle Miocene hiatuses on the shelf.

The eastern Mediterranean continental margin is a mature passive margin, and the subsidence during most of the Miocene times followed a smooth path characteristic of thermal subsidence (Tibor et al. 1992). However, the close proximity of the plate boundary to the African-Eurasian collision zone in the northwest and the Dead Sea Transform to the east, may have resulted in deviation from a smooth subsidence path. Tectonic uplifting of the shelf area during Early Miocene and Middle Miocene (Serravalian) times may, therefore, account for the interrupted succession of marine sedimentation during the Miocene epoch.

The Serravalian sea-level curve of the Mediterranean may have departed from that of the global sea level. The indication of salinity stratification at that time (Van der Zwaan and Gudjonssen, 1986) may indicate a departure from global sea level. The occurrence of Serravalian evaporites in some Mediterranean localities (Esteban, this volume) may also indicate restriction of the inflow/outflow exchange with the Atlantic, when the connection with the Indian Ocean and the Paratethys was already severed because of plate convergence (Buchbinder, this volume). This may also explain the termination of the Ziqlag reefal phase in early Serravalian (N10) and the absence of N10-N13 sediments in the shelf area.

## ACKNOWLEDGMENTS

I thank C. G. Adams of the British Museum (Natural History) for identifying the larger foraminifera, G. M. Martinotti and R. Siman-Tov of the Israel Geological Survey for establishing the planktic foraminifera biostratigraphy, and C. Jordan, M. Colgan, M. Esteban, E. Franseen and C. Mankiewicz for their constructive reviews.

## REFERENCES


AIGNER, T., 1982, Calcareous tempestites storm-dominated stratification in Upper Muschelkalk limestones (Middle Trias, SW Germany), *in* Einsele, G. and Seilacher, A., eds., Cyclic and Event Stratification: Berlin, Springer-Verlag, p. 180-198.

BIZON, G., 1981, Variation climatique dans le domaine méditerranéen au cours de Néogène. Quelques remarques deduites de la répartition des fossiles marines: Annales Géologiques Des Pays Helleniques, Hors Serie, v. 4, p. 194-198.

BUCHBINDER, B., 1975, Lithogenesis of Miocene reef limestones in Israel with particular reference to the significance of the red algae: Jerusalem, Israel Geological Survey, Report OD/3/75, 173 p.

BUCHBINDER, B., 1977a, Systematics and paleoenvironments of the calcareous algae from the Miocene (Tortonian) Ziqlag Formation, Israel: Micropaleontology, v. 23, p. 415-435.

BUCHBINDER, B., 1977b, Different responses to diagensis of various coral groups in the Miocene Ziqlag Formation, Israel: Orléans, Bureau Recherches Geologiques et Minieres, Mémoire 89, p. 26-32.

BUCHBINDER, B., 1979, Facies and environments of Miocene reef limestone in Israel: Journal of Sedimentary Petrology, v. 49, p. 1323-1344.

BUCHBINDER, B., MARTINOTTI, G. M., SIMAN-TOV, R., and ZILBERMAN E., 1993, Temporal and spatial relationships in Miocene reef carbonates in Israel: Palaeogeography, Palaeoclimatology, Palaeoecology, v. 101, p. 97-116.

CARANNATE, G., ESTEBAN, M., MILLIMAN, J. D., AND SIMON, L., 1988, Carbonate lithofacies as paleoaltitude indicators: problems and limitations: Sedimentary Geology, v. 60, p. 333-346.

CHEVALIER, J. P., 1977, Apercu sur la faune corallienne recifale du Néogène: Orléans, Bureau Recherches Geologiques et Minieres, Mémoire 89, p. 359-366.

COHEN, A., 1987, Subaqueous origin of the Mavqiim anhydrites (Messinian): Jerusalem, Israel Geological Survey, Report GSI/15/87, 81 p.

COHEN, A., 1993, Halite-clay interplay in the Israeli Messinian: Sedimentary Geology, v. 86, p. 211-228.

COHEN, A. and PARCHAMOVSKY, S., 1986, The Mavqiim Formation of Israel - a series of onlapping evaporites of Messinian age: Jerusalem, Israel Geological Survey, Report GSI/19/86, 31 p.

DEMARCQ, G., 1989, Nemesis and the Serravallian crisis in the Mediterranean: Annales Géologiques Des Pays Helleniques, v. 34, p. 1-8.

DERIN, B. and REISS, Z., 1973, Revision of marine Neogene stratigraphy in Israel: Israel Journal of Earth Science, v. 22, p. 199-210.

DOTT, R. H. and BOURGEOIS, J., 1982, Hummocky stratification: significance of its variable bedding sequences: Bulletin of the Geological Society of America, v. 93, p. 663-680.

DRUCKMAN, Y., MARTINOTTI, G. M. and SIMAN-TOV, R., 1983, The Late Cretaceous-Neogene stratigraphy of the Beeri west-1 well borehole: Jerusalem, Israel Geological Survey, Report S/8/83, p. 1-12.

FRANSEEN, E. K. and MANKIEWICZ, C., 1991, Depositional sequences and correlation of middle to late Miocene carbonate complexes, Las Negras and Nijar areas, southeastern Spain: Sedimentology, v. 38, p. 871-897.

GELBERMAN, E. and GROSSOWICZ, Y., 1990, Seismic, structural, and isopach maps: central, southern, and offshore Israel: Holon, Israel Petroleum and Infrastructure Corporation, Document 3395a.

GOLDHAMMER, R. K., 1990, Depositional cycles, composite sea-level changes, cycle stacking patterns and the hierarchy of stratigraphic forcing - examples from Alpine Triassic platform carbonates: Bulletin of Geological Society of America, v. 102, p. 535-562.

GOLDSMITH, N. F., HIRSCH, F., FRIEDMAN, G. M., TCHERNOV, E., DERIN B., GERRY, E., HOROWITZ, A., and WEINBERG, G., 1988, Rotem mammals and Yeroham Crassostreids: stratigraphy of the Hazeva Formation (Israel) and the paleogeography of Miocene Africa: Newsletters on Stratigraphy, v. 20, p.73-90.

GVIRTZMAN, G., 1970, The Saqiye Group (Late Eocene to Early Pleistocene) in the Coastal Plain and Hashephela regions, Israel: Jerusalem, Israel Geological Survey, Report OD/5/67, 170 p.

GVIRZTMAN, G. and BUCHBINDER, B., 1969, Outcrops of Neogene formations in the central and southern Coastal Plain, Hashephela and Beer Sheva regions, Israel: Jerusalem, Israel Geological Survey, Bulletin, no. 50, 73 p.

GVIRTZMAN, G. and BUCHBINDER, B., 1978, The late Tertiary of the Coastal Plain and continental shelf of Israel and its bearing on the history of the eastern Mediterranean, *in* Ross, D. A. and Neprochnov, Y. P., eds., Initial Reports: Deep Sea Drilling Project, v. 42B, p. 1195-1222.

HAMBLIN, N. A. and WALKER, R. G., 1979, Storm-dominated shallow marine deposits: the Fernie-Kootenay (Jurassic) transition, southern Rocky Mountains: Canadian Journal of Earth Sciences, v. 16, p. 1673-1690.

HAQ, B. V., HARDENBOL, J., and VAIL, P. R., 1987, Chronology and fluctuating sea levels since the Triassic (250 millions years ago to present): Science, v. 235, p. 1156-1167.

HAQ, B. H., HARDENBOL, J., and VAIL, P. R., 1988, Mesozoic and Cenozoic chronostratigraphy and cycles of sea-level change, *in* Wilgus, C. K., Ross, C. A., Posamentier, H., Wagoner, J. van, and Kendall, G. St. C., eds., Sea-level Changes: An Integrated Approach: Tulsa, Society of Economic Paleontologists and Mineralogists Special Publication 42, p. 71-108.

HOROWITZ, A. and DERIN, B., 1987, Palynological correlations of continental and marine Neogene sequences in Israel: Annales Instituti Geologici Publici Hungarici, v. 60, p. 503-507.

JENKINS, D. A., 1990, North and central Sinai, *in* Said, R., ed. The Geology of Egypt: Rotterdam, A. A. Balkema, p. 361-380.

KREISA, R. D., 1981, Storm-generated sedimentary structures in subtidal marine facies with examples from the Middle and Upper Ordovician of southwestern Virginia: Journal of Sedimentary Petrology, v. 51, p. 823-848.

KUMAR, N. and SANDERS, J. E., 1976, Characteristics of shoreface storm deposits: Modern and ancient examples: Journal of Sedimentary Petrology., v. 46, p. 145-162.

LOUTIT, T. S., HARDENBOL, J., VAIL, P. R., and BAUM, G. R., 1988, Condensed sections: the key to age dating and correlation of continental margin sequences, *in* Wilgus, C. K., Ross, C. A., Posamentier, H., and Kendall, C. G. St. C., eds., Sea-level Changes: An Integrated Approach: Tulsa, Society of Economic Paleontologists and Mineralogists, Special Publication 42 p.183-216.

MAGARITZ, M., 1973, Carbon and oxygen isotopic composition of some carbonate rocks from Israel: Unpublished Ph.D. Dissertation, The Weizmann Institute of Science, Rehovot, 158 p.

MARTINOTTI, G. M., 1981, Biostratigraphy and planktonic foraminifera of the Late Eocene to ?Pleistocene sequence in the Ashqelon 2 well (southern coastal plain, Israel): Revista Espanola Micropaleotologia, v. 12, p. 343-381.

MARTINOTTI, G. M., 1989, The last occurrence of *Classigerinella chipolensis* in the Mediterranean region: Journal of Foraminiferal Research, v. 19, p. 180-184.

MARTINOTTI, G. M., 1990, The stratigraphic significance of *Globigerinoides ruber* and *Globigerinoides obliquus obliquus* in the Mediterranean Middle Miocene: Micropaleontology, v. 36, p. 96-101.

MARTINOTTI, G. M., GVIRTZMAN, G., and BUCHBINDER, B., 1978, The Late Miocene marine transgression in the Beer Sheva area: Israel Journal of Earth Sciences, v. 27, p. 72-82.

MCCROY, V. L. C. and WALKER, R. G., 1986, A storm and tidally-influenced prograding shoreline-Upper Cretaceous Milk River Formation of southern Alberta, Canada. Sedimentology, v. 33, p.47-60.

MOSHKOVITZ, S. and EHRLICH, A., 1982, Biostratigraphical problems of the Middle Miocene calcareous nannofossils and the paleoecological significance of the braarudosphaerids in the Coastal Plain and offshore of Israel: Jerusalem, Israel Geological Survey, Current Research, 1981, p. 43-47.

OSWALD, E. J., MEYERS, W. J., and POMAR, L., 1990, Dolomitization of an Upper Miocene reef complex, Mallorca, Spain: evidence for Messinian dolomitizing sea: American Association of Petroleum Geologists Bulletin, v. 74, p. 735.

POMAR, L., 1991, Reef geometries, erosion surfaces and high-frequency sea-level changes, Upper Miocene reef complex, Mallorca, Spain: Sedimentology, v. 38, p. 243-269.

POMAR, L., and RODRIGUEZ-PEREA, 1983, El Neogeno inferior de Mallorca:

Randa, *in* Pomar, L., Obrador, A., Fornós, J., and Rodriguez-Perea, A., eds., El Terciario de la Baleares (Mallorca-Menorca): Guia de las Excursiones del X Congreso Nacional Sedimentologia, Instituto Estatal Balearics and Universidad Palma de Mallorca, p. 115-137.

REISS, Z. and GVIRTZMAN, G., 1966, *Borelis* from Israel: Eclogae Geologicae Helvetae, v. 59, p. 438-447.

ROUCHY, J. M. and SAINT MARTIN, J. P., 1992, Late Miocene events in the Mediterranean as recorded by carbonate evaporite relations: Geology, v. 20, p. 629-632.

SAINT-MARTIN, J. P. and ROUCHY, J. M., 1990, Les plate-formes carbonatées messiniennes en Méditerranée occidentale: leur importance pour la reconstitution des variations du niveau marin au Miocene terminal: Bulletin Géologique de France, v. 6, p. 83-94.

SCHWARZACHER, W., 1987, The analysis and interpretation of stratification cycles: Paleooceanography, v. 2, p, 79-95.

TIBOR, G., BEN-AVRAHAM, Z. and FLIGELMAN, H., 1992, Late Tertiary subsidence history of the southern Levant margin, eastern Mediterranean Sea, and its implications to the understanding of the Messinian event: Journal of Geophysical Research, v. 97, p. 17,593-17,614.

Van der ZWAAN, G. J. and GUDJONSSON, L., 1986, Middle Miocene-Pliocene stable isotope stratigraphy and paleoceanography of the Mediterranean: Marine Micropaleontolgy, v. 10, p. 71-90.

WANLESS, H., R., 1991, Observational foundation for sequence modeling, *in* Franseen, E., Lynn Watney, W., Kendal, C. G. St. C., and Ross, W., eds., Sedimentary modeling: computer simulations and methods for improved parameter definition: Kansas Geological Survey Bulletin, v. 233, p. 43-62.

# MIOCENE REEFS OF THE NORTHWEST RED SEA

BRUCE H. PURSER, JEAN-CLAUDE PLAZIAT,
*University of Paris Sud, Bât. 504, F-91405 Orsay, France*
AND
BRIAN R. ROSEN
*Natural History Museum, Cromwell Road, London 5W7 5BD, United Kingdom*

ABSTRACT: Numerous, Early to Middle Miocene reefs cropping out along the northwest (Egyptian) coast of the Red Sea have variable geometries which were determined mainly by tectonics and relative sea-level changes. Distributed between Abu Ghusun in the south to Esh Mellaha in the north, these Burdigalian-Langhian reefs are frequently multiple, most being located near the crests of structural blocks. Several have grown on submarine fans. Although one reef complex (Sharm el Qibli) slopes to the west, most reefs face towards the rift axis. Stacking of individual reef bodies within any given complex may be essentially lateral, the reef at Sharm el Luli exhibiting clear down-stepping relating to structural uplift. Others (Zug el Bohar) show marked lateral accretion while the spectacular complex at Abu Shaar el Qibli is mainly vertical. Because of repeated Miocene faulting, certain reef bodies are preserved mainly in the form of olistoliths (Abu Ghusun) or as coarse reef debris deposited on the periplatform talus, as on the eastern flank of Abu Shaar el Qibli.

Reef distribution appears to have changed during rift evolution. Early Miocene reefs are located on structural blocks separated from the western periphery of the rift. However, with progressive deepening of the Red Sea rift, blocks and associated reefs have been drowned although less subsidence within the Gulf of Suez has favoured the local survival of offshore reefs. Elsewhere, modern reefs fringe much of the continental shoreline.

## INTRODUCTION

Early Miocene reefs of the northwest Red Sea occur both at outcrop and in the adjacent subsurface where they are locally oil productive. This study is based on a series of spectacular exposures whose quality enable a three dimensional study of most reef bodies. Individual reefs, although generally small, are of economic importance, especially when associated with adjacent platform sediments. Their geometry is highly variable being conditioned mainly by the pre- and synreef tectonic movements. Of the 10 reefs studied, the authors have selected 6 to demonstrate the major variations.

### *Previous Work*

Following the pioneer regional studies of Gregory (1906) which included the first descriptions of corals from the Abu Shaar-Esh Mellaha reefs, the synrift Neogene sediments of the Gulf of Suez and the northwest Red Sea have been described briefly by Heybroek (1965), Rouchy (1982), Rouchy et al. (1983) and by Haddad et al. (1984). These initial studies concerned the general organization of the platform and did not consider the diagenetic or petrophysical properties of the reef bodies and related platform carbonates. The first detailed descriptions were those of Aissaoui et al. (1986), Coniglio et al. (1988), James et al. (1988) and Monty et al. (1987) who associated both the ecological and diagenetic evolution of the Miocene platform at Abu Shaar el Qibli (Fig. 1). While the present authors do not necessarily agree with certain aspects of these contributions (see subsequent paragraphs), nevertheless they are the only relatively synthetic studies of a Miocene reef complex in the northern Red Sea region. However, regional studies of the Egyptian northwest Red Sea coast (Montenat et al. 1986) have shown that many reefs occur between the Gulf of Suez and Sudan. Some of these have also been discussed briefly by Purser et al. (1990) and by Plaziat et al. (1990a) who emphasised the variable composition and geometry of these bodies

The present contribution evaluates a series of reef bodies, including two which have never been described, stressing the relationships between sedimentation and structural framework.

### *Location*

The reefs discussed (Fig. 1) are kilometer-sized bodies distributed between the southern Gulf of Suez and Abu Ghusun, some 350 km to the south. Readily accessible by the coastal road, these reefs and associated Neogene synrift sediments form a narrow band (1 to 5 km) of low hills (50 to 150 meters) situated on the edge of higher (1000 to 1500 m) Late Precambrian-Paleozoic basement relief. Similar bodies occur offshore where they have been buried during subsidence of the axial parts of the rift system.

### *Stratigraphic Sequence*

The Miocene reefs are a characteristic element of the syn-rift sequence (Fig. 2). Following Late Oligocene, proto-rift continental sedimentation (Orszag-Sperber and Plaziat, 1990, Plaziat et al., 1990c), the Early Miocene transgression penetrated depressions in a marked Miocene morphology whose various structural blocks, to a large extent, determined the local stratigraphic succession (Plaziat et al., 1990a). In structural lows (generally half-grabens), early Miocene marine sedimentation was terrigenous. Reefs developed during limited episodes, occasionally on the crests of the detrital submarine fans and more typically on structural highs. Reefs rarely occur directly on

Models for Carbonate Stratigraphy from Miocene Reef Complexes of Mediterranean Regions,
SEPM Concepts in Sedimentology and Paleontology #5, Copyright © 1996,
SEPM (Society for Sedimentary Geology), ISBN 1-56576-033-6, p. 347-365.

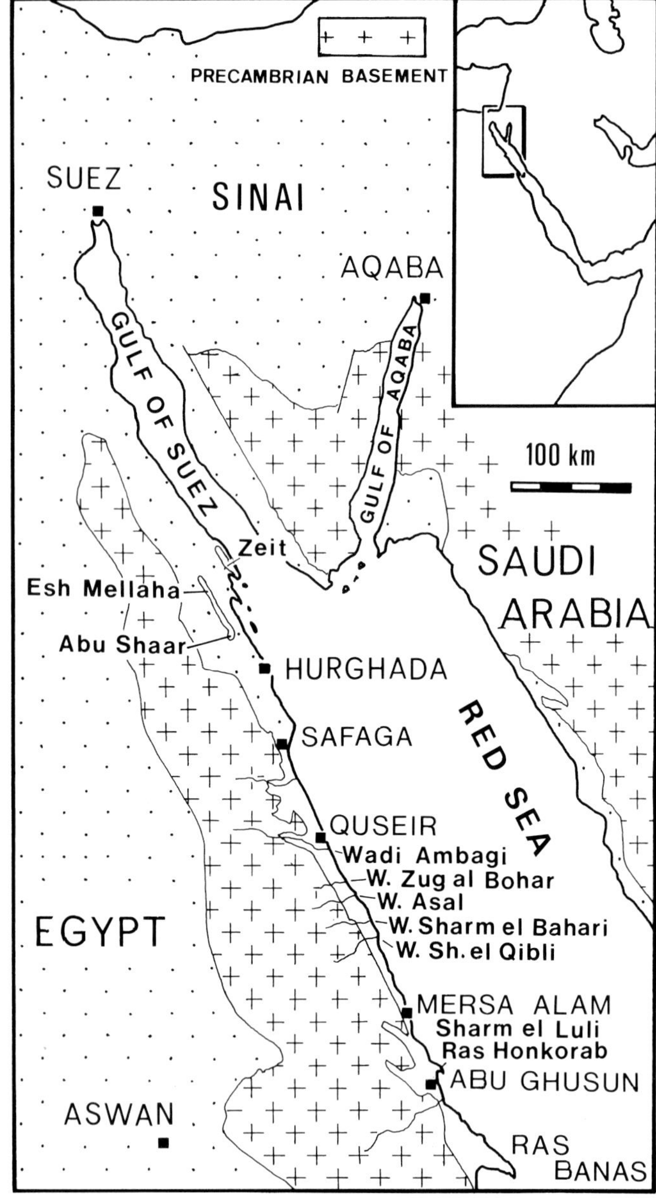


FIG. 1.—Map of the northwest Red Sea region showing the distribution of reefs discussed in this contribution.

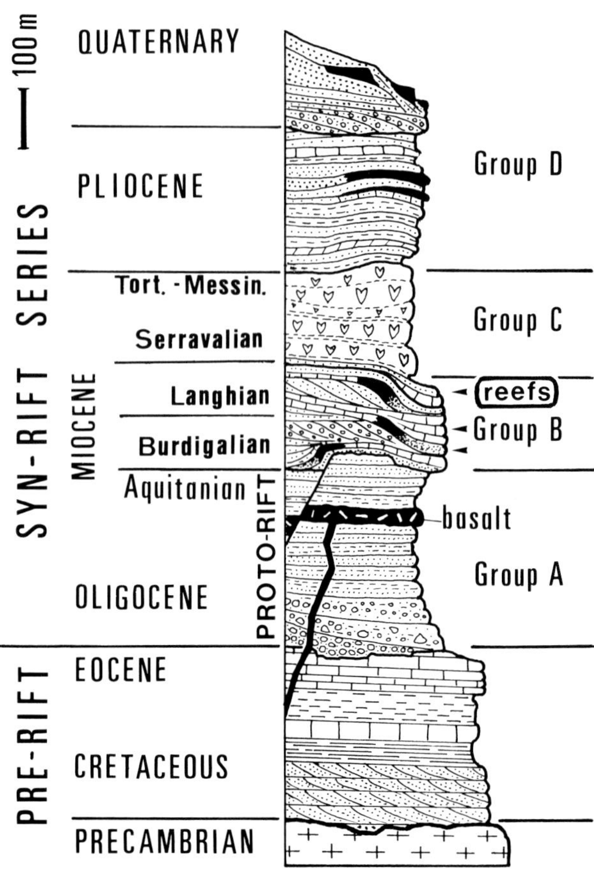


FIG. 2.—Generalized stratigraphic section showing the location of reefs within the synrift sediments of the northwest Red Sea; the present study concerns only those of Burdigalian-Langhian age.

Paleozoic basement but on relatively thin (1 to 20 m) transitional sands or conglomerates. In both low and high structural settings, reefs and related carbonates generally are overlain by a thick (10-50 m) series of irregularly laminated dolomicrites. This widespread formation, described elsewhere by Rouchy et al. (1983) and by Monty et al. (1987), and probably of microbial origin, expresses Middle to Late Miocene restriction immediately preceeding the onset of evaporite sedimentation. This evaporite series attains a thickness of about 2000 m in structural lows (Heybroek, 1965, Hassan and El Dashlouty, 1970, Richardson and Arthur, 1988), while the evaporites lie discordantly against the talus of the more peripheral Miocene marine platforms. The marine Miocene series and associated reefs belonged to the Mediterranean domain. The evaporitic formation passes upwards into Plio-Quaternary marine carbonates and siliciclastics, locally exceeding 1000 m in thickness, recording both major peripheral uplift and the opening of the Red Sea to the Indian Ocean.

### *Geologic Age of the Reefs*

According to nomenclature employed by oil companies working in the Gulf of Suez region, the reefs are referred to as the Lower Rudeis Formation. However, along the Red Sea coast where the sequence is somewhat different, they belong to the Group B of Montenat et al. (1986a) or to the Um Mahara Formation of the NSSC (1974). Their Burdigalian to Langhian age is established indirectly: underlying marine clays near Ras Honkorab have been dated as Burdigalian by Montenat et al. (1986) and as Early Langhian by Philobbos (oral commun.). Platform sediments associated with reefs at Abu Shaar contain the alveolinid *Borelis melo*, a benthic foraminifera ranging from Burdigalian to Late Miocene (Magné, 1978). The rareness of marine shales unaffected by calcium sulfate replacement and the multiple diagenesis of most carbonates, limit paleontologic and radiometric dating.

### *Regional Geologic Framework*

The lithostratigraphic sequence in the northwest Red Sea is related closely to the Neogene tectonic framework whose elements were generated during the early stages of rifting. Multiphased, essentially distensive tectonics have created a series of tilted blocks whose existence strongly influenced reef distribution. Following a Late Oligocene phase of strike-slip movement (Thiriet et al., 1985, Jarrige et al., 1986) affecting the proto-rift continental sediments, a second, Early Miocene distensional phase has produced multiple fault blocks. These deformational phases have produced multiple fault-blocks and structural depressions whose major axes are generally oriented parallel to the axis of the rift (*clysmic* direction of Hume, 1921), they are frequently limited by oblique (N10° or N20°) grabens such as at Wadi Ambagi (Quseir), Ras Honkorab and Abu Ghusun, some of which have favoured transport of detrital discharges towards the axial depression (Philobbos and El Haddad, 1983) otherwise obstructed by the predominant NW-SE (clysmic) block system.

Most Miocene reefs are located along positive structural blocks. These are tilted either towards the periphery or towards the axis of the rift, depending on the region. In addition, the main axis of any given block often plunges and the block thus has a complex asymmetry (transverse and longitudinal) whose vectors strongly influenced contemporaneous sedimentation and reef geometry.

## REEFS AND ASSOCIATED SEDIMENTS

### *Data Base*

As already noted, this study is based almost exclusively on field observations. These were strongly oriented towards reef geometries and their relationships with the structural framework. To date, little attention has been given to detailed petrographic or mineralogic analyses except at Abu Shaar, although this is currently being undertaken. The ecological aspects of Miocene reefs have been studied briefly and new coral identifications (B. R. Rosen) are included.

### *Depositional Facies and their Distribution*

This study involves a series of six reefs which will be treated successively. Although most have sedimentary characteristics which are peculiar to a given reef, nevertheless, there are certain features common to most, if not all reefs.

In general, autochthonous reef bodies are relatively small, being only 100-350 m in width and usually less than 10 m in thickness. The volume of the reef is considerably smaller than that of the adjacent peri-reefal talus, perhaps the most characteristic element of Red Sea Miocene reef complexes (Figs. 3A, B). This is due to the fact that most (but not all) reef bodies are located on structural highs.

Although most are located on structural blocks, certain reefs have formed on highs created by sedimentary processes. Thus, at Wadi Sharm el Bahari (Fig. 4), a coral-algal reef is situated on the crest of a submarine cone, and the reef talus is relatively thin and widespread. Virtually all reefs may be regarded as fringing reefs bordering structural highs, the single exception being the platform-edge barrier at Abu Shaar.

The six examples are presented according to their tectonic setting. One reef (Sharm el Qibli) formed relatively early in rift evolution and is not oriented toward the rift axis. However, most reefs formed somewhat later when the axis of the rift was beginning to deepen and when terrigenous input from the west generally limited reef growth. Thus, reefs were developed preferentially on the axial (northeastern) sides of tilted blocks. On a local scale, certain prereef blocks were positioned close to or above Miocene sea-level and have risen during reef growth whereas others have subsided. These structural parameters, although certainly not the only factors, are the main elements controlling reef geometry.

*Wadi Sharm el Qibli.—*

Situated on the south side of the wadi, some 3 km from the sea, this lenticular body (Figs. 4B, C) is poorly preserved. Nevertheless, this reef is unusual, not only because it is possibly one of the oldest, but also because of its geometry and ecology.

*Reef substrate.*—Basement rocks here consist of proto-rift, continental playa muds and sands which dip at about 20° to the southwest (i.e., towards the periphery of the rift). Together with a Late Oligocene basalt flow, these continental sediments have been tilted, the crest of this proto-rift relief being transgressed by Early Miocene carbonates within which are developed three lensoid reef bodies (Fig. 4B).

*Reef-core facies.*—The three individual reef bodies have formed on surfaces inclined toward the southwest, the inclinations tending to diminish toward the uppermost reef whose base is approximately horizontal. Each lense comprises both massive and branched corals, the latter being dominated by *Stylophora* (Fig. 5) especially well developed within the second reef where their vertical growth position clearly confirms the pre-reef age of substrate tilting. The principal components include *Stylophora regulata* (Chevalier, 1962), with subsidiary *Tarbellastraea* spp., *Montastrea alloiteaui* (Chevalier, 1954), *Porites* sp., and solitary mussids. The predominance of *Stylophora* suggests that this westward-facing reef developed under somewhat protected conditions.

*Off-reef facies.*—The outcrop, limited to the west by a northeast-southwest trending fault, consists essentially of reef core material. Locally, this core is flanked by a debris flow rich in molluscs and red algae. This detrital material also dips toward the southwest suggesting somewhat deeper water between the reef and the western periphery of the rift.

*Post-reef facies.*—The individual reef lenses are separated by several meters of skeletal wackestone, the uppermost unit being overlain by about 30 m of mixed carbonate-siliciclastic detritus.

*Tectonic setting.*—The southwest inclination of the reef base

FIG. 3.—General field characteristics of Miocene reefs: (A.) The Zug al Bohar complex; synsedimentary faulting of the first reef (1) is represented by a planar surface with borings (arrow). The second reef (2') is a narrow, multilayered lense located on the seaward margin of terrigenous slope deposits (see Fig. 10C). The third reef (3) is covered by oolitic sands. (B.)Reef talus at Esh Mellaha which measures about 80 m in height. (C) and (D) Coral colonies (*Favites*) in vertical growth position within the slope deposits (dip 25°) at Esh Mellaha (see photo B). (E) *Porites* in growth position affected by *Lithophaga* borings, at Abu Ghusun .

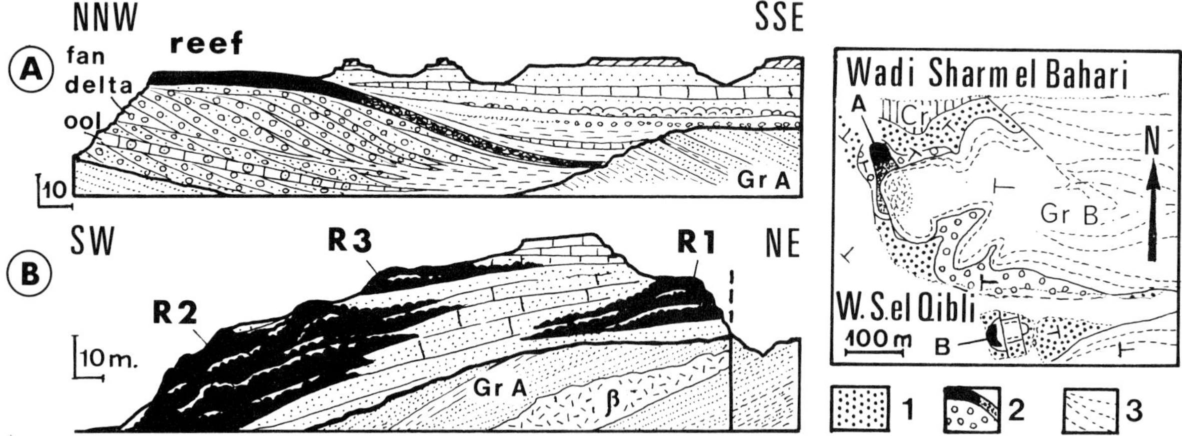


FIG. 4.—Various settings of Early Miocene reefs. The reef at Sharm el Bahari (A) developed on a submarine conglomeratic cone sloping to the south-southeast. On the other hand, that at Wadi Sharm el Qibli (B) is located on a paleo-relief consisting of tilted, Group A, continental deposits; sediments deposited in the adjacent depression (left bank of the wadi) have no reefs. In the location map, B= basalt ; 1= Oligocene-Early Miocene Gr. A ; 2= Early-Middle Miocene cone and reefs ; 3= post-reef Group B Early to Middle Miocene sediments.

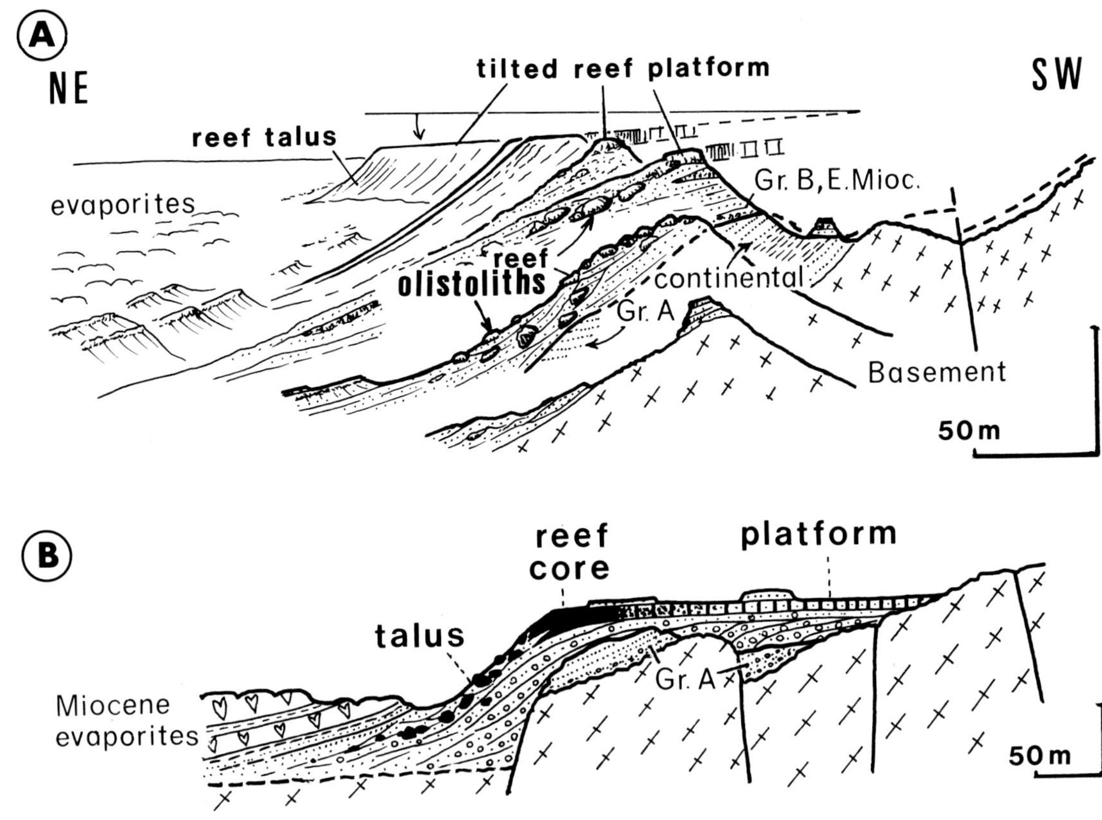


FIG. 6.—Reef platform and talus at Abu Ghusun. The general slope of the platform results from differential uplift which has induced erosion of the N parts of the reef core which is recorded exclusively by reef olistoliths (see Fig. 7). (A)= series of sections along the E border of the platform showing lateral variations; (B)= schematic reconstitution of the platform.

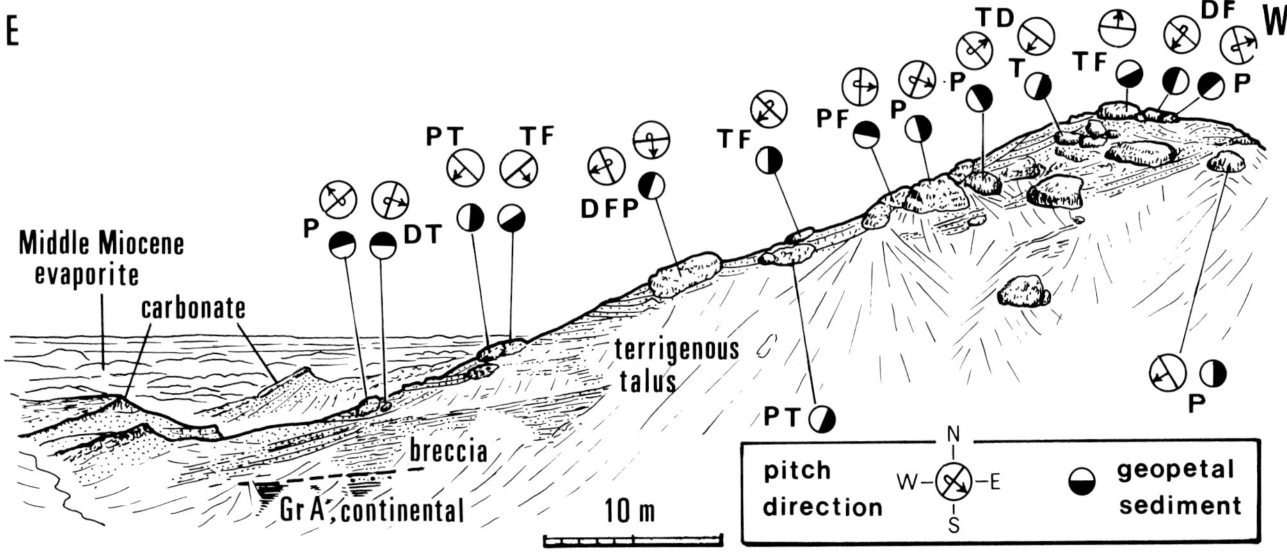


FIG. 7.—Distribution of reef olistoliths in the terrigenous talus at Abu Ghusun. The olistoliths are expressed in terms of coral content: P = *Porites*; D= *Diploastraea*; T = *Tarbellastraea*; F = *Favites*. The direction and normal to overturned orientation of corals in growth position are indicated by arrows.

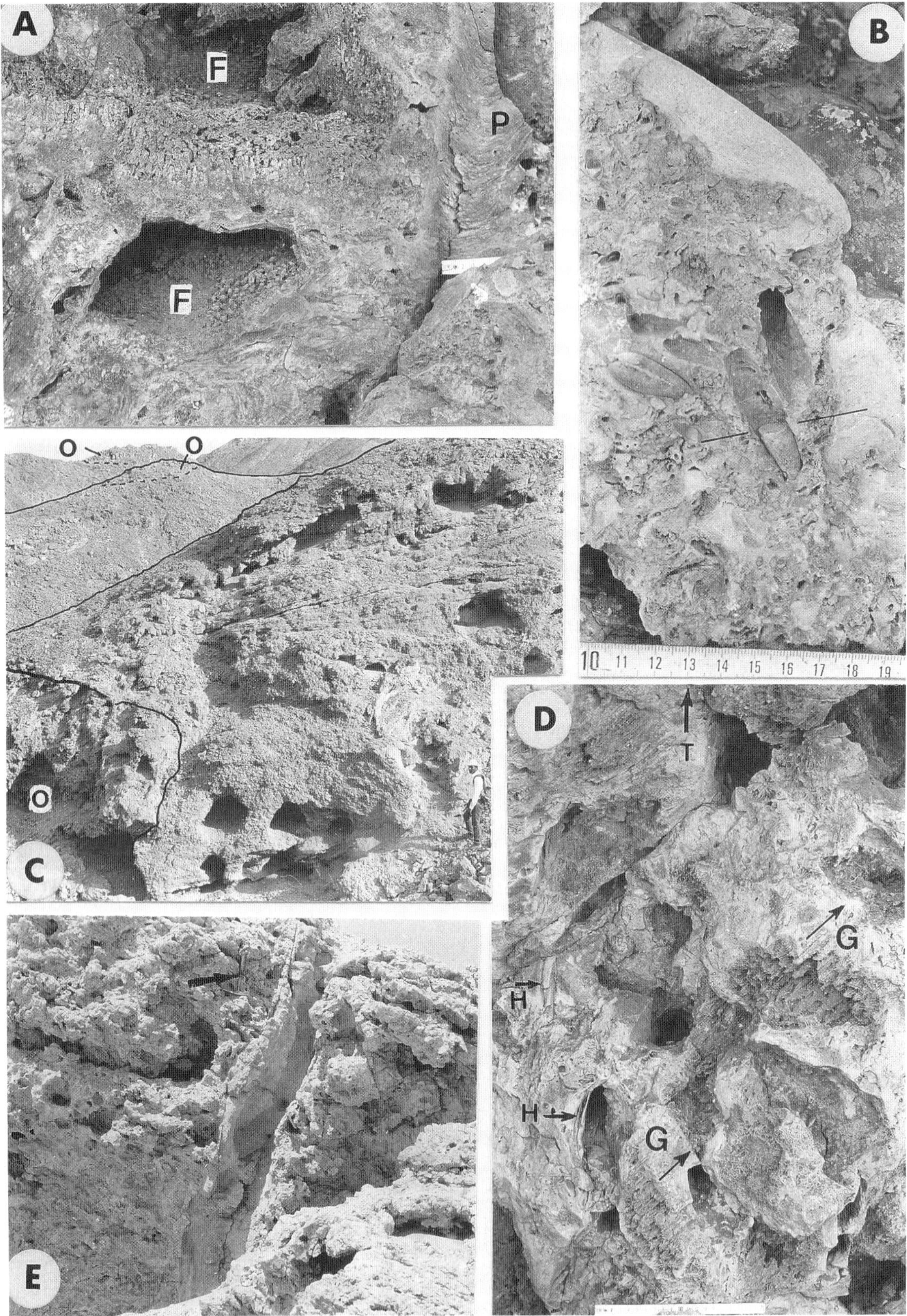


FIG. 8.—Detail of various reef talus deposits: (A) Overturned colonies of *Porites* (P) and *Favites* (F) within a reef olistolith at Abu Ghusun. (B) Inclined internal sediment and geopetal filling in *Lithophaga* boring within gently tilted (12°) reef at Zug al Bohar (scale in cm). (C) Terrigenous talus deposits with reef olistoliths (O) at Abu Ghusun. (D) Growth direction of corals (G) and tilting of internal sediment (H) within an olistolith at Abu Ghusun (scale 20 cm; T = present top). (E) Sand-filled neptunian dike within the R2 reef at Sharm el Luli (arrow = hammer).

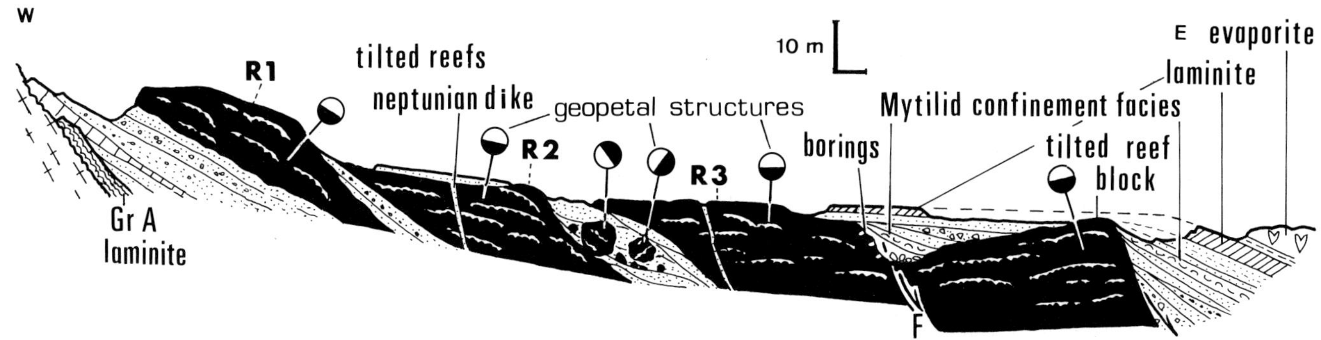


FIG. 9.—The Sharm el Luli reef complex: progressive decrease in slope suggests synsedimentary uplift of R1 and R2 reefs followed by tilting of part of the last (R3) reef. The resulting structural depression (fault-plane bored by molluscs) is filled with beach boulders, and monospecific mytilid lamellibranchs, capped by pre-evaporite microbial laminite.

and erosion and the existence of a reef is recorded only by the preservation of its slope detritus. Reef destruction is not surprising in view of the proximity of a major fault system situated immediately (500 m) to the southwest.

*Sharm el Luli.—*

Located about 500 m from the present shoreline, 6 km northwest of Ras Honkorab, the well preserved reef complex (Fig. 9) at Sharm el Luli is located along the northeast flank of a major structural block. The reef clearly records progressive uplift with respect to sea level, this adjustment resulting in down-stepping of the reef complex.

*Pre-reef facies.*—A steeply inclined (35°) basement surface is overlain by about 30 m of mixed siliciclastic-carbonate sand and gravel enriched in molluscs and small rhodoliths. These beds are also inclined at 20° toward the northeast. They include a large (10 m) reef olistolith, the prereef unit being topped by an erosional surface on which is developed the first autochtonous reef core.

*Reef-core facies.*—Possibly one of the most interesting Miocene reefs of the northwest Red Sea, this complex consists of three individual prograding reef bodies, the downstepping indicating fall of relative sea-level.

Reef R1 is composed mainly of small (30-cm-high) massive coral colonies of the genera *Porites, Diploastraea, Tarbellastraea* and *Favites* forming a coral wackestone to floatstone deposit. Many colonies are in growth position. Faint traces of inclined stratification indicate progradation toward the northeast. The top of the reef is inclined at about 10° to the northeast. That this surface was originally horizontal is indicated by the disposition of geopetal sediments filling small cavities within the underlying reef. These fillings, approximately parallel to the upper surface, are also tilted to the northeast indicating tectonic adjustment. The front of reef R1 is an inclined surface (20°) overlain by fine siliciclastic sands and bioclasts of slope deposits that gradually decrease in dip before passing into the following reef.

Reef R2, located about 25 m to the northeast and some 10 m lower than its predecessor, contains abundant massive corals in a fine carbonate matrix. Geopetal fillings, together with the surface limiting the reef, are gently inclined at 5°. However, the presence of several large (3 m) olistoliths in the frontal part of the body, together with subvertical fractures (neptunian dikes) filled with marine sands (Fig. 8E), suggest continual tectonic instability. The top of reef R2 is bordered to the NE by a low (1-2 m) bioconstructed ridge suggesting a weakly developed barrier. Behind this ridge (to the southwest), 3 m of well-bedded siliciclastic sands with molluscs overlie both reefs R1 and R2 suggesting a slight rise in relative sea level.The sharply defined surface marking the front of reef 2 is overlain by lenses (3 m) of siliciclastic sand and massive carbonate, the latter highly deformed during slumping.

Reef R3, rich in lamellar corals, overlies the spectacular slump structure. Geopetal fillings within this reef core are subhorizontal and the upper surface is overlain by several meters of bedded carbonate deposited during and subsequent to reef R3.

*Post-reef facies.*—The front of reef R3 is marked by a bored surface cut locally by synsedimentary fault-planes sealed by clays and, subsequently, by ooid, intraclastic grainstone beds with dips of 35°. The discontinuity marking the top of this youngest reef is overlain by 2 m of cryptalgal laminite marking somewhat more restricted conditions preceding evaporite deposition.

The front of the reef mass is overlain by 5 m of inclined, densely cemented carbonate rich in small, monospecific mytilids indicating conditions of increasing restriction which terminated reef development.

*Tectonic setting.*—The reef complex at Sharm el Luli is located on the northeast flank of a clearly defined, linear horst whose continual uplift during reef sedimentation is recorded by numerous discontinuities, olistoliths, slump structures, and the overall geometry of the reef complex. The degree of uplift is far less than at Abu Ghusun, as is suggested both by the in situ preservation of reefs and the moderate difference in altitude (about 20 m) between reefs R1 and R3.

*Zug al Bohar.—*

Between wadis Asal and Zug al Bohar, some 5 km to the north, two Neogene fault blocks whose main axes are approximately parallel to the Red Sea (Fig. 10), are composed of granite and prerift Nubia Sandstone probably of Mesozoic age. Present

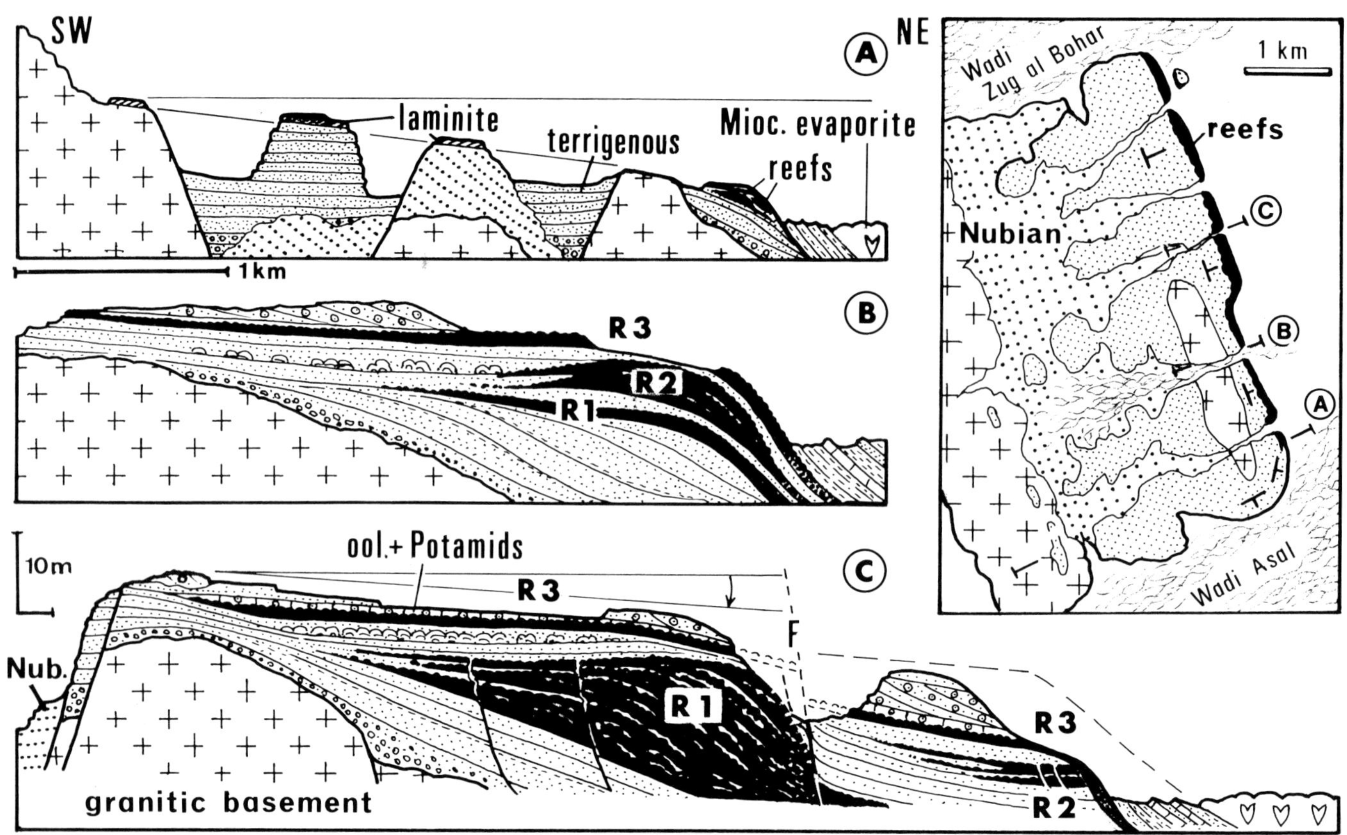


FIG. 10.—Reef complex at Zug al Bohar. Section A shows localization of reefs along the northeastern edge of a series of basement blocks composed of Precambrian granites and Cretaceous Nubia Sandstone. Sections B and C illustrate details of the reefs which have formed on the margin of terrigenous platform deposits which were tilted subsequently. The restricted back-reef deposits include stromatolites, potamid and mytilid molluscs and oolite. Synsedimentary tectonics and relative sea-level changes explain the variable reef geometries and settings. F = fault.

during the Early Miocene transgression, these blocks have favored the development of an elongate (3 km) reef complex (Figs. 3A, 10, 11) situated on the northeast front of the outer (eastern) block. The three reef masses, affected by synsedimentary tectonics, are overlain by open marine carbonates and siliciclastics indicating tectonic or eustatic deepening.

*Pre-reef facies.*—Pre-rift basement is overlain by subautochtonous boulders and pebbles filling local depressions. These are covered by 10-20 m of mixed siliciclastic-carbonate sands and conglomerates whose sedimentary dips are inclined at about 10° to the northeast. The pre-reef clastics are locally rich in rhodoliths, notably in their upper parts, which grade into the overlying reef-core.

*Reef-core facies.*—In common with the reefs at Sharm el Luli, the three reefs (Fig. 11) are located along the northeastern side of the adjacent basement block whose top probably emerged during growth of reef R1.

Reef R1 (Figs. 10B, C) is frame-supported with a diverse coral biota including *Porites, Tarbellastraea, Heliastraea, Diploastraea, Favites* and mussids. The lenticular core facies has a vague stratification indicating its propagation towards the northeast giving a total width of about 350 m. The front of the reef is a steeply inclined (65°), flat surface perforated by numerous *Lithophaga* (Fig. 10C) which can be traced laterally for at least 1 km. Dragging of overlying talus deposits suggests that it was a persistently active, synsedimentary fault plane. Its tilting effect is demonstrated by the presence of a shallow back-depression which evolved into a lagoon containing oolitic layers with potamid gastropods and stromatolites.

Reef R2 is located in front of and somewhat below its predecessor (R1) on the seaward margin of sands and gravels which bury the fault plane. R2 is composed of diverse massive corals and large branching colonies of *Porites* in vertical growth position. Small in size, this body is discontinuous. Its front is also limited by a second fault plane (Fig. 10C) overlain by steeply dipping (40°), essentially siliciclastic slope deposits which pass

FIG. 11.—Typical Miocene reef-core facies: (A) Dissolved corals whose form is defined by a micrite coating, possibly of microbial origin. Inter-colony voids are partially filled with internal sediment; scale indicated by hammer head (Zug al Bohar); (B) Branching *Stylophora* colony; middle reef at Sharm el Qibli (scale 5 cm). (C) Polyphased geopetal fillings within a reworked coral colony. Multidirectional inclinations of the fillings show that the colony has moved at least twice; Wadi Sharm el Luli, reef R3 (scale 1 cm); (D) Slightly tilted geopetal fillings within *Porites* in growth position; Wadi Sharm el Luli, reef R2 (scale 5 cm). (E) Massively bedded reef sediments including "cowrie" shells and a coral; reef platform, Abu Ghusun (scale in mm.). (F) *Lithophaga* borings in massive (infilled) reef facies; bored fault-plane cutting reef R1 at Zug al Bohar (scale = 2 mm).

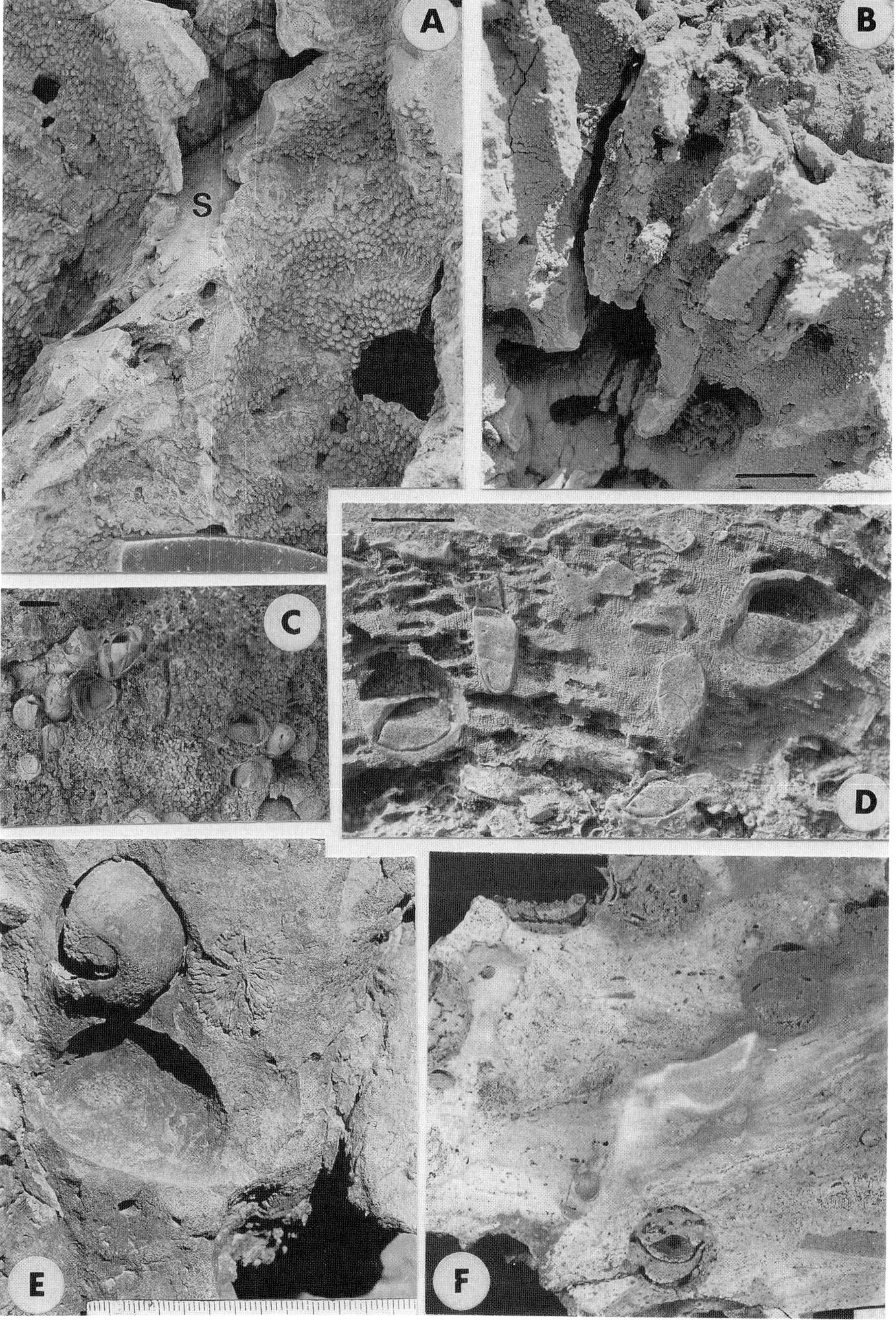
A
S
B
C
D
E
F

laterally over the reef top.

Reef R3, a thin (1 m) biostromal blanket containing scattered massive colonies, which covers the preceding reefs, rises and thins progressively towards the culmination of the block. This uppermost body is overlain by about 5 m of cross-bedded oolitic grainstone which has prograded across the underlying reef.

The three reef bodies, of which the initial reef (R1) is by far the most important (Fig. 10C), extend along the northeastern, frontal slopes of the basement block. Nowhere do they overlie the culmination of this structural block. In contrast with the reef complex at Sharm el Luli, the three units, at least in part, are superimposed.

*Tectonic setting.*—This geometry, together with the vertical stacking of associated marine detritus, suggests initial lateral accretion (reef R1) followed by repeated abrupt subsidence. These movements are mainly the result of synsedimentary faults which limit reefs R1 and R2. On a somewhat larger scale, it is interesting to compare this northeast flank of the structural block, characterised by reef growth, with that on the opposing southwestern side. Here, the adjacent structural depression is filled with cross-bedded, marine detritus including coarse conglomerates (Fig. 10A). There are no reefs. The preferential development of reefs along the northeastern flank probably also reflects more open marine conditions related to the deepening of the rift.

*Off-reef detritus.*—Zug al Bohar is one of the few Miocene localities where the nature and geometry of contemporaneous detritus may be mapped. The structural block measures about 3 Km in length and 1 Km in width. Reefs are best developed near the northeastern extremity of the block, and these pass along strike into cross-bedded bioclastic sands which have been deposited on the southeastern extremity of the block (Fig. 10).

*Abu Shaar.*—

Situated near the entrance to the Gulf of Suez some 25 km north of Hurghada, this complex (Figs. 3, 12) is readily accessible. It has received much attention since its initial discovery by Hume (1921). Most studies have concerned only Abu Shaar in spite of the fact that this platform is a direct extension of the Esh Mellaha block whose eastern flank supports spectacular reef talus.

In contrast with the preceding examples, the reefs at Abu Shaar are essentially a vertically stacked complex some 50 m in thickness. Together with associated shallow platform carbonates, this 100-m-thick sequence clearly suggests considerable relative sea-level rise during sedimentation. The present authors will treat only certain aspects of the Abu Shaar complex comparing it with the preceding examples. This more general approach may help in appreciating the variable nature of Miocene reef geometries and their relationships with their structural framework.

*Pre-reef facies.*—These vary according to the position of the reef with respect to the underlying basement which, although tilted to the west, also plunges to the southeast. On the structurally higher parts of this block, notably towards the northern end of Abu Shaar, a thin reef body has developed directly on crystalline basement (N. Wadi Bali). However, in the more southern parts, reefs are developed on a coarse conglomerate comprising decimeter-sized basement blocks locally overlying finer beach conglomerates.

*Reef-core facies.*—The organic composition of the reef is varied, the reader being referred to the zonation proposed by James et al. (1988) and work in progress by C. Perrin, B. Rosen and D. Bosence. It should be stressed, however, that the massive carbonates concentrated along the eastern margins of the block are only locally bioconstructed. There occur several, relatively thick (5-10 m) members composed of bedded bioclastic material rich in red algae detritus, notably in the vicinity of Bir Abu Shaar.

The overall geometry of the reef and interbedded sediment varies from north to south, mainly as a function of synsedimentary erosion relating to reactivation of the adjacent fault system which limits the eastern side of the platform (Burchette, 1988). Although well preserved locally, reef development is confined mainly to the SE corner and, especially, to the southern flank of the platform, at Wadi Kharaza (Fig. 12) in areas slightly more removed from the synsedimenary fault system.

Thickness of the reef also is related to the geometry of the underlying basement which plunges toward the south. In the north where basement attains altitudes of about 150 m, only reef talus fringes the eastern flanks of the platform. Farther south, in the vicinity of Wadi Bali, the coral-rich facies attains a thickness of about 20 m and forms a weakly defined ridge along the eastern margins of the platform (Fig. 12). As the basement block plunges to the south, it has clearly been situated well below Miocene sea level. It has supported a 5-km-wide carbonate platform, the eastern and southern fringes of which are bordered by a narrow barrier reef whose cumulated thickness may attain 50 m. Thus, the reef complex at Abu Shaar, too often discussed only in terms of east-west profiles, in fact exhibits northwest-southeast variations both in thickness (Fig. 12) and in composition, the importance of which tends to be underestimated.

*Tectonic framework.*—The localization and geometry of the reef complex is closely related to the morphology of the underlying structural block. This block is tilted to the west, and its highest point, located along the eastern periphery, coincides with the major reef development. Because many Miocene reefs of the northwestern Red Sea are preferentially located along the eastern flanks of structural highs (see previous examples), one should consider possible regional effects relating to the presence of deeper waters along the axial parts of the rift.

As already noted, basement rock also plunges to the southeast. The more northerly Esh Mellaha basement range was situated above sea level and the elongate Miocene island was lined by a narrow fringing reef and talus. Farther to the south, however, basement rock plunged below sea level and the reef body has bordered a 5-km-wide submarine platform with few basement islets in the form of a narrow (350 m) barrier. The lateral transition between reef and back-reef facies is subvertical (Fig.

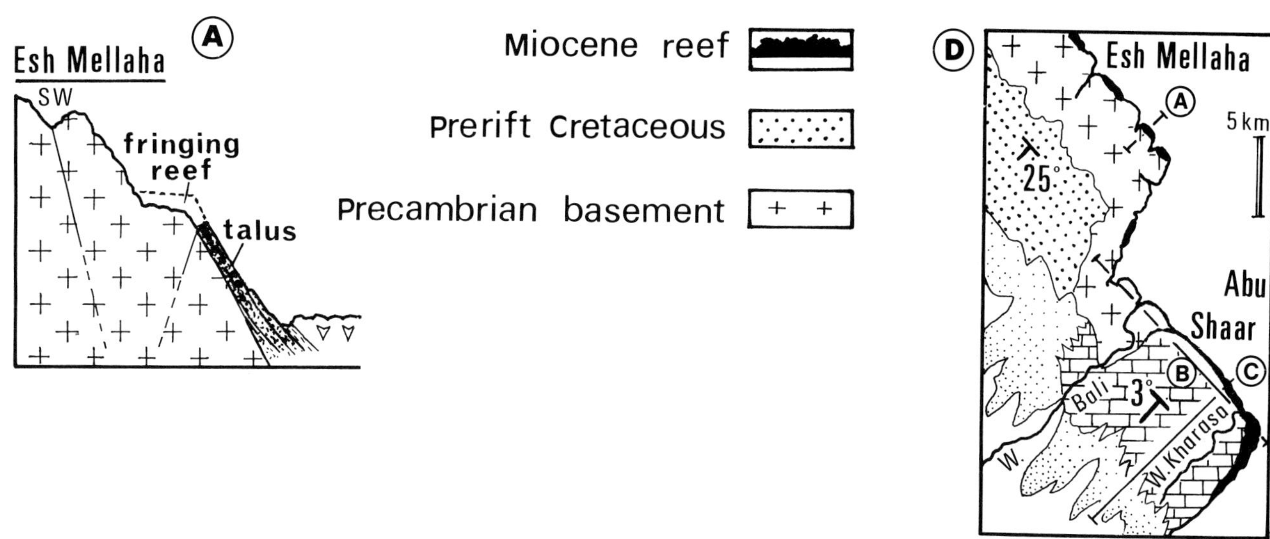


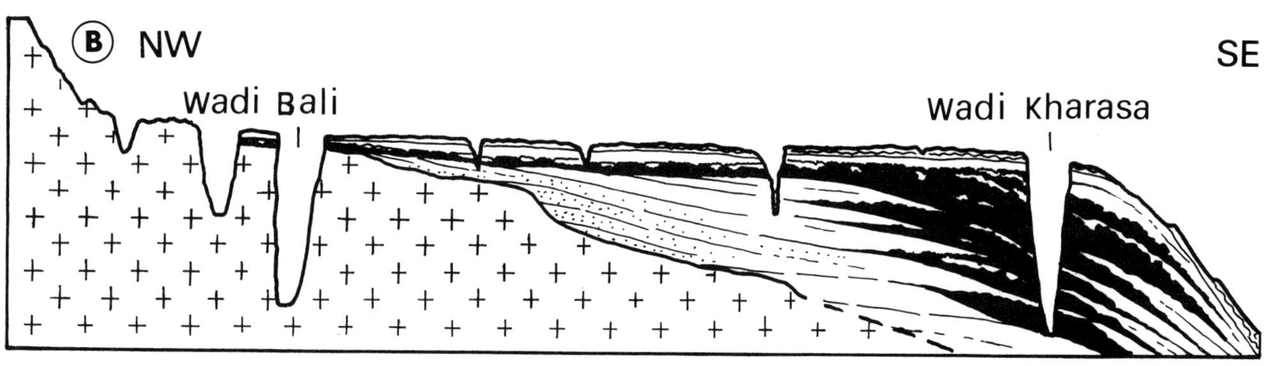


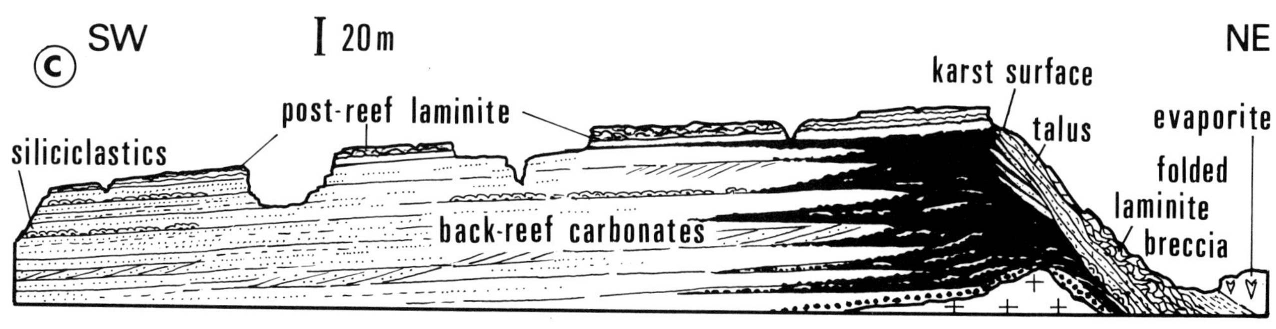


FIG. 12.—Generalized profiles showing the geometry of the reefs at Esh Mellaha and Abu Shaar. Along the Esh Mellaha block (A) reefs, represented only by their talus, fringed a Miocene island while reefs at Abu Shaar (B and C) formed a barrier which thickens to the southeast. This barrier protected a 5- to 10-km-wide platform whose somewhat restricted carbonates grade westwards into mixed terrigenous facies.

12), suggesting that the rates of reef growth have been approximately equal to the rates of subsidence or relative sea-level rise.

*Off-reef facies.*—The platform and barrier reef complex at Abu Shaar has been described by Rouchy (1982), Haddad et al. (1984), Prat et al. (1986), Burchete (1988), Monty et al. (1987) and by James et al. (1988). It is flanked to the east (i.e., towards the axis of the rift) by spectacular talus deposits (Fig. 3), affected locally by synsedimentary faults. The height of these slope deposits indicates that water depths in front of the barrier exceeded 100 m. To the west (i.e., towards the platform interior) the reef passes rapidly into bedded carbonate sands composed of pellets, foraminifera and molluscan debris. Several stromatolitic levels are a characteristic element of these environments. The back-reef carbonates at Wadi Kharasa grade progressively to the north (i.e., laterally, along the platform) into mixed carbonate-feldspathic sands (Wadi Bali) whose presence reflects the proximity of the Esh Mellaha paleo-island.

The reef complex is topped by a weakly developed karst surface and the entire complex, including the back-reef facies, is overlain by 10-15 m of laminated dolomicrite with scattered gypsum pseudomorphs indicating increasing salinities prior to the onset of regional evaporite sedimentation. This laminite is folded and brecciated, both on the fore-reef slope and on the subhorizontal platform. This spectacular deformation has been interpreted (Plaziat et al., 1990b) as being caused by major synsedimentary earthquakes.

## DIAGENETIC OVERPRINT

With the exception of the complex at Abu Shaar, no systematic study has been made of the diagenetic processes affecting these Miocene reefs. However, this research is in progress. Initial results based on several reef complexes indicate that the general paragenetic sequence is as follows: *submarine cementation-dissolution and sparitisation-dolomitization-sulfate replacement.* Certain phases, notably dolomitization, may be multiple.

### *Submarine Cementation*

As already noted by Aissaoui et al. (1986) and Purser et al. (1988), the eastern flanks of Abu Shaar platform are strongly cemented by fibrous, isopachous carbonate (now dolomite, Fig. 13), while both at Abu Shaar and at Zug al Bohar bored surfaces confirm early marine lithification (Fig. 11F). This early marine diagenesis has resulted in a hard carapace which envelopes the eastern, basinwards peripheries of these platforms. Within the reef-core itself, laminated micrite coatings of coral framework predate internal sediments. The laminated nature of the micrite suggests a microbial origin which could have contributed to early cementation.

### *Dissolution and Sparitization*

All reefs examined have been strongly affected by dissolution; most coral colonies and back-reef platform sands are dissolved (Fig. 13 B, C) and identification is often based on molds. These molds are generally well preserved, especially in the case of certain branching corals which are protected externally by a coating of laminated micrite. At Abu Shaar (Wadi Bali), internal dissolution of large (5 m) coral colonies has resulted in a cavernous structure which is lined by a second phase of fibrous submarine cement (Fig. 13C).

In spite of this intense dissolution, reefs and associated sediments are little affected by sparitic cement; sparitic calcite is rare! It would seem that great quantities of dissolved carbonate either have been incorporated into the dolomite or have been flushed out of the system. The reefs examined occur near the periphery of the rift system and have been buried to depths rarely exceeding 100 m. Peripheral uplift probably has stimulated active hydrodynamic flow (see following paragraph) favoring removal of carbonate.

### *Dolomitization*

Most Miocene carbonates, including reefs, are dolomitized; in many localities dolomitization is multiphased. Petrographic study shows that the dolomite is generally very fine and is fabric-preserving (Fig. 13). Its oxygen and carbon isotopic properties, although generally negative, are highly variable, 0 ranging from +1.2 to -8.5‰ PDB (Aissaoui et al., 1986, Coniglio et al., 1988). The general predominance of negative oxygen and carbon values suggests dolomitization from non-marine fluids. Influx of dolomitizing meteoric groundwaters is a distinct possibility in view of the proximity (10 km) of the elevated (2000 m) rift periphery. Theoretically, negative oxygen values may reflect abnormal geothermal conditions typical of rifts, but this hypothesis is difficult to accept for the following reasons:

1. The negative oxygen values (-8‰) would require relatively elevated temperatures (about 70°C), incompatible with the very shallow (less than 200 m) burial history of the platform which is situated near the periphery of the rift;
2. Dolomitization is a regional phenomenon and is not related geographically to the major fault systems, and;
3. The petrographic attributes of most of the dolomite are not those normally associated with high-temperature dolomites.

The exceptionally negative carbon isotopes (-13‰) may reflect the influence of organic carbon. Finally, selective sampling (Aissaoui et al., 1986) has revealed the grouping of distinct isotopic values suggesting that dolomitization is multiphased.

FIG. 13.—Typical diagenetic fabrics within Miocene reefs and associated sediments: (A) General view of the northeastern margin of Abu Shaar platform at Wadi Bali. The reef complex (arrow) is flanked to the right by talus and to the left by a weakly developed lagoonal depression. The reef complex is developed on Precambrian basement. (B) Back-reef packstone whose skeletal components have been dissolved; notice absence of sparitic cement. Within this pure dolomite, porosity is a function of sedimentary fabric and early dissolution (scale = 1 mm). (C) Dissolution cavity within a faviid coral colony. The vug is lined with fibrous marine cement (arrow) exhibiting characteristic botryoidal morphology. Both coral and cement are now dolomite; Wadi Bali, scale (coin) = 3cm. (D) Polished surface showing multiphased diagenesis. All is dolomite. However, the lower, light grey parts (a) are pseudomorphs of celestite which are overlain by an isopachous layer of fibrous cement (b). The upper parts (c) are dolomitised internal sediment whose lithoclasts have been dissolved, Wadi Asal. (E) Karst cavity in reef-core lined with an isopachous layer of fibrous cement (arrow) overlain by laminated internal sediment. All is polyphased dolomite; N Wadi Bali, Abu Shaar (scale 1cm). (F) Fore-reef talus with pisolites, a characteristic facies along the upper parts of slope deposits on the eastern flanks of Abu Shaar. Pisolites are highly cemented by fibrous carbonate, both sediment and early marine cements being replaced by mimetic dolomite. This dense fabric is typical of the platform periphery (scale 0.5 mm).

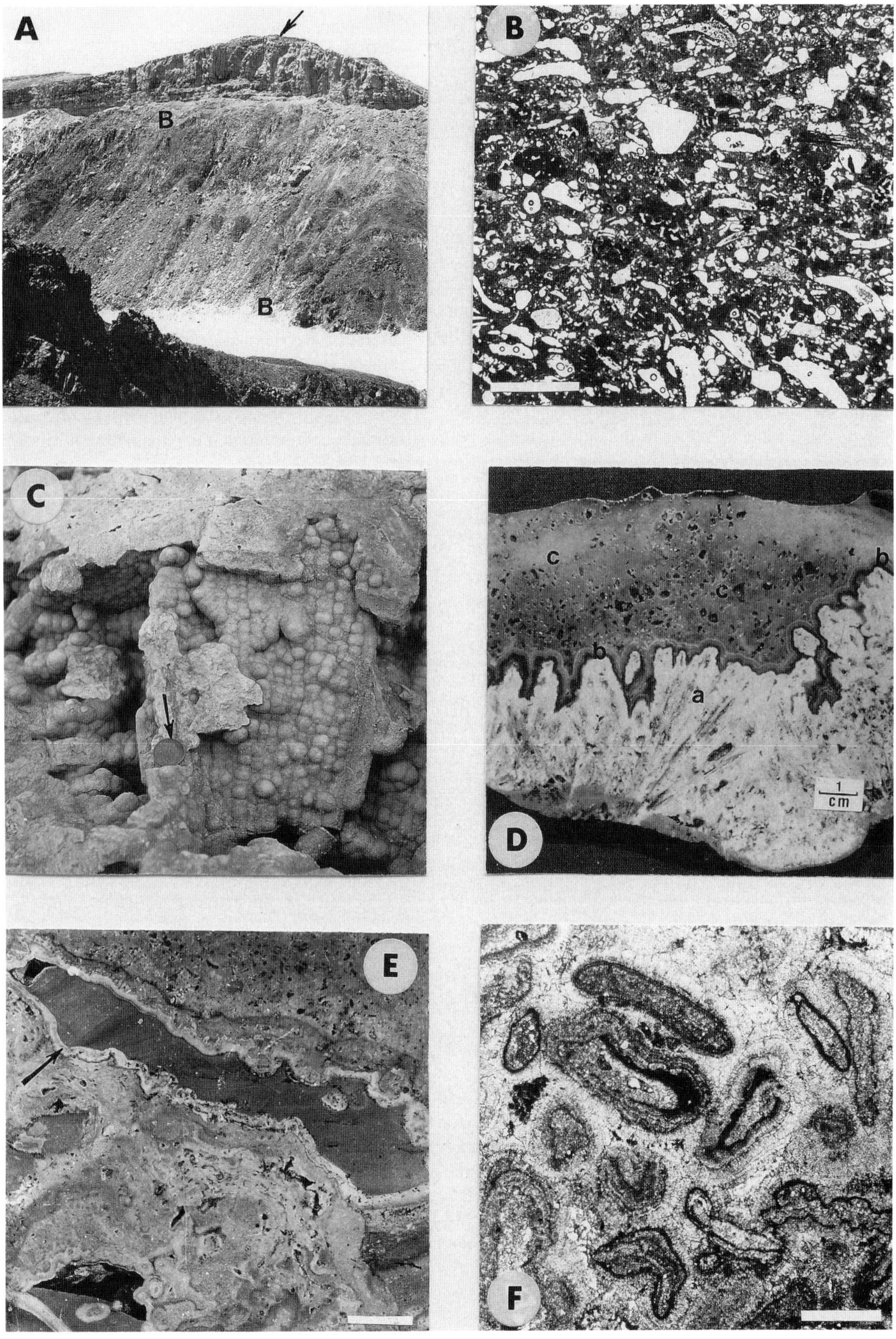
A
B
B
B
C
c
c
b
b
a
1
cm
D
E
F

Clearly, the precise diagenetic history of these carbonates remains to be determined.

Porosity of the Miocene dolomites (see following paragraph) is generally high being inherited from the sedimentary and early diagenetic textures (Purser et al., 1994; Figs. 13B, F).

### *Sulfate Replacement*

Miocene dolomites, including parts of the reef complex at Sharm el Luli, are replaced by secondary calcium sulfate (Fig. 14). As noted by Orszag-Sperber et al. (1986), this generally occurs where subhorizontal Miocene evaporites onlap inclined reef or talus deposits. Sulfate-rich waters have penetrated into the permeable carbonates which they replace or cement; corals and molluscs often are replaced by gypsum (Fig. 14). The precise age of this process may vary. It has occurred during or immediately subsequent to the deposition of the Miocene evaporites, as in the region of Um Gheig (Orszag-Sperber et al., 1986). However, gypsification of Miocene dolomite can also be a relatively recent process notably where sulfate-rich waters flow across exposed carbonates.

### *Diagenetic Model*

The Miocene reefs and associated carbonates of the northwest Red Sea are affected by multiple phases of early diagenesis whose exceptional intensity may be conditioned by particular morphologies relating to the structural framework (Purser et al., 1987b, 1990). Rifting has created a series of blocks upon which reefs and other platform carbonates have accumulated and has also resulted in the uplift of peripheral areas and subsidence of the axial zone. The resulting tectonically-controlled morphologies with numerous inclined submarine surfaces and considerable vertical relief, both on local and on regional scales, must favour active hydrodynamic systems. Within this context, early diagenesis is stimulated by the following factors:

1. Penetration of sea water into inclined substrates via surface currents and waves may favour submarine cementation, notably along the platform periphery;
2. Interstitial flow relating to regional hydrostatic disequilibrium causing dissolution and possibly dolomitization;
3. Local vertical influx of meteoric waters during repeated emergence of platforms during tectonic uplift or eustatic lowering of sea-level provoking dissolution and/or dolomitization, and;
4. Density-driven displacement (reflux) relating to varying salinities developed within separate water bodies formed in the multiple structural depressions favouring replacement of calcium carbonate by calcium sulfate.

Speculating, a high degree of diagenetic activity may well be an important attribute of rifts and, as such, could be a determining factor in reservoir potential within rift systems. However, this could be true mainly for peripheral situations, carbonates buried to greater depths within the more axial parts of the rift possibly being less affected.

## POROSITY ASSOCIATED WITH REEFS

Most reefs and associated sediments examined are very porous, and their reservoir potential is excellent. Porosity (Fig. 13B,C) has two origins.

### *Primary intergranular porosity*

In spite of pervasive dolomitization, porosity, both in reefs and in back-reef sediments, is essentially a function of primary sedimentary texture. In the reef core, primary, decimeter-sized cavities in the framework generally remain open in spite of partial filling by sediment. Carbonate sands on the platform are both porous and permeable (Fig. 13B), although precise measurements have not been made. Here, intergranular sparitic cements are only weakly developed. However, sands and coarse fore-reef debris located on talus-slopes, on the contrary, are often affected by intense submarine cementation and, in spite of subsequent dolomitization, remain dense (Fig. 13F).

### *Secondary, Vuggy Porosity*

Both aragonitic and calcitic constituents in reefs and perireefal sediments have been dissolved giving spectacular vuggy porosity (Fig. 13B, C). Dissolution, however, may be selective. Certain corals, notably the massive favids, have not been totally dissolved probably as the result of their denser skeletal structure. Branching *Porites*, on the contrary, are frequently dissolved. Because the sandy or muddy matrix within the reef is also porous, the overall reef body, with the exception of its seawards (eastern) periphery, is an excellent reservoir. As already noted, there is very little sparitic calcite cement . Furthermore, dolomitization has only minor effects on the petrophysical properties of these carbonates, porosity in the dolomite being inherited from the sedimentary texture or from the predolomite diagenetic fabric (Figs. 13B, F).

## OIL AND GAS CONSIDERATIONS

This study is based on outcrop. However, similar carbonate bodies, presumably including reefs, constitute important reservoirs within several fields in the nearby Gulf of Suez. This is the case, for example, in Zeit Bay Field, briefly described by Hassan and Swidan (1990) where Miocene reservoir dolomites form a lenticular mass situated near the top of a tilted basement block. Porosity appears to be due mainly to the dissolution of corals and molluscs within the dolomite.

## DISCUSSION AND CONCLUSIONS

The size of the Lower to Middle Miocene reefs of the northwest Red Sea is very modest: rarely do they exceed 350 m in width and several kilometers in length, with maximum cumulated thickness on the order of 50 m (at Abu Shaar). However, when considered together with their detritus, they constitute

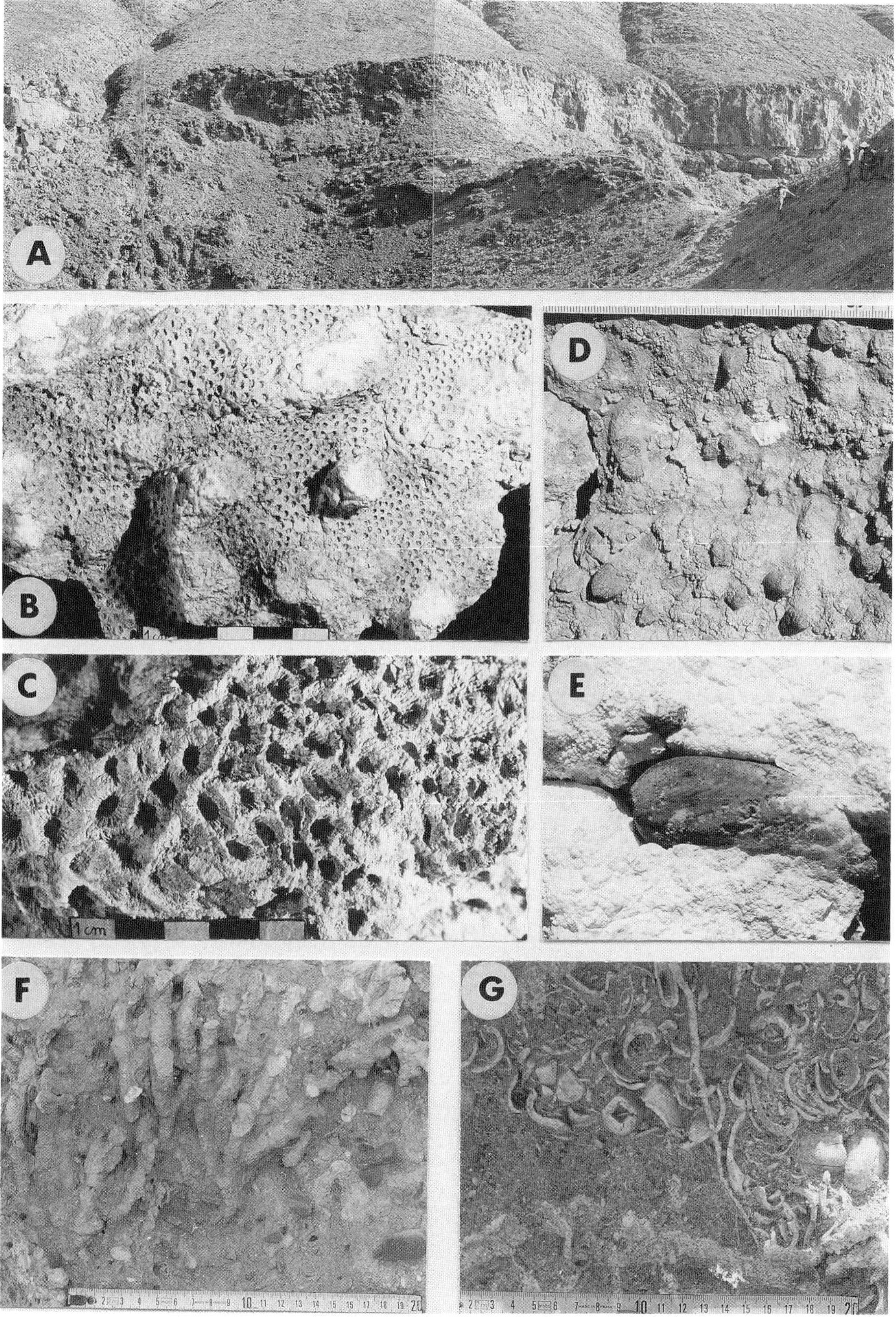


FIG. 14.—Various diagenetic phenomena: (A) Lateral replacement of talus beds composed of marl and dolomitized bioclasts (right) by secondary sulfate (left), south Wadi Asal. (B and C) Colonies of *Tarbellastraea* and *Favites* replaced by calcium sulfate, north Wadi Ambagi. (D and E) *Botula* borings filled with dolomitized sediment within a coral colony which is replaced by calcium sulfate, Wadi Ambagi. (F and G) Selective replacement of branching corals and molluscs within unaltered terrigenous sands and gravels, north Wadi Ambagi.

porous carbonate masses of economic dimensions.

Their reservoir properties are exceptionally favourable due mainly to intense dissolution and the virtual absence of sparitic cements; excess carbonate seems to have been flushed out of the system, possibly due to active water circulation relating to tectonically induced relief. Virtually all reefs and associated sediments are dolomite with petrophysical properties inherited from the original primary or early diagenetic secondary porosities.

In several cases (Abu Ghusun, Abu Shaar and Esh Mellaha), most of the reef core has been eroded during or immediately following its formation. This marine collapse of the reef mass appears to have been stimulated by reactivation of major faults which border each platform. In the case of Abu Shaar, a significant part of the eastern edge of the platform, together with its associated reef, has been downfaulted (or otherwise collapsed), with only the youngest parts of the Miocene platform and reef being preserved. Those segments of the platform which have been eroded must exist within the adjacent subsurface, probably in the form of coarse breccias and olistoliths. This reef-associated debris is sealed by overlying Miocene evaporites which onlap the platform. As such, these periplatform breccias constitute potential oil traps comparable to those associated with oil producing Tertiary reef breccias in the Philippines, described by Longman (1985).

Each Miocene reef in the northwest Red Sea has its own peculiarities which are expressed mainly in terms of thickness and geometry. Some (Abu Ghusun, Esh Mellaha) are recorded mainly by the products of their destruction, others, including Sharm el Luli, by downstepping, prograding geometry, while at Zug el Bohar and southern Abu Shaar, the principal reef masses are stacked vertically. These marked variations, indicating changes with respect to relative sea level, are the result of local structural movement the amplitude and frequence of which have varied from platform to platform. Although these reefs are not necessarily contemporaneous, there being no precise dating, all have comparable coral faunas. With the exception of the reefs at Sharm el Qibli, they all occupy the same general stratigraphic position with respect to the overlying evaporites.

On any given platform reefs are multiple, individual reef bodies being limited by erosional or diagenetic discontinuities some of which indicate paleoemergence. However, it is not possible to correlate these surfaces from one platform to another, their origins probably being related to local structural movement. Nevertheless, the Early Miocene series of the northwest Red Sea records a major marine transgression following deposition of the Late Oligocene, protorift, continental series. This general rise in relative sea level probably was eustatic. However, its effects on reef geometry are masked by those resulting from local tectonic instability.

Because the localisation and geometry of individual reefs are closely related to the tectonic framework (the main subject of this contribution), their prediction, based mainly on seismic methods, is possible. However, the precise relationship between reef distribution and tectonic frame seems to have changed with rift evolution. Reefs which developed relatively early may be located on either side of a given structural block, as at Sharm el Qibli-Sharm el Bahari. However, those formed somewhat later are generally located preferentially on the eastern flanks of highs, probably because of more favourable open marine conditions created during the deepening of the rift.

Nearly all reefs are located on structural blocks and, as such, tend to be isolated from terrigenous influx derived from the periphery of the rift. This same situation occured during the Pliocene when certain reef bodies were located on the top of halokinetic domes. Finally, with Plio-Quaternary deepening and the overall drowning of most offshore highs in the northwest Red Sea, reefs, favoured by warm arid climates, migrated to the periphery of the rift where they fringe much of the present continental shoreline.

## REFERENCES


AISSAOUI, D. M., CONIGLIO, M., JAMES, N .P., AND PURSER, B. H., 1986, Diagenesis of a Miocene reef platform: Jebel Abu Shaar, Gulf of Suez, Egypt, *in* Schroeder, J. H. and Purser, B. H. eds., Reef Diagenesis: Heidelberg, Springer-Verlag, p. 112-131.

BURCHETTE, T . P. 1988, Tectonic control on carbonate platform facies distribution and sequence development: Miocene, Gulf of Suez: Sedimentary Geology, v. 59, p. 179-204.

CHEVALIER, J-P., 1954, Contribution à la révision des Polypiers du genre *Heliastraea:* Annals Hebert et Haug, t. 8, p. 105-190.

CHEVALIER, J-P., 1962, Recherches sur les madréporaires et les formations récifales de la Méditerranée occidentale: Mémoire Société Géologique France, v. 40, 562 p.

CONIGLIO, M., JAMES, N. P., AND AISSAOUI, D. M., 1988, Dolomitization of Miocene carbonates, Gulf of Suez: Journal Sedimentary Petrology, v. 58, p. 110-119.

GREGORY, J. W., 1906, On a collection of fossil corals from Eastern Egypt; Abu Roash and Sinai: Geological Magasine, London, v. 3, p. 50-58.

HADDAD, E. A., AISSAOUI, D. M., AND SOLIMAN, M., 1984, Mixed carbonate-siliciclastic sedimentation on a Miocene fault-block, Gulf of Suez, Egypt: Sedimentary Geology, v. 37, p. 185-202.

HASSAN, F. AND EL-DASHLOUTY, S., 1970, Miocene evaporites in the Gulf of Suez region and their significance: American Association Petroleum Geologists Bulletin, v. 54, p. 1686-1696.

HASSAN, M. AND SWIDAN, N., 1990, Carbonate facies and reservoir heterogeneity in Zeit Bay field: Cairo, Tenth Petroleum Exploration and Production Conference, p. 20.

HEYBROEK, F., 1965, The Red Sea Miocene evaporite basin: London, Salt Basins Around Africa, London Institute of Petroleum, p. 14-40.

HUME, W. F., 1921, Relations of the northern Red Sea and its associated gulf areas to the "rift" theory: Quarterly Journal Geological Society London, v. 77, p. 96-101.

JAMES, N. P., CONIGLIO, M., AISSAOUI, D. M., AND PURSER, B. H., 1988, Facies and geological history of an exposed Miocene rift-margin carbonate platform. Gulf of Suez, Egypt: American Association Petroleum Geologists Bulletin, v. 72, p. 555-572.

JARRIGE, J. J., OTT D'ESTEVOU, P., BUROLLET, P. F., THIRIET, J. P., ICART, J. C., RICHERT, J. P., SEHANS, P., MONTENAT, C., AND PRAT, P., 1986, Inherited discontinuities and Neogene structure: the Gulf of Suez and NW edge of the Red Sea: Philosophical Transactions Royal Society London, v. A317, p. 129-139.

LONGMAN, M. W., 1985, Fracture porosity in proximal reef talus of a Tertiary pinnacle reef, the Philippines, *in* Roehl P. O. and Choquette P. W., eds., Carbonate Petroleum Reservoirs: New York, Springer-Verlag, p. 122-128.

MAGNE, J., 1978, Etudes stratigraphiques sur la Néogène de la

Méditerranée nord-occidentale, 1. Les bassins néogènes catalans: Toulouse, Centre National des Recherches Scientifiques, 265 p.

MONTENAT, C., OTT D'ESTEVOU, P., AND PURSER, B. H., 1986a, Tectonic and sedimentary evolution of the Gulf of Suez and the northwestern Red Sea: a review: Paris, Documents et Travaux de l'IGAL, n° 10, p. 7-18.

MONTENAT, C., ORSZAG-SPERBER, F., OTT D'ESTEVOU, P., PHILOBBOS, E., PURSER, B. H., and RICHERT, J. P., 1986b, Etude d'une transversale de la marge occidentale de la Mer Rouge: le secteur de Ras Honkorab - Abu Ghusun: Paris, Documents et Travaux de l'IGAL, n° 10, p. 145-170.

MONTY, C. L., ROUCHY, J. M., MAURIN, A., BERNE-ROLLANDE, M. C., AND PERTHUISOT, J. P., 1987, Reef-stromatolite-evaporite relationships from Middle Miocene: examples of the Gulf of Suez and the Red Sea, *in* Peryt, T. M., ed., Evaporite Basins: Heidelberg, Springer-Verlag, p. 75-86.

NATIONAL STRATIGRAPHIC SUB-COMMITTEE OF GEOLOGICAL SCIENCES OF EGYPT, 1974, Miocene rock stratigraphy of Egypt: Egyptian Journal Geology, v. 18, 69 p.

ORSZAG-SPERBER, F., FREYTET, P., MONTENAT, C., and OTT D'ESTEVOU, P., 1986, Métasomatose sulfatée de dépôts carbonatés marins miocènes sur la marge occidentale de la Mer Rouge: Compte Rendu de l'Académie des Sciences de Paris, v. 302, p. 1079-1984.

ORSZAG-SPERBER, F. AND PLAZIAT, J-C., 1990, La sédimentation continentale (Oligo-Miocène) des fossés du proto-rift du NW de la Mer Rouge: Bulletin de la Société Géologique de France, v. VI, p. 385-396.

PHILOBBOS, E. R., BEKIR, R., EL HADDAD, A. A., LUGER, P., AND MAHRAN, T., 1992, Facies and age of syn-rift Miocene Sediments around fault-blocks in the Abu Ghusun-Wadi el Gemal area, Red Sea, Egypt: International Symposium on Sedimentation and Rifting in the Red Sea and Gulf of Aden, Abstract, p. 37.

PHILOBBOS, E. R. and EL HADDAD, A. A., 1983, Tectonic control of Neogene sedimentation along the Egyptian coast of the Red Sea: Cairo, Proceedings Fifth International Conference on Basement Tectonics, p. 307.

PLAZIAT, J.-C., PURSER, B. H., AND SOLIMAN, M., 1990a, Les rapports entre l'organisation des dépôts marins (Miocène inférieur et moyen) et la tectonique précoce sur la bordure NW du rift de Mer Rouge: Bulletin Société Géologique de France, v. VI, p. 397-418.

PLAZIAT, J-C., PURSER, B. H., AND PHILOBBOS, E., 1990b, Seismic deformation structures (seismites) in the syn-rift sediments of the NW Red Sea (Egypt): Bulletin Société Géologique de France, v. VI, p. 419-434.

PLAZIAT, J-C., MONTENAT, C., ORSZAG-SPERBER, F., PHILOBBOS, E., AND PURSER, B. H., 1990c, Geodynamic significance of continental sedimentation during initiation of the NW Red Sea rift (Egypt): Journal African Earth Sciences, v. 10, p. 335-360.

PRAT, P., MONTENAT, C., OTT D'ESTEVOU, P., AND BOLZE, J., 1986, La marge occidentale du Golfe de Suez, d'après l'étude des Gebels Zeit et Mellaha: Paris, Documents et Travaux de l'IGAL, n° 10, p. 45-74.

PURSER, B. H., AISSAOUI, D. M., AND ORSZAG-SPERBER, F., 1987b, Diagenèse et rifting: évolution post-sédimentaire des sédiments carbonatés miocènes sur la bordure NW de la Mer Rouge: Paris, Notes et Mémoires de la Compagnie Francaise des Pétroles, n° 21, p. 145-166.

PURSER, B. H., BROWN, A., AND AISSAOUI, D. M., 1994, Nature, origins and evolution of porosity in dolomite, *in* Purser, B. H., Tucker, M., and Zenger, D., eds., Dolomite: Oxford, International Association of Sedimentologists Special Volume 21, Blackwell Scientific Press, p. 283-309.

PURSER, B. H. AND HOTZL, H., 1988, The sedimentary evolution of the Red Sea rift: a comparison of the NW (Egyptian) and NE (Saudi Arabian) margins of the Red Sea and Gulf of Suez: Tectonophysics, v.153, p.193-208.

PURSER, B. H., ORSZAG-SPERBER, F., AND PLAZIAT, J-C., 1987a, Sédimentation et rifting: les séries néogènes de la marge nord-occidentale de la Mer Rouge (Egypte): Paris, Notes et Mémoires Compagnie Francaise des Pétroles, n° 21, p. 111-114.

PURSER, B. H., PHILOBBOS, E., AND SOLIMAN, M., 1990, Sedimentation and rifting in the NW parts of the Red Sea : a review: Bulletin Société Géologique de France, v. VI, p. 371-384.

RICHARDSON, M. AND ARTHUR, M. A., 1988, The Gulf of Suez- northern Red Sea Neogene rift: a quantitative basin analysis: Marine and Petroleum Geology, v. 5, p. 247-270.

ROUCHY, J. M., 1982, La genèse des évaporites messiniennes de Méditerranée: Unpublished Ph.D. Dissertation, University of Paris, Paris, 292 p.

ROUCHY, J. M., BERNET-ROLLANDE, M. C., MAURIN, A. F., AND MONTY, C., 1983, Signification sédimentologique et paléogéographique de divers types de carbonates bioconstruits associés aux évaporites du Miocène moyen près du Gebel Esh Mellaha (Egypte): Compte Rendu de l'Académie des Sciences de Paris, v. 296, p. 457-462.

THIRIET, J. P., BUROLLET, P. F., GUIRAUD, R., ICART, J. C., JARRIGE, J. J., MONTENAT, C., AND OTT D'ESTEVOU, P., 1985, Sur l'existence de jeux décrochants transpressifs dans la structuration précoce du Golfe de Suez et de la Mer Rouge; l'exemple de Port Safaga (Egypte): Compte Rendu de l'Académie des Sciences de Paris, v. 301, p. 207-212.

# ABU SHAAR COMPLEX (MIOCENE) GULF OF SUEZ, EGYPT: DEPOSITION AND DIAGENESIS IN AN ACTIVE RIFT SETTING

MARIO CONIGLIO,
*Department of Earth Sciences, University of Waterloo, Waterloo, Ontario N2L 3G1, Canada*
NOEL P. JAMES, AND
*Department of Geological Sciences, Queen's University, Kingston, Ontario K7L 3N6, Canada*
DJAFAR M. AÏSSAOUI
*CNRS-UA 723, Département de Géologie, Bâtiment 504, Université Paris-Sud, 91405 Orsay, France*

ABSTRACT: Abu Shaar is a pervasively dolomitized Miocene carbonate platform that veneers tilted Precambrian basement blocks on the western margin of the Gulf of Suez. This reef-rimmed complex developed in an active margin setting with facies composition and distribution controlled by eustasy and local tectonics. The first stage of platform development is recorded by the Kharasa Member which resulted from initial marine onlap as aprons of mixed coarse-grained siliciclastics and carbonate. These sediments pass gradually upward into open-platform carbonates and a reef-dominated platform margin. A second phase of platform development is recorded by the complex stratigraphy of the Esh el Mellaha Member, which also included prolific reef growth in platform margin and back reef areas. Following a brief period of subaerial exposure, a third phase of platform development, the Bali'h Member, saw only minor reef development and soon became dominated by restricted carbonate and evaporite facies. The fourth and final stage of development is recorded on the platform margin by an enigmatic Chaotic Breccia Member, a unit composed of the insoluble material of a formerly extensive evaporite sequence. Microfabrics of allochems that were originally Mg-calcite or calcite in composition were generally well preserved whereas originally aragonitic allochems were replaced by dolomite with significant loss of microfabric or dissolved resulting in moldic pores.

These rocks were probably dolomitized by marine to hypersaline fluids. These fluids could have refluxed into the complex from overlying shallow subtidal or sabkha environments. Alternatively, fluids could have originated as hypersaline brines while the platform lay exposed during an extensive evaporative phase accompanying Middle to Late Miocene sea-level lowstand. These fluids were subsequently introduced into the carbonate complex during transgression, causing dissolution of aragonite and dolomitization. In order to explain the wide ranging and negative $\delta^{18}O$ data, these sediments are thought to have later recrystallized, involving meteoric or more likely hydrothermal waters.

Sedimentary facies and diagenesis of the carbonates at Abu Shaar are strikingly similar to other Miocene reefs described from elsewhere in the Red Sea and western Mediterranean areas. The reasons for this, although unclear, likely include regional tectonic effects operating against a backdrop of fluctuating sea levels.

## INTRODUCTION

Deeply-incised wadis and extensive bedding plane exposures at Abu Shaar el Quibli provide a rare opportunity to trace the spatial and temporal stratigraphic development of a completely dolomitized Miocene reef-rimmed platform. Recent studies of these carbonates, one of several isolated, undeformed, Miocene platforms veneering Precambrian basement fault blocks in the Gulf of Suez area, have focused on facies development, carbonate-siliciclastic interaction and platform response to extensional tectonics in this classic active rift setting (Haddad et al., 1984; Cofer et al., 1984; Burchette, 1988; and James et al., 1988a). Reefs, stromatolites and related sediments were also examined by Rouchy et al. (1983), Rouchy (1986), Monty et al. (1987) and Purser et al. (1988a; also see citations within these sources). The diagenesis of these carbonates has been recently discussed by Aïssaoui et al. (1986), Purser et al. (1988b) and Coniglio et al. (1988). More regional studies of sedimentation and tectonics incorporating the sequence at Abu Shaar include Purser and Hötzl (1988), Purser et al. (1990), Plaziat et al. (1990) and Purser et al. (this volume). In this paper, we summarize the stratigraphic and diagenetic history of this complex based primarily on James et al. (1988a) and Coniglio et al. (1988).

The Abu Shaar sequence continues to attract attention for several reasons. This reef-rimmed platform sequence developed in an early rift setting on an active fault block, and so this study contributes to our knowledge of reefs and platforms in such tectonically active regions. In addition, Abu Shaar seems to be one of the few places where the facies, timing and style of margin collapse, and nature of platform recovery can be observed. The combination of active tectonic setting, eustatic sea-level fluctuations, evaporation, shallow-burial history, and possible hydrothermal influences during the post-depositional history of these carbonates produced dolomite whose outcrop and petrographic attributes are strikingly similar to those of meteorically-altered limestones reported from the Pleistocene units of such areas as Bermuda, Barbados, and southern Florida (Aïssaoui et al., 1986; Coniglio et al., 1988). The petrographic attributes of the dolomite in this sequence are also similar to Cenozoic dolomite studied from the Bahamas, Bonaire, Grand Cayman, Jamaica, San Salvador and the western Mediterranean (Land, 1973; Supko, 1977; Sibley, 1980; Kaldi and Gidman, 1982, Rouchy, 1982; Jones et al., 1989). The carbonates at Abu Shaar, therefore, provide further insight into possible mechanisms of pervasive dolomitization.

Knowledge of the stratigraphy and diagenesis of these carbonates and how they control the nature and extent of porosity also has considerable economic importance. Comparable platform

Models for Carbonate Stratigraphy from Miocene Reef Complexes of Mediterranean Regions,
SEPM Concepts in Sedimentology and Paleontology #5, Copyright © 1996,
SEPM (Society for Sedimentary Geology), ISBN 1-56576-033-6, p. 367-384.

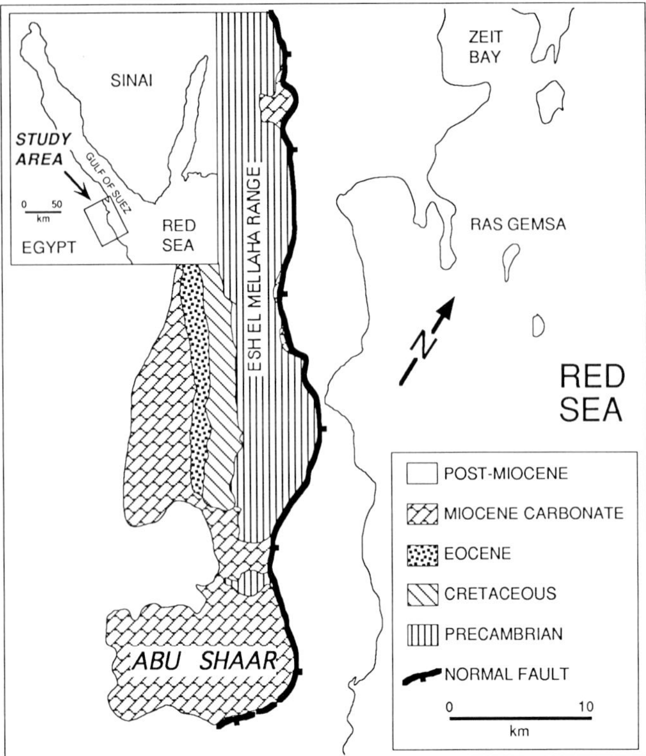


FIG. 1.—Location and geologic setting of study area. Fault trace is from Thiébaud and Robson (1979). Extension into Abu Shaar area is from Burchette (1988). Modified from Coniglio et al. (1988).

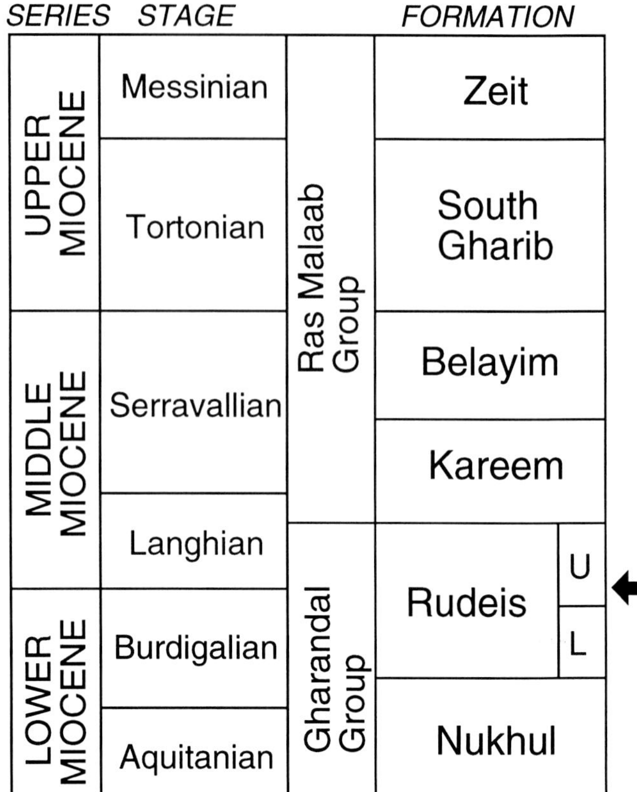


FIG. 2.—Generalized stratigraphic column of major Miocene units in the Gulf of Suez region (after Scott and Govean, 1985; James et al., 1988a). Arrow indicates approximate position of the Abu Shaar platform in the basal part of the Upper Rudeis Formation.

carbonates, developed on basement horsts and sealed by thick evaporites, are hydrocarbon reservoirs in the subsurface of the Red Sea area (Thiébaud and Robson, 1979; Kulke, 1982).

## GEOLOGIC SETTING

The Gulf of Suez is a 60-80 km wide, normal fault-bounded rift basin branching from the Red Sea rift. Evolution of the Suez rift and the early Red Sea resulted from separation of the Sinabian (Sinai and Arabia) and Nubian (Africa) plates (Girdler and Southren, 1987). Sediments in the Suez rift zone are 3 to 4 km thick and range from Oligocene to Holocene in age (Purser et al., 1990). The stratigraphy of the region is outlined in Sellwood and Netherwood (1984) and Scott and Govean (1985), with their work based mainly on subsurface data of the eastern and northern Gulf. The relationship between sedimentation and rifting is reviewed in Purser et al. (1990).

Abu Shaar is located approximately 8 km from the western shore of Gulf of Suez, at the southern end of the Esh el Mellaha block, one of four fault-bounded and tilted basement blocks that border the Gulf of Suez (Fig. 1). These fault blocks are the main structural features of the Gulf of Suez area and are defined by the intersection of NW-SE and NE-SW to E-W fault trends. The margin of each block adjacent to the Gulf of Suez is a listric normal master fault (Burchette, 1988). The blocks were active during Late Eocene time at the onset of rifting and later during Oligocene or Lower Miocene time (Haddad et al., 1984). The gentle southwest dip of the basement blocks is locally offset by an irregular erosional topography on the Precambrian surface (Haddad et al., 1984).

Miocene deposits veneer parts of these blocks and fill synclinal depressions (Purser et al., this volume). The Miocene record in the Gulf of Suez region is subdivided into a lower Gharandal Group comprising predominantly siliciclastic rocks of the Nukhul and Rudeis Formations, and an upper Ras Malaab Group containing widespread evaporites (Fig. 2; Sellwood and Netherwood, 1984; Scott and Govean, 1985).

The carbonate sequence exposed at Abu Shaar covers an area of approximately 100 km$^2$ and forms a nearly flat plateau which stands some 200 m above the surrounding coastal plain (Fig. 3). Strata on the platform dip 2 to 5° to the southwest into a sand and gravel plain formed by Quaternary alluvium from the Red Sea Hills. In contrast, strata on the steep eastern and southern margins may reach dips up to 40° (James et al., 1988a).

A Miocene age for the Abu Shaar sequence has been known since the early part of the century based largely on the coral faunas (Gregory, 1906; Hume, 1916; Madgwick et al., 1920). The precise age of these carbonates, however, is unresolved largely due to dolomitization. A probable age of latest Early Miocene (latest Burdigalian) or early Mid Miocene (Langhian)

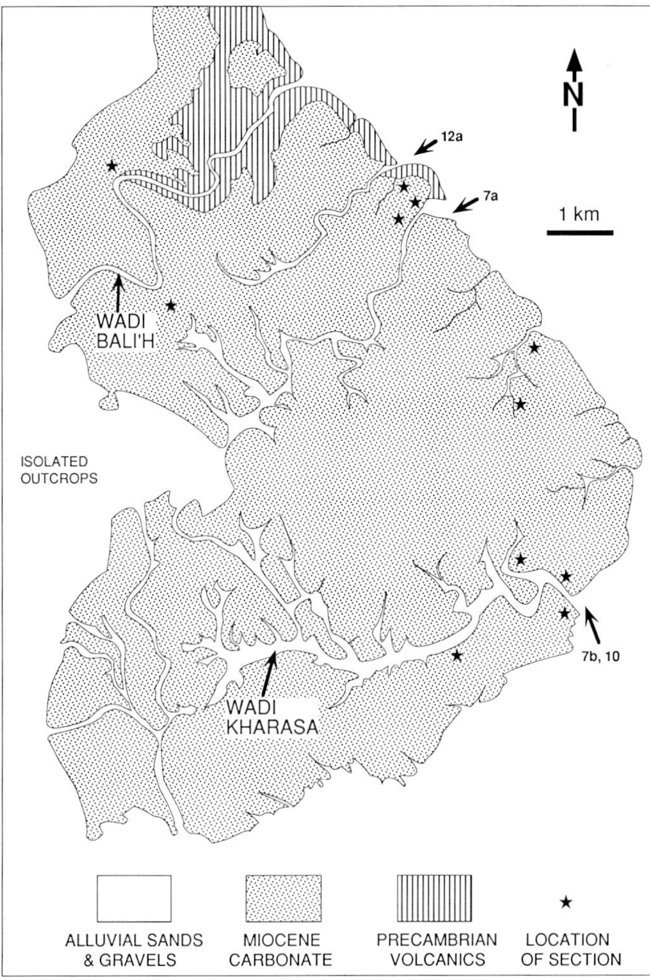


FIG. 3.—Map of Abu Shaar showing major wadis and locations of measured sections. Localities of outcrops in Figures 7A, 7B, 10 and 12A are also shown. Modified from James et al. (1988a).

is indicated by calcareous nannoplankton ages (NN4 nannoplankton zone) recovered from basinal carbonates in subsurface sections in the Zeit Bay area approximately 50 km north of Abu Shaar (Cofer et al., 1984). A Langhian age for the Abu Shaar sequence is consistent with that of similar reefal carbonates in the Middle East (Buchbinder and Martinotti, this volume), Paratethys (Pisera, this volume) and the western Mediterranean (Calvet et al., this volume; Esteban et al., this volume). James et al. (1988a) suggest that these sediments lie in the basal part of the Upper Rudeis Formation of the Gharandal Group.

The Rudeis Formation is predominantly composed of basinal marine globigerinid marls and mudstones hundreds of meters thick. Black organic-rich shales in the lower part of the formation occur under the Gulf and grade laterally into conglomerates and sandstones toward the rift margins. Deposition of the Upper Rudeis Formation was preceded by major uplift, known as the 16.5 Ma mid-Rudeis Event (or Mid-Clysmic Event), that occurred at the N7-N8 planktonic foraminiferal zone boundary (Blow, 1969, cited in Smale et al., 1988). This disturbance was manifested through increased fault block reactivation and rotation of tilt blocks and was coincident with a global sea-level lowstand (Smale et al., 1988). The stratigraphic record comprises local unconformities and extensive conglomerate beds that grade into calcarenites away from the faults and then pass laterally into reef limestones. This general setting is similar to that along the present-day margins of the Gulf of Suez (Purser et al., 1987).

The Abu Shaar complex is bordered to the east by younger Miocene evaporites that apparently were also deposited over the sequence but have since been eroded. These evaporites dip 10-20° to the east and pinch out against the sloping strata of the complex (Monty et al., 1987).

## LITHOFACIES

The Abu Shaar complex is composed of four discrete stratigraphic units — Kharasa, Esh el Mellaha, Bali'h, and Chaotic Breccia — here described as informal local members of the Rudeis Formation (James et al., 1988a). Reefal sediments are a conspicuous component of the platform margin in the Kharasa and Esh el Mellaha Members and, to a lesser extent, the Bali'h Member. Reefs are continuous to patchy, approximately 100 m in width with coral abundances typically ranging from 30-50%. A comprehensive description and discussion of facies compositions and distribution are provided in James et al. (1988a) and the reader should consult this source for systematic coverage.

### *Kharasa Member - Stage 1*

The Kharasa Member records the transition from siliciclastic-dominated transgressive sediments to the establishment of a reef-rimmed carbonate platform (Fig. 4a). The early phase of Middle Miocene (Langhian, latest Burdigalian) sea-level rise resulted in flooding of the tilted blocks of Precambrian basement initially from the south and west (also see Burchette, 1988). The emergent southern end of the Esh el Mellaha range was a persistent source of boulder conglomerates and terrigenous sands that were shed into the adjacent shallow marine environment as halos around basement highs and in topographic depressions on the Precambrian basement (Fig. 5a). Bioclastic grainstones became increasingly more important away from the coarse siliciclastic halos.

Overlying thinner bedded strata are dominated by sand- to granule-sized siliciclastics with normal grading, trough cross bedding and layers of bivalve coquina, rhodolites, and burrowed wackestone (Fig. 5b). The range of mixed carbonate and siliciclastic sediments suggest depositional conditions varied from high-energy beaches to quiet-water lagoons in which coarse storm beds were periodically introduced.

Continued sea-level rise forced a northward retreat of the shoreline, significantly reducing but not eliminating the siliciclastic component in the sediment. The upper half of the Kharasa Member consists of: (1) open-platform carbonates dominated by variably bioturbated rudstones, grainstones,

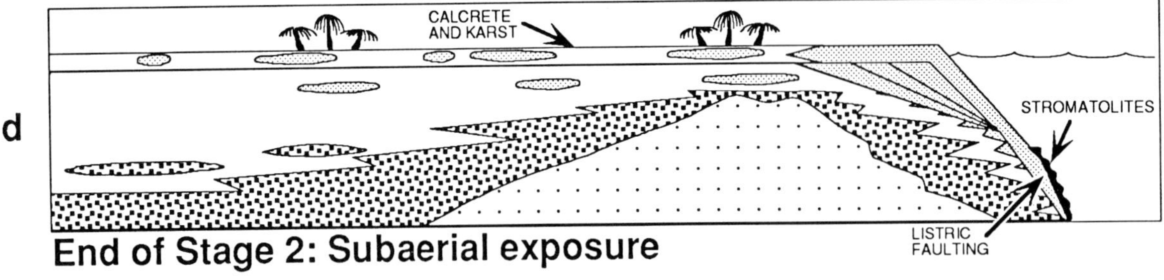


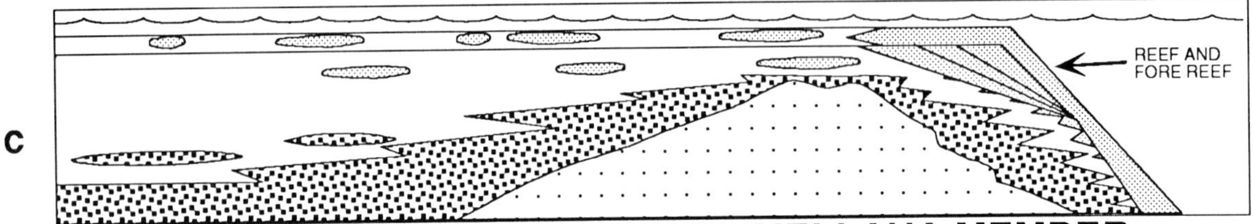


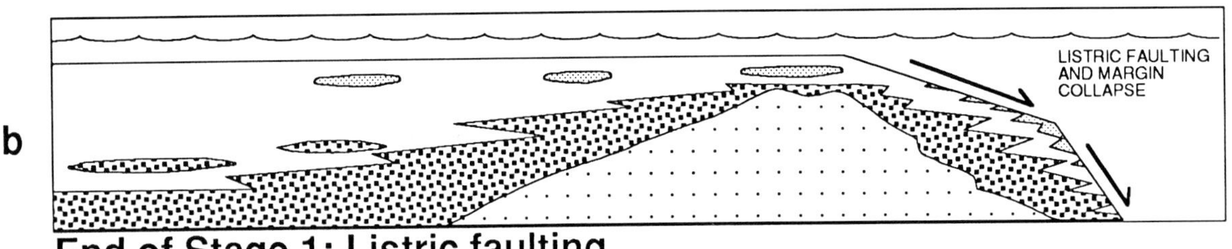


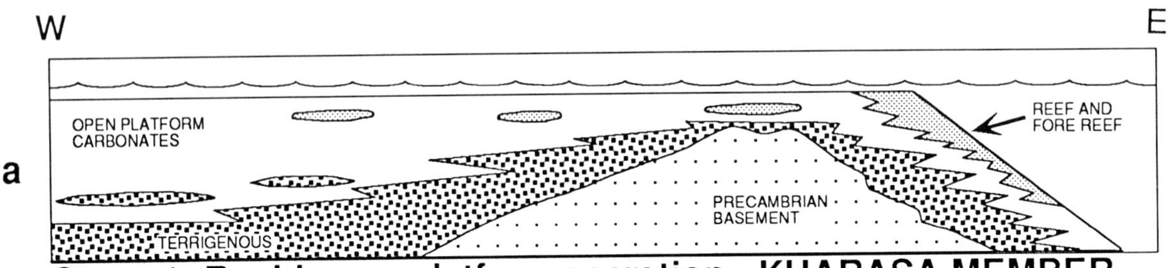


FIG. 4.—Schematic summary of stratigraphic development of the Abu Shaar complex. See text for details. Platform and basement depicted in frame "g" is approximately 8 km wide and 200 m thick (from James et al., 1988a).

packstones and wackestones, with minor reefs and (2) platform-margin reefs. The diverse fauna of the open platform includes corals, bivalves, gastropods, bryozoans, echinoids, and benthic foraminifera. Mud-free, unbioturbated sediments indicate deposition in relatively high-energy settings, whereas bioturbated bioclastic grainstones to wackestones were deposited in lower-energy settings, likely a grass-covered sea bottom. Rhodolite rudstones (Fig. 5c) locally grade laterally into small patch reefs. Pavements of encrusting coralline algae also occur. Interbedded bivalve coquinas probably represent deposition from periodic storms (Fig. 5d).

Reefs are not common on the open platform. Small bioherms 2-3 m across may be composed almost exclusively of the branching coral *Stylophora*. One particularly notable bioherm in Wadi Bali'h is estimated to be 6-8 m high and 25-30 m across and is composed of massive *Montastrea* colonies up to 1 m high containing numerous pholad borings. Bioherms are typically surrounded by a halo of coarse bioclastic sediment, usually rhodolite rudstones and/or bivalve coquinas. Throughout the Abu Shaar sequence, corals are commonly preserved as molds.

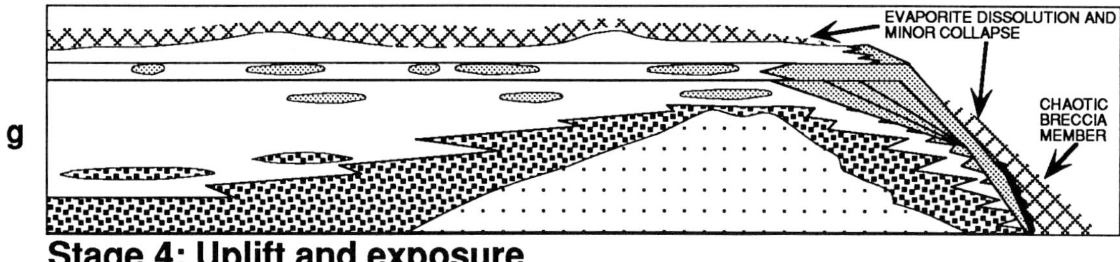


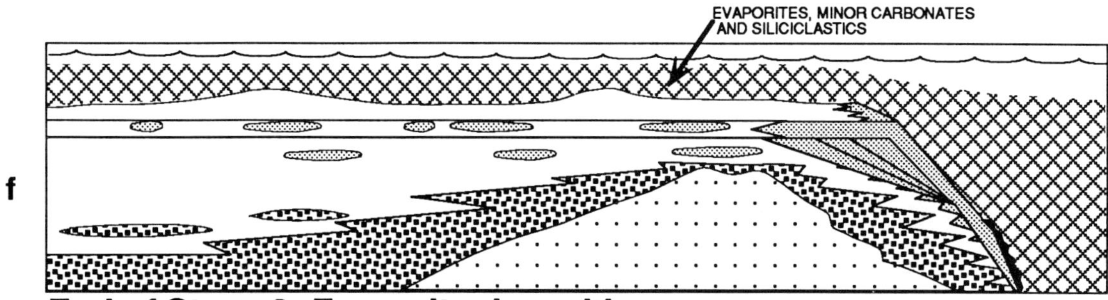


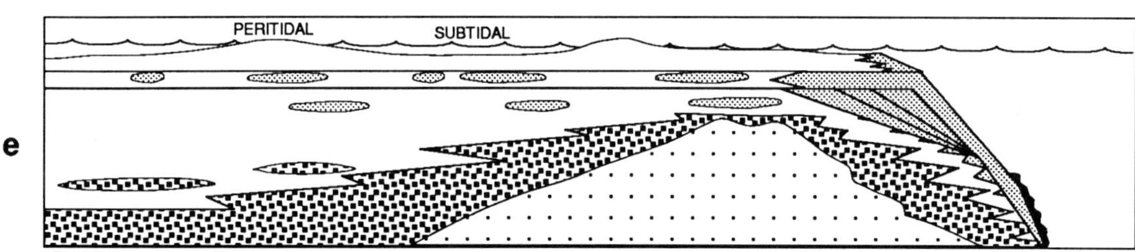


FIG. 4.—Continued.

Many of the large nondescript vugs encountered in outcrop are likely to be solution-enhanced coral molds.

Reef facies at the platform margin comprise patch reefs, semicontinuous barrier reefs and biostromes containing a variety of corals (James et al., 1988b). The outer 10 m or so is generally a framestone composed of large columnar *Porites* and numerous 10-30 cm-diameter faviids (Figs. 6a, b). Corals occupy approximately 30% of the rock, with the interframework sediment typically being a bioclastic mudstone. The upper parts of some reefs are dominated by faviids, especially *Montastrea* and *Tarbellastrea*. Corals occurring here and throughout the Abu Shaar succession are commonly encrusted by coralline algae, bored by endolithic molluscs, and more rarely coated by marine cement. These massively bedded reef carbonates grade shelfward into the more conspicuously bedded, open-platform facies described earlier.

Not all of the platform-margin setting, however, is dominated by massive coral framestone. At the southern margin of the complex, it is clear that bioclastic sand shoals are locally an important component of the platform-margin facies mosaic. On the eastern margin, 4- to 6-m-thick biostromal beds of branching *Stylophora* and *Caulastrea*(?), and minor faviids are interbedded with rhodolite rudstones, bioclastic grainstones, and bivalve coquina (Figs. 6c, d).

This first stage of platform accretion was interrupted by normal listric faulting along the eastern edge, which truncated some of the platform-margin facies and likely redeposited them in deep water (Fig. 4b; discussed further below). These redeposited sediments are presumably now buried beneath the adjacent evaporite plain to the east.

### *Esh el Mellaha Member - Stage 2*

Deposition resumed as a series of narrow, fringing reefs and coralline algal flats initially along the margin and then later across much of the platform as well as the steep front of the platform (Fig. 4c). Along the eastern margin most of the earliest reef facies form an upper slope progradational "wedge" up to 50

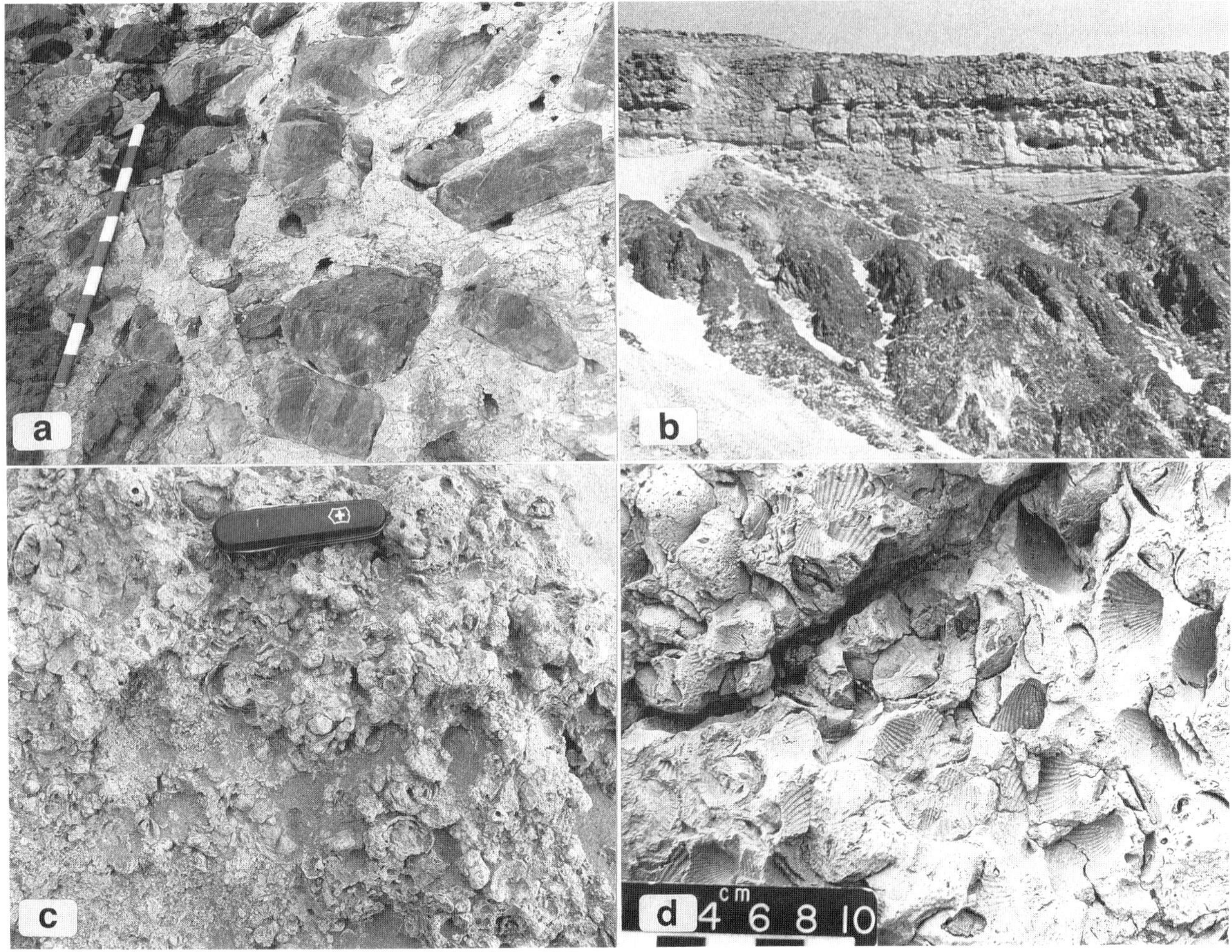


FIG. 5.—Outcrop photographs of Kharasa Member open-platform facies. (a) Basal conglomerate composed of large angular clasts in bioclastic carbonate matrix. Scale divisions are 10 cm. (b) Siliciclastic-rich facies approximately 30 m thick overlying dark-colored Precambrian basement. (c) Rhodolite rudstone. Pocket knife for scale. (d) Bivalve rudstone in which all bivalves are leached, yielding biomoldic porosity. Scale divisions are 2 cm.

m thick deposited on the truncated Kharasa platform margin (Fig. 7a). The wedge is composed of smaller stacked wedges which taper basinward. Sediments in the wedge include coral-rich rhodolite rudstones with a grainstone matrix and local layers of bivalve coquina and massive coral framestone composed predominantly of faviids and mussids. Steeply-dipping fore-reef strata downslope from the wedge are slope-parallel and consist of bivalve-rich grainstones, bivalve-coral-coralline algal rudstones and coral rudstones.

Overlying these sediments is a continuous, 2- to 4-m-thick "reef veneer" that extends from the basal fore reef to the reef crest at the top of the wedge and then as intermittent biostromes and patch reefs over the open-platform facies of the Kharasa Member (Fig. 8a). This unit dips 35-40° from the platform margin and forms the modern surface slope. Thickets of columnar *Porites* or branching *Caulastrea* occur at the reef crest, which is also the modern break in slope (Figs. 8b, c). These grade downslope into branching *Stylophora* thickets or massive faviid framestone or floatstone in the reef front (Fig. 8d) and then into skeletal grainstones containing varying amounts of coral, coralline algae, bivalves, benthic foraminifera and *Halimeda*. Locally sand- to granule-sized pisoids occur (Fig. 9a). Although their origin remains unclear, the fibrous carbonate comprising the cortex of the pisoids bridges between particles in some cases and suggests *in situ* growth on the slope. Detailed discussion on these pisoids is provided in Aïssaoui et al. (1986), Monty et al. (1987), James et al. (1988a), and Coniglio et al. (1988).

In contrast to the eastern margin described above, discernible facies differentiation is not present along the southern margin. Here, reef facies comprise a massive 100-m-thick prograding unit plastered to the front of the Kharasa reef margin (Figs. 7b, 10). Bedding is obvious and dips southward at 25-30°. This unit is interpreted as a prograding reef and fore-reef complex. The upper, more massive part is mostly a framestone composed of

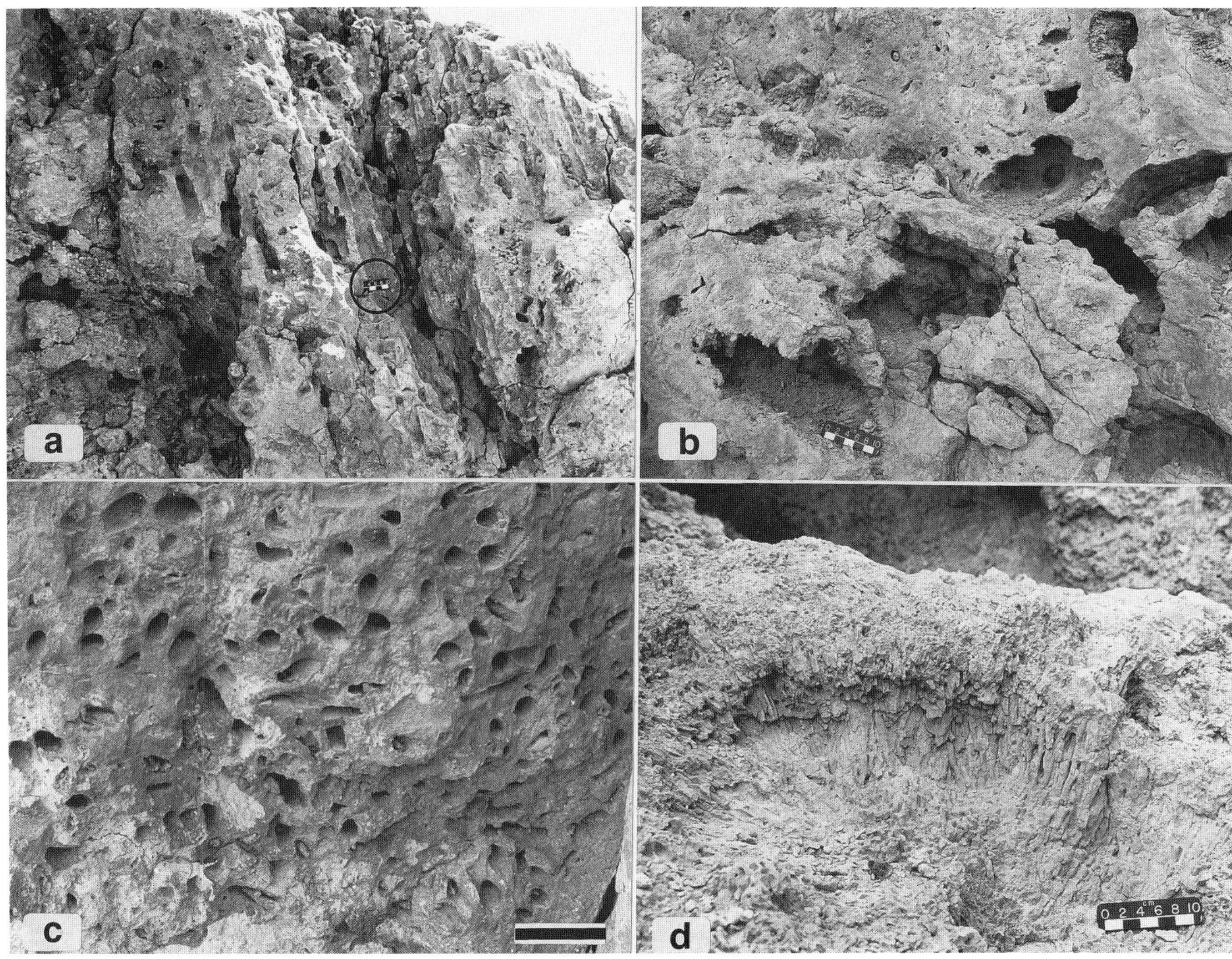


FIG. 6.—Outcrop photographs of Kharasa Member platform-margin facies. (a) Massive *Porites* framestone at platform margin with leached corals forming tubular pores. Scale (circled) divisions are 2 cm (near center of photograph). (b) Massive faviid framestone near platform margin in which dissolved corals form large vugs. Scale divisions are 2 cm. (c) *Stylophora* (leached) floatstone approximately 50 m from truncated margin. Scale bar is 10 cm. (d) Well-preserved *Caulastrea* approximately 500 m from platform margin. Scale divisions are 2 cm.

approximately equal proportions of meter-sized *Porites* and faviid colonies, many in growth position. Framestones grade downslope into bedded fore-reef bioclastic grainstones to wackestones with scattered faviids and thickets of *Caulastrea* encrusted by coralline algae.

The youngest part of the Esh el Mellaha Member is a widespread 1 m-thick stromatolite unit (Rouchy, 1982) exposed at the base of the present-day physiographic slope (Figs. 7a, b, 9b). Haddad et al. (1984) and James et al. (1988a) suggested a deep-water origin for these stromatolites based on the following observations: (1) their superposition over deep fore-reef sediments or deep-water ahermatypic corals; (2) possible ahermatypic corals within the stromatolites; and (3) absence of evidence of evaporites or shallow-water desiccation features. In addition, their unusual petrographic characteristics suggest they may be analogous to hardgrounds and crusts presently forming today on the lower slopes of carbonate platforms (James et al., 1988a; Coniglio et al., 1988).

Following deposition of the Esh el Mellaha sediments, a period of subaerial exposure and attendant meteoric diagenesis resulted from eustasy or tectonic tilting. This caused the sporadic development of a calcrete (now dolomitized) and karst surface at the top of the flat lying Esh el Mellaha Member (Fig. 4d) and at the top of the Kharasa Member where the Esh el Mellaha Member was not deposited. Evidence for subaerial exposure is only recognizable in the outer parts of the platform, along the southern and eastern sides. During this exposure event, numerous corals were leached, creating secondary pores that later, during subsequent transgression, became partly filled with marine cement and internal sediment.

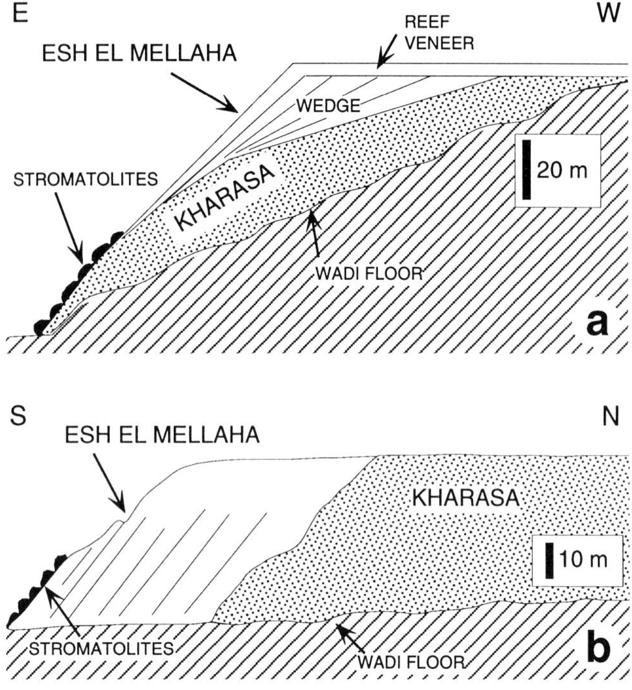


FIG. 7.—Cross sections through Esh el Mellaha Member. Vertical scales are approximate. (a) Upper slope wedge and reef veneer as viewed towards south along eastern margin. Width is approximately 1 km. (b) Prograding reef and fore-reef complex as viewed towards west along southern margin at Wadi Kharasa. Width is approximately 300 m. See Figure 10.

### *Bali'h Member and Chaotic Breccia Member - Stages 3 and 4*

Reflooding of the platform was accompanied by a gradual change in the depositional style from carbonate to progressively more evaporitic, foreshadowing the eminent demise of the reef and open-platform environment (Figs. 4e, f). The Bali'h Member consists of poorly developed, low-energy, peritidal, shallowing-upward sequences with open-marine sediments grading upward through stromatolitic carbonates into evaporites, now present as collapse horizons or chert. These in turn are capped by laminated carbonate mudstones. This unit is up to 50 m thick in the northwest and is eroded to 20 m along the eastern margin. Rapid lateral and vertical facies changes characterize this unit with terrigenous siliciclastics being more important towards the north. The generally good local but poor regional correlation of units suggests that both shoreline and island tidal flats were important elements in the paleoenvironmental mosaic.

The lower half of the Bali'h Member includes: (1) rippled and burrowed, occasionally oolitic, bioclastic grainstones, (2) faviid coral rudstones and framestones with minor branching *Stylophora* biostromes, (3) rudstones, packstones, and wackestones containing bivalves, gastropods (particularly cerithids), or rhodolites, and (4) mud-cracked mudstones. These deposits represent a variety of shallow-marine, low to high energy, subtidal to strandline environments. The monospecific to low diversity molluscan fauna in some wackestones and mudstones suggest that restricted, probably metahaline conditions were locally developed (Cofer et al., 1984). Even though platform-margin facies have largely been removed by modern erosion, the northern part of the Abu Shaar complex contains a coral framestone to rudstone dominated by *Acanthastrea*, *Montastrea* and other miscellaneous faviids, with accessory rhodolites and bivalves. James et al. (1988a) interpreted this feature to be part of a series of patch reefs or a remnant of more continuous reefs on the platform margin.

The sediments in the upper half of the Bali'h Member are similar to those in the lower half but also include stromatolites (Monty et al., 1987), dololaminites and evaporite solution horizons (Figs. 11a-c). The assemblage of normal-marine bioclastic and stromatolitic carbonates, dololaminites, teepees, mudcracks and evaporites led James et al. (1988a) to interpret the upper Bali'h Member as representing deposition on arid tidal flats bordering a normal marine setting, much like the present Gulf of Suez.

Irregular large domes, up to 20 m across, cap the Miocene sequence (Fig. 11d). These domes are poorly bedded and locally brecciated, possibly resulting from evaporite solution or modern subaerial exposure (James et al., 1988a). Alternatively, Purser et al. (1990) interpreted these structures to have resulted from periodic seismicity resulting from earthquakes. Similar megadomes occur at the top of Messinian reef facies of the western Mediterranean and appear to be a common facies during or after evaporite formation (Esteban, 1979).

The adjacent basin was eventually filled with subaqueous Middle to Upper Miocene evaporites, comprising mainly anhydrite, gypsum and halite, with some shale, dololaminite and limestone (end of Stage 3; Fig. 4f). This largely evaporitic sequence averages approximately 100 m in thickness along most of the northwest coast of the Red Sea but may thicken to 2000 m in some wells due to axial subsidence prior to or during evaporite deposition (Purser et al., 1990). The extent of burial of the Abu Shaar platform by evaporites is unknown; however, it is thought to be a few hundred meters or less based on the lack of pressure solution features, poor definition of bedding planes and lack of grain fracturing. These evaporites were later removed and in the study area are now represented by the Chaotic Breccia Member ("subaerial talus" of Haddad et al., 1984), a unit of massive collapse breccia on the platform margin (Stage 4; Fig. 4g). The degree of lithification and porosity development in this 20-m-thick dolomite unit is variable. Well-lithified portions range from dolomite with numerous gypsum crystal molds to very porous tufa-like rocks to dolomite-cemented sandstone. James et al. (1988a) interpreted this unit to be the insoluble material within and overlying the formerly extensive evaporite.

## TRUNCATION OF PLATFORM MARGIN

The Abu Shaar complex contains several truncation surfaces in various locations along the seaward margin (Figs. 7a, 12a, b). These surfaces trend parallel to the attitude of the major faults

FIG. 8.—Outcrop photographs of corals from the Esh el Mellaha Member. (a) Faviid biostrome forming prominent bench at top of reef veneer. (b) Massive *Porites* at the reef crest and present-day break in slope. Columnar corals are approximately 1.5 m high. Hammer (circled) for scale. (c) Phaceloid *Caulastrea* in reef-crest facies. (d) Faviid floatstone in which most corals are dissolved. Scale divisions are 2 cm.

and are discussed fully in James et al. (1988a) and Burchette (1988). A series of these surfaces are clearly displayed in an unnamed wadi south of Wadi Bali'h (referred to as S. Wadi Bali'h by Burchette, 1988; Fig. 12a). Here, a gently dipping (ca. 20°) slightly concave upper surface cuts the platform-margin facies of the Kharasa Member. A steeply dipping (ca. 35°) middle surface truncates both the Kharasa Member and Precambrian basement. The above two surfaces predate deposition of the Esh el Mellaha Member. A very steeply-dipping (47-52°) lower surface cuts the lower facies of the Esh el Mellaha Member and basement but is covered by younger facies of the unit. Numerous observations indicate that there are no slickensides on these surfaces, which cut carbonate, conglomerate and basement smoothly.

James et al. (1988a) interpreted the truncation surfaces to be the lower detachment planes of high-angle collapse structures, the result of synsedimentary normal faulting related to intermittent activation of basement faults. This explanation is consistent with the continuity of the truncation surface through both relatively soft Miocene and hard Precambrian strata, their occurrence along the axis-facing side of the rift and the presence of local synsedimentary vertical faults up to 10 m high that cut Kharasa conglomerates but are buried by Esh el Mellaha fore-reef carbonates. Presumably, the rocks of the hanging wall moved and collapsed into the adjacent basin, either as huge olistoliths or as debris flows.

## DIAGENESIS

### *Petrography*

*Dolomite crystal characteristics.—*

Haddad et al. (1984) were the first to recognize that these carbonates are entirely dolomite. Under high magnification in transmitted light, dolomite crystals, whether replacement or pore-filling, are anhedral to subhedral with relatively sharp

FIG. 9.—Outcrop photographs of Esh el Mellaha facies. (a) Plan view of pisoids on reef front. (b) Cross section of stromatolites (S) on Precambrian basement (B) at West Gemsa, approximately 40 km north of Abu Shaar. Scale bar is 30 cm.

FIG. 10.—West wall of Wadi Kharasa showing massively bedded Kharasa reef margin (K) overlain by steeply-dipping Esh el Mellaha beds (M). The entire sequence is capped by the Bali'h Member (B). Approximately 80 m of section is shown.

extinction. Replacement and cement crystals are commonly 10 µm or smaller although some mosaics are dominated by 50- to 100-µm-sized crystals. In a given sample, replacement dolomite is usually more finely crystalline than dolomite cement. Dolomite cement commonly occurs in primary inter- and intraparticle pores and secondary vuggy and moldic pores. Equant dolomite cement lines pores which remain incompletely cemented or is followed by a later phase of calcite or evaporite cement.

FIG. 11.—Outcrop photographs of evaporitic carbonates of Bali'h Member. (a) Stromatolites under hammer (circled) overlain by carbonate with disrupted bedding and extensive chert replacement. (b) Disrupted bedding. (c) Evaporite crystal molds (white arrows) in hard white dolomite at the top of the Bali'h Member. (d) Convex structure of large domes at the top of the Bali'h Member. Person (circled) for scale.

Most dolomite crystals have cloudy, inclusion-rich cores and clear, relatively inclusion-free rims. Their cathodoluminescence is a relatively uniform, dull to moderately bright orange-red color with faint, if any, zoning. Cloudy cores are more brightly luminescent than their inclusion-free rims. Both cement and replacement dolomite form porous, sucrosic aggregates or tightly intergrown mosaics. Replacement dolomite preserves microfabric mainly by crystal size differences and slight differences in luminescence that help differentiate interparticle matrix from allochem.

The most distinctive characteristic of both cement and replacement dolomite crystals is their non-planar surfaces, seen clearly either by focusing through larger crystals in thin sections or SEM observation of fractured samples and polished thin sections (Figs. 13a-c). Many crystals are anhedral and have smooth surfaces. Other crystals have complex surfaces, the result of peculiar growth features and dissolution. This small-scale surface relief or microtopography is shown by all dolomite crystals, except in some resistant-weathering dolomite at the top of the Bali'h Member where crystals are multi-faceted polyhedra. Many cement and replacement dolomite crystals also show intensive intracrystalline leaching, resulting in moldic zones and dissolved crystal cores (Fig. 13d). These microfabrics are further discussed in Aïssaoui et al. (1986) and Coniglio et al. (1988). Similar moldic zones and anhedral or non-planar crystals were described in finely crystalline dolomite in dolomitized reef complexes in the western Mediterranean by Oswald et al. (1991a).

*Allochems.—*

The microfabrics of Mg-calcite allochems, such as coralline algae, benthic foraminifera, and echinoderms and less common calcite allochems, such as ostreids and pectenids, are generally well preserved, commonly with mimic replacement of original optical c-axis orientations in the replacement dolomite (cf. Bullen and Sibley, 1984; Fig. 14a). Despite preservation of Mg-

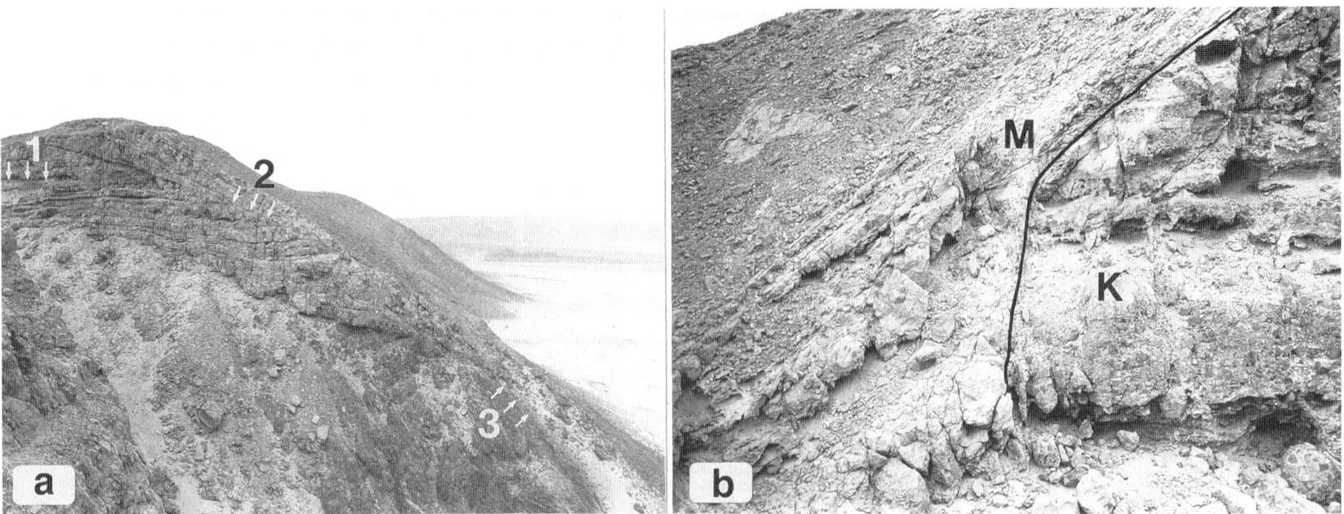


FIG. 12.—Truncation surfaces. (a) Kharasa Member is truncated by surface 1. Surface 2 truncates Kharasa Member as well as Precambrian basement. These surfaces are overlain by the Esh el Mellaha wedge and reef-front sediments. Surface 3 truncates basement and is overlain by Esh el Mellaha fore-reef sediments. Total section is approximately 20 m thick. (b) Horizontally-bedded Kharasa conglomeratic carbonates (K) are cut by inclined and near-vertical truncation surfaces (faults) and overlain by steeply dipping beds of the Esh el Mellaha (M) Member. Section is approximately 15 m thick.

calcite and calcite allochems, there is no evidence of any dolomitized low-Mg calcite cements.

Originally aragonitic allochems are poorly preserved by dolomicrospar or dissolved (Fig. 14b). The molds are either empty or partially to completely filled with later dolomite, post-dolomite calcite, or evaporite cement. Solution enlargement of various biomolds is thought to account for the origin of the majority of non-descript, millimeter- to decimeter-sized vugs in these rocks. Some coral molds are filled with fibrous marine cement and geopetal sediment, indicating that porosity development in some instances was a relatively early diagenetic event.

*Fibrous and Micritic Cements.—*

A volumetrically minor but widespread pore-filling, dolomitized, fibrous cement occurs in primary and secondary pores in the reef-front and fore-reef sediments of the Esh el Mellaha Member and in secondary pores in patch reefs of the Kharasa Member. In the Esh el Mellaha Member, the fibrous cement may be overlain by or interlaminated with dolomitized micritic internal sediment, some of which is demonstrably marine in origin (Aïssaoui et al., 1986; Fig. 15a). Dolomitized, fibrous calcite also occurs in the fibrous cortex of pisoids and in the peculiar fore-reef stromatolites of the Esh el Mellaha Member. Fibrous cement fringes are isopachous to irregular, typically 10-500 μm in width and rarely as thick as 1 cm. Most crystals are approximately 100 μm long and a few micrometers wide. The high degree of fabric preservation similar to dolomitized Mg-calcite allochems indicates these cements were originally precipitated as Mg-calcite marine cement (Coniglio et al., 1988).

Minor amounts of micrite cement are now represented by equant dolomite crystals 1-15 μm in size. The cement occurs as circumgranular isopachous crusts, circumgranular irregular crusts, and intergranular micritic clots or peloids. Isopachous crusts reach 25 μm in width and are most clearly observed in well-sorted grainstones (Fig. 15b). The isopachous distribution, as well as the small crystal size is reminiscent of modern micritic marine cements (e.g., James and Choquette, 1990). The original mineralogy of the micrite cement may have been either Mg-calcite or aragonite. Circumgranular irregular crusts are rare and restricted to grainstones adjacent to subaerial surface horizons or which contain other microfabrics suggestive of calcite precipitated within calcretes. The intergranular micrite clots are typically silt-sized, vary considerably in their definition and are most readily seen in well-sorted grainstones (Fig. 15c). They are similar to peloidal Mg-calcite cements described from modern reefal sediments (Macintyre, 1985).

*Postdolomite Phases.—*

Calcite cement, pore-filling evaporite, and authigenic quartz are ubiquitous although they are of minor importance overall. Their relative timing of precipitation is difficult to determine, but all clearly postdate dolomitization. Calcite crystals are considered to be vadose precipitates based on the occurrence of microstalactites and are commonly associated with red brown iron oxide or hydroxide crusts and impregnations. Gypsum or anhydrite overlies calcite cements and iron oxides and locally fills porosity as the last precipitate. Authigenic quartz occurs mainly as 1- to 15-cm-sized black chert nodules associated with stromatolites and related sediments in the Bali'h Member. Chert always postdates dolomitization as indicated by entombed crystals of dolomite.

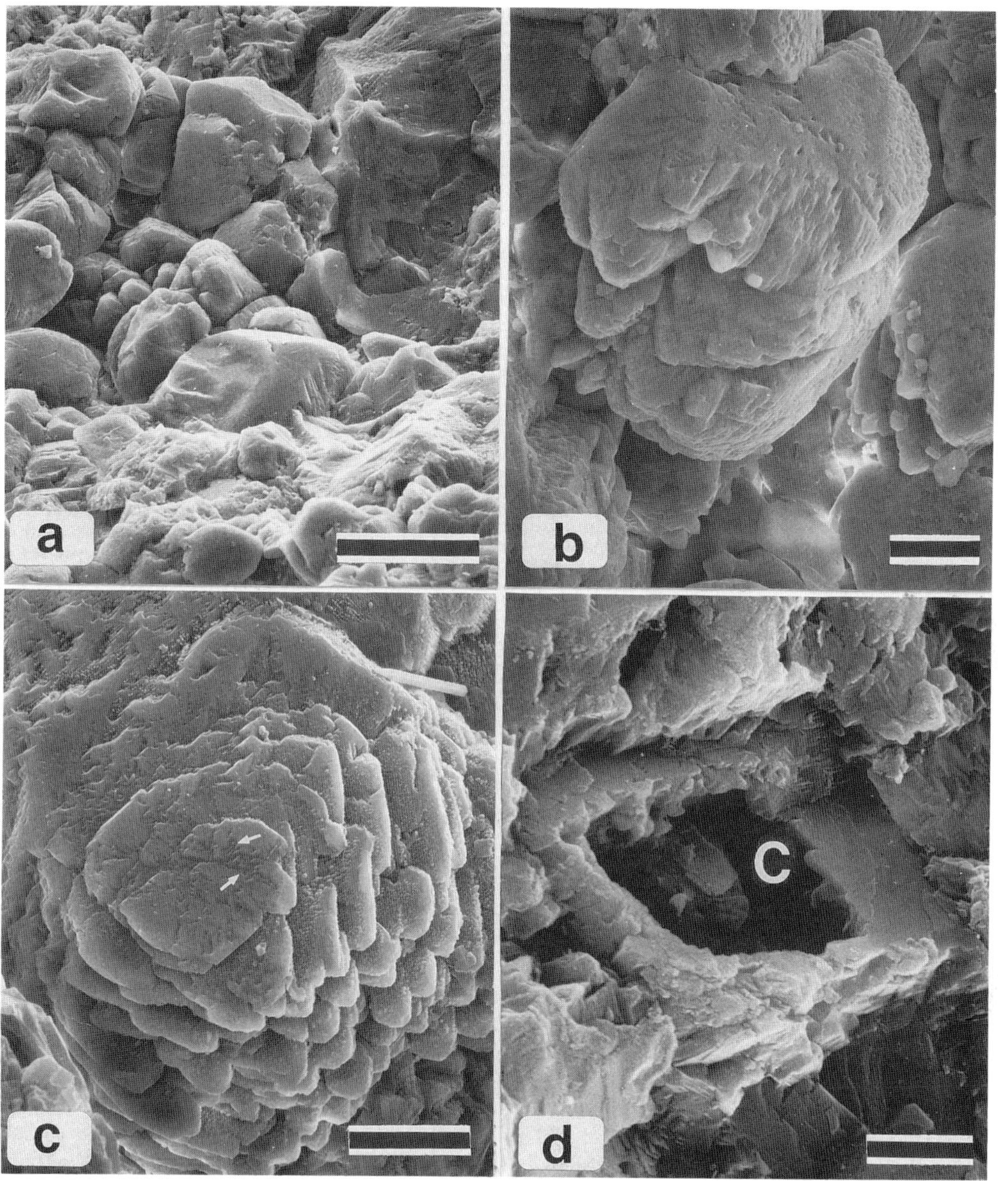


FIG. 13.—SEM photomicrographs. (a) Porous mosaic of anhedral dolomite with smooth crystal surfaces; Kharasa Member. Scale bar is 20 μm. (b) Dolomite crystal with cleavage-controlled etching; Kharasa Member. Scale bar is 10 μm. (c) Dolomite crystal with cleavage-controlled etching and dissolution microchannels (arrows); Kharasa Member. Scale bar is 10 μm. (d) Dolomite crystal with dissolved core (C); Kharasa Member. Scale bar is 3 μm.

*Dolomite Geochemistry*

Ninety-nine samples of dolomite were analysed for carbon and oxygen stable isotope compositions. Results are reported in parts per mil (‰) versus the PDB-1 standard. Untreated samples were prepared by reaction in 100% anhydrous phosphoric acid at 25° for 4 days. $\delta^{18}O$ values were left uncorrected for acid fractionation effects relative to calcite. Precision for both $\delta^{13}C$ and $\delta^{18}O$ is ±0.2‰. Samples include whole rock (grains and matrix), homogeneous mudstones, cements, pisoids, and fossils.

Considering all isotope analyses, there is a wide spread of approximately 12‰ in both $\delta^{13}C$ and $\delta^{18}O$ values, with $\delta^{13}C$ concentrated between -2 and +2‰, and $\delta^{18}O$ ranging mostly between +1 and -7‰ (Fig. 16). Each of the three stratigraphic members has wide-ranging $\delta^{13}C$ and $\delta^{18}O$ values, and no stratigraphic or geographic trends are discernible. Individual components, such as fibrous cement, may show significant differences in $\delta^{18}O$ and, to a lesser extent, $\delta^{13}C$ among different samples taken from the same slab. There are also nonsystematic variations in $\delta^{13}C$ and/or $\delta^{18}O$ between different components, such as fibrous cement and pisoids.

Microprobe analyses (n = 53) indicate that Abu Shaar dolo-

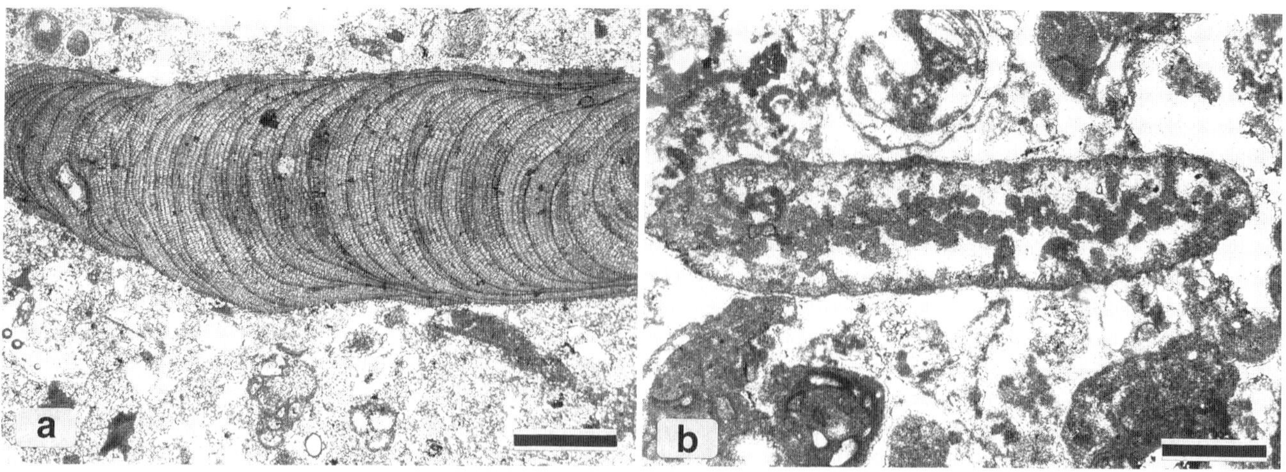


FIG. 14.—Thin-section photomicrographs in plane light of preservation styles of various dolomitized allochems. Scale bars are 500 µm. (a) Coralline algae exhibits excellent microstructure retention in porous, sucrosic dolomite mosaic; Esh el Mellaha Member. (b) *Halimeda* plate in bioclastic grainstone is mostly leached but still preserved by micrite envelope and micritic filling of utricles. A portion of the moldic porosity is filled with dolomite cement; Bali'h Member.

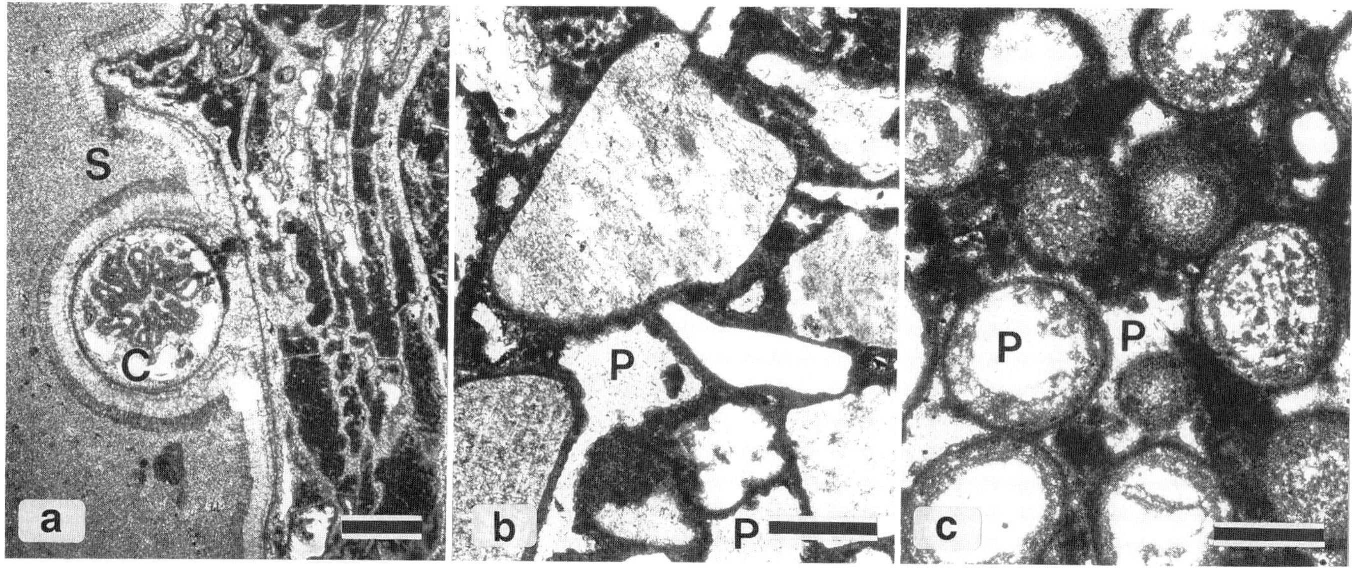


FIG. 15.—Thin-section photomicrographs in plane light of dolomitized marine cements. (a) Isopachous fringe of multi-generation marine cement which was precipitated directly on coral (C) and is overlain by micritic internal sediment (S); Esh el Mellaha Member. Scale bar is 1 mm. (b) Isopachous, micritic, marine cement coats sand-sized grains of quartz (clear) and feldspar (turbid). Interparticle porosity lined by marine cement is indicated by P; Bali'h Member. Scale bar is 250 µm. (c) Interparticle pores in well-sorted, but poorly-preserved ooid grainstone, contain peloidal micrite, interpreted to be marine cement. Interparticle and oomoldic porosity is indicated by P; Bali'h Member. Scale bar is 250 µm.

mite is stoichiometric. From inductively-coupled argon plasma (ICAP) emission spectroscopy, Sr ranges from 70 to 440 ppm, averaging 150 ppm (n = 16). Additional details concerning isotopic and elemental analyses are reported in Coniglio et al. (1988).

## DISCUSSION

### *Interpretation of Diagenesis*

The $\delta^{18}O$ values of Abu Shaar dolomite suggest the presence of isotopically light or warm waters rather than seawater or concentrated $^{18}O$-rich seawater-derived brines (cf. Oswald et al., 1990). The more negative $\delta^{18}O$ values are unlikely to be the result of burial as constrained by the geological history of the area. The presence of: (1) a karst surface and dolomitized calcrete, (2) secondary pores filled by later marine internal sediment, and (3) evidence for dissolution of evaporites in the upper part of the Bali'h Member indicate that subaerial exposure and some meteoric water alteration did occur during and after development of the platform. In light of this, a possible origin for these rocks is a single-stage mixing zone dolomitization of previously undolomitized sediment. Petrographically similar

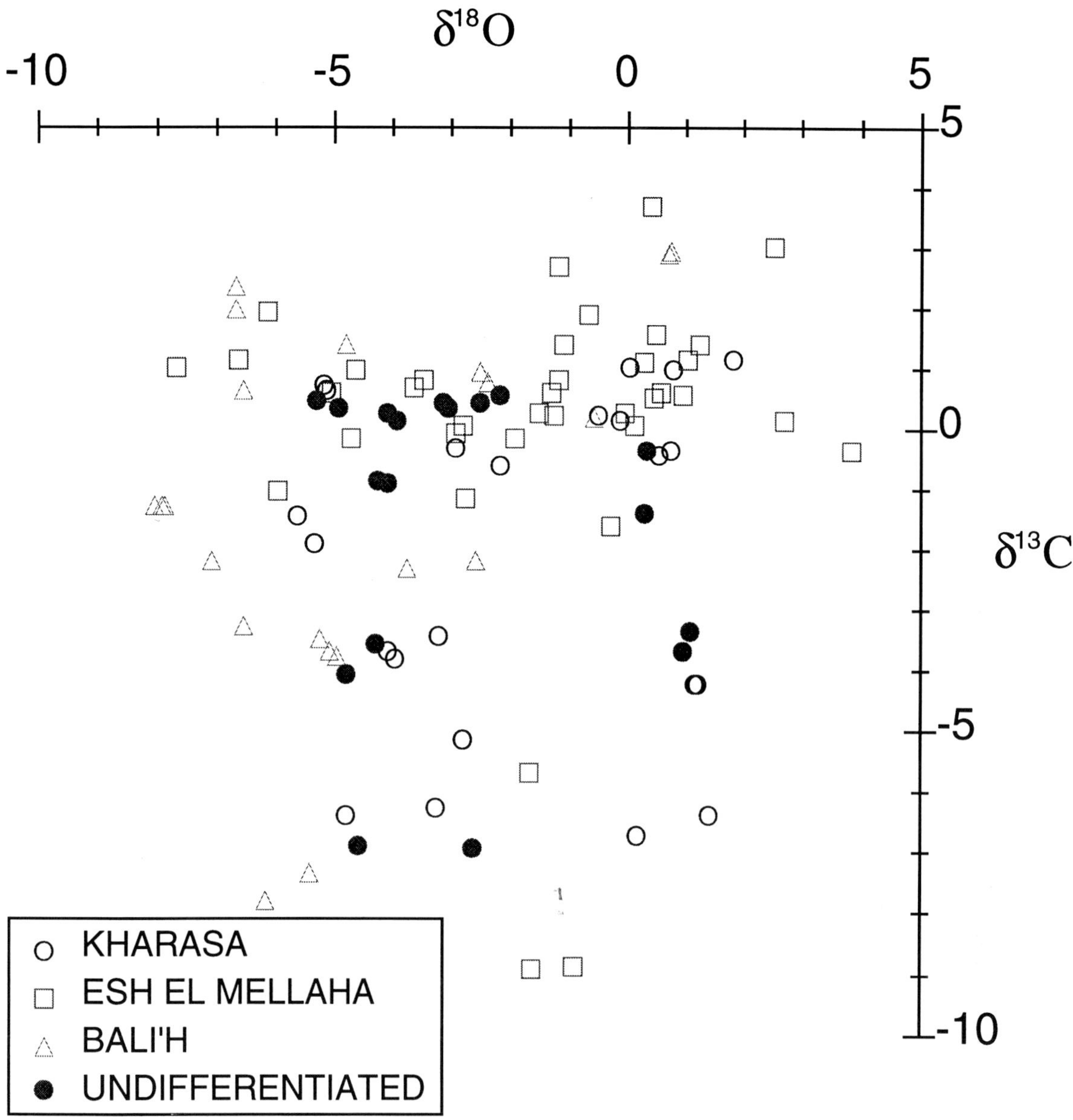


FIG. 16.—Plot of $\delta^{13}C$ (PDB) versus $\delta^{18}O$ (PDB) for all samples considered in this study (modified from Coniglio et al., 1988).

Cenozoic dolomites are regionally widespread and have been previously ascribed to a mixing zone origin (e.g., San Salvador — Supko, 1977; Bonaire — Sibley, 1980; the Bahamas — Kaldi and Gidman, 1982; Barbados — Humphrey and Quinn, 1989). Recent critical appraisals of the field, petrographic and geochemical data supporting mixing zone dolomitization, however, have concluded that pervasively dolomitized sequences, like that at Abu Shaar, are very unlikely to result from this mechanism (Machel and Mountjoy, 1986; Hardie, 1987). In addition, the association of dolomite and evaporites in the Abu Shaar area suggests an arid climatic setting, and thus the formation of significant amounts of fresh groundwater would have been unlikely (e.g., Sun, 1992).

A preferred explanation for dolomitization at Abu Shaar involves recrystallization and a two-stage dolomitization history. In the first stage, pervasive dolomite could have been formed by shallow-subtidal reflux of marine or hypersaline water (Simms, 1984; Machel and Mountjoy, 1986). Alternatively, initial dolomite precipitation may have occurred below a sabkha. We infer that the original dolomite isotopic composition would have been the low positive values of $\delta^{18}O$ characteristic of marine or hypersaline environments. An especially attractive dolomitization mechanism is one that was used to explain dolomitization in Miocene carbonates from Mallorca and other areas in southern Spain (Oswald et al., 1990, 1991a, b; Sun, 1992). Like Abu Shaar, the Spanish carbonates were pervasively dolomitized and associated with extensive evaporite. In addition, meteoric-water cements were apparently absent and originally aragonitic allochems were dissolved. During sea-level fall, seawater-derived brines became undersaturated with respect to aragonite. With subsequent transgression, these brines mixed with seawater and were introduced into the platformal carbonates, causing aragonite dissolution and incorporating released $Ca^{2+}$ into dolomite cement and replacive dolomite. The mechanism appears equally applicable to the Abu Shaar carbonates. Dolomitization could also occur if these evolved, transgressing brines mixed with continental waters (Oswald et al., 1991b).

The first stage thus explains the pervasive nature of dolomite at Abu Shaar. In order to explain the stable isotopic compositions, a subsequent stage of recrystallization is inferred to have taken place either in a meteoric water-dominated fluid or perhaps more likely, in a solution in which there was a significant hydrothermal component. The initial rifting between the Sinabian and Nubian plates was accompanied by volcanism, dike injection and, if the present-day metalliferous hot brines in the Red Sea are a guide, considerable hydrothermal activity (see Thisse et al., 1983). Basement faults may have acted as conduits for these hot fluids. The influence of hydrothermal fluids, which could have a marine subsurface brine as well as meteoric origin, is therefore a reasonable possibility although supporting field or petrographic evidence is absent.

Explanations for the problematical wide scatter of $\delta^{13}C$ and $\delta^{18}O$ values are speculative and include variations in fluid mixing, rock composition and/or degree of water-rock interaction. In addition, the wide range in the $\delta^{18}O$ values could be explained by a variable temperature influence during the postulated recrystallization or, alternatively, bulk-solution disequilibrium (e.g., Veizer, 1983). The variable morphology and irregular microtopography of most dolomite crystals suggest collectively that dissolution rather than precipitation was the last phase of crystal-water interaction. This lends support to both dolomitization hypotheses. If other studies of mixing-zone dolomites are a guide, then dolomitization occurred during a sea-level drop, forcing the coastal aquifer and mixing zone down through the sediments. The mixing-zone waters, while initially saturated with respect to dolomite, probably became undersaturated as the $Mg^{2+}$ source — seawater — became progressively diluted. The leached zones and centers of the dolomite crystals may have resulted from the preferential dissolution of a calcian or trace element-rich phase during progressive dilution while the more stoichiometric dolomite remained stable in these mixed waters.

Dissolution in Abu Shaar dolomite crystals can also be explained in the recrystallization hypothesis if pore fluids following recrystallization later became undersaturated with respect to dolomite. An extrinsic control, however, such as the dropping sea level in the mixing-zone hypothesis, is not apparent.

### *Timing of Dolomitization*

Considering the mixing-zone hypothesis, dolomitization at Abu Shaar could have occurred after Pliocene uplift, during which the enveloping, younger evaporites were removed by solution in fresh groundwaters and possibly also erosion (James et al., 1988a). As relative sea level continued to drop, either through additional uplift or real lowering of sea level, the dolomitizing mixing zone migrated downward through the Abu Shaar complex. The occurrence of dolomite with cloudy or leached cores, relatively clear rims, but severely etched crystal surfaces attests to the progressive freshening of pore waters through time and eventual undersaturation with respect to dolomite.

In the preferred recrystallization hypothesis, the first stage of dolomitization could have resulted from gradual restriction of the platform, possibly during the Mid to Late Miocene evaporitic phases. Marine or hypersaline fluids, driven downward by density contrast with the underlying pore waters or pumped downward by wind or waves, dolomitized the sediments. Alternatively, subsequent transgression following an evaporitic phase may have forced hypersaline waters to enter the carbonate complex, dissolve aragonite and precipitate dolomite. If later recrystallization occurred in a meteoric-water dominated system, this could have been related to Pliocene uplift and telogenesis during which the enveloping, younger evaporites were removed. On the other hand, we have no basis to suggest timing of recrystallization if it involved hydrothermal fluids.

## CONCLUDING REMARKS: IMPLICATIONS FOR POROSITY DEVELOPMENT

Overall, the Abu Shaar dolomite is quite porous. Point count estimates of porosity range from 10 to 30%, averaging 22% in open-platform sediments and 14% in platform-margin strata (James et al., 1988a). Pore types observed include primary interparticle and intraparticle pores as well as vuggy, moldic, and intercrystalline types. The above porosity estimates do not include pore types that are too small to be resolved with the light microscope, such as the intracrystalline and intercrystalline pores discernible only with SEM, nor do they include pores that are too large to be sampled in a thin section. Vugs are the most commonly encountered pore type.

A strong facies control on the distribution of porosity in the Abu Shaar sequence is a function of: (1) original sediment composition and (2) synsedimentary cementation and internal sedimentation. Open-platform facies are the most porous, based on matrix porosity from point count data, although reefal facies are clearly very macroporous. Aragonitic allochems such as corals, gastropods and bivalves were commonly dissolved and are now preserved as molds or cement-reduced molds; whereas, components that were originally Mg-calcite are typically well preserved. Consequently, facies that were initially aragonite-rich tend to be more porous than those dominated by Mg-calcite which have mainly interparticle porosity. On a bedding scale this explains the presence of interbedded porous and less porous horizons. Position on the platform margin is also an important porosity control. Because platform-margin strata were commonly cemented by what is interpreted to have originally been Mg-calcite marine cement and these strata also contain abundant internal sediments, fore-reef and reef-front strata tend to be less porous than open-platform facies.

The relatively high porosity throughout the complex and additional microporosity occurring in intercrystalline and intracrystalline areas makes reef-rimmed platforms such as Abu Shaar attractive targets providing adequate seals are available. Similar platforms that occur in the offshore of the Gulf of Suez region are significant hydrocarbon reservoirs. An example of such a buried platform is the K84 structure offshore Ras Gharib, approximately 120 km NNE from Abu Shaar (Kulke, 1982). This 7-km-long and 1-km-wide, fault-bounded horst block has a 200-m-thick cap of porous dolomitic limestone and dolomite. Although the age of this structure is uncertain, there are significant similarities with the Abu Shaar sequence. The carbonates are mostly bioclastic packstones and rare coral-algal boundstones with abundant vuggy and biomoldic porosity. They contain the same skeletal elements and their diagenetic history also includes intensive dissolution, dolomitization and anhydrite precipitation in voids.

Whichever of the dolomitization hypotheses proposed above is the most realistic also has significant implications for exploration. If the recrystallization/hydrothermal explanation is correct, the potential for any carbonate veneer over active fault blocks in the region to have developed into a reservoir rock is high. On the other hand, if dolomitization is related to uplift, as would be likely if meteoric waters were involved, then subsurface platforms may still be largely limestone in composition, less porous and permeable, and less likely to form an important reservoir rock.

## ACKNOWLEDGMENTS

We are grateful to M. Esteban, E. Oswald and T. Simo for providing numerous constructive comments on earlier versions of this manuscript. This research was financially supported by Marathon Oil Company and the Natural Sciences and Engineering Research Council of Canada (NSERC). We are particularly indebted to J. L. Wray, formerly of Marathon Research Center (Denver), and J. D. Denton of Marathon Petroleum Egypt for the opportunity to examine these sediments. C. Cofer, A. Morrisy, M. Atta, and A. Muktar (all Marathon Petroleum Egypt) assisted with various aspects of our study while in Egypt. B. R. Rosen of the British Museum, Natural History identified numerous representative coral specimens. Thin sections, stable isotope, and wet chemical analyses were provided by Marathon Research Center. We also thank the AAPG and SEPM for permission to use many of the figures in this contribution.

## REFERENCES

Aïssaoui, D. M., Coniglio, M., James, N. P., and Purser, B. H., 1986, Diagenesis of a Miocene reef-platform, Jebel Abu Shaar, Gulf of Suez, Egypt, *in* Schroeder, J. H. and Purser, B. H., eds., Reef Diagenesis: Berlin, Springer-Verlag, p. 112-131.

Blow, W. H., 1969, Late middle Eocene to Recent planktonic foraminiferal biostratigraphy, *in* Bronnimann, R. and Renz, H. H., eds., Proceedings of the First International Conference on Planktonic Microfossils: Leiden, E. J. Brill, v. 1, p. 199-421.

Bullen, S. B. and Sibley, D. F., 1984, Dolomite selectivity and mimic replacement: Geology, v. 12, p. 655-658.

Burchette, T. P., 1988, Tectonic control on carbonate platform facies distribution and sequence development: Miocene, Gulf of Suez: Sedimentary Geology, v. 59, p. 179-204.

Cofer, C. R., Lee, K. D., and Wray, J. L., 1984, Miocene carbonate microfacies, Esh el Mellaha Range, Gulf of Suez: Cairo, Egyptian General Petroleum Corporation Seventh Exploration Seminar, 36 p.

Coniglio, M., James, N. P., and Aïssaoui, D. M., 1988, Dolomitization of Miocene carbonates, Gulf of Suez, Egypt: Journal of Sedimentary Petrology, v. 58, p. 100-119.

Esteban, M., 1979, Significance of the upper Miocene coral reefs of the western Mediterranean: Paleogeography, Paleoclimatology, Paleoecology, v. 29, p. 169-188.

Girdler, R. W. and Southren, T. C., 1987, Structure and evolution of the northern Red Sea: Nature, v. 330, p. 716-721.

Gregory, J. W., 1906, Fossil coral from Egypt: Geological Magazine, v. 3, p. 50-58.

Haddad, A. El., Aïssaoui, D. M., and Soliman, M. A., 1984, Mixed carbonate-siliciclastic sedimentation on a Miocene fault-block, Gulf of Suez: Sedimentary Geology, v. 37, p. 185-202.

Hardie, L. A., 1987, Dolomitization: a critical view of some current views: Journal of Sedimentary Petrology, v. 57, p. 166-183.

Hume, W. F., 1916, Report on the oilfield region of Egypt, with a geological map of the region from surveys by Dr. John Ball: Cairo, Egypt Survey Department, 130 p.

Humphrey, J. D. and Quinn, T. M., 1989, Coastal mixing zone dolomite, forward modeling, and massive dolomitization of platform-margin carbonates: Journal of Sedimentary Petrology, v. 59, p. 438-454.

James, N. P. and Choquette, P. W., 1990, Limestones — the sea-floor diagenetic environment, *in* McIlreath, I. A. and Morrow, D. W., eds., Diagenesis: St. John's, Geoscience Canada Reprint Series 4, p. 13-34.

James, N. P., Coniglio, M., Aïssaoui, D. M., and Purser, B. H., 1988a, Facies and geologic history of an exposed Miocene rift-margin carbonate platform: Gulf of Suez, Egypt: American Association of Petroleum Geologists Bulletin, v. 72, p. 555-572.

James, N. P., Rosen, B., and Coniglio, M., 1988b, Miocene platform-margin reefs, Gulf of Suez, Egypt (abs.): American Association of Petroleum Geologists Bulletin, v. 72, p. 200-201.

Jones, B., Pleydell, S. M., Ng, K.-C., and Longstaffe, F. J., 1989, Formation of poikilotopic calcite-dolomite fabrics in the Oligocene-Miocene Bluff Formation of Grand Cayman, British West Indes: Bulletin of Canadian Petroleum Geology, v. 37, p. 255-265.

Kaldi, J. and Gidman, J., 1982, Early diagenetic dolomite cements: examples from the Permian Lower Magnesian Limestone of England and the Pleistocene carbonates of the Bahamas: Journal of Sedimentary Petrology, v. 52, p. 1073-1085.

Kulke, H., 1982, A Miocene carbonate and anhydrite sequence in the Gulf of Suez as a complex oil reservoir: Cairo, Egyptian General Petroleum Corporation Fifth Exploration Seminar, 16 p.

Land, L. S., 1973, Contemporaneous dolomitization of middle Pleistocene reefs by meteoric water, North Jamaica: Bulletin of Marine Science, v. 23, p. 64-92.

Machel, H. G. and Mountjoy, E. W., 1986, Chemistry and environments of dolomitization - a reappraisal: Earth Science Reviews, v. 23, p. 175-222.

Macintyre, I. G., 1985, Submarine cements — the peloidal question, *in* Schneidermann, N. and Harris, P. M., eds., Carbonate Cements: Tulsa, Society of Economic Paleontologists and Mineralogists Special Publication 36, p. 109-116.

MADGWICK, T. G., MOON, F. W., AND SADED, H., 1920, Preliminary geological report of the Abu Shaar El Quibli (Black Hill) district: Cairo, Petroleum Research Bulletin, Government Press, v. 6, 11 p.

MONTY, C. L. V., ROUCHY, J. M., MAURIN, A., BERNET-ROLLANDE, M. C., AND PERTHUISOT, J. P., 1987, Reef-stromatolites-evaporites facies relationships from Middle Miocene examples of the Gulf of Suez and the Red Sea, *in* Peryt, T. M., ed., Evaporite Basins, Lecture Notes in Earth Sciences: Berlin, Springer-Verlag, p. 133-188.

OSWALD, E. J., FRANSEEN, E. K., AND MEYERS, W. J., 1991a, Similarities in the dolomitization of Upper Miocene reef complexes in Mallorca and the Las Negras area, Spain: possible evidence for a Mediterranean dolomitizing event during the Messinian (abs): American Association of Petroleum Geologists Bulletin, v. 75, p. 649.

OSWALD, E. J., MEYERS, W. J., AND POMAR, L., 1990, Dolomitization of an Upper Miocene reef complex, Mallorca, Spain: evidence for a Messinian dolomitizing Mediterranean Sea (abs): American Association of Petroleum Geologists Bulletin, v. 74, p. 735.

OSWALD, E. J., SCHOONEN, M. A. A., AND MEYERS, W. J., 1991b, Dolomitizing seas in evaporitic basins: a model for pervasive dolomitization of Upper Miocene reefal carbonates in the western Mediterranean (abs): American Association of Petroleum Geologists Bulletin, v. 75, p. 649.

PLAZIAT, J.-C., PURSER, B. H., AND SOLIMAN, M., 1990, Les rapports entre l'organisation de dépôts marins (Miocène inférieur et moyen) et la tectonique précoce sur la bordure NW du rift de mer Rouge: Bulletin de la Societé géologique de France, v. 8, p. 397-418.

PURSER, B. H., AÏSSAOUI, D. M., AND ORSZAG-SPERBER, F., 1988b, Diagenèse et rifting: évolution post-sédimentaire de sédiments carbonatés Miocènes sur la bordure NW de la mer Rouge: Paris, Notes et Mémoires, Total Compagnie Française des Pétroles, no. 21, p. 145-166.

PURSER, B. H. AND HÖTZL, H., 1988, The sedimentary evolution of the Red Sea rift: a comparison of the northwest (Egyptian) and northeast (Saudi Arabian) margins: Tectonophysics, v. 153, p. 193-208.

PURSER, B. H., ORSZAG-SPERBER, F., AND PLAZIAT, J.-C., 1988a, Sédimentation et rifting: les séries Néogène de la marge nord-occidentale de la mer Rouge (Egypte): Paris, Notes et Mémoires, Total Compagnie Française des Pétroles, no. 21, p. 111-144.

PURSER, B. H., PHILOBBOS, E. R., AND SOLIMAN, M., 1990, Sedimentation and rifting in the NW parts of the Red Sea: a review: Bulletin de la Societé géologique de France, v. 8, p. 371-384.

PURSER, B. H., SOLIMAN, M., AND M'RHABET, A., 1987, Carbonate, evaporite, siliciclastic transitions in Quaternary rift sediments of the northwestern Red Sea: Sedimentary Geology, v. 53, p. 247-267.

ROUCHY, J. M., 1982, La genèse des évaporites messiniennes de Méditerranée. Mémoires du Muséum national d'Histoire naturelle, Paris, C, 50, 267 p.

ROUCHY, J. M., 1986, Les évaporites miocène de la Méditerranée et de la mer Rouge et leurs enseignements pour l'interpretation des grandes accumulations évaporitiques d'origine marine. Bulletin de la Societé géologique de France, v. 8, p. 511-520.

ROUCHY, J. M., BERNET-ROLLANDE, M.-C., MAURIN, A. F., AND MONTY, C., 1983, Signification sédimentologique et paléogéographique des divers types de carbonates bioconstruits associés aux évaporites du Miocène Moyen près du Gebel Esh Mellaha (Égypte): Paris, Compte Rend Academie Science, Série II, v. 296, p. 457-462.

SCOTT, R. W. AND GOVEAN, F. M., 1985, Early depositional history of a rift basin: Miocene in western Sinai: Paleogeography, Paleoclimatology, Paleoecology, v. 52, p. 143-158.

SELLWOOD, B. W. AND NETHERWOOD, R. E., 1984, Facies evolution in the Gulf of Suez area: sedimentation history as an indicator of rift initiation and development: Modern Geology, v. 9, p. 43-69.

SIBLEY, D. F., 1980, Climatic control of dolomitization, Seroe Domi Formation (Pliocene), Bonaire, N.A, *in* Zenger, D. H., Dunham, J. B., and Ethington, R. L., eds., Concepts and Models of Dolomitization: Tulsa, Society of Economic Paleontologists and Mineralogists Special Publication 28, p. 247-258.

SIMMS, M., 1984, Dolomitization by groundwater - flow systems in carbonate platforms: Transactions of the Gulf Coast Association of Geological Societies, v. XXXIV, p. 411-420.

SMALE, J. L., THUNELL, R. C., AND SCHAMEL, S., 1988, Sedimentological evidence for early Miocene fault reactivation in the Gulf of Suez: Geology, v. 16, p. 113-116.

SUN, S. Q., 1992, Skeletal aragonite dissolution from hypersaline seawater: a hypothesis: Sedimentary Geology, v. 77, p. 249-257.

SUPKO, P. R., 1977, Subsurface dolomites, San Salvador, Bahamas: Journal of Sedimentary Petrology, v. 47, p. 1063-1077.

THIÉBAUD, C. E. AND ROBSON, D. A., 1979, The geology of the area between Wadi Wardan and Wadi Gharandal, East Clysmic Rift, Sinai, Egypt: Journal of Petroleum Geology, v. 1, p. 63-75.

THISSE, Y., GUENNOC, P., POUIT, G., AND NAWAB, Z., 1983, The Red Sea: a natural geodynamic and metallogenic laboratory: Episodes, v. 1983, p. 3-9.

VEIZER, J., 1983, Chemical diagenesis of carbonates: theory and application of trace element technique, *in* Arthur, M. A., Anderson, T. F., Kaplan, I. R., Veizer, J, and Land, L. S., eds., Stable Isotopes in Sedimentary Geology: Tulsa, Society of Economic Paleontologists and Mineralogists Short Course 10, p. 3-1—3-100.

# INDEX

## D



## E



## F



## G



## H



## I



## K



## L



## M

## N



## O



## P



## R

## S



## T

**U**



**V**



**W**